# International Conference on Aerospace, Aerodynamics and Mechatronics Engineering (AAME 2022)



Journal of Physics: Conference Series Volume 2364

Singapore
22-24 July 2022

ISBN: 978-1-7138-6472-1
ISSN: 1742-6588

**Printed from e-media with permission by:**

Curran Associates, Inc.
57 Morehouse Lane
Red Hook, NY 12571

**Some format issues inherent in the e-media version may also appear in this print version.**



Institute of Physics
Dirac House, Temple Back
Bristol BS1 6BE UK

Phone: 44 1 17 929 7481
Fax: 44 1 17 920 0979

techtracking@iop.org

**Additional copies of this publication are available from:**

Curran Associates, Inc.
57 Morehouse Lane
Red Hook, NY 12571 USA
Phone: 845-758-0400
Fax: 845-758-2633
Email: curran@proceedings.com
Web: www.proceedings.com

# TABLE OF CONTENTS

# Preface

The International Conference on Aerospace, Aerodynamics and Mechatronics Engineering (AAME 2022) was successfully held on July 29, 2022 in the form of online conference. AAME 2022, mainly focusing on the latest research on "aerospace engineering" and "mechatronics engineering", gathered scholars, researchers and related practitioners in this field to share research results, explore current hot news, and exchange new ideas, providing a face-to-face communication for all participants in the struggling period of COVID-19 pandemic.

We were delighted and honored to invite Prof. Ian R. McAndrew from Capitol Technology University, USA to serve as our Conference Chair. There were 90 individuals and enterprises all over the world attending the conference. Divided into three parts, the conference agenda covered keynote speeches, oral presentations, and online Q&A discussion. Firstly, keynote speakers were each allocated 30-45 minutes to hold their speeches. Then in the oral presentations, the excellent papers we had selected were presented by their authors by turns.

During the conference, four sophisticated professors were invited to address keynote speeches. Among them, Prof. Bing Li from Northwestern Polytechnical University, China delivered a speech on the topic: Extraordinary Elastic-Wave Control Based on Metamaterials and Metasurfaces. In this report, a series of elastic metamaterials and metasurfaces used for broadband vibration-suppression and extraordinary wavefront manipulation were introduced, a variety of novel elastic-wave manipulations were realized, and the proposed designs paved feasible ways for noise and vibration control in elastodynamics. Additionally, Prof. Ian R McAndrew from Capitol Technology University, USA gave a report on Cybersecurity Issues in the Aerospace Industry. He presented to us the findings from several aligned research projects that focused on the cyber issues, challenges and risks involved at present and in the future. The subjects of safety and risk were presented with unique findings and recommendations on what was the need to prioritize. The wonderful and dramatic speeches of each keynote speaker had triggered heated discussion. Moreover, every participant praised this conference for disseminating useful and insightful knowledge.

After months of well preparation and hard work, the proceedings of AAME 2022 covering a bunch of papers is smoothly published. These papers feature but are not limited to the following areas: Rocket Theory and Design, Communication Systems and Technologies, Interoperability, Synthetic Aperture Radar Technology, etc. All the papers have been checked through rigorous review and processes to meet the requirements of publication.

On behalf of the conference organizing committee, we would like to express our heartfelt appreciation to all the keynote speakers, peer reviewers, and all the participants. In particular, we would like to acknowledge the Journal of Physics: Conference Series, for the hard work of all the colleagues in publishing this paper volume. We firmly believe that all the attendees have had fruitful discussions and gained valuable knowledge, and will enjoy the opportunity for future collaborations.

Committee of AAME 2022

# Committee member

**Conference Chair**
Professor Ian R. McAndrew, Dean of doctoral programs at Capitol Technology University, USA

**Technical Program Committee Chair**
Prof. George Zhu, Academic Director of Research Commons, Office of Vice-President Research and Innovation, York University, Canada
Prof. P. K. Dash, Professor and Head, Dept. of Aeronautical Engineering, Nitte Meenakshi Institute of Technology, Bangalore, Karnataka, India

**Publication Chair**
Prof. Bing Li, Department of Aircraft Design Engineering, School of Aeronautics, Northwestern Polytechnical University, China

**Technical Program Committee Members**
Prof. Bing Li, School of Aeronautics, Northwestern Polytechnical University, China
Prof. Kun Li, ChongQing University, China
Prof. Hamid Bahai,Queen Mary College, England
Prof. Dr. Murat Tolga OZKAN, Gazi University, Turkey
Prof. Alexander Lukin, Western-Caucasus Research Center, Russia
Prof. Ozgur Koray, Biruni University, Turkey
Prof. Sandeep Saxena, IMS Unison University, India
Prof. Hossein Hosseinkhani, Innovation Center for Advanced technology, Matrix, Inc. Canada
Prof. Mingwei Qin, Southwest University of Science and Technology, China
Dr. Siting Chen, South China University of Technology, China
Assoc Prof. Muhyiddine Jradi, University of Southern Denmark, Denmark
Assoc Prof. Wei Wei, BeiHang University, China
Assoc Prof. SANDIP KUNAR, Aditya Engineering College, India
Assoc Prof. Chao Chen, Taiyuan University of Technology, China

# Peer Review Statement

All papers published in this volume have been reviewed through processes administered by the Editors. Reviews were conducted by expert referees to the professional and scientific standards expected of a proceedings journal published by IOP Publishing Publishing.

- **Type of peer review:** Single Anonymous
- **Conference submission management system:** Morressier
- **Number of submissions received:** 146
- **Number of submissions sent for review:** 131
- **Number of submissions accepted:** 68
- **Acceptance Rate (Submissions Accepted / Submissions Received × 100):** 46.6
- **Average number of reviews per paper:** 2
- **Total number of reviewers involved:** 40
- **Contact person for queries:**
  **Name:** Xuexia Ye
  **Email:** xx.ye@keoaeic.org
  **Affiliation:** AEIC Academic Exchange Information Centre

# Research on space construction and selection method of space-based information support equipment

**Qing Chang[a], Desheng Liu[b*] and Yang Yang[c]**

Science and Technology on Complex Electronic System Simulation Laboratory, Space Engineering University, Beijing, 101416, China

[a]email: 936500617@qq.com,[b]email: liudsnudt@126.com,[c]email:1783249960@qq.com

[*]Corresponding author's e-mail: liudsnudt@126.com

**Abstract.** With the rapid development of information technology and space technology, the important position of space-based information support equipment is increasingly prominent. However, the increasing types and quantities of space-based information support equipment also put forward higher requirements for the formulation of combat plans, decision-making and deployment of combat equipment. In this paper, based on the seven-step method, an ontology construction method suitable for space-based information support equipment is proposed, and the ontology construction tool Protégé is used to realize the ontology modeling of space-based information support equipment. By acquiring open source satellite data, the knowledge map construction tool Neo4j is used to construct the knowledge map of space-based information support equipment. Finally, the space-based information support equipment conforming to the combat mission can be obtained to provide assistance for the formulation of the combat plan and the decision of equipment utilization.

## 1. Introduction

The data acquisition of space-based information support equipment is difficult and intuitionistic, and it is difficult in management and application. Knowledge graph has achieved good results in fields such as medicine and e-commerce. Therefore, the method of constructing knowledge graph can be used to realize space construction of space-based information support equipment.

According to the characteristics of weapon equipment system and the seven-step method, Li Xiao [1] combined with the seven-step method, proposed an ontology construction method suitable for weapon equipment system and constructed the knowledge graph of weapon equipment system. Cheng Jianbo [2] constructed the ontology model of command information system from two aspects: entity and relationship. Ding Junyi [3] adopted the method of artificial modeling and applied Protégé software to complete the construction of ontology in the field of weapons and equipment .Han Jinan [4] built a knowledge graph system based on space equipment data for space equipment data.

At the same time, the application plan formulation process of space-based information support equipment should not only rely on equipment knowledge and data, but also need to make comprehensive judgment combined with mission constraints. Analytical Graphics' STK (Satellite Toolbox) has been leading in the field of simulation analysis of aerospace equipment [5]. Miao Jisong [6] used MATLAB and STK co-simulation to achieve the optimal design of satellite orbit, sensor, satellite formation and other parameters in Walker constellation. Guan Qingbo [7] proposed the relevant principles and processes of combat simulation design for space-based information support. Xu Jun [8] designed three different

Walker constellation schemes, calculated the coverage of the three schemes on a target area, and evalu ated the three different Walker constellation schemes with coverage time, coverage gap and other indic ators.

## 2. Object description

Space-based information support equipment can be divided into reconnaissance satellites, early warnin g satellites, communication satellites, navigation and positioning satellites, etc., which mainly provide combat operations supported by battlefield situation, communication relay, navigation and positioning information for the battlefield [9].

Among them [10], Reconnaissance satellites can use reconnaissance equipment to obtain visible, ra dar or electronic information of ground targets from orbit in space and transmit the information back to the ground station. Early warning satellites can detect and track ballistic missile launches in time and give early warning of incoming missiles. Communication satellites can realize the communication bet ween the ground and spacecraft or between the ground and the ground. Navigation and positioning sate llites can provide navigation and positioning information for combat units.

When making equipment operation plan for space-based information support equipment, it is neces sary to consider the operation parameters of equipment, such as operation state, time and space locatio n, and conduct scene simulation analysis.

## 3. Ontology-based space-based information support equipment space construction method

### *3.1. Ontology construction of space-based information support equipment*

In 1993, Gruber of Stanford University provided a formal definition of ontology in the field of informa tion science, namely, "Ontology is a clear specification of conceptualization", which is the most widel y recognized concept definition of ontology at present [11]. Ontology is a standardized description of t he hierarchical structure of concepts in a specific domain, which can be regarded as people's unified un derstanding of concepts in this domain [12].

Ontology is defined as $D$, which mainly includes five parts:

$$D = \{C, A, R, I, M\} \quad (1)$$

Where, $C$ is concept class; $A$ is attribute set; $R$ is relation; $I$ is the instance set; $M$ is the mapping rela tionship between instances and concept classes. Using Protégé ontology construction tool and referring to the seven-step method, this paper puts forward the ontology construction steps of space-based infor mation support equipment. The specific process is as follows:

(1) Determine the subject field and category of ontology. The object of this research is Space Infor mation Supporting Equipment (SISE).

(2) List important terms in the ontology. The terms in the Chinese People's Liberation Army Militar y Language and other related books were used as the basis for modeling.

(3) Define concept classes and their hierarchical relationships. First, it should confirm whether ther e are existing ontologies available for reuse, and then divide the criteria in a top-down hierarchy, from general concepts down to examples.

Combined with the terms listed in step (2), the Protégé ontology architecture tool is used to define t he concept class hierarchy, and the result is shown in Figure 1.

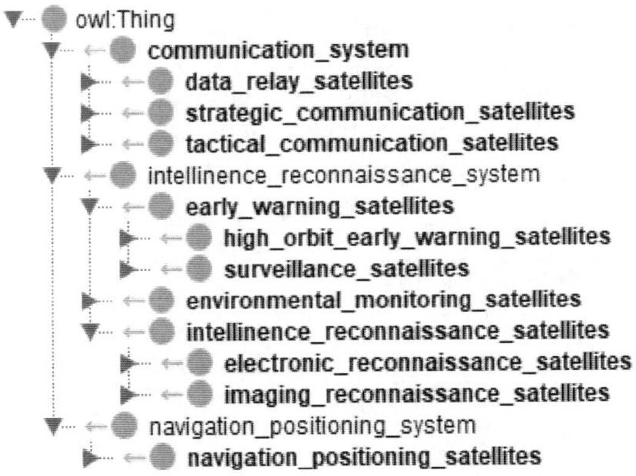


Figure 1. Concept class construction results.

(4) Defining conceptual class relations.

The previous step defined parent-child relationships for concept classes. In addition to the parent-child relationship, some space-based information support equipment concept classes have cooperative relations, and the relationship between concept classes needs to be constructed. The concept class relationship and name of space-based information support equipment are shown in Table 1, and the schematic diagram of concept class relationship is shown in Figure 2.

Table 1. Part of the conceptual class relationship.

| Relationship | Concept of class |
|---|---|
| Support information | Communication system→ Intelligence reconnaissance system |
| Support information | Communication system→ navigation positioning system |
| Support of information relay | Data relay satellites→ Strategic communication satellites |
| Support of information relay | Data relay satellites→ Tactical communication satellites |
| Warning relay | High orbit early warning satellites→ Surveillance satellites |
| ...... | ...... |

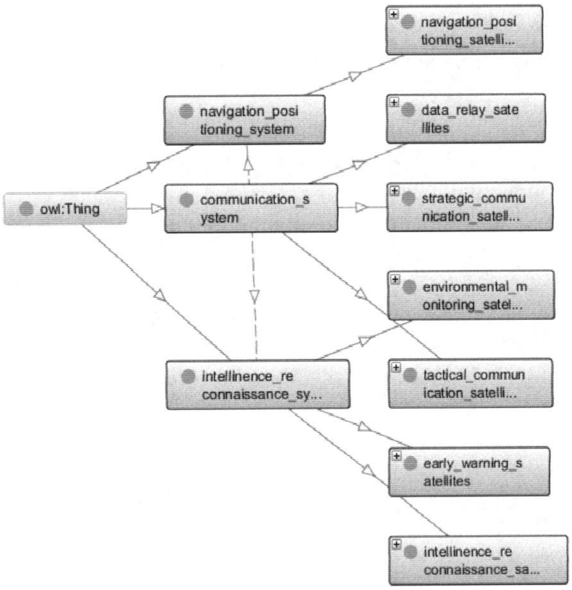


Figure 2. Schematic diagram of conceptual class relations.

(5) Define the attribute of concept class.

Concept class attributes refer to the set of attributes that can represent the characteristics of the conc ept itself. The process of defining conceptual class properties should also follow a top-down order. Att ributes will be inherited by subclass concept classes and instances. Taking optical imaging reconnaissa nce satellite as an example, the attributes of this type of satellite include detection range, image resolut ion, processing delay, etc. Some of the concept class attributes are shown in Table 2..

Table 2. Concept class attributes.

| Attributes | Concept class | Attribute types |
|---|---|---|
| ID<br>Date of Launch<br>Country/Organization<br>Service life<br>Orbital parameter<br>...... | Space-based information support equipment | INT<br>Array<br>String<br>INT<br>Vector<br>...... |
| Timing accuracy<br>Dynamic positioning accuracy<br>Static positioning accuracy<br>Information acquisition time<br>Probability of anti-interference<br>...... | Navigation positioning system | INT<br>INT<br>INT<br>INT<br>INT<br>...... |
| Coverage<br>Processing delay<br>Detection probability<br>Resolution<br>...... | Intelligence reconnaissance system | INT<br>INT<br>INT<br>INT<br>...... |
| ...... | ...... | ...... |

(6) Inspection.

The reasonability of the model construction is tested by expert review or pre-construction. If the m odel passes the test, the final ontology is formed. On the contrary, it is necessary to repeat steps 3, 4 an d 5, and make corresponding modifications until passing the test.

*3.2. Space-based information support equipment space construction*

Neo4j database supports three knowledge data import methods [13], which respectively use the CREA TE statement in Cypher language, load. CSV statement and the official Neo4J-admin import tool. In th is paper, the second knowledge data import method is selected. The nodes constructed in this time incl ude concept nodes and entity nodes, defined as shown in Table 3.

Table 3. Knowledge graph node definition.

| | Node types | ID | Implication |
|---|---|---|---|
| Concept nodes | Space-based informatio n support equipment | SISE | The concept classes defined in the ontology represent the weapon equipment system itself and constitute the whol e network as the central node. |
| | Equipment types | type | The concept class defined in the ontology, for example: i ntelligence reconnaissance satellites. |
| | Equipment functions | function | Concept classes defined in the ontology, such as missile early warning satellites. |
| | Equipment models | model | The smallest concept class defined in the ontology, such as BEIDOU_G. |
| Entity no des | Equipment individual | Individual | The instance corresponding to the ontology, such as SBI RS_HEO-2. |

The knowledge graph of space-based information support equipment constructed finally has a total of 1689 equipment nodes, and each node corresponds to an equipment instance individual. Only part o f the nodes are shown here, as shown in Figure 3.

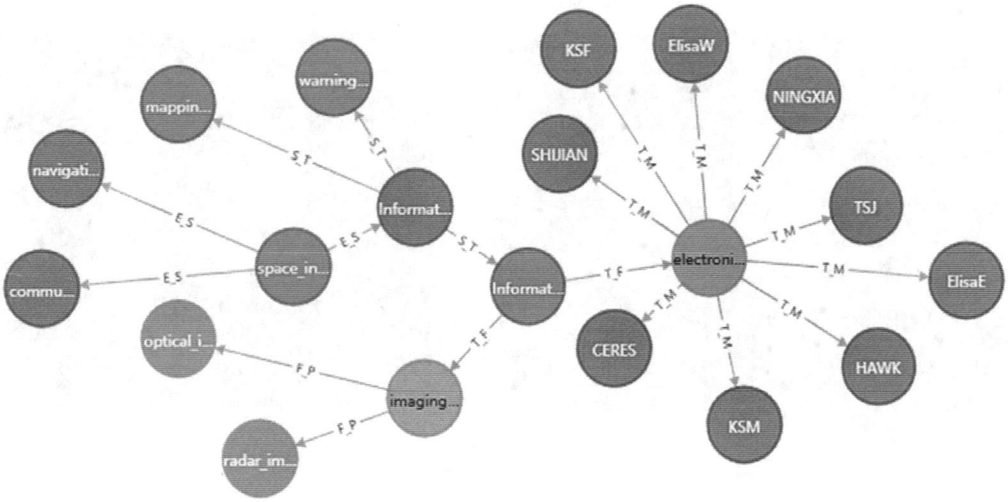


Figure 3. Partial knowledge graph.

## 4. Space-based Information support equipment screening

When making the equipment application plan of space-based information support equipment, it is nece ssary to consider the equipment capability, mission constraints and other factors. The task constraints o f space-based information support equipment are analyzed through simulation experiments to achieve t he screening of space-based information support equipment.

### *4.1. Simulation scene construction*

In order to realize the task constraint screening of space-based information support equipment, differen t simulation scenarios should be constructed according to the corresponding mission characteristics of t he satellites, and corresponding satellites should be constructed in the simulation scenarios for analysis according to the entity nodes of the knowledge graph.

Taking the visible light imaging reconnaissance satellite as an example, the simulation scene constr uction process is introduced:

(1) Create a scene named “Optical Imaging Satellite”, and set the simulation scene time as “1 Jul 20 20 06:00:00. 000” and “1 Jul 2020 09:00:00. 000”.

(2) Create an AreaTarget object, name it “AreaTarget1”, set the “AreaTarget” type as “Pattern”, an d set the longitude and latitude coordinates of the four vertices of the rectangular “AreaTarget”, as sho wn in Table 4 and Figure 4.

Table 4. AreaTarget vertices.

| Vertices number | Vertices coordinates |
|---|---|
| 1 | 20,120 |
| 2 | 25,120 |
| 3 | 25,125 |
| 4 | 20,125 |

For the other types of satellites, the simplified model adopts the regional target as the combat objec t.

(3) Input the visible light imaging reconnaissance satellite orbit data stored in the knowledge graph into the simulation scene in the form of TLE data to generate satellite objects. A sensor is added for ea ch satellite object. This experiment is a simplified model, and the sensor is uniformly set as a rectangu lar field of view with a side swing Angle of 45°. The simulation scenario is shown in Figure 5.

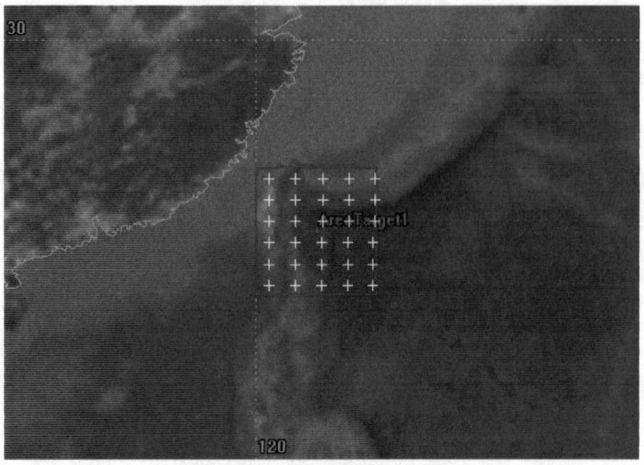


Figure 4. AreaTarget.

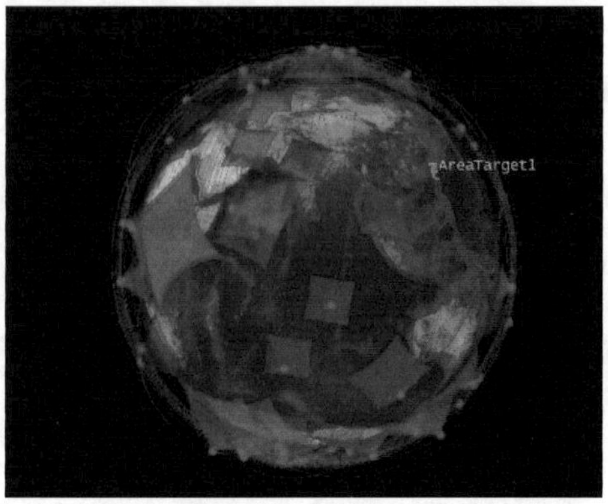


Figure 5. 3D graphic.

*4.2. Selection of available equipment based on mission constraints*

Taking the visible light imaging reconnaissance satellite as an example, the screening based on mission constraints is performed as follows.

(1) Set the type of analysis parameters according to the working characteristics of the satellite.

According to the working characteristics of the visible light imaging reconnaissance satellite, the coverage ratio of the satellite to the regional target can be calculated by the coverage analysis module, so as to judge the degree of fit between the satellite and the combat mission. The mission constraint parameters of other kinds of satellites are shown in Table 5.

Table 5. Simulation scenario Parameters.

| Simulation scenario | Calculating parameter | Report |
|---|---|---|
| Optical imaging reconnaissance satellites | Custom Region- number of accesses | Percent Coverage |
| Radar imaging reconnaissance satellites | Custom Region- number of accesses | Percent Coverage |
| Electronic reconnaissance satellites | Chain-Compute Accesses | Complete Chain Access |
| Environmental monitoring satellites | Custom Region- number of accesses | Percent Coverage |
| High orbit early warning satellites | Custom Region- coverage time | Grid Stats |
| Surveillance satellites | Chain-Compute Accesses | Complete Chain Access |
| Navigation positioning satellites | Custom Region- Dilution Of Precision | Percent Coverage |
| Strategic communication satellites | Chain-Compute Accesses | Complete Chain Access |
| Tactical communication satellites | Chain-Compute Accesses | Complete Chain Access |
| Data relay satellites | Chain-Compute Accesses | Complete Chain Access |

(2) Activate a single satellite and sensor in sequence, calculate Coverage parameters, and calculate "CoverageDefinition-Percent Coverage" to obtain the Coverage ratio P of satellite sensors to the target range during the simulation duration. The formula for calculating the coverage ratio P is:

$$P = S_i / S, i = 1, 2, \ldots, n \tag{2}$$

Where, $S_i$ is the coverage area of the ith satellite to the target area, $S$ is the total number of network points in the target area, and $n$ is the number of satellites.

The satellites with coverage ratio equal to 100 are recorded as the equipment output that meets the r equirements of combat missions. The satellites with coverage ratio of 100% are shown in Table 6. The chart of "Percent Coverage" of the satellite "Yaogan9C_36415" is shown in Figure 6.

Table 6. Candidate equipment.

| Satellite number | Satellite name | Coverage ratio | Satellite number | Satellite name | Coverage ratio |
|---|---|---|---|---|---|
| 13 | Coms1_36744 | 100 | 90 | Yaogan8_361 21 | 100 |
| 55 | Saudisat1C_276 07 | 100 | 91 | Yaogan9A_36 413 | 100 |
| 78 | Yaogan15_3835 4 | 100 | 92 | Yaogan9B_36 414 | 100 |
| 87 | Yaogan4_33446 | 100 | 93 | Yaogan9C_36 415 | 100 |

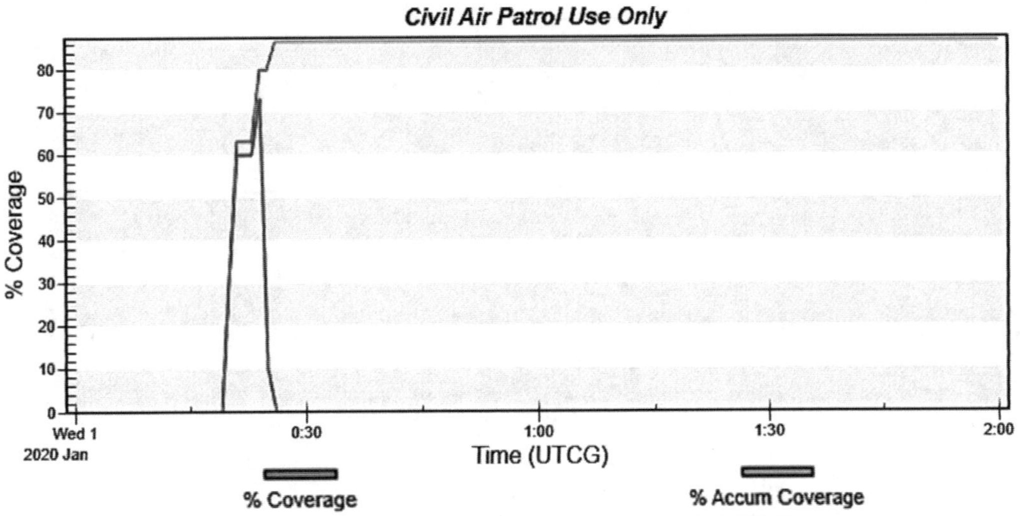


Figure 6. "Percent Coverage" of "Yaogan9C_36415".

Next, taking the tracking and surveillance satellite as an example, the screening based on mission c onstraints is carried out, and the detailed steps are omitted due to the length. Through visibility analysi s, the calculated parameter is "Compute Accesses", and the output report selects "Complete Chain Acc ess" to screen out the satellites with visibility to the target.

The combat object of the tracking and monitoring satellite is ballistic Missile. The Missile object w as created as shown in Figure 7 and named "Missile1". The parameters are shown in Table 7.

Table 7. Missile object parameters.

| Parameter name | Parameter values |
|---|---|
| Launch point coordinates | 20,120 |
| impact point coordinates | 40,120 |
| Fixed Apogee Alt | 1876 km |

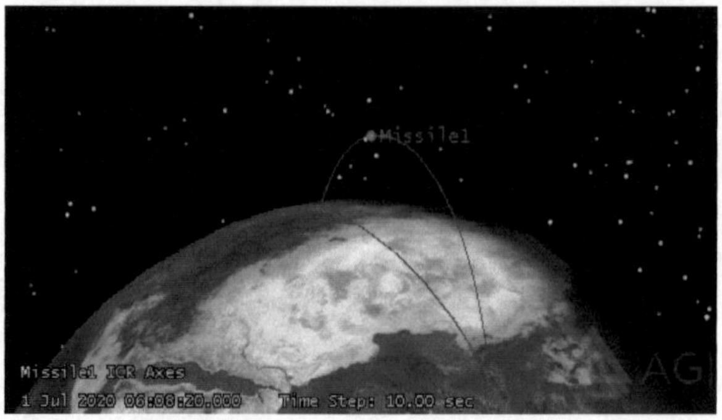

Figure 7. Missile object graphic.

The visible satellites are recorded as shown in Table 8. The Individual Object Access is shown in Figure 8.

Table 8. Missile object parameters.

| Satellite number | Satellite name | Object Access | Satellite number | Satellite name | Object Access |
|---|---|---|---|---|---|
| 3 | STSS12 | 275.318 | 14 | STSS42 | 1297.465 |
| 4 | STSS13 | 1188.188 | 17 | STSS51 | 991.909 |
| 7 | STSS23 | 1413.067 | 18 | STSS52 | 1434.592 |
| 8 | STSS24 | 417.5090 | 21 | STSS61 | 1463.183 |
| 10 | STSS32 | 447.327 | 22 | STSS62 | 905.026 |
| 11 | STSS33 | 836.422 | 24 | STSS64 | 723.743 |

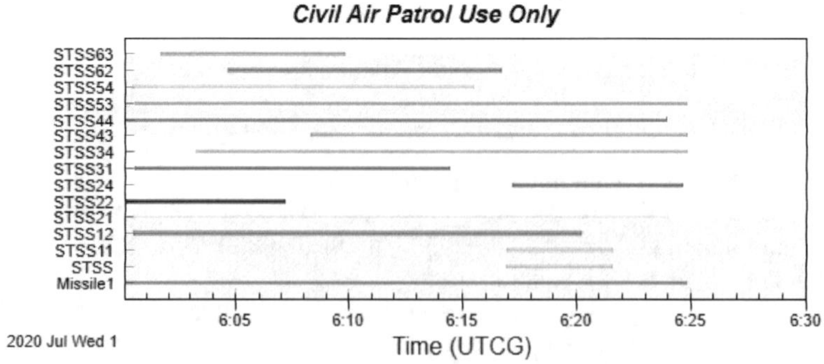

Figure 8. Individual Object Access.

### *4.3. Analysis of conclusion*

Conclusions can be drawn from the selection process of imaging reconnaissance satellite mission constraints as follows:

(1) A total of 96 imaging reconnaissance satellites are obtained from the space of space-based information support equipment based on knowledge Atlas. Most of these satellites are located in LEO orbit with an earth-orbiting period of about 96 minutes, so only a small number of satellites can achieve full imaging coverage of the target area. In this simulation experiment, 8 satellites with coverage ratio of 100 were selected, accounting for 8. 3% of the total number of satellites, which is in line with the reality.

(2) The satellite screening results based on mission constraints are mainly affected by the location of the target area, simulation time, the resolution of the coverage grid and the size of the field of view of

the satellite sensor. In this simulation experiment, the size of the field of view Angle of the satellite se nsor is uniformly set as 45°×45° to simplify the model. Although the size of the field of view Angle of the sensor of the satellite equipment is different in the actual combat process, it does not affect the rati onality of the proposed method.

(3) At present, the simulation modeling and screening method of space-based information support e quipment in this study is only based on satellite functions for classification modeling, but many constra ints involved in practice are not considered due to simplified model factors, and subsequent research n eeds to be further strengthened.

## 5. Conclusion

This paper adopts the top-down construction method and takes the seven-step method as the reference, proposes an ontology construction method suitable for space-based information support equipment, an d uses the ontology construction tool Protege to realize the ontology modeling of space-based informat ion support equipment. On the basis of ontology construction, the knowledge map of space-based infor mation support equipment was constructed by acquiring open source satellite data and using the knowl edge map construction tool Neo4j.

In view of the equipment used in the decision-making process equipment selection problem, put for ward a method of space-based information support and equipment selection using STK satellite toolbo x to build simulation scenarios, by analyzing the task constraints of space-based information support e quipment selection, eventually get in line with the mission of space-based information support equipm ent, and equipped for battle plan of using provide auxiliary decision-making.

## References

[1] Li Xiao, Liu Desheng. Ontology Construction for Weapon Equipment System Knowledge Gr aph [J]. Ordnance Automation, 222,41(03):25-30.

[2] Cheng Jianbo, Liu Desheng. Command information system modeling based on ontology [J]. Com mand control and simulation,2017,39(06):82-88.

[3] Ding Junyi, Zhao Qingsong, Xia Boyuan, Zou Zhigang. Research on weapon equipment knowled ge graph construction method based on open source data [J]. Command control and simulati on,2018,40(02):22-26.

[4] Han Jinan, Yuan Hongyi, Zheng Kuizhi, Zhu Chenlai, Zhang Bing. Aerospace equipment large da ta oriented knowledge map system [J]. Journal of national defense science and technology, 2 021, and (5) : 40-46. DOI: 10. 13943 / j. i ssn1671-4547. 2021. 05. 07.

[5] Wang Xin, Jin Hu. Communication satellite Scene Modeling Based on STK/MATLAB [J]. Elect ronic Test and Measurement,2009(04):1-5+41.

[6] Miao Jisong, Shao Qiongling, Ren Yuan, Lv Yongshen, Wang Chen. MATLAB/STK joint simula tion method to study and GUI design [J]. Computer measurement and control, 2017, 25 (8) : 199-202. The DOI: 10. 16526 / j. carol carroll nki. 11-4762 / tp. 2017. 08. 051.

[7] Guan Qingbo, Cao Yuhua, Zhuang Jinshan. Space-based information support system simulation c ase design [J]. Journal of system simulation technology, 2014, 10 (02) : 165-170. The DOI: 10. 16812 / j. carol carroll nki cn31-1945. 2014. 02. 015.

[8] Xu Jun, Bai Xiangfeng, Zhou Xiaohua. Simulation analysis of reconnaissance satellite coverage b ased on STK [J]. Software guide,2016,15(08):156-158.

[9] Zhang Hanfeng, Liu Xianmin, Liu Qiang. Research on the Requirements of Space-based Informat ion Support for Air Defense Operations [M]. Beijing: National Defense Industry Press,2018.

[10] Su Jinhai. Military Information System [M]. Publishing House of Electronics Industry, 2010.

[11] Gruber T R . A translation approach to portable ontology specifications [J]. Knowledge Acquisit ion, 1993, 5( 2):199-220.

[12] Huang Zhiduan. Exploration of Constructing New Media Domain Ontology by Protégé [D]. Wuh an University of Technology,2013.

[13] Cao Haowei, Xu Jianliang, Dou Fangkun. Based on secondary biological medicine [J]. Journal of knowledge map building computer age, 2020 (6) : 35-38. DOI: 10. 16644 / j. carol carroll n ki cn33-1094 / tp. 2020. 06. 010.

# Study on the relationship between the length of mixing layer and the characteristic scale of mixing layer in multi-strut ejector

**Letian Zhou[1,a], Wanwu Xu[1,b*] and Wei Ye[1,c]**

[1] Science and Technology on Scramjet Laboratory, College of Aerospace Science and Engineering, National University of Defense Technology, Changsha, Hunan, 410073, China

[a]email: zhouletian20@nudt.edu.cn, [c]email: yewei@nudt.edu.cn

[b*]Corresponding author's e-mail: cfdxww@nudt.edu.cn

**Abstract.** The flow structure of the multi-strut ejector based on jet segmentation has an obvious relationship with the transverse characteristic scale of the jet. Therefore, this paper studies the relationship between jet transverse characteristic size and mixing layer. Through the simulation of multi-strut ejectors with different number of struts, the development of mixing layer in the mixing process of jets with different transverse characteristic sizes is compared under the same boundary conditions. It is concluded that when there is no obvious coupling effect between the mixing layer and the shock & expansion wave, there is a linear relationship between the transverse characteristic scale of the jet and the mixing length of the flow direction, and the jet mixing with different transverse characteristic scales is consistent. When the shockwaves and expansion waves are strongly coupled with the mixing layer, the linear relationship between the jet transverse characteristic scale and the mixing layer length is not strong.

## 1. Introduction

The ejection mixing layer[1] is an important mechanism of mass, momentum and energy exchange in the process of ejector active flow and passive flow mixing. Industrial refrigeration[2], propulsion system[3], high-energy laser pressure recovery system[4], hypersonic ground test equipment[5], and other research focus. The development and mixing efficiency of jet mixing layer have always been the focus of the research on ejection mixing. It is found that the larger the mixing area of active and passive flow, the shorter the length of the mixing diffuser[6]. Based on this principle, the concept of multi-jet ejector is proposed. Such as multi-nozzle ejector[7], multi-strut ejector (MSE) [8], differentuional ejector[10].

In the mixing chamber of multi-jet injector, the mixing area is divided into several mixing units by multi-jet, and a excellent mixing effect is obtained. In particular, MSE [8] has excellent clamping and supercharging performance and can achieve high performance sub-super ejecting mode and super-super ejecting mode. Preliminary study shows that the length of jet direction is proportional to the lateral characteristic scale of jet[11]. However, due to the influence of scale effect, the nonlinear relationship between mixing length and characteristic scale is not clear. Therefore, based on two-strut ejector, three-strut ejector and four-strut ejector, the development of mixing layers at different jet level characteristic scales was compared to determine the relationship between the two.

Considering that most cases of ejection mixing are mixing under pressure mismatch. If the velocity or momentum thickness is used to describe the mixing layer, it is easy to be affected by the expansion and compression of the primary flow and secondary flow, resulting in noticeable errors. Therefore, the definition of scalar mixing layer thickness is used for description[12][13], and the Pitot pressure thickness is used in this study.

## 2. Numerical study

### *2.1. Calculation model and boundary conditions*

The four-strut ejector configuration studied in this paper adopts the configuration used by Ye Wei[8][9], the two-strut ejector adopts the configuration used by Ye We[11], and the three-strut ejector is constructed by the authors using the scaling method. The ejector mixing chamber is rectangular section, width 280mm, height 120mm, water diameter D=168mm. Figure 1. shows the schematic diagram of the diffuser system of multi-strut ejector. This study mainly analyzes the relationship between the non-mixing length (non-mixing length is the length from the position where the secondary flow is completely covered by the mixing layer to the nozzle exit end face) and the transverse characteristic scale of the jet. This part of the flow characteristics mainly occurs in the initial development stage of the ejector mixing layer, so the straight mixing section of the ejector is mainly simulated. The simulation length of the straight section is set to 500 mm, and the length-diameter ratio L/D is 2.976. This configuration is a two-dimensional configuration, and the development of the mixing layer on the central symmetry plane is the focus of this study. The main work is to study the relationship between jet transverse characteristic size and mixing length. Therefore, this study mainly uses two-dimensional simulation method to simulate the central symmetric surface flow in the straight section of the multi-strut ejector system.

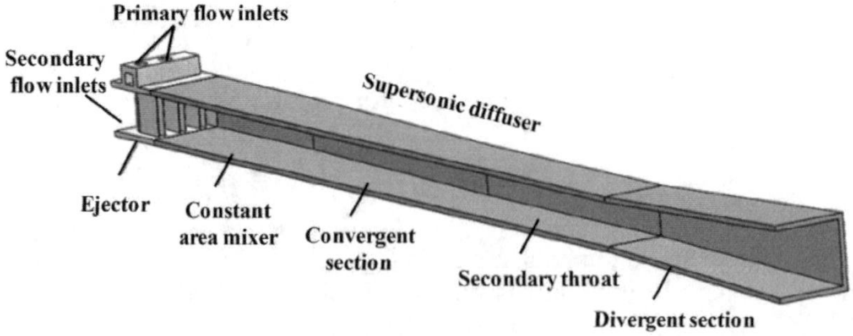


Figure 1. Schematic of the MSE–diffuser system [8].

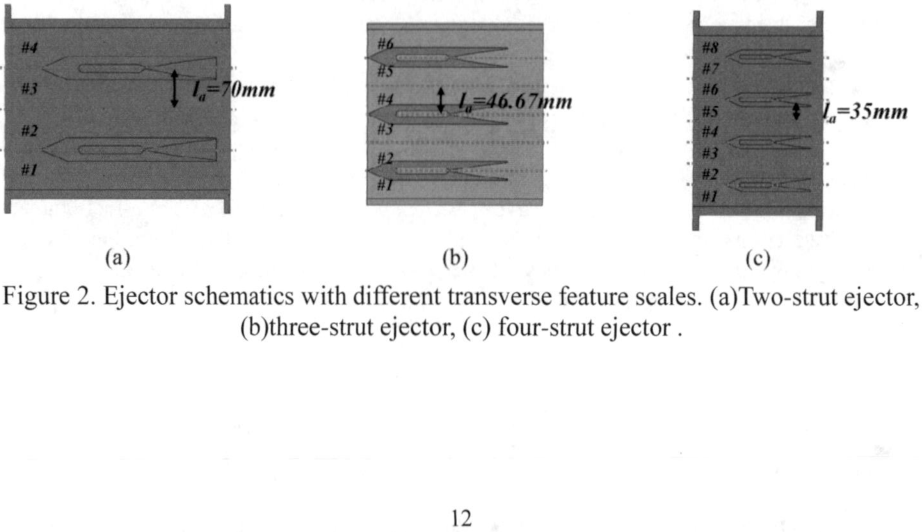


Figure 2. Ejector schematics with different transverse feature scales. (a)Two-strut ejector, (b)three-strut ejector, (c) four-strut ejector .

The ejector section in Figure 1 can be replaced by three ejectors in Figure 2 to complete the replacement of ejectors with different number of struts, so as to obtain jet mixing with different transverse characteristic scales. The transverse characteristic scale is the transverse distance between the primary flow center line and the secondary flow center line, as shown in Figure 2(a), the transverse characteristic scale of the two-strut ejector is la=70 mm.

In order to facilitate comparative analysis, the same boundary conditions are set for primary and secondary flow of different ejectors. Since the secondary throat will isolate the back pressure through the pseudo-shock wave structure, the equal straight section is almost unaffected by the back pressure, so the outlet pressure Pout is set to 0 Pa. The detailed parameters of other specific flow field boundary conditions are shown in Table 1.

Table 1. Summary of the boundary conditions.

| | | $P_0$(pa) | $P$(Pa) | $\dot{m}$ (kg/s) | $T$(K) | $P_{out}$(pa) | $ER$ |
|---|---|---|---|---|---|---|---|
| Two-strut ejector | Primary flow(*Air*) | $3.13\times10^6$ | $3.1\times10^6$ | 5.2099 | 300 | 0 | 0.1513 |
| | Secondary flow(*Air*) | $17\times10^3$ | $13\times10^3$ | 0.7882 | | | |
| Three-strut ejector | Primary flow(*Air*) | $3.13\times10^6$ | $3.1\times10^6$ | 5.9112 | | | 0.1222 |
| | Secondary flow(*Air*) | $17\times10^3$ | $13\times10^3$ | 0.7225 | | | |
| Four-strut ejector | Primary flow(*Air*) | $3.13\times10^6$ | $3.1\times10^6$ | 5.1018 | | | 0.1452 |
| | Secondary flow(*Air*) | $17\times10^3$ | $13\times10^3$ | 0.7406 | | | |

### *2.2. Numerical method*

The calculation of ejector has the characteristics of large computational domain and complex flow structure. If LES or DNS is used for calculation, the cost is too high, so the RANS method with higher cost performance is generally used in previous studies. Turbulent model should be selected for CFD calculation, and the model of $k$-$\varepsilon$ and $k$-$\omega$ is used to solve the overall performance of the ejector, such as flow rate and ejector ratio, with good results [14]. By comparing six different turbulence models, Bartosiewicz and Aidoun [15] found that SST model was the best model to predict the mixing phenomenon of two-stream. Giorgio Besagni and S. Croquer [16] concluded that SST model had excellent performance in predicting global and local flow phenomena. Therefore, this paper uses SST model for simulation calculation.

### *2.3. Meshing generation and grid sensitivity analysis*

Structured grids are used to mesh the near-wall and active-passive mixing center area, $y^+ \approx 8.5$. The two-strut ejector is divided into 644,424 quadrilateral cells, the three-strut ejector is divided into 546,140 quadrilateral cells, and the four-strut ejector is divided into 584,169 quadrilateral cells. Figure 3. describes the grid division of the three ejectors.

Taking the four-strut ejector as an example, the flow field is solved by three sets of grids G1 (494,64 ), G2 (584,169) and G3 (706,377) under the same boundary conditions, and the flow field data under different spatial discretization are obtained. The velocity magnitude of the flow field centerline is extracted, as shown in Figure 4. It is found that the development trend of resultant velocity on the centerline is basically the same, and the difference between the results is small, and important structures of the flow field can be captured, with the maximum error not exceeding 7%. Therefore, considering the calculation cost and accuracy, G2 (584, 169) grid is used in the calculation of the following four-strut ejectors.

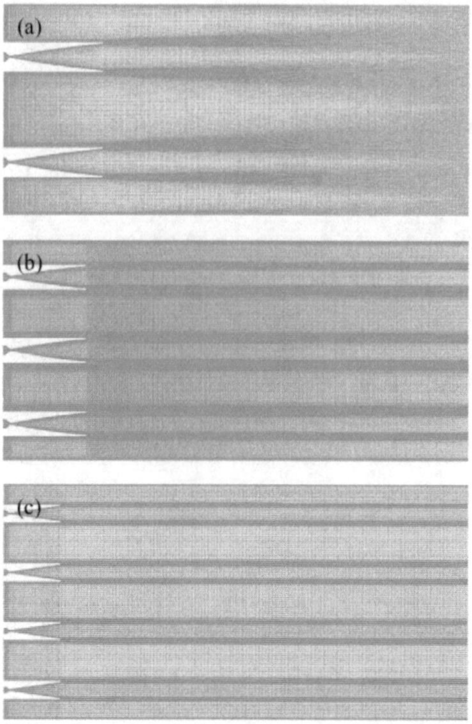


Figure 3. Meshing of the 500 mm straight mixing section of the multi-strut ejector. (a) Two-strut ejector, (b) three-strut ejector, (c) four-strut ejector.

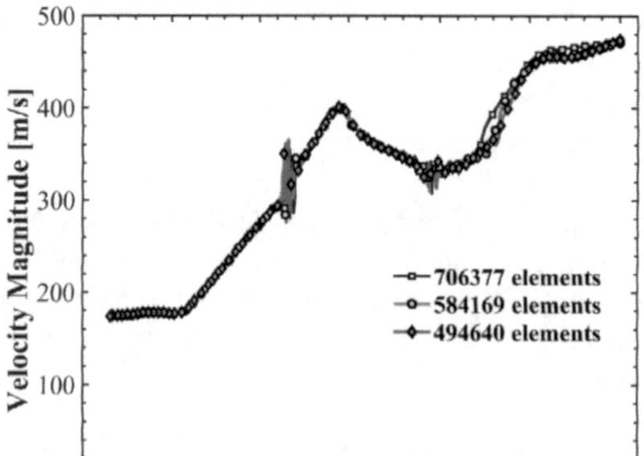


Figure 4. The velocity magnitude curves on the central line of the four-strut ejector with different grid elements.

## 3. Result and Discussion

The flow field structures of the three ejectors in the straight section are calculated in Figure 5, which is a typical multi-jet mixing process, and there is a coupling process of strong expansion, strong shock wave and mixing layer. Whether such flow field structure still maintains a simple linear relationship is worth further analysis. The compression strength in Figure 5. (a), (c), (e) is $U/a\cdot(\nabla p/\|\nabla \rho\|)$.

Considering that the wall will affect the development of the mixing layer, the development characteristics of free shear are mainly concerned. Therefore, the results are analyzed only by averaging the submixing regions of #2 to #3, #2 to #5 and #2 to #7 of the two-strut, three-strut and four-strut ejectors. In this study, the Pitot pressure thickness is used to evaluate the mixing layer, that is, Papamoschou et al. [17] defined the width of the Pitot pressure profile with the free flow difference of 5% to 95% as the Pitot pressure thickness $\delta_{pitot}$ .

As shown in Figure 5., the primary and secondary flows of the multi-strut ejector are divided into multiple jets by the strut for mixing, which has obvious matrix characteristics. The ejector mixing is a mixture under the condition of pressure mismatch in confined space. In addition to the strong shear mixing between jets, there is a strong coupling between shock wave and expansion wave in the flow field. The internal flow field structure of two-strut ejector, three-strut ejector and four-strut ejector is very complex. Therefore, the relationship between the transverse characteristic size of the jet and the mixing length of the flow direction of the jet does not have a clear linear relationship as that of the free jet, and it should be disturbed by the scale effect and the shock wave & expansion wave of the flow field, with obvious nonlinear characteristics.

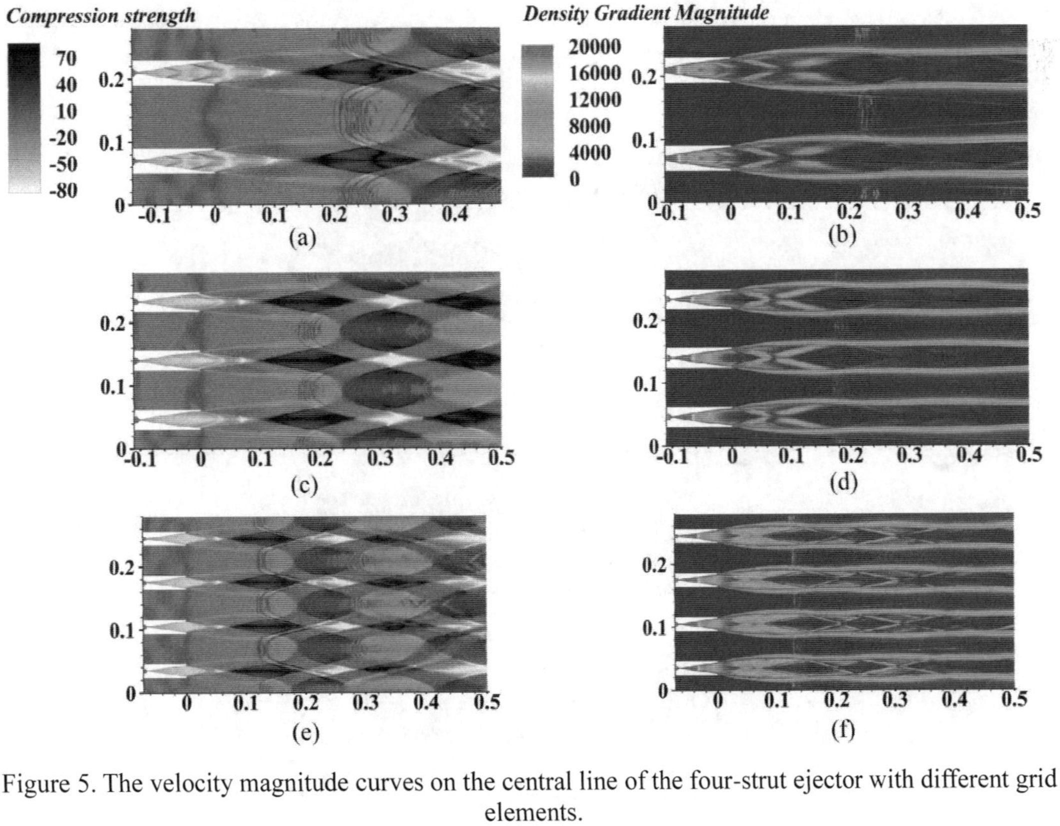


Figure 5. The velocity magnitude curves on the central line of the four-strut ejector with different grid elements.

In this paper, the mixing layer is marked in the flow field through the identification method of the mixing layer of the Pitot pressure thickness, and obtained the Figure 6. Combined with Figure 5., it can be confirmed that jet mixing with different transverse characteristics has certain similarity in macroscopic development. By comparing (a), (b) and (c) in Figure 6., it is found that the jet mixing with the transverse characteristic size develops faster under the same mixing chamber size, and the mixing layer of the two-strut ejector can be regarded as the initial stage of the three-strut ejector and the four-strut ejector. In particular, the flow is relatively simple in the 0–0.2 m section of the straight section of the two-strut ejector, the 0–0.16 m section of the straight section of the three-strut ejector and the 0–0.12 m section of the straight section of the four-strut ejector (expansion only once, occupation of the secondary flow channel, resulting in the compression and growth of the secondary flow). It can be clearly found that the three are linear.

The thickness $\delta_{pitot}$ of the mixing layer and the mixing length of the flow direction L were standardized by the transverse characteristic size la of each jet, and the Figure7. At the mixing position of about 4 times la, the mixing of three jets with different transverse characteristic sizes does have strong consistency. The 4 times la flow direction mixing position corresponds to the 0–0.28 m section of the mixing chamber of the two-strut ejector, the 0–0.187 m section of the mixing chamber of the three-strut ejector and the 0–0.14 m section of the four-strut ejector respectively, and this position is the position where the aerodynamic congestion is completed. After the airflow passes through the choke point, the oblique shock wave begins to enter the mixing layer, resulting in an obvious coupling effect between the shock wave and the mixing layer. Due to the scale effect of jets with different transverse characteristic scales under the same boundary conditions, the shock wave intensity and expansion wave intensity with different intensities are caused, resulting in different coupling effects of mixing layer and shock wave expansion. Therefore, the curve development is obviously different after 4 times A flowing to the mixing position point. Such results are basically consistent with the previous analysis. Therefore, it can be concluded that the jet transverse characteristic scale is consistent with the mixing layer thickness in the mixing process of multi-strut ejector without strong expansion compression coupling effect. There is a linear relationship between jet transverse characteristic scale and mixing length.

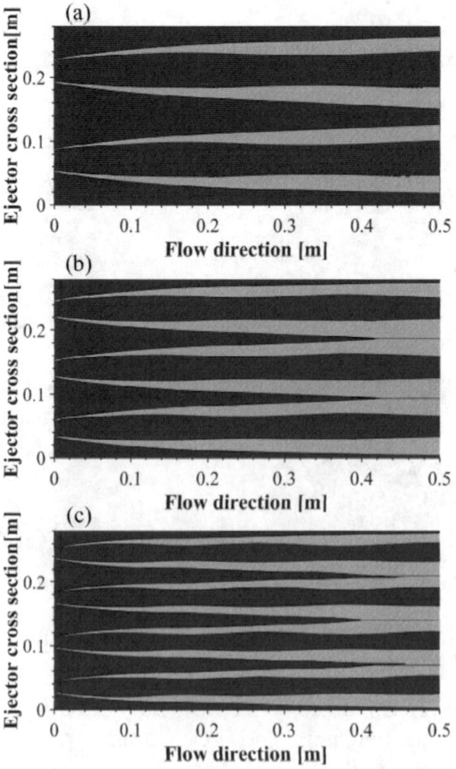


Figure 6. Development of mixing layer in straight section of multi-strut ejector. (a) Two-strut ejector, (b) three-strut ejector, (c) four-strut ejector.

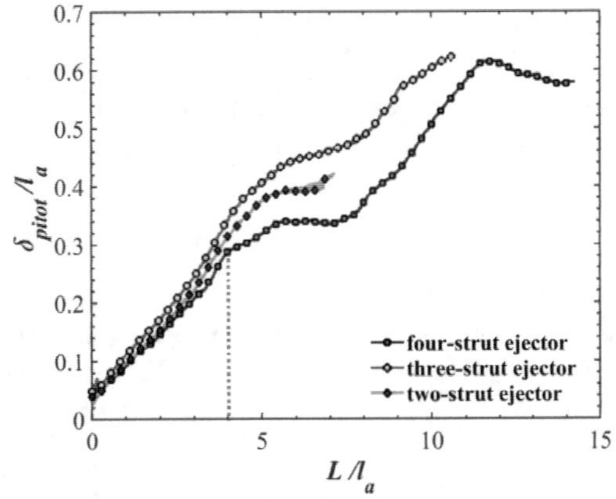


Figure 7. The standardized growth curve of the mixing layer thickness of multi-plate ejector.

## 4. Conclusions

Based on the jet segmentation method, the multi-strut ejector uses the strut to divide the jet into multiple jets with small characteristic sizes for mixing. There is an obvious relationship between the flow structure inside the multi-strut ejector and the number of struts, that is, the transverse characteristic scale of the jet. Since the multi-strut ejector is a new type of ejector, there are relatively few studies on the internal flow phenomenon of multi-strut at present. Therefore, the relationship between jet transverse characteristic size and mixing layer was studied in this study. Through the simulation of two-strut ejector, three-strut ejector and four-strut ejector, the development of mixing layer in the process of jet mixing with different transverse characteristic sizes is compared under the same boundary conditions. The following conclusions are drawn:

- There is no mixing section with obvious coupling effect of mixing layer with shock wave and expansion wave, and there is a linear relationship between jet transverse characteristic scale and mixing length of flow direction.
- In the case of no obvious coupling effect of mixing layer with shock wave and expansion wave in the mixing section, the mixing of jets with different transverse characteristic scales is consistent.
- In the case of no obvious mixing layer coupling effect with shock wave and expansion wave in the mixing section, there is a linear relationship between the transverse characteristic scale of the jet and the mixing length of the flow direction.

## References

[1] Tashtoush, B. M., Al-Nimr, M. D. A., Khasawneh, M. A. (2019) A comprehensive review of ejector design, performance, and applications. Appl. Energy, 240: 138-172.

[2] Eames, . W., Ablwaifa, I, A. E., Petrenko, V. (2007) Results of an experimental study of an advanced jet-pump refrigerator operating with R245fa. Appl. Therm. Eng., 27: 2833-2840.

[3] Etele, J., Sislian, J., Parent, B. (2004) Variable area ejectors for increasing the compression factor of an RBCC engine. AIAA Paper, 2004-3651.

[4] Boreisho, A. S., Khailov, V., Malkov, V., Savin, A. V. (2001) Pressure recovery systems for high-power gas flow chemical lasers. Proc.SPIE, 4184: 401-405.

[5] Reubush, D., Puster, R., Kelly, H. (1987) Modification to the Langley 8-foot high temperature tunnel for hypersonic propulsion testing. AIAA Paper, 87-1887.

[6] Liu, H. (2009) Development of Numerical Method and Investigation on Performances of Supersonic Ejectors. Ph.D. thesis.China Aerodynamic Research and Development Center Graduate School, Mianyang.

[7] Yisheng, Y., Xiqiang, Y., Hequan, J., Zongzheng, L. (2020) Experimental Study on Ejector Vibration Characteristics Face for Advanced Engine Technology. IEEE 7th ICIEA, 674-677.
[8] Ye, W., Xu, W., Li, P., Zhang, F., Zhang, J. (2019) Experimental investigation on the performance of a multi-strut mixing-enhancement ejector. Appl. Therm. Eng., 154: 650-656.
[9] Ye, W., Zhang, F., Xu, W., Zhang, Z. (2020) Numerical investigation on the flow structures of the multi-strut mixing enhancement ejector. Appl. Therm. Eng., 179: 115653.
[10] Urukov, B. A. (1963) Theory of' differentuional ejector. Appl. Mech. Thech. Phys. 5.
[11] Ye, W. (2021) Studies on the Multi-strut ejector designed by jet-segment method. Ph.D. thesis. National University of Defense Technology, Changsha.
[12] Wang, Y., Liang, J., Cai, X., Mahmoudi, Y. (2022) Generation mechanism of a new type of unburnt gas pocket and its influences on the detonation-wave/boundary-layer interaction. Phys. Fluids, 34: 041704.
[13] Cai, X., Deiterding, R., Liang, J., Sun, M., Mahmoudi, Y. (2018) Diffusion and mixing effects in hot jet initiation and propagation of hydrogen detonations. J. Fluid Mech., 836: 324-351.
[14] Tashtoush, B.M., Al-Nimr, M.D.A., Khasawneh, M.A. (2019) A comprehensive review of ejector design, performance, and applications. Appl. Energy, 240: 138-172.
[15] Bartosiewicz, Y., Aidoun, Y., Desevaux, P., Mercadier, Y. (2003) CFD-experiments integration in the evaluation of six turbulence models for supersonic ejectors modeling. Integrating CFD and Experiments Conference.
[16] Besagni, G., Inzoli, F. (2017) Computational fluid-dynamics modeling of supersonic ejectors: Screening of turbulence modeling approaches. Appl. Therm. Eng., 117: 122-144.
[17] Papamoschou, D., Roshko, A. (1988) The compressible turbulent shear layer: an experimental study. J. Fluid Mech., 197: 453-477.

# Fault Detection and Diagnosis of Gear Train Based on Motor Current Signature Analysis

**Junjie Diao[1*], Yuan Tian[1], Yuelong Zhang[1], Shisheng Huang[2], Yijun Xin[1] and Yulong Guo[1]**

[1] School of Mechanical and Electrical Engineering, Shenyang Aerospace University, Shenyang, Liaoning, 110136, China

[2] School of Safety Engineering, Shenyang Aerospace University, Shenyang, Liaoning, 110136, China

*Corresponding author's e-mail: 10422@email.sau.edu.cn

**Abstract.** Gear fault diagnosis generally uses oscillation signals to extract fault features, but the oscillation identify technique is inconvenient to set up sensors and is easily affected by the surrounding and noise. The motor itself has the property of a sensor, and signals such as motor current is able to reflect the change of the load torque. This paper proposes a gear fault detection approach, which applies the Motor Current Signature Analysis (MCSA) to exploit spectral characteristics to recognize malfunction in servo motor drive gears. First, we calculate the nominal revolutions per minute (rpm) to explore frequencies of interest. Second, frequency bands where faulty signals would be appear are constructed. Next, we extract power spectral density (PSD) feature data. Finally, the paper computes spectral metrics for the frequency band of interest. By using two spectral metrics, gear health and failure data are clearly grouped in different areas of the scatter chart. The experimental results give evidence of the effectiveness of analyzing current characteristics of servo motors to classify fault and health data.

## 1. Introduction

Gear transmission mechanism is accurate and reliable, and can transmit a wide range of power and speed, so it has been widely used in modern industry [1]. However, affected by the running state and environment and its complex structure, the gear is easy to fail, resulting in major production accidents and even endangering personal safety. Therefore, the malfunction diagnosis of the gear is highly considerable. At present, the mainstream diagnosis method is mainly vibration diagnosis method, that is, install vibration sensors on the gearbox to collect vibration signals, extract fault features of vibration signals, and realize fault diagnosis [2-6].

Although the vibration diagnosis method developed earlier and the theory is relatively mature, its disadvantages are becoming more and more obvious. The installation of vibration sensor is inconvenient, and the installation position of sensor will have a great impact on the acquisition of vibration signal, and then affect the diagnosis effect. The vibration signal may also be disturbed by mechanical resonance and industrial noise. Moreover, the gearbox will inevitably generate heat in practical application, which also requires considering the temperature application range of the vibration sensor. In contrast, noninvasive detection has more advantages [7], and motor current signature analysis (MCSA) is a good representative [8]. MCSA is a beneficial means for the distinguish of malfunction that cause torque and speed undulations, and has been certificated to be superior for motor malfunction pattern recognition. Gear malfunction diagnosis utilizing conventional vibration instruments can be challenging, especially where instruments with accelerometers or other vibration sensors are not easily accessible to the gear train, such as in the inner workings of those high-risk equipment systems. The motor itself has the characteristics of sensor. The electrical characteristics of the motor can be used to diagnose the faults of the motor body and load [9-11]. In recent years, some scholars also began to use the motor current signal to detect the malfunction of the transmission mechanism. Chen and Yang et al [12,13] realized the fault detection of gear transmission system based on current signal respectively, which is mainly aimed at the broken tooth fault . In 2015, Hou and others [14] realized the extraction of fault features such as misalignment and imbalance of rotor system by using current signal. Aiming at the broken teeth and wear faults in gear transmission system, this paper takes the current signal as the starting point, analyzes the fault characteristics, and realizes fault diagnosis.

## 2. Servo Gear Train

The Futaba S3003 servo includes four groups of nylon gears as exhibited in the Figure 1 below. The figure 2 below summarises the output speed at each gear mesh, the gear mesh frequency and the number of teeth and theoretical values of the cumulative gear reduction. When meshing transmission occurs in the gear system, even if the input torque of the involute gear mechanism remains constant, the output torque must have this certain fluctuation. The fluctuation of the output torque is closely related to the sliding friction coefficient between teeth, manufacturing accuracy and the number of teeth. When the gear fails in the transmission process, it will affect the load torque of the motor. Specifically, the broken tooth fault of the gear will introduce an impact torque; The gear wear will increase the torque fluctuation during gear meshing. When the rotor system operates normally, only one phase current characteristic can be examined according to the three-phase symmetry of the power supply. A total of 10 health were collected prior to the introduction of malfunction into step gears G2 and G3. A total of 10 fault were recorded after the introduction of the malfunction in gears G2 and G3.

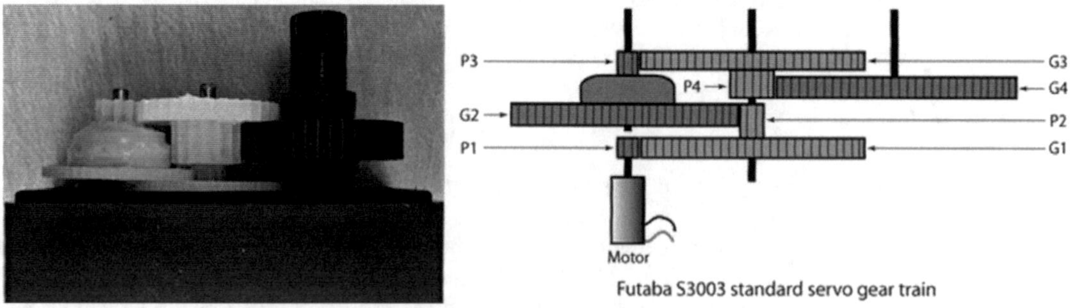


Figure 1 The mechanical schematic of Futaba S3003 servo.

| Pinion | Gear | Pinion Teeth | Gear Teeth | Output Speed (RPM) | Output Speed (Hz) | Gear Mesh Frequency (Hz) | Cumulative Gear Reduction |
|---|---|---|---|---|---|---|---|
| P1 | | 10 | | 13901.6 | 231.69 | | 1 |
| P2 | G1 | 10 | 62 | 2242.2 | 37.37 | 2316.9 | 6.2 |
| P3 | G2 | 10 | 50 | 448.5 | 7.47 | 373.7 | 31 |
| P4 | G3 | 16 | 35 | 128.1 | 2.14 | 74.7 | 108.5 |
| | G4 | | 41 | 50 | 0.83 | 34.2 | 278 |

Theoretical values calculated from tooth count assuming 50 RPM at the output shaft

Figure 2 The tooth count and theoretical values of output speed.

## 3. Simulation Experimental Results and Analysis

All simulation and analysis are based on the MATLAB@R2021a environment. The first column of servoData covers a schedule of 10 health data, however, the second pillar is comprised of a schedule of 10 failure information. Each dataset is comprised of approximately eleven seconds of material sampled at 1500 Hz. Each timetable in servoData involves a dataset, the first column is the tachometer pulses data and the second pillar is the present signal. As shown in Figure 3, the outcome spline has 8 pulses per rotation, corresponding to the 8 caves on the annulus.

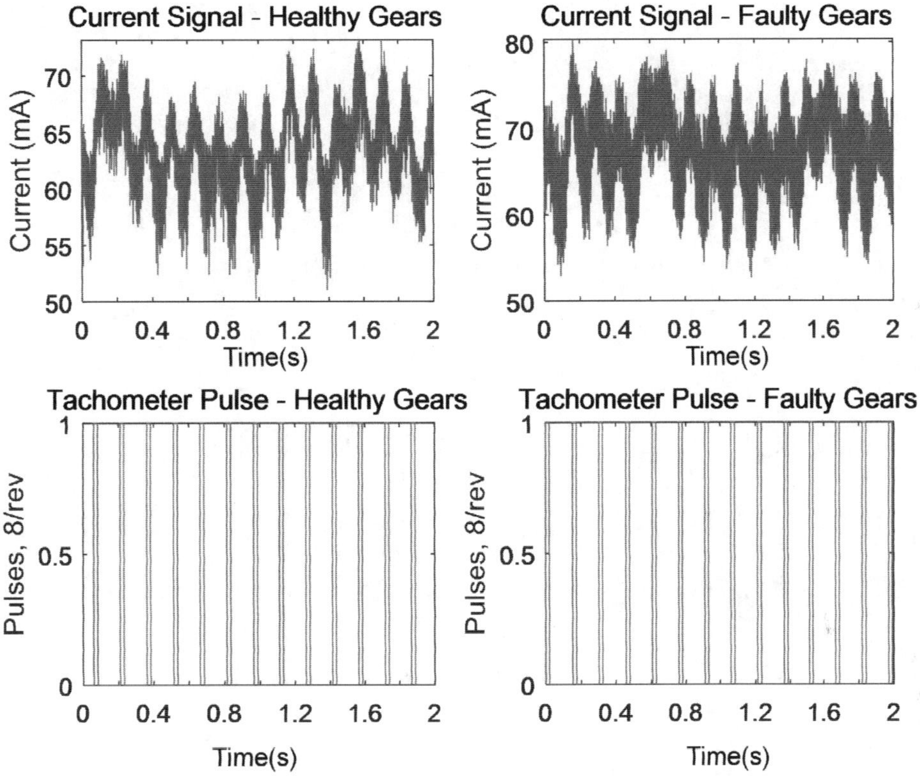


Figure 3 Visualize the current signal and tachometer pulses.

### *3.1. Compute Nominal Rpm*

We calculate the nominal speed to discover the interest frequency in the gear system and match it properly to the frequency on the power spectrum. Utilizing the sampling periodicity cost of 1500 Hz, we display the tachometer semaphore and count the nominal rpm taking advantage of 'tachorpm' function, as shown in Figure 4. Among them, rpm_Healthy = 49.8550; rpm_Faulty = 49.5267. We notice which there is a extremely tiny discrepancy in the output shaft velocity between the normal and abnormal information. The practical nominal rpm measures are still familiar to the theoretical measurements of 50 rpm. Accordingly, this paper premeditates the same measurement of 49.9 rpm for both the healthy and malfunction data resolution.

### *3.2. Establish Frequency Bands*

Establishing frequency bands is a significant prerequisite for figuring the spectral indicator information. Taking advantage of the tooth count of the drive gears in the gear train and the nominal rpm, we primarily estimate the frequencies of interest.

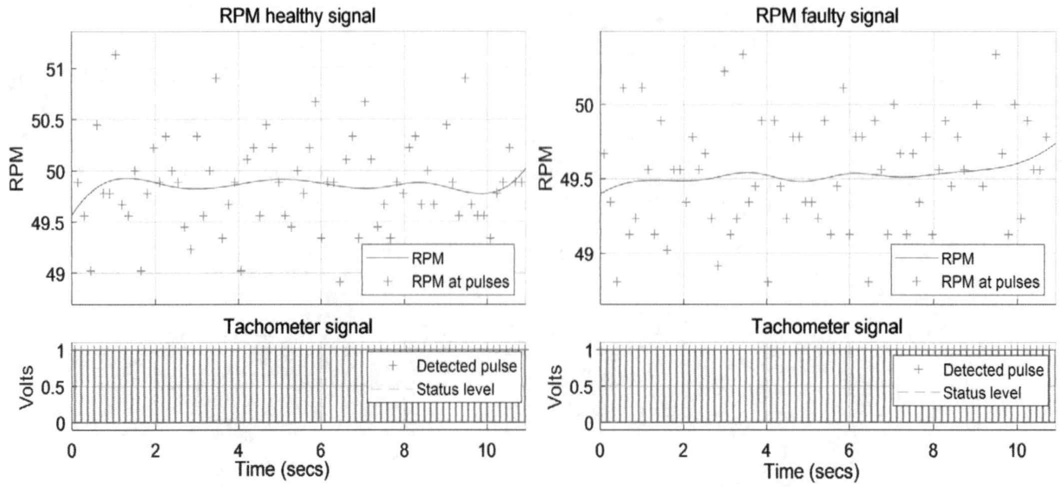


Figure 4 Visualize the current signal and tachometer pulses.

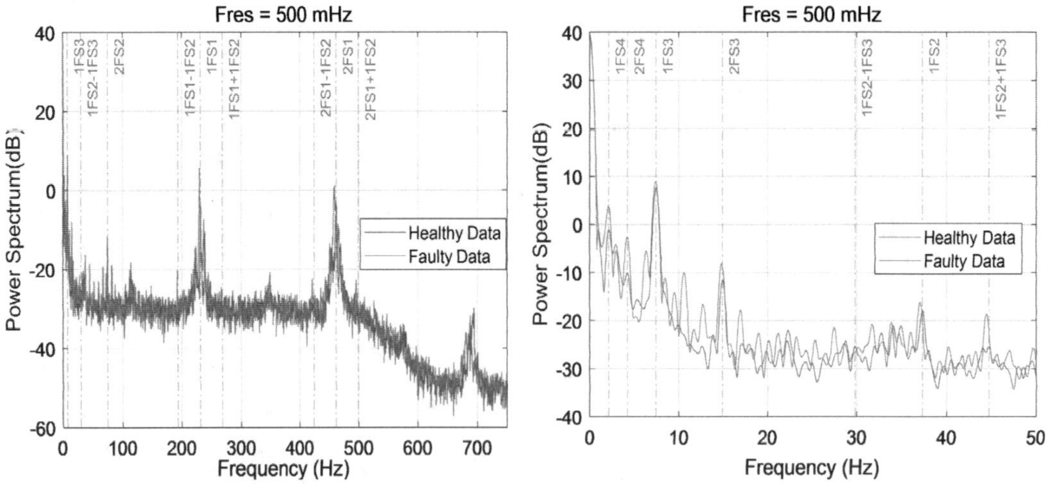


Figure 5 Revealing the power spectrum characteristic of the healthy and malfunction.

The frequency of interest is realistic output velocity value, which is close to the theoretical value enumerated in Figure 2. G4 = 41; G3 equals 35; G2 equals 50; G1 equals 62; P4 is equal to 16; P3 is equal to 10; P2 is equal to 10; P1 is equal to 10. FS5 is avg (rpm) divide by 60; FS4 is G4/P4*FS5; FS3 is G3/P3*FS4; FS1 is equal to G1/P1*FS2; FS is equal to [FS1,FS2,FS3,FS4,FS5] is equal to [231.0207, 37.2614, 7.4523, 2.1292, 0.8309]. Next, we build the frequency bands for the every the output which involve the following frequencies of interest applying 'fault_Bands' function.

### *3.3. Extract Power Spectral Density*

We calculate and exhibit the power spectrum of the normal and fault. In addition, we scheme the frequencies of interest on the spectrum picture by making use of the knowledge in the structure information, as Figure 5. The drawing in blue symbolizes the normal spectrum while the drawing in red represents the spectrum of the malfunction signal. From the painting, we can pay attention the enhancement in oscillations of malfunction frequencies. In other words, the left side of Figure 5 shows the PSD distribution of normal and malfunction information at sampling frequencies from 0 to 700 Hz. It can be seen that most of them overlap and are difficult to distinguish. The right side of Figure 5 shows the PSD distribution of the sampling frequency in a small range of 0-50 Hz. Although its normal data and malfunction data are highly coincident at 8HZ, the cross-fluctuation is obvious in the 20-30 HZ interval, which is beneficial to extract its differentiated features and perform the next pattern classification.

### *3.4. Compute Spectral Metrics*

In this section, using the band and PSD data, we utilize the 'faultBandMetrics' function to compute spectral metrics for health and fault data. The output is a 49×3 metric table of frequency ranges in FB. The elementary column is the metrics (PeakAmplitude-1~16, PeakFrequency-1~16, BandPowe-1~16), the second column contains metrics for health data, and the third column contains metrics for fault data. We investigate that the following coefficients have fairly larger numerical values for the malfunction signal than for the normal signal. PeakAmplitude-2 for the frequency at 231 Hz with a discrepancy of 3.1842 units. Total Band Power with a diversity of 6.01 units. Consequently, we employ The above two categories index to generate a scatter chart to distinguish the malfunction and wholesome signal respectively, as Figure 6.

**Table 1.** Spectral metrics of frequency ranges in FB.

| Spectral Metrics | Healthy Data | Faulty Data |
|---|---|---|
| PeakAmplitude-1 | 0.0020712 | 0.009804 |
| PeakFrequency-1 | 193.75 | 192.44 |
| BandPower-1 | 0.010881 | 0.017916 |
| PeakAmplitude-2 | 0.50813 | 3.6921 |
| PeakFrequency-2 | 231.06 | 229.44 |
| BandPower-2 | 0.46652 | 2.9975 |
| PeakAmplitude-3 | 0.0019579 | 0.0035204 |
| PeakFrequency-3 | 272.5 | 266.44 |
| BandPower-3 | 0.011841 | 0.015639 |
| ... | ... | ... |
| PeakAmplitude-16 | 0.09849 | 0.55167 |
| PeakFrequency-16 | 4.25 | 4.25 |
| BandPower-16 | 0.01806 | 0.10029 |
| Total Band Power | 3.9737 | 9.9838 |

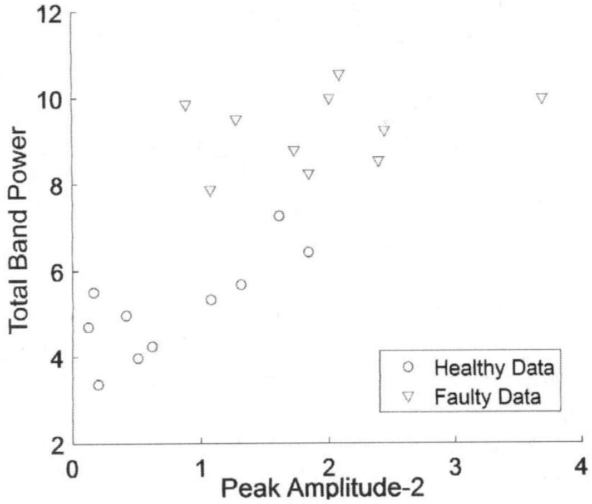


Figure 6 The gear failure and health data classification result.

## 4. Conclusion

We proposes a approach for applying the Motor Current Signature Analysis (MCSA) to extract spectral characteristics to distinguish malfunction in servo motor gears. This paper calculates the nominal rotational speed, constructs the frequency bands where potential malfunction data may exist, extracts the power spectral density, and calculates the spectral metric for the frequency band of interest. By using two spectral characteristic metrics (i.e., PeakAmplitude-2 and Total Band Power), gear health and failure data are clearly grouped in different regions of the scatter plot. The experimental results further prove the effectiveness of analyzing the current characteristics of the servo motor for gear fault diagnosis and identification.

## References

[1] H. Henao, S. H. Kia and G. Capolino, Torsional-Vibration Assessment and Gear-Fault Diagnosis in Railway Traction System, in IEEE Transactions on Industrial Electronics, vol. 58, no. 5, pp. 1707-1717, 2011, doi: 10.1109/TIE.2011.2106094.

[2] Guo, Y., Yan, W., Bao, Z., Gear fault diagnosis of wind turbine based on discrete wavelet transform. Proceedings of the World Congress on Intelligent Control and Automation (WCICA), no. 5554606, pp. 5804-5808, 2010.

[3] Z. Li, J. Zuo, J. Wu, Z. Liu, Y. Song, and S. Geng. (2021) Application of Decision Tree Integrated Hybrid Classifier in Feature-Fused Robot Big Data. In Proceedings of the 2021 International Conference on Control and Intelligent Robotics (ICCIR 2021). ACM, pp. 681-686.

[4] L. Han, Gear fault detection and diagnosis based on Hilbert Huang transform. Proceedings 2010 3rd International Congress on Image and Signal Processing, CISP 2010, vol. 7, no. 5647578, pp. 3323-3326, 2010.

[5] Shufeng, A., Hui, L., Application of order cepstrum and neural network to gear fault detection. IMACS Multiconference on "Computational Engineering in Systems Applications", CESA, no. 4105675, pp. 1822-1827, 2006.

[6] Z. Li, J. Zuo, J. Wu, Y. Song, S. Geng and J. Li, Deep Convolution and Gated Recurrent Unit Network for Robot Perceptual Intelligent Recognition, 2021 6th International Conference on Intelligent Computing and Signal Processing (ICSP), 2021, pp. 792-795.

[7] S. H. Kia, H. Henao, G. A. Capolino. A comparative study of acoustic, vibration and stator current signatures for gear tooth fault diagnosis. Proceedings- 2012 20th International Conference on Electrical Machines, ICEM 2012, no. 6350079, pp. 1514-1519, 2012.

[8] Jung, J.-H., Lee, J.-J., Kwon, B.-H. Online diagnosis of induction motors using MCSA. IEEE Transactions on Industrial Electronics, vol.53, no. 6, pp:1842-1852, 2006.

[9] K. K. Huh , R. D. Lorenz, N. J. Nagel. Gear fault diagnostics integrated in the motion servo drive for electromechanical actuators. IEEE Transactions on Industry Applications, vol. 48, no.1, pp. 142-150, 2012.

[10] C. M. Wolf, K. M. Hanson, R. D. Lorenz and M. A. Valenzuela, "Using the Traction Drive as the Sensor to Evaluate and Track Deterioration in Electrified Vehicle Gearboxes," in IEEE Transactions on Industry Applications, vol. 49, pp. 2610-2618, 2013.

[11] C. M. Wolf and R. D. Lorenz, "Using the Motor Drive as a Sensor to Extract Spatially Dependent Information for Motion Control Applications," in IEEE Transactions on Industry Applications, vol. 47, no. 3, pp. 1344-1351, 2011, doi: 10.1109/TIA.2011.2126014.

[12] Z. Chen, T. Wang, F. Gu, et al. Gear transmission fault diagnosis based on the bispectrum analysis of induction motor current signature. Journal of Mechanical Engineering, vol. 48 no. 21, pp. 84-90, 2012.

[13] M. Yang , G. Li , C. Dong , et al. Gear fault diagnosis method based on MCSA. Journal of Beijing Jiaotong University, vol. 39, no. 5, pp. 87-91, 2015.

[14] J. Hou, H. Ding, Z. Yang. Fault diagnosis of rotor bearing system based on stator current. Coal Technology, vol. 34, no. 7, pp. 268-271, 2015.

Journal of Physics: Conference Series **2364** (2022) 012004 doi:10.1088/1742-6596/2364/1/012004

# Noise simulation of wake field of solid rocket motor

**Yuhang Gao**[1], **Jian Zheng**[1*]

[1] School of Mechanical Engineering, Nanjing University of Science and Technology, Nanjing, Jiangsu, 210094, China

[*]Corresponding author's e-mail: zhengjian@njust.edu.cn

**Abstract.** When the rocket engine works, it will produce high-temperature and high-speed gas, which will be rapidly mixed with the surrounding medium, and will produce huge jet noise. This paper studies jet flow field and sound field of rocket engine using Fluent. The broadband noise simulation shows that the noise distribution is related to the nozzle jet. The sound pressure data of transient FW-H observation points show that the noise frequency distribution is wide. At the same time, the size of the noise will increase with the increase of the inlet pressure, and the influence range will become larger with the increase of area expansion ratio.

## 1. Introduction

When the solid rocket engine works, the high-temperature and high-pressure gas flow will be compressed and expanded into the atmosphere through the Laval nozzle at the rear of the engine, and the jet will generate a strong thrust to propel the rocket. When the high-temperature and high-pressure gas enters the air, it will be violently mixed with the surrounding atmosphere at the end of the jet velocity[1]. The air vibrates in the collision between the gas and the outside world, and the formed physical wave continuously propagates backward in the air, thereby producing a strong jet noise. The site sound pressure level of the jet at the end of the rocket can reach an intensity of 180dB. Such intensity of noise will not only cause irreversible damage to the human ear. Moreover, the strong noise will continuously impact the rocket body through vibration, causing damage to the rocket body and the important electronic components carried inside [2]. The noise will also interfere with the precision measuring instruments of the ground console or monitoring station, making it impossible to obtain accurate rocket launch data. In noise research. Korean scholars Zhenglie Li and Zheng Li have investigated the supersonic jet of the engine, and designed and evaluated a SRM noise reduction design and manufacturing technology. Michael Muhlestein,Kent L Gee et al. made a prediction of noise propagation based on generalized Burgers equation for the nonlinearity of solid rocket motor noise [3]. At present, the noise source of the engine mainly comes from the high temperature and high-pressure gas jet ejected by the nozzle in the outer flow field. Therefore, use the simulation of supersonic jet to study the characteristics of noise propagation in the external flow field and the fundamental factors of noise generation, so as to replace complicated and expensive engineering experiments. The validity and success rate of the experiment can be guaranteed, it is of great significance to study the distribution of the external flow field and sound field of the rocket engine for the control of the flame jet during the rocket launch and to reduce the impact of the high temperature and high-pressure gas noise of the jet on the console and personnel around the rocket launch.

## 2. Calculation method

*2.1 Computational fluid dynamics*

In this paper, the fluent commercial software is used to solve the governing equations, continuous equations, momentum equations, and energy equations of fluid flow. Calculate the basic physical quantities of the flow field. Summary of Unsteady 3D Compressible Viscous Flow Equations:

1. Continuity equation

$$\frac{\partial \rho}{\partial t}+\nabla\cdot(\rho\vec{V})=0 \tag{1}$$

2. Momentum equation

$$\begin{aligned}
\frac{\partial(\rho u)}{\partial t}+\nabla\cdot(\rho uV)&=-\frac{\partial P}{\partial x}+\frac{\partial\tau_{xx}}{\partial x}+\frac{\partial\tau_{yx}}{\partial_y}+\frac{\partial\tau_{zx}}{\partial_z}+\rho f_x\\
\frac{\partial(\rho v)}{\partial t}+\nabla\cdot(\rho vV)&=-\frac{\partial P}{\partial y}+\frac{\partial\tau_{xy}}{\partial x}+\frac{\partial\tau_{yy}}{\partial_y}+\frac{\partial\tau_{zy}}{\partial_z}+\rho f_y\\
\frac{\partial(\rho w)}{\partial t}+\nabla\cdot(\rho wV)&=-\frac{\partial P}{\partial z}+\frac{\partial\tau_{xz}}{\partial x}+\frac{\partial\tau_{yz}}{\partial_y}+\frac{\partial\tau_{zz}}{\partial_z}+\rho f_z
\end{aligned} \tag{2}$$

3. Energy equation

$$\begin{aligned}
&\frac{\partial \rho e}{\partial t}+\nabla\cdot(\rho eV)=P\dot{q}-\left(\frac{\partial}{\partial x}\left(-k\frac{\partial T}{\partial x}\right)+\frac{\partial}{\partial y}\left(-k\frac{\partial T}{\partial y}\right)+\frac{\partial}{\partial z}\left(-k\frac{\partial T}{\partial z}\right)\right)\\
&+\tau_{xx}\frac{\partial u}{\partial x}+\tau_{yx}\frac{\partial u}{\partial y}+\tau_{zx}\frac{\partial u}{\partial z}-P_x\frac{\partial u}{\partial x}\\
&+\tau_{yy}\frac{\partial v}{\partial y}+\tau_{xy}\frac{\partial v}{\partial x}+\tau_{zy}\frac{\partial v}{\partial z}-P_y\frac{\partial v}{\partial y}\\
&+\tau_{zz}\frac{\partial w}{\partial z}+\tau_{yz}\frac{\partial w}{\partial y}+\tau_{xz}\frac{\partial w}{\partial x}-P_z\frac{\partial w}{\partial z}
\end{aligned} \tag{3}$$

The calculation adopts the Standard k-epsilon model:

The standard model was proposed by Launder and Spalding. The standard model obtains the solutions of k and e by solving the turbulent kinetic energy (k) equation and the turbulent dissipation rate (e) equation, and then uses the values of k and e to calculate the turbulent viscosity, the solution of Reynolds stress is obtained by the Boussinesq assumption[4].

$$\begin{aligned}
&\frac{\partial}{\partial t}(\rho k)+\frac{\partial}{\partial x_i}(\rho k u_i)=\frac{\partial}{\partial x_j}\left[\left(\mu+\frac{\mu_t}{\sigma_k}\right)\frac{\partial k}{\partial x_j}\right]+G_K+G_b-\rho\varepsilon-Y_M+S_K\\
&\frac{\partial}{\partial t}(\rho\varepsilon)+\frac{\partial}{\partial x_i}(\rho\varepsilon u_i)=\frac{\partial}{\partial x_j}\left[\left(\mu+\frac{\mu_t}{\sigma_\varepsilon}\right)\frac{\partial\varepsilon}{\partial x_j}\right]+G_{1\varepsilon}\frac{\varepsilon}{k}(G_k+G_bC_{3\varepsilon})-C_{2\varepsilon}\rho\frac{\varepsilon^2}{k}+S_\varepsilon
\end{aligned} \tag{4}$$

Where $G_k$ is Turbulent kinetic energy produced by laminar flow, $G_b$ is Turbulent kinetic energy produced by buoyancy, $Y_m$ is A wave produced by the diffusion of a transition occurring in a compressible flow, $C_1C_2C_3$ is constant, $\sigma_k\sigma_e$ is the turbulent Prandtl number for the k and e equations.

*2.2 Broadband Noise*

The generation of fluid noise is explained by the Lighthill equation. The Lighthill equation is derived by solving the flow and continuity equations of a fluid. Its first term represents the propagation of noise due to a change in density. The second two terms of the equation are the sources of turbulent noise, most of the noise comes from the four-level sub-noise, which is caused by the interaction of jet and air, and a small part comes from the dipole noise. In fluid flow, dipole noise is caused by the interaction of the flow forces on the surface of the object and the pressure fluctuations caused by the reaction on the surface of the object[5].

$$\frac{\partial^2}{\partial t^2}\rho - c\nabla^2\rho = \frac{\partial}{\partial x_i}\frac{\partial}{\partial_{xj}}\rho v_i v_j - \frac{\partial}{\partial x_i}F_i \tag{5}$$

*2.3 FW-H noise model*

Jet noise during a rocket launch is caused by where the gas jet meets the air. The main source of noise is from the vortex flow generated at the junction position, where the interaction force is due. The sound field calculation is based on the FW-H equation and requires a transient calculation, In the first step, the results of steady-state flow calculation are converged by the conventional turbulence model, Use the initial data of the steady-state cloud map after the convergence and stabilization to perform the transient calculation again. The FW-H method needs to first select the receiving position of the sound field in the external flow field. In the transient calculation, the effective sound field parameter information is continuously obtained with the progress of the time step.

The FW-H equation is a non-homogeneous wave equation derived from the continuity equation and N-S equations[6].

$$\begin{aligned}\frac{1}{a_0^2}\frac{\partial^2 p'}{\partial t^2} - \nabla^2 P' &= \frac{\partial^2}{\partial x_i \partial x_j}\left[T_{ij}H(f)\right] - \frac{\partial}{\partial x_i}\left\{\left[p_{ij}n_j + \rho u_i(u_n - v_n)\right]\delta(f)\right\} \\ &+ \frac{\partial}{\partial t}\left\{\left[\rho_0 v_n + \rho(u_n - v_n)\right]\delta(f)\right\}\end{aligned} \tag{6}$$

Where $u_i$ is the velocity component in the direction, $u_n$ is normal component of fluid velocity in the $f = 0$ plane, $v_i$ is The velocity component above the $x_i$ direction, $v_n$ is the normal component of the velocity above the plane, $\delta(f)$ is the $\delta$ function, $H(f)$ is the far-field pressure, $p'$ is the far-field pressure, $p' = p - p_0$ ,The $n_i$ is out-of-plane $(f > 0)$ direction vector for pointing,

$a_0$ is far-field sound velocity, $T_{ij}$ is Lighthill stress tensor, $p_{ij}$ is the compressible stress tensor.

$$\begin{aligned} T_{ij} &= \rho u_i u_j + p_{ij} - a_0^2(\rho - \rho_0)\delta_{ij} \\ p_{ij} &= p\delta_{ij} - \mu\left[\frac{\partial u_i}{\partial x_j} + \frac{\partial u_j}{\partial x_i} - \frac{2}{3}\frac{\partial u_k}{\partial x_i}\delta_{ij}\right]\end{aligned} \tag{7}$$

## 3. Analysis of three-dimensional nozzle flow field

*3.1 computational model*

The nozzle model and boundary conditions are shown in Figure 1 and Table 1. In the selection of the external flow field, considering the broadness of the sound field propagation range, 16 times the length of the nozzle is selected for the length of the external flow field.,Ensure that the sound field information can be clearly displayed. 19 times the nozzle inlet radius was chosen in terms of the width of the outer flow field.

Table 1. Nozzle and external flow field parameters(unit:mm)

| Inlet radius | nozzle throat | Outlet radius | Nozzle total length | Diameter of flow field | total length |
|---|---|---|---|---|---|
| 97.50 | 46 | 115.96 | 502.79 | 1095.75 | 8000 |

In this paper, the 3D nozzle and its external flow field are meshed by ICEM,Using a structured grid, the grid is divided as shown in Figure 2 below.

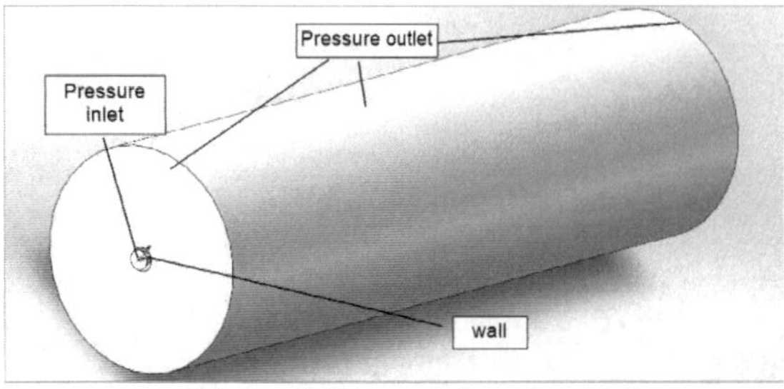


Figure 1 Computational model

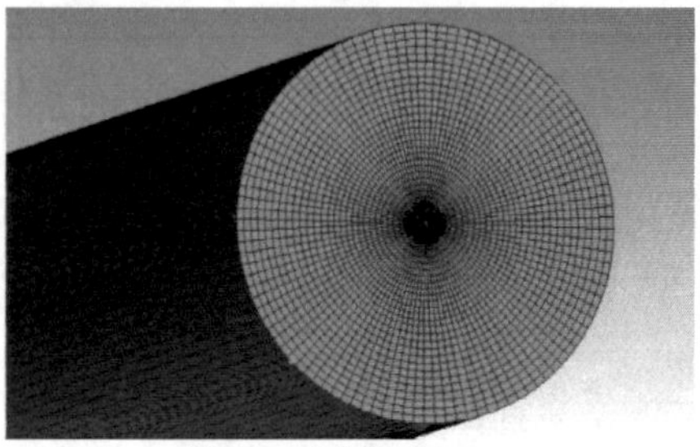

Figure 2 Meshing

*3.2 Broadband Noise Analysis of Flow Field Outside Nozzle*

The velocity cloud of the external flow field is shown in Figure 3 below,The maximum speed of the nozzle outlet can reach 3023m/s,The powerful jet will propagate backward along the axis from the center of the nozzle, which is in line with the actual working status of the flow field outside the nozzle.

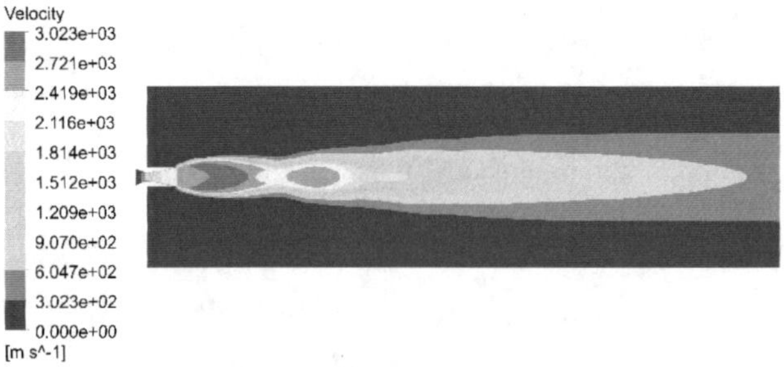


Figure 3 Nozzle velocity contour

Sound pressure is commonly used to measure the size of a sound,Sound pressure is the ratio of the change in actual sound field pressure to a defined fixed reference pressure. is a relative magnitude.

$$SPL = 10 * \log(\text{p}/\text{pref})^2 = 20\log(\text{p}/\text{pref}) \tag{8}$$

In this paper,select the sound pressure reference value is pref=2e-5.

As shown in Figure 4 below,The strong turbulent jet can bring many vortices and produce four-order sub-noise,You can see it in the cloud,The broadband noise is generated along the direction of 45° from the exit position of the jet and continues to propagate backwards.It can be seen that the broadband noise at the end of the wake jet reaches a peak of 145dB and surrounds the tail flame in the 3D simulation, This is where the turbulence of the jet interacts most strongly with the air. It is also found that the decibel region of the small sound field at the jet apex is caused by the relatively small turbulent kinetic energy at this point. The fourth-order subnoise propagation source is shown in figure 5 below,The turbulent noise is mainly concentrated in the place where the four-level sub-noise is generated. The place where the difference between the high-speed jet and the low-speed jet is the largest will produce strong vibration, These are the four levels of noise generated by turbulence.

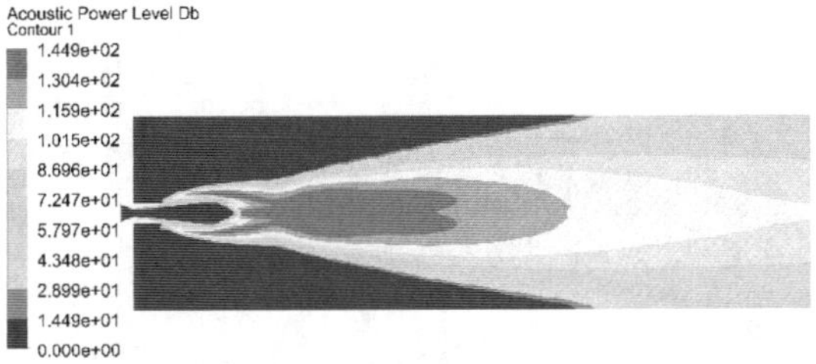


Figure 4 Outer flow field broadband noise

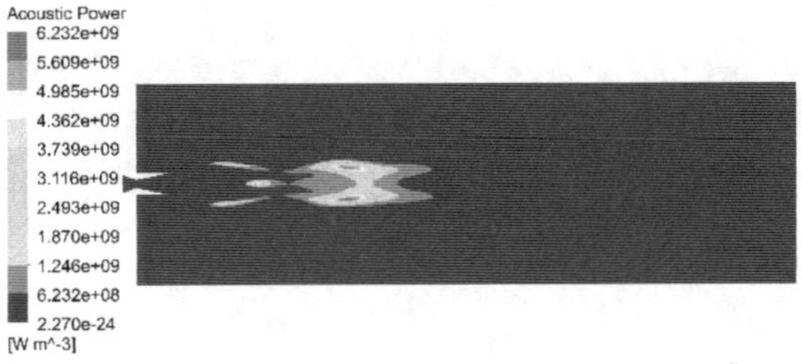


Figure 5 Fourth-order sub-noise

*3.3 FW-H sound pressure and frequency*

In the three-dimensional model, the transient calculation is carried out by using the convergent steady state as the initial condition. When setting the transient calculation, set the time step to 1e-5s. The maximum frequency of 25000HZ can be obtained. In the transient calculation of sound field, it is necessary to set the receiving position of the sound field first. Through figure 6, set the position of receiving the sound field signal as shown in the figure, and select the special point of the sound pressure position for analysis. Select the position of the central axis and the 45° angle propagation position as the analysis positions of sound pressure and frequency.

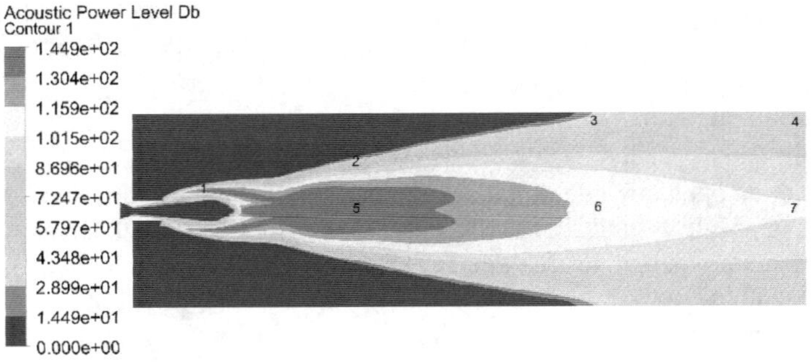


Figure 6 Measurement point location

As shown in Figure 7 below, the relationship between sound pressure and frequency along the 45° angle of propagation is shown in the figure,The results show that the maximum sound pressure in the 45° propagation direction is concentrated in the high frequency region of 0Hz-10000Hz. At the same time, in the frequency range of 10000Hz-20000Hz, the sound pressure value is similar, and there is no big difference with the change of distance. The frequency range of 0Hz-7000Hz and higher than 18000Hz at the position of point 1 presents a high decibel noise different from the other three points. A sound pressure range smaller than the reference sound pressure appears in the four-point portion. The remaining three points overall reflect the high sound pressure level noise in the range of frequencies below 10,000 Hz and above 20,000 Hz. The sound pressure at the same frequency is consistent with the result of wide-frequency noise decreasing along the direction of 45° sound field.

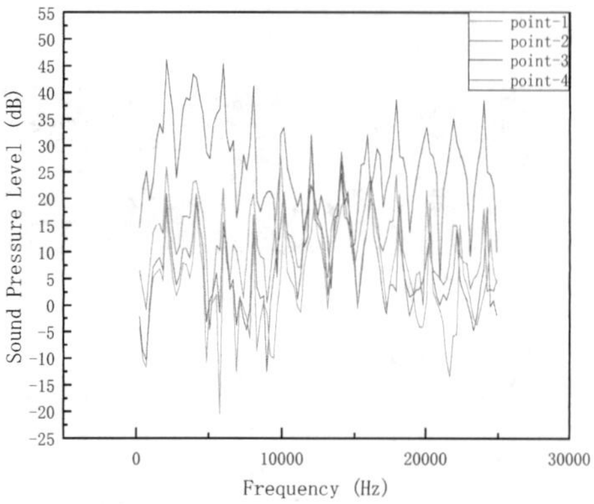


Figure 7 45° angle measuring point

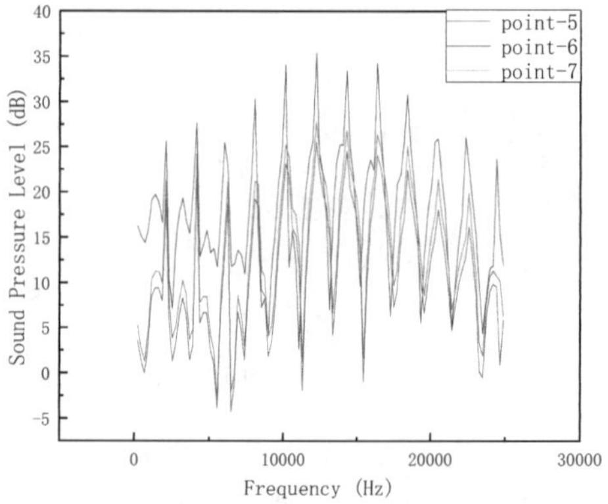


Figure 8 Central axis measurement point

As shown in Figure 8 above, the relationship between sound pressure and frequency along the axis of propagation is shown in the figure. The results show that the sound pressure on the same axis line has similarities in the range of frequency and sound pressure. The whole shows a consistent relationship of frequency and sound pressure,At 10,000 Hz-20,000 Hz the sound pressure is high, and at positions below 10,000 Hz and above 2,000 Hz the sound pressure is low. At the same time, the relationship between frequency and sound pressure also satisfies the broadband noise distribution law that decreases with the increase of the axis distance.

## 4. Effect of nozzle parameters on acoustic field

### *4.1 Effect of inlet pressure on external flow field noise*

By continuously reducing the inlet pressure, Figure 9 shows the broadband analysis of the noise of the flow field outside the nozzle with 6MPa, 8MPa, and 10MPa working pressure. As can be seen from the figure, with the continuous increase of the working pressure, The range occupied by high-decibel sound fields is increasing, and the range of external flow field radiation is also increasing. At the nozzle outlet position, the sound pressure of the nozzles with 6MPa and 8MPa working pressures is significantly lower than that of 10MPa working pressures.

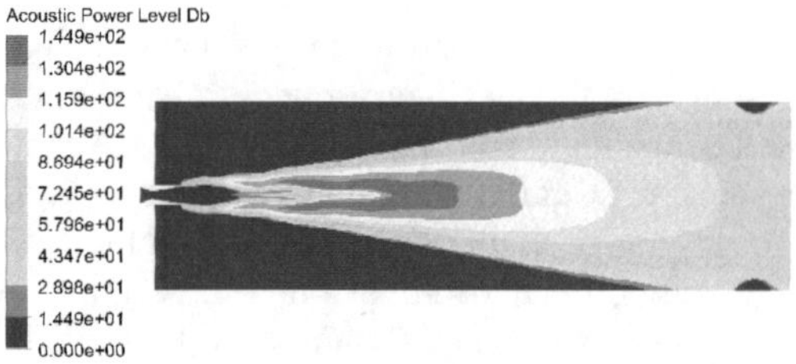


(a) 6MPa

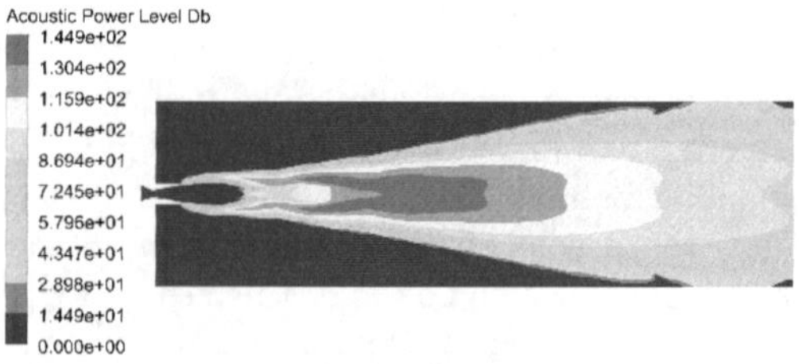


(b) 8MPa

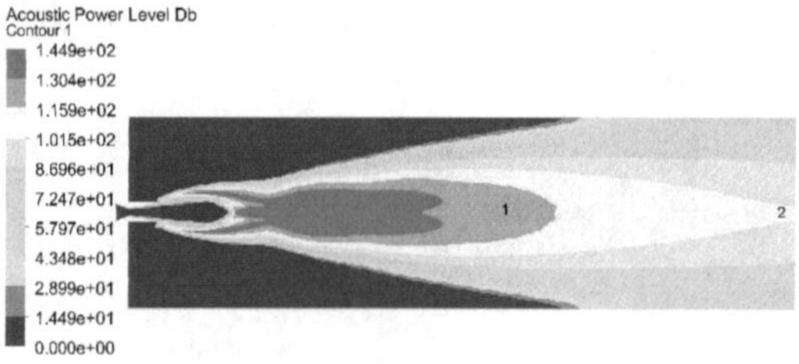


(c) 10MPa

Figure 9 Broadband noise at different working pressures

Sound pressure frequency analysis for the same position of three nozzles,The selected points are shown in Figure 9(c) above, which are located at 3m and 7.8m from the exit of the axial nozzle respectively. The relationship between sound pressure and frequency at two points is shown in Figure 10. It can be seen from the figure that the frequency and sound pressure distribution of different inlet pressures at the same position are similar,It is more obvious at point 2. At the same time, at point one, the sound pressure value of the 10MPa working pressure nozzle is significantly greater than that of 6MPa and 8MPa.

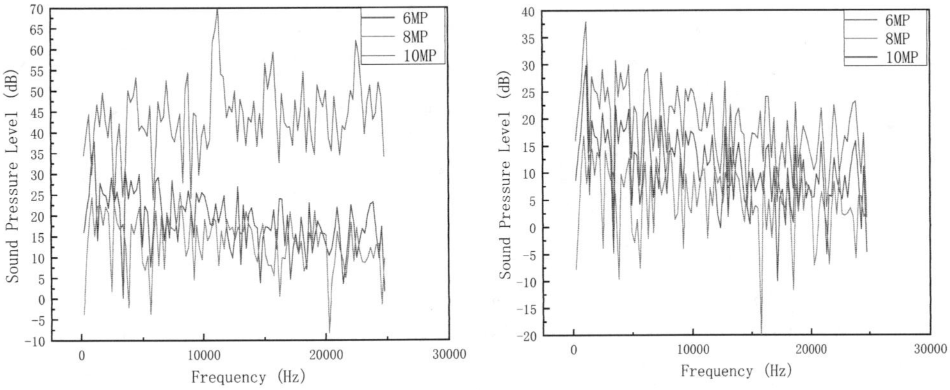


（a） point-1 （b） point-2

Figure 10 Measurement point sound pressure frequency comparison

*4.2 Effect of nozzle outlet area on external flow field noise*

The throat area and the total length of the nozzle remain constant. By changing the nozzle outlet area, change the size of the nozzle expansion ratio from 3.83 to 6.83. It can be clearly seen from the Figure 11 that as the expansion ratio increases, the influence range of high-decibel broadband noise continues to expand. The 5.83 expansion ratio is compared with the nozzles with the 3.83 and 6.83 expansion ratios, and the 100dB noise influence range is larger. Even reaching the end of the wake field. At the same time, as the expansion ratio continues to increase, the influence angle of broadband noise also decreases.

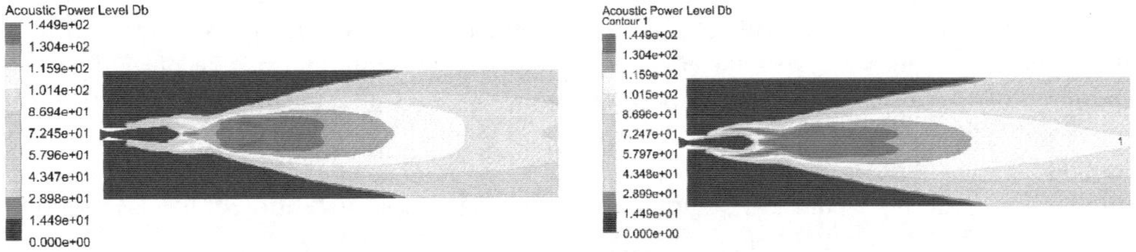


(a) 3.83 (b) 5.83

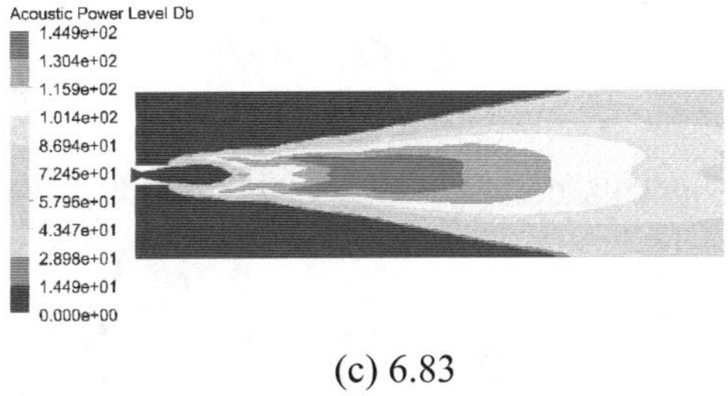


(c) 6.83

Figure 11 Broadband noise with different expansion ratios

Considering the different distribution of broadband noise at the end,The sound pressure frequency is measured and compared at 7.8m from the nozzle outlet, and the position of the measurement point is shown in Figure 11(b). Point-1 sound pressure frequency relationship is shown in Figure 12.It can be seen that the sound pressure frequency of 5.83 expansion ratio is higher than the sound pressure frequency of 3.83 and 6.83 expansion ratio,Overall sound pressure distribution High sound pressure is concentrated in the low frequency region of 0-10000Hz. As the frequency increases, the sound pressure value gradually decreases. At the same time, the sound pressure frequency distributions of different expansion ratios are similar.

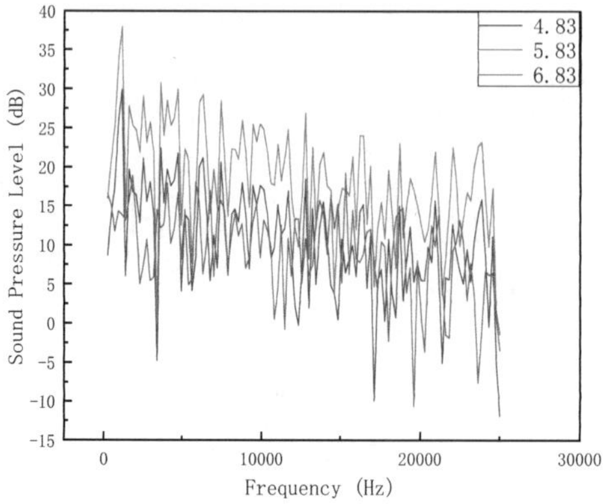


Figure 12 Point-1 sound pressure frequency relationship

## 5. Conclusion

1. The overall propagation of the sound field intensity of the outer flow field of the nozzle is related to the flame of the outer flow field,Specifically, it is related to the four-level sub-noise generated by the turbulent flow at the nozzle exit. At the outermost end of the nozzle flame is the maximum value of the sound field, which is about 144 dB.At the same horizontal or vertical position, the farther away from the nozzle, the smaller the sound pressure.

2. In the frequency graph, it can also seen that the noise frequency is broadband. Point-1 on the 45° propagation line shows that the sound pressure value of 10000-2000Hz is lower than the sound pressure of the frequencies on both sides.The remaining points at 45° and on the central axis show a sound pressure value of 10000-20000Hz that is higher than the sound pressure of the frequencies on both sides.

3. The noise distribution under different nozzle parameters is analyzed. The sound pressure value of broadband noise will increase with the increase of working pressure and nozzle outlet area. But the distribution of noise frequency and sound pressure at the same position is similar.

## References

[1] PEN wenjun,(2019) Numerical Simulation of Gas Jet Noise in Solid Rocket Motor[J].Digital Ocean & Underwater Warfare, 2(02):56-59.
[2] LIU xinqiang,HUANG wenchao,LI hongli,(2019) Simulation calculation of jet aerodynamic noise [J]. Science Technology and Engineering,338-342.
[3] Michael Muhlestein,Kent L. Gee,Tracianne B. Neilsen,Derek C. Thomas.(2013) Prediction of nonlinear propagation of noise from a solid rocket motor[J]. Proceedings of Meetings on Acoustics ,18(1).
[4] ZHOU zhipeng,WANG hongping,WANG shizhao.(2020) HE guowei,FW-H Integral Quadrupole Acoustic Source Correction Model Based on Correlation Velocity[J]. Acta Aerodynamica Sinica,38(06):1129-1135.
[5] CHENG xiuyan,RONG jili,A nisu,et al.Research on aerodynamic noise characteristics of supersonic overexpanded jet of rocket engine [J]. Journal of Astronautics,2020,41(09):1204-1211.

[6] ZHOU junjie. (2010) FLUENT Engineering Technology and Case Analysis [M]. Beijing Publishing House of China water conservancy and hydropower,Beijing.

# Technical Comparison Report on Aerospace Plane Propulsion System

**Siwei Li**

College of Aerospace Science and Engineering, National University of Defense Technology, Changsha, China

lisiwei19@nudt.edn.cn

**Abstract**—The aerospace plane is a new type of aircraft used for both aviation and space flight. This paper analyzes the performance requirements of the propulsion system of space airplanes and focuses on the introduction and comparison of the propulsion systems used in three typical aerospace planes. Then, some suggestions are put forward for the subsequent development of the related technology.

## 1. Introduction

The aerospace plane is an aircraft that highly integrates aviation technology with aerospace technology. It is capable of both aviation and aerospace, pushing space development to a new stage[1].

The aerospace plane is a future plane. It takes off horizontally like an ordinary plane and then flies in the atmosphere at a hypersonic speed of 16,000 to 30,000 kilometers per hour. At an altitude of 30 to 100 kilometers, the flight speed is 12 to 25 times the speed of sound. It can be directly accelerated into the earth's orbit to become a spacecraft. After returning to the atmosphere, it will land at the airport like an airplane. The aerospace plane becomes a means of freely traveling between the sky and the earth.

Before the appearance of the aerospace plane, aeronautics and aerospace were two different technical fields. Aircraft and spacecraft operated in and out of the atmosphere respectively. The air transport system was reused, and the space transport system generally could not be reused. Furthermore, the aerospace plane can achieve the purpose of complete reuse and significantly reduce the cost of space transportation.

The transportation cost of the aerospace plane is only one-tenth of that of the space shuttle, and it does not require a large-scale and complex space launch site; only one airport is needed. After an aerospace plane completes one flight, it can take off again after a week of maintenance. One can travel the universe by space airplane, just like taking a normal plane. Even if you do not go to space, it is very convenient to take it to see friends across the ocean. Moreover, it can also be converted into aircraft such as aerospace fighter jets, aerospace bombers, and aerospace transport aircraft. The aerospace plane has strong penetrating power, which can easily break through the enemy's defense system and attack the enemy, destroying the enemy's equipment and facilities and having high military value.

The United States, Russia, China, Japan, and Germany are researching aerospace planes, but none has actually succeeded. Most plans still rely on disposable launch vehicles to lift off, similar to the old-fashioned space shuttle.

## 2. Analysis of Task Requirements

Aerospace planes fly from inside to outside the atmosphere at speeds from 0 to Ma=25. Such a large speed span and changes in the working environment are beyond the capabilities of all single types of engines now.

Air-breathing engines need to take in air in the atmosphere without carrying oxidants, but they cannot work outside the atmosphere and have a lower practical speed; The rocket engine has its oxidant, which can work outside the atmosphere and has a wide range of practical speeds, but the oxidant carried is bulkier and has a small specific impulse.

It is imperative to develop a combined engine with a multi-cycle working system and stable performance so that the aerospace aircraft can work effectively within the flight range from zero speed at take-off to acceleration at the time of orbit (up to Mach 25).

## 3. Introduction to Propulsion Systems

Three engines are generally installed in an aerospace plane: turbine engines, ramjets, and rocket engines. The turbojet engine can make the space plane take off horizontally. When the speed exceeds 2,400 kilometers, the ramjet is used, making the space plane fly at a speed of nearly 30,000 kilometers per hour in the atmosphere 60 kilometers above the ground. And then, if a rocket engine accelerates it, the aerospace plane will burst out of the atmosphere, like the space shuttle, and go directly into space.

The propulsion systems of three typical space planes are selected as follows.

### *3.1. X-37B*

The X-37B Orbital Demonstration Vehicle (also known as Unmanned Aerospace Vehicle) is one of the secret development projects of the U.S. military. It is technically and financially supported by the Defense Advanced Research Projects Agency (DARPA) of the Defense Department. The U.S. Air Force is responsible for program management. Boeing, the second-largest arms dealer, is the prime contractor responsible for technology research and development[2].

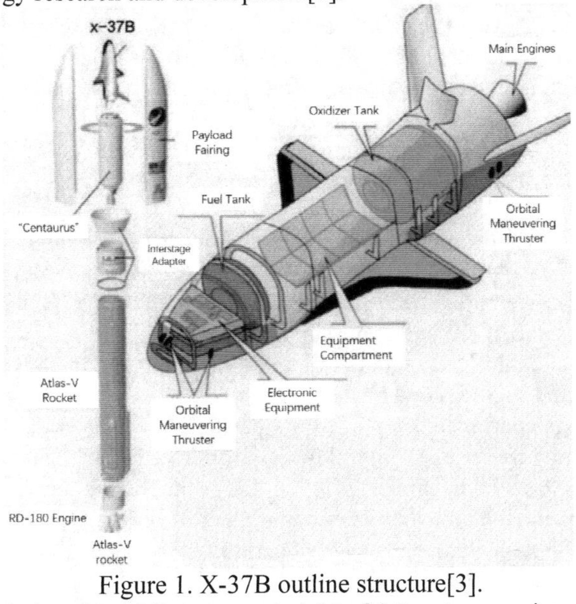


Figure 1. X-37B outline structure[3].

The X-37B has a body length of 8.9 meters, a height of 2.9 meters, a wingspan of 4.5 meters, and a launch mass of 4990 kg. The load compartment is 2.13 meters long, 1.22 meters wide, and has a payload of 227-300 kg. The cargo space and capacity are relatively small.

Since the X-37B is a highly classified flight test program in the United States, after the previous tests, the U.S. Air Force rarely disclosed the details and only introduced the test situation in general; more information comes from the reports of authoritative foreign media and the observations and analysis of astronomy enthusiasts.

When the X-37B took off, it used a two-stage launch system, and it needed to be transported to suborbital rockets before it could cruise. The launch vehicle used for the first four flight tests of the X-37B was the Atlas-V, and the launch vehicle used for the fifth flight test was the Falcon 9.

The Atlas-V launch vehicle adopts a two-stage or two-stage-semi structure, and the standard modules used in each model are a 3.81m diameter Atlas core stage and a 3.05m diameter Centaur upper stage. The Atlas core stage uses an RD-180 liquid oxygen/kerosene rocket engine. This engine adopts a staged combustion cycle and dual combustion chambers and has 47% and 100% throttling capacity. The vacuum thrust reaches 422t, and the vacuum specific impulse is 3304m/s (337s) when working without throttling. The upper stage of the Centaur uses one or two RL10A-4-2 liquid hydrogen/liquid oxygen rocket engines of Pratt & Whitney Group, its vacuum thrust reaches 10t/unit, and the specific impulse is 4415 m/s (451s)[4].

The Falcon 9 launch vehicle is the world's first reusable orbital stage rocket. It adopts a two-stage structure, uses a Merlin engine, and is fueled by liquid oxygen and kerosene. The height is 70 meters, and the diameter is 3.7 meters. The first-stage rocket has nine Merlin engines and a fuel tank made of Al-Li alloy. The sea-level thrust of a single rocket is 854kN (87.1 tons), the sea-level specific impulse is 2695m/s (275s), and the vacuum thrust is 914kN (93.3 tons). At the moment of the rocket launch, the Falcon 9 generates more than 7607kN of thrust. As the altitude increases, the rocket thrust will become larger and larger, and 8227kN of thrust is generated in a vacuum. The second stage rocket is powered by a single Merlin vacuum engine with a vacuum thrust of 981kN (100 tons) and a vacuum specific impulse of 3439m/s (351s). The two-stage rocket engine will fire a few seconds after the first and second stages are separated and can fire multiple times to ensure that multiple loads are sent to different orbits[5].

The launching success rate of both rockets is very high. Falcon 9 wins in reusability, and it is more economical.

In the fourth flight test, the X-37B carried Hall thrusters as on-orbit thrusters to test the modified characteristics of Hall thrusters in space.

In orbit changing stage, it is envisaged that aeroassisted orbital transfer technology with a restartable and variable thrust engine will be used on the X-37B. In step 1, a slight reverse boost is applied to the main rocket engine, which causes the spacecraft to lower its initial orbit and intersect the upper atmosphere at perigee; In the second step, the spacecraft on the elliptical maneuvering orbit rolls the wings, and generates a lift vector with the help of aerodynamic force to change the direction of the flight orbit; The third step is to use the small thrust generated by the main rocket engine to boost, and return to the original height with a new orbital inclination[6].

However, the reality is that the orbit becomes unstable when it encounters atmospheric disturbances and resistance migration energy from the orbits, which can cause catastrophic damage to the orbit and potentially endanger the spacecraft. NASA and the U.S. Navy have investigated a method for stabilizing the orbit, which utilizes thrust from the orbital maneuvering main engine during an aerodynamic maneuver by designing a closed-loop regulator that uses inertial velocity and position data as feedback measurements to adjust the engine's thrust.

Since the orbital energy consumed from the orbit by atmospheric drag is replaced by the engine thrust, there is no danger of orbital collapse. All lateral forces are derived from the aerodynamic lift vector, and the engine is only used to maintain the stability of the orbit. During an aerodynamic maneuver, the propellant required to maintain orbital stability is significantly less than that required to perform a comparable number of orbital plane changes using thrust alone. The process requires the engine to reliably perform multiple starts capability and the ability to adjust thrust with high precision.

Aerodynamically assisted maneuvers require the ability to have real-time feedback to ensure that the spacecraft has sufficient energy to compensate for the energy lost in the atmosphere due to drag. There is uncertainty in the atmospheric crossing of a spacecraft, so the thrust required for each crossing cannot be accurately calculated in advance, and the thrust and thrust duration required for each successive atmospheric crossing is different from the previous one. The aircraft needs to be equipped with variable thrust engines for these reasons. Furthermore, since multiple traversals of the atmosphere are required to accomplish a predetermined plane change, the propulsion system must also have the capability to restart multiple times.

Propellant consumed during aerodynamic maneuvers, for an initial parked orbit of 350km, a 10° change in the aerodynamic plane can save roughly 1400kg of propellant compared with an equivalent propulsion-only maneuver, saving over 200% of fuel. However, this technology has not yet been verified in X-37B flight tests[6].

### *3.2. Skylon*

Britain's Skylon aerospace plane is a hybrid, reusable, horizontal take-off and landing single-stage orbital vehicle. It is currently the only aerospace plane globally that uses a pre-cooled combined engine.

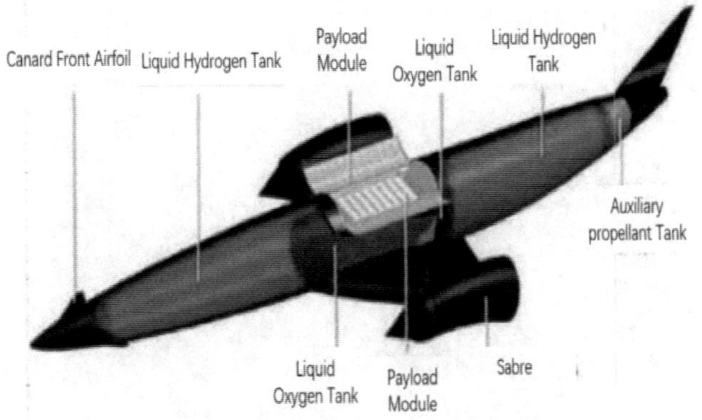


Figure 2. Skylon layout schematic

The Skylon uses a new "Sabre" dual-mode engine. The Sabre has two modes of operation: air-breathing mode and rocket mode. It has a pre-cooled turbo compressor that supplies high-pressure air to the combustor, enabling it to take off from the runway at Mach number 0.46 in air-breathing mode during the ascent phase and accelerate to Mach number 5.5 at 25 km altitude. The air is cooled by a pre-cooler during this period before entering the combustor. The hydrogen fuel is the coolant for the closed-cycle helium circuit before entering the combustor. Compared with traditional rockets, the air-breathing mode can significantly reduce the oxidant consumption and the aircraft's total weight (the X-30 in the NASP plan carries 70% of the total weight of propellant when it takes off). After flying at Mach number 5.5, the engine switches to pure rocket mode and starts to work with its liquid hydrogen/liquid oxygen propellants so that the Sky Tower can reach the orbital speed of Mach number 25[7].

The air-breathing/rocket hybrid engine uses the same combustor, exhaust dust, and pump, reducing the increase in mass caused by adding an air-breathing engine and eliminating the aerodynamic resistance induced by the overload of the vehicle during the air-breathing ascent stage[8].

The following table shows the main parameters of the Skylon：

Table 1. Main parameters of the Skylon.

| Parameters | | indicators |
|---|---|---|
| Size | Length of Body/m | 83.3 |
| | Diameter of Body /m | 6.75 |
| | Wingspan/m | 25.4 |
| Mass | Net Mass /t | 53 |
| | Mass of Propellant/t | 277 |
| | Maximum Theoretical Take-off Mass /t | 345 |
| Carry Capacity | An Orbit 300km High Above the Equator/t | 15 |
| | "International Space Station" (Orbital Altitude 408km, Inclination 51.6°)/t | 9 |
| | Polar orbit (Orbit Altitude 250km, Inclination 98°)/t | 4.8 |
| Engine | Maximum Thrust of Air-breathing Engine /kN | 2×1350 |

| | | |
|---|---|---|
| | Specific Impulse of Air-breathing Engine /(m/s) | 35000 |
| | Maximum Thrust of Rocket Engine/kN | 2×1800 |
| | Specific Impulse of Rocket Engine/(m/s) | 4500 |
| Service Life/times | | 200 |

The core of the engine shutdown technology is the design and validation of the pre-cooler and the exhaust duct. The function of the pre-cooler is to ensure that the air entering the engine can be cooled from 1 000 °C to -150 °C within 0.01s. At the same time, the anti-icing control technology is used to prevent the water vapor in the inhaled air from freezing at low temperatures[9]. As shown in Figure 3, after a series of ground tests at the B9 jet engine power test site, the high-speed cooling capability of the pre-cooler has been verified.

Figure 3. Ground test of pre-cooler.

In the design process of the nozzle, the exhaust duct must be adapted to all atmospheric pressure environments from the ground to the space, and there should be no excessive thermal expansion and contraction to affect the performance. The University of Bristol led the completion of two experimental works[10]: the static test of expansion and rotation nozzle (STERN) and the static test of a rocket with a cooled thrust chamber. The first experiment demonstrated that the flow in the expansion and rotation nozzle is stable over a wide range of pressure ratio conditions. The second experiment investigated the high-speed stability of the jet and the heat flux law of the engine wall, extended the working time through a water-cooled system, and tested the performance of the heat exchanger inside the nozzle. At the same time, the selection work was carried out for the nozzle, and the engine power was increased through the optimization of the nozzle shape, which increased the carrying capacity by 10%.

### *3.3. Tengyun Project*

China's "Tengyun Project" is a combined horizontal take-off and landing space-to-ground shuttle system consisting of a large carrier (first stage) and a small orbiting spacecraft (second stage), sometimes called a two-stage orbiting TBCC sky aircraft. It realizes low-cost round-trip transportation between the space and the earth through two-stage-to-orbit[16].

In September 2018, the "Tengyun Project" realized the first domestic flight verification of combined engine mode conversion[11]. The first flight test is planned for 2025.

Judging from the model exhibited at the Zhuhai airshow in 2021, the first-level aircraft of the "Tengyun Project" carries the second-level aerospace plane on its back, takes off from the ground level, and turns on the TBCC turbine-based combined cycle power system. The gas turbine power is used at low speed, and the scramjet power is used at high speed to accelerate the combination of primary and secondary aircraft to above Mach number 6 to 7. The second-level aerospace plane is separated from the ground at about 30 kilometers. The second-level aerospace plane turns on the rocket engine and directly enters the orbit. At the same time, the first-level aircraft flies through the powered section, glides without power, and finally returns to the launch site, getting ready for the next flight.

Figure 4. Tengyun engineering model.

Turbine, scramjet, and rocket engines cooperate, and the three-pronged approach maximizes the speed so that the fastest speed can be achieved within the limited fuel to circle the earth. Furthermore, the aerospace plane can be reused many times, which dramatically reduces the cost of launching.

## 4. Comparison of Performance Parameters and Technical Characteristics

Since most aerospace planes are currently in the program design and test verification stage and have not yet been used, the key technologies and specific performance parameters of aerospace planes in various countries have not been disclosed. Most of the information comes from reports from authoritative foreign media and observations and analyses by astronomers. Therefore, this article mainly compares typical aerospace plane propulsion systems from the technical characteristics combined with rough parameters.

Table 2. Comparison of propulsion schemes for aerospace plane.

| | X-37B | Skylon | Tengyun Project |
|---|---|---|---|
| Injection Method | Two-stage | Single-stage | Two-stage |
| Propulsion Mode | rocket engine | Pre-cooling Combined Engine（Air-breathing Mode and Rocket Mode） | Turbo-based Combined Cycle Power、Rocket engine |
| Engine Model | Falcon 9 and etc. | Sabre dual-mode engine | Unpublished |
| Can be Reused | Yes | Yes | Yes |

From the perspective of the injection method, compared with the single-stage orbiting vehicle, the two-stage orbiting vehicle is relatively less technically tricky and easy to implement. It does not need to bring all the structural mass into the orbit, so it has a higher payload carrying coefficient[12].

The single-stage orbital space transportation system is simple in structure, reliable, and low in cost. However, it has great technological barriers and requires an integral combined engine (such as an air-breathing/rocket engine). The combined engine first works in the air-breathing mode so that the speed of the single-stage orbiting space transportation system reaches the conversion Mach number and then starts the rocket engine to enter the space orbit. A common nozzle and fuel pump can be used in this case. The use of a common nozzle can reduce the launch area of the vehicle, increase the thrust, and increase the slenderness of the flame[13].

From the view of propulsion mode, the aerospace plane with a combined engine is more portable. The "X-37B" aerospace plane of the United States cannot be launched autonomously, and it relies on rockets to drive the launch when flying. The "Tengyun Project" can take off directly from the airport without relying on external force and can independently complete the tasks of manning and carrying objects, which dramatically reduces the cost.

In addition, the thrust and equivalent specific impulse of the combined cycle engine is more than twice that of the rocket engine, the efficiency of launching into orbit is higher, and the fuel cost is lower.

The pre-cooling combination engine adopted in the U.K. can improve the performance of the turbine engine and greatly reduce the consumption of oxidants and the total mass of the aircraft.

China's aerospace plane has broken through the critical technology of scramjet and realized the combination with gas turbine power, so it has good flight performance whether at low Mach number or high Mach number.

## 5. Suggestions for Future Development

### *5.1. Multi-mode Combined Power Engine*

The aerospace plane has a wide flight range and a large Mach number span. If multiple engines are installed, each engine can only work within a specific range, and the rest of the engines will become dead weight, which is not conducive to increasing the payload. Therefore, it is necessary to develop a combined cycle engine with multiple functions[14]. Combined power can fully utilize oxygen in the atmosphere, reduce propellant consumption, and improve the economy.

Fan Zhang from Tsinghua University proposed a dual rocket-based combined cycle (DRBCC) scheme[15]. The DRBCC consists of a liquid oxygen/methane staged combustion cycle liquid rocket engine and a dual-mode ramjet with a distributed ejector in the inner flow channel. There are five operating modes: hybrid mode, pure ejector mode, sub-ramjet mode, scramjet mode, and pure rocket mode. This scheme can break through several technical bottlenecks in the RBCC propulsion system: The thrust gain of the ejector mode in the take-off phase is not significant; The embedding of the ejector rocket into the airflow channel brings severe technical challenges to the development of the rocket, such as miniaturization, thermal protection, and large-scale thrust change; The fuels used by scramjets flying at Mach 6-8 and 8-10 are incompatible.

### *5.2. Pre-cooling Engine*

The pre-cooling engine adds a pre-cooler based on the traditional turbine engine to cool the turbine engine, thereby improving its power and performance.

The Precooling Power Technology Center of 31st Inst. of CASIC has completed the Ma5 thermal state test of the "Yunlong" engine pre-cooler, achieving a temperature drop of nearly 1000 °C in milliseconds, reaching the expected goal.

In the future, the pre-cooling engine can be combined with a ramjet or a rocket engine to obtain the power of the aerospace aircraft with better performance.

### *5.3. Single-stage Orbital Aerospace Plane*

A single-stage orbiting aerospace plane can propel the plane from take-off to orbit with a single integrated engine combination. This injection is more consistent with people's original idea of aerospace planes: it integrates flight vehicles, space vehicles and spacecraft, and the structure is more integrated.

This method has high requirements on the engine, which may bring problems such as complex structure and poor controllability. It also requires a large number of theoretical research and experimental verification. The injection method adopted by most countries is still based on two stages. It is hoped that there will be a breakthrough in single-stage orbital entry in the future.

### *5.4. Aeroassisted Orbital Transfer Technology*

Aeroassisted orbital transfer technology can utilize the resources of the earth's atmosphere, take advantage of the good aerodynamic shape of aerospace planes, and use aerodynamic force to realize orbit transfer by entering and exiting the atmosphere. It can significantly reduce propellant consumption, increase the payload, and have broad application prospects.

A worthy research direction is to combine aeroassisted orbital transfer technology with a variable thrust engine for aerospace plane orbit transfer.

## 6. Conclusion

To sum up, the technical key to aerospace plane propulsion systems lies in combined engine technology. Judging from the current research situation in various countries, the combined engines used in the aerospace plane are still in the proposal and test verification stage.

We look forward to the gradual improvement of the technical level of various countries in this regard and the development of combined engines with more stable performance.

Maybe we will all have the opportunity to travel to space by aerospace plane in the future.

## References

[1] Baidu baike. (2013) aerospace plane. https://mbd.baidu.com/ug_share/mbox/4a83aa9e65/share?product=smartapp&tk=0218032b50f955efd56263c5c4dbac44&share_url=https%3A%2F%2Fvhsagj.smartapps.baidu.com%2Fpages%2Flemma%2Flemma%3F_swebfr%3D1%26lemmaTitle%3D%25E7%25A9%25BA%25E5%25A4%25A9%25E9%25A3%259E%25E6%259C%25BA%26lemmaId%3D799948%26from%3DbottomBarShare%26_swebFromHost%3Dbaiduboxapp&domain=mbd.baidu.com

[2] Wen Li. (2015) Competing for space superiority——Analysis of the U.S. military X-37B aerospace plane. China Economic and Trade Tribune, 18:77-79.

[3] Terigele, Nannan Wang, Yuan Yao, Xuan Chen. (2020) A brief analysis of the development of the X-37B in the United States. Aerospace China, 04:37-40.

[4] Shu Zhang. (2008) The Present Situation and Future Development of Atlas 5 Series Launch Vehicles. Missiles and Space Vehicles , 02:56-60.

[5] Ninety-nine star cat. (2019) Introduction of Falcon 9. https://www.bilibili.com/read/cv2395546.

[6] Sun Zongxiang, Tang Zhigong, ChenXilan, Zhong Ping, Liu Xiaobo,(2015) Review of the state-of-art and aerodynamic technology of X-37B. Journal of Experiments in Fluid Mechanics,29(01):1-14+24

[7] Fan Deng, Lin Chen, Wei Song, Tong Chen, Zhuxi Wang. (2017) The overall scheme design and task demand analysis of the Skylon Aerospace Plane of British. Aerodynamic Missile Journa,02:41-45.

[8] Wenjie Shan, Junpeng Hui , Shiyong Huang.(2012) The secret of the British " Skylon " aerospace plane. Space International,01:6-12.

[9] Varvill R, Bond A . (2003) A comparison of propulsion concepts for SSTO reusable launcher. Journal of the British Interplanetary Society. 56:3-4.

[10] Taylor N, Steelant J, Bond R. (2011) Experimental comparison of dual bell and expansion deflection nozzles. AIAA , 5688.

[11] Xinhua News Agency. (2018) The six major commercial aerospace projects of CASIC have achieved phased results. https://baijiahao.baidu.com/s?id=1612743426873478001&wfr=spider&for=pc

[12] M. A. Brock, M. E. Franke. (2004) Two - stage - to - orbit reusable launch vehicle propulsion performance study. AIAA, 3903.

[13] Xuewu Si. (1990) Comparison of single-stage and two-stage orbital aerospace plane with horizontal take-off. The International Missile and Spacecraft,02:23-26.
[14] XIE Rui-xuan. (2018) Research on a New Type of Liquid Air Cycle Engine for SSTO Spaceplane. Internal Combustion Engine & Parts, 10
[15] Fan Zhang , Huiqiang Zhang , Bing Wang. (2016) Preliminary design and analysis of DRBCC carrier.In: The 37th Technical Exchange Conference of China Aerospace Third Professional Information Network and the First Joint Conference on Aerospace Power .XI'an. 189-195.
[16] Ling Xiao's seven-color flower. (2022) There is great good news for China's aerospace plane, the combined engine has successfully made its first flight, and the round-trip to space is just around the corner. https://3g.163.com/dy/article/H5PRIODE05529M6C.html

# Analysis of the Attachment Process of Bionic Spines Based on the Coupling of Discrete Elements and Multi-body Dynamics

**Ran Liu**[1,a]**, Yu Xin Li**[2]**, Haitao Luo**[3*]**, Miao Yu**[4]

[1]School of Mechanical Engineering,Shenyang Ligong University, Shenyang, China

[2]State Key Laboratory of Robotics, Shenyang Institute of Automation Chinese Academy of Sciences,Shenyang, China

[3]State Key Laboratory of Robotics,Shenyang Institute of Automation Chinese Academy of Sciences Shenyang, China, Institutes for Robotics and Intelligent Manufacturing, Chinese Academy of Sciences, Shenyang, China

[4]State Key Laboratory of Robotics, Shenyang Institute of Automation Chinese Academy of Sciences, Shenyang, China, Institutes for Robotics and Intelligent Manufacturing, Chinese Academy of Sciences,Shenyang, China

[a]liuran@sia.cn; [*] Corresponding author: luohaitao@sia.cn

**Abstract**—As the only natural satellite of the Earth, the Moon has profound significance for deep space research, and the undifferentiated lunar rock core plays a key role in the investigation of the origin of the Moon. In recent years, there have been numerous deep space sampling techniques, but all of them are at the cost of destroying the sample, while drilling can retain the sample in its original state to the maximum extent. In this paper, based on the reference of the claw spine attachment device at home and abroad, the theoretical modeling of the claw spine piece which plays the role of grasping and attaching is carried out, and the simulation analysis is carried out by the coupling method of discrete element and multi-body dynamics, which provides important guidance for the subsequent experi-ments and the optimized design.

## 1. INTRODUCTION

Chang'e 5 landed safely in Inner Mongolia in the early morning of December 17, 2020. Chang'e 5 brought back 1731 grams of lunar soil, which is of great significance for China to understand the compositional characteristics of the sampling area (rock types, mineral components, chemical composition, etc.), the interaction between lunar surface materials and the solar wind, and the study of the soil-forming mechanism of lunar soil. The unweathered lunar cores are also of great reference value for studying the origin of the Moon and calculating the age of the Earth. In the future, China will send astronauts to the Moon to carry out lunar rock drilling and sampling missions.How to eliminate the recoil force generated during drilling and how to drill lunar rock cores in a stable manner, which poses new challenges to the way of lunar rock gripping and attaching.

Insects are distributed in all corners of the world and have evolved different forms to adapt to their environment, among which the claw-spin grip is prevalent in insects, especially adapted to rough surfaces. Insects either use their claws to form mechanical locks with the micro-valleys and convex peaks of the crawling surface, or directly pierce the soft crawling surface to generate greater adhesion and ensure that they can stay stable and move freely on the surface they are crawling on. The moon rock surface is porous due to its long differentiation, which is its main characteristic, which makes it possible to use the claw sting to grasp and steadily drill the moon rock core.

Boston Dynamics developed the RISE robot [1-3] in 2006, which can move on walls, rocks, and horizontal surfaces; the Jet Propulsion Laboratory developed the Silver Fox IIB [4] in 2013, which can move on vertical, overhanging, and reversed rock surfaces; the University of Science and Technology of China developed a tracked pair of gripping foot robots in 2015, which can move on rough avoided, ceiling motion; Nanjing University of Posts and Telecommunications developed a barb-and-claw grasping wall-climbing robot in 2017, which can move on rough wall surfaces, and all the above mentioned robots use claw-spike grasping attachment, which provides experience in completing sampling tasks.

In this paper, we will first establish a theoretical mechanical model of the bionic claw-spike sheet, then perform a joint discrete element and multi-body dynamics simulation, followed by an explanation of the problems arising from it, and finally an outlook for the claw-spike grasping attachment method.

## 2. THEORETICAL MODELING

Bionic claw stinger single piece attachment mechanism schematic diagram is shown in Fig. 1, the orange square represents the position of the claw stinger piece in the attachment device simplified diagram, with the blue flexible rod indicates the single claw stinger piece, the number 1 ~ 4 represents the claw stinger piece movement process of four positions, green represents the wall, the black dot represents the bump, single claw stinger piece attachment process is as follows.

First, the claw spike sheet from position 1 with a certain horizontal speed and vertical speed close to the wall, movement to position 2, at this time the claw spike sheet just touch the wall and not under force; then, the claw spike sheet from position 2 with vertical speed downward, in position 3 to capture the protrusion on the wall; finally, the claw spike from position 3 to position 4, in this process, the flexible rod deformation elongation, resulting in a normal force perpendicular to the wall and tangential force parallel to the wall [5].

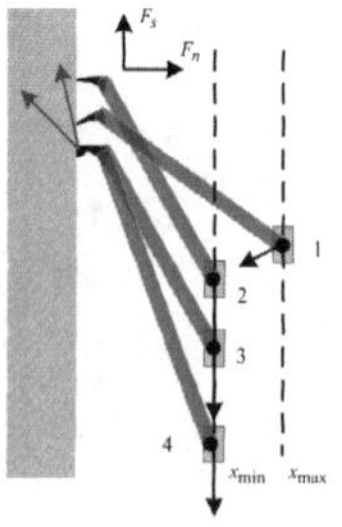


Figure 1. Schematic diagram of the adhesion mechanism

The above mechanism can be simplified to a flexible rod model.

The flexible rod mechanics model is shown in Fig. 2.

TABLE 1. LIST OF PARAMETERS OF FLEXIBLE ROD MODEL

| Variables | Description | Variables | Description |
|---|---|---|---|
| $F_r$ (N) | Axial force | $F_\theta$ (N) | Turning force |
| $F_s$ (N) | Tangential force | $F_n$ (N) | Normal force |
| $k_r$ (N/m) | Tensile stiffness | $r_0$ (m) | Initial length |
| $k_\theta$ (N/rad) | Rotational stiffness | $\theta_0$ (rad) | Initial angle |

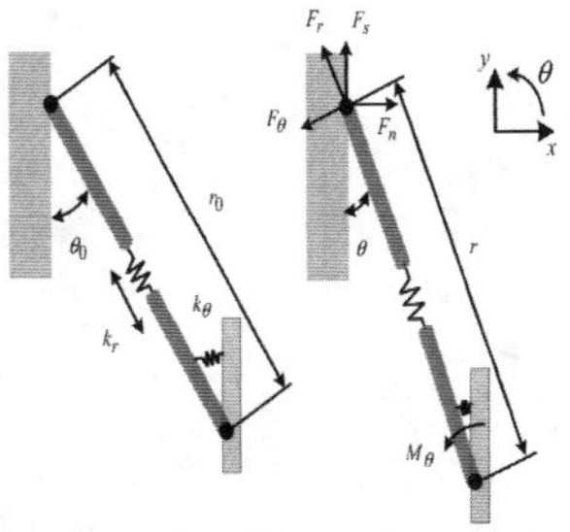

Figure 2. Flexible rod model

The force analysis and orthogonal decomposition of the (1)

$$F_r = (r - r_0)k_r \tag{1}$$

$$F_\theta = \frac{(\theta - \theta_0)k_\theta}{r} \tag{2}$$

Can be launched

$$F_n = -F_r \sin\theta - F_\theta \cos\theta \tag{3}$$

$$F_s = F_r \cos\theta - F_\theta \sin\theta \tag{4}$$

Substituting (1) (2) into (3) (4) and expressing it in matrix form

$$\begin{bmatrix} F_n \\ F_s \end{bmatrix} = \begin{bmatrix} -\sin\theta & -1/r\cos\theta \\ \cos\theta & -1/r\sin\theta \end{bmatrix} \begin{pmatrix} k_r(r-r_0) \\ k_\theta(\theta-\theta_0) \end{pmatrix} \tag{5}$$

## 3. Simulation Modeling

A 3D model of the actual claw barb sheet is created in Solidworks as shown in Fig. 3, with the yellow part being the flexible material and the green part being the rigid material [6]. The actual way of working is shown in Fig. 4, the frame moves with the attachment device, and the bionic claw barb sheet in the process of moving with the frame, the fish hook captures the surface bump or pit to be grasped for attachment, when the fish hook captures the surface, as the frame is still moving, it makes the positioning pin move backward, at this time, the tensile force is generated due to the tensile deformation of the flexible part, which can be decomposed into horizontal and vertical forces, and the force in the vertical direction, which is the required the required sampling adhesion force. When the positioning pin moves to the position of the rigid part a will cause the hook attached to the surface to detach from the surface.

The frame has three locating pins, which protect the flexible part from stretching too long and breaking beyond the material stretching limit, and give the rigid part at the back a certain degree of freedom of rotation to adapt to different contact surfaces.

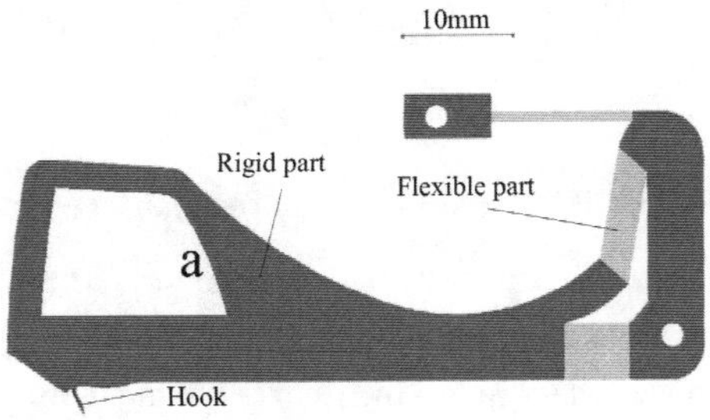


Figure 3. Single spike piece model

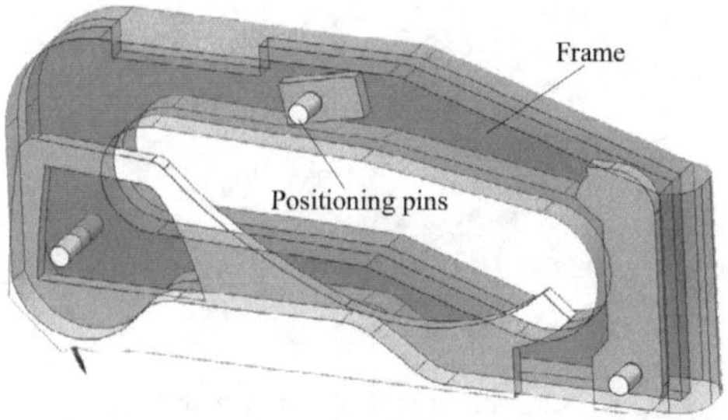


Figure 4. Single stinger model of the placement frame

To shorten the simulation process, the flexible part of the model is replaced by a rigid spring, and the model shown in Fig. 5 is established in the multi-body dynamics simulation software RecurDyn, in which the lower five springs all have a stiffness coefficient of 0.08 N/mm, a damping coefficient of 0.03 N·s/mm, a free length of 6 mm and a preload tension of 0.04 N; the middle three springs all have a stiffness coefficient of 0.06N/mm, a damping coefficient of 0.03N·s/mm, and a free length of 10mm; the upper spring has a stiffness coefficient of 0.15N/mm, a damping coefficient of 0.15N·s/mm, a free length of 19mm, and a preload tension of 0.1N. The frame is given motion along the horizontal X-axis with a motion speed of 2.5mm/s.

The volcanic rock surface was established in the discrete element software EDEM, and four kinds of particles were taken for filling, respectively, a single sphere model with a radius of 1 mm, a double sphere model with a radius of 0.9 mm, a three-sphere model with a radius of 0.6 mm and a four-sphere model with spherical particles with radii of 0.4 mm, 0.5 mm, 0.6 mm and 0.7 mm for simulation, and the radius values were different because it was hoped that by this way to the surface porous state is simulated as much as possible, and the established surface is shown in Fig. 6. After setting the parameters for the coupling simulation, the simulation time is 12s.

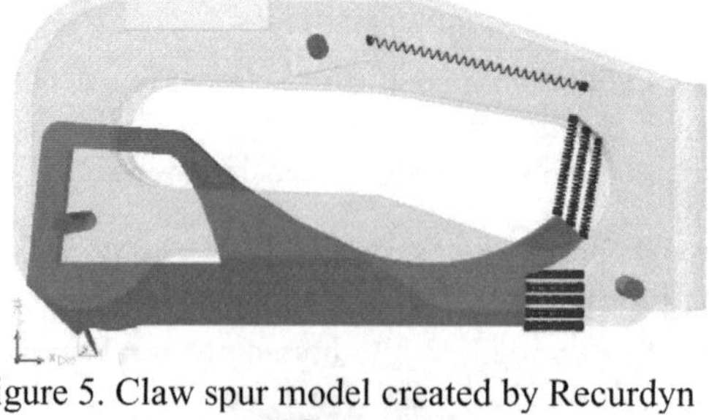

Figure 5. Claw spur model created by Recurdyn

Figure 6. EDEM-created volcanic rock surface model

## 4. Simulation Results and Analysis

Fig. 7 shows the comparison of the simulation and theoretical results obtained under the above conditions, according to (5), in accordance with the relevant parameters of the coupling simulation. The tangential force $F_s$ is significantly larger than the normal force $F_n$ in the Fig., and the two values keep rising from 0.8s to 6s. The reason for the fluctuation of the simulation data on Fig. 7 is that the coupling simulation process is the process of contact and collision between the fish hook (rigid body) and the spherical particle (rigid body), and vibration will be generated when the fish hook is in contact with the particle surface, and fluctuation will be generated when the spring is stretched. According to Hooke's law, the force is proportional to the deformation of the spring, so it is more appropriate to replace the fluctuation section of the coupled simulation with a theoretical linear section, and the error between the coupled simulation and the theoretical model is less than 10% in numerical terms.

Fig. 8 shows the attachment state diagram of the single claw stinger piece grasping the surface of the particle, the hook did not capture the surface particle before 0.8s, in 0.8s to 6s, the hook captures the surface particle, the spring stretches, when the spring stretches a certain length to contact the positioning pin, the force will appear to increase rapidly, after increasing to 23N, the hook detaches from the surface of the particle, the moment of detachment, producing a dramatic change in the value at the end of Fig. 8 .

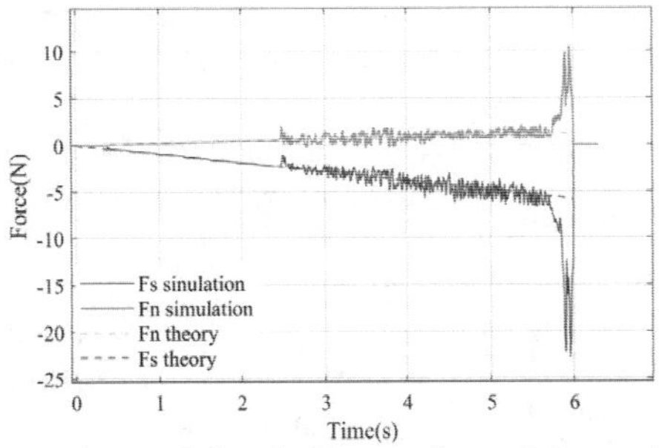


Figure 7. Comparison of simulation results and theoretical results

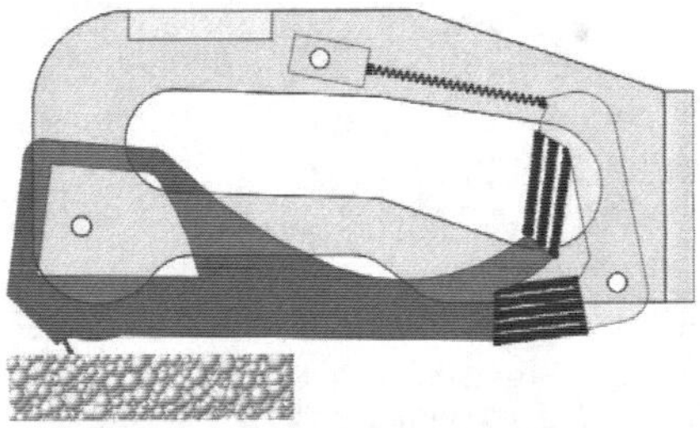

(a) During the attachment process

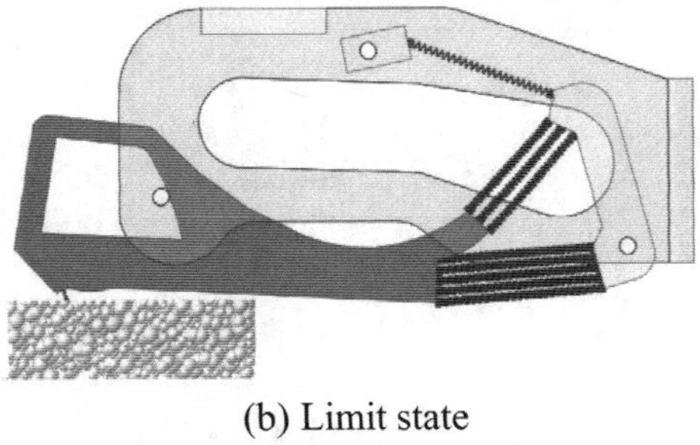

(b) Limit state

Figure 8. Claw stab piece deformed by force stretching

As shown in Fig. 9, the barb does not reach the peak tension during the first grip, and the barb falls off the particle surface at some point during its movement; in the second grip, the barb firmly grips the particle surface until the protective effect of the positioning pin is triggered, at which point the tension reaches its peak; in the third grip, as in the first grip, the tension does not reach its peak, and a small force is generated before it breaks away from the particle surface.

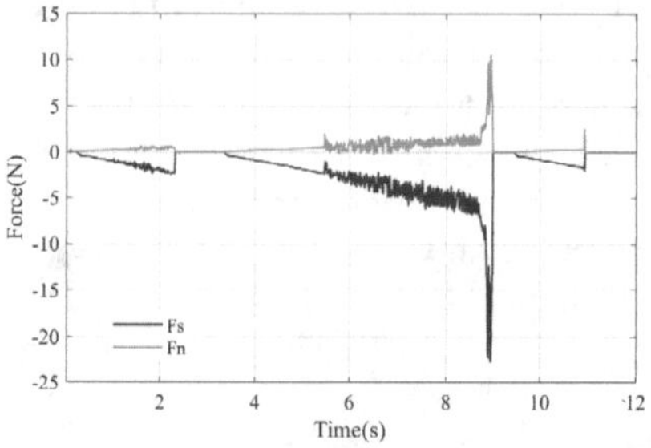


Figure 9. Single piece coupling simulation grip attached to the force diagram

## 5. Conclusions

The error between the simulation results and the theoretical model is controlled within 10%, which shows that the coupling simulation is in basic agreement with the theory, which provides the possibility of subsequent tests and stinger optimization, shortens the optimization test cycle and test cost, and saves manpower. In this paper, although the coupling simulation is basically consistent with the theory, there is a lack of experimental data. In the future, we will conduct several experimental studies to compare with the simulation results, find out the rules and optimize the stab structure by simulation method.

## Acknowledgment

This research was supported by the Development Fund of Space Automation Technology Laboratory, SIA, CAS (E155380104); National Natural Science Foundation of China (No.51975567); Liao Ning Revitalization Talents Program (XLYC1907152); Natural Science Foundation of Liaoning Province (2019-MS-347); State Key Laboratory of Robotics (2022-Z01); Chinese Academy of Science Youth Innovation Promotion Association (No.2018237) and Jiang Xin-song Innovation Fund (No.20180504).

## References

[1] Autumn, Kellar, et al. "Robotics in scansorial environments." Unmanned ground vehicle technology VII. Vol. 5804. SPIE, 2005.
[2] J. Spenko, Matthew, et al. "Biologically inspired climbing with a hexapedal robot." Journal of field robotics, 2008, pp. 223-242.
[3] G. C. Haynes, A. Khripin, G. Lynch, et al. Rapid pole climbing with a quadrupedal robot, Proceeding of the 2009 IEEE International Conference on Robotics and Automation(ICRA). Piscataway, USA: IEEE, 2009, pp. 2762-2772.
[4] A. Parness, M. Frost, N. Thatte, et al. Gravity-independent rock-climbing robot and a sample acquisition tool with microspine grippers. Journal of Field Robotics, 2013, vol. 30, pp. 897-915.
[5] A. T. Asbeck, M. R. Cutkosky. Designing compliant spine mechanisms for climbing[J]. 2012:2-4.
[6] A. Parness, M. Frost, N. Thatte, et al. Gravity-independent mobility and drilling on natural rock using microspines, 2012 IEEE International Conference on Robotics and Automation. IEEE, 2012, pp. 3437-3442.

# Research on construction method of modular missile family general components based on self-organizing mapping

**Feiran Guo[1a*], Jianqiao Yu[1b*]**

[1]School of Aerospace Engineering, Beijing Institute of Technology, Beijing 100081, China

[a*]745896011@qq.com, [b*]jianqiao@bit.edu.cn

**Abstract**—As far as the missile family is concerned, the general components are an important part of the missile types in the missile family. At the same time, they are also a major advantage and feature of collaborative design of multiple missile types to form the missile family. In view of the poor universality of modules in traditional missile design, this paper proposes a general components construction method based on self-organizing mapping, which aggregates the components with similar structure and little difference in performance parameters as general components, so as to maximize the proportion of general components in the component library under the condition of meeting the performance index, and then improve the efficiency of missile equipment system while reducing the design cost.

## 1. Introduction

With the rapid development of weapon and equipment technology, facing the military requirements of joint operation under the condition of future information, the weapon systems constituting the equipment system need to cooperate under certain capability conditions, so as to give full play to the efficiency of the equipment system to the greatest extent. In order to meet the requirements of various combat tasks in modern war, the missile family collaborative design including multiple missile types has become the trend and direction of missile equipment system design. According to the structure and function composition of the missile, in the process of forming different types of general-purpose component libraries, there will be several general-purpose components with different attributes and related parameter values in each type of component library. In this way, according to different combat mission requirements, different functional modules meeting corresponding mission requirements can be matched in real time and assembled into corresponding missile types in real time. Thus, the operational flexibility of missiles under different mission requirements in real-time battlefield environment is greatly improved. The composition of the general component library requires that the components in it are basically the same in composition and structure, and the difference in corresponding performance parameters should be within an acceptable range.As an unsupervised learning algorithm for clustering and high-dimensional visualization, self-organizing mapping (SOM) algorithm is an artificial neural network[1,2] developed by simulating the characteristics of human brain in signal processing. The model was proposed by Teuvo Kohonen[3], Professor of Helsinki University in Finland in 1981. Now it has become the most widely used self-organizing neural network method. WTA (winner takes all) competition mechanism[4] reflects the most fundamental characteristics of self-organizing learning. SOM algorithm is an unsupervised clustering method. Since it was proposed 20 years ago, many researchers have done a lot of work in the theoretical and application fields of pattern recognition, signal processing, data mining and so on, and achieved a lot of research results[5,6,7]. These achievements are largely due to the simplicity and practicability of SOM algorithm itself. According to the clustering characteristic of self-organizing mapping, this paper realizes the construction of modular general component library through self-organizing mapping neural network algorithm based on genetic algorithm.

## 2. Self-organizing Mapping Network Structure

SOM network structure is shown in Figure 1. It is composed of input layer and competition layer (output layer). The number of neurons in the input layer is one-dimensional or two-dimensional planar array composed of neurons in the competition layer. The network is fully connected, that is, each input node is connected with all output nodes.

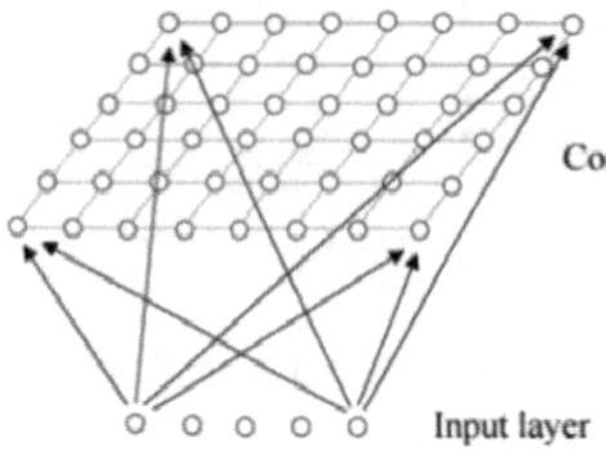


Fig.1 SOM network structure diagram

SOM network can map any dimensional input mode into one-dimensional or two-dimensional graphics in the output layer, and keep its topology unchanged. Through the repeated learning of the input mode, the network can make the probability distribution of the weight vector space and the input mode tend to be consistent, that is, probability retention. Each neuron in the competition layer of the network competes for the response opportunity to the input mode, and the weights related to the winning neuron are adjusted in the direction more conducive to its competition, that is, taking the winning neuron as the center, it shows exciting side feedback to the nearest neighbor's neurons, while inhibitory side feedback to the far neighbor's neurons. The nearest neighbors encourage each other and the far neighbors inhibit each other. Generally speaking, the nearest neighbor refers to the neuron that sends out the signal as the center of the circle, with a radius of about 50 μm-500 μm neurons; Far neighbor refers to a radius of 200 μm-2 mm neurons. Neurons farther from distant neighbors showed weak excitation, as shown in Figure 2.

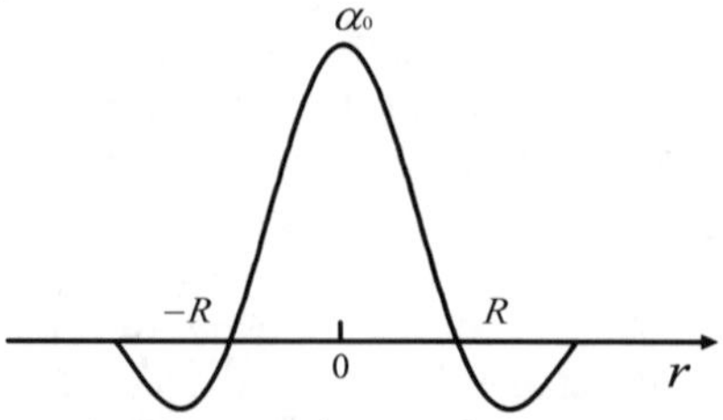


Fig.2 Neuron interaction pattern

Note that the set composed of all output neurons is $\delta$, and the connection weight vector between neurons and input layer neurons is $W_c$. As an unsupervised method, SOM algorithm can theoretically map the input mode into one-dimensional, two-dimensional or even higher-dimensional discrete graphics in the output layer, and keep its topology unchanged. With the continuous progress of learning, the learning rate will continue to decrease, and the neighborhood will continue to shrink. The ownership vectors will be separated from each other in the input vector space, each representing a kind of pattern in the input space, which is the clustering function of automatic feature recognition of SOM network.The algorithm of self-organizing mapping neural network is mainly based on genetic algorithm[8]. Due to the good search characteristics of genetic algorithm, combined with the continuous learning process of each neuron in the self-organizing mapping network structure, with the continuous iteration and population update of genetic algorithm, the learning rate of each neuron will continue to decline, and the neighborhood radius of learning will continue to shrink. Finally, the clustering result when the learning rate decays to 0 will be obtained.

## 3. General Components Construction Based on SOM

According to the characteristics of self-organizing mapping neural network algorithm, it can be applied to the construction of general components. The input vector is the $m$ dimensions parameter vector of different index attributes of various types of components, corresponding to $n$ different component nodes. After the network training of genetic algorithm, when the number of iterations reaches the given value and the learning rate decays to 0, the component clustering result in the final output layer can be obtained. The construction process of general components is shown in Figure 3.

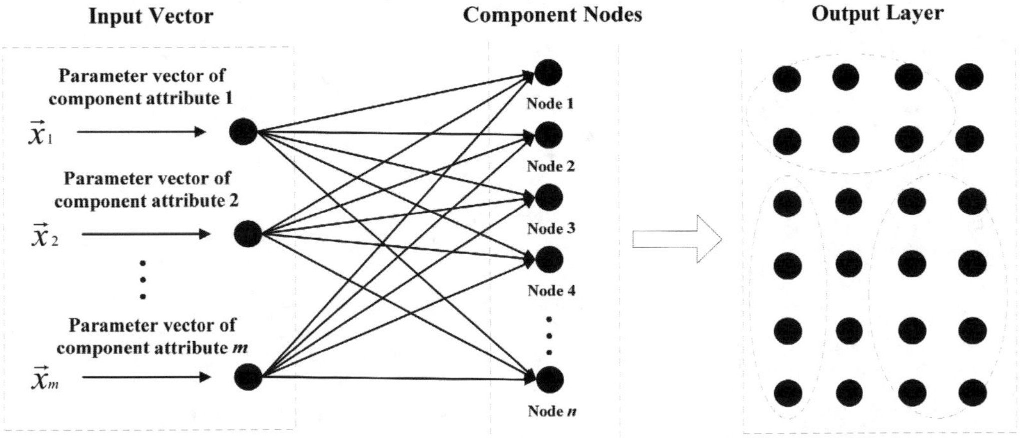


Fig. 3 Schematic diagram of general component construction process

Referring to the common components of various types of missiles, the index attributes to be clustered of various types of components are given, as shown in Table 1.

Table 1 Index attributes of various types of components to be clustered

| Components types | Index attributes |
|---|---|
| Seeker | Effective action distance/(km) |
| IMU | Attitude accuracy/(deg/h)<br>Position accuracy/(m/h) |
| Steering gear | Maximum rudder deflection angle/(deg)<br>Maximum rudder yaw rate/(deg/s) |
| Engine | Thrust/(kN)<br>Total impulse/(kN·s) |
| Warhead | Mass/(kg) |

The initial parameter setting of self-organizing mapping neural network algorithm is shown in Table 2.

Table 2 Initial parameters of self-organizing mapping neural network algorithm

| Initial parameters | Set value |
|---|---|
| Learning rate $\eta$ | 0.8 |
| Effective width $\sigma$ | 2 |
| Stride $\alpha$ | 0.2 |
| Power of activation function | 10 |
| Minimum evolutionary step | $10^{-5}$ |
| Maximum genetic algebra | 100 |

| | |
|---|---|
| Crossover probability $P_c$ | 0.1 |
| Variation probability $P_m$ | 0.01 |
| Diversity evaluation parameters | 0.95 |

According to the algorithm flow of self-organizing mapping network based on genetic algorithm, the eight index attributes of five types of components are clustered and calculated respectively, and the corresponding clustering results are obtained.

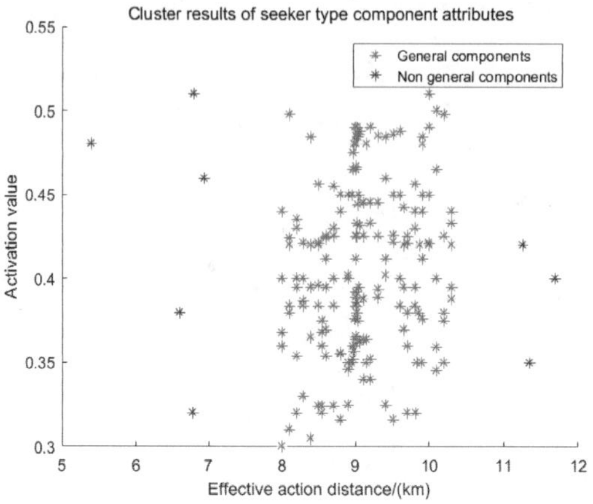


Fig. 4 Cluster results of seeker type components

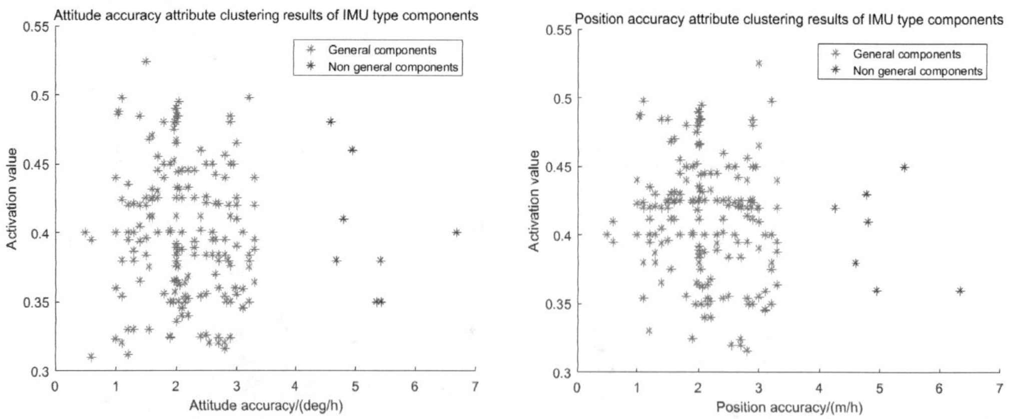


(a)Attitude accuracy attribute clustering results of IMU type components

(b)Position accuracy attribute clustering results of IMU type components

Fig. 5 Cluster results of IMU type components

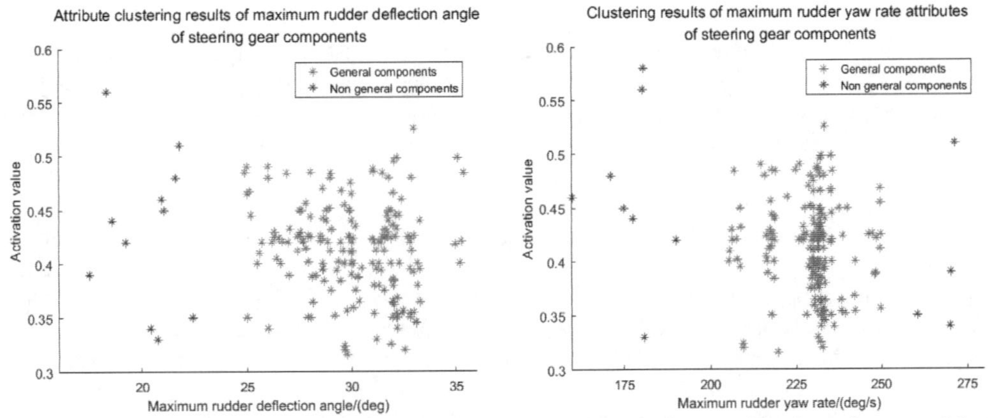


(a)Attribute clustering results of maximum rudder deflection angle of steering gear components
(b)Clustering results of maximum rudder yaw rate attributes of steering gear components

Fig. 6 Cluster results of steering gear type components

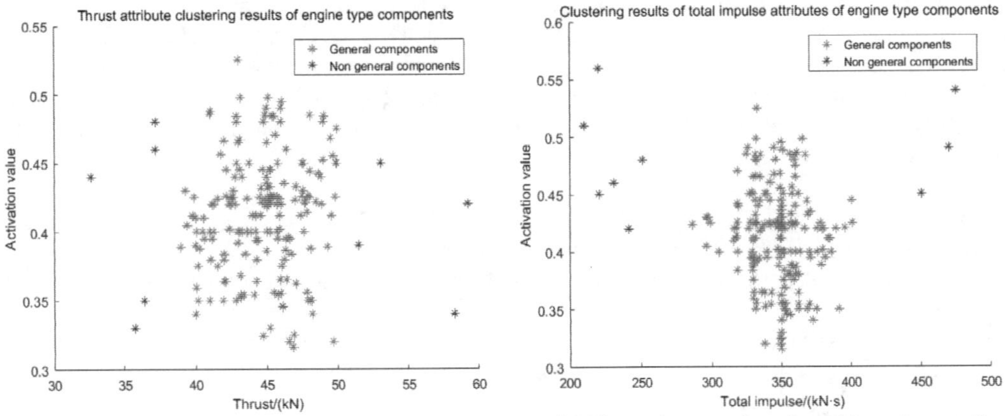


(a)Thrust attribute clustering results of engine type components
(b)Clustering results of total impulse attributes of engine type components

Fig. 7 Cluster results of engine type components

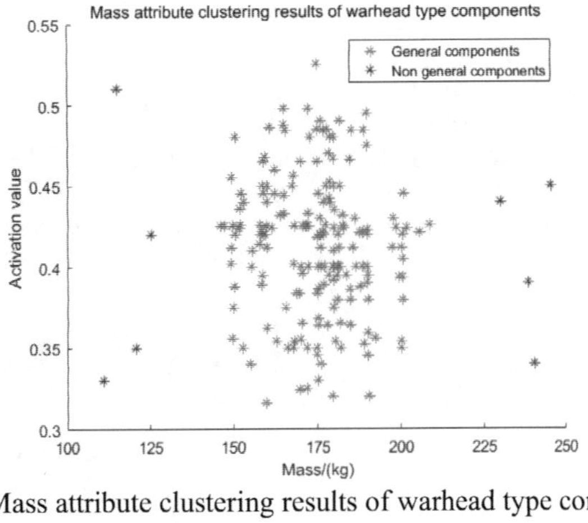


Fig. 8 Mass attribute clustering results of warhead type components

## 4. Conclusion

It is not difficult to see from Figures 4 to 8 that the obvious clustering effects of the corresponding attributes of various types of components can be obtained by using the self-organizing mapping neural network algorithm based on genetic algorithm. For the convenience of expression, the cluster value of effective operating distance attribute of seeker type components is rounded to 8km, 9km and 10km; The cluster values of attitude accuracy attributes of IMU type components are rounded to 1deg/h, 2deg/h and 3deg/h, and the cluster values of position accuracy attributes are rounded to 1m/h, 2m/h and 3m/h; The maximum rudder deflection angle attribute cluster values of steering gear type components are rounded to 25deg, 30deg and 35deg, and the maximum rudder yaw rate attribute cluster values are rounded to 200deg/s, 225deg/s and 250deg/s; The thrust attribute cluster values of engine type components are rounded to 40kN, 45kN and 50kN, and the total impulse attribute cluster values are rounded to 300kN·s, 350kN·s and 400kN·s; The cluster values of mass attributes of warhead type components are rounded to 150kg, 175kg and 200kg.

Based on the results and discussions presented above, the conclusions are obtained that the general component construction method based on self-organizing mapping neural network proposed in this paper can realize the construction of general component library, and the composition structure of each component is roughly the same, and the difference in corresponding performance parameters is within an acceptable range. The cluster analysis of the effective operating distance attribute of the seeker type components, the attitude accuracy attribute and position accuracy attribute of the IMU type components, the maximum rudder deflection angle attribute and maximum rudder yaw rate attribute of the steering gear type components, the thrust attribute and total impulse attribute of the engine type components and the mass attribute of the warhead type components proves the effectiveness of the method proposed in this paper.

## References

[1] Giordano Laura,Gliozzi Valentina,Theseider DuprÉ Daniele. A conditional, a fuzzy and a probabilistic interpretation of self-organizing maps[J]. Journal of Logic and Computation,2022,32(2).

[2] Ferles Christos,Papanikolaou Yannis,Savaidis Stylianos P.,Mitilineos Stelios A.. Deep Self-Organizing Map of Convolutional Layers for Clustering and Visualizing Image Data[J]. Machine Learning and Knowledge Extraction,2021,3(4).

[3] Teuvo Kohonen. Essentials of the self-organizing map[J]. Neural Networks,2013,37.

[4] Yan Yuduo. Research on weapon target allocation for army unit level CGF[D]. University of Defense Science and technology, 2016.

[5] Song Xinping, LV Guodong, Shen Yan, Wang Xiao. Research on competitor identification method based on SOM neural network in big data environment [J/OL] Information theory and practice: 1-13 [2022-04-06].

[6] Wu Hongying. Research on bank customer segmentation based on improved SOM[D]. Changchun University of technology, 2021.

[7] Xu Lili. Research on identification method and applicability of deterioration state of rolling bearing of urban rail train based on feature screening[D]. Beijing Jiaotong University, 2021.

[8] Li Pengsong, Liu Xin, Li Junda. Application research of self-organizing mapping network based on immune genetic algorithm[J]. Journal of Northeast Electric Power University, 2018,38 (06): 82-85.

# Analysis of the influence of wave parameters on helicopter floating characteristics

**Ting Jiang**[*1,2]**, Sheng-zhe Shi**[1,2]**, Xiao-long Zheng**[1,2]

[1]Aviation Industry Corporation of China Special Vehicle Research Institute, Hubei·Jingmen448035,P.R.China

[2]Key Aviation Scientific and Technological Laboratory of High-Speed Hydrodynamic, Jingmen 448035, China

[*]Corresponding author Ting Jiang's e-mail: hshl_jiangt@163.com

**Abstract.** Foreign countries have accumulated rich experience in the research of helicopter floating characteristics. From the model test in the 1950s to the current numerical simulation and extrapolation evaluation, its technical maturity is significantly higher than that in domestic related fields. In order to fully study the influence of wave parameters on helicopter floating characteristics, sort out and refine the model test methods suitable for the study of helicopter floating characteristics, this paper takes a helicopter model as the research object, and carries out the research on the helicopter model test technology with emergency buoys, focusing on the influence range of wave parameters and wave direction on helicopter test results. The results show that the motion response of the helicopter under regular wave conditions presents typical periodic changes. When the helicopter encounters the influence of transverse wave, the rolling amplitude decreases with the increase of wavelength, and the acceleration amplitude of the nose and in the aircraft first increases and then decreases with the increase of wavelength; Under other wave direction conditions, the rolling angle, the acceleration amplitude of the nose and in the machine first increase and then decrease with the increase of the wavelength.

## 1. Introduction

The floating characteristics of helicopter are the comprehensive embodiment of its ability to float stably on the wavy water surface, anti sinking performance, personnel emergency evacuation and so on. Excellent floating characteristics are the prerequisite for helicopter ditching / landing on water, and it is also a necessary condition for the helicopter to have stable floating to ensure the emergency escape of passengers in the aircraft and the smooth recovery of body devices. For any helicopter flying across the sea, it is necessary to carry out a comprehensive analysis and verification study on the floating characteristics of forced landing on water, so as to determine the stable floating sea conditions and the limit overturning boundary of the helicopter [1].

Aiming to solve the floating characteristics test technology of helicopter in China, this paper base on the target helicopter, and carries out a series of regular wave model tests with variable wave parameters, focusing on the motion law and deviation correction ability of the helicopter floating on the water surface. Based on the model motion, the influence law of wave direction and wavelength parameters on the helicopter floating characteristics is preliminarily explored.

## 2. Test method

The model test of the floating characteristics of a certain type of helicopter was carried out in the Key Laboratory of high-speed hydrodynamic aviation science and technology of China Institute of special aircraft.The whole test device is installed at the lower part of the test support, the auxiliary protection device required for the floating test is fixed on the model, and the head and tail navigation pieces of the model are installed on the navigation rod of the support bottom plate. During installation, ensure that the center line of the navigation piece is along the vertical direction to limit the yaw movement of the model. After installation, insert the auxiliary device into the navigation piece.

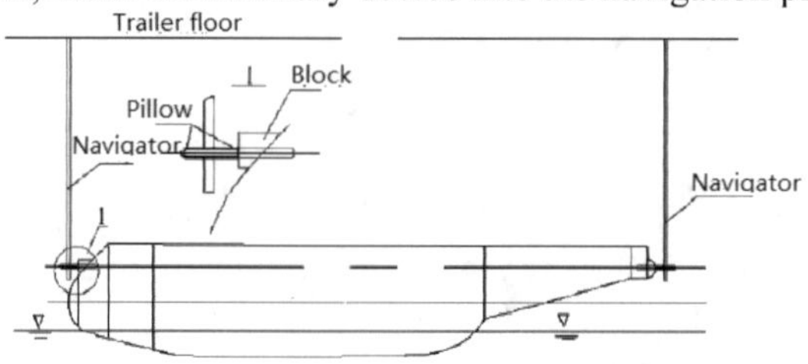


Fig 1 Helicopter floating test scheme

The helicopter floating characteristic test is mainly based on the floating stability test device of the long towing pool [2] The configuration and installation method of the test device are detailed in document 2.

The whole test process is mainly divided into the following steps:

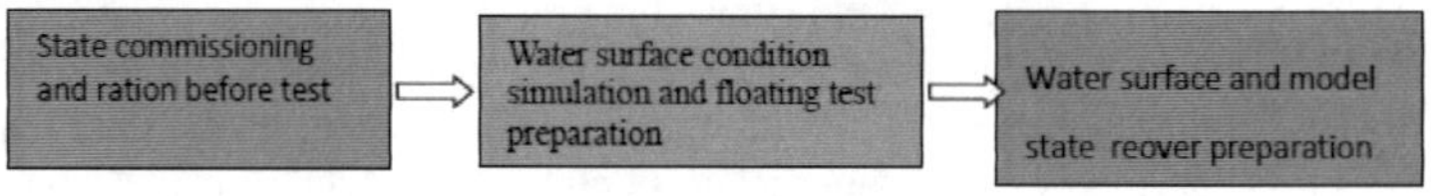


Fig 2 Test procedure

## 3. Test result and analysis

For this test, the floating characteristics test mainly focuses on the influence of wave direction and wavelength on the floating characteristics of helicopters.

Based on the research background, this paper tests and analyzes the floating motion characteristics of the helicopter based on the variable wavelength parameters and variable wave direction parameters under the same wave height.

### *3.1. Floating characteristic in regular wave*

Based on the principle of seakeeping analysis, the rolling motion of a sailing body on a regular wave surface can be divided into three parameters: roll, heave and pitch. However, the helicopter's surface floating characteristics focus on it's lateral stability and balanced floating ability, this paper will focus on the analysis of the change law of roll angle and corresponding acceleration.

Figure 1 shows the roll angle of the helicopter under regular wave conditions $\varphi$、 Variation curve of nose acceleration $A_{fz}$ and middle fuselage acceleration $a_{cz}$ with time. From the above data curve, it can be seen that the motion of the helicopter in the regular wave surface shows a significant periodic change law, that is, due to the influence of regular wave disturbance, the helicopter makes a continuous forced sway motion [3-6].During data processing, the amplitude is extracted based on the time history curve of rolling angle obtained from the test, and the influence law of wave parameters on helicopter motion is determined by comparing the magnitude and range of amplitude.

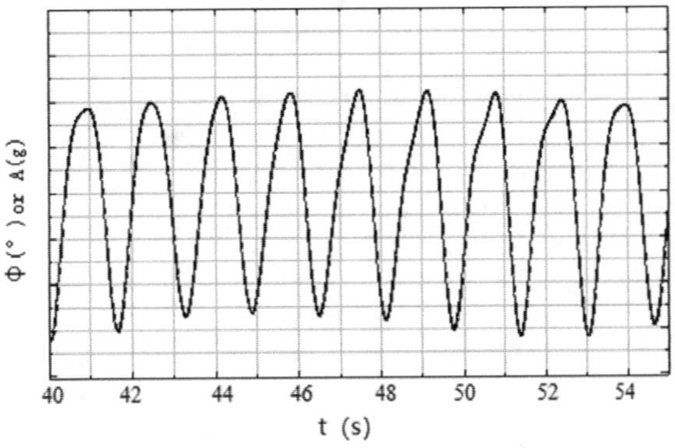


Fig 3 Time domain motion curve of helicopter in waves

In order to fully study the influence of wave direction on the helicopter's floating attitude, taking the target helicopter as the test model, The division of wave direction is as follows [7-9].

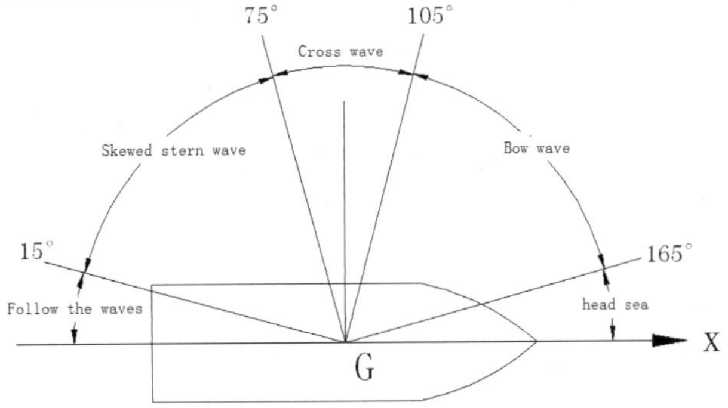


Fig 4 Wave direction classification

*3.2. Analysis of the influence of wave parameters on helicopter floating characteristics*

Figure 3 shows the test curve of helicopter equipped with flexible buoyancy bladder under free floating characteristics test conditions based on different wave direction analysis.

When the helicopter equipped with flexible buoyancy bladder floats freely on the water, its course will change significantly. When it is disturbed by slight waves or external forces, a large yaw occurs to get rid of the most unfavorable "horizontal" state (the fuselage is facing the waves laterally).

When the helicopter is in different wave directions, there are certain differences in the motion response values of the model: among them, the rolling double amplitude value φ The change is the most obvious. Under the cross wave condition, the rolling double amplitude value φ Maximum, and with the change of wave direction, this value decreases significantly; However, for the nose acceleration $A_{fz}$ and the acceleration $a_{cz}$ in the middle of the fuselage, the change of wave direction will not cause a significant change in this parameter, which is mainly manifested in that the nose acceleration $A_{fz}$ in the transverse wave state in the small wavelength range is slightly larger than that in other wave directions, and the acceleration $a_{cz}$ in the middle of the fuselage in other wave directions is slightly larger than that in transverse wave. Then, when the wavelength continues to increase, the amplitude parameters of both the nose acceleration Afz and the middle fuselage acceleration $a_{cz}$ in other wave directions are greater than the corresponding values of the transverse wave state.

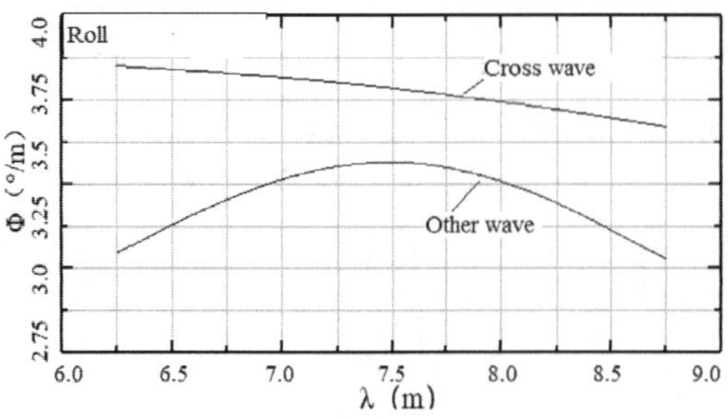


a) Roll angle comparison

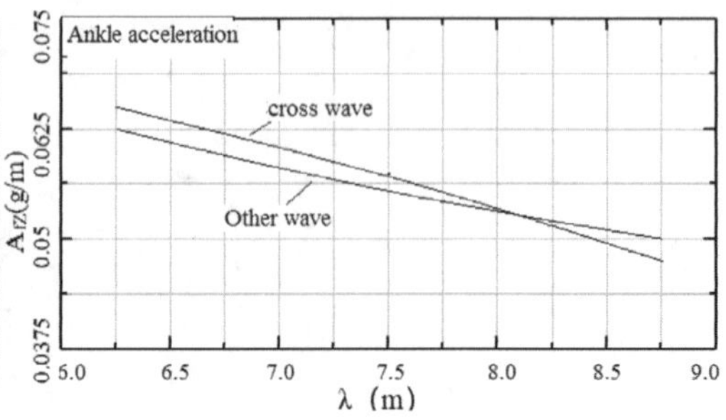


b) Ankle acceleration comparison

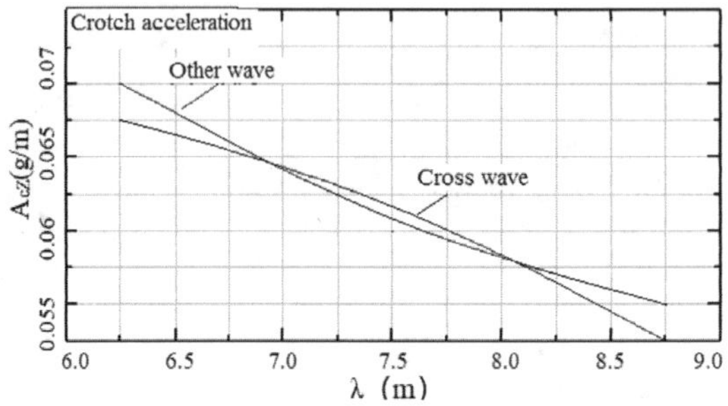


c) Crotch acceleration comparison

Fig 5 Comparison curves with different wave direction

## 4. Conclusion

Based on the regular wave surface conditions, this paper studies the helicopter floating response characteristics with different wave parameters and wave direction ranges. Through the analysis of various test data, the following conclusions can be drawn:

a) When the helicopter floats on the water, the rolling angle and acceleration both decrease with the increase of wavelength - the longer the wavelength, the smaller the motion amplitude;

b) For different wave directions, the extreme span of motion response changes greatly, among which the transverse wave time range is the largest, and the other wave direction angle time range span is small, that is, the swaying of the helicopter is the most intense under the transverse wave condition.

## References

[1] China's Civil Aviation Regulations "Part 29 Provisions on Airworthiness of Transport Rotor Aircraft"[M]. General Administration of Civil Aviation of China.2002..

[2] Jiang Ting, Jiao Jun, Tang Binbin et.Research on test technology of helicopter floating characteristics[J].2020.34（4）：30--35.

[3] Wang Zhengzhong，Chen Lixia,et.Test Research on Helicopter Ditching Load[J]. Journal of Nanjing University of Aeronautics and Astronautics 2017.2（49）：258-263.

[4] Jiang Ting，Wu Bin，et. Capillary Water Migration Effects of Modified Loess in Vertical Direction[J].2015.15（28）：200-204.

[5] Duan Guohe, Lu Jiang. Time Domain Analysis Method and Verification of Helicopter Lateral Floating Stability [J], Helicopter Technology Time Domain Analysis Method and Verification of Helicopter Lateral Floating Stability [J], Helicopter Technology.

[6] Huang Xinfeng，Stability Analysis of the Helicopter with Flotation Bags[D]，Nanjing University of Aeronautics and Astronautics，2011.

[7] Huang Miao, Wu Bin, Jiang Rong, et. An Experimental Study about Motion Response of a Seaplane on Waves[J], Journal of Experiments in Fluid Mechanics, 2015.29（3）：41-46
[8] EASA. Study on Helicopter Ditching and Crashworthiness.2007.
[9] Sheng zhen-bang, Principle of ship[M].shanghai:shanghai Jiaotong University Press,2003.

Journal of Physics: Conference Series **2364** (2022) 012009 doi:10.1088/1742-6596/2364/1/012009

# Optimization Design of Propeller for Ultra-high-altitude Unmanned Aerial Vehicle

**Yuan Yao[1,a], Dongli Ma[1,b], Liang zhang[1,*], Xiaopeng Yang[1,c]**

[1]School of Aeronautic Science and Engineering, Beihang University, Beijing, China, 100184

[a]email: 228452403@qq.com; [b]email: madongli@buaa.edu.cn

*email: microbeast@126.com; [c]email: xiaopengyang@buaa.edu.cn

**Abstract**—Ultra-high-altitude unmanned aerial vehicle raises a high demand for the efficiency of propeller at low Reynolds numbers. An optimization method for low Reynolds number propeller is established by iterative optimizations of the sectional airfoils and the chord, pitch distributions in this paper. Different from the traditional method, each sectional airfoil is continuously optimized under the specific Reynolds number and lift coefficient which are updated with the iteration in the proposed method. Adopting this method, the propeller for an ultra-high-altitude unmanned aerial vehicle is optimized. And the result shows that the optimized propeller can achieve 3.6% higher efficiency than the propeller designed by the traditional Betz method at the cruising condition ($Re \approx 4.0 \times 10^4$).

## 1. INTRODUCTION

Ultra-high-altitude unmanned aerial vehicle has great superiorities in flight height and endurance [1], which has become a research hotspot in recent years. Ultra-high-altitude unmanned aerial vehicles use propeller, which is powered by batteries and solar panels, to provide thrust. The efficiency of the propeller is the essential factor for ultra-high-altitude unmanned aerial vehicle to achieve permanent cruise, because of the aircraft's weak balance of energy [2]. However, the working Reynolds number of such propeller is very lower, due to the low air density in the near space and its low cruising velocity. Thus, laminar separation bubbles can easily generate on the blades, resulting in a significant decrease in propeller efficiency. If the influence of low Reynolds number is not considered in the design of such propeller, the propeller performance may be far less than expected.

The efficiency of propeller can be increased by improving the performance of the sectional airfoil or reducing the induced loss of the blade [3]. Adkins et al. [4] proposed a fast design method of propeller based on the Betz optimum condition and the blade element momentum theory (BEMT). When the rotational velocity, diameter of the propeller, and required thrust were given, a propeller with minimum induced loss could be quickly obtained. García-Gutiérrez et al. [5] optimized a propeller for the stratospheric airship and achieved 5% improvement in efficiency compared with the traditional method by take the variation of Reynolds number along the radial direction into consideration. However, these methods could not realize targeted optimization of each sectional airfoil under the specific working conditions and lift coefficient determined by the cruise state of the propeller. Because in the traditional non-iterative method, the design parameters and working status of the propeller are unknow before the optimization of propeller is completed. Therefore, the above methods could be further improved.

In this study, an optimization method is established for the propeller of ultra-high-altitude unmanned aerial vehicle under low Reynolds numbers. In Section II, the aerodynamic analysis methods of sectional airfoil and propeller are introduced. In Section III, the proposed iterative optimization strategy for low Reynolds number propeller is described, which can iteratively carry out the optimization of sectional airfoils and the optimization of chord and pitch distributions. In Section IV, the feasibility of the proposed method is validated by case study. In Section V, the aerodynamic characteristics of the optimized propeller is compared with that of the initial propeller designed by Betz method. Finally, the study is summarized in Section VI.

## 2. Aerodynamic analysis methods

In this study, the computational fluid dynamics (CFD) method based on $\gamma$-$Re_\theta$ model, which is proposed by Menter and Langtry [6, 7], is used to realize the high-fidelity calculation of airfoil as well as to analyze the performance of optimized propellers. Although CFD method has been widely used in aerodynamic optimization design of propeller, it is ineffective for the calculation of three-dimensional propeller in the propeller optimization. Because thousands of performance calculations for propellers need to be completed in the optimization process and each calculation can take several hours.

Thus, a faster method is needed for evaluating the performance of propeller during the optimization. BEMT [8] maybe a better choice. McCrink and Gregory [9] proposed a corrected BEMT model with corrections for tip losses, Mach effects, three-dimensional flow components, and Reynolds scaling. And XFOIL is used to generate the aerodynamic coefficients of airfoils and consider the Reynolds scaling in his study. In this paper, the corrected BEMT model is furtherly improved by adopting CFD data of airfoils under low Reynolds number instead of XFOIL. The procedure of BEMT method is summarized as follows [3]:

1) Firstly, the parameters such as free-stream velocity ($V_\infty$), propeller radius ($R$), propeller rotational velocity in revolutions per second ($n_s$), blade number ($N_B$), blade chord ($c$), and pitch ($\theta$) distributions must be known.

2) The blade is divided into several blade elements along the radial direction, and the aerodynamic coefficients of each sectional airfoil at corresponding angles of attack and Reynolds numbers are calculated by CFD method. Then, based on the CFD data, the response surface of the sectional airfoils is constructed by interpolation. The Mach correction is applied to the lift coefficient and drag coefficient of airfoils in this step.

3) Since the thrust on the blade element can be expressed by the sum of the lift and drag components or by the momentum theorem, the equation of the induced angle ($\beta$) of each blade element can be established. Thus, $\beta$ can be solved by Newton iteration method.

4) Then, the inflow angle ($\varphi$) and the induced velocity coefficient ($a_0$, $a_1$) for each blade element can be obtained. Further, the corrections for tip losses are implemented in this step.

5) Finally, the thrust ($\mathrm{d}T$) and torque ($\mathrm{d}Q$) of each blade element are calculated, and the total thrust ($T$) and torque ($M$) of the propeller can be obtained by summing $\mathrm{d}T$ and $\mathrm{d}Q$ of all the blade elements. Further, the efficiency of the propeller ($\eta$) can be calculated.

In order to validate the precision of BEMT method, the propeller in the reference [10] is selected as the test model (see Fig. 1), and the installation pitch of the propeller is 12°. The aerodynamic forces of the propeller at different advance ratio were calculated by the CFD and BEMT method. As shown in Fig. 2, the results by BEMT are in good agreement with that by CFD method and experiment data. Therefore, BMET mothed is used in the propeller optimization, and the CFD method is used to analyze the optimization results.

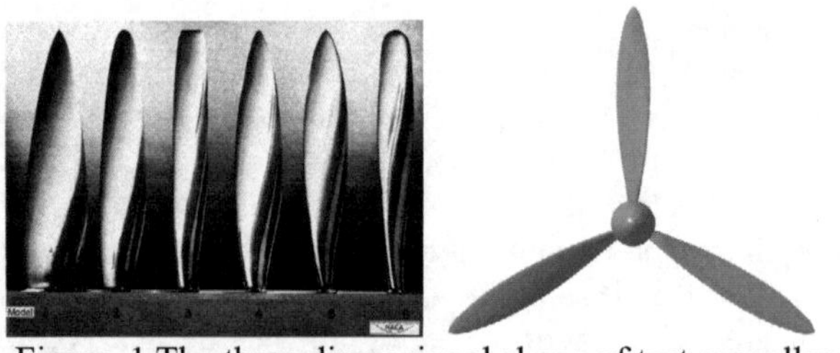

Figure. 1 The three-dimensional shape of test propeller.

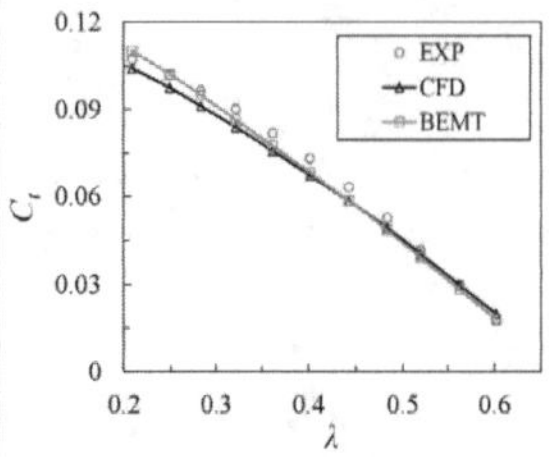


(a) Thrust coefficient vs. advance ratio

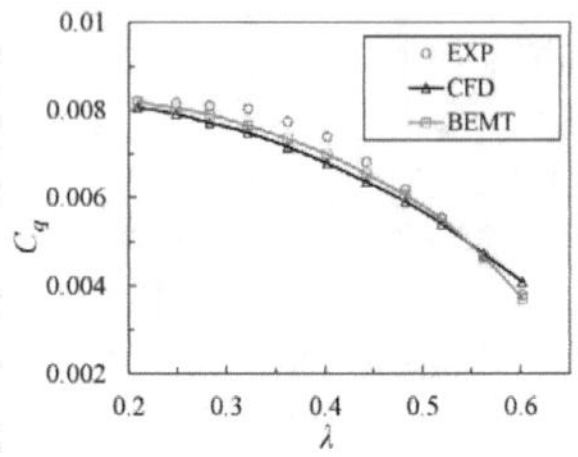


(b) Torque coefficient vs. advance ratio

Figure. 2 The experimental data and the results of CFD method and BEMT.

## 3. Iterative optimization method of low Reynolds number propeller

In this section, an optimization method for propeller working at low Reynolds number is established, including the optimizations of sectional airfoil and of chord and pitch distributions.

### *3.1. Optimization Model of Sectional Airfoil*

As shown in Fig. 3, two 5th order class-shape transformation (CST) [11] curves are used to parameterize the airfoil, which are controlled by twelve parameters ($a_0$-$a_5$ and $b_0$-$b_5$, respectively).

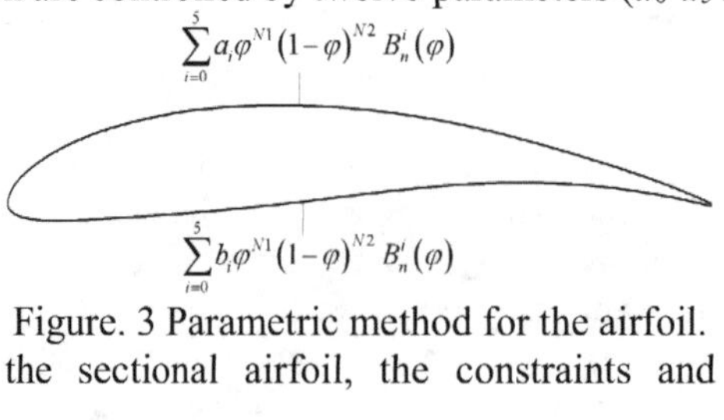


Figure. 3 Parametric method for the airfoil.

In the optimization of the sectional airfoil, the constraints and optimization objective are constructed as follows:

(a) Constraint of relative thickness

The variation in the relative thickness distribution of blade usually depends on the structural strength requirements. Generally, the relative thickness of wooden propeller is 7-8% at the tip and 20-30% near the hub [3]. In this paper, the relative thickness of the blade ($t_{\text{target}}$) is given as follows:

$$t_{\text{target}} = 0.48(\frac{r}{R})^4 - 1.62(\frac{r}{R})^3 + 2.04(\frac{r}{R})^2 - 1.14\frac{r}{R} + 03 \quad (1)$$

where $r$ is the radial position of the blade element.

(b) Constraint of lift coefficient

In order to improve the lift-drag ratio of the airfoil under the required lift coefficient, the working lift coefficient of sectional airfoils should be consistent with the actual working state of the propeller.

In the iterative optimizations, the lift coefficient distribution ($C_{l\text{-target}}$) can be calculated by BEMT method after the propeller optimization. But for the first optimization of sectional airfoils in the first iteration, the lift coefficient distribution needs to be calculated by Betz method [4] as follows.

Firstly, according to the circulation equations, the circulation distribution of the blade can be expressed as follows:

$$\Gamma = \frac{V_\infty^{\ 2}\zeta Fx\cos\varphi\sin\varphi}{N_{\mathrm{B}}n_s} \tag{2}$$

where $\Gamma$ is the circulation distribution along the radial direction of the blade, $x$ is defined as $x=2\pi n_s r/V_\infty$, $r$ is the radial position of blade element, $\zeta$ is a constant for an optimum design, $F$ is the momentum loss factor and can be estimated as follows [3, 4]:

$$\begin{cases} F = (2/\pi)\arccos(e^{-f}) \\ f = (N_{\mathrm{B}}/2)(1-r/R)/\sin\varphi_{\mathrm{t}} \end{cases} \tag{3}$$

Then, the lift of each blade element can be calculated by

$$\mathrm{d}L = N_{\mathrm{B}}\Gamma\,\rho\cdot V\mathrm{d}r \tag{4}$$

where $V$ is the actual velocity of blade element (see Fig. 4).

In addition, the lift of each blade element can be also written as

$$\mathrm{d}L = \frac{1}{2}N_{\mathrm{B}}c\rho V^2\cdot C_l\mathrm{d}r \tag{5}$$

According to Eqs. (6) and (7), the expected lift coefficient distribution ($C_{l\text{-target}}$) can be expressed [4] as

$$C_{l-\mathrm{target}} = \frac{2\Gamma}{cV} \tag{6}$$

(c) Constraint of Reynolds number

The working Reynolds number of sectional airfoil ($\mathrm{Re}_{\mathrm{target}}$), which is determined by the working status of propeller, can be calculated by

$$\mathrm{Re}_{\mathrm{target}} = \frac{\rho cV}{\mu} \tag{7}$$

where $\mu$ is the dynamic viscosity coefficient.

(d) Objective function of the sectional airfoil optimization

The efficiency of blade element is used as the optimization objective function in the optimization of corresponding sectional airfoil, which can be calculated by

$$\eta_{\mathrm{element}} = \frac{\mathrm{d}T\cdot V_\infty}{\mathrm{d}F\cdot 2\pi n_s r} \tag{8}$$

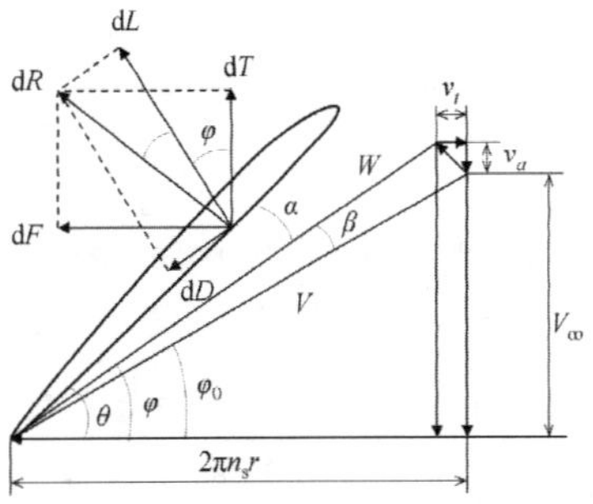


Figure. 4 Force diagram for a blade element.

As shown in Fig. 4, the efficiency of each blade element can be written as

$$\eta_{\text{element}} = \frac{\cos\varphi - \gamma\sin\varphi}{\sin\varphi + \gamma\cos\varphi} \cdot \tan\varphi_0 \tag{9}$$

where $\gamma$ is drag-to-lift ratio.

The induced angle ($\beta$) is ignored in the initial optimization, which means that the efficiency of each blade element can be written as

$$\eta_{\text{element}} = \frac{\sin\varphi_0(1-\gamma\tan\varphi_0)}{\sin\varphi_0 + \gamma\cos\varphi_0} \tag{10}$$

Thus, the optimization model of the sectional airfoil can be constructed as follows:

$$\begin{aligned} &\text{find} && \boldsymbol{x} = [\, a_0, \cdots, a_5, b_0, \cdots, b_5] \\ &\max && \eta_{\text{element}} \\ &\text{s.t.} && \mathrm{Re} = \mathrm{Re}_{\text{target}} \\ &&& C_l = C_{l\text{-target}} \\ &&& t = t_{\text{target}} \end{aligned} \tag{11}$$

### *3.2. Optimization Model of Chord and Pitch Distributions*

After finishing the optimization of sectional airfoils, the chord and pitch distributions of the propeller can be optimized.

The chord and pitch distributions of the blade are parameterized by two 4th order CST curves and are controlled by parameters $c_0$-$c_4$ and $d_0$-$d_4$, respectively. To avoid unreasonable design, the pitch is limited in the range of 0° to 70°. Then, similar to the airfoil optimization, the constraints and objective function for propeller optimization are determined as follows:

(a) Constraint of required thrust

The efficiency corresponding to the required thrust ($T_{\text{req}}$) of the propeller is the key factor that affects the energy consumption of aircraft in cruising. Newton iterative method is used to solve the propeller rotational velocity and aerodynamic efficiency at the required thrust. The detailed steps are as follows:

Based on the BEMT method, the thrust $T(n_s)$ at the rotational velocity $n_s$ can be calculated. Then, the function $F(n_s) = T(n_s) - T_{\text{req}}$ is constructed. The value of function at an initial rotational velocity $n_{s0}$ is expressed as $F(n_{s0})$, and the function value at $n_{s0}+\Delta n_s$ is expressed as $F(n_{s0}+\Delta n_s)$. As shown in Fig. 5, the equation of the straight line through these two points can be expressed as follows:

$$\frac{F(n_{s0}+\Delta n_s) - F(n_{s0})}{\Delta n_s} = \frac{F - F(n_{s0})}{n_s - n_{s0}} \tag{12}$$

The intersection of the straight line and the zero axis ($F(n_s) = 0$) is defined as $n_{s1}$, and its value can be calculated as follows:

$$n_{s1} = n_{s0} + \frac{\Delta n_s \cdot F(n_{s0})}{F(n_{s0}) - F(n_{s0}+\Delta n_s)} \tag{13}$$

Then, the next intersection $n_{s2}$ can be calculated based on $n_{s1}$ using the same method (see Fig. 5). The same operation will be executed $i$ times until the function value $F(n_{si})$ corresponding to $n_{si}$ becomes sufficiently close to 0, which means that the propeller thrust $T(n_i)$ is close enough to $T_{\text{req}}$. Therefore, $n_{si}$ can be considered as the propeller rotational velocity $n_{s\text{-req}}$ corresponding to $T_{\text{req}}$ when the convergence condition $F(n_{si}) < 1\times10^{-3}$ is satisfied.

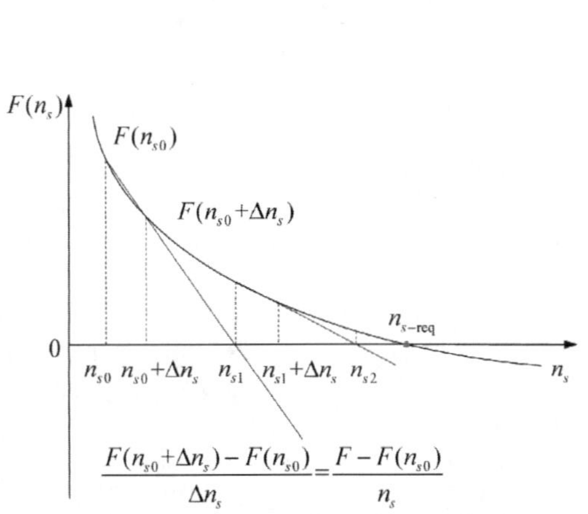


Figure. 5 Newton iteration method

(b) Objective function for the optimization

The aim of the optimization is to maximize the efficiency ($\eta_{\text{req}}$) at the required thrust ($T_{\text{req}}$), which can be expressed by

$$\eta_{\text{req}}=\frac{T_{\text{req}}V_{\infty}}{2\pi n_{s\text{-req}}Q_{\text{req}}} \tag{14}$$

where $Q_{\text{req}}$ is the propeller torque corresponding to $T_{\text{req}}$.

In summary, the optimization model of chord and pitch distributions can be described as follows:

$$\begin{aligned} &\text{find} \quad \boldsymbol{x}=[\,c_0,\cdots,c_4,d_0,\cdots,d_4] \\ &\max \quad \eta_{\text{req}} \\ &\text{s.t.} \quad T=T_{\text{req}} \end{aligned} \tag{15}$$

*3.3. Iterative Optimization Strategy for Low Reynolds Number Propeller*

The main idea of iterative optimization strategy is to iteratively optimize the sectional airfoils as well as the chord and pitch distributions under low Reynolds number until the convergence condition is satisfied. The latter optimization is conducted based on the former optimization results, and the former optimization can be conducted again according to the results of the latter. Therefore, in each iteration, all the sectional airfoils can be optimized under specific Reynolds number and lift coefficient determined by the cruising state of the propeller. After several iterations, a satisfactory design can be obtained. The detailed optimization method is introduced as follows:

Step 1. Input the initial design parameters

Firstly, free-stream velocity ($V_{\infty}$), propeller radius ($R$), blade number ($N_B$), required thrust ($T_{\text{req}}$), propeller rotation velocity ($n_s$) and the aerodynamic data of the initial sectional airfoils are initialized. Notice that the rotational velocity is just empirically given in this step and will became $n_{s\text{-req}}$ (the rotational velocity corresponding to the required thrust) after the optimization of the propeller in Step 5.

Step 2. Design the initial propeller by Betz method

Using the known parameters, an initial propeller is designed by Betz method to get the initial chord and pitch distributions.

Step 3. Optimize the sectional airfoils.

All the sectional airfoils in the radial range of 0.2$R$-1.0$R$ on the blade are optimized according to the optimization model of sectional airfoil and genetic algorithm. And the constraints of lift coefficient ($C_{l\text{-target}}$) and working Reynolds number ($\text{Re}_{\text{target}}$) are calculated according to the optimization result in Step 2 in the first iteration. And in following iterations, they are determined by the optimization result in Step 5.

Step 4. Construct the aerodynamic response surface of optimal sectional airfoils.

For each sectional airfoil, the aerodynamic response surface is constructed by polynomial interpolation of CFD data.

Step 5. Optimize the chord and pitch distributions.

The chord and pitch distributions of the propeller are optimized according to the optimization model of the chord and pitch distributions and genetic algorithm.

Step 6. Obtain the final optimal solution.

Step 3, Step 4 and Step 5 are repeated until the relative difference in maximum $\eta_{\mathrm{req}}$ in Step 5 between two adjacent iterations is less than $\varepsilon$. Once the convergence criterion is satisfied, the whole optimization is finished and the propeller with maximum efficiency $\eta_{\mathrm{req}}$ in the last iteration is considered as the final optimal solution.

The optimization procedure is illustrated in Fig. 6.

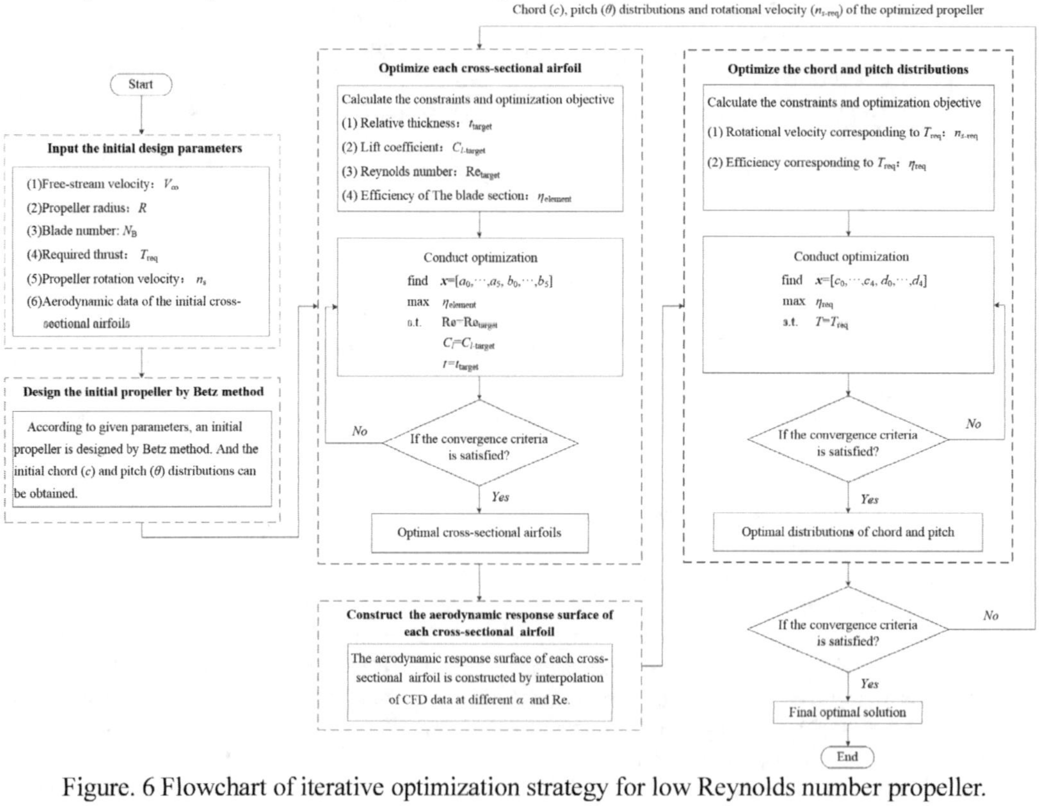


Figure. 6 Flowchart of iterative optimization strategy for low Reynolds number propeller.

## 4. Case Study

### 4.1. Design Input and Conditions

The propeller for an ultra-high-altitude unmanned aerial vehicle working at the altitude of 20 km is optimized to test the proposed method. The design requirements and cruise conditions for the propeller are given in Table 1.

TABLE 1. PARAMETERS OF PROPELLER AND CRUISE CONDITIONS.

| $N_B$ | D (m) | $V_\infty$ (m/s) | $\rho$ (kg/m$^3$) | $\mu$ (Ns/m$^2$) | $T_{\mathrm{req}}$ (N) |
|---|---|---|---|---|---|
| 2 | 2 | 32 | 0.0889 | $1.422\times10^{-5}$ | 20 |

### 4.2. Optimization Results

The optimization of the propeller for the ultra-high-altitude unmanned aerial vehicle is finished after several iterative optimizations of sectional airfoils and chord, pitch distributions. Fig. 7 displays the convergence history of the propeller efficiency in the final iteration. Besides, an initial propeller is designed by Betz method adopting given initial sectional airfoils in the second step of the optimization procedure.

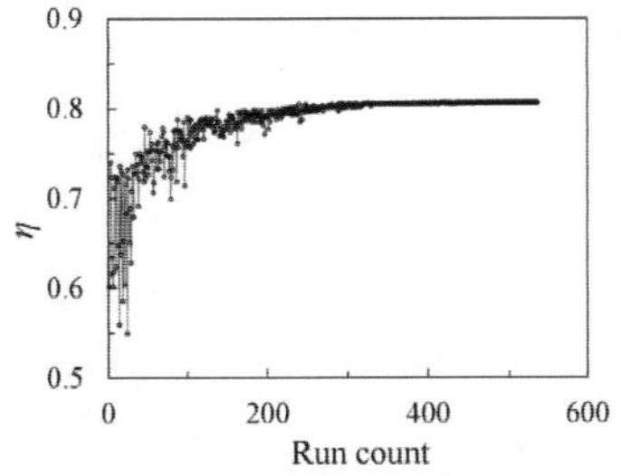


Figure. 7 Convergence history of the optimization for the regular propeller.

(a) Prop A

(b) Prop B

Figure. 8 Geometries of two propellers

For convenience, the propeller initially designed by Betz method is denoted by Prop A, the optimized propeller by the iterative optimization is denoted by Prop B in the following. Fig. 8 displays the geometries of two propellers, and Fig. 9 and Fig. 10 show the chord and pitch distributions of them, respectively. It can be seen that Prop B has larger chord and pitch than Prop A, which implies that Prop B have lower rotational velocities at the same thrust.

As shown in Table 2, the errors of two propellers' efficiencies at cruise state obtained by the BEMT in the optimization are all within 1% compared with that by the CFD method. And the optimized propeller can achieve 3.6% higher efficiency than the initial propeller designed by the Betz method at the cruising altitude of 20 km at $Re_{average} \approx 4.0\times10^4$ (see Table 2). Thus, the iterative optimization is effective.

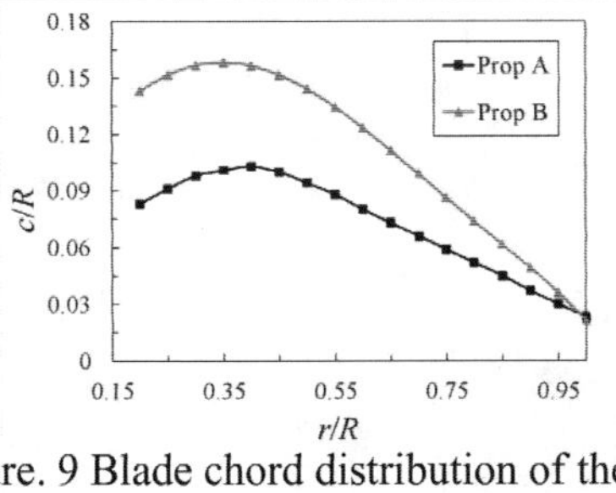


Figure. 9 Blade chord distribution of the two propellers.

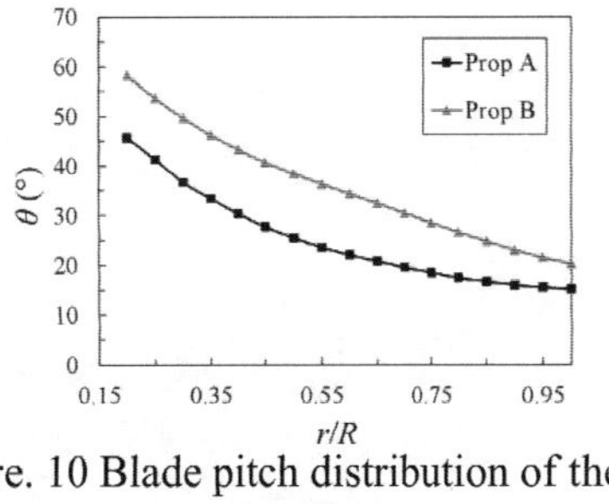


Figure. 10 Blade pitch distribution of the two propellers.

TABLE 2. COMPARISON BETWEEN PROP A AND PROP B AT THE CRUISE STATE.

| Propeller | $T_{req}$(N) | $n$ (rpm) | $\lambda$ | $\eta$ | | $Re_{average}$ |
|---|---|---|---|---|---|---|
| | | | | BEMT | CFD | |
| Prop A | 20 | 1515 | 0.6354 | 76.5% | 77.0% | $4.03\times10^4$ |
| Prop B | 20 | 978 | 0.9816 | 80.0% | 80.6% | $4.23\times10^4$ |

## 5. Mechanism analysis

### *5.1. Aerodynamic analysis of two-dimensional sectional airfoil*

The sectional airfoils at 0.7R on Prop A and Prop B are analyzed in this subsection, which are respectively denoted by Airfoil A and Airfoil B for convenience. Fig. 11 displays the geometries of the two sectional airfoils.

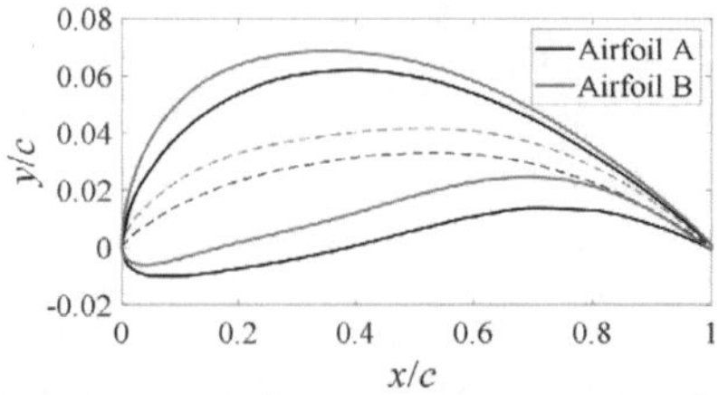


Figure. 11 Geometries of Airfoil A and Airfoil B.

The aerodynamic performances of the two sectional airfoils at different Reynolds numbers (from $Re = 3.4\times10^4$ to $Re = 2.05\times10^5$) are displayed in Fig. 12, which are calculated by CFD method based on $\gamma$-$Re_\theta$ model. It can be seen that the Airfoil B has higher lift coefficient and larger maximum lift-to-drag ratio than Airfoil A. This is because the optimization under specific Reynolds number and lift coefficient helps Airfoil B in obtaining higher lift coefficients by larger camber to overcome the lift loss under low Reynolds number and achieve greater lift-to-drag ratio.

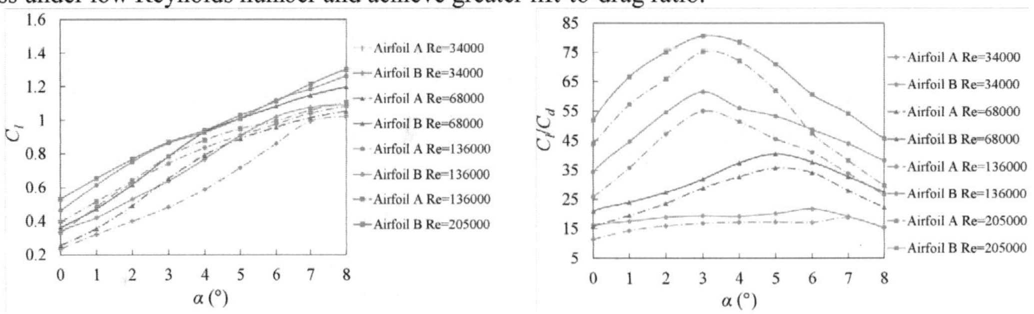


(a) Lift coefficient vs. angle of attack (b) Lift-to-drag ratio vs. angle of attack

Figure. 12 Aerodynamic performance of Airfoil A and Airfoil B at different Reynolds numbers.

### *5.2. Aerodynamic Analysis of Three-dimensional Blade*

The three-dimensional flow characteristics of the two propellers under $Re_{average} \approx 4.0\times10^4$ are analyzed in this subsection. The computational grids of the two propellers are displayed in Fig. 13. The pressure coefficient distributions on the blades and streamlines on different cross-sections of the two propellers are displayed in Fig. 14.

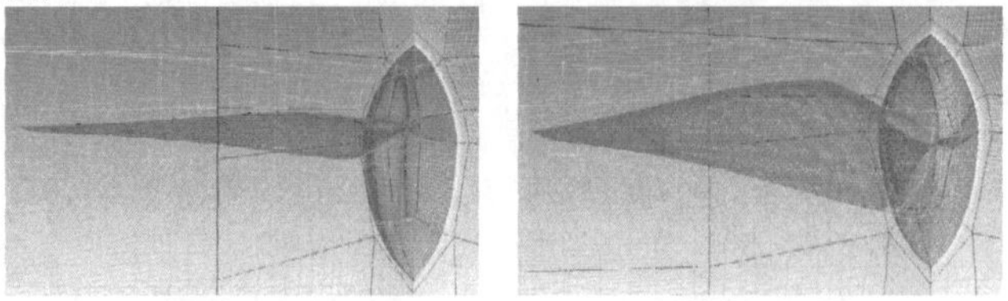

(a) Prop A (b) Prop B

Figure. 13 Computational grid of the two propellers.

It can be seen that there are obvious laminar separation bubbles on the surface of two propellers at $Re_{average} \approx 4.0\times10^4$. At this Reynolds number, Prop B has higher maximum efficiency than Prop A. Because the sectional airfoils of Prop B obtained better the aerodynamic performances under low Reynolds number after the iterative optimization under specific Reynolds number and lift coefficient, and form a better combination with the propeller's chord and pitch distributions. The streamlines of two propellers show that the laminar separation bubbles on Prop B are smaller than that on Prop A, which implies that the laminar separation bubbles are suppressed effectively after the optimization.

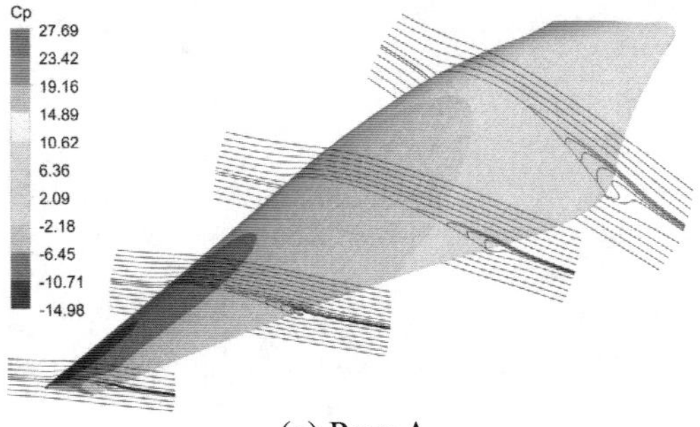


(a) Prop A

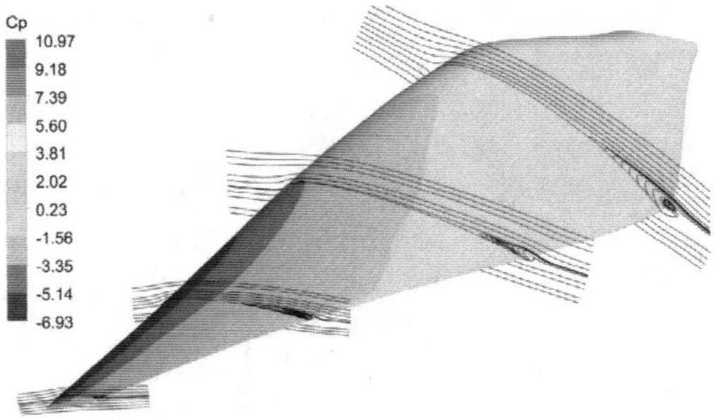


(b) Prop B

Figure. 14 Pressure coefficient distribution and streamlines on different cross-sections of two propellers at $Re_{average} \approx 4.0\times10^4$.

## 6. Conclusion

This paper proposed an optimization method of the propeller on ultra-high-altitude unmanned aerial vehicle, which can iteratively optimize the sectional airfoils and the chord, pitch distributions. The feasibility and effectiveness of the proposed methodology were examined by a case study. The main conclusions are summarized as follows:

(1) The laminar separation bubble was the main reason for the deterioration of the propeller efficiency under very low Reynolds number. The key to design such propeller is to overcome the adverse effects of laminar separation bubbles on the blade.

(2) Properly increasing the camber of the sectional airfoil can help improve the aerodynamic efficiency of the low Reynolds number propeller.

(3) The propeller optimized by proposed method can achieve 3.6% higher efficiency than the initial propeller designed by the Betz method at the cruising altitude of 20 km (Reynolds number was near $4.0\times10^4$). Because after the iterative optimizations, the aerodynamic performances of the sectional airfoils of the propeller are significantly improved and form a better combination with the chord and pitch distributions.

## Acknowledgment

This research was funded by the National Key Project GJXM92579.

## References

[1] G. Romeo, G. Frulla, and E. Cestino, "Design of solar high altitude long endurance aircraft for multi payload & operations," *Aerosp. Sci. Technol.*, vol. 10, no. 6, pp. 541-550, Sep. 2006.

[2] R. Ma, B. Zhong, P. Liu. "Optimization design study of low-Reynolds-number high-lift airfoils for the high-efficiency propeller of low-dynamic vehicles in stratosphere," *Sci. China Technol. Sci.*, vol. 53, no.10, pp. 2792–2807, Oct. 2010.

[3] P. Liu. *Air propeller theory and its application*. Beijing, China: Beihang University Press; 2006. pp. 75-99.

[4] C. N. Adkins, and R. H. Liebeck, "Design of optimum propellers," *Journal of Propulsion & Power*, vol. 10, no. 5, pp. 676-682, Oct. 1994.

[5] A. García-Gutiérrez, J. Gonzalo, D. Domínguez, D. López, and A. Escapa, "Aerodynamic optimization of propellers for High Altitude Pseudo-Satellites," *Aerosp. Sci. Technol.*, vol. 96, p. 105562, Nov. 2019.

[6] F. R. Menter, R. B. Langtry, S. R. Likki, Y. B. Suzen, P. G. Huang, and S. Volker, "A Correlation-Based Transition Model Using Local Variables-Part I: Model Formulation," *J. Turbomach.*, vol. 128, no. 3, pp. 413-422, 2006.

[7] R. B. Langtry, F. R. Menter, S. R. Likki, Y. B. Suzen, P. G. Huang, and S. VöLker, "A Correlation-Based Transition Model Using Local Variables: Part II — Test Cases and Industrial Applications." *J. Turbomach.*, vol. 128, no. 3, pp. 433-434, 2006.

[8] L. D. Koch. "Design and performance calculations of a propeller for very high altitude flight," Degree awarded by Case Western Univ 1998.

[9] M. McCrink, H. Matthew and J. W. Gregory, "Blade element momentum modeling of low-Re small UAS electric propulsion systems," in *Proc. 33rd AIAA Applied Aerodynamics Conference*, June. 2015, p. 3296.

[10] E. G. Reid, "The influence of blade-width distribution on propeller characteristics," *technical report archive & image library,* 1949.

[11] B. M. Kulfan, "Universal parametric geometry representation method," *J. Aircr.*, vol. 45, no. 1, pp. 142-158, Jan. 2008.

[12] Y. Yao, D. Ma, L. Zhang, X. Yang, and Y. Yu, "Aerodynamic Optimization and Analysis of Low Reynolds Number Propeller with Gurney Flap for Ultra-High-Altitude Unmanned Aerial Vehicle," *Applied Sciences,* vol. 12, p. 3195, 2022.

# 3D Mapping and Positioning Technology of Lunar Environment Based on LiDAR

**Shi Biao Du[1,2a], Xin Zhao[3,4b], Hong Yong Fu[5,6c*]**

[1]Technology and Engineering Center for Space Utilization, Chinese Academy of Sciences, Bei Jing, China

[2]University of Chinese Academy of Sciences, Bei Jing, China

[3]Technology and Engineering Center for Space Utilization, Chinese Academy of Sciences, Bei Jing, China

[4]University of Chinese Academy of Sciences, Bei Jing, China

[5]Technology and Engineering Center for Space Utilization, Chinese Academy of Sciences, Bei Jing, China

[6]University of Chinese Academy of Sciences, Bei Jing, China

[a]e-mail: 15712596378@163.com; [b]e-mail: zhaoxin19@csu.ac.cn

[*]Corresponding author: [c]fuhongyong@csu.ac.cn

**Abstract**—Using intelligent guidance system for lunar sample collection, Simultaneous lunar environment 3D mapping and localization technology (SLAM) is a key problem to be solved urgently. The lunar illumination changes dramatically and the terrain is complex. And limited computing power of the intelligent guidance system on the moon will affect the positioning accuracy of the algorithm. In view of the low positioning accuracy of LiDAR in the current lunar environment. This paper use an algorithm with higher positioning accuracy, which integrates the front-end feature extraction algorithm, the back-end optimization algorithm based on factor graph and the Scan Context loop closure detection algorithm. The simulation platform of lunar environment is built by Airsim to test the algorithm, and the robustness and accuracy of the algorithm are verified.

## 1. Introduction

Accurate positioning is crucial when collecting samples on the lunar surface using an intelligent guidance system with a 3D lidar. Many scholars at home and abroad have studied the lunar surface positioning technology. The main methods are: dead reckoning and positioning technology, visual navigation and positioning technology, astronomical navigation and positioning technology, radio navigation and positioning technology, combined navigation and positioning technology, etc. Among them, the navigation and positioning based on the vision of the lunar rover is gradually becoming the development trend of the navigation and positioning of the lunar rover [1].

The important reason why the lunar environment is more difficult than the earth's environment is the complexity and uncertainty of the lunar environment itself. At present, most earth's positioning technologies are not suitable for the positioning on the moon [2]. In recent years, some research institutions have tested the SLAM technology in the simulation lunar environment. In reference 3 a Lidar and IMU fusion method is proposed, By incremental optimization algorithm based on surface location capability estimation and salient feature cloud extraction to improve the robustness [3]. However, the backend adopts a filtering-based method, and the update efficiency will decrease as the running time increases. In reference 4 a camera and IMU fusion method is proposed, By an optical flow tracing algorithm based on quaternion tree to improve the accuracy of relative posture estimation between keyframes[4]. However, when the lunar surface illumination changes too drastically, the camera cannot estimate its own pose well. Reference 5 study the consistency of the SLAM algorithm of Mars rover based on EKF frame-work. The effec-tiveness and accuracy of the navigation system of Mars rover are proved by the ground tests and the analysis results [5]. The disadvantage is also that the filter-based method will reduce the efficiency over time. Therefore, in order to solve the above problem, the algorithm of this paper is as follows.

The lunar illumination changes dramatically, which has a great impact on the camera imaging effect. The terrain of the lunar surface is complex, and meteorite impact craters account for more than 10% of the lunar surface area. And the computing power of intelligent guidance system is limited. So this paper use a Lidar SLAM algorithm which is not affected by illumination. The front end of SLAM algorithm calculates the curvature of point cloud in each frame, and selects feature points with large curvature to replace the overall point cloud, which greatly reduces the computational cost of the algorithm. And the back-end of SLAM algorithm adopts the back-end optimization method based on factor graph. It is suitable for positioning and mapping of large scenes. In the process of positioning and mapping, the old point cloud is discarded in batches and new point cloud is added, which not only ensures the robustness of the algorithm in large-scale scenes, but also ensures the positioning accuracy.

This paper uses UE4 and Airsim to build a simulated lunar environment. The algorithm is tested, and the results show that the average positioning accuracy of the three axes of x, y, and z is below 1.3m.

## 2. Lidar Odometer

The LOAM algorithm was proposed in 2014[7]. Later, based on this paper, some scholars proposed LEGO-LOAM , L IO-SAM, LV-LOAM and so on[8-10]. The algorithm mainly consists of two parts, one part performs odometry at a high frequency but low stability to estimate motion of the Lidar. Another part runs at a low frequency for fine matching and registration of the point cloud. Combining the two algorithms can complete real-time mapping. Next, we will introduce from two aspects of feature extraction and feature matching..

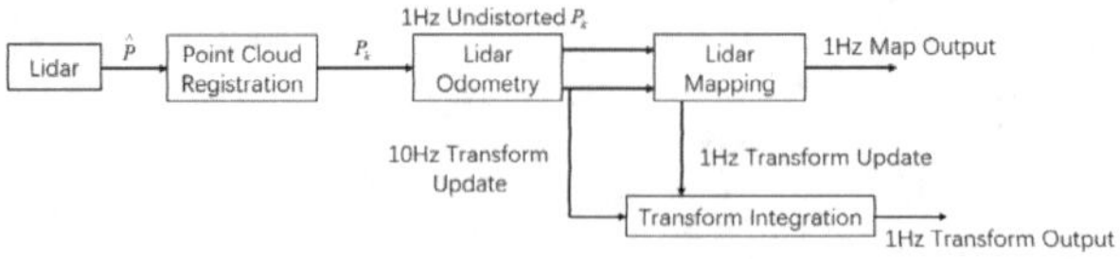


Fig. 1 Block diagram of the lidar odometry and mapping software system

$\hat{p}$ be the points received in a laser scan, and $p_k$ represents the point cloud perceived during sweep k. The lidar odometry takes the point cloud and computes the motion of the lidar between two consecutive sweeps to correct distortion in $p_k$ . Then, the $p_k$ is sent to the Lidar mapping node at a frequency of 1 Hz. The mapping node also outputs the mapping result at a frequency of 1 Hz. The $p_k$ is sent to the transform integration node at a frequency of 10 Hz, and the mapping node sends the mapping result to the transform integration node at a frequency of 1 Hz. The transform integration node uses the above two parts of information to output the pose of the Lidar relative to the map at a frequency of 10Hz, that is the final odometer information.

*2.1.Feature extraction*

LOAM innovatively proposes a feature point extraction method. By calculating the smoothness of the points, edge points and planar points are extracted, and these two types of points are used to replace the overall points to reduce the computation.

$$c = \frac{1}{|S| \left\| X^L_{(k,i)} \right\|} \sum_{j \in S, j \neq i} \left\| X^L_{(k,i)} - X^L_{(k,j)} \right\| \tag{1}$$

$S$ be the set of continuous points returned by the laser scanner in the same scan, $X^L_{(k,i)}$ be the coordinates of a point $i \in P_k$ in the laser coordinate system L. The c value represents the smoothness of the point in the set S. Determines whether it is a planar point or an edge point by setting a threshold. In order to make the feature points evenly distributed in the environment. Separate a scan into 4 segments, each providing 2 edge points and 4 planar points.

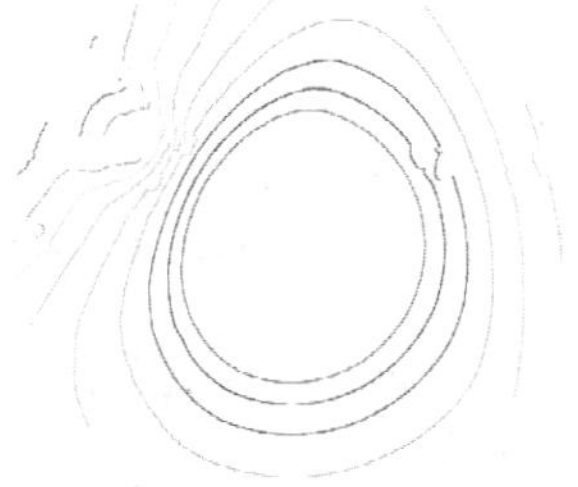

Fig. 2 Lidar point cloud

Fig. 3 Point cloud after feature extraction

At the same time, in order to improve the accuracy, point clouds with the following two characteristics cannot be selected.

- A local plane roughly parallel to the laser beam
- The surrounded points have been selected as a planar points or edge points.
- The points at boundary of occluded regions

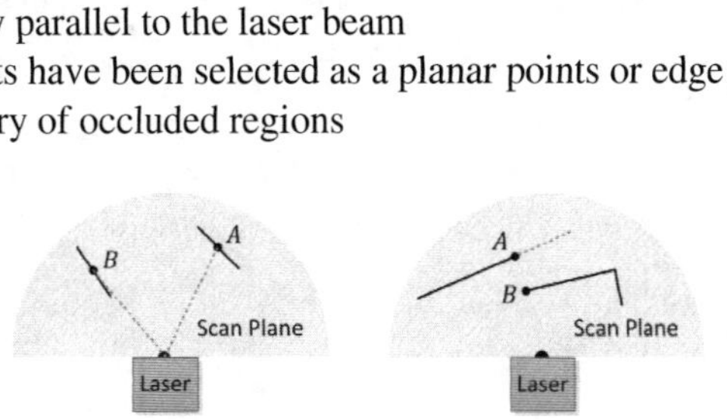


Fig. 4 remove special points

### 2.2. Feature matching

After feature extraction, we need to perform feature matching. First, to project the two frames of point clouds to the same timestamp. The $t_k$ be the starting time of a sweep k. $t_{k+1}$ be the ending time of a sweep k , which is also the starting time of a sweep k+1. Then the point cloud of the sweep k are projected to time $t_{k+1}$ to match.

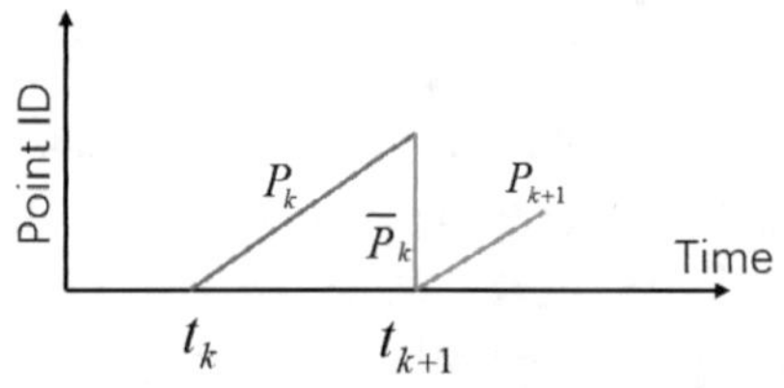


Fig. 5 Point cloud projection

Let $P_k$ be the points cloud of sweep k, $\overline{P}_k$ be the reprojected point cloud. Let $E_k$ and $H_k$ denote the set of edge points and planar points, respectively. Similarly, Let $P_{k+1}$ be the points cloud of sweep k+1. Let $E_{k+1}$ and $H_{k+1}$ denote the set of edge points and planar points, respectively. It is required to solve the pose transformation relationship between two frames. It only needs to solve the pose relationship between the plane point and the edge point in two frames. Reprojecting $E_{k+1}$ and $H_{k+1}$ to the staring time of the sweep as $E_{k+1}$ and $H_{k+1}$

With the above reprojecting, you can construct a function and iteratively recover the lidar motion. During edge point matching, Finding an edge line in $E_k$ as the correspondence for an edge point in $E_{k+1}$. By minimizing Point-to-Line distance to solving lidar motion.

The point in $E_{k+1}$ to line in $E_k$ distance can be computed as

$$d_\varepsilon = \frac{\left|\left(X^L_{(k+1,i)} - \overline{X}^L_{(k,j)}\right) \times \left|\left(X^L_{(k+1,i)} - \overline{X}^L_{(k,l)}\right)\right.\right.}{\left|\overline{X}^L_{(k,j)} - \overline{X}^L_{(k,l)}\right|} \tag{2}$$

The numerator of this formula represents the area of a triangle formed by the three points. The denominator represents the distance between two points in $E_k$. By minimizing the $d_\varepsilon$ recover the lidar motion.

### 2.3. Lidar Odometry Algorithm

The overall Lidar Odometry Algorithm is as follows :

- Enter the result obtained from the previous iteration $\overline{P}_k, P_{k+1}, T^L_{k+1}$
- $T^L_{k+1}$ is set to zero, while a new sweep is started.
- Extracts feature points $E_{k+1}$ and $H_{k+1}$ from $P_{k+1}$
- For each feature point $E_{k+1}$ and $H_{k+1}$ we find its correspondence in $\overline{P}_k$
- Iterative updates $T^L_{k+1}$ make the distance function trend zero.
- Iterating terminates when the convergence is found or exceeded maximum number of iterations.
- When the algorithm reaches the end of a sweep, $P_{k+1}$ reprojected to $t_{k+2}$, which is used to estimate the motion of the next sweep. $\overline{P}_{k+1}$ and $T^L_{k+1}$ are returned, otherwise only the pose transform $T^L_{k+1}$ is returned.

## 3. BACK-END OPTIMIZATION

A factor graph is an undirected graph consisting of two types of nodes: variable nodes, which represent optimization variables, and factor nodes, which represent factors. This part of the basic theory comes from the reference[11,12].

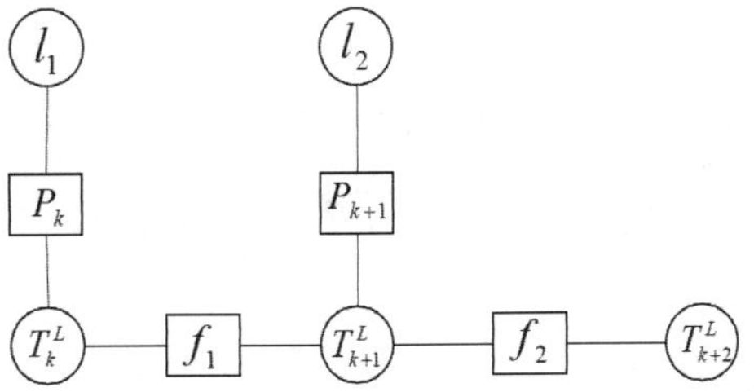


Fig. 6 factor graph

where and $T_k^L$ are variables, be the Lidar poses. $P_k$ is observation. And $l_i$ is landmark.

$$T_k^L = [x, y, z, yaw, pitch, roll] \tag{3}$$

$f$ be the transformation between two poses, commonly known as the equation of motion.

$$T_{k+1}^L = f(T_k^L, u_k) + w_k \tag{4}$$

$w_k$ is noise and obeys a Gaussian distribution with a mean of 0. $w_k = N(0, R_k)$ Then the following formula is satisfied.

$$P(T_{k+1}^L \mid T_k^L) = N(T_k, u_k, R_k) \tag{5}$$

Similarly for the observation data there are:

$$P(P_k \mid T_k^L, l_k) = N(h(T_{k-1}^L, l_k), Q_{kj}) \tag{6}$$

The Lidar pose and map can be optimized by maximizing the posterior probability. In order to facilitate the solution, the negative logarithm is used to transform it into a least squares problem, which is solved by GN or LM[13].

As the map is constantly accumulating, the nodes are constantly increasing. It is obviously unreasonable to recalculate all nodes every time while a node is added. Based on the reference [11,12]. A novel Bayesian tree is proposed to deal with sparse matrices. As shown in the figure below, when nodes are continuously added, the affected nodes can be regarded as all nodes in the region from the beginning of the loop closure to the current frame, so that we do not need to optimize the entire graph. Significantly saves computation.

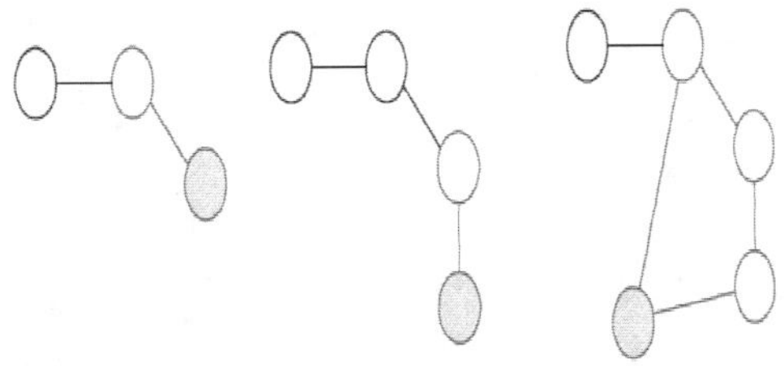

Fig. 7 Incremental processing nodes

## 4. LOOP CLOSURE DETECTION BASED ON SCAN CONTEXT

Scan context was proposed by Giseop Kim and Ayoung Kim from KAIST University in Korea in 2018[14]. Mainly used for relocation and scene recognition based on 3D point cloud. It can also be used for loop closure detection of Lidar slam. The main process is as follows.

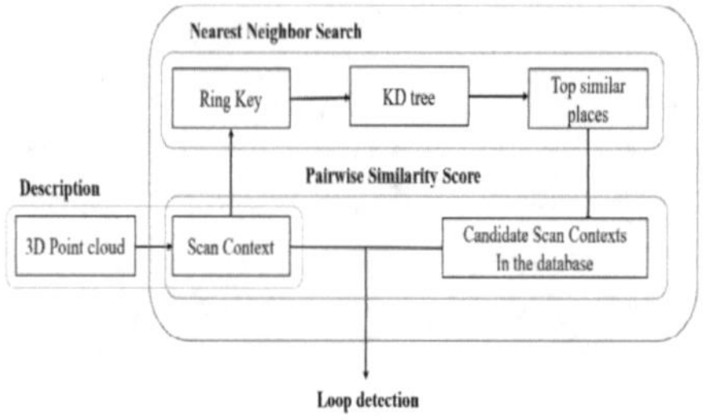


Fig.8 Loop closure detection based on scan - context

First, the point cloud of each sweep is segmented.

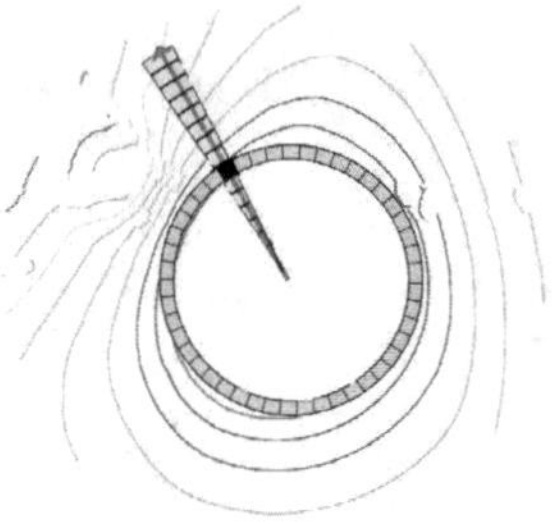

Fig.9 point cloud segmentation

The value of each grid takes the maximum z value of the point cloud of the grid as the value of the grid.

$$\phi(P_k) = \max_{X^L_{(k,i)} \in P_k} z(X^L_{(k,i)}) \tag{7}$$

In order to overcome the rotation-invariant, a rotation invariant descriptor needs to be constructed to match the point clouds of two sweep.

Ring - Key descriptor is a rotation-invariant descriptor. Specifically represented as a one-dimensional array $k$, each element is as follows:

$$k = (\varphi(r_1), ..., \varphi(r_n)) \tag{8}$$

$$\varphi(r_i) = \frac{\|r_i\|_0}{N_s} \tag{9}$$

where $\|r_i\|$ is the number of non-zero numbers in $r_i$. $r_i$ is a circle of segmented point cloud.

Ring-Key is to compressing and reducing the dimension of data, and since Ring-Key is a rotation-invariant descriptor, the similarity of two sweep of point cloud can be directly compared use Ring-Key.

## 5. SIMULATION EXPERIMENT

### *5.1. Environment construction and mapping results*

This paper uses Airsim and UE4 to build a lunar simulation environment. Simulate an intelligent guidance system carrying a 3D lidar to perform a mission on the lunar surface. The lunar environment is as follows:

Journal of Physics: Conference Series **2364** (2022) 012010 doi:10.1088/1742-6596/2364/1/012010

Fig.10 Simulation of the lunar environment

Lidar data was collected in lunar simulation environment. The results show that both methods can realize the point cloud mapping function of the lunar surface. The result is shown in Fig 11 and Fig 12.

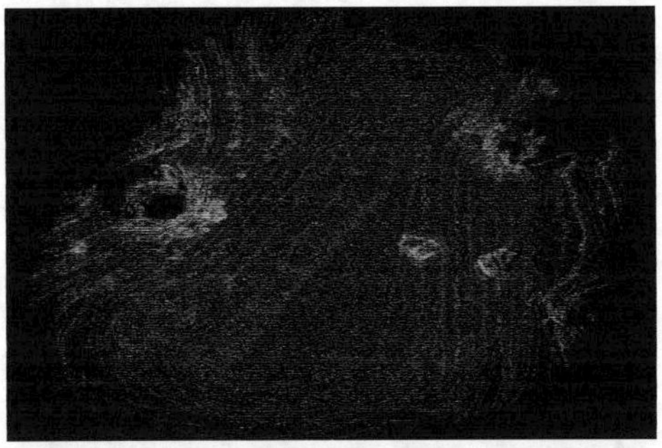

Fig.11 Original algorithm mapping results

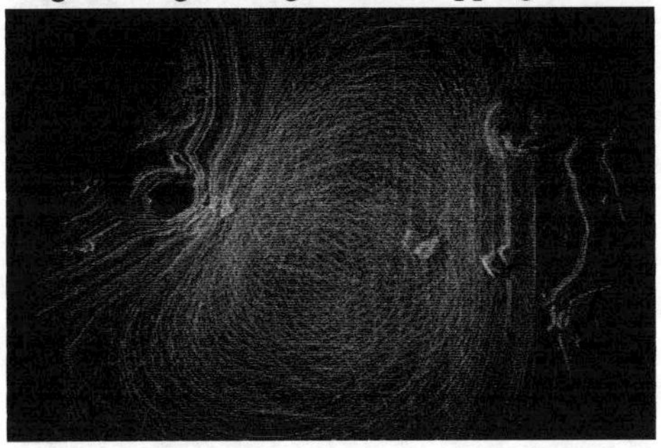

Fig.12 Improved version mapping result

*5.2.Positioning results*

The positioning error results of the two methods in the x-axis, y-axis, and z-axis are shown in Figure 13. It can be seen from the figure that after adding back-end optimization and loopback detection, the positioning performance of x-axis and y-axis has been significantly improved.

The comparison between the true trajectory and the trajectory calculated by the two algorithms is shown in Figure 14

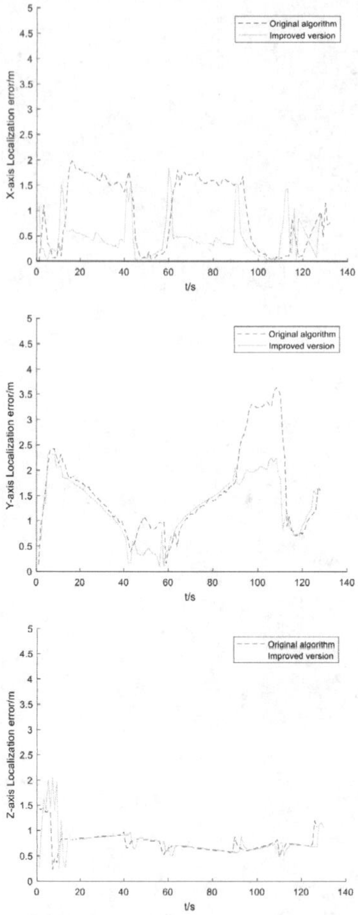


Fig.13 X, Y, Z axis positioning error

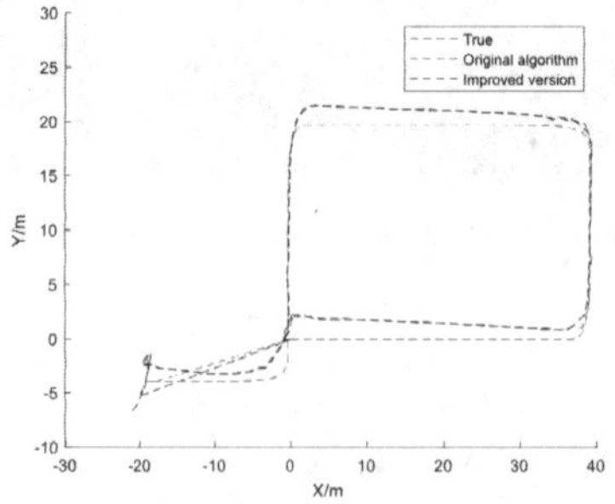


Fig.14 track comparison

TABLE 1. POSITIONING ERROR COMPARISON

| | ***Original algorithm*** | ***Improved version*** |
|---|---|---|
| X-axis/m | 0.9618 | 0.4617 |
| Y-axis/m | 1.5434 | 1.2698 |
| Z-axis/m | 0.7659 | 0.7720 |

## 6. Conclusion

This paper builds a lunar surface simulation environment through Airsim and UE4, and verifies the mapping and positioning effect of Lidar SLAM in this environment. In view of the sparse feature points on the lunar surface and the limited computing power, this paper uses the front-end odometer of feature point extraction to reduce the cost of computation, and adds back-end optimization based on factor diagram and loopback detection algorithms to reduce the accumulative error during positioning and mapping. After the algorithm is improved, the positioning accuracy of X and Y axes is increased from 0.9618m and 1.5434m to 0.4617m and 1.2698m respectively. The positioning accuracy of X and Y axes has been significantly improved.

## References

[1] Yue Fuzhan, Cui Pingyuan, Cui Hutao. (2006) Research on positioning technology of lunar patrol probe [J]. Robot, 235-240.

[2] Wang Yiqiao, Zhang Wei. (2018) Research progress of SLAM technology based on unstructured lunar complex environment [J]. Journal of Zhengzhou University (Engineering Edition), 39(03):45-50.

[3] Shang Tianxiang, Wang Jingchuan, Dong Lingfeng, Chen Weidong. (2021) Three-dimensional laser SLAM technology for lunar environment [J]. Journal of Aeronautics and Astronautics, 42(01): 296-304.

[4] Xie Hongle, Chen Weidong, Fan Yaxian, Wang Jingchuan. (2021) Visual-inertial SLAM method in lunar feature sparse environment [J]. Journal of Aeronautics and Astronautics, 42(01): 305-315.

[5] Zheng Bo. (2020) Research on SLAM method for Mars rover [D]. Harbin Institute of Technology.

[6] Shital Shah et al. (2017) "AirSim: High-Fidelity Visual and Physical Simulation for Autonomous Vehicles" field and service robotics.

[7] ZHANG J, SINGH S. (2017) LOAM: Lidar Odometry and Mapping in Real-time[C] Robotics: Science and Systems. 2(9)

[8] Shan, T. , & Englot, B. . (2019). LeGO-LOAM: Lightweight and Ground-Optimized Lidar Odometry and Mapping on Variable Terrain. IEEE/RSJ International Conference on Intelligent Robots and Systems (IROS). IEEE.

[9] Tixiao Shan et al. (2020) LIO-SAM: Tightly-coupled Lidar Inertial Odometry via Smoothing and Mapping ,intelligent robots and systems.

[10] Ji Zhang and Sanjiv Singh. Ji, Z. , & Singh, S. . (2015). Visual-lidar odometry and mapping: low-drift, robust, and fast. IEEE International Conference on Robotics & Automation. IEEE.

[11] Kaess, M. , Ranganathan, A. , & Dellaert, F. . (2007). iSAM: Fast Incremental Smoothing and Mapping with Efficient Data Association. IEEE International Conference on Robotics & Automation, 1365–1378.

[12] Kaess, Johannsson, Roberts, Ila, Leonard, & JJ, et al. (2012). Isam2: incremental smoothing and mapping using the bayes tree. INT J ROBOT RES, 2012,31(2)(-), 216-235

[13] Madsen K , Nielsen HB , Tingleff O . (2004) Methods for Non-Linear Least Squares Problems (2nd ed.)[J]. society for industrial & applied mathematics.

[14] Kim G , Kim A . (2018) Scan Context: Egocentric Spatial Descriptor for Place Recognition Within 3D Point Cloud Map[C]// 2018 IEEE/RSJ International Conference on Intelligent Robots and Systems (IROS).

Journal of Physics: Conference Series **2364** (2022) 012011 doi:10.1088/1742-6596/2364/1/012011

# The Satellite Coverage Analysis is Based on a Combination of STK and MATLAB

**Cheng Zhang [1,a], Xuegang Wang [2,b], Yunzheng Liu [1,c]**

[1] Aerospace Command College, Space Engineering University, Beijing, China

[2] Department of Aerospace Security, Space Engineering University, Beijing, China

[a]email:616980682@qq.com,[b]email:598456920@qq.com,[c]email:2486829152@qq.com

**Abstract:** In maritime military operations, reconnaissance satellites can provide all-around and efficient reconnaissance and surveillance of target areas in established orbits, regardless of territorial, airspace, and territorial sea restrictions, and are a beneficial means of leadership in today's military. STK can construct satellite motion models, and MATLAB can invoke the STK function module to achieve complex scene generation. The two are combined, respectively, with Walk Constellation grouping and single-satellite simulation of specific target areas at sea, resulting in coverage analysis reports with good results.

## 1. Introduction

In modern technological warfare, satellite surveillance has become an important way to obtain information [1]. Reconnaissance satellites can be independent of territory, airspace, territorial waters, regional ground targets, and maritime targets for the efficient survey to overcome the shortcomings of traditional means of surveillance. Currently, reconnaissance satellites are developing from single to multi-satellite networks, shortening the time interval between targets, increasing the scope of targets by satellites, and enhancing the performance of reconnaissance coverage [2]. STK software is a powerful tool for satellite mission simulation, providing 2D and 3D simulation scenarios as well as analysis reports and various types of accurate graphical data, and MATLAB software provides many advanced mathematical calculation tools, as well as MATLAB software, provides many advanced mathematical calculation tools and calls to STK's supporting function packets, significantly enhancing the computational and analytical capabilities of STK software. This paper uses the Satellite Simulation Toolkit (STK) software interfaced with MATLAB software to simulate the performance of individual reconnaissance satellites and the walk constellation network for target coverage in specific sea areas, comparing the results of the analysis and drawing reliable conclusions for operational planning.

## 2. Status of reconnaissance satellite development

With satellite information supporting maritime military operations, weapons platforms can perform precision and long-range strike capabilities better. The first US reconnaissance satellite, Discovery, was launched in February 1952, and since then, it has developed rapidly, becoming an important channel for countries to obtain military information. Since then, reconnaissance satellites have grown quickly and become an important channel for governments to get military information. [3] According to some data, over 70% of military intelligence in Russia comes from reconnaissance satellites and up to 95% in the US. At present, the United States has successfully launched the "lock-eye" series, "lacrosse" series of reconnaissance satellites, and the White Cloud ocean surveillance satellite, forming an imaging reconnaissance satellite, signal intelligence reconnaissance satellite, ocean surveillance satellite, three major space reconnaissance system.

Reconnaissance satellites are divided into optical imaging reconnaissance satellites and electronic reconnaissance satellites. Optical imaging reconnaissance satellites orbit at an altitude of several hundred kilometers and are mainly used for intelligence gathering and tactical reconnaissance in the military. Electronic reconnaissance satellites receive electromagnetic signals from target radar and other equipment to determine the type of target parameters as well as location and additional information. They can be divided into active electronic reconnaissance satellites (also called synthetic aperture radar satellites, using onboard radar equipment to actively release electromagnetic signals based on the reflected signals to determine, and analyze the target) and passive electronic reconnaissance satellites (which can be seen as a radio, directly intercepting the reconnaissance of enemy targets electromagnetic signals). The orbit of the electronic reconnaissance satellite is circular or near circular in order to increase the monitoring area of the Earth, reduce the drag, extend the life and take into account the needs of positioning accuracy; the orbit plane inclination is large, the detection area is a generally circular area, the radius can be more than 2000 km. Satellite orbit, this paper focuses on the simulation of electronic reconnaissance satellite coverage of the target area [4].

## 3. Reconnaissance satellite coverage analysis

Satellite coverage of the Earth refers to the effective area of the Earth's surface, a spacecraft's instrument or antenna at a given moment or over a longer period of time. Satellite coverage of an area is generally to use grid point analysis, where an area of the Earth's surface is first divided into grid points, the special simulation time, and the grid points covered by onboard sensors are counted as a basis for coverage characteristics assessment. The grid point analysis method has a high simulation accuracy and applies to any orbit and shape of coverage field of view, and has become the most dominant method for coverage solutions. The simulation software STK provides a grid method of sampling points, dividing the Earth's surface latitude to longitude into parallel and equal grids to perform coverage calculations [5].

### *3.1 Coverage performance indicators*

For reconnaissance missions, coverage capability is the most important performance of a reconnaissance constellation [6]. There are five common satellite ground coverage performance metrics, namely coverage percentage. Maximum revisit time, average revisit time, time-averaged gap, and average response time. Coverage percentage $t_{pc}$ is the percentage of coverage time at any point on the ground and is equal to the total satellite coverage time $t_c$ divided by the total simulation time t, i.e.

$$t_{pc} = \frac{t_c}{T} = \frac{1}{T}\sum_{i=1}^{n} t_{ci} . \tag{1}$$

The maximum revisit time is the longest interval of time that any point on the simulation grid is not covered by a satellite, this statistical property gives the worst-case information about the target point, and it is equal to the longest response time. Mean revisit time is the total length of the coverage gap at the ground point divided by the number of coverage gaps. The number of coverage gaps is the number of times in simulation time t that the point is not covered by a satellite, and the total length of the coverage gap is the total time that the point is not covered by a satellite. This statistical characteristic gives the average length of time that coverage is interrupted at a given grid point. The time-averaged gap is the average of the gap lengths, and the metric is numerically equal to the sum of the squares of the n coverage gap lengths divided by the simulation time t. The average response time is the average waiting time from the time a random request to observe a point is received by the ground station receiving equipment until the point can be observed [7]. It has the advantage of handling the delay in communication, the delay in response, and the request for data. This performance metric is the best criterion for evaluating coverage and takes into account both the statistical characteristics of the coverage gap and the statistical characteristics of the coverage.

*3.2 Reconnaissance satellite orbit*

Satellite orbits are divided into three categories according to their altitude, namely high orbit (GEO), medium orbit (MEO), and low orbit (LEO); high orbit orbits are located above 20,000km, medium orbit orbits are located between 2000-20000km and low orbit orbits are located between 200 -2000km [8]. The common orbit for electronic reconnaissance satellites is the critical inclination orbit, with an orbital height of 500km and an orbital inclination of 63.4°. As the Earth itself is an irregular sphere with uneven mass and shape, its polar radius is smaller than the equatorial radius, and the equator is similar to an ellipse, resulting in the generation of regenerative forces [9]. The use of a critical inclination orbit reduces the impact of regenerative forces on the satellite and ensures a stable orbit. In addition, the long return period of a critical inclination orbit allows satellites to conduct reconnaissance from different positions above the Earth, which is one of the most common orbits used by electronic reconnaissance satellites [10]. The simulation uses the critical inclination circular orbit of the Walk constellation to analyze the coverage of the target area and to derive the best coverage performance constellation by different grouping methods.

## 4. Overlay simulation and analysis based on STK and MATLAB

The STK call module encapsulates all operations such as setting satellites, orbits, coverage performance analysis, etc. Using the connect command enables MATLAB to interconnect the STK software and call the program director for operations [11]. First, open the STK software, create a new simulation scenario, create a satellite and Walk constellation, create a sub-object sensor on the satellite, use the sensor to cover the target, then open the MATLAB software and enter the "stkInit" command to establish the interconnection, enter the relevant Astrogator Connect command, to set the parameters of the satellite orbit, Walk constellation, sensor, and finally get the simulation scenario and simulation report of different scenarios.

Assume a 24-hour satellite reconnaissance of a specific sea target area. The target area range is set as (N15°00′00″, E110°00′00″), (N15°00′00″, E115°00′00″), (N8°00′00″, E110°00′00″), (N8°00′00″, E115°00′00″), (N8°00′00″, E115°00′00″). The extent of the area is shown in Figure 1.

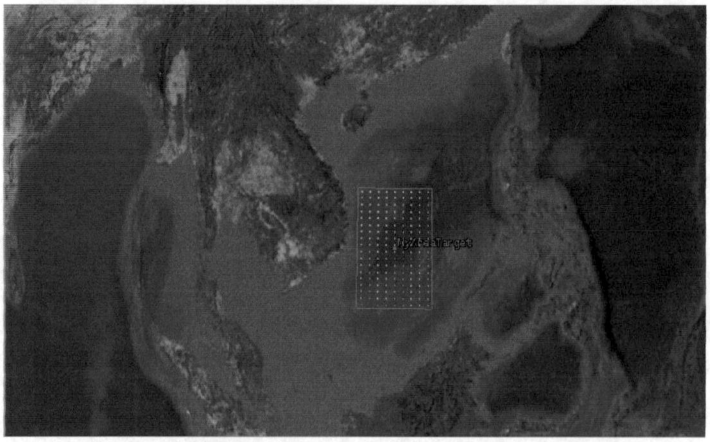

Fig. 1 Map of the reconnaissance target area

For better coverage of the target area, electronic reconnaissance satellites are usually located in satellite orbits with a large inclination angle, giving a wider coverage area. The simulation uses a circular orbit with an inclination of 63.4° in combination with the Walk constellation for coverage analysis of a specific area. A single satellite is first used for the coverage analysis of the target area. Assume an orbit altitude of 500km, an orbit inclination of 63.4°, an eccentricity of 0 due to the circular orbit, a sensor setting of a simple vertebral sensor, an observation angle of 50°, and a grid constraint of 0.5°. The coverage was simulated for 24 hours from May 26, 2022, at 04:00 to May 27, 2022, at 04:00. The simulated 3D scene is shown in Figure 2.

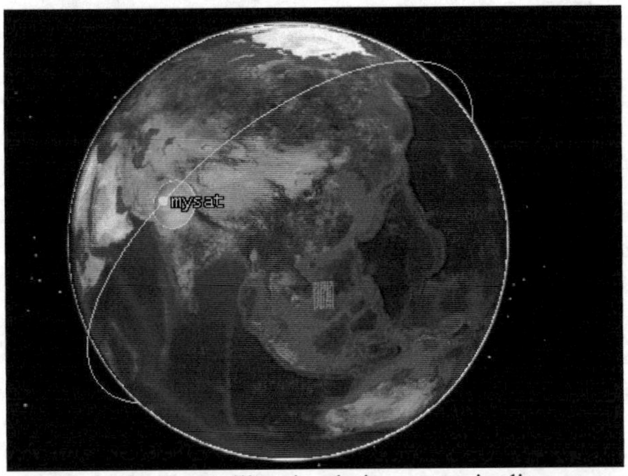


Fig. 2 Single satellite simulation scenario diagram

When a single satellite simulation is used, the percentage coverage of the set target area is shown in Figure 3: the coverage time of the set target area by a single satellite is shown in Figure 4. The step size is set to 10 seconds. From Figure 3 and Figure 4, it can be seen that: the maximum coverage of the target area by a single satellite is 100%, the average coverage percentage is 2.7%, and the time to reach the 100% coverage percentage is 10:57:30 on May 26, 2022.

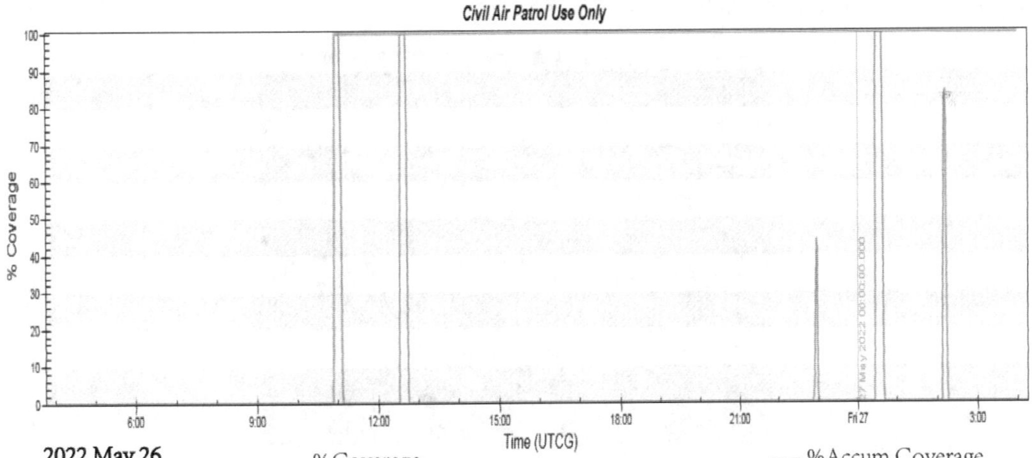


%Coverage %Accum Coverage

Fig. 3 Percentage of time a single satellite covers a target area.

| | | | |
|---|---|---|---|
| | 27 May 2022 03:43:00.000 | 0.00 | 100.00 |
| | 27 May 2022 03:44:00.000 | 0.00 | 100.00 |
| | 27 May 2022 03:45:00.000 | 0.00 | 100.00 |
| | 27 May 2022 03:46:00.000 | 0.00 | 100.00 |
| | 27 May 2022 03:47:00.000 | 0.00 | 100.00 |
| | 27 May 2022 03:48:00.000 | 0.00 | 100.00 |
| | 27 May 2022 03:49:00.000 | 0.00 | 100.00 |
| | 27 May 2022 03:50:00.000 | 0.00 | 100.00 |
| | 27 May 2022 03:51:00.000 | 0.00 | 100.00 |
| | 27 May 2022 03:52:00.000 | 0.00 | 100.00 |
| | 27 May 2022 03:53:00.000 | 0.00 | 100.00 |
| | 27 May 2022 03:54:00.000 | 0.00 | 100.00 |
| | 27 May 2022 03:55:00.000 | 0.00 | 100.00 |
| | 27 May 2022 03:56:00.000 | 0.00 | 100.00 |
| | 27 May 2022 03:57:00.000 | 0.00 | 100.00 |
| | 27 May 2022 03:58:00.000 | 0.00 | 100.00 |
| | 27 May 2022 03:59:00.000 | 0.00 | 100.00 |
| | 27 May 2022 04:00:00.000 | 0.00 | 100.00 |
| Global Statistics | | | |
| ----------------- | | | |
| Min % Coverage | 26 May 2022 04:00:00.000 | 0.00 | 0.00 |
| Max % Coverage | 26 May 2022 10:57:30.000 | 100.00 | 100.00 |
| Mean % Coverage | | 2.70 | |

Fig. 4 Time to target area coverage by a single satellite

Option 1: 4-star, 4-track Walk constellation option.

The simulation uses a Walk scheme with a uniform distribution of satellites, a total of 4 satellites, four orbits, 1 star per orbit, an orbital inclination of 63.4°, and an orbital altitude of 500km, other simulation conditions and times are the same as the single star scheme simulation, at this time the satellite coverage 3D scene is shown in Figure 5.

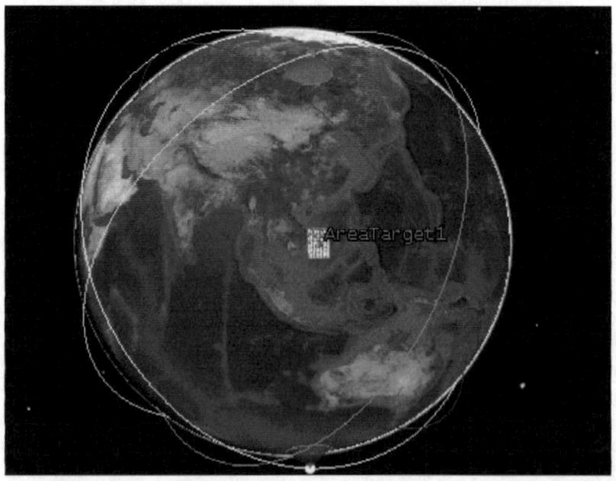


Fig. 5 3D scene of the simulation of the four stars of the Walk constellation

When the coverage simulation of the target area by 4 satellites in 4 orbital planes of the Walk constellation is used, the step size is set to 10 seconds, and the coverage percentage of the set target area is shown in Figures 6; Figures 6 and 7, it can be seen that the maximum coverage percentage of the target area by 4 satellites in 4 orbital planes of the Walk constellation is 100%, the average coverage percentage is 8.9% and the time to reach 100% coverage percentage is May 26, 2022 at 05:03:30.

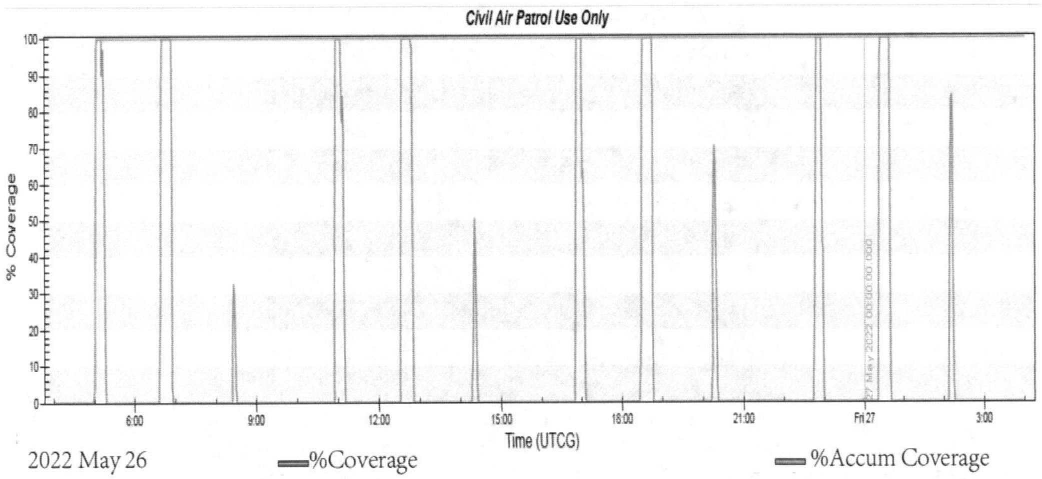


Fig. 6 Percentage of time that the four stars of the Walk constellation cover the target area

| | | | |
|---|---|---|---|
| | 27 May 2022 03:44:00.000 | 0.00 | 100.00 |
| | 27 May 2022 03:45:00.000 | 0.00 | 100.00 |
| | 27 May 2022 03:46:00.000 | 0.00 | 100.00 |
| | 27 May 2022 03:47:00.000 | 0.00 | 100.00 |
| | 27 May 2022 03:48:00.000 | 0.00 | 100.00 |
| | 27 May 2022 03:49:00.000 | 0.00 | 100.00 |
| | 27 May 2022 03:50:00.000 | 0.00 | 100.00 |
| | 27 May 2022 03:51:00.000 | 0.00 | 100.00 |
| | 27 May 2022 03:52:00.000 | 0.00 | 100.00 |
| | 27 May 2022 03:53:00.000 | 0.00 | 100.00 |
| | 27 May 2022 03:54:00.000 | 0.00 | 100.00 |
| | 27 May 2022 03:55:00.000 | 0.00 | 100.00 |
| | 27 May 2022 03:56:00.000 | 0.00 | 100.00 |
| | 27 May 2022 03:57:00.000 | 0.00 | 100.00 |
| | 27 May 2022 03:58:00.000 | 0.00 | 100.00 |
| | 27 May 2022 03:59:00.000 | 0.00 | 100.00 |
| | 27 May 2022 04:00:00.000 | 0.00 | 100.00 |
| Global Statistics | | | |
| ----------------- | | | |
| Min % Coverage | 26 May 2022 04:00:00.000 | 0.00 | 0.00 |
| Max % Coverage | 26 May 2022 05:03:30.000 | 100.00 | 100.00 |
| Mean % Coverage | | 8.90 | |

Fig. 7 Coverage time of the target area by four stars of the Walk constellation

Option 2: 12-star, 4-track Walk constellation option.

The simulation uses a Walk scheme with a uniform distribution of satellites, a total of 12 satellites, four orbits, three satellites per orbit, an orbital inclination of 63.4°, and an orbital altitude of 500km; other simulation conditions and times are the same as the single star scheme simulation, the satellite coverage 3D scene is shown in Figure 8.

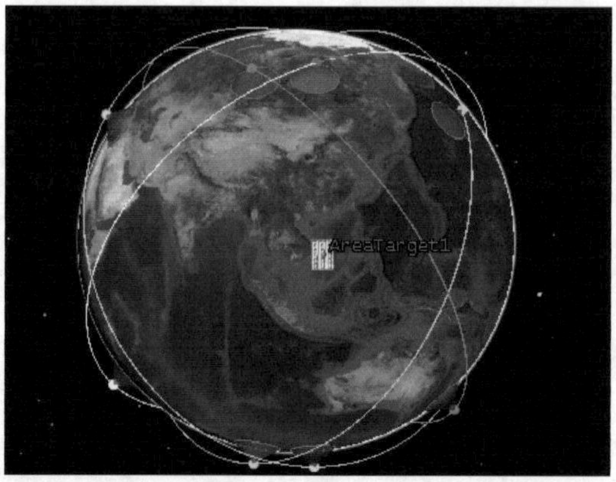


Fig. 8 3D scene of the 12 stars of the Walk constellation simulation

When the coverage simulation of the target area by 12 satellites in 4 orbital planes of the Walk constellation is used, the step size is set to 10 seconds, and its coverage percentage of the set target area is shown in Figure 9; from Figure 9 and Figure 10, it can be seen that the maximum coverage percentage of the target area by 12 satellites in 4 orbital planes of the Walk constellation is 100%, and the average coverage percentage is 26.81%, reaching 100% coverage The time to reach 100% coverage was 04:33:30 on May 26, 2022.

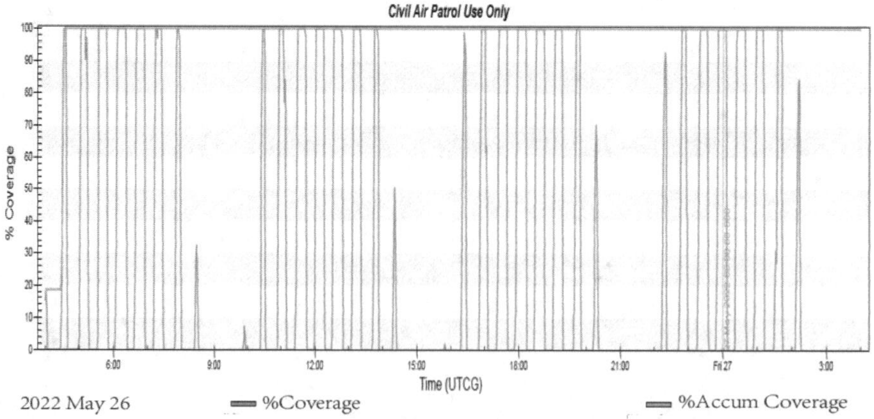


Fig. 9 Percentage of time the 12 stars of the Walk constellation cover the target area

```
27 May 2022 03:44:00.000        0.00        100.00
27 May 2022 03:45:00.000        0.00        100.00
27 May 2022 03:46:00.000        0.00        100.00
27 May 2022 03:47:00.000        0.00        100.00
27 May 2022 03:48:00.000        0.00        100.00
27 May 2022 03:49:00.000        0.00        100.00
27 May 2022 03:50:00.000        0.00        100.00
27 May 2022 03:51:00.000        0.00        100.00
27 May 2022 03:52:00.000        0.00        100.00
27 May 2022 03:53:00.000        0.00        100.00
27 May 2022 03:54:00.000        0.00        100.00
27 May 2022 03:55:00.000        0.00        100.00
27 May 2022 03:56:00.000        0.00        100.00
27 May 2022 03:57:00.000        0.00        100.00
27 May 2022 03:58:00.000        0.00        100.00
27 May 2022 03:59:00.000        0.00        100.00
27 May 2022 04:00:00.000        0.00        100.00

Global Statistics
-----------------
Min % Coverage    26 May 2022 04:04:30.000        0.00         18.38
Max % Coverage    26 May 2022 04:33:30.000      100.00        100.00
Mean % Coverage                                  26.81
```

Fig. 10 Coverage time of the target area by the 12 stars of the Walk constellation

Option 3: The 24-star, 4-track Walk constellation option.

The simulation uses a Walk scheme with a uniform distribution of satellites, a total of 24 satellites, four orbits set, six satellites per orbit, an orbital inclination of 63.4°, and an orbital altitude of 500km, other simulation conditions and times are the same as the single-star scheme simulation, at this time the satellite coverage 3D scene is shown in Figure 11.

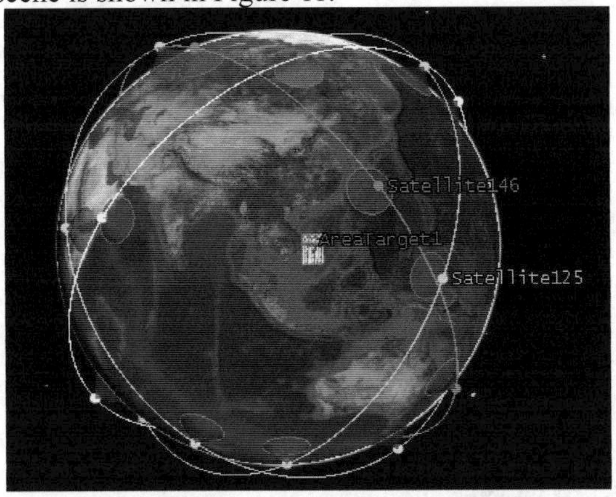


Fig. 11 Simulated 3D scene of the 24 stars of the Walk constellation in 4 orbits

When the coverage simulation of the target area by 24 satellites in 4 orbital planes of the Walk constellation is used, the step size is set to 10 seconds, and its coverage percentage of the set target area is shown in Figure 12; from Figure 12 and Figure 13, it can be seen that the maximum coverage percentage of the target area by 24 satellites in 4 orbital planes of the Walk constellation is 100% and the average coverage percentage is 46.43%, reaching 100% coverage The time to reach 100% coverage was 04:36:30 on May 26, 2022.

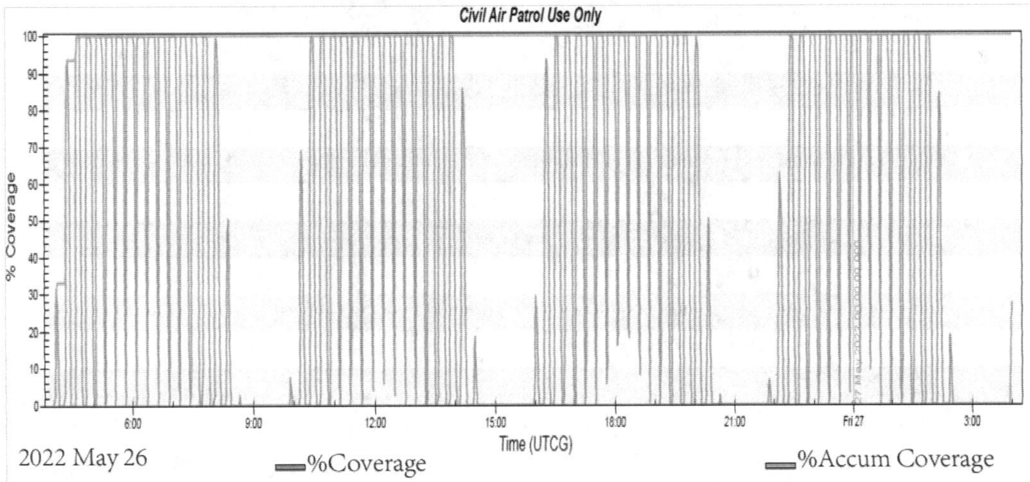


Fig. 12 Percentage of coverage time of the target area by 24 stars in 4 orbits of the Walk constellation

| | | |
|---|---|---|
| 27 May 2022 03:44:00.000 | 0.00 | 100.00 |
| 27 May 2022 03:45:00.000 | 0.00 | 100.00 |
| 27 May 2022 03:46:00.000 | 0.00 | 100.00 |
| 27 May 2022 03:47:00.000 | 0.00 | 100.00 |
| 27 May 2022 03:48:00.000 | 0.00 | 100.00 |
| 27 May 2022 03:49:00.000 | 0.00 | 100.00 |
| 27 May 2022 03:50:00.000 | 0.00 | 100.00 |
| 27 May 2022 03:51:00.000 | 0.00 | 100.00 |
| 27 May 2022 03:52:00.000 | 0.00 | 100.00 |
| 27 May 2022 03:53:00.000 | 0.00 | 100.00 |
| 27 May 2022 03:54:00.000 | 0.00 | 100.00 |
| 27 May 2022 03:55:00.000 | 0.00 | 100.00 |
| 27 May 2022 03:56:00.000 | 0.00 | 100.00 |
| 27 May 2022 03:57:00.000 | 1.83 | 100.00 |
| 27 May 2022 03:58:00.000 | 12.22 | 100.00 |
| 27 May 2022 03:59:00.000 | 23.28 | 100.00 |
| 27 May 2022 04:00:00.000 | 27.61 | 100.00 |

Global Statistics
-----------------

| | | | |
|---|---|---|---|
| Min % Coverage | 26 May 2022 04:00:00.000 | 0.00 | 0.00 |
| Max % Coverage | 26 May 2022 04:36:30.000 | 100.00 | 100.00 |
| Mean % Coverage | | 46.43 | |

Fig. 13 Coverage time of the target area by 24 stars in 4 orbits of the Walk constellation

Option 4: 24-star, 6-track Walk constellation option.

The simulation uses a Walk scheme with a uniform distribution of satellites, a total of 24 satellites, six orbits, four satellites per orbit, an orbital inclination of 63.4°, and an orbital altitude of 500km. Other simulation conditions and times are the same as the single star scheme simulation. The satellite coverage 3D scene is shown in Figure 14.

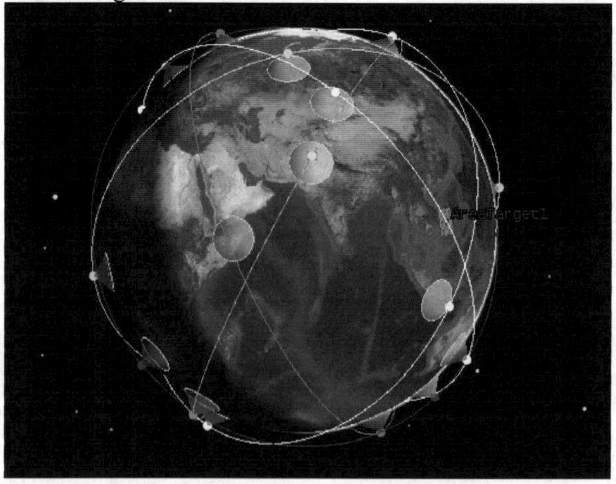

Fig. 14 Simulated 3D scene of the 24 stars of the Walk constellation with six orbits

While the total number of 24 satellites remains the same, the number of orbits is increased. When the simulation of coverage of the target area by 24 satellites on six orbital planes of the Walk constellation is used, the step size is set to 10 seconds, and the coverage percentage of the set target area is shown in Figure 15; from Figure 15 and Figure 16, it can be seen that the maximum coverage percentage of the target area by 24 satellites on six orbital planes of the Walk constellation is 100%. The average coverage percentage is 56.35%, and the time to reach 100% coverage percentage is 04:00:00 on May 26, 2022.

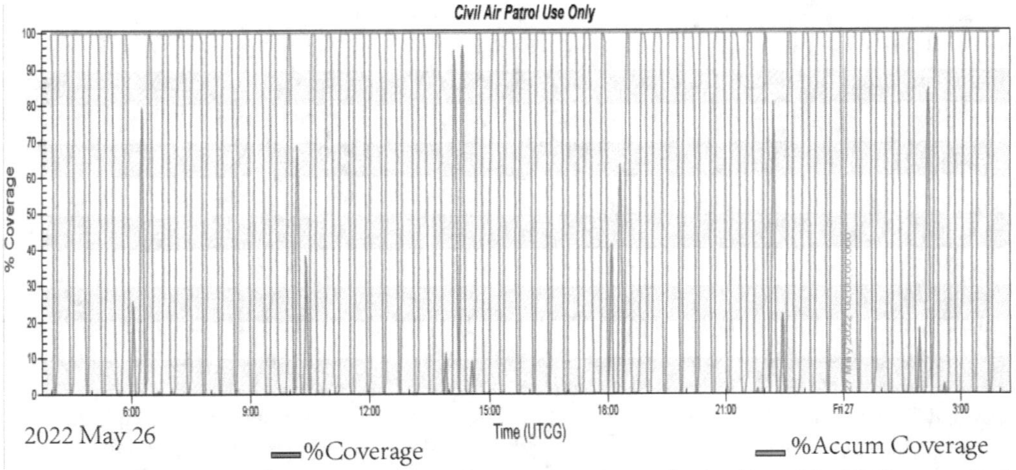


Fig. 15 Percentage of coverage time of the target area by 24 stars in 6 orbits of the Walk constellation

| | | | |
|---|---|---|---|
| | 27 May 2022 03:43:00.000 | 0.00 | 100.00 |
| | 27 May 2022 03:44:00.000 | 0.00 | 100.00 |
| | 27 May 2022 03:45:00.000 | 0.00 | 100.00 |
| | 27 May 2022 03:46:00.000 | 0.00 | 100.00 |
| | 27 May 2022 03:47:00.000 | 3.74 | 100.00 |
| | 27 May 2022 03:48:00.000 | 50.31 | 100.00 |
| | 27 May 2022 03:49:00.000 | 98.17 | 100.00 |
| | 27 May 2022 03:50:00.000 | 100.00 | 100.00 |
| | 27 May 2022 03:51:00.000 | 100.00 | 100.00 |
| | 27 May 2022 03:52:00.000 | 100.00 | 100.00 |
| | 27 May 2022 03:53:00.000 | 100.00 | 100.00 |
| | 27 May 2022 03:54:00.000 | 100.00 | 100.00 |
| | 27 May 2022 03:55:00.000 | 100.00 | 100.00 |
| | 27 May 2022 03:56:00.000 | 100.00 | 100.00 |
| | 27 May 2022 03:57:00.000 | 100.00 | 100.00 |
| | 27 May 2022 03:58:00.000 | 100.00 | 100.00 |
| | 27 May 2022 03:59:00.000 | 100.00 | 100.00 |
| | 27 May 2022 04:00:00.000 | 100.00 | 100.00 |
| Global Statistics | | | |
| ----------------- | | | |
| Min % Coverage | 26 May 2022 04:04:30.000 | 0.00 | 100.00 |
| Max % Coverage | 26 May 2022 04:00:00.000 | 100.00 | 100.00 |
| Mean % Coverage | | 56.35 | |

Fig. 16 Coverage time of the target area by 24 stars in 6 orbits of the Walk constellation

The results of the coverage analysis of several simulation scenarios show that in a set scenario with an orbital inclination of 63.4, an orbital altitude of 500km, a sensor observation angle set at 50°, a step size of 60 seconds, a target area of (N15°00′00″, E110°00′00″), (N15°00′00″, E115°00′00″), (N8°00′00″, E110°00′00″), (N8°00′00″, E115°00′00″), with a grid constraint of 0.5° in a set scenario for a 24-hour simulation from 0400 hours on May 26, 2022, to 0400 hours on May 27, 2022.

The Single Star coverage scenario for the target area and Scenarios 1, 2, 3, and 4 all achieve 100% coverage of the target area. As can be seen in Figures 4, 7, 10, 13, and 16, the maximum percentage coverage of the target area is achieved by the single star scenario, Scenario 1, Scenario 2, Scenario 3, and Scenario 4 at 10:57:30 on May 26, 2022, 05:03:30 on May 26, 2022, 04:33:30 on May 26, 2022, 04:33:30 on May 26, 2022, 04:36:30 on May 26, 2022, and May 26, 2022, at 04:36:30, and May 26, 2022, at 04:00:00; the average percentage of coverage of the target area by the single star scenario, Scenario 1, Scenario 2, Scenario 3 and Scenario 4 is 2.7%, 8.9%, 26.81%, 46.43%, and 56.35% respectively. It can be seen that the single-star coverage scenario takes the longest time to reach 100% coverage percentage of the target area, has the lowest timeliness and the lowest average coverage percentage; Scenario 4 of the 6-track 24-star Walk constellation network has the shortest time to reach 100% coverage percentage of the target area, has the highest timeliness and the highest average coverage percentage. From the coverage percentage results of the target area in Figure 3, Figure 6, Figure 9, Figure 12, and Figure 15, the single-star coverage scheme, Scheme 1 and Scheme 2 have poor coverage performance of the target area; Scheme 3 and Scheme 4 are both 24-star network coverage schemes, and Scheme 4, on the basis of Scheme 3, keeps the total number of satellites unchanged and adopts the method of increasing the orbital plane to cover the target area for simulation, which basically It can achieve continuous coverage of the target area within 24 hours and meet the demand of electronic reconnaissance satellites to achieve continuous coverage of the enemy target area.

## 5. Conclusion

In this paper, a combination of MATLAB and STK software is used to simulate the coverage performance of electronic reconnaissance satellites over the target area. Five scenarios, namely single-satellite coverage, Walk constellation with four orbital planes and four, 12, 24, and six orbital planes and 24 satellites, are simulated and analyzed, respectively. According to the simulation results, it can be seen that as the number of satellites increases, the coverage performance of the target area becomes better; among the Walk constellation network coverage scenarios, scenario 4, while keeping the total number of satellites unchanged, adopts the method of increasing the orbital planes to improve the coverage performance and basically achieves the purpose of real-time coverage of the target area by the Walk constellation network, meeting the electronic reconnaissance satellite's demand for maritime target area It has an important reference value and engineering significance.

## Reference.

[1] Su Jianwei, Song Yuan, Xu Linzhou. Analysis of electronic reconnaissance effectiveness of ocean surveillance satellites on surface ships[j]. Ship Electronic Countermeasures, 2009, Vol. 32(4): 51-53

[2] Yu Hang-Hua, Liu Lei. Design of simulation analysis software for reconnaissance satellite constellation coverage performance[j]. Aerospace Electronic Countermeasures, 2019, Vol. 35(2): 33-37

[3] Chen Qiuqiong. Satellite visibility analysis of targets based on the combination of STK and MATLAB [J]. Electronic Design Engineering, 2020, vol. 28(1): 180-183, 188

[4] Gao Jing , Liu Xiaomeng . Simulation and analysis of hybrid constellation performance based on STK [J]. Journal of Shandong Agricultural University (Natural Science Edition), 2016, Vol. 47(5): 773-778

[5] Wu Hao, Wang Yu. Simulation and coverage analysis of STK-based MEO satellite communication system [J]. Electronic Design Engineering, 2017, Vol. 25(22): 120-123,127

[6] Wang Hao, Zhang Zhanyue, Zhang Haitao, Jiang Ping. Performance evaluation study of small satellite reconnaissance constellation[j]. Chinese space science and technology, 2020, vol. 40(6): 68-76

[7] He Jing , Liu Minwei , Song Guolin , Wang Dong-Yu , Wang Quan . Design and performance analysis of satellite AIS constellation [j]. Electronic Design Engineering, 2020, vol. 28(4): 100-104

[8] Ma Xiaopan, Kong Nianping, Yu Shaobo. Satellite constellation design and analysis with coverage simulation applied to the Internet of Things [j]. Digital Communication World, 2021, (8): 21-25

[9] Tan Zhenhua. The functional design of satellite orbit scanning coverage simulation program [d]. China University of Geosciences (Beijing), 2013

[10] Xu Jun, Bai Xiangfeng, Zhou Xiaohua. Simulation analysis of reconnaissance satellite coverage based on STK [j]. Software Guide, 2016, Vol. 15(8): 156-158

[11] Yu Hanghua, Liu Lei. Design of simulation analysis software for reconnaissance satellite constellation coverage performance [j]. Aerospace Electronic Countermeasures, 2019, vol. 35(2): 33-37

# Study on the Influencing Factors of Modal Transformation of Continuous Rotating Detonation Wave Propagation

**Chen Shu [1], Wei Lin [1], Chenglong Yan [1], Qiang Shi [1*]**

[1] Department of Aerospace Science and Technology, Space Engineering University, Huairou, Beijing, 101416, China

[*]Corresponding author's e-mail: shuchenhgd@163.com

**Abstract:** In order to study the propagation mode of rotating detonation wave, the detailed reaction mechanism of hydrogen-oxygen with 9 components and 19 steps is used, and the finite rate chemical reaction model is used for numerical calculation. After ignition, there is a stable collision period in the flow field for a certain period, and the collision extinguishing and new detonation wave generation occur during the mode conversion. By adding nitrogen or argon to the hydrogen-oxygen mixture for dilution, when the dilution ratio increases, the activity of the gas mixture decreases, the height of the combustible gas layer increases, and the temperature of the combustion product after the detonation wave decreases away from the head of the detonation wave, which will reduce the mixture that is burned in advance. There are different degrees of reflected shock at the outlet of the combustion chamber, and the upward detonation wave will cause the height change of the detonation wave head and the instability of the flow.

## 1. Introduction

A detonation is a form of self-pressurized combustion that couples detonation waves with combustion waves with high thermal cycling efficiency and heat release rate [1]. The detonation engine concept has become a new trend in aerospace propulsion research. The rotating detonation engine can be continuously operated with only one ignition, with a high operating frequency and stability, making it the most valuable for engineering applications. Accurately understanding the formation of the collision model of the detonation wave in the combustion chamber and its transformation mechanism, and realizing the control of the wave propagation mode of the detonation wave, are the prerequisites for the future realization of rotating detonation propulsion.

The annular and empty barrel combustion chamber test and numerical calculation simulation study found different degrees of detonation wave collision phenomenon. The U.S. Air Force Institute of Technology Shank[2, 3] conducted experimental studies in a 6-inch diameter rotating detonation model engine and captured forward, reverse, and two-wave collision modes of blast wave propagation through high-speed photography. The U.S. Air Force Laboratory Rankin [4,5] conducted a parameter study on the flow field structure of the rotating detonation wave: the structure of the detonation wave was carefully studied through experiments and by changing the parameters such as air quality flow rate, equivalent ratio, and air injection area, the detonation tube was used, and the single-wave transition to two-wave collision mode appeared under some specific operating conditions, which was considered to be an uncertain mode caused by the poor mixing effect of fuel and oxidant. Ohio State University's Schocki[6] experimented with the effect of air quality flow on the number of detonation waves. As the mass flow increases, it changes from single-wave to two-wave collision mode. The collision of detonation waves can produce three different results: co-directional two-wave mode, transition to single-wave mode, and two-blast detonation wave extinguishing. It was found that the two-wave model that appeared before stabilization was uncertain by changing the flow rate. Tae-Hyeong Yi[7] numerical simulations of the two-dimensional rotating detonation current field found that two reverse-directional detonation waves are generated during the initial ignition process, and the flow field will experience a collision of detonation waves before continuing to develop.

In this study, the rotating detonation explosion, modal conversion, and stable propagation are studied by changing the dilution degree of the oxidant to flow deeply understand the flow field characteristics after detonation wave collision, understand the transformation process and mechanism of the propagation mode after the detonation wave collision.

## 2. Physical models and numerical methods

### *2.1 Numerical calculation models*

For better convection field observations and to ignore the effect of curvature, a two-dimensional rectangular combustion chamber after the ring chamber is expanded is used to simulate the propagation of the rotating detonation wave [8,9], as shown in Fig 1. Fig. 2 shows the calculation area of the combustion chamber with a circumferential length of 400 mm and an axial length of 70 mm, with the upper-end face as the mass flow inlet and the lower-end face as the pressure outlet. Among them, the inlet adopts the method of interval injection; 80 nozzles are distributed, the ratio of nozzles to wall area is 2:3, and the nozzles enter the premixed gas. The left and right sides are periodic boundary conditions.

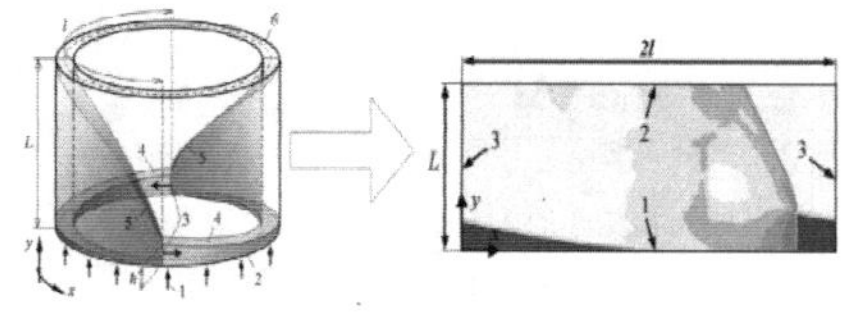

Figure. 1 2D flow-field expanded along the circumference

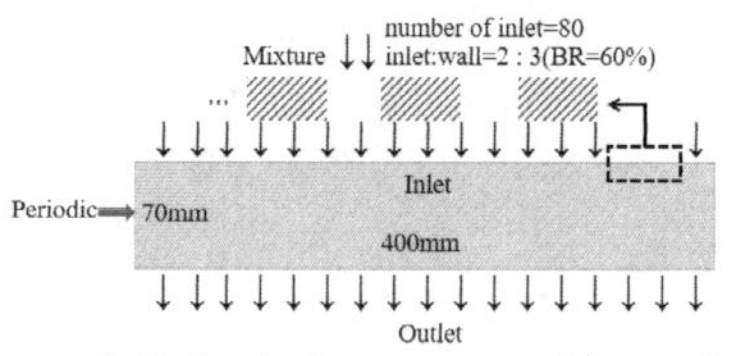


Figure.2 Calculation zone and boundary conditions

### *2.2 Numerical calculation methods*

Based on the density based explicit solver in this study, the chemical reaction model adopts a 9-component 19-step detailed mechanism of hydrogen and oxygen[10] finite rate model. Arrhenius' law calculates the reaction rate constant. Since the scale of the chemical reaction of the detonation wave surface is much smaller than the diffusion scale, using the two-dimensional Euler equation. The gas mixture ignores viscosity, heat transfer, and diffusion and uses the ideal gas. The physical flux is decomposed in the commonly used Roe-FDS (Flux-Difference Splitting) format, and the convection terms are separated in second-order Upwind (SOU) format:

$$\phi_{f,SOU} = \phi + \nabla\phi \cdot \vec{r} \tag{1}$$

$\phi$ And $\nabla\phi$ are the center values and gradients within the upstream cell, respectively. The governing equation is as follows:

$$\frac{\partial U}{\partial t} + \frac{\partial E}{\partial x} + \frac{\partial F}{\partial y} = Q \tag{2}$$

$$U = \begin{bmatrix} \rho \\ \rho u \\ \rho v \\ \rho e \\ \rho_1 \\ \cdots \\ \rho_{N-1} \end{bmatrix}, \quad E = \begin{bmatrix} \rho v \\ \rho uv \\ \rho v^2 + p \\ v(\rho e + p) \\ \rho_1 u \\ \cdots \\ \rho_{N-1} u \end{bmatrix}, \quad F = \begin{bmatrix} \rho v \\ \rho uv \\ \rho v^2 + p \\ v(\rho e + p) \\ \rho_1 v \\ \cdots \\ \rho_{N-1} v \end{bmatrix}, \quad Q = \begin{bmatrix} 0 \\ 0 \\ 0 \\ 0 \\ \sigma_1 \\ \cdots \\ \sigma_{N-1} \end{bmatrix} \tag{3}$$

where Q is the chemical reaction source term, $\rho$ is the density, $u$ , $v$ is the velocity in the X direction and Y direction, respectively, p is the pressure, e the total energy per unit volume, and N the serial number of the component. Ideal gas equation:

$$p = \rho RT \tag{4}$$

*2.3 Mesh independence verification and calculation scenarios*

First, the reaction mechanism accuracy and grid independence are verified.The grid scales are 0.16 mm, 0.20 mm, and 0.4 mm, respectively. Table 1 compares the velocity calculation results in the detonation tube using the the detailed chemical reaction. The theoretical value is the CJ detonation velocity under the same calculation conditions by NASA CEA software. Fig 3 shows the pressure and temperature curves during the detonation wave propagation extracted by the overall reaction and the detailed chemical reaction at the 0.2 mm grid scale.

Table 1 Comparison of the calculated values between different grids and mechanisms in the detonation tube with the theoretical values

| | Δx/(mm) | Detonation Velocity/(m/s) | Reaction | Error |
|---|---|---|---|---|
| Case1 | 0.4 | 2002.01 | 19-step | 1.45% |
| Case2 | 0.2 | 1989.67 | 19-step | 0.83% |
| Case3 | 0.16 | 1989.08 | 19-step | 0.80% |
| Case4 | 0.2 | 1993.72 | 7-step | 1.09% |
| Case5 | 0.2 | 2000.80 | One step | 1.39% |
| Theoretical | | 1973.3 | $H_2$/air | |

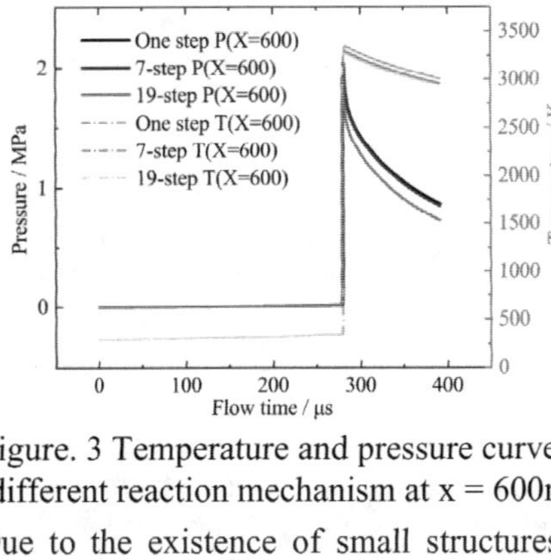


Figure. 3 Temperature and pressure curves of different reaction mechanism at x = 600mm

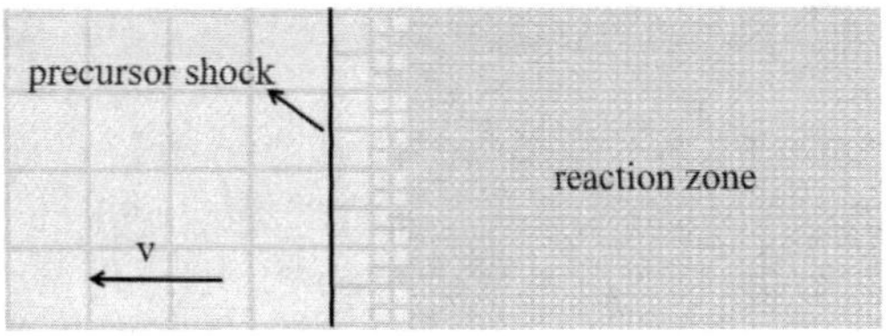


Figure. 4 Schematic diagram of mesh adaptation at detonation wave surface

Due to the existence of small structures such as induction zones and three-wave points on the detonation wave surface, in order to accurately calculate the detonation flow field, the adaptive grid technology is used to modify the density of the grid or the direction of the layout grid according to the calculation results in the flow field. The maximum number of mesh refinement layers set is three layers, the base mesh size is 0.2 mm, and the minimum size of the mesh after three layers is 0.025 mm.

## 3. Effect of gas dilution ratio on detonation combustion mode

In this study, the total inlet flow is kept unchanged, different types of gases are added to the hydrogen-oxygen mixture for dilution, and the dilution ratio is changed. The types of dilution gases were selected from nitrogen present in the air and argon commonly used in laboratories, where the nitrogen dilution ratio increased from 25% to 56%, the argon dilution ratio gradually increased from 40% to 70%, and the test showed that 70% argon could produce stable detonation cell [11]. The calculation conditions are shown in Table 2

Table 2 Characteristics of different gas mixtures

| | $m_{Total}$ /(g/s) | $S_d$ | $R_d$ /% | $v_{Total}$ /(m/s) | $m_d$ /(g/s) |
|---|---|---|---|---|---|
| Case1 | 560 | Ar | 40 | 486.0 | 385.9 |
| Case2 | 560 | Ar | 47 | 455.0 | 417.3 |
| Case3 | 560 | Ar | 56 | 422.0 | 451.6 |
| Case4 | 560 | Ar | 70 | 376.1 | 496.1 |
| Case5 | 560 | $N_2$ | 25 | 657.4 | 244.7 |
| Case6 | 560 | $N_2$ | 40 | 594.4 | 339.4 |
| Case7 | 560 | $N_2$ | 56 | 540.6 | 417.3 |

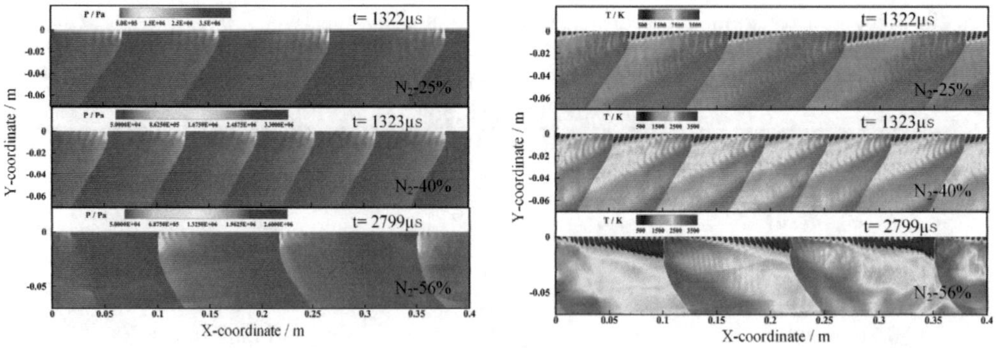


(a) Pressure cloud diagram of different dilution ratio stabilization stages

(b) Temperature cloud map of different dilution ratio stabilization stages

Figure 5 Pressure and temperature cloud diagram of the stabilization stage under different nitrogen dilution ratios

Fig 5 shows a pressure and temperature cloud diagram of the stabilization phase under different nitrogen dilution ratios is 25% at the stabilization stage. Four detonation wave heads propagate in the same direction to the right in the sound stage. The local peak pressure reaches 3.5 MPa, the peak temperature reaches 3358 K, and most of the flow field is in a high-temperature state. Increase the dilution ratio to = 40%, the number of heads in the stabilization stage increases to six, and the height of the detonation wave surface decreases, which is due to the increase in the dilution ratio, and the total volume of the mixture is reduced first, the injection speed of the mixture is reduced, and the fresh mixture has more time to stay near the injection end, which is easy to induce new detonation heads. Secondly, due to the reduction of hydrogen and oxygen content in the mixture, less total heat is released after combustion, the local peak pressure is reduced, and the overall temperature in the flow field after the detonation wave is also reduced. Continue to increase the dilution ratio to 56%, and the local peak pressure is further reduced, although the injection speed becomes slower; at this time, because the activity of the mixture is reduced, it is not enough to appear more detonation hotspots, the number of wave heads is reduced to three, and the height of the detonation wave surface is increased. Due to the increase in the cumulative height of the fresh mixture and the increase in the distance between the two blast waves, the flow after the detonation wave becomes less regular. In addition, under the condition of =56%, a weak upward reflective shock can be seen at the outlet of the combustion chamber, which will also have a particular impact on the flow of the post-wave product.

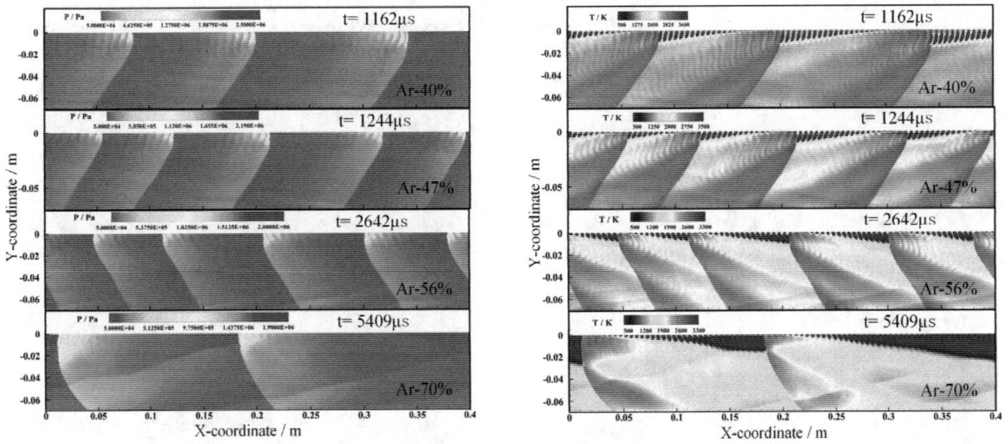


(a) Pressure cloud diagram of different dilution ratio stabilization stages

(b) Temperature cloud map of different dilution ratio stabilization stages

Figure. 6 Pressure and temperature cloud diagram of the stabilization stage under different argon dilution ratios

Figure 6 shows a cloud of pressure and temperature in the stabilization phase under different argon dilution ratios. At = 40%, there are three detonation wave heads in the flow field, the same as when the nitrogen is diluted; at this time, the flow field pressure peak is 2.5 MPa, and the temperature peak is 3616.4K, which are the highest values under four operating conditions. The overall temperature of the flow field is at a high level. When the dilution ratio is increased to 47% and 56%, five detonation wave heads are propagating in the same direction in the flow field, and the detonation wave height decreases. With the dilution ratio increase, the flow field's peak pressure gradually decreases, and the overall temperature of the flow field decreases. At 70%, due to the further decrease in the activity of the gas mixture, only two detonation wave heads are present in the flow field. The temperature of the post-wave flow field drops sharply. At a distance from the wave head, the temperature is only about 1300 K. Although the presence of reflected shocks in the flow field will increase the temperature of the flow field, the intensity is still unable to induce new detonation hot spots, so the mixture can accumulate to a higher height, which is one of the main reasons for the small number of detonation wave in the combustion chamber under high dilution specific conditions.

Combined with different nitrogen dilution ratios, there is no reflected detonation wave at the outlet at low dilution ratios. As the dilution ratio increases, the angle of the outlet reflected shock increases and the distance upstream in the combustion chamber increases. When diluted with 56% nitrogen and 70% argon, the upstream reflected wave will affect the fresh mixture accumulation layer so that the post-wave product flow and the mixture layer are irregular, resulting in changes in the height of the wave head and instability in the flow of the post-wave product.

Fig 7 shows the pressure curve of the process under different dilution gas transition stages at the monitoring point. As the dilution ratio of argon is gradually increased in Fig 7(a), the transition phase time gradually increases. However, at 70%, there is a low-pressure state at the monitoring point for a long time, and no detonation wave passes through the monitoring point for a considerable period. From Fig 7(a), it can be seen that there is a stable collision period for a certain period after ignition. This is because there has been a collision extinguishing of the detonation wave and the regeneration of the new wave head during the transformation process. The intensity of the detonation wave in the flow field that travels in opposite directions is similar, showing a symmetrical two-wave collision form. This situation occurs because at a high dilution ratio, the injection rate of fresh mixture is slow, and the activity is low. After the collision of two detonators with little difference in intensity, the intensity of the transmitted detonation wave is brutal to induce the mixed gas layer with a relatively small height after the wave to form a detonation hot spot. Near the left and proper boundaries of the combustion chamber, the mixture accumulation layer reaches a certain height, and new detonation waves of opposite motion will develop. New detonation waves will interact with the transmitted detonation waves so that the transmitted detonation waves are swallowed up and thus cycle.

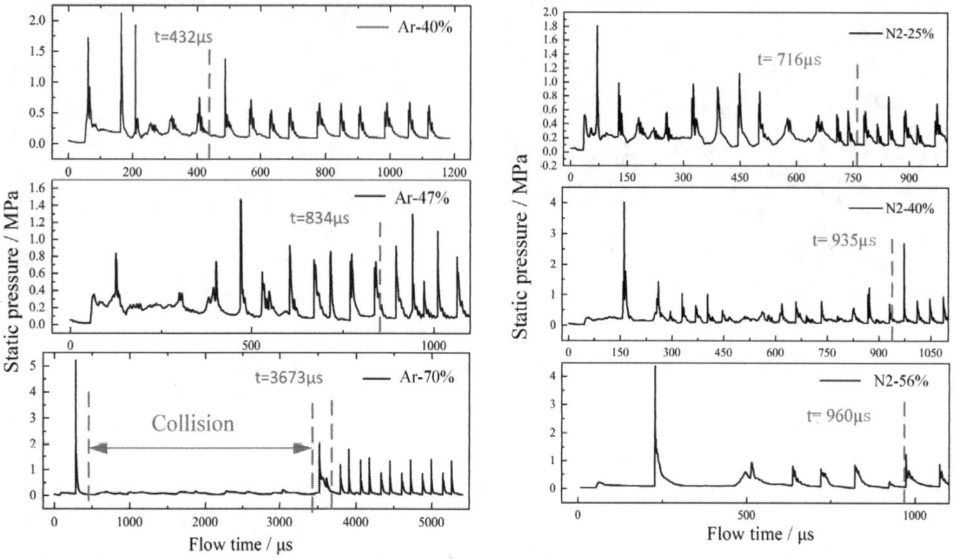


(a) Transformation stage curves for different argon dilution ratios

(b) Transition stage curves for different nitrogen dilution ratios

Figure. 7 Pressure curve of the process under different dilution gas transition stages at monitoring point (0,-5).

When the nitrogen dilution ratio is gradually increased in Fig 7(b), the transition stage time also shows a gradual increase trend. Moreover, from the number of curve pressure peaks can be found that the lower the dilution ratio, the greater the number of pressure peaks in the transition stage; that is, because of the high gas mixing activity and injection speed in the flow field, it is easier to appear new *detonation* hot spots, and the internal chaos of the flow field is more chaotic.

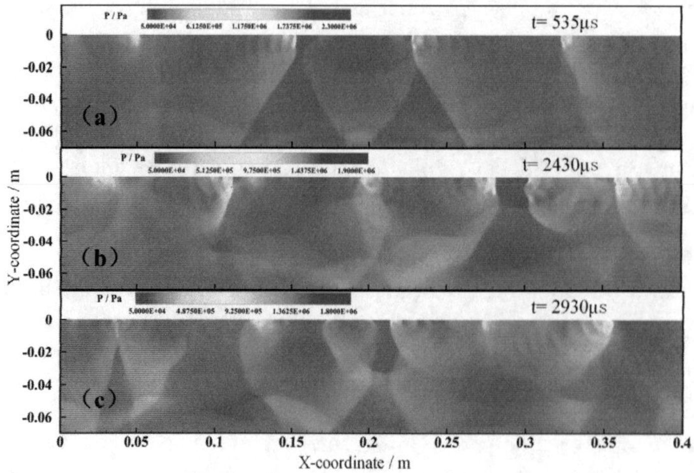


Figure 8 Pressure cloud diagram of the flow field transition phase during 70% argon dilution

When the accumulated height of the fuel layer in the flow field is insufficient, or there is a disturbance such as a vortex, the strength of the detonation wave on both sides changes, which will break the cycle. The two-wave collision in Figure 8(b) is no longer symmetrical. In contrast, the number of detonation waves propagating to the left in Figure 8(c) is already more significant than the number of detonation waves propagating to the right. The cyclic collision mode is broken, eventually forming the flowfield of the two heads in the same direction, as shown in the stabilization stage.

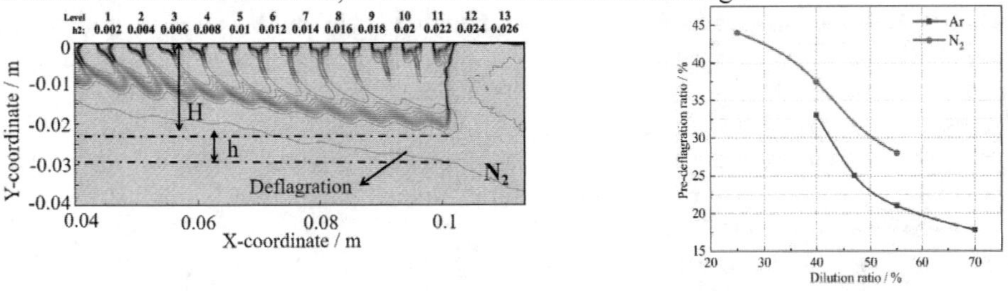


（a）Schematic diagram of deflagration （b）Deflagration curves with different dilution ratios

Figure 9 Case7 hydrogen contour plot at the stabilization stage and deflagration curve at different dilution ratios

Figure 9 shows the hydrogen contour plot at case7 and the deflagration curve at different dilution ratios. H in Figure 9(a) is the height of the fresh mixture accumulation layer, i.e., the height of the detonation wave surface. The contour plot shows that some hydrogen has been consumed before the blast wave surface reaches the fresh mixture layer. These consumed hydrogens are due to deflagration on the contact surface with the high-temperature and high-pressure products after the previous wave. The deflagration phenomenon consumes the mixture in advance, reduces the height of the mixture, affects the strength of the detonation wave, and adversely affects the propagation of the detonation wave. In the figure, h is the height of the hydrogen consumed in advance; we define the deflagration rate, H = 23 mm for Case7, h = 6.5 mm, in this case, = 28.3%. Figure 9(b) plots the deflagration rate under different argon and nitrogen dilution ratios. As the dilution ratio increases, the deflagration rate shows a downward trend in nitrogen and argon. First, when the dilution ratio increases, the activity of the gas mixture decreases, and the height of the combustible gas layer increases; Secondly, when the dilution ratio is increased, the temperature of the post-detonation combustion product decreases away from the detonation head, which will reduce the mixture that burns early.

## 4. Conclusion

In this study, by changing the dilution ratio of the flow mixed gas, the influence on the controlled collision process of the detonation wave is explored, and the following conclusions are mainly obtained: Change the type and dilution ratio of the dilution gas of the incoming mixture, and find that with the increase of the dilution ratio, the number of detonation waves in the transition stage increases first and then decreases. The angle of the reflected light wave gradually increases, and the deflagration rate decreases. At nitrogen mass fractions of 25%, 40%, and 56%, the propagation modes of 4, 6, and 3 rotating detonation waves were obtained, respectively. The propagation modes of 3, 5, 5, and 2 detonation waves were obtained at argon mass fractions of 40%, 47%, 56, and 70%, respectively. There is no reflected shock at the outlet at low dilution ratios. As the dilution ratio increases, the angle of the outlet reflected shock increases and the distance upstream in the combustion chamber increases. The lower the dilution ratio, the greater the number of pressure peaks in the transition stage; that is, because of the high gas mixing activity and injection velocity in the flow field, new detonation hot spots are more likely to appear, and the internal chaos of the flow field is more chaotic.

**Acknowledgments**

The authors gratefully acknowledge support from the National Natural Science Foundation of China (Grant No. 11902361).

**Reference**

[1] Wang Jianping,Zhou Rui,Wu Dan.(2015) Progress of continuously rotating detonation engine research. Journal of Experiments in Fluid Mechanics,29(04): 12-25.
[2] Shank J, King P, Karnesky J et al.(2012) Development and Testing of a Modular Rotating Detonation Engine. 50th AIAA Aerospace Sciences Meeting Including the New Horizons Forum and Aerospace Exposition.
[3] Shank J.(2012) Development and Testing of a Rotating Detonation Engine Run on Hydrogen and Air [PhD Thesis]. Ohio:Wright-Patterson Air Force Base,Air Force Institute of Technology.
[4] Rankin B, Richardson D, Caswell A et al.( 2017) Chemiluminescence imaging of an optically accessible non-premixed rotating detonation engine. Combustion and Flame, 176: 12-22.
[5] Rankin B, Richardson D, Caswell A et al.(2015) Imaging of OH* Chemiluminescence in an Optically Accessible Nonpremixed Rotating Detonation Engine. 53rd AIAA Aerospace Sciences Meeting.AIAA 2015-1604.
[6] Suchocki J, Yu S-T, Hoke J et al.( 2012) Rotating Detonation Engine Operation . 50th AIAA Aerospace Sciences Meeting Including the New Horizons Forum and Aerospace Exposition.
[7] Yi T-H, Lou J, Turangan CK et al.(2020) Numerical Study of Detonation Processes in Rotating Detonation Engine and its Propulsive Performance. Transactions on Aerospace Research, 2020(3): 30-48.
[8] Xia Z-J, Luan M-Y, Liu X-Y et al.(2020) Numerical simulation of wave mode transition in rotating detonation engine with OpenFOAM. International Journal of Hydrogen Energy, 45(38): 19989-19995.
[9] Jiang Z-L.(2020) Physics of gas detonation and its unified framework theory. Beijing:Science Press.Beijing
[10] Hishida M, Hayashi AK. (1992) Numerical simulation of the interaction between a flowfield and chemical reaction on premixed pulsed jet combustion. Computational Fluid Dynamics '92.
[11] Neumann V.( 1942) Progress report on "Theory of detonation waves"[R]. National Defense Research Committee Div. , OSRD. Rept.

# Mars Ascent Trajectory Optimization Based on Convex Optimization

**Yu Lan**[1, a], **Ji Li**[2,b*]

[1]Beijing Institute of Control Engineering, Beijing, 100089, China

[2]Beijing Institute of Control Engineering, Beijing, 100089, China

[a]email: lanyu1328@163.com,

[*b]Corresponding author's e-mail: Jerem-lee@sohu.com

**Abstract.** In this paper, a Mars ascent trajectory optimization algorithm based on convex optimization is designed for the Mars ascent trajectory optimization problem. The main accomplishment of this paper is the trajectory optimization of the second-stage ignition based on the convex optimization. For the first-stage ignition, open-loop guidance is used. In this stage, the ascender flies using the maximum thrust to track the attitude profile, thus completing the ascent flight control. In the presence of errors in the first-stage ignition, the experiment generates multiple sets of first-stage shutdown point data and completes simulation verification experiments based on them. The simulation verifies that the second-stage ignition trajectory optimization based on convex optimization algorithm can achieve high accuracy into orbit despite the large starting point deviation.

## 1. Introduction

Since ancient times, human beings have been full of infinite reveries about the vastness of the universe. With the breakthrough and development of a series of technologies such as artificial earth satellites, manned spacecraft, artificial lunar satellites, space stations, manned moon landings, Mars exploration, etc., we are unveiling the mystery of the universe step by step. Now the depth of exploration and the breadth of exploration targets are expanding. Exploration requirements are no longer macroscopic or contoured detection. Sample return and other detection have become a new demand for scientific development.

The desire for Mars exploration is due to the fact that Mars is seen as the most likely planet for humans to move to. But the exploration of Mars is more difficult compared to the Moon. Earth is farther away from Mars and Mars has a more complex atmosphere[1]. Only a very small number of countries have the capability to explore Mars. By 2020, only the United States has successfully completed seven Mars landing probes[2][3][4]. China launched the Tianwen-1 Mars probe on July 23, 2020, which reached Mars in February 2021 and was successfully captured by Mars. These successful explorations and subsequent Mars exploration missions by various countries will further unveil the mystery of Mars for us.

In future exploration missions, all countries will definitely enhance their understanding and exploration of Mars. At present, the Mars sample return mission is still in the pre-programming demonstration stage. The methods for Mars ascent guidance include "Apollo" type nominal trajectory tracking guidance[5] and predictive correction guidance[6], but neither of them can meet the requirements of high-precision orbital entry.

The advantages of convex optimization algorithms in terms of mathematical principles and computations make them particularly suitable for online trajectory optimization applications. The lossless convexification method was proposed by Açıkmeşe and Ploen et al. for the Mars power descent guidance problem[7].

To complete the Mars high-precision orbit entry design, this paper designs a Mars second-stage ignition ascent trajectory optimization algorithm based on convex optimization. First-stage ascent guidance This paper adopts open-loop guidance. In this stage, the ascender flies with maximum thrust and the control system tracks the attitude profile so that the ascender can reach an altitude at a certain height from the ground when the first stage is shut down. Since the first-stage flight is in the atmosphere, the thrust magnitude and rotation angle will have some deviation, so there will be some error in the shutdown point in this stage. The position, velocity, and horizontal angle of the final first-stage shutdown point will fluctuate within a range. In this paper, the initial conditions are mainly based on several different first-stage shutdown points with atmospheric drag error, ascent thrust error, and rotation control error. A convex optimization algorithm is used to complete the simulation experiments for the second-stage trajectory optimization.

## 2. Model of Mars Rising Problem

### *2.1. Kinematic model of the ascender*

This paper is based on theoretical research. The ascender studied has no specific name and model. The Mars ascent guidance method studied in this paper can be applied to all kinds of Mars ascenders that can achieve secondary ignition. The ascender is viewed as a rigid body. Due to the relatively thin atmosphere on the surface of Mars, the effects of the atmosphere are ignored during the secondary ignition ascent stage.

The 3--degree-of-freedom model of the Mars ascender can be described as:

$$\begin{cases} \dot{m}(t) = -\alpha_{\dot{m}} \left\| \boldsymbol{T}(t) \right\|_2 \\ \dot{\boldsymbol{r}}(t) = \boldsymbol{v}(t) \\ \dot{\boldsymbol{v}}(t) = \dfrac{1}{m(t)} \boldsymbol{T}(t) - \dfrac{\mu_m}{\left\| \boldsymbol{r} \right\|^3} \boldsymbol{r} \end{cases} \tag{1}$$

In the above equation: $\alpha_{\dot{m}} \triangleq \dfrac{1}{I_{sp} g_e}$ is the mass of the ascender. $I_{sp}$ is the specific impulse of the reaction thrust engine, and $g_e$ is the gravitational acceleration of the Earth, both of which are constant, so $\alpha_{\dot{m}}$ is also a constant. $\boldsymbol{T}(t)$ is the thrust vector. $\boldsymbol{r}(t) \in R^3, \boldsymbol{v}(t) \in R^3$ are the position and velocity vectors. $\mu_m = 4.285 \times 10^{13} \mathrm{m}^3/\mathrm{s}^2$ is the gravitational constant of Mars. $\boldsymbol{r} = r_m + h$ , $r_m = 3398000\mathrm{m}$ is the radius of Mars. $h$ is the altitude of the ascender to the surface of Mars.

### *2.2. Description of the trajectory optimization problem for the ascender*

#### *2.2.1. Conditions*

Specifying that the engine can be shut down, then the constraint on the magnitude of the thrust can be described as:

$$0 = T_{\min} \leqslant \left\| \boldsymbol{T}(t) \right\|_2 \leqslant T_{\max} \tag{2}$$

The initial and terminal moment constraints are:

$$\begin{aligned} &\boldsymbol{r}(0)=\boldsymbol{r}_i, \boldsymbol{v}(0)=\boldsymbol{v}_i, m(0)=m_{wet} \\ &\boldsymbol{r}(t_f)=\boldsymbol{r}_f, \boldsymbol{v}(t_f)=\boldsymbol{v}_f \end{aligned} \tag{3}$$

Since the fuel carried by the ascender is finite, the fuel constraint can be described as:

$$0 < m_{dry} \leqslant m(t) \tag{4}$$

*2.2.2. Objective function*
Since the fuel carried by the ascender is finite, the goal of optimization is to consume the least amount of fuel during the entire flight:

$$\max\left(m\left(t_f\right)\right)=\min \int_0^{t_f} \|\boldsymbol{T}\| dt \tag{5}$$

*2.3. Convexification transformation*
Since the lower limit of thrust can be zero, there is no non-convex constraint in the thrust constraint. However, mass values appear in the denominator of the state equation, so a convex transformation is also required.

The relaxation variables are introduced to change the thrust constraint to the following form:

$$\begin{aligned} &0 < T_{\min} \leqslant \Gamma(t) \leqslant T_{\max} \\ &\|\boldsymbol{T}(t)\|_2 \leqslant \Gamma(t) \end{aligned} \tag{6}$$

Perform variable substitution:

$$\sigma \triangleq \frac{\Gamma}{m}, u \triangleq \frac{T}{m}, z \triangleq \ln m \tag{7}$$

The equation of state constraint can be redescribed as:

$$\begin{cases} \dot{z}(t)=-\alpha_{\dot{m}}\sigma(t) \\ \dot{\boldsymbol{r}}(t)=\boldsymbol{v}(t) \\ \dot{\boldsymbol{v}}(t)=u(t)-\dfrac{\mu_m}{\|\boldsymbol{r}\|^3}\boldsymbol{r} \end{cases} \tag{8}$$

The fuel-optimal objective function can be rewritten as:

$$\min \int_0^{t_f} \sigma(\tau) d\tau \tag{9}$$

The thrust constraint can be redescribed as:

$$\begin{aligned} &\| u(t) \| \leqslant \sigma(t) \\ &T_{\min} e^{-z_0}\left[1-(z-z_0)+\frac{(z-z_0)^2}{2}\right] \leqslant \sigma(t) \leqslant T_{\max} e^{-z_0}\left[1-(z-z_0)\right] \end{aligned} \tag{10}$$

The final continuous convex optimization problem is described as follows:

$$\begin{aligned} &\min \quad (9) \\ &\text{subject to} \quad (8)(10)(3)(4) \end{aligned} \tag{11}$$

Above, the description of the continuous problem is completed. In order to be able to solve the problem on a computer, the continuous problem needs to be discretized. The discretization method used for the problem in this paper is the same as in the literature[7].

## 3. Simulation Verification

### *3.1. Simulation parameter setting*

The simulation case chosen for this section is the Mars ascender starting from the surface of Mars and reaching the near-Mars point at an altitude of 200 km in the circum-Mars orbit. The terminal position velocity of the second-stage guidance can then be determined. In the first-stage ascent guidance, open-loop guidance is used in this paper. The ascender flies for 120s using the maximum thrust, and the control system tracks the attitude profile so that the ascender finally rises to an altitude 55.39km high from the start point. It is necessary to ensure that the velocity of the ascender can reach $1.75\ \mathrm{km/s^2}$, which is at an angle of 30 degrees with the horizontal plane. Since the first-stage flight is in the atmosphere, the thrust magnitude and rotation angle will have some deviation, so there will be some error in the shutdown point. The position, speed and horizontal angle of the final first-stage shutdown point will fluctuate within a range. In this paper, five different sets of first-stage shutdown points with atmospheric drag, ascending thrust error and rotation control error are used as initial conditions to complete the second-stage trajectory optimization simulation experiments.

Table 1. Parameter setting.

| Parameter | Value | Unit |
|---|---|---|
| $g_e$ | 9.81 | m/s$^2$ |
| $m_{wet}$ | 1000 | kg |
| $m_{wet}$ | 200 | kg |
| $T_{\min}$ | 0 | N |
| $T_{\max}$ | 6490 | N |
| $I_{sp}$ | 600 | /s |
| $tf$ | 600 | s |
| $\boldsymbol{r}_i$ | $\begin{bmatrix} r_x & r_y & r_z \end{bmatrix}^T$ | m |
| $\boldsymbol{r}_f$ | $\begin{bmatrix} 3116000 & -179000 & 0 \end{bmatrix}^T$ | m |
| $\boldsymbol{v}_i$ | $\begin{bmatrix} v_x & v_y & v_z \end{bmatrix}^T$ | m/s |
| $\boldsymbol{v}_f$ | $\begin{bmatrix} -1725.5 & -2988.7 & 0 \end{bmatrix}^T$ | m/s |
| $K$ | 1001 | -- |

### *3.2. Simulation results*

Table 2. Comparison of secondary guidance simulation with different starting points.

| No. | **First stage shutdown point data** | | **Secondary optimization results** | | |
|---|---|---|---|---|---|
| | Position (m) | Velocity (m/s) | Quality Consumption (kg) | Position Error (m) | Velocity Error (m/s) |
| 1 | $\begin{bmatrix} 55930 & 31744 & 0 \end{bmatrix}^T$ | $\begin{bmatrix} 873.5 & 822.8 & 0 \end{bmatrix}^T$ | 639.2 | $2.68\times10^{-4}$ | $9.51\times10^{-6}$ |

| | | | | | |
|---|---|---|---|---|---|
| 2 | $[54778\ \ 31217\ \ 0]^T$ | $[863.1\ \ 813.8\ \ 0]^T$ | 649.2 | $9.85\times10^{-5}$ | $9.82\times10^{-6}$ |
| 3 | $[54769\ \ 31178\ \ 0]^T$ | $[863.3\ \ 813.2\ \ 0]^T$ | 650.1 | 0.0031 | $4.40\times10^{-6}$ |
| 4 | $[54856\ \ 31097\ \ 0]^T$ | $[864.7\ \ 811.8\ \ 0]^T$ | 652.3 | 0.0049 | $1.09\times10^{-6}$ |
| 5 | $[54765\ \ 31104\ \ 0]^T$ | $[863.0\ \ 811.6\ \ 0]^T$ | 652.6 | $1.98\times10^{-7}$ | $1.00\times10^{-5}$ |

For intuitive, a simulation diagram based on the first set of simulation data in the table is given below. The data in this set are the nominal parameters for the first stage of shutdown, the next four sets are all shutdown parameters with deviations.

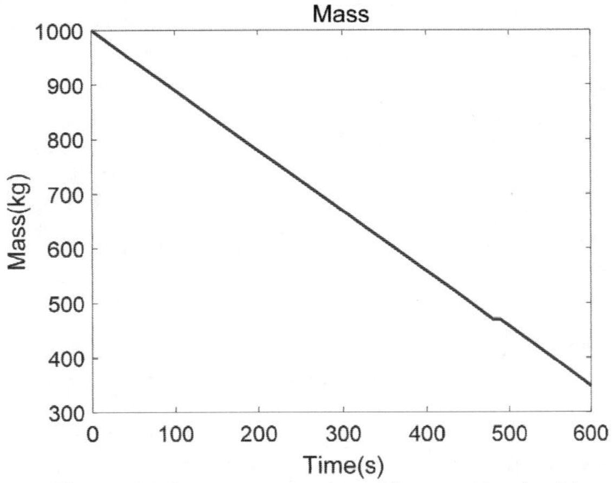

Figure 1 Mass consumption of secondary ignition

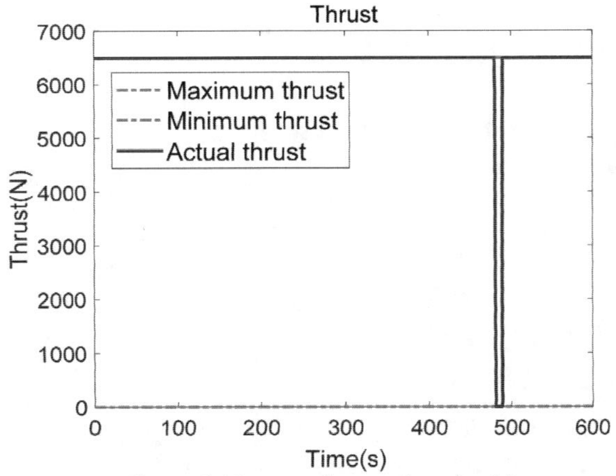

Figure 2 Thrust of secondary ignition

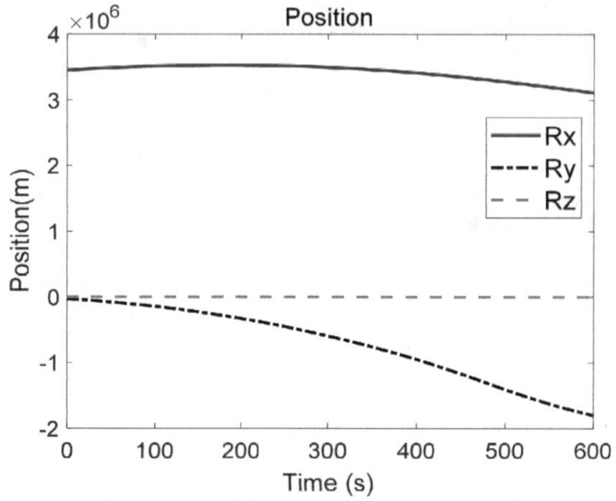

Figure 3 Position of secondary ignition

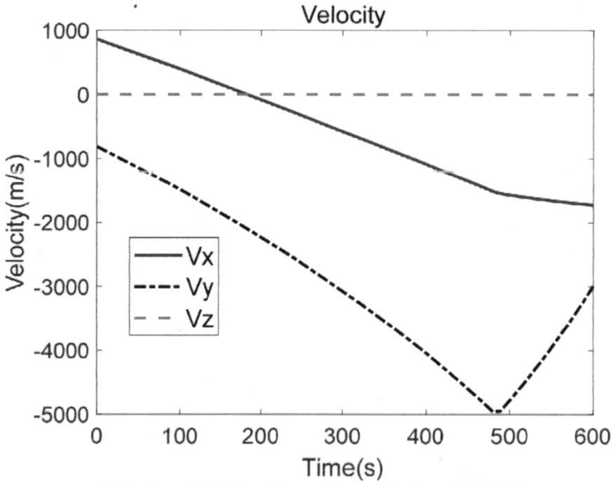

Figure 4 Velocity of secondary ignition

According to the simulation data and the simulation result graph, it can be concluded that the convex optimization algorithm can be used for the second-stage ignition of the Mars ascender to complete the optimization of the ascent guidance. For different first-level shutdown point data, the second-level optimization algorithm can complete the trajectory optimization with high precision. The thrust and mass of the ascender show a three-stage trend, which is in line with the optimal theoretical design. The two-stage ignition ascent guidance method is feasible.

## 4. Conclusion

This paper focuses on a two-stage ignition trajectory optimisation based on convex optimisation. Simulation experiments have demonstrated that the ascender can be placed in orbit with high accuracy at different starting points using this method. The algorithm ensures that the mass consumption is around 650kg, so that it reaches the near-Mars point in the circum-Mars orbit before running out of fuel. The trajectory optimisation algorithm for the second stage design is capable of high precision ascent guidance optimisation. The second stage is able to reprogram the orbit with high accuracy for different first stage shutdown point positions and velocities caused by various errors during the first stage flight.

## References

[1] Cui Y P, Yu Z S, Zhu S Y. (2013)Research Progress and Prospect of Autonomous Navigation Techniques for Mars Entry Phase . Journal of Astronautics,34(04):447-456.
[2] Li Y F. (2018) Research on the Key Technologies in Manned Deep Space Exploration. Missiles and Space Vehicles, 4(1): 24.
[3] Yu D Y, Sun Z Z, Meng L Z, et al. (2016) The development process and prospects for Mars exploration. Journal of Deep Space Exploration, 3(2): 108-113.
[4] Braun R D, Manning R M. (2007) Mars Exploration Entry, Descent, and Landing Challenges. Journal of Spacecraft and Rockets, 44(2): 310-323.
[5] Queen E M. (2000) Guidance Concept for a Mars Ascent Vehicle First Stage. NASA Langley Technical Report Server.
[6] McCormick B, Comptopn J. 1986 Mars sample return ascent vehicle conceptual guidance algorithm. Guidance, Navigation and Control Conference. 1: 429-434.
[7] Açıkmeşe B, Ploen S R. (2007) Convex Programming Approach to Powered Descent Guidance for Mars Landing. Journal of Guidance, Control, and Dynamics, 30(5): 1353-1366.

Journal of Physics: Conference Series **2364** (2022) 012014 doi:10.1088/1742-6596/2364/1/012014

# The dynamic model of the high-thrust orbital maneuvering engine with storable propellant

**Qingdong Su[1,2], Jinjin Wang[1*], Mingxia Yan[1,2], Zhensheng Sun[1], Bailin Zha[3] and Weifeng Huang[2]**

[1] Missile Engineering College, Rocket Force University of Engineering, Xi'an 710025, China

[2] Department of Mechanical Engineering, Tsinghua University, Beijing 100084, China

[3] Project Management Center, PLA Rocket Force Equipment Department, Beijing 100085, China

*Corresponding author's e-mail: suqd19@mails.tsinghua.edu.cn

**Abstract.** The storable propellant rocket engine has the advantages of long-term storage and rapid use. A new design of a large-thrust orbital maneuvering engine is proposed for the first time and the dynamic model for working process is established by the multi-disciplinary dynamics simulation software AMEsim. The accurate model parameters of hydraulic and combustion components are obtained through actual unit tests. The dynamic characteristics of the engine are calculated and evaluated in detail, and the results show that the system has good robustness to disturbance parameters. The fluctuation of the propellant tank pressure has little influence on operating parameters. The peak pressure of the water hammer, especially at the tank outlet pipe, is about 17 times higher than the stable state. The starting sequence influences working conditions of the mixing ratio regulator and the thrust chamber for less than 0.1s, and has no influence on the stable working state of the engine. The research provides an effective method for designing a high-thrust orbital maneuvering engine system, which has broad prospects in low-Earth orbit transportation and military missions.

## 1. Introduction

The orbital maneuvering engine is a kind of high-thrust upper stage rocket engine system, which works under unsteady conditions most of the time. This kind of engine provides main power for orbital maneuvering of spacecraft. Therein, an engine with storable propellant has outstanding advantages of long-term preservation and rapid use, which has been paid attention and widely studied by various countries. But there exists some serious problems such as the fluctuation of the propellant tank pressure, water hammer phenomenon and starting sequence in the propellant supply system when the thruster is frequently started, shut down, or changed working conditions[1]. They will damage the pipeline valves of the propulsion system, even affect the dynamic performance of the system coupled through the hydraulic circuit. In addition, the system works primarily at high altitude, resulting in high cost, long term and poor safety of thermal test. Thence, as an economical, safe and efficient method, the numerical simulation has been widely used in the design of various rocket power systems, which can be used to study the influence of single factors on engine performance. It is of great significance for avoiding the design defects, optimizing the performance, shortening the development cycle, and reducing the costs of the new rocket engine system[2].

The early research on system characteristics was mainly through the establishment of ordinary differential equations or algebraic equations by simple forms such as component mathematical models, conservation laws and empirical fitting formulas, to form special simulation programs. Typical examples include that the calculation of the start-up process of the Agena rocket engine of the US Air Force Rocket Propulsion Laboratory[3], the start-up characteristic neural network program of the Space Shuttle Main Engine (SSME) of the Lewis Research Center[4], LE-5, LE-7 engine dynamic process simulation program[5] and so on. In the process of developing high-thrust liquid oxygen/kerosene engines, some Chinese researchers developed special simulation programs to solve rapid iterative design problems[6]. Zhang and Xu compiled the steady-state/dynamic working model of the storable propellant rocket engine based on Matlab, and analyzed the fault mode of the engine by using BP neural network[7]. However, these special simulation programs depend on the explicit data format and transport relationship, which can not meet the requirement of rapid system simulation tasks.

Up to now, general simulation programs have become the main development trend of software engineering. In terms of system characteristics simulation platform, Beijing Aerospace Power Research Institute[8,9], China Aerodynamics Research and Development Center[10], North-western Polytechnical University[11], National University of Defense Technology[12] carried out dynamic characteristics researches on various power systems such as expansion cycle liquid oxygen methane engine, low temperature wind tunnel liquid nitrogen supply system, and electric pump pressure liquid oxygen kerosene engine based on the AMEsim platform. These researches show that the simulation method and AMEsim platform are reliable, which provide extremely favorable conditions for solving many system simulation analysis problems of storable propellant orbital maneuvering engines.

The study establishes a system nonlinear mathematical AMEsim model suitable for the storable propellant orbital maneuvering rocket engine with the modular modeling method. Detailed models are carried out in the aspects of cooling and heat exchange structure, key regulating valve components, thrust chamber, and pipeline distribution parameters. Moreover, the influence of key issues such as tank pressure fluctuation, pipe network water hammer, and thrust chamber working sequence on the characteristics of the engine system are analyzed.

## 2. Mission profile and system structure

The integrated orbital maneuvering engine is significant for material transportation in low-Earth orbit, space station and rapid deployment of large satellites. Attitude control and orbit control functions are integrated, and the typical working process includes multiple stages such as start-up, rated operating conditions, throttling operating conditions and shutdown.

A typical engine system uses Unsymmetrical dimethylhydrazine (UDMH) as fuel and dinitrogen tetroxide (NTO) as oxidant. The thrust of a single nozzle is as high as about 7000N, and a combination of 4 combustion chambers is usually used to generate vector thrust. Throttling conditions are typically 50% of rated value.

The system structure diagram is shown in Figure 1, which is divided into the propellant supply system and the thrust chamber. The propellant supply system includes propellant tanks, pumps, regulating valves and other components. The thrust chamber is composed of a combustion chamber, a nozzle and a regenerative cooling channel.

## 3. System components model

### *3.1. Piping model*

During the working process of the engine system, the low-pressure air-entrained filling of the pipeline has a certain influence on the start-up process. The AMEsim model simplifies the part of the air-containing filling model, and mainly verifies the pipeline transport characteristics of NTO and UDMH media at higher pressures. According to the low-frequency analysis target of the system, the actual pipeline geometry parameters are imported, and the corresponding C, IR or R pipeline units considering capacitive, inertial and resistive properties are established respectively[13,14].

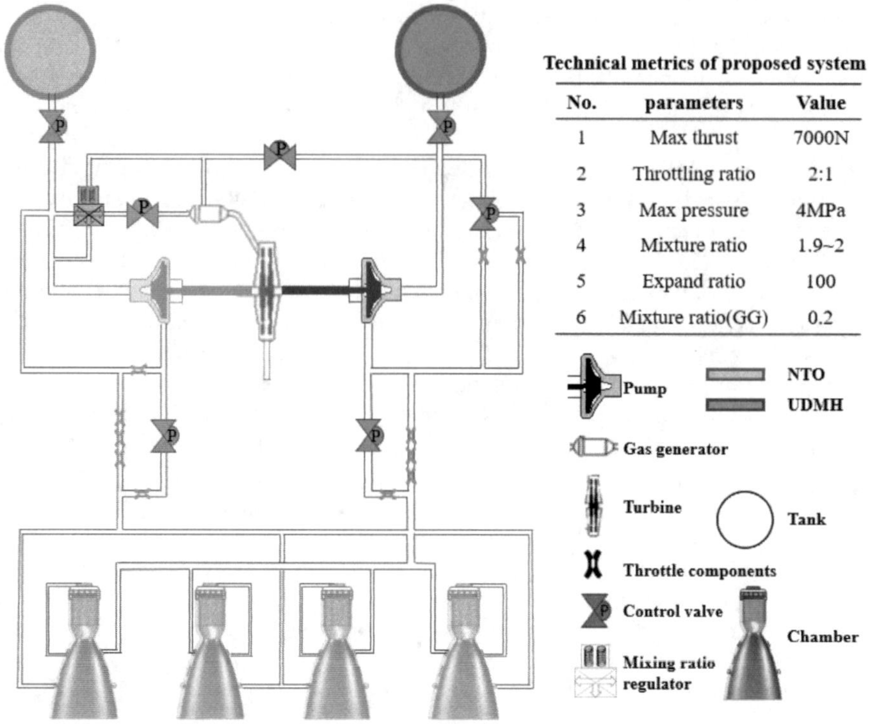


| No. | parameters | Value |
|---|---|---|
| 1 | Max thrust | 7000N |
| 2 | Throttling ratio | 2:1 |
| 3 | Max pressure | 4MPa |
| 4 | Mixture ratio | 1.9~2 |
| 5 | Expand ratio | 100 |
| 6 | Mixture ratio(GG) | 0.2 |

Figure 1. Schematic diagram of engine system structure

### *3.2. Orifice model*

The mass flow rate of the liquid orifice is calculated as follows:

$$\dot{m} = c_q A \sqrt{2\rho_{up}\Delta p} \tag{1}$$

where, $\dot{m}$ mass flow rate, $c_q$ flow coefficient, $A$ fluid passage area, $\rho_{up}$ calculated fluid density under upstream conditions, $\Delta p$ upstream and downstream pressure drop.

The enthalpy flow of the orifice is calculated according to the upwind style.

$$dmh = \dot{m} \cdot h\left(P_{up}, T_{up}\right) \tag{2}$$

For gas orifice, the flow algorithm is:

$$\dot{m} = C_q \cdot A \cdot \Psi \cdot \sqrt{\frac{2 \cdot P_{up} \cdot \rho_{up}}{k_{dp}}} \tag{3}$$

where $k_{dp}$ is the pressure loss coefficient and $\Psi$ is the flow function.

When the nozzle is not clogged, the above equation degenerates to equation (1). When the nozzle is clogged, it is a one-dimensional compressible flow, which is related to the pressure ratio as follows:

$$\Psi = \begin{cases} \sqrt{\dfrac{1}{1-\gamma_s}} \cdot \sqrt{\eta^{2\gamma_s} - \eta^{1+\gamma_s}} & \eta > \eta_{cr} \\ \sqrt{\dfrac{\gamma_s}{1+\gamma_s}} \cdot \left(\dfrac{2\gamma_s}{\gamma_s+1}\right)^{\frac{\gamma_s}{1-\gamma_s}} & \eta \le \eta_{cr} \end{cases} \quad (4)$$

In the formula, $\gamma_s$ is the non-ideal gas state isentropic heat factor, $\eta$ is the current pressure ratio, $\eta_{cr}$ is the critical pressure ratio.

### *3.3. Pump model*

The pump model is a centralized inertial flow element (IR element) on the fluid and a torque source on the machine. Considering the inertia of the pump, the rate of change of the pump volume flow is calculated as follows:

$$\frac{dQ}{dt} = \frac{\rho g H - \left(p_{up} - p_{down}\right)}{I} \quad (5)$$

Where $\rho$ is average density of pump (average of inlet and outlet), $H$ is head, $I$ is fluid inertia, $p$ is fluid Pressure. The fluid inertia is calculated as follows:

$$I = \frac{\rho_R L_R}{A_R} \quad (6)$$

In the formula, $\rho_R$ is the fluid density at the design point, $L_R$ is the length of the design point along the fluid path of the pump, and $A_R$ is the design flow cross-sectional area of the pump.

The pump head $H$ and torque $N$ are calculated using the Suter pump model. The dimensionless parameters of the Suter pump are as follows:

$$v = \frac{Q}{Q_R}, \quad \alpha = \frac{N}{N_R}, \quad h = \frac{H}{H_R}, \quad \beta = \frac{T}{T_R} \quad (7)$$

In the formula, $Q$ is volume flow, $N$ is speed, $H$ is head, $T$ is torque, subscript $R$ represents design point parameters. The detailed parameter definitions refer to the Reference [15]. The dimensionless head characteristics and efficiency characteristics can be expressed as functions of Suter angle and specific speed:

$$\begin{aligned} WH &= WH\left(N_s, \theta\right) = \frac{h}{\alpha^2 + v^2} \\ WT &= WT\left(N_s, \theta\right) = \frac{\beta}{\alpha^2 + v^2} \end{aligned} \quad (8)$$

From this, the characteristics of the pump are calculated as follows:

$$\begin{aligned} H &= H_R \cdot WH \cdot \left(\alpha^2 + v^2\right) \\ T &= T_R \cdot WT \cdot \left(\alpha^2 + v^2\right) \end{aligned} \quad (9)$$

Among them, the design point torque of the pump is:

$$T_R = \frac{H_R \cdot Q_R \cdot \rho \cdot g}{N_R \cdot \eta_R} \quad (10)$$

*3.4. Turbine model*

The turbine model is a concentrated inertial flow element (IR element) on the fluid and a torque source on the machine. The calculation method of the actual mass flow of the turbine is similar to the above orifice model.

To avoid the cyclic solution problem caused by implicit variables, the turbine mass flow is treated as an inertial model:

$$\frac{d\dot{m}_t}{dt} = \frac{\dot{m}_t - \dot{m}_{t-1}}{\tau} \tag{11}$$

The isentropic expansion speed $c_0$ is calculated as follows:

$$c_0 = \sqrt{2(h_{0,up} - h_{0,s,down})} \tag{12}$$

In the formula, $h$ is specific enthalpy, subscript 0 is total enthalpy, $s$ is isentropic flow.

The key parameter speed ratio is defined as follows:

$$\frac{U}{c_0} = \frac{\omega \cdot R}{\sqrt{2\left(h_{0,up} - h_{0,s,down}\right)}} \tag{13}$$

Turbine efficiency is the ratio of the actual enthalpy drop to the isentropic enthalpy drop:

$$\eta = \frac{h_{0,up} - h_{0,down}}{h_{0,up} - h_{0,s,down}} \tag{14}$$

The torque and turbine power calculation models are:

$$T = \frac{\eta \dot{m}\left(h_{0,up} - h_{0,s,down}\right)}{\omega} \tag{15}$$

$$P_{mech} = \omega \cdot T \tag{16}$$

Among them, the turbine output power is essentially the reduction of the enthalpy flow.

*3.5. Combustor and Nozzle Models*

The combustion chamber is the main area where the reaction processes such as atomization, evaporation, mixing and combustion occur. In this paper, without considering the adiabatic uniformity of liquid volume, a zero-dimensional premixed combustion time-delay model is used to study its low-frequency dynamics[14].

The nozzle model also considers the wall free shock separation problem of the gas jet. According to the nozzle flow separation standard in the literature [16], the separation position is calculated to improve the accuracy of thrust calculation. In the cooling jacket part, the Bartz method [15] is used to calculate the heat transfer problem on the gas side, and the heat transfer tube flow [17] model is used to calculate the heat transfer on the wall of the thrust chamber and the heat transfer problem with the cooling water. The radiative cooling expansion section is calculated based on the heat conduction and radiative heat transfer models in the literature[18].

## 4. Calculation method

*4.1. AMEsim shetch*

Based on the idea of modular modeling, a dynamic simulation model of the AMEsim system is established, and the structure is shown in Figure 2. The main components are oxidant source, fuel source, mixing ratio control valve, gas generator, turbine, oxidant pump, fuel pump, injector and thrust chamber.

The main simulation sequence and variable setting values of the system are controlled by prefabricated piecewise functions. The whole system contains 431 state variables, corresponding to 431 dynamic equations, which are solved by a variable-step integral solver, The method effectively avoids the traditional single Runge-Kutta Method Solving dilemmas in solving high frequency oscillations and discontinuities of system equations.

## *4.2. Model details*

In the process of engine regeneration and cooling, solid heat conduction, convective heat transfer between fluid and solid, and heat radiation between fluid/solid and the environment are carried out simultaneously. The working principle of the thrust chamber with staged regenerative cooling designed and calculated in this paper is shown in Figure 3. The heat exchange section near the entrance of the combustion chamber is cooled by liquid UDMH, the middle section is cooled by liquid NTO, and the heat exchange section closest to the nozzle is used for cooling. The third section of the outlet uses radiative cooling.

In order to further ensure the simulation accuracy, the geometric dimensions of the other components of the system are consistent with the actual parameters, and the pipeline specifications, valve flow coefficient and other data are calibrated according to the rated nominal parameters. The elastic modulus of the tube wall and the thermal conductivity of the material, which are not easily obtained by direct experiments, were selected by referring to the AMEsim material library and the chemical kinetics database (National Institute of Standards and Technology, NIST).

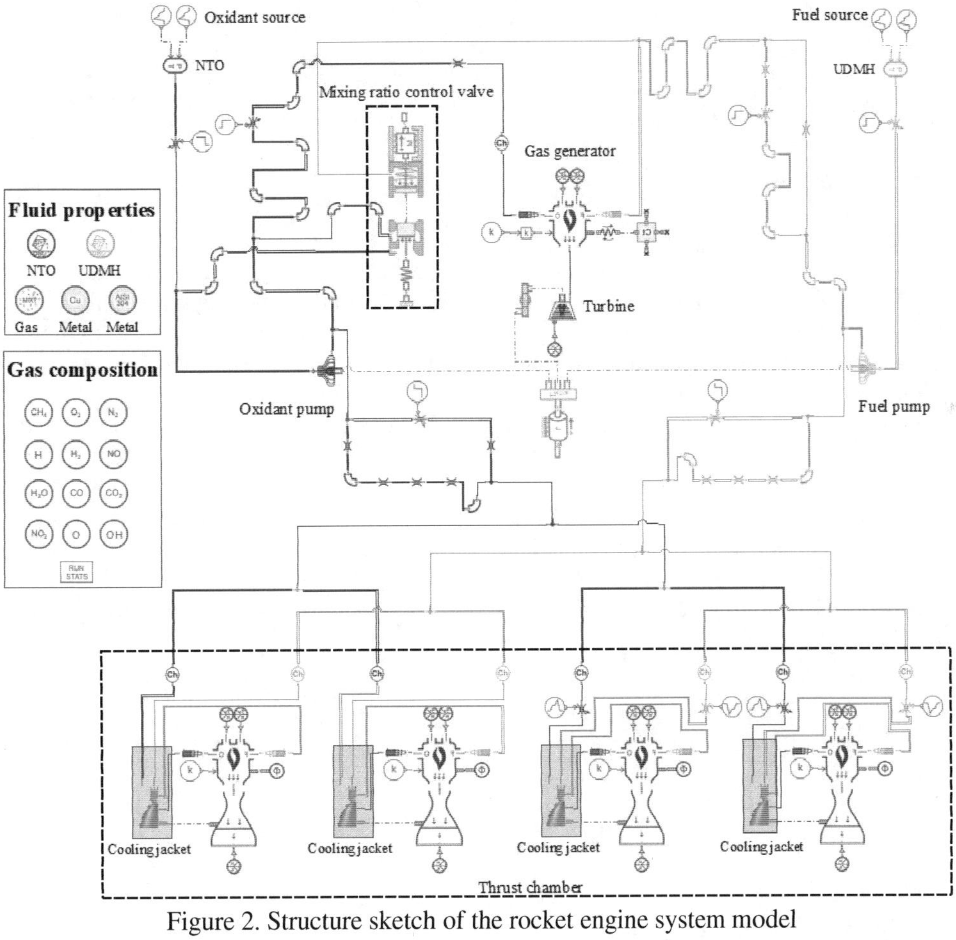


Figure 2. Structure sketch of the rocket engine system model

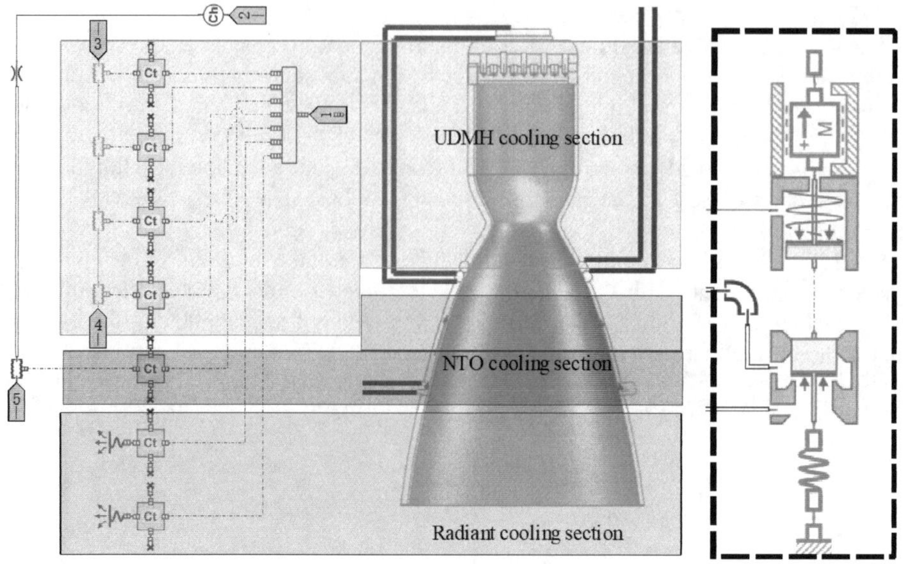


Figure 3. Structure of thrust chamber for staged regenerative cooling and mixing ratio control valve

## 5. Results and discussion

### *5.1. Disturbance effects of tank pressure fluctuations*

Under the main working condition, the response characteristics of the system working process with different fluctuation frequencies, amplitudes and forms of the oxidant/fuel tank pressure of the whole system are simulated. The results show that using the automatic batch operation function of the software, six different tank pressure fluctuation frequencies ($1\times10^{-6}$Hz, 0.1Hz, 0.5Hz, 1Hz, 5Hz, 10Hz) and the fluctuation range of 10% of the stable working pressure are simulated. In Figure 4(d), the blue curve represents the pressure fluctuation of the oxidant tank. It can be seen that the displacement change of the mixed ratio control spool and the tank pressure show synchronous anti-phase fluctuation, which means the spool displacement reaches the minimum value when the tank pressure reaches the maximum fluctuation value. According to the working principle of mixed ratio control valve, the inlet pressure and valve opening increase while the spool displacement decreases under the condition that the control pressure is basically unchanged. With the increase of the frequency of tank pressure fluctuations, the displacement of the spool shows a stable fluctuation within a certain threshold range, and the fluctuation amplitude is about 0.008mm, which has great flow stability.

Journal of Physics: Conference Series **2364** (2022) 012014 doi:10.1088/1742-6596/2364/1/012014

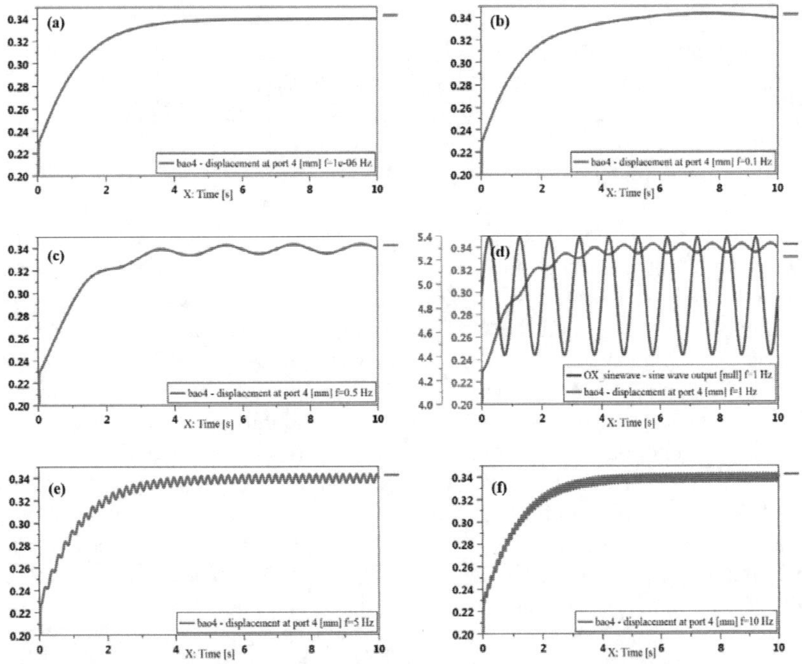


Figure 4. Displacement curve of mixing ratio regulating valve spool under different tank pressure fluctuation frequencies

Figure 5 shows the variation of nozzle thrust under different tank pressure fluctuation frequencies. The partial enlargement of the thrust rising to the stable section at 6 different frequencies is depicted in Figure 5(a), and the nozzle thrust is stable. Under different fluctuation frequencies, the range of nozzle thrust fluctuation is very small, only about 1N, which is consistent with the tank pressure fluctuation frequency. This dynamic response law is consistent with the response characteristics of the control valve.

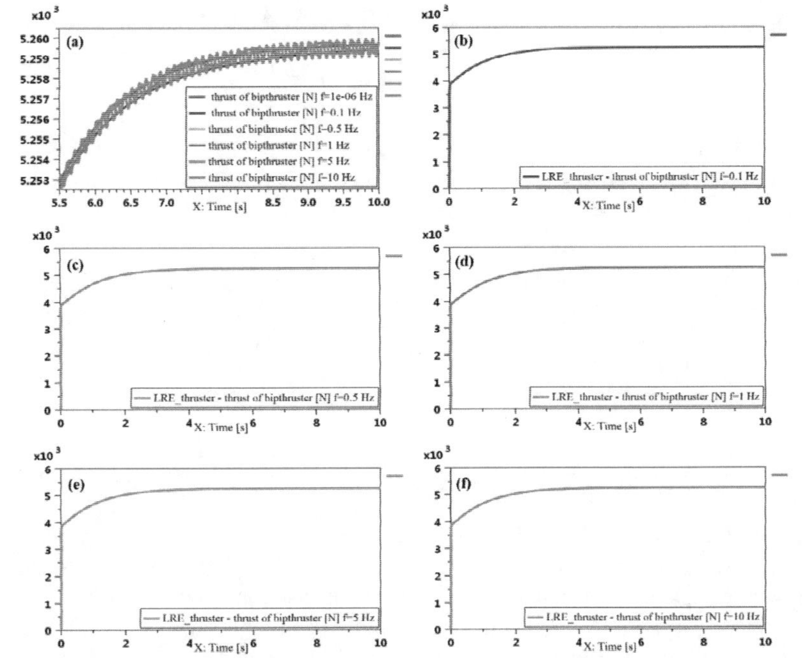


Figure 5. Variation of nozzle thrust under different tank pressure fluctuation amplitudes

*5.2. Water hammer effect during start-up*

The water hammer problem mainly considers the fluid inertia, and the segmented pipeline model is used in the simulation model to study the water hammer characteristics of the pipeline.

The AMEsim software is used to build a dynamic system simulation model, and the water hammer characteristics of the outlet pipeline of the oxidant tank and the fuel tank are analyzed. The pipeline submodel adopts HL0040 model, which is an inertial pipeline model with distributed parameters (C- IR - C-IR).

As shown in the Figure 6, when the four thrust chambers are started at the same time, the outlet pipeline of the oxidant tank has a large pressure fluctuation within 0.1s of the start-up. For pipeline node 1, the peak water hammer pressure peaks at 0.002s, and the peak pressure is 1.67 MPa, then begins to approximate damped oscillation, and the oscillation amplitude decreases rapidly. After about 0.3s, it gradually recovers to 0.489MPa, the peak pressure is 3.4 times the stable pressure, and there is no obvious water hammer in the outlet pipeline. The flow reaches its peak value at 0.004s, and the peak flow is 1.96L/s. After about 0.3s, the oscillation ends and gradually rises and stabilizes to 2.32L/s.

The fuel tank outlet pipeline has large pressure fluctuations within 0.1s of start-up. For pipeline node 1, the peak water hammer pressure appears at 0.0067s, and the peak pressure is 4.57MPa. Then the oscillation amplitude decreases rapidly, and gradually recovers to 0.25MPa after about 0.1s. The peak pressure is 18.28 times to the steady pressure, and there is a relatively obvious water hammer phenomenon in the outlet pipeline of the fuel tank. The fuel flow reaches a peak value of 4.34L/s at 0.004s, and the oscillation ends after about 0.1s, then rises gently and stabilizes to 3.32L/s.

Figure 6 shows the chamber pressure change of the thrust chamber before 0.1s during the system start-up process. The pressure fluctuates with the pipeline pressure and flow, and there are slight oscillations during the rising process. In general, the water hammer phenomenon during start-up has little effect on the combustion pressure of the thrust chamber. Furthermore, the engine thrust is also maintained stable.

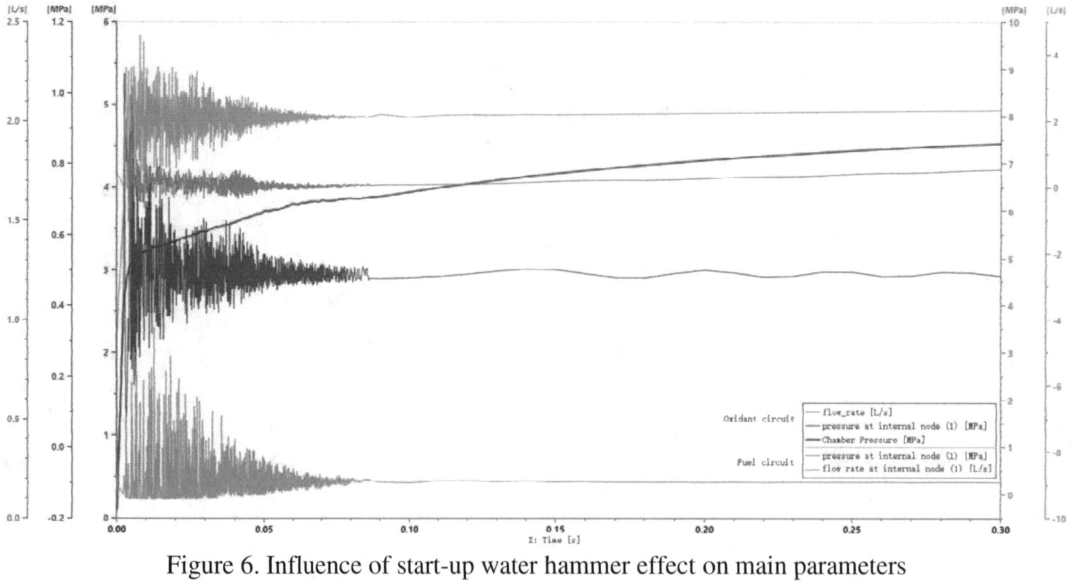


Figure 6. Influence of start-up water hammer effect on main parameters

*5.3. Staggered start-up optimization strategy of thrust chambers*

Under the main stage condition, the dynamic simulation of the four thrust chambers starting at staggered time is carried out. Considering the balance of the system thrust, the diagonal thrust chambers starts first and the remaining two staggered starts within 0.5s. The starting sequence is realized based on the timing control valve at the end of the propellant pipeline in the thrust chamber.

Figure 7 shows the effect of staggered start-up on the outlet line pressure of the oxidant tank. The blue curve in the figure is the staggered starting condition, and the red curve is the normal starting condition. The normal starting pressure fluctuation range is slightly larger, indicating that the staggered starting of the two thrust chambers slightly reduces the starting water hammer effect.

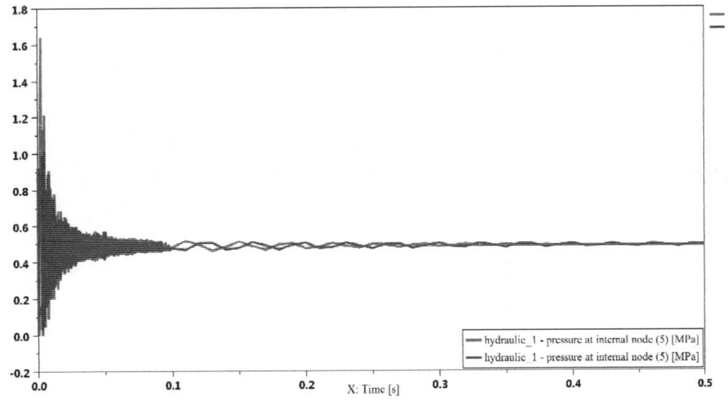


Figure 7. Influence of staggered start-up on oxidant tank outlet pipeline pressure

Figure 8 is the thrust change when the four thrust chambers are staggered, the orange curve is the case where the four thrust chambers started at the same time, and the red and blue curves are the thrust changes when the thrust chambers are started at different time. Before 0.5s, the thrust of the first two thrust chambers rose rapidly to 7604N at 0.5s, the remaining two thrust chambers started, and the thrust instantly increased to 6915N. In a short period of time, the thrust of the four thrust chambers recovered to about 6997N. The partial magnification of the curve before 0.2s shows that even in the presence of water hammer, the thrust changes slightly and the system is stable.

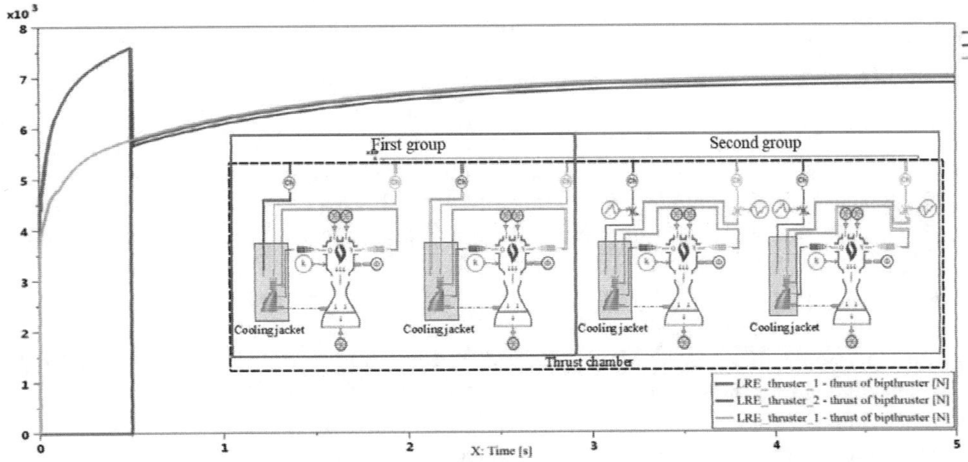


Figure 8. Comparison of nozzle thrust under staggered starting conditions

## 6. Conclusions

In the paper, a mathematical model of the working process of the storable propellant orbital maneuvering engine system is established on the basis of the unit tests and real measurement calculations of the components such as the piping, mixing ratio control valve, etc.

The integrated system scheme evaluation dynamic characteristics calculation is realized by the multi-disciplinary dynamics simulation software AMEsim. The influence law of factors such as the response characteristics of the tank pressure parameter fluctuation system, the water hammer of the pipeline network, and the staring sequence of the power system are analyzed. The results show that the simulation model and method components established are flexible and versatile, and conform to the trend of advanced modular modeling.

All in all, the designed system has good robustness to disturbance parameters. The controllable water hammer problem exists, especially the peak water hammer pressure of the fuel tank outlet pipe is about 18 times the stable pressure, and it has little effect on the operating parameters of the system. The starting sequence of the engine system has a short-term effect on the working conditions of the mixing ratio regulating valve and thrust chamber within less than 0.1s, and the overall working parameters change little. The starting sequence can be controlled immediately according to the thrust demand, proving the system is stable and reliable.The research provides a reliable analysis method for the further structure design and test of integrated storable propellant orbital maneuvering engine system.

**Acknowledgments**
The support of the Chinese Program for the National Natural Science Foundation of China (No. U1737209) and the National Basic Research Program of China (Grant No. 91952110) are greatly appreciated.

**References**
[1] Zhu N.C. et al. (1994) Design of liquid rocket engine (upper and lower) [M]. Beijing: China Aerospace Publishing House.
[2] Zhang Y.L., Liu K., Cheng M.S. (2005) Dynamics Theory and Application of Liquid Rocket Motor[M]. Beijing: Science Press.
[3] Powell M, Butler P, Beck L. (1975) Ground test simulation of rocket engine start transients[C]// 11th Propulsion Conference.
[4] Meyer C M, Maul W A. (1991) The application of neural networks to the SSME start-up transient. Presented at the 27th Joint Propulsion Conference, Sacramento, CA, 24-27 Jun. 1991; sponsored by AIAA, SAE, ASME, and American Society for Electrical Engineers.
[5] Kanmuri A., Kanda T., Wakamatsu Y., et al. (1989) Transient analysis of $LOX/LH_2$ rocket engine (LE-7)[C]// 25th Joint Propulsion Conference.
[6] Xu H. H. (2003) Research on start-up process of liquid oxygen kerosene supplementary combustion engine ［D］. Xi'an: The Eleventh Research Institute of the Sixth Research Institute of China Aero-space Science and Technology Corporation.
[7] Zhang W., Tian Q., Xu Z.G. (2014) Analysis and Diagnosis of Liquid Missile Engine Failure Char-acteristics [M]. National Defense Industry Press.
[8] Li J.J. (2012) A fast method for dynamic simulation of low temperature engine system [J]. Missile and Space Launch Technology, (01):13-17.
[9] Zheng D.Y., Wang H.Y., Hu J. (2021) Research on transient characteristics of high-thrust hydro-gen-oxygen engine [J]. Propulsion Technology, 42(8):9.
[10] Zhang W., Sun D.W., Chen W.H. et al. (2022) Simulation of water hammer in liquid nitrogen supplying system based on AMEsim [J]. Journal of Aerospace Power, 37（2）: 366-374.
[11] Liu Y., Fu B.S., Yang J.G., et al. (2019) Modeling and Simulation of Electric Pump-Pressure Liquid Rocket Engine System [J]. Manned Aerospace, 25(1):9.
[12] Cui P., Song J., Li Q.L., et al. (2022) Dynamics Modeling and Simulation Analysis of Electric Pump-Pressure Liquid Oxygen Kerosene Variable Thrust Rocket Engine: Part I-Single-Point Condition Analysis [J]. Chinese Journal of Aeronautics and Astronautics, 43 (1): 15.
[13] Glickman. (1995) Automatic adjustment of liquid rocket motor [M]. Translated by Gu Mingchu. Beijing: Aerospace Press.
[14] Zhai Y.F. Research on dynamic simulation of expansion cycle engine thrust regulation process [D]. China Aerospace Science and Technology Corporation.
[15] Siemens. Simcenter Amesim 2020.1 Liquid Propulsion Library User's Guide ［R］. Brussels: Siemens Industry Software NV, 2020.

[16] Stark R. Flow Separation in Rocket Nozzles, a Simple Criteria[C]//41th Joint Propulsion Conference & Exhibit. DLR, 2005.
[17] Liang T., Cui P., Li Q. L., et al. (2022) Dynamic Characteristics of Electric Expander Cycle System for LOX/$LCH_4$ Liquid Variable Thrust Liquid Rocket Engine［J/OL］. Journal of Propulsion Technology, 1-14: 210365.
[18] Siemens. Simcenter Amesim 2020.1 Thermal Hydraulic Component Design library User's Guide［R］. Brussels：Siemens Industry Software NV, 2020.

# Structural Optimization of Single Crystal Turbine Blade Based on Recrystallization Critical Stress

**Shifeng Li[1*], Suling Chen[2], Siyuan Ren[1]**

[1]China Aerodynamics Research and Development Center, Aerospace Technology Institute, Mianyang Sichuan, 621000, China

[2]Department of Material Science and Engineering, Sichuan Engineering Technical College, Deyang, 618000, Sichuan, China

*E-mail: 275400376@qq.com

**Abstract.** In view of the problem that the surface structural recrystallization (SRR) is very easy to occur in the splitting gap area of the single crystal blade. A method to control the residual stress is proposed that based on the critical stress of single crystal recrystallization, and the structural details are optimized by combined with the numerical calculation and experimental verification. To reduce the residual stress level and change the residual stress distribution in the single crystal blade splitting gap. So as to control the level of residual stress, which is no more than the recrystallization critical stress. Finally, the experimental verification is completed by the optimized blade. The verification results show that the maximum residual stress is reduced by about 10% in the tail splitting region, and the recrystallization structure is completely eliminated in the original structure. It is proved that this method is a great effectiveness to control the structural integrity and surface structure connectivity of the single crystal blade.

## 1. Introduction

Turbine blades work under harsh conditions such as high temperature, high pressure, high speed and strong alternating load. They are not only subjected to vibration stress caused by centrifugal force, aerodynamic force and vibration, but also to comprehensive effects such as thermal shock, thermal stress and thermal corrosion of gas. They are the most stringent parts in aero-engine working environment [1-3]. At present, with the continuous improvement of performance for aero-engine, the gas temperature is getting higher and higher in front of turbine. So it is the single crystal blades that are widely used in advanced aero-engines at home and abroad, and which has the higher temperature resistance. However, the recrystallization problems often occur in domestic single crystal blades at this stage, which seriously affects the structural integrity and continuity of turbine blades [4, 5].

From the research and engineering application at home and abroad, there are two main types of surface recrystallization structures locally appearing of the single crystal blades: Surface Plastic Recrystallization (SPR) and Surface Structure Recrystallization (SSR), and the failure probability of blade is more than 30% which caused by recrystallization. Up to now, around the SPR problem, there are lots of research in the research institutions and industrial departments of domestic and foreign, which force on the aspects of alloy optimization and process optimization and control quality, etc., so the SPR phenomenon has been basically eliminated [6-12]. However, SSR is not only related to alloys and processes, but also mainly restricted by blade structure. Especially for single crystal blades with complex hollow thin-walled structure, the wall thickness is about 2.50-0.40mm, the cooling structure is very complicated, and the section of blade body changes suddenly. Therefore, to completely eliminate the SSR, which caused by large residual stress of structure, the most feasible and key method are to control the structural residual stress of single crystal blades. So it is an aim that does not exceed the critical stress when SSR occurs[13]. That is reduced by optimizing the details of blade structure, which include the stress level and the stress distribution forms, and the residual stress level of local structure is mostly lower than the critical stress where recrystallization occurs in single crystal alloy. In this way, SSR can be completely eliminated in single crystal blade.

Through the above research and understanding, this paper takes the tail splits groove as the research object of a single crystal blade. And based on the critical stress of DD6 single crystal alloy recrystallization, the optimum design was done by combining numerical simulation and experimental verification for the blade structure under multiple constraints , which is of great significance to eliminate SSR. Meanwhile, it can ensure the structural integrity and continuity of single crystal blade.

## 2. Optimizing model establishment

### *2.1. Optimizing mathematical model*

In this study, the low-pressure turbine blade of an aeroengine is taken as the research object. It is a thin-walled and hollow and complex structure, which the tail splits groove often occurs SSR of a single crystal blade. So it is selected as the optimization design objective, based on the critical stress boundary [13, 14] of recrystallization of DD6 alloy, and with the requirement of the cold air flow area and strength reserve, the optimum design method is adopted. And the maximum residual stress ($\sigma_{max}$)is taken as the research object which is no more than the critical stress ($\sigma_c$)of single crystal recrystallization. At last, the relationship is optimized between the maximum residual stress （$\sigma_{max}$） and design parameters (D, H, R, $\alpha$） . Optimized mathematical models such as formulas (1) and (2).

Objective function:

$$\sigma_{max} = \phi(R_i, D_j, H_n, L_m, \alpha) \tag{1}$$

Constraints:

$$\begin{cases} \sigma_{max} = \phi(R_i, D_j, H_n, L_m, \alpha) \\ \sigma_{max} \le \sigma_c = F(T) \\ k = \sigma_{max}/[\sigma] \ge 1.25 \end{cases} \tag{2}$$

In equation: $\sigma_{max}$ as the maximum residual stress, unit MPa; $\sigma_c$ as the recrystallization critical stress, unit MPa; $D_j$ as the variable width, unit mm; $H_n$ as the variable depth, unit mm; $R_i$ as the variable radius, unit mm; $\alpha$ as the inclination angle, unit degrees; $F(T)$ as the constraint function.

### *2.2. Design variables and constraints*

When optimizing design of the blade tail splitting groove structure, the restrictions are imposed such as the radial total height（$D_j + L_n$）, and the pneumatic shape, and the cold air flow area ($A_0$) and the strength reserve of blade body. At last, the overall structure of tail splitting groove can only be designed by the width（$D_j$）and the radius($R_i$)of the groove, which considering the convenience of wall thickness control and the ceramic positioning and the process operability, especially the "adjustability" of curve configuration and corner line. As shown in Figure 1, the specific design variables and constraints are expressed in Table 1.

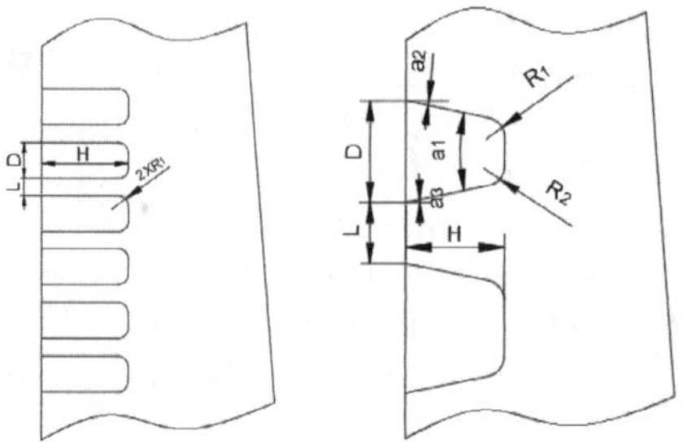

Figure 1. Schematic diagram of half-split groove structure

Table 1. Variables for optimized design of tail splitting grooves

| design variable | constraint condition | optimizing objectives |
|---|---|---|
| $D_j$、$R_i$<br>$D_j \geq 3.0$<br>$R_i \geq 0.3$ | $D_j = H_n$; $L = D_j/2$<br>$\alpha_2 = \alpha_3 = \alpha_1 / 2$<br>$\tan\left(\alpha_1/2\right) = 1 - \frac{A_0}{D_j^2}$<br>$A_0 = D_j \cdot H_n - H_n^2 \cdot \tan\left(\frac{\alpha}{2}\right)$ | $\sigma_{max} < \sigma_c$ |

In order to ensure the maximum residual stress ( $\sigma_{max}$ ) , which does not exceed the recrystallization critical stress ( $\sigma_c$ ) of DD6 alloy in the tail splitting slot area. According to the work hardening rate theory, and considering the material nonlinearity, the relationship with the stress-strain are obtained between hardening rate ($\sigma - \varepsilon$) and strain ($\ln\theta - \varepsilon$) curves of nickel-based single crystal material, which based on the tensile stress-strain curves of the standard test rods, and the thin-walled circular tubes and the thin plate samples. On this basis, the critical characteristic of recrystallization to the corresponding temperature and critical stress [13, 15-18] are given. Finally, the relationship model between critical stress ($\sigma_c$) and temperature (T) is obtained by fitting the curve of temperature equation and critical stress ($\sigma_c$). It is shown in the following formula (3), and as shown in Figure 2.

$$\begin{cases} \sigma_c = 47454.8 \cdot \exp\left(-1.02 \cdot T/233.5\right) - 48.2 \\ 1173\,K \leq T \leq 1550\,K \end{cases} \tag{3}$$

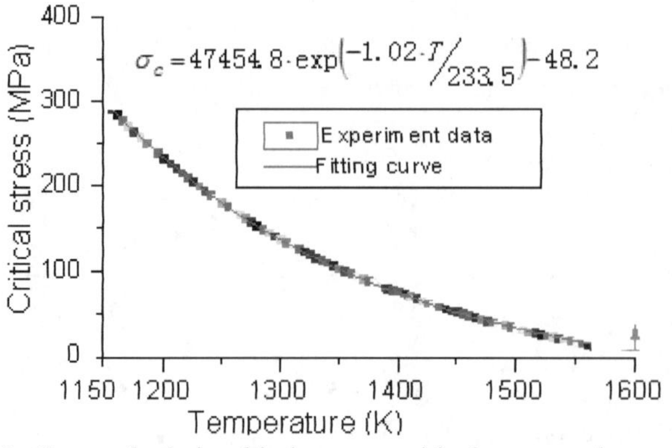

Figure 2. Curve of relationship between critical stress and temperature

## 3. Analysis of optimization results

The low-pressure turbine blade is a shrouded blade. The adjacent shrouds are in normal contact through the working surface of the shrouds to form a cyclic symmetrical model. Its grid model is shown in Figure 2, with a total of 450000 units and about 700000 nodes. Under the conditions of specific aerodynamic load and the speed and temperature load, the strength of blade tail slit is compared and analyzed before and after optimization.

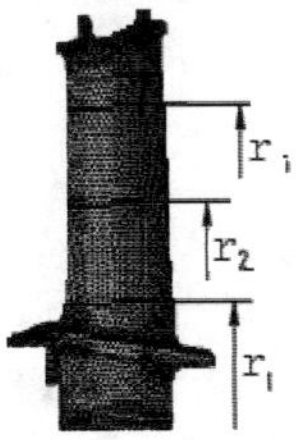


Figure 3. Schematic diagram of low-pressure vane grid model

### *3.1. Strength comparison and analysis*

Two sections with radial height of R1 and R2 , which are selected for lasting strength analysis. And the R1 is root section, the R2 is the maximum temperature section. The equivalent stress distribution of each section is shown in Figure 4. Compared with before optimization, it can be seen from the figure that the overall stress level is significantly reduced after the optimization of the blade body. Especially in the exhaust edge area, the maximum equivalent stress is reduced by about 15%, and the stress distribution is more uniform.

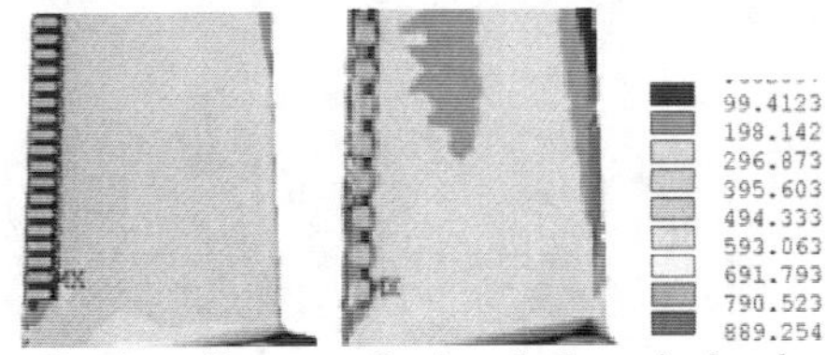


Figure 4. Equivalent stress distribution diagram (before left optimization and after right optimization)

At the same time, the R1 and R2 sections are selected for comparative analysis of section stress and strength reserve, as shown in Figure 5 and 6.

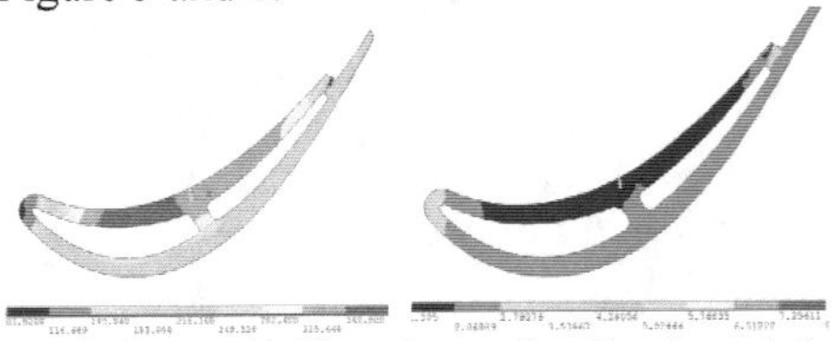

Figure 5. R2 section (left equal effect force distribution, right strength reserve)

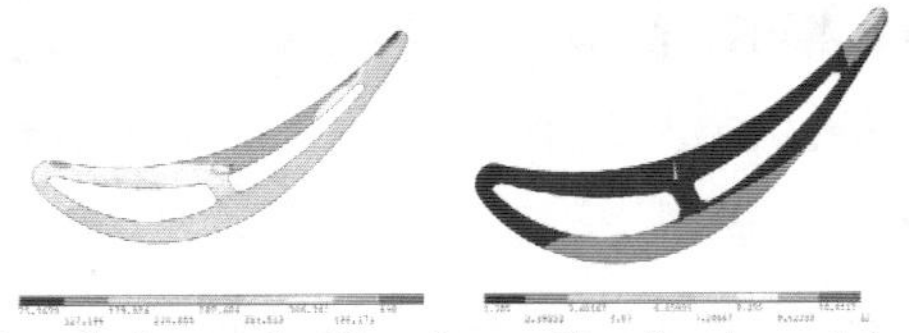

Figure 6. R1 section (left equal effect force distribution, right strength reserve)

After the optimization of the tail split joint structure, it can be seen from the figure that the maximum stress is between 119 MPa and 463 MPa in the groove area, and the overall stress level is low. The maximum stress appears in the rib area between the groove and the groove, the stress value is 463 MPa, and the strength reserve is greater than 1.35 at the corresponding position, the design requirements that the strength reserve is greater than 1.25. The maximum residual stress is shown in Table 2 with the inlet window area under different optimization parameters.

Table 2. Maximum residual stress under different optimization parameters (MPa)

| Optimization parameter | residual stress（MPa） | | | |
|---|---|---|---|---|
| | $R_i$=0.3 | $R_i$=0.5 | $R_i$=0.8 | $R_i$=1.0 |
| $D_j$=3.0 | 463 | 402 | 342 | 286 |
| $D_j$=3.5 | 421 | 343 | 274 | 177 |
| $D_j$=4.0 | 383 | 306 | 228 | 151 |
| $D_j$=4.5 | 336 | 268 | 154 | 119 |

*3.2. Relationship model between residual stress and variables*

For the data in Table 2, the surface fitting function is established by means of binary multiple linear regression [15], as shown in formula (4). In order to ensure the fitting accuracy of the surface, it is necessary to determine the m and n values of the polynomial when fitting the surface. Meanwhile, for ensuring the accuracy of the fitting function, setting x as the curvature radius ($R_i$)of the corner at the groove bottom, setting y as the groove width( $D_j$) and z as the maximum residual stress（$\sigma_{\max}$）. The fitting of the surface function is realized; the values are shown in Figure 7 which is the determination coefficients with corresponding to different exponents in the function.

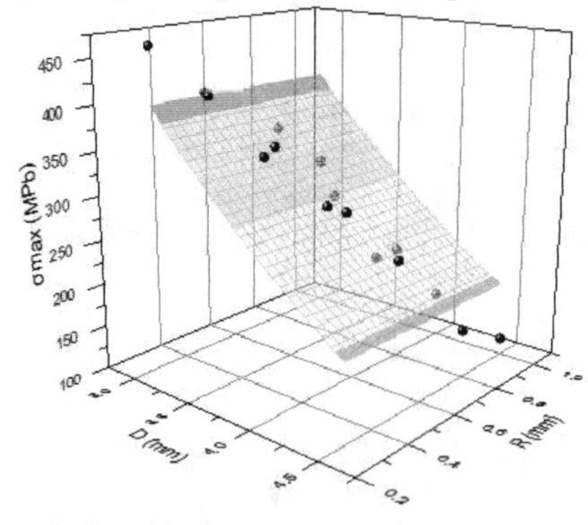


Figure 7. Curve fitting of the relationship between maximum residual stress and design parameters chart

$$\sigma_{\max} = 1460.9 \cdot \exp\left(-\frac{R}{3.1} - \frac{D}{352.5}\right) - 155.4 \qquad (4)$$

Combined with formula (3), the critical stress ($\sigma_c$) of formula (3) is equal to the maximum residual stress ($\sigma_{\max}$) of formula (4) as the threshold value, so as to obtain the relationship criterion between the maximum residual stress ($\sigma_{\max}$) and structural optimization parameters (R, D) and temperature (T), as shown in formula (5).

$$\begin{cases} \sigma_{\max} = 730.45 \cdot \exp\left(-\frac{R}{3.1} - \frac{D}{352.5}\right) + \\ 23727.4 \cdot \exp\left(-\frac{1.02 \cdot T}{233.5}\right) - 101.8 \\ 1173K \le T \le 1550K \end{cases} \qquad (5)$$

## 4. Test verification

In order to verify the effectiveness of this research method, the optimized tail split groove is verified by engineering. The single crystal casting blade is inspected to recrystallized after solution heat treatment. The inspection results show that there is no recrystallization in the groove area of the tail slit on the single crystal blade. Compared with the structure before optimization, the optimization effect is remarkable, and the recrystallized grains are completely eliminated in the original structure area, as shown in Figure 8.

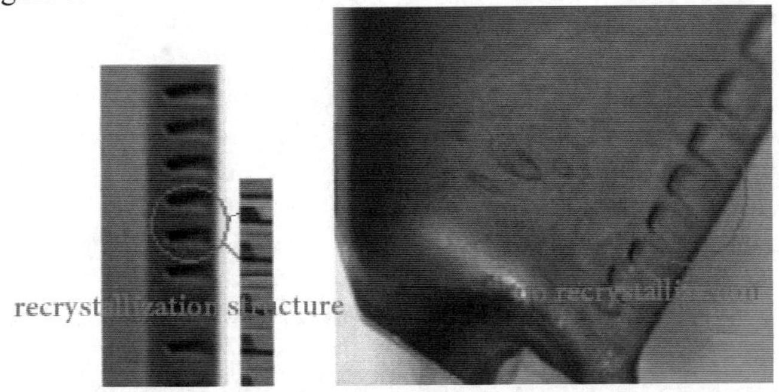

a）before - optimization b）after -optimization

Figure 8. Comparison of verification before and after optimization of exhaust edge half split groove

The verification shows that this method can realize to the reasonable regulation and redistribution of residual stress in complex structure. Reducing the residual stress level and the residual stress distribution form in the complex area of blade. Therefore, this method can effectively control the structural integrity and the structural continuity of single crystal blade.

## 5. Conclusion

(1) Based on the critical stress characteristics of single crystal recrystallization, the level and the distribution of residual stress are significantly improved by optimizing in tail splitting structure, and the maximum residual stress is reduced by about 15%. At last, the residual stress is controlled below the critical stress of recrystallization.

(2) The experimental results show that this method can control the level of structural residual stress to be under the safe area of recrystallization critical stress, and realizing the reasonable regulation of structural residual stress, Then, the surface SSR is completely controlled for the single crystal blade.

## Reference

[1] Bu Kun, Qiu Fei, Wang Zhihong. (2016)Research Progress on manufacturing technology and recrystallization of nickel based single crystal blade. Aviation manufacturing technology, 21:34-39.

[2] Chen suling, Jiang Wenjuan, Du Juan. (2020)Research progress of surface recrystallization control technology of nickel base single crystal superalloy. Journal of Chongqing University, 8: 74-79.

[3] Ma Jian, Li Shifeng. (2012)Study on crack and recrystallization of single crystal turbine blade edge plate. Journal of Aeronautical dynamics, 5: 1069-1073.

[4] Qiu Fei, bu Kun. (2017)Structural optimization of single crystal blade tenon inlet window based on residual stress. Journal of Harbin Engineering University, 2: 235-240.

[5] Li Shifeng, Wei Gang. (2015)Structural detail optimization of single crystal turbine blade with geometric and dynamic constraints. Journal of Aeronautical dynamics, 1: 90-95.

[6] Cox D.C, Roebuck B. (2003)Recrystallization of single crystal super alloy CMSX24. Materials Science and Technology, 19: 440-446.

[7] Jo C.Y, Cho H.Y, Kim HM. (2003)Effect of recrystallization on microstructural evolution and mechanical properties of single crystal nickel base superalloy CMSX22. Materials Science and Technology, 19: 1665-1670.

[8] Xiong Jichun, Li Jiarong. (2014)Recrystallization structure of Single Crystal Superalloy DD6 and its effect on rupture properties. Journal of metals, 6: 737-743.
[9] Okazaki M, Ohera I. (2003)Prevention of fatigue life reduction relating to cellular microstructure in single crystal Ni-base superalloy CMSX-4. In: Warrendale, pp. 63-73.
[10] Shi Zhenxue, Liu Shizhong. (2020)Receystallization Behavior of Single Crystal Superalloy at As-cast State. Journal of Materials Science & Engineering, 3: 345-354.
[11] Lei Yi. (2020)Study on surface recrystallization behavior of directionally solidified nickel base high temperature alloys dd483 and pwa1483. Xi'an University of technology, Xi'an.
[12] Han Mei, Xie Hongji, Li Jiarong. (2019)Effect of recrystallization on axial high cycle fatigue properties of DD6 single crystal superalloy. Materials engine, 6:161-168.
[13] Li Shifeng, Wei Gang. (2014)Mapping model between critical stress of single crystal recrystallization and structural design parameters of turbine blade. Journal of Nanjing University of Aeronautics and Astronautics, 6: 909-914.
[14] Wang Jian, Cheng Gengdong. (2003)Topology optimization design of continuum structure under multi condition stress constraints. Mechanical strength, 1: 55-57.
[15] Zhao Yang. (2005)Recrystallization behavior of Co-based superalloy at elevated temperature. Northeast University, Shenyang.
[16] Liu Pengfei, Liu Dong. (2009)Hot deformation behavior and dynamic recrystallization model of GH761 alloy. Rare metal materials and engineering, 2: 275-278.
[17] Huang Guangjie, Qian Baohua. (2007)Study on critical conditions of initial dynamic recrystallization of AZ31 magnesium alloy. Rare metal materials and engineering, 12: 1080-1085.
[18] Ding Hanlin. (2007) Experimental study and numerical simulation of high temperature deformation behavior of AZ91 magnesium alloy. Shanghai Jiao Tong University, Shanghai.

# Influence of brake-disc friction characteristics on anti-skid braking control system of aircraft

**Xiaoqiong Du[1], Bin Li[1*] and Linyin Luo[2]**

[1]School of AeronauticsDepartment, NorthwesternPolytechnical University, Xi'an, Shaanxi, 710072, People's Republic of China

[2]AVIC General Huanan Aircraft Industry Co.Ltd., Zhuhai, Guangdong, 519040, People's Republic of China

[*]Corresponding author's e-mail:leebin@nwpu.edu.cn

**Abstract.** The low-frequency landing gear walk of aircraft during braking processis studied. Firstly, by virtual of LMS Virtual.Lab motion software, the multi-body dynamic model of the landing gear is established, the motion pair and internal and external loads are applied, and the torsion spring is set at the connection between the landing gear and the fuselage to simulate the flexibility of the landing gear. Then, the variation of friction coefficient of brake discis considered in the model of braking pressure and torque. When the braking control system works, if the braking pressure is less than the limit value, the gear walk is less affected by the friction characteristics of the brake disc. If the braking pressure reaches the limit value, the law of gear walk is the same as that of the pilot's manual control of the braking pressure. At this time, the gear walk is only affected by the positive and negative slope of the friction coefficient of the brake disc. If the friction coefficient curve of the brake disc presents a negative slope, it may cause gear walk, and the greater the absolute value of the slope, the more obvious the gear walk is. Improving the friction characteristics of brake discs can provide a new strategy forgear walk control.

## 1. Introduction

Brake-induced vibrationof landing gear is a complex multi-disciplinary coupling problem[1]. When the aircraft brakes, the huge kinetic energy generated by the wheel deceleration is transformed into heat energy, which causes different vibration of landing gear[2], such as chatter, squeal and gear walk. Chatter and squeal are medium and high frequency vibration with small amplitude respectively. Gear walk is a kind of low-frequency vibration with frequency of 10~20Hz, defined as the periodic longitudinal fore-and-aft vibration of the landing gear strut. In terms of harmfulness, gear walk causes fatigue damage of landing gear and even damage of brake parts. There are many factors causing braking vibration [3]. The mismatch of braking control law, the friction characteristics of brake disc, the fluctuation of braking pressure and assembly clearance may cause the instability of braking torque, which are closely related to the gear walk.

Previous studies focused on the impact of the braking control law on gear walk [4-8], assuming that the friction coefficient of brake disc is constant. Few researchers pay attention to the influence of dynamic friction characteristics of brake disc materials on gear walk. Based on the multi-body dynamics model of landing gear, firstly, a series of curves between the friction coefficient of brake disc and the relative velocity of rotors and stators are constructed. Then, thecurve slope of friction coefficient of brake disc on gear walk with and without anti-skidPID braking control systemis studied.

## 2. Multi-body dynamics model of landing gear

The multi-body dynamic model of landing gear is established by LMS Virual.Lab motion software, and the motion pair and external force are applied to the model according to the connection relationship between components, as shown in Fig. 1. External force mainly refers to the buffer axial force and tire force. The buffer axial force includes air spring force, oil damping force and structural limiting force [9]. The vertical force of the tire is calculated based on complex tire model of software [10].

The adhesion coefficient $\mu$ between the wheel and the ground is related to the wheel slip rate, which can be expressed as [11]:

$$\mu=0.8\sin(1.5377\times\arctan(14.0326\sigma)) \quad (1)$$

$$\sigma=(V_c-R_g\omega)/V_c \quad (2)$$

where $\sigma$ is the slip rate of wheel, $V_c$ is the longitudinal velocity of the wheel, $R_g$ denotes the rolling radius, $\omega$ is the angular velocity of the wheel.

Gear walk is a kind of longitudinal dynamic problem. It is necessary to set a torsion spring at the connection between the outer cylinder and the fuselage to reflect the flexibility of the landing gear. The longitudinal stiffness and damping of the landing gear can be expressed by the stiffness and damping of torsion spring.

Fig.1 Multi-body dynamic model of landing gear

The relationship between braking pressure and torque[7] is described as follows,

$$M_b=\mu_m(P_b-P_0)N_m\frac{R+r}{2} \quad (3)$$

Where $P_0$ is the brake pressure loss; $N_m$ is the number of friction surfaces; $P_b$ is the brake pressure; $R$ is the outer radius of the brake stators; $r$ is the inner radius of brake rotors; $\mu_m$ is the friction coefficient of brake disc.

The braking torque is related to the friction characteristics of brake disc materials. The real friction coefficient of brake disc is not constant, $\mu_m$ can be expressed as a function of the relative velocity of brake stators and rotors, as shown in Fig.2. Fig.2(a) and Fig.2(b) reflect the positive and negative slope characteristics of the curve composed by the friction coefficient of the brake disc and the relative velocity of brake stators and rotors respectively. The value range of $\mu_m$ is 0.15 ~ 0.35.

There are two ways to control the braking pressure in this paper: automatic control of the anti-skid PID control system and manual control of the pilot. The longitudinal vibration of the landing gear when the braking pressure is output in the above two ways and the friction coefficient of the brake disc changes according to the curve shown in Fig2 will be studied in the following sections.

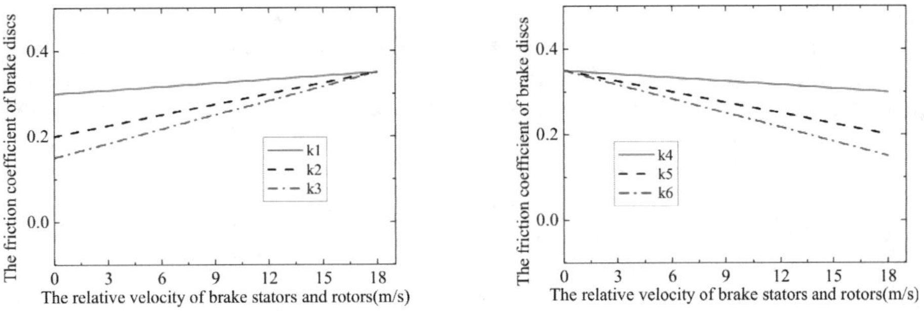


(1) Positive slope (2) Negative slope

Fig.2 The curve between the friction coefficient of brake discs and the relative velocity of rotors and stators

## 3. Anti-skid PID control system

By adjusting the control parameters, the slip rate of the anti-skid PID control system can always close to the optimal slip rate, and the corresponding adhesion coefficient between the wheel and the ground reaches the maximum value, which can make the aircraft achieve the highest braking efficiency. Fig.3 is the schematic diagram of the control system.

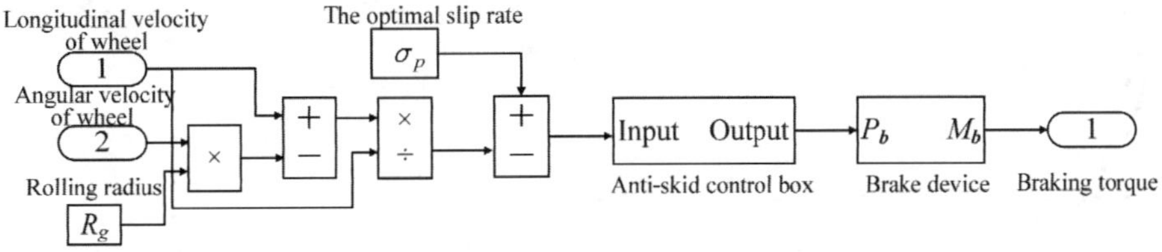


Fig.3 Schematic diagram of PID anti-skid braking control system

$$P_b = K_p \cdot \Delta\sigma + K_i \cdot \int(\Delta\sigma)dt + K_d \cdot \frac{d(\Delta\sigma)}{dt} \tag{4}$$

$$\Delta\sigma = \sigma_p - \sigma \tag{5}$$

Where$K_p$, $K_i$ and $K_d$ are the proportional, integral and differential coefficients of the control box, respectively. $\sigma_p$ is the optimal slip rate， $\sigma$ is the wheel slip rate, $\Delta\sigma$ is the difference between $\sigma_p$ and $\sigma$. The input and output of the anti-skid control box is $\Delta\sigma$ and brake pressure $P_b$. The wheel decelerates under the action of the braking torque output by the brake device. The initial velocity of the aircraft is 180km/h, and the anti-skid braking control system works after the wheel touches the ground for 1s. The friction coefficient of brake disc does not change, decrease or increase with the relative velocity of rotors and stators. Compare the variation laws of braking torque, the longitudinal displacement of wheel axle, slip rate and braking pressure.

### *3.1. Constant friction coefficient of brake disc*

When the wheel touches the ground, the ground adhesion force makes the velocity of wheel quickly consistent with the initial velocity of aircraft, resulting in the transient gear walk, however, this vibration converges quickly. The wheel enters the pure rolling state from the static state, and the slip rate changes from 1 to 0. The anti-skid control system starts working after 1 s of landing， only when the braking torque output by the anti-skid system changes as shown in Fig. 4 (b), can the slip rate of wheel maintain the optimal value 0.12, as shown in Fig. 4 (d). The braking torque is fully controlled by the braking pressure, and the longitudinal displacement of wheel axle in Fig. 4 (c) is closely related to braking torques, which shows that the gear walk does not occur.

### *3.2. Friction coefficient of brake disc decreases with decreasingvelocity*

The friction coefficient$\mu_m$ of the brake disc are taken as shown in Fig. 2 (a). When the curve slope of$\mu_m$ is k1, as$\mu_m$ decreases, the braking system needs to output a braking torque similar to that in Fig. 4 (b) to maintain the wheel at the optimal slip rate, therefore the overall value of the braking pressure of Fig.5(a) increases compared with that of Fig. 4 (a).

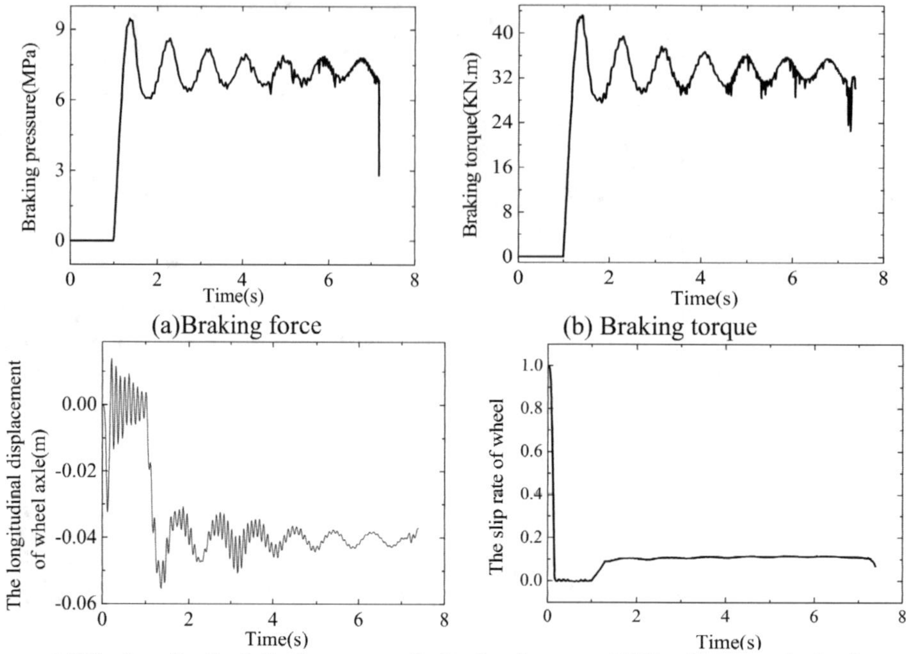


(a)Braking force (b) Braking torque

(c)The longitudinal displacement of wheel axle (d)Theslip rate of wheel

Fig.4 The friction coefficient of brake disc is constant

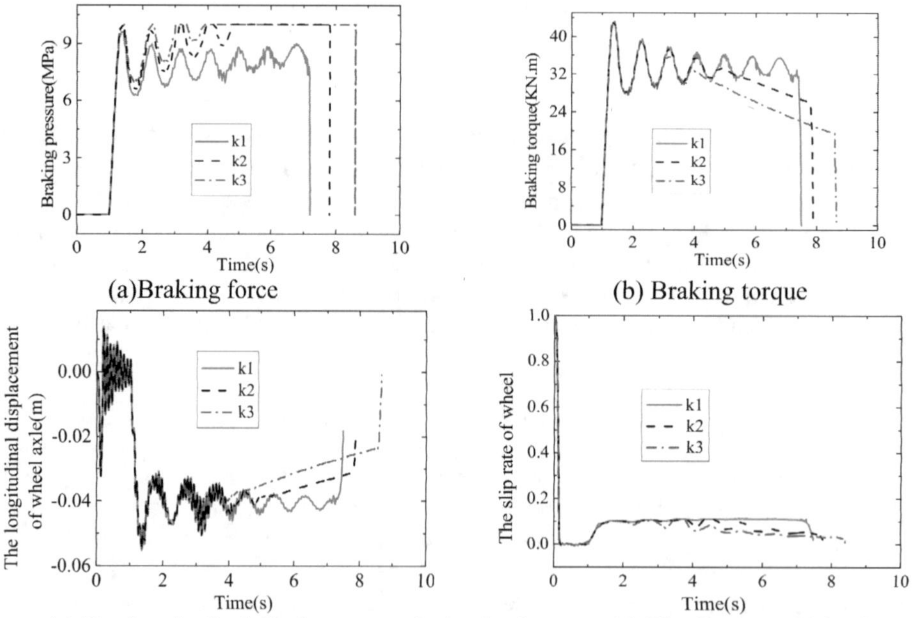


(a)Braking force (b) Braking torque

(c) The longitudinal displacement of wheel axle (d) Theslip rate of wheel

Fig.5 The friction coefficient of brake disc decreases with decreasing velocity

It can be seen from Fig. 5 (a) that the braking pressure with curve slope k2 and k3 of$\mu_m$ reach the limit value 10MPa at 4.74s, and 3.81s respectively. When the braking pressure is less than the limit value, the control of the slip rateof wheel is not affected. The braking pressure is controlled by the anti-skid system, and the shape of the braking torque curve is consistent with Fig. 4(a). However, when the braking pressure reaches the limit value, the braking torque is only affected by$\mu_m$ and decreases linearly, and the slip rate cannot be controlled at the optimal value, which decreases continuously until the aircraft stops.

During the whole braking process, the trend of longitudinal displacement of wheel axle is consistent with that of braking torque, as shown in Fig. 5 (b) and Fig. 5 (c). The positive slope of the friction coefficient of the brake disc does not cause the gear walk.

### *3.3. Friction coefficient of brake disc increases with decreasingvelocity*

The friction coefficient$\mu_m$ of the brake disc are taken as shown in Fig. 2 (b). When the curve slope of$\mu_m$ is k4, as the friction coefficient increases, the overall value of the braking pressure in Fig.6(a) decreases compared with Fig. 4 (a).

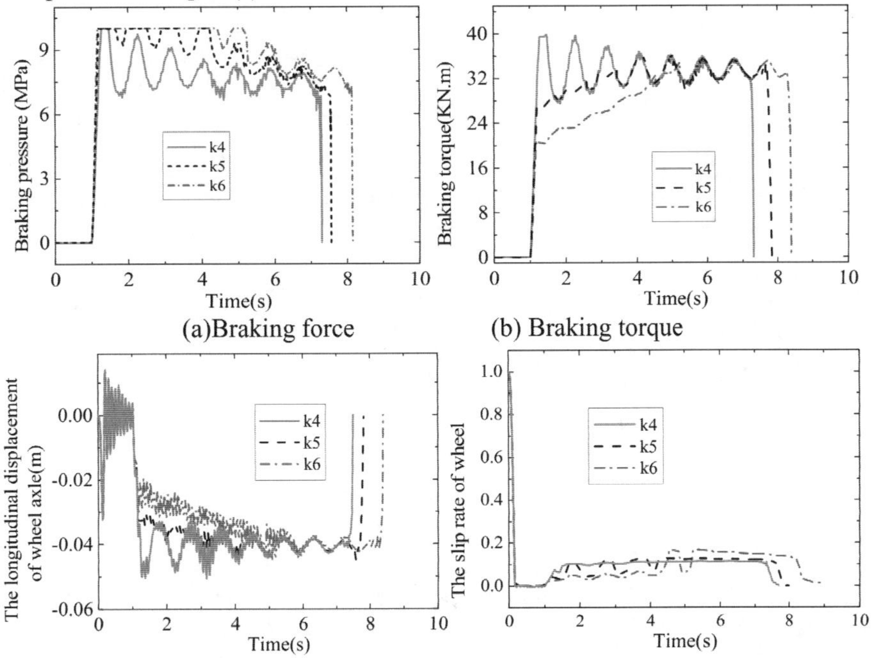


(a)Braking force (b) Braking torque

(c)The longitudinal displacement of wheel axle (d)Theslip rate of wheel

Fig.6 The friction coefficient of brake disc increases with decreasing velocity

As shown in Fig. 6, when the curve slope of the friction coefficient of the brake disc are taken ask5 and k6, at the initial braking stage , $\mu_m$is small, the braking pressure reaches the limit value, the corresponding braking torque increases linearly, and the slip rate cannot always maintain the optimal value. The landing gear shows low-frequency vibration with slope k6 of $\mu_m$, and the maximum amplitude range of the longitudinal displacement of wheel axle is 11.07mm.

As the aircraftvelocity decreases, the friction coefficient$\mu_m$increases, braking pressure decreases gradually until automatic control is restored. When the curve slope of$\mu_m$ is k5 and k6, the braking pressure resume automatic control at 4.27s and 5.12s. After that, the braking torque is consistent with that ofFig. 4 (b). The gear walk decreasesgradually until it disappears, the slip rate of wheel gradually returns to the optimal value.

## 4. Constant braking pressure

When the anti-skid braking system does not work, the braking pressure is controlled by the pilot stepping on the pedal. Assuming that the braking pressure is a constant value, the influence of the positive and negative slope characteristics of the braking torque caused by the friction coefficient of the brake disc on gear walk is studied.

### *4.1. Positive slope of braking torque*

When the curve slope of the friction coefficient$\mu_m$ of the brake disc is taken according to Fig. 2(a), the correspondingsimulation results of the braking torque, the slip rate of wheel, the longitudinal displacement and velocity of wheel axle are shown in Fig. 7.

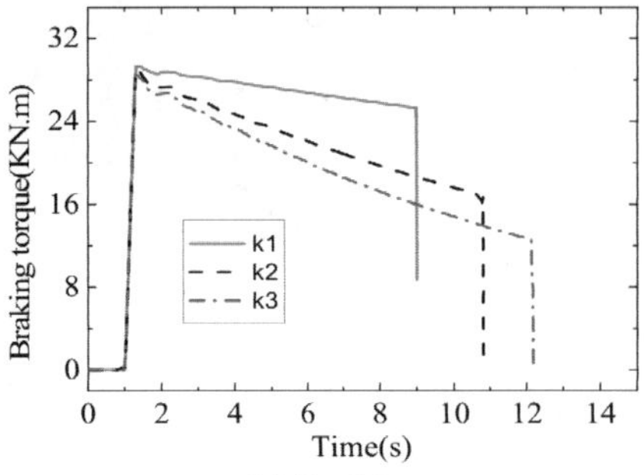


(a) Braking torque

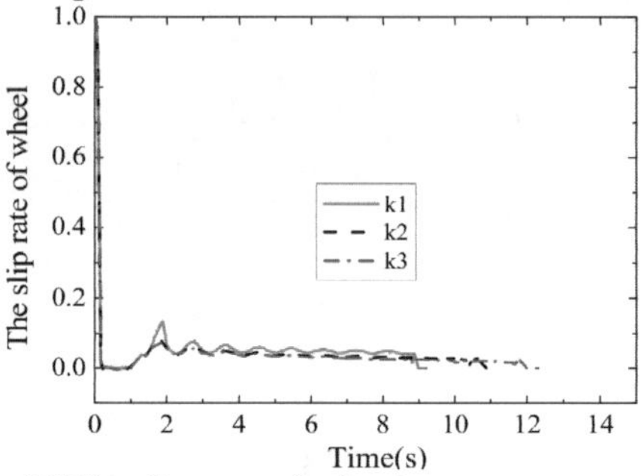


(b)The slip rate of wheel

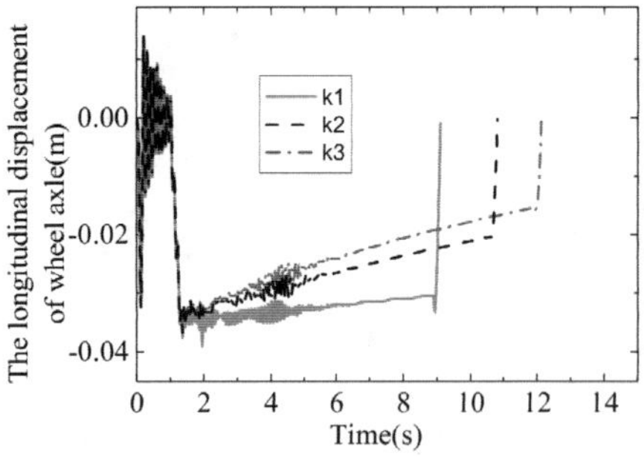


(c)The longitudinal displacement of wheel axle

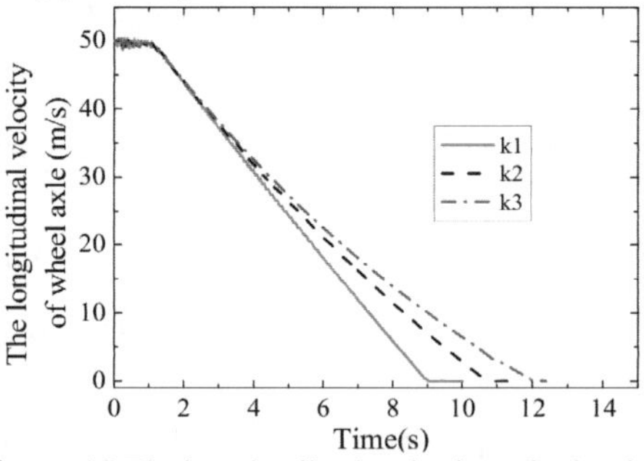


(d)Thelongitudinal velocity of wheel axle

Fig.7 Positiveslopeof braking torque

The braking torque curve in Fig. 7 (a) shows positive slope characteristics. The change trend of the longitudinal displacement of wheel axle is consistent with that of braking torque. The slip rate of wheel remains at a small value, and the landing gear walk does not occur. Fig. 7(d) shows that larger curve slope of $\mu_m$ causes a longer braking time and affects the braking efficiency.

### *4.2. Negative slope of braking torque*

When the curve slope of $\mu_m$are taken according to Fig. 2(b), the correspondingsimulation results are shown in Fig. 8.

The landing gear walk is not obvious with curve slope k4 of $\mu_m$. When the slope of $\mu_m$ is k5, the longitudinal displacement of wheel axle shows slight low-frequency vibration with small amplitude. As the absolute value of the curve slope of $\mu_m$ continues to increase to k6, the gear walk is obvious andthe maximum vibration range of the longitudinal displacement of wheel axle is 34.11mm. The braking torque curve oscillates slightly and is coupled with the wheel axle vibration. The slip rateare all at a low value, which indicates that the gear walk caused by the negative slope of braking torque is not induced by continuous braking application and release.

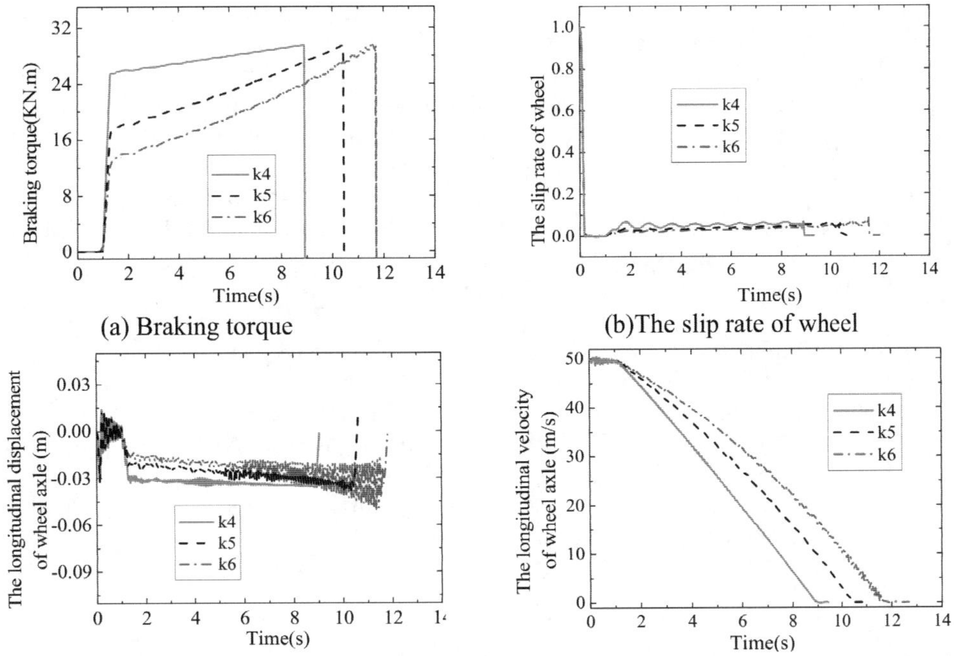


(a) Braking torque (b)The slip rate of wheel

(c)The longitudinal displacement of wheel axle (d)Thelongitudinal velocity of wheel axle

Fig.8 Negative slope of braking torque

## 5. Conclusion

Based on the multi-body dynamic model of landing gear, the influence of friction coefficient of brake disc on gear walk is studied when the braking pressure is automatically controlled by the anti-skid control system and manually controlled by the pilot. The results show that:

(1) The anti-skid control system always tries to make the wheel maintain the optimal slip rate by adjusting the braking pressureto make the braking torque increases and decreases periodically. When the braking pressure is less than the limit value, it can realize automatic adjustment, so that the wheel slip rate reaches the optimal value, and the landing gear walk does not occur or is in a weakened state.

(2) When the braking pressure under the action of the anti-skid control system reaches the limit value, the control of the optimal slip rate of the wheel cannot be realized. The influencing factors of gear walk are the same as that corresponding to the constant braking pressure discussed in Section 3.The braking torque only shows the positive and negative slope characteristics with the friction coefficient of the brake disc. Positive slope ofthe braking torque will not cause gear walk. The negative braking torque slope may cause gear walk. The greater the absolute value of the negative slope, the more obvious the longitudinal vibration of the wheel axle. Gear walk affects passenger comfort and braking efficiency. Improving the friction characteristics of the brake disc can effectively suppress gear walk.

## Acknowledgments

The supports from the National Natural Science Foundation of China (Grant No. 11872312) and the Prediction Project of Civil Aircraft (2018F14) are much appreciated.

## References

[1] Pritchard,J. (2001) Overview of Landing Gear Dynamics. Journal of Aircraft, 38(1),130-137.

[2] Krüger, W.R., and Morandini, M. (2011)Recent Developments at the Numerical Simulation of Landing Gear Dynamics. CEAS Aeronautical Journal ,1(1-4): 55-68.

[3] Wu,H.W.,Chen,T.F., Hu,C.K., et al. (2012) Solutions of the resonance caused by aircraft braking. Aviation Precision manufacturing technology,48 (2): 50-53.
[4] Lernbeiss, R., and Pliöchl,M.(2007) Simulation Model of an Aircraft Landing Gear Considering Elastic Properties of the Shock Absorber. Journal of Multi-body Dynamics.221:77-86.
[5] Khapane, P.D.(2006) Gear Walk Instability Studies Using Flexible Multi-body Dynamic Simulation Methods in SIMPACK. Aerospace Science and Technology,10: 19-25.
[6] D'Avico, L., Tanelli, M., Savaresi, S .M., et al.(2017) A Deceleration-based Algorithm for Anti-Skid Control of Aircraft. International Federation of Automatic Control, 50(1): 14168-14173.
[7] Yin, Q. Z.,Nie, H., Zhang, M., et al.(2016) Analysis of the Influence of Braking Control Laws on Gear Walk. Journal of Vibration Engineering, 18(3):1723-1741.
[8] Du , X.Q., Li,B., and Luo, L. Y.(2022) Study on braking vibration behavior of high strut landing gear of amphibious aircraft. ActaAeronauticaetAstronauticaSinica, 43(6):526199.
[9] Yin, Q. Z. ,Nie, H., and Zhang, M.(2016) Vibration Characteristics of Gear Walk on Half-axle Landing Gear.Journal of Vibration Engineering,29(6):954-962.
[10] Yu, T.X., Zhang,Y.G., Wang,X.F., et al(2017) LMS Virtual.Lab Motion Advanced Tutorial. Northwestern Polytechnical University Press, Xi 'an .
[11] Wang, J.S.,and He,C.A.(2000)Determination of tire-runway friction coefficient, Northwestern Polytechnical University, 18(4): 569–571.

# Application of LiDAR Survey Technology in Space Launch Sites

**Zhenzhen Ren, Qiang Wang, Yuchao Wang, Gaoyang Wu, Kang Wang, Wen'an Zhong***

Xichang Satellite Launch Center, Xichang, Sichuan, 615000, China

*Corresponding author: zhongwenan888@163.com

**Abstract:** LiDAR has the advantages of small size, high imaging accuracy, high resolution, all-day survey, and strong adaptability to complex environments. With the continuous improvement of technical performance, LiDAR is widely used in aerospace, mapping, military, and civil fields. In this study, the principle of LiDAR survey technology was introduced, and the key technology and application method of using LiDAR to measure the initial section of rocket launch were discussed in detail; a LiDAR survey system was designed on this basis. Afterward, the feasibility and rationality of the survey system and survey method were verified by rocket launch tests. The results show that compared with the traditional survey methods, LiDAR survey technology has obvious advantages in terms of survey accuracy, automation, and real-time visualization of survey data, and it can greatly reduce labor intensity and improves work efficiency, which provides an important reference for the design and application mode of LiDAR survey system in space launch sites.

## 1. Introduction

The active photoelectric beam sensor emits light waves to interact with long-distance targets, and the scattered light is then sensed by the receiving system to obtain target distance, velocity, vibration, reflected light intensity, phase, polarization, and angular displacement information. Compared with a passive photoelectric beam sensor, it not only achieves 2D imaging at longer distances, but also accomplishes near-photographic-quality 2D imaging, 3D imaging, and other forms of imaging discrimination when the target is obscured by camouflage nets, smoke, or weather conditions. Based on this rich target image information, it can enhance a range of engineering applications. Similar to the radio radar system, the active photoelectric beam sensor also has the advantage of a controllable target illumination range, and works in the visible and infrared regions of the electromagnetic spectrum; due to the shorter wavelengths of infrared visible light, it allows for higher range accuracy, angular resolution, and comprehensible information [1].

In active photoelectric beam sensors, some detection means are often named LiDAR (Light Detection and Ranging). LiDAR is widely used in various fields such as aerospace, mapping, military, and civil applications. At present, lots of international companies and research institutions have invested a lot of human and financial resources in the research and development of related technologies and systems. LiDAR shows the advantages of high imaging accuracy, high resolution, all-day survey, and strong adaptability to complex environments. Among them, 3D mapping is the main application direction of LiDAR, which can pass through other obstacles to obtain real-time high-precision 3D scene data information that is difficult to be provided by other traditional photographic survey technologies. As a basic data layer integrating all types of geographic distribution information, 3D data sets support the needs of terrain mapping, weapons guidance, and spacecraft in-orbit services. In addition, LiDAR surveys enable highly sensitive vibration detection, which is used in many applications to determine the state of a target for fault diagnosis and feature recognition [1-7].

This study explores the feasibility and rationality of ground-based LiDAR survey in space launch site applications, and traces and summarizes its research progress. Compared with the traditional survey method, the proposed method proves its obvious advantages in terms of survey accuracy, automation, and real-time visualization of survey data, which provides an important reference for the subsequent application mode of the LiDAR survey system in space launch sites.

## 2. Principle of LiDAR Scanning Survey

Fig. 1 shows the block diagram of a typical transceiver split direct-detection scanning LiDAR. The laser first generates a laser pulse, reduces its divergence angle by shaping and expanding the beam, and then the scanning mechanism points the laser beam at the imaging area; the target scattered light signal is collected by an optical antenna such as a telescope and focused to a photosensitive unit for photoelectric conversion; after filtering and denoising the resulting electrical signal, the arrival time of the echo signal can be determined by analyzing the electrical signal; the pulse round trip time can determine the relative distance from the detector to the target irradiated point; at this time, the scanning mechanism controls the laser to change the irradiated point of the target, and the 3D imaging process is a continuous repetition of the above steps [8].

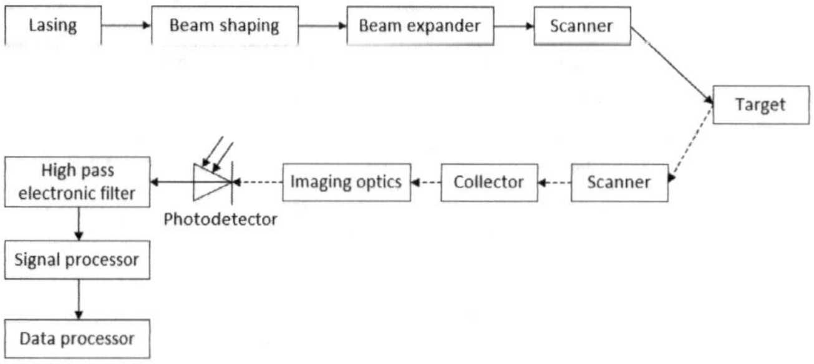


Fig. 1 Block Diagram of Typical Transceiver Split Direct-Detection Scanning LiDAR

The schematic diagram of the LiDAR scanning survey system space is shown in Fig. 2. In the ground LiDAR survey system, if the scanning mechanism follows a linear scan, the resulting data can be combined to generate a TV image, which consists of two angular and one distance information. These data are described in a spherical coordinate system with the origin at the detector and reflect the geometry of the target, which can be converted to other coordinate forms such as the Cartesian coordinate system by trigonometric transformation. If the position and orientation of the detector are known, the 3D coordinate data of the target surface can be obtained for various calculations or 3D models

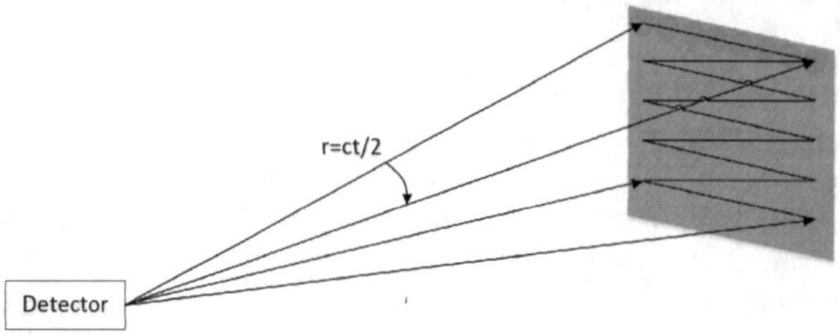


Fig. 2 Schematic Diagram of The LiDAR Scanning Survey System Space

## 3. LiDAR Survey System

### *3.1 Composition of Survey System*

LiDAR survey system is a new active photoelectric detection survey system, which has been widely used in mapping, imaging, and velocity survey. Among them, the ground-based LiDAR uses a tracking frame as its observation platform and a laser scanning distance survey system as its sensing device. The system can obtain the 3D spatial information of the target surface in real-time, measure the parameters of velocity, displacement, and acceleration of the target, and realize the acquisition of vibration information of the target. Therefore, the LiDAR survey system is one of the contemporary high-tech means of acquiring spatial information about targets.

The survey system mainly consists of LiDAR, tracking frame, mounting platform, vibration isolation platform, controller, and data processing system. A schematic diagram of the composition of the LiDAR survey system is shown in Fig. 3. Using the high accuracy and good tracking performance of the tracking frame, the LiDAR is mounted on the tracking frame through the mounting platform, which drives the LiDAR to track and scan the fixed position of the target with high accuracy and obtains the attitude change and vibration information of the target in real-time. The tracking frame is mounted on a vibration isolation platform to isolate the effect of large vibration. The controller is used to control the tracking frame with high precision, and the data processing system completes the real-time processing of laser point cloud data.

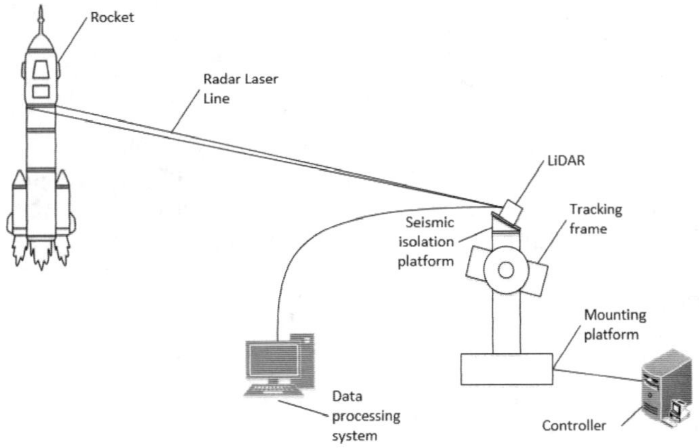


Fig. 3 Schematic Diagram of the Composition of the LiDAR Survey System

### *3.2 Survey Process*

The survey process is as follows: While the rocket is stationary, the LiDAR continuously scans to acquire static laser cloud data. The mounting platform is adjusted to the appropriate position. During the rocket takeoff, the tracking frame tracks the rocket with high precision and drives the LiDAR to scan the target point positions on the rocket continuously. The data processing system receives the laser point cloud data and processes it in real-time to determine information such as the rocket takeoff drift, attitude change, or vibration change.

## 4. Practical Application and Development Direction of LiDAR in Space Launch Sites

### *4.1 Long-range, High-Precision 3D Imaging*

In the initial stage of the space launch, the high frame frequency live scenes are mainly recorded by the high-speed TV survey instrument, but there are many shortcomings, mainly including two points. First, influenced by the survey system, it has the disadvantages of many scattered and fixed points, difficult securing, high coordination requirements, and poor mobility. Second, due to the influence of light conditions and target background contrast characteristics, passive detection leads to “two tone colors, bright light interference saturation and night detection performance degradation, and other phenomena. As a supplement to the existing survey means, the new survey system using laser 3D imaging radar breaks through the traditional imaging concept and mode, and images dynamic targets without distortion; therefore, it has the advantages of high imaging accuracy, long detection distance, fast 3D image acquisition, and strong adaptability to complex environments [9-10]. The application of LiDAR in the initial take-off stage of a rocket launch can achieve the following features:

1) Single laser 3D imaging radar for emergency mobile launch mission rapid response surveying capability. As a type of mobile equipment with strong mobile emergency capability, the laser 3D imaging radar can be deployed at each point in a maneuverable manner, but also be rapidly deployed in non-scheduled temporary sites for emergency maneuvers to achieve flexible networking and rapid response, and complete tasks such as 3D imaging of the initial stage of emergency mobile launch.

2) All-day, all-weather fast, and distortion-free acquisition of 3D images of the target function. Compared with passive detection optoelectronic equipment, it has the features of acquiring accurate 3D images of rockets instead of 2D images and being independent of light conditions and the contrast characteristics of the target background. Compared with millimeter-wave radar, it has the advantages of high resolution, good anti-interference, unaffected by ground clutter, and can realize the function of all-day, all-weather acquiring 3D images of the target quickly without distortion.

3) The ability to provide high-precision 3D images for real-time, automated tracking. High-resolution 3D imaging, quasi-real-time processing, and automatic target recognition algorithms make automatic target acquisition and tracking more accurate.

### *4.2 Survey of Attitude at the Vertical Takeoff Stage*

The real-time attitude survey data of the vertical takeoff stage of a rocket is of great significance in analyzing the orbit, aerodynamic parameters, and flight control performance. However, there are a lot of disturbances such as dense wake flames, ground vibrations, and ground clutter signals during the takeoff, and it is difficult for the traditional attitude survey methods to survey the attitude changes in this stage with high accuracy in real-time. At present, there is no external real-time attitude survey data in the vertical takeoff stage, so it is urgent to fill the data gap in this stage with a new survey method to provide a real-time, high-precision attitude data source for the safe launch of the rocket.

Using the LiDAR-based attitude survey method at the rocket's vertical takeoff stage, LiDAR continuously scans the rocket to obtain the laser point cloud data, including the horizontal angle, vertical angle, and distance data centered on the LiDAR. The spatial coordinates under the radar coordinate system can be calculated by correction, and then the spatial 3D coordinates in the rocket launch coordinate system can be solved by the coordinate system conversion method. The multi-ellipse center-fitted mid-axis algorithm is used to calculate and analyze the attitude survey value with high accuracy in real-time.

According to the results of a satellite launch center's pre-rocket launch test, the real-time dynamic attitude survey accuracy is about five times higher than the optical survey equipment's post-attitude survey data, thus proving the high accuracy and reliability of the LiDAR survey method, providing a real-time high-precision attitude data source for the safety control equipment, and ensuring the safety of the rocket launch process; therefore, this has important engineering application value[11-12].

### *4.3 Survey of Active Drift*

The takeoff drift of the rocket refers to the deviation of the centroid relative to the reference ballistic trajectory caused by various disturbances during the takeoff of the carrier rocket. Before the rocket takes off and leaves the launch tower, the rocket inevitably swings, where the vertical takeoff stage is the lateral drift. Currently, the takeoff drift is surveyed by at least three high-speed TV survey instruments. Due to considering both field of view coverage and ballistic survey outside the initial stage, the survey accuracy of rocket takeoff drift is low, only in the order of decimeters. This cannot meet the requirements of some specific rockets, especially the new model rockets, for the takeoff drift survey accuracy. In addition, this method has the defects of post-event image interpretation error, long data processing time, and failure in real-time acquisition of the drift. In the new era, therefore, a more accurate rocket takeoff drift survey system is needed for space launch sites, so as to make up for the shortage of the existing survey system.

A LiDAR-based active scanning method is used to measure the rocket takeoff drift in real-time. A 2D precision turntable drives the LiDAR to continuously scan the rocket with high accuracy and obtains laser point cloud data. The ellipse center fitting algorithm is used to calculate and analyze the real-time drift of the rocket during the takeoff stage. The survey of all-day rocket takeoff drift can greatly improve the survey accuracy and realize the data output of takeoff drift with high real-time and high accuracy.

According to the results of a satellite launch center's initial rocket launch test, a high real-time, high-precision takeoff drift survey can be achieved. Compared with the high-speed TV alternate survey method, the real-time survey's accuracy is raised from the decimeter level to the centimeter level[13-14]. Meanwhile, the real-time output drift provides real-time high-precision rocket flight status for the safety control equipment to ensure the safety of the rocket launch process and provide an important basis for rocket flight control performance, launcher design, and safety control, which has critical engineering application value.

*4.4 Survey of Engine Vibration Monitoring*

A rocket costs hundreds of millions or even billions. Rockets also carry expensive spacecraft that would be disastrous if they failed. The core of any rocket is the launcher. The stable working condition of the launcher directly determines the success or failure of the rocket launch. The fuel is docked above the rocket engine, and the gas is discharged below. Most of the traditional vibration surveys are contact surveys, and the additional mass and mass load will cause large errors. Therefore, it is urgent to accurately survey engine vibrations through a non-contact method, assess the engine operating status, provide real-time, high-precision engine operating status, and ensure the safety of the launch process. The laser vibration survey also provides a basis for the commissioning and improvement of new rocket engines, which has important engineering application value[15-17].

Currently, the Doppler vibration survey instrument is the most common commercial laser vibration survey technology. Compared with traditional sensors such as accelerometers, it has the advantages of the long-distance survey, non-invasive property, high spatial resolution, short survey time, wide response band, and high-velocity resolution. The laser vibration survey instrument is mainly of single-point laser Doppler vibrometer. With the development of calibration technology, scanning laser Doppler vibrometer, 3D laser Doppler vibrometer, microscopic laser Doppler vibrometer, differential laser Doppler vibrometer, etc., have been widely used. The physical principle of laser Doppler vibrometry is that the reflected light from a moving object will carry the vibration characteristics of the moving object itself, i.e., Doppler frequency shift.

$$\Delta f_D = 2V / \lambda \tag{1}$$

where, $\Delta f_D$ is the Doppler shift of the laser after reflection from a vibrating object, $V$ is the velocity of the object's motion, and $\lambda$ is the laser wavelength.

The principle of the engine vibration survey is given in Fig. 4. A laser beam with frequency $f$ is emitted from the laser and incident on the surface of the engine through a beam splitter. Due to the vibration of the engine surface, the reflected light produces a Doppler frequency shift $\Delta f_D$, the reference beam with frequency $f + f_r$ and the reflected light with frequency $f + f_D$. Through the reflection of the reflector, they are projected to the photodetector for mixing, and the difference frequency signals between them can generate a photocurrent with a sinusoidal change of frequency Y. After reflecting from the reflector, the reflected light is projected onto the photodetector for mixing, and the difference signal between them can generate a sinusoidally varying photocurrent with frequency $f_D - f_r$. The output signal of the laser Doppler vibrometer is analyzed and processed to obtain the vibration signal of the engine [5]. The output signal of the photodetector is mixed with a reference beam of known direction and frequency, so that the information reflecting the vibration characteristics of the engine itself, such as the direction, amplitude, and frequency of motion, can be distinguished, which can be used for engine condition monitoring and fault diagnosis.

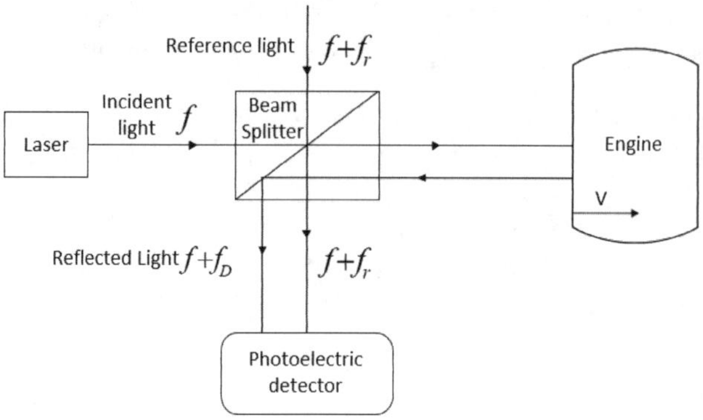


Fig. 4 Vibration Survey Principle of Laser Doppler Vibrometer

## 5. Conclusion

Through the analysis of existing problems such as 3D imaging, attitude survey, drift survey, and engine vibration survey of the space launch sites at the initial take-off stage, the importance and necessity of the application of LiDAR survey technology in space launch survey is fully realized. By virtue of LiDAR's small size, high imaging accuracy, high resolution, all-day survey, and strong adaptability to complex environments, this study explores the working principle of LiDAR survey technology and explores the feasibility and rationality of applying LiDAR survey system and surveying methods at space launch sites. The comparison with the traditional survey method shows that LiDAR survey technology has obvious advantages in survey accuracy, automation, and real-time visualization of survey data, and can greatly reduce labor intensity and improves work efficiency. Therefore, this study can provide an important reference for making full use of LiDAR survey technology and bringing out its application value in space launch sites.

## References

[1] Du X. P., Song Y. S., Zhao J. G. LiDAR-active photoelectric detection progress and development opportunities[M]. National Defense Industry Press, 2020
[2] Bu Y. M., Du X. P., Zeng C. Y. Research progress and trend analysis of non-scanning laser 3D imaging radar [J]. China Optics, 2018,11(5):711-727
[3] Li Q. Q., Li B. J. & Chen J. Research on LiDAR measurement technology and its application [J]. Journal of Wuhan University of Surveying and Mapping, 2000, 25(5): 387-392
[4] Lv G. P., Liao C. R. Application of LiDAR technology in monitoring of mine ecological environment[J]. Journal of Ecology and Rural Environment, 2017, 33(7): 577-585
[5] Xiang J., Hao W., Liu F. Research on LiDAR measurement technology and its application[J]. Science and Technology Innovation Herald, 2019,10(49):49-51
[6] Ma H. C. Several applications of LiDAR measurement technology in geology[J]. Journal of China University of Geosciences, 2011, 36(2): 347-354
[7] Zhang L., Zhang Y., Tu J. L. The key technology of LiDAR and its application in unmanned vehicles[J]. Modern Industrial Economy and Information, 2021, 12: 156-158
[8] Zhan Q. M., Liang Y. B. LiDAR data processing, information extraction and application[J]. Geomatics World, 2011, 4(2): 38-40
[9] Wang S., Sun H. Y., Guo H. C. Development and current situation of space-based LiDAR 3D

imaging[J]. Laser and Infrared, 2018,48(9):1073-1081
[10] Yuan X. X., Wang X. Q., Dou A. X. 3D modeling analysis of surface rupture based on ground LiDAR Yushu earthquake[J]. Seismology and Geology, 2012(1)
[11] Qiu J. W., Zhang Z., Yu S. F. Research progress of 1.5-micron atmospheric detection LiDAR (special invitation) [J]. Infrared and Laser Engineering, 2021, 50(3): 20210079. doi: 10.3788/IRLA20210079
[12] Shi H., Gao X., Li X. Y. Attitude measurement technology of rocket vertical take-off stage based on LiDAR[J]. Acta Photonica Sinica, 2022
[13] Shi H., Gao X., Li X. Y. Rocket active drift measurement technology based on LiDAR[J]. Infrared and Laser Engineering, 2021
[14] Zhang Y., Wang H., Huang C. Research on multi-constraint take-off drift control of a new generation of large launch vehicles[J]. Missile and Space Vehicle Technology, 2021(4):7-10
[15] Liu B. Vibration measurement and analysis method of the aero-engine complete machine[J]. Computer Measurement and Control, 2019, 27(5): 22-25, 43
[16] Lv H. S., Liu B. The latest development of laser Doppler vibration measurement technology [J]. Laser Technology, 2005,4,29(2):176-179
[17] Liu J. K., Ma X. S., Ma X. Review of laser Doppler vibrometer research[J]. Laser Journal, 2014, 12(35): 1-5

# Comprehensive Risk Assessment of Satellite on-Orbit Operation

**Ru Feng[1], Heng Zheng[1], Bo Zhou[1], Zongsheng Xie[1], Xuemei Hu[2], Xia Shang[3] and Lei Chen[3*]**

[1]China Aerospace Standardization Research Institute, Beijing, 100071, China

[2]China Academy of Space Technology, Beijing, 100094, China

[3]Beijing Institute of Tracking and Telecommunication Technology, Beijing, 100094, China

*Corresponding author's e-mail: shanpao@sjtu.edu.cn

**Abstract:** Aiming at the lack of research on quantitative evaluation of satellite on-orbit operation risk, this paper from two aspects: the special features and general quality characteristics analysis and identify factors that influence the satellite provide services, and on the basis build satellite on-orbit operation risk comprehensive evaluation index system. By combining analytic hierarchy process (AHP) and satellite availability evaluation model, this paper complete the research on comprehensive evaluation of satellite on-orbit operation risk. The results of the case show that the satellite is in good operating condition and capable of providing services normally. Thus, this method can realize the comprehensive quantitative evaluation of the actual operational risks of satellites on-orbit, find out the weak points of risks according to the evaluation results, and provide effective support for the operational health management of satellites on-orbit.

## 1. Introduction

Satellite is an important space resource, providing remote sensing, navigation, communication, weather and other services for the ground. It is widely used in all aspects of the national economy, and is an indispensable space information infrastructure in current society. Due to the high requirements of the overall performance of the satellite system, it is necessary to provide high continuity and high stability services during the on-orbit operation. However, in the actual operation of the satellite, it is inevitable that the health status of the satellite system will gradually appear or suddenly deteriorate due to various risk factors, resulting in the interruption of satellite service performance. How to carry out the research on the risk management of satellite on-orbit operation and effectively evaluate the risk of satellite on-orbit operation are the key technologies that need to be broken through in the current research.

Foreign research and methods in the field of aerospace risk management can be divided into two stages: from the qualitative and quantitative approach to the popularity of Probabilistic Risk Assessment (PRA). The risk management of domestic aerospace projects started late. Reliability block diagram analysis, fault tree analysis and failure mode and impact analysis are mainly adopted, and the CPRA method was proposed based on PRA[1,2]. Literature[3,4] describes the development and typical application of health management technology for similar complex equipment systems. Literature[5] realizes the model and professional operation of satellite on-orbit management through hierarchical management of on-orbit organization, hierarchical disposal of abnormal problems, hierarchical monitoring of telemetry parameters, and classified management and control of support system. Literature[6] takes the on-orbit satellite control system as the research object and proposes a health status evaluation method based on variable weight comprehensive AHP method, which can solve the health status evaluation problems of key components under variation, abnormality and failure that cannot be effectively solved by the constant weight AHP method. Literature[7] proposes an on-orbit real-time health evaluation method for satellite attitude control system based on multi-level fuzzy comprehensive evaluation framework, and adopts fuzzy comprehensive evaluation algorithm to achieve comprehensive evaluation of the overall health performance of the attitude control system. Literature[8] identifies and evaluates the risk items of the satellite integrated electronic subsystem by using fault mode and impact analysis, and then identifies the weak links of the technology by using AHP for the evaluated key risk items, so as to formulate corresponding countermeasures. Literature[9] analyzes the key factors affecting the success or failure of the mission from the aspects of satellite service performance and on-orbit safe and stable operation, and adopts the parameter tradeoff method to carry out quantitative calculation of the constructed navigation satellite health management index system.

At present, most studies at home and abroad focus on the health evaluation of satellite subsystems, but there are few studies on the comprehensive evaluation of the risk of on-orbit satellite operation, lacking standard evaluation indicators and clear evaluation mechanism, and most satellite indicators are difficult to quantify. Therefore, this paper takes on-orbit navigation satellite as the research object and uses analytic AHP to carry out on-orbit risk comprehensive assessment of navigation satellite. By analyzing the assessment results, risk prevention and control measures can be effectively formulated, thus providing theoretical basis and guidance for practical engineering applications.

## 2. Materials and Methods

This paper first introduces the basic concept and steps of analytic hierarchy process (AHP), then analyzes and identifies the factors that affect the on-orbit service delivery of navigation satellites, and then establishes the on-orbit risk comprehensive evaluation index system of navigation satellites. On this basis, the scoring criteria and corresponding weights of each index are determined.

### *2.1. Analytic Hierarchy Process*

The Analytic Hierarchy Process (AHP), as a relatively mature decision-making method at present, is a systematic and hierarchical analysis method combining qualitative and quantitative analysis[10]. It can quantify the qualitative factors, examine and reduce the influence of subjective factors to some extent, and make the evaluation more scientific. The main steps are as follows:

(1) A hierarchical structure is constructed to analyze the risk factors that affect the services provided by satellites in the process of on-orbit operation by taking navigation satellite operation risks as the target. A comprehensive evaluation index system for on-orbit operation risks of satellites is constructed from top to bottom according to the thinking of target layer, criterion layer and index layer, as shown in Figure 1.

(2) Do dimensionless processing for the index layer and explain the scoring criteria;

(3) Determine the relative weight of elements in each layer, construct a judgment matrix by pairwise comparison of indicators in each layer, and then determine the relative weight of each indicator through matrix operation;

1) Construct the judgment matrix

Assume that the criterion B1 has p ($0<p<m$) indexs, using 1-9 scale method, as shown in Table 1.

$a_{ij}$ is used to represent the comparison result of index i with index j, then:

$$a_{ij} = \frac{1}{a_{ji}}. \quad (1)$$

Thus, the pair-judgment matrix of p indicators against criterion B1 can be constructed, as shown below:

$$A=\left(a_{ij}\right)_{p\times p}=\begin{pmatrix} a_{11} & \cdots & a_{1p} \\ \vdots & \ddots & \vdots \\ a_{p1} & \cdots & a_{pp} \end{pmatrix}. \tag{2}$$

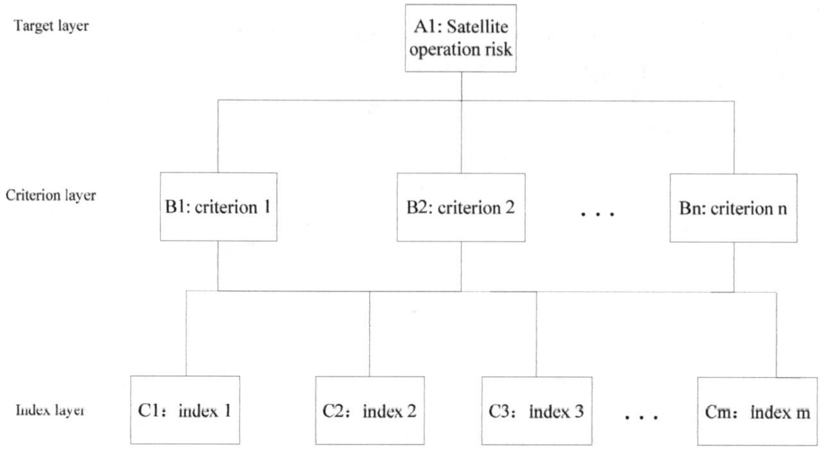

Fig. 1 Risk evaluation hierarchy.

Table 1 Comparison matrix scale of AHP.

| Scale | Definitions |
|---|---|
| 1 | The influence of index i and index j is the same |
| 3 | The influence of index i is slightly stronger than that of index j |
| 5 | The influence of index i is stronger than that of index j |
| 7 | The influence of index i is stronger than that of index j |
| 9 | The influence of index i is absolutely stronger than that of index j |
| 2,4,6,8 | The influence of index i to index j is between the two adjacent grades |

2) Calculate the weight
Multiply the matrix A by rows to obtain:

$$r_i=\prod_{j=1}^{p} a_{ij}\left(i=1,2,\cdots,p\right). \tag{3}$$

Then the matrix A becomes a p×1 matrix, and then raise each row of elements to the p power to obtain the matrix $A_1$:

$$A_1=\begin{pmatrix} \sqrt[p]{r_1} \\ \vdots \\ \sqrt[p]{r_p} \end{pmatrix}_{p\times 1}. \tag{4}$$

Let $w_i$ be the weight of index i, then for index i, its weight to criterion B1 can be expressed as:

$$w_i=\frac{\sqrt[p]{r_i}}{\sum_{i=1}^{p}\sqrt[p]{r_i}}\left(i=1,2,\cdots,p\right). \tag{5}$$

The weight calculation process of other index is as above. Finally, the weight vector of p indexs to criterion B1 can be expressed as $\{w_1, w_2, \cdots, w_p\}$.

3) Consistency test

For the constructed judgment matrix, consistency index CI, random consistency index RI and consistency ratio CR should be used to conduct consistency test on the basis of calculating the maximum eigenvalue and its corresponding feature vector. If and only if CR<0.1, it is considered that the inconsistency degree of the judgment matrix is within the allowable range, and the calculation result of its weight vector is reliable. The calculation process of consistency index CI, random consistency index RI and consistency ratio CR is as follows:

$$CI = \frac{\lambda_{max} - p}{p-1}. \tag{6}$$

$$CR = \frac{CI}{RI}. \tag{7}$$

Where, $\lambda_{max}$ is the maximum eigenvalue of the judgment matrix, and RI can be obtained by searching relevant data.

iv. Calculate the results and make decisions.

For target A1, its comprehensive evaluation mode is as follows:

$$A1 = \sum_{i=1}^{n} w_{Bi} Bi = \sum_{i=1}^{n} \left( w_{Bi} \cdot \sum_{i=1}^{m_i} w_{Cj} Cj \right). \tag{8}$$

Where, $w_{Bi}$ is the weight of criterion i relative to target, Bi is the weight of criterion i, $w_{Cj}$ is the weight of index j relative to criterion i, and Cj is the weight of index j.

### *2.2. Identification of risk fac*tors

From the perspective of affecting the normal operation performance of satellites, this paper summarizes the evaluation of satellite system operation risk at home and abroad, and takes the satellite operation risk as the target, and mainly identifies the risk factors of satellite on-orbit operation risk from the following two aspects.

#### *2.2.1. Specific features of satellite*

For a navigation satellite, the special features mainly refer to the basic support capability of the whole satellite to provide services. Through searching relevant information, it can be found that the services provided by the satellite include power supply subsystem, attitude and orbit control subsystem, uplink and downlink .

The power supply subsystem provides power support for the whole operation process of the satellite, mainly relying on solar array and battery to complete the storage and generation of power supply. During the period of solar shadow, solar panels are mainly used to complete photoelectric conversion and power supply for various components on the satellite to make them work normally. During the period of earth shadow, due to the loss of solar illumination, solar panels on the satellite have no power output, and the satellite needs to rely on battery power supply. The attitude and orbit control subsystem is responsible for the timely adjustment of satellite attitude so as to keep the satellite on-orbit. It mainly relies on the wheel, gyroscale, earth sensor and star sensor to complete the measurement and adjustment of satellite orbit position, so as to ensure the accuracy of satellite attitude data. Uplink and downlink can ensure the transmission of relevant command information and telemetry data during satellite operation. In the uplink, mainly relies on the responder 1, synthetic shunt, diplexer to complete the signal injection and transmission process; In the downlink, mainly relies on TWTA, responder 2 to complete the downlink signal broadcast and transmission process. The health status of each key product above will affect the work completion of the subsystem, and then affect the special features of the satellite. Therefore, the telemetry parameters representing the health status of the product were selected as monitoring indicators, and the collation results are shown as follows.

Table 2 product and telemetry information table.

| product | Telemetry parameters |
|---|---|
| Solar array | Voltage, current |
| Battery | Discharge voltage |
| Gyroscale | Motor current, angular rate |
| Wheel | Current, speed, friction moment |
| Star sensor | Temperature, angular rate |
| Earth sensor | Temperature, pitch angle |
| Responder 1 | Voltage, power |
| Synthetic shunt | state |
| Diplexer | shell temperature |
| TWTA | Current, power |
| Responder 2 | Voltage, power |

*2.2.2. General quality characteristics of satellite*

General quality characteristics can be used to measure the relationship between specific features and time, including reliability, maintainability, supportability, testability, safety and environmental adaptability, namely six characteristics. In the process of satellite operation, the most important concern of users is the long-term service availability of satellite. Accurate measurement of service availability can better reflect the capability of providing services and provide guidance for making preventive decisions. Therefore, on the basis of traditional General quality characteristics, this paper selects service availability as satellite universal quality characteristics index. Availability is mainly calculated from short-term and long-term unplanned outages caused by satellite failures or anomalies, as well as short-term and long-term planned outages for maintenance purposes, and each type of outages includes average interval time and recovery time.

*2.3. Establish a satellite operation risk comprehensive evaluation index system*

Based on the above analysis, this paper takes the special features and general quality characteristics as the two factors affecting the risk level of satellite operation, and divides them into criterion layer. According to the above description of special features and general quality characteristics, the subcriteria layer, product layer and index layer are successively divided. Thus, the satellite operation risk comprehensive evaluation index system is constructed, as shown below.

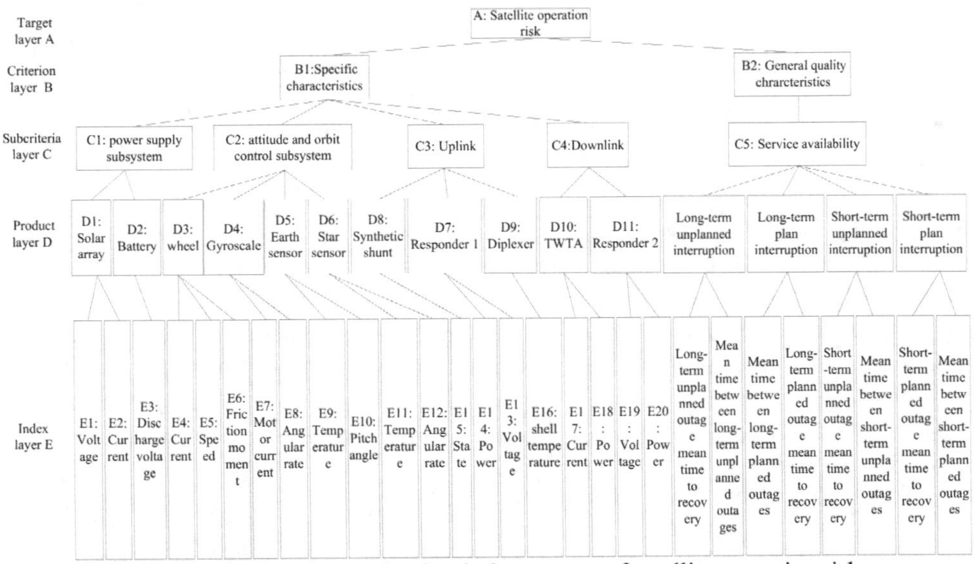


Fig. 2 Comprehensive evaluation index system of satellite operation risk.

### *2.3.1. Weight calculation of indicator layer*

(1) Specific features index

For each indicator under the specific features, it is the telemetry parameter representing the health status of the corresponding single machine. This article selects the telemetry parameters of product both qualitative index (state), also including the quantitative index (voltage, current, power, etc.), collected from every product's telemetry parameter values, analysis of the volatility of telemetry data, scope of the normal range and the envelope with the corresponding file (can be based on the satellite remote sensing parameters), The scores of qualitative and quantitative indicators are shown in the table below.

Table3 Corresponding table for scores of special characteristic qualitative indicators.

| Score interval | Qualitative indicators |
|---|---|
| 0 | Incorrect on-orbit status and telemetry |
| 60-100 | On-orbit status is correct. Score based on telemetry fluctuation range |
| 100 | The on-orbit state and telemetry are correct |

Table 4 Corresponding table for scores of special characteristics quantitative indicators.

| Score interval | Telemetry data condition | The case of product |
|---|---|---|
| 0 | Out of normal range | product failure |
| 60 | Above and below the normal range | product normal |
| 60-100 | Outside the envelope, within the normal range | product normal |
| 100 | within the envelope | product normal |

(2) General quality characteristic index

Satellite availability is the only index under general quality characteristics, and the state transition model of Markov chain can be used to evaluate the availability of navigation satellites. As shown in the figure 3, state 1 represents the normal operation of the satellite, state 0 represents the functional failure state of the satellite caused by fault or interference, $\lambda$ represents the failure rate of the satellite, $\mu$ represents the repair rate of the satellite.

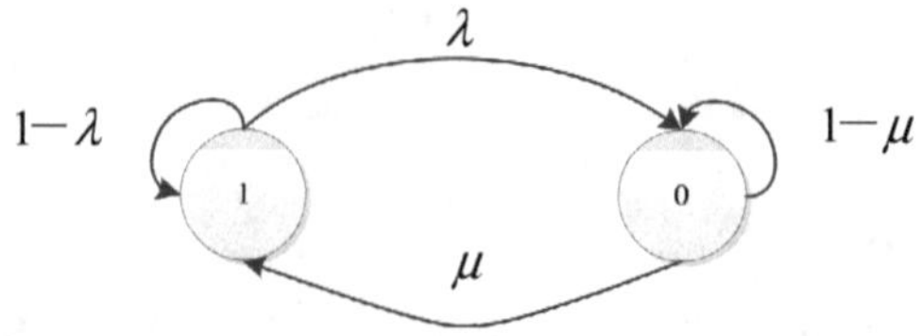


Fig. 3 Markov state transition process of satellite.

Assuming that the navigation satellite has only available state (State 1) and unavailable state (state 0), the transition from available state to unavailable state is caused by four different interrupt types, as shown in the figure 4.

In the figure 4, $\lambda_{STU}$、$\lambda_{STS}$、$\lambda_{LTS}$ and $\lambda_{LTU}$ represent the failure rate of short-term planned interruption, short-term unplanned interruption, long-term unplanned interruption and long-term planned interruption; $\mu_{STU}$、$\mu_{STS}$、$\mu_{LTS}$ and $\mu_{LTU}$ indicate the recovery rate of short-term planned interruption, short-term unplanned interruption, long-term unplanned interruption, and long-term interruption respectively.

Then, the calculation formula of satellite service availability is:

$$A_s = A_s(\infty) = \left(1 + \frac{\lambda_{STS}}{\mu_{STS}} + \frac{\lambda_{STU}}{\mu_{STU}} + \frac{\lambda_{LTS}}{\mu_{LTS}} + \frac{\lambda_{LTU}}{\mu_{LTU}}\right)^{-1} \quad (9)$$

Since the value of availability is [0,1], in order to facilitate the unification of scoring results with the dedicated feature index, the calculation formula of score of satellite service availability index (C5) is as follows:

$$C5 = A_s \times 100. \quad (10)$$

*2.3.2. Determining weights*

Using the 1-9 scale method, combined with the expert experience and the running status of satellite on-orbit history can be indicators for single layer and single layer for child rule, sub rule layer for the rule layer, as well as the criteria for the target layer two judgment matrix, and generation into the consistency check formula (7), the final weight can be set for each layer element, as shown Table 5, Table 6, Table 7, and Table 8.

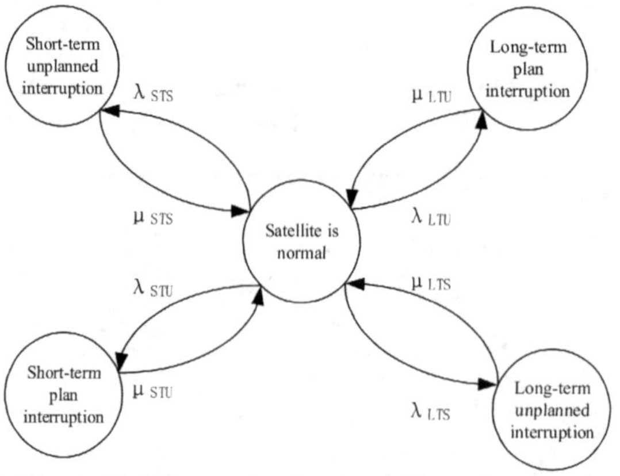


Fig. 4 Satellite availability evaluation model based on Markov process.

Table 5 Index layer weight value.

| Index layer | weight | Index layer | weight | Index layer | weight | Index layer | weight |
|---|---|---|---|---|---|---|---|
| E1 | 0.5 | E6 | 0.4 | E11 | 0.5 | E16 | 1 |

| Index layer | weight | Index layer | weight | Index layer | weight | Index layer | weight |
|---|---|---|---|---|---|---|---|
| E2 | 0.5 | E7 | 0.5 | E12 | 0.5 | E17 | 0.5 |
| E3 | 1 | E8 | 0.5 | E13 | 0.5 | E18 | 0.5 |
| E4 | 0.3 | E9 | 0.5 | E14 | 0.5 | E19 | 0.5 |
| E5 | 0.3 | E10 | 0.5 | E15 | 1 | E20 | 0.5 |

Table 6 Product layer weight value

| Product layer | weight | Product layer | weight | Product layer | weight |
|---|---|---|---|---|---|
| D1 | 0.5 | D5 | 0.3 | D9 | 0.25 |
| D2 | 0.5 | D6 | 0.3 | D10 | 0.25 |
| D3 | 0.25 | D7 | 0.4 | D11 | 0.3 |
| D4 | 0.25 | D8 | 0.3 | | |

Table 7 Subcriteria layer weight value

| Subcriteria layer | weight | Subcriteria layer | weight |
|---|---|---|---|
| C1 | 0.25 | C3 | 0.25 |
| C2 | 0.25 | C4 | 0.25 |

Table 8 Criterion layer weight value

| Rule layer | weight |
|---|---|
| B1 | 0.5 |
| B2 | 0.5 |

## 3. Results & Discussion

Take a certain navigation satellite named X-1 as an example. It has been in operation for three years and nine months since it was launched at 9:00 a.m. on July 18, 2018. Based on the comprehensive evaluation index system of operational risk constructed in Figure 2, comprehensive risk assessment is carried out for satellite operation during 2021.09.01-2021.09.29. The corresponding table of risk assessment results is described as follows.

Table 9 Description of risk assessment of satellite on-orbit operation.

| Satellite on-orbit operation risk assessment score | instructions |
|---|---|
| 95-100 | The satellite is functioning very well and is capable of providing normal service and maintaining regular monitoring. |
| 85-95 | The satellite is in good condition and capable of providing normal services. It can pay attention to the weak points of risks according to the actual situation. |
| 75-85 | The satellite is in good condition and has the ability to provide services normally. It is necessary to find out the weak points of risks and formulate corresponding risk plans. |
| 60-75 | The satellite is in poor condition, and there is a risk that it cannot provide services. Measures should be taken to reduce the risk. |
| Below 60 | The satellite was in poor condition and could not provide normal service. |

*3.1. Calculate the score of each index*

Take solar cell array current and voltage as an example. In this satellite, the normal range of solar cell array current is 38-52A, and the envelope value is 42-48A. By collecting and sorting the telemetry data of solar cell array current, the relationship between the solar cell array current and the normal range and envelope value range is shown in the figure 5.

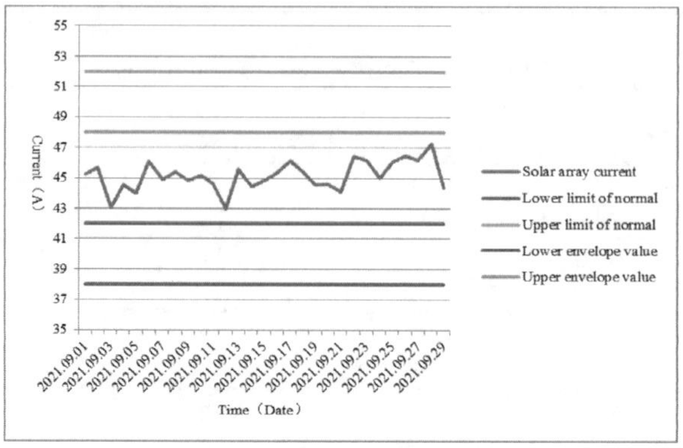


Fig. 5 Solar array current telemetry.

It can be seen from the figure 5 that the solar cell array current fluctuates up and down within the envelope value, and the current score of the solar cell array can be obtained as 90 points according to the linear interpolation method.

The normal range of solar cell array voltage is 31.5-60V, and the envelope value is 33-50V. By collecting and sorting the telemetry data of solar cell array voltage, its relationship with the normal range and envelope value range is shown in the figure 6. As can be seen from the figure 6, the voltage of solar cell array fluctuates up and down within the envelope value range. According to the linear interpolation method, the voltage score of solar cell array is 100 points.

The calculation formula of solar array score is:

$$D1 = E1 \times w_{E1} + E2 \times w_{E2}. \quad (11)$$

Thus, the solar array scored 95 points. Other product scores can be obtained according to the above method.

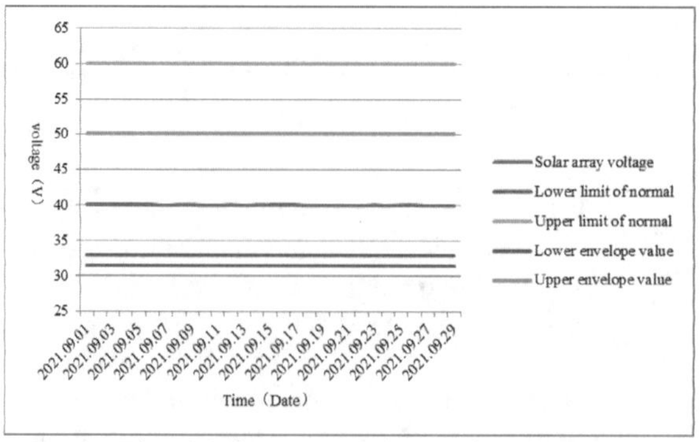


Fig. 6 Solar array voltage telemetry.

By collecting relevant materials, the satellite experienced once short-term unplanned interruption, twice short-term planned interruptions, none long-term unplanned interruption and none long-term planned interruption during 2021.09.01 to 2021.09.29. The specific failure rate and repair rate are as follow:

$$\lambda_{STS} = 0.0001; \lambda_{STU} = 0.0001; \lambda_{LTS} = 0.000001; \lambda_{LTU} = 0.000002;$$

$\mu_{STS} = 0.01; \mu_{STU} = 0.01; \mu_{LTS} = 0.05; \mu_{LTU} = 0.02.$

By substituting the above data into Equation (9), it can be calculated that the satellite service availability is 0.9802 and the score is 98.02.

### *3.2. Comprehensive risk assessment of on-orbit satellite operation*

By substituting the above analysis and calculation results into Equation (8), the on-orbit operation risk of the satellite and the score of each layer index can be obtained, as shown in figure 7.

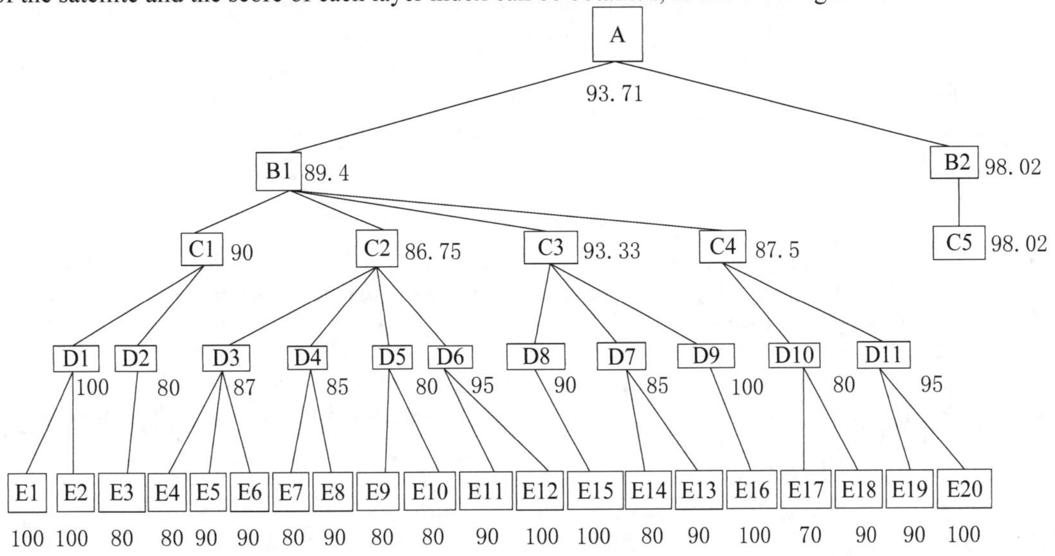


Fig. 7 Satellite operational risk indicator score.

It can be seen from the above table that the comprehensive assessment result of satellite operation risk is 93.71 points, the satellite is in good operating condition and has the ability to provide services normally. Among them, the satellite attitude and orbit control subsystem and downlink scored 86.75 points and 87.5 points, respectively, which are relatively low, and the gyroscope and earth sensor in the attitude and orbit control subsystem scored 85 and 80 points, and the downlink scored 80 points. Combining the weights of these three types of products, it can be seen that these three types of products are the objects that need to be paid attention to at present. We should always do a good job of monitoring the operation situation, and formulate plans and preventive measures in advance, so as to improve the overall operation service capability of the satellite. provide assurance.

## 4. Conclusion

From the perspective of the normal service capability of navigation satellites, this paper proposes a quantitative evaluation method for satellite on-orbit operation risk. Based on the principle of ensuring the normal operation of satellites and the maximum availability of services, the risk factors affecting satellite services are analyzed, and a comprehensive evaluation index system for satellite on-orbit operation risk is constructed. This paper firstly analyzes the variation range of telemetry data that characterizes the health status of a single satellite, and obtains the evaluation value of each parameter according to the proposed scoring standard. At the same time, using the availability evaluation method, combined with the outage data during on-orbit operation, the satellite service availability score is obtained. Finally, a comprehensive evaluation of the application cases of satellite on-orbit operation risks is carried out by using the Analytic Hierarchy Process (AHP). According to the evaluation results, the current satellite operation situation can be directly explained, and the weak links in the score are pointed out as the low-risk management of satellites, which is beneficial to the relevant departments. Take appropriate measures in a timely manner to reduce the risk of satellite operations, thereby improving the ability of satellites to provide services. In the future, the idea of constructing an index system based on this method can also be applied to the operational risk assessment of other types of satellites and spacecraft.

**References**

[1] Baron J, Hershey J.C, Kunreuther H. (2004) Determinants of Priority for Risk Reduction: the Role of Worry. Risk Analysis., 4:413-427.
[2] Meulbroek L.K. (2002)The Rromise and Challenge of Integrated Risk Management.Risk Management and Insurance Review., 4:413-427.
[3] Wang SP. (2014) Large aircraft airborne subsystem prediction and health management shutdown technoligy. Journal of Aeronautics., 6:1459～1472.
[4] Luo RY, Sun B, Zhang L et al.. (2013) Research on spacecraft prediction and health management technology. Spacecraft engineering., 4:95～101.
[5] Ma L, Yuan LF, Wang L et al.. (2017) On orbit operation analysis and management of navigation satellite. Spacecraft engineering., 5:121～125.
[6] Li JC, Li GH, Yang TS, et al. (2012) An Evaluation method for health status of orbiting satellites . In: 2012 2nd International Conference on Aerospace Engineering and Information Technology. Hong Kong. pp. 368-373.
[7] Song F, Qin SV. (2019) On-orbit real-time health evaluation of satellite attitude control system. Journal of Beijing university of aeronautics and astronautics., 11:1581-1588.
[8] Liu ZC, Li JX. (2019) Risk management of satellite integrated electronic subsystem based on AHP and FMEA. Computer age., 5:16-22.
[9] Pan YQ, Feng WJ,Zhang G et al.. (2021) Construction method of navigation satellite independent health management index system. Spacecraft engineering., 4:8～14.
[10] Ni HH, Chen A, Chen N. (2010) Some extensions on risk matrix approach. Safety Science., 48:1269.

Journal of Physics: Conference Series **2364** (2022) 012019 doi:10.1088/1742-6596/2364/1/012019

# General compensation control method of flexible manipulator driven by tendon-sheath mechanism

**Cunxiang Zhang**

Shandong Vocational College of Light Industry, Zibo, Shandong, China

41195986@qq.com

**Abstract.** Flexible manipulator has been widely used because of its flexibility. In addition, the tendon-sheath mechanism has the advantages of compact structure and strong flexibility, and it can be used as the driving transmission mode of the flexible manipulator. However, the nonlinear and hysteretic characteristics of the tendon-sheath mechanism directly affect the motion accuracy of the manipulator. Firstly, the kinematic and static models of the flexible manipulator are analyzed considering the influence of multi section coupling. In addition, the mechanical transmission process of the tendon-sheath mechanism is analyzed. On this basis, a general feedforward compensation method for the flexible manipulator is established. The proposed model-based compensation control method can be applied to all flexible manipulators, and it is of great significance to promote the practical application of the manipulator.

## 1. Introduction

Compared with the rigid manipulator, the flexible manipulator reaches the target position through its own deformation, resulting in better flexibility [1, 2]. It can be used in the laparoscopic surgery, nasal surgery, or other surgical procedures. In addition, it can also be used in the industrial field. For example, it can be used to detect unstructured space, including aeroengine and other structures [3].

Flexible manipulator includes pneumatic, hydraulic drive, tendon drive and other drive modes. Compared with other driving modes, the tendon driven mode is easier to miniaturize. The tendon driven manipulator usually controls its bending accurately by controlling the tendon length [4]. In order to further simplify the structure of the manipulator system, the tendon-sheath mechanism is widely used. However, the utilize of tendon-sheath mechanism brings new problems. For example, the backlash of the system and the deformation of the tendon results in the poor accuracy [5, 6]. Many scholars have studied the transmission model, and try to apply it to the control of flexible manipulator. However, it is only for specific structures, or additional sensors are required to measure the driving force. For example, according to the input force measured by the sensors in [7], a feedforward compensation method for manipulator is proposed. But this method is not suitable for situations where additional sensors cannot be added. A general feedforward compensation method is proposed in this paper, which is based on the static model of the manipulator, combined with the tendon-sheath mechanism model. The proposed method can improve the motion accuracy of the system

The arrangement of this paper is as follows: The section II introduces the manipulator structure. The section III and section IV are the models of the manipulator and the tendon-sheath mechanism system respectively, followed by the compensation method and experiment. The last section is the conclusion.

## 2. System structure introduction

In this paper, the most commonly used flexible manipulator is used as the object for analysis [8, 9]. As shown in Fig. 1, in order to improve the flexibility of the manipulator, it can include multiple sections according to the application requirements. The lower left of Fig. 1 shows four bending sections, while the right represents the flexible manipulator with two bending sections. The center of the flexible manipulator adopts hyperelastic nitinol alloy as its skeleton, and the driving tendon is the cable of stainless steel. The tendon-sheath mechanism is used to separate the manipulator from the driving system, which increases the distance between the drive unit and the manipulator.

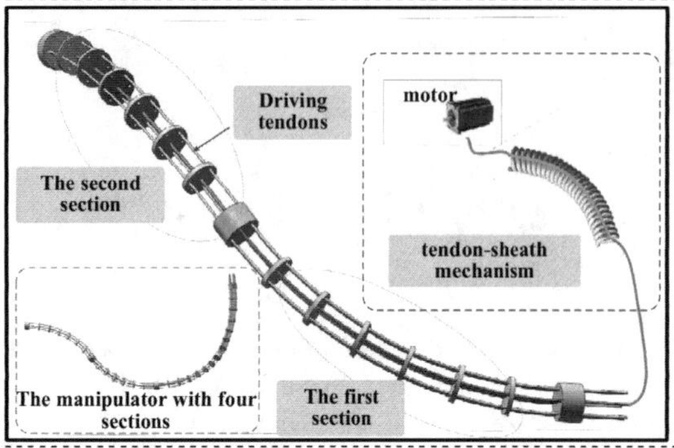


Figure 1. Structure of flexible manipulator and the tendon-sheath mechanism

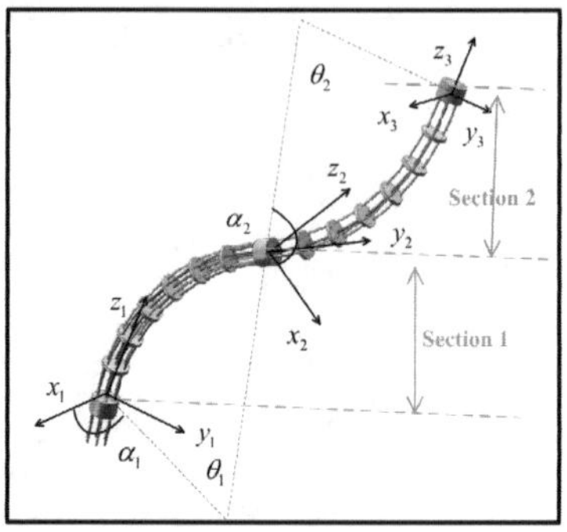


Figure 2. The coordinate system and bending angle diagram of the flexible manipulator

## 3. Modeling of flexible manipulator

Precise modeling of flexible manipulator is the basis of its precise control. In this paper, a flexible manipulator with two sections is taken as an example for modeling and analysis. The manipulator is driven by tendons, and the kinematics and statics of the manipulator are analyzed respectively.

### 3.1. Kinematic analysis

It is assumed that the curvature of each bending section is constant [10]. The kinematics of the manipulator is to describe the relationship between the three spaces: 1. Driving space $Q=[l_{s,i}]^T (s=1,2,i=1,2,3)$ represents the length of the driving tendons, where s is the sth section and i indicates the ith driving tendon. 2. Configuration space $\psi=[\theta_s \quad \alpha_s]^T (s=1,2)$ indicates the bending direction and bending angle of each section. 3. Workspace $P=[x_s \quad y_s \quad z_s]^T (s=1,2)$ indicates the position of the end center of each section.

The definition of coordinate system and parameter representation are shown in Fig. 2, and the $x$ axis passes through the intersection of the first tendon and the guiding disk. There is a coupling relationship between the lengths of the driving tendons for the two sections.

$$l_{s,i}=\sum_{p=1}^{s}\left[L_p - r\theta_p \cos((i-1)\alpha+(s-1)\beta-\alpha_p)\right] \quad (s=1,2,i=1,2,3) \tag{1}$$

where $L_p$ represents the length of the center line of each section, $r$ indicates the distance between the driving tendon and the center line, which is 3 mm in this paper, $\alpha=120°$ represents the angle between the tendons of the first section, while $\beta=60°$ is angle between the driving tendons of the second section and the first section. Based on homogeneous coordinate transformation, the transformation relationship between the configuration space and the workspace space can be obtained. The transformation relationship can be expressed as

$$^{s-1}_{s}R = Rot(z_{s-1},\alpha_s)Rot(y_1,\theta_s)Rot(z_2,-\alpha_s) \tag{2}$$

$$ {}^{s-1}_{s}P = trans(\frac{L_p}{\theta_s} c\alpha_s (1-c\theta_s), \frac{L_p}{\theta_s} s\alpha_s (1-c\theta_s), \frac{L_p}{\theta_s} s\theta_s) \tag{3} $$

According to the transformation between the driving space and the parameter space, its instantaneous kinematics can be carried out. It will not be described in detail due to space limitation.

### *3.2. Static analysis*

It is assumed that the external force on the manipulator is $F_e = \left[ f_e^T \quad m_e^T \right]^T$, and the displacement under external force can be expressed as $\Delta x$. The change of driving tendon and driving force are expressed as $\Delta Q = \left[ \Delta l_1 \quad \Delta l_2 \quad \Delta l_3 \right]^T$ and $\tau = \left[ \tau_1 \quad \tau_2 \quad \tau_3 \right]^T$. The change of system potential energy is recorded as $\Delta E$. It can be obtained according to the principle of virtual work [11, 12]

$$ F_e^T \Delta x + \tau^T \Delta Q = \Delta E \tag{4} $$

The general solution can be expressed as $\tau = \left( J_{Q\psi}^T \right)^+ \left( \nabla E - J_{P\psi}^T F_e \right) + \lambda \left( I - \left( J_{Q\psi}^T \right)^+ J_{Q\psi}^T \right) \eta$. In particular, when the external force is 0, the driving force can be expressed as $\tau = \left( J_{Q\psi}^T \right)^+ \left( \nabla E \right)$.

Because the driving tendon is stainless steel, the strain energy is neglected in static analysis, and the strain energy of the hyperelastic central skeleton needs to be considered, which can be expressed as

$$ w = \int_0^{L_p} \frac{EI}{2} \left( \frac{d\theta}{ds} \right)^2 ds = \frac{EI\theta^2}{2L_p} \tag{5} $$

The coupling effect should be considered in the modeling of manipulator with two sections [13]. First, the coupling moment of the second section and the first one can be calculated by $M = \dfrac{EI\theta}{L_p}$. The direction of the moment can be solved by $C = A \cdot B$, where $A$ is the transformation matrix of section 1, and $B = R(z, \alpha_2)$. The direction of coupling torque is $\left[ C(1,2) \quad C(2,2) \quad C(3,2) \right]^T$. According to the magnitude and direction of the coupling torque, the driving force considering the coupling of two sections can be calculated by substituting it into equation (4).

## 4. Modeling of tendon-sheath mechanism

The mechanical transfer model of the tendon-sheath system should be studied [14]. The micro segment is selected for analysis, as shown in Fig. 3. The mechanical equilibrium equation is

$$ \begin{cases} F(c,t)\cos\dfrac{d\varphi}{2} = \mathrm{sign}(\dot{\varepsilon}) f(c,t) + F(c+dc,t)\cos\dfrac{d\varphi}{2} \\ F(c,t)\sin\dfrac{d\varphi}{2} = N(c,t) - F(c+dc,t)\sin\dfrac{d\varphi}{2} \end{cases} \tag{6} $$

where $\mathrm{sign}(\dot{\varepsilon})$ is used to describe the motion direction. The difference in tension between the two sides can be expressed as $dF(c,t) = F(c+dc,t) - F(c,t)$. The value of $\varphi$ is small enough, and $dF(c,t)\sin\dfrac{d\varphi}{2}$ is negligible. It is assumed that

$$F'(c,t)=\frac{dF(c,t)}{dc}=\frac{-\text{sign}(\dot{\varepsilon})f(c,t)}{dc} \tag{7}$$

It can be obtained by integration

$$\int_0^c \frac{F'(c,t)}{F(c,t)}dc=\int_0^c -\text{sign}(\dot{\varepsilon})\mu k(c,t)dc \tag{8}$$

The solution of the differential equation is

$$F(c,t)=F(0,t)\exp\left[-\text{sign}(\dot{\varepsilon})\mu\int_0^c k(\sigma,t)d\sigma\right] \tag{9}$$

And the overall elongation of the tendon in the tendon-sheath system is

$$\Delta L=\int_0^L \frac{F(c,t)}{EA}dc=\frac{F(0,t)}{EA}\int_0^L \exp\left[-\text{sign}(\dot{\varepsilon})\mu\int_0^c k(\sigma,t)d\sigma\right]dc \tag{10}$$

Furthermore, the mechanical transmission relationship and elongation of the tendon are

$$F_{out}=\begin{cases} F_{in}\exp\left[-\text{sign}(\dot{\varepsilon})\mu\int_0^L k(\sigma,t)d\sigma\right] & \dot{\varepsilon}\neq 0 \\ F_{out}(t^-) & \dot{\varepsilon}=0 \end{cases} \tag{11}$$

$$\Delta L=\begin{cases} \frac{F(0,t)}{EA}\int_0^L \exp\left[-\text{sign}(\dot{\varepsilon})\mu\int_0^c k(\sigma,t)d\sigma\right]dc & \dot{\varepsilon}\neq 0 \\ \Delta L(t^-) & \dot{\varepsilon}=0 \end{cases} \tag{12}$$

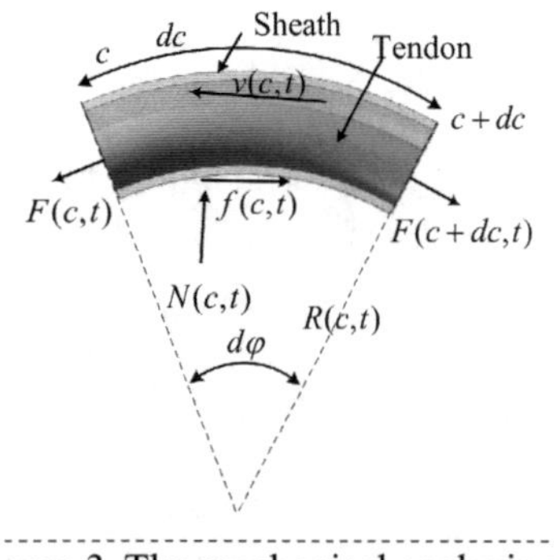


Figure 3. The mechanical analysis of the tendon-sheath system segment.

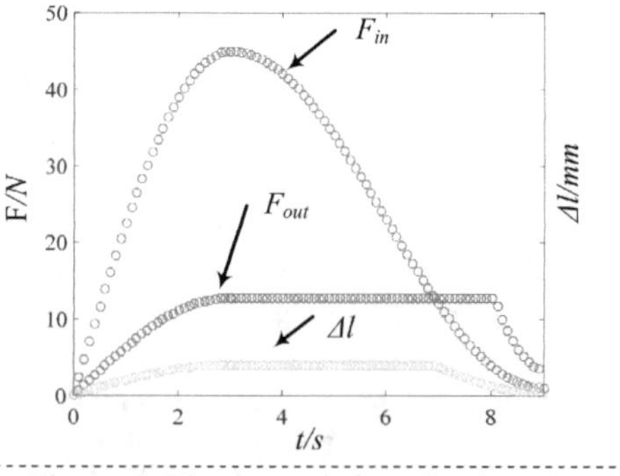


Figure 4. The force and tendon length variation of the tendon-sheath system.

It is assumed that the length of the tendon-sheath system is 1000 mm, and the bending angle is 180 degrees. The friction coefficient is 0.4. The diameter of the tendon is 0.2 mm, and the elastic modulus is 200GPa. Then the input and output force and the stretch length of tendon can be obtained and shown in Fig. 4. It can be seen that the tendon-sheath system can cause the loss of force and the elongation of the tendon.

## 5. Compensation method and experiment

### *5.1. Compensation method*

To improve the motion accuracy of the manipulator, a general feedforward compensation controller is proposed according to the mechanical and kinematic models of the manipulator and the tendon-sheath system. The output force of the tendon-sheath system is the bending driving force of the manipulator.

In order to obtain accurate position control result, the corrected input tendon length can be expressed as

$$Q_m(t) = Q_d(t) + \zeta(L_{sheath}, R_{sheath}, \varphi_{sheath}, F_{tendon}) \tag{13}$$

where $Q_d(t)$ and $Q_m(t)$ are the ideal kinematic movement of the driving tendon and its correction result. $\zeta$ is the elongation calculated by equation (12), and $F_{tendon}$ is the driving force, which is not measured by sensors, but through static model. This method is generally applicable, and can be easily applied to various types of flexible manipulator.

### *5.2. Experimental verification*

In order to verify the feedforward compensation method of the manipulator, an experimental platform is set up, as shown in Fig. 5. To simplify the experimental process, it is supposed that the manipulator is driven by a single tendon and the manipulator has only one section. The system driving force is transmitted by the tendon-sheath system, and the single driving tendon is driven by the steering engine.

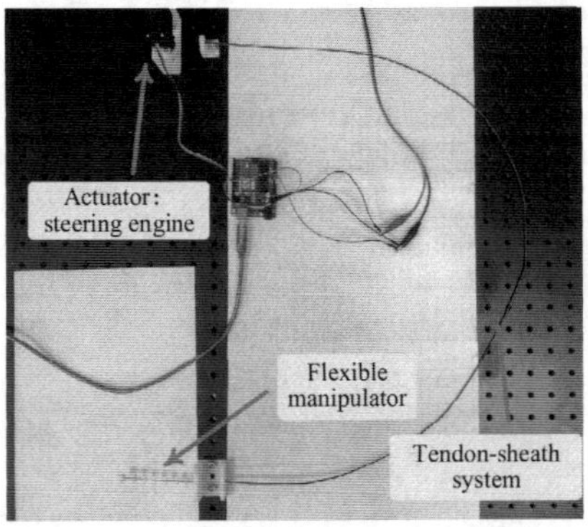


Figure 5. The establishment of experimental platform.

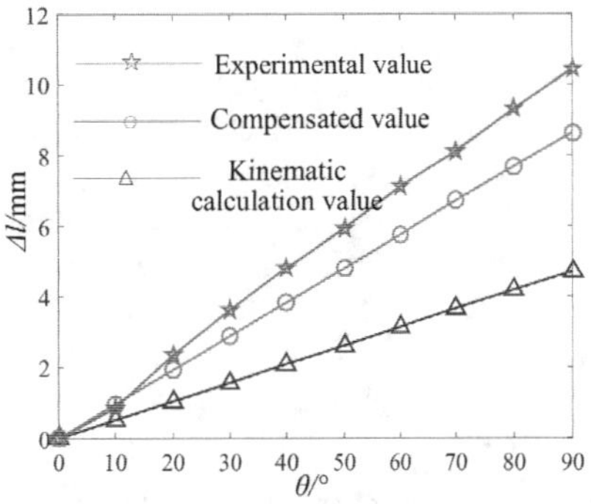


Figure 6. Driving tendon length corresponding to different bending angles.

The driving tendon is made of stainless steel with the outer diameter of 0.2 mm, while the sheath is stainless steel spring with outer diameter of 2 mm and inner diameter of 1.4mm. The manipulator consists of 1 section and 7 segments. The total length is 80 mm and the outer diameter is 8 mm. The bending angle of the tendon-sheath system is 180 degrees. The elastic modulus of hyperelastic central skeleton is 70 GPa. During the experiment, the manipulator was driven by a single driving tendon and bent from 0 degrees to 90 degrees at a distance of 10 degrees. First, the driving force of the system without the tendon-sheath system is measured and the mechanical model is modified and multiplied by a scale factor of 1.34, which is identified by experimental data.

The change of driving tendon length corresponding to different bending angle is recorded. It is compared with that of the manipulator with and without compensation, and the results are shown in Fig. 6. After compensation, the average error of driving tendon length decreases from 53.8% to 15.1%. It can be seen that the proposed compensation method greatly reduces the error.

Therefore, the proposed compensation method can greatly improve the motion accuracy of the manipulator, and the effectiveness of the proposed compensation method is verified. The error of the system may be caused by the assembly error and the clearance of the tendon-sheath system.

## 6. Conclusions

In order to maximize the performance of the manipulator, the tendon-sheath transmission system is used to separate the manipulator from the driving part. In this paper, the flexible manipulator and the transmission system are modeled and analyzed, and a feedforward compensation method based on the manipulator model is proposed. Finally, the effectiveness of the proposed compensation method is verified by experiment.

This paper ignores the friction between the guiding disk and the driving tendon of the manipulator, and the constant curvature assumption is used for kinematic analysis. These are not accurate enough, resulting in some errors. However, the manipulator can generally adopt the master-slave control mode, and the requirement for accuracy is not very high, So the compensation method proposed in this paper is feasible.

Next, the feedforward compensation effect of the manipulator will be studied when a single driving tendon is continuously stretched and released, in addition, it is necessary to further verify the movement effect of the manipulator when there are multiple driving tendons and the manipulator contains multiple sections. In order to further improve the motion accuracy of the manipulator, it is necessary to study the influence of the internal friction and external contact force of the manipulator on the motion of the system.

**References**

[1] Zhong, G., Peng, B., Dou, W. Kinematics analysis and trajectory planning of a continuum manipulator. (2022) INTERNATIONAL JOURNAL OF MECHANICAL SCIENCES, 222: 107206.

[2] Wang, C., Geng, S., Branson, D T., Yang, C., Dai, J., Kang, R. (2019) Task space-based orientability analysis and optimization of a wire-driven continuum robot. Proceedings of the Institution of Mechanical Engineers, Part C: Journal of Mechanical Engineering Science, 233(23-24): 7658-7668.

[3] Wang, M., Dong, X., Ba, W., Mohammad, A., Axinte, D., Norton, A. (2021) Design, modelling and validation of a novel extra slender continuum robot for in-situ inspection and repair in aeroengine. Robotics and Computer-Integrated Manufacturing, 67: 102054.

[4] Webster, III R J., Jones, B. (2010) Design and kinematic modeling of constant curvature continuum robots: A review. The International Journal of Robotics Research, 29(13): 1661-1683.

[5] Lau, K., Leung, E., Chiu, P., Yam, Y., Lau, J., Poon, C. (2016) A flexible surgical robotic system for removal of early-stage gastrointestinal cancers by endoscopic submucosal dissection. IEEE Transactions on Industrial Informatics, 12(6): 2365-2374.

[6] Sun, Z., Wang, Z., Phee, S. (2014) Elongation Modeling and Compensation for the Flexible Tendon--Sheath System. IEEE/ASME Transactions on Mechatronics, 19(4):1243-1250.

[7] Xu, W., Poon, C., Yam, Y., Chiu, P. (2016) Motion compensated controller for a tendon-sheath-driven flexible endoscopic robot. International Journal of Medical Robotics and Computer Assisted Surgery, 13(1): e1747.

[8] Simaan, N. (2005) Snake-Like Units Using Flexible Backbones and Actuation Redundancy for Enhanced Miniaturization.In: IEEE International Conference on Robotics & Automation. IEEE, pp. 3012-3017.

[9] Xu, K., Zhao, J., Zheng, X. (2015) Configuration comparison among kinematically optimized continuum manipulators for robotic surgeries through a single access port. Robotica, 33(10):2025-2044.

[10] Jones, B., Walker, I. Kinematics for multisection continuum robots. (2006) IEEE Transactions on Robotics, 22(1): 43-55.

[11] Rone, W., Ben-Tzvi, P. (2013) Continuum robot dynamics utilizing the principle of virtual power. IEEE Transactions on Robotics, 30(1): 275-287.

[12] Wei, X., Zhang, Y., Ju, F., Guo, H., Chen, B., Wu, H. (2021) Design and analysis of a continuum robot for transnasal skull base surgery. The International Journal of Medical Robotics and Computer Assisted Surgery, 17(6): e2328.

[13] Roy, R., Long, W., Simaan, N. (2017) Modeling and Estimation of Friction, Extension, and Coupling Effects in Multisegment Continuum Robots. IEEE/ASME Transactions on Mechatronics, 22(2):909-920.

[14] Wu, Q., Wang, X., Du, F., Chen, L. (2015) Torque Transmission Characteristics and Friction Compensation for a Tendon-sheath Actuation System in Pull-pull Configuration. JOURNAL OF MECHANICAL ENGINEERING, 51(5): 22-29. (In Chinese)

# Design and optimization of booster gliding model guided rocket scheme

**1st Author: Yifan Zhang[1, a], Co-1st Author: Meng Zhao[1, b], 2nd Author: Zhengxin Tao[1, c], 3rd Author: Wabu Mao[1, d], 4th Author: Shijin Luo[1, e], 5th Author: Xibin Bai[1, f], Ke Peng[1,*]**

[1]College of Aerospace Science and Engineering, National University of Defense Technology, Changsha,Hunan, China

[a]e-mail: 1228613509@qq.com; [b]e-mail: 1290185302@qq.com;

[c]e-mail: 872679253@qq.com; [d]e-mail: 3470583740@qq.com;

[e]e-mail: 1458364901@qq.com; [f]e-mail: baixibin@hotmail.com

[*]e-mail: pengke.nudt@nudt.edu.cn

**Abstract**—In order to meet the needs of rocket innovation practice learning that takes into account both low-cost and high-level engineering practical training, the design of flight mission of the model guided rocket is proposed and the technical index analysis is carried out in the present study. The design of the rocket thrust system is completed using two paralleled YSR-32 motor, whose total engine impulse is 296.14$\boldsymbol{N \cdot s}$; The mathematical model of the integrated aerodynamic / ballistic design optimization problem of rocket is constructed, adopting the aerodynamic configuration of side strip wing and x-shaped tail rudder and using Boost-Glide trajectory. Take the farthest range as the objective function, the influence of 8 variables on it is analyzed; By using Astrolab, a rapid overall design optimization platform of the rocket, the integrated simulation process of the aerodynamic / ballistic performance of the rocket is established; The maximum range of the optimization results is 1.29 $\boldsymbol{km}$, the takeoff mass is 2.8 $\boldsymbol{kg}$, the speed at the shutdown point is 93.6 $\boldsymbol{m/s}$, and the maximum flight time is 28.4s; The single use cost of the guided rocket is not exceed 1300 yuan.

## 1. Introduction

Innovative practices in the field of rockets play an important role in improving the comprehensive ability of aerospace students. Nowadays, the rockets used in rocket clubs in universities at home and abroad are mainly small uncontrolled rockets whose difficulty of production and flight tests is relatively low. It mainly involves two aspects of professional knowledge such as aerodynamic and ballistic design, which can be completed by using open source software such as OpenRocket. This kind of rockets can stimulate students' interest in aerospace, but can not fully improve the comprehensive ability; some colleges and universities have carried out medium and large rocket innovation practice. The theoretical knowledge and hands-on ability of the participants are improved. However, for the long implementation cycle, it is hard to conduct more flight tests and make continuous iterative improvement. The economy is low for the high development and launch costs. Therefore, it is urgent to develop a low-cost model guided rocket to meet the needs of rocket practical learning. The design optimization of the boost-gliding model guided rocket scheme is carried out in the present study, which provides support for the development of low-cost test rockets that meet the needs of high-order engineering practical training of the whole process and all elements.

## 2. TASK DESIGNING AND ANALYSIS OF TECHNICAL INDICATORS

Aiming at the in-depth scientific research practice requirements and economic requirements of model rockets, and at the same time provides a low-cost reusable verification platform for cutting-edge scientific research related to aerospace, a modular model guided rocket with the advantages of high technology, good economy, high flight accuracy, safety and reliability is designed in this paper. The range of the model rocket should be in the range of 600-1000m and can be controlled in three channels. The overall rocket specifications can be summarized as low cost, modular and reusable, and the rocket should meet the following requirements:

1. Mainly use products that are widely sold and 3D-printed components to reduce the cost of production and better balance the economic and practical needs.
2. The assembly process is simple and can be retrofitted according to the needs of different tasks.
3. It should be recyclable in order to shorten the cycle of scientific research and learning practice projects.

In order to enable the model guided rocket to be put into the scientific research practice of the relevant students who major in aerospace, promote the deep learning of aerospace knowledge, and can be used as a technical verification platform to verify the relevant technology, the corresponding mission of the rocket should be set. Fig. 1 shows the diagram of the mission.

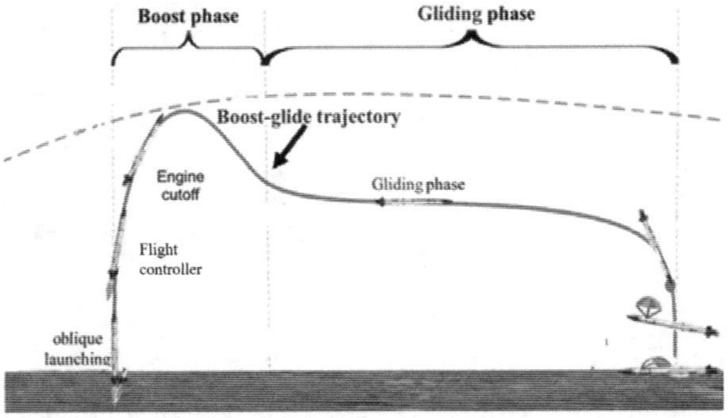


Figure 1 Schematic of the flight mission

According to the needs of the mission and the requirements of practice, the value of the model guided rocket indicators are shown in TABLE 1.

TABLE 1 TECHNICAL INDICATORS OF MODEL GUIDED ROCKETS

| Name of rocket technical indicator | Value of rocket technical indicator |
|---|---|
| **total length(mm)** | ≯1500 |
| **Maximum outer diameter(mm)** | 90 |
| **Take-off mass(kg)** | ≯4 |
| **Range(m)** | 600-1000 |
| **Speed of the shutdown point(m/s)** | ≮90 |
| **Maximum flight time(s)** | ≮20 |
| **Cost per use(yuan)** | ≯2000 |
| **Recycle performance** | Recyclable and reusable |

## 3. MATERIALS AND METHODS

Astrolab, a rocket overall rapid design optimization platform independently developed by College of Aerospace Science and Engineering, National University of Defence Technology, was used to optimize the design of rocket programs. The simulation modules called ACP (Aerodynamic Characteristics Prediction) and NTS (Normal Trajectory Simulation) are involved.

*3.1.Engineering calculation method of aerodynamic characteristics*

ACP (Aerodynamic Characteristics Prediction) software V2.4 is used to calculate the Aerodynamic Characteristics of multi-stage model rocket. The ACP software can calculate the Aerodynamic Characteristics of multi-stage model rocket according to the parameters of the shape size, flight state and centre of mass of the rocket. It can quickly calculate the axial force coefficient, normal force coefficient, lateral force coefficient and pitching moment coefficient, yaw moment coefficient, roll moment coefficient, drag coefficient, lift coefficient, lateral force coefficient, lift-to-drag ratio, static stability, maneuverability, hinge moment characteristics and static dynamic derivative calculation of the rocket, satisfy the requirement of rocket aerodynamic characteristics accurately and fast calculation. The software interface is shown in Fig. 2.

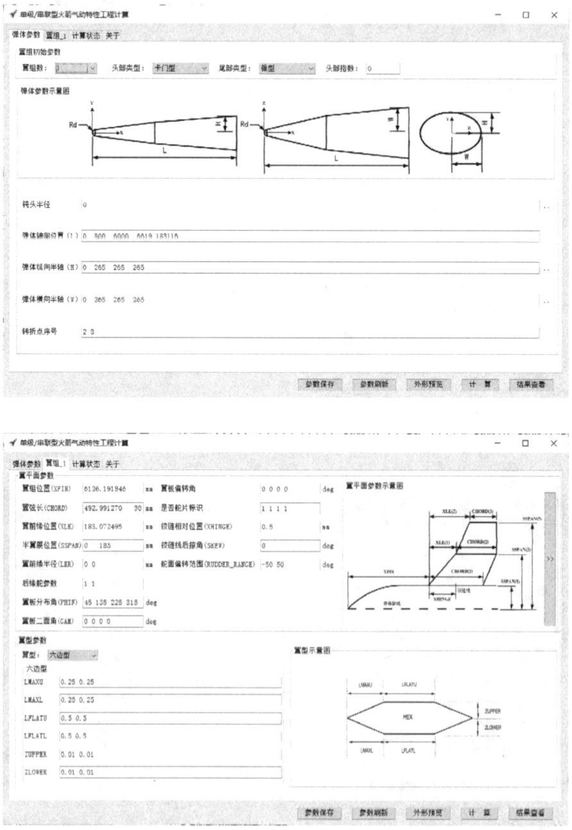

Figure 2 ACP Software Interface

*3.2.Trajectory simulation method*

Aerodynamic characteristics of multi-stage model rockets are calculated by using NTS (Normal Trajectory Simulation)V4.1 software. The NTS software can realize the calculation of three-degree of freedom ballistic parameters of rockets according to the parameters of rocket engine, aerodynamics, mass characteristics, launch elements, flight program, and earth model, etc. The software interface is shown in Fig. 3.

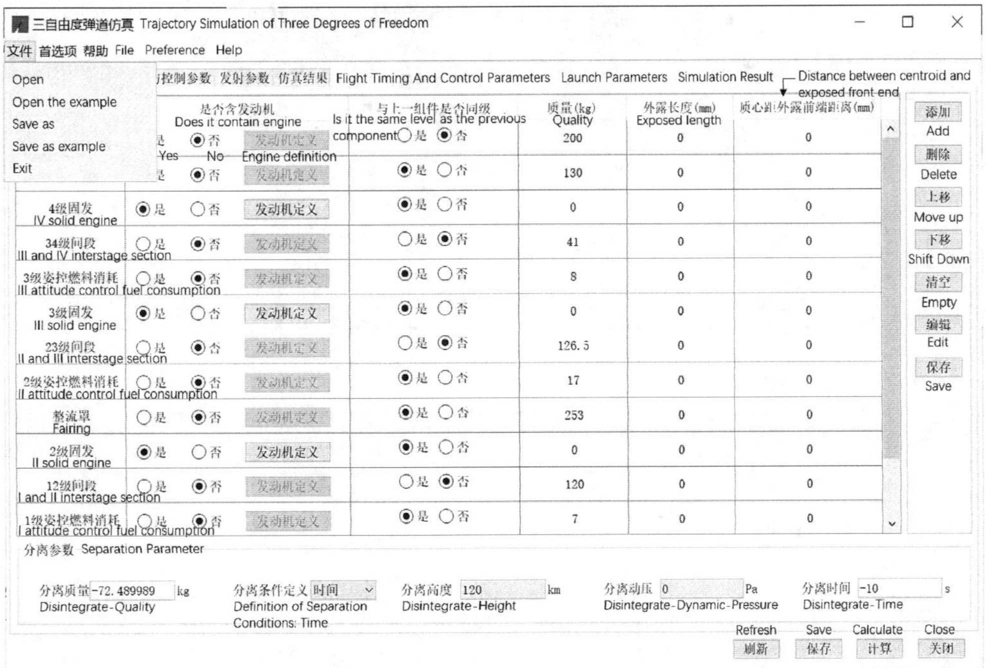


Figure 3 NTS Software Interface

## 4. DESIGN AND OPTIMIZATION OF ROCKET PROGRAM

### *4.1.Rocket design optimization*

The rocket is divided into head cone module, navigation module, control module, load module, recovery module, actuator module and power module. The basic structure layout and shape of the rocket are shown in Fig. 4.

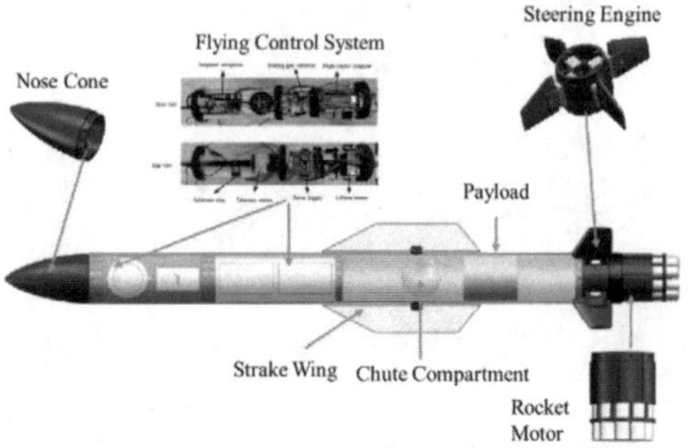


Figure 4 Basic structural layout of Model guided rocket

Two YSR-32 engines in parallel are used as the power of the model rocket. Fig. 5 shows the engine compartment.

Figure 5 Diagram of the engine compartment

The thrust characteristics and thrust curves of the scheme are shown in TABLE 2 and Fig. 6.

TABLE 2 THRUST CHARACTERISTICS OF THE POWER SCHEME

| Power scheme | Explosive load(N·S g) | Total impulse (N·S) | Maximum thrust(N) |
|---|---|---|---|
| 2*YSR32 | 140.0 | 296.14 | 248.17 |

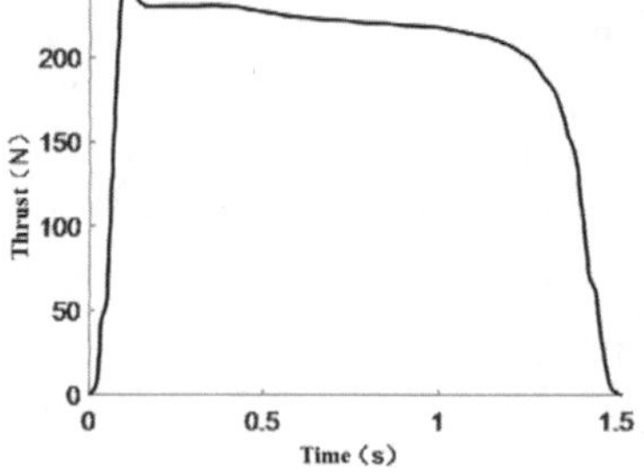


Figure 6 Thrust curves of Two YSR-32 in parallel

*4.2.Design and optimization*

*1) Design variables*

The rocket ballistic design variables are rocket aerodynamic discipline variables and ballistic discipline variables.

Since there are many rocket aerodynamic design variables, this paper studies a set of aerodynamic discipline design variables that are uncorrelated and can clearly characterize the aerodynamic shape (Fig. 7). There are 8 variables, which are the position and sizes of the strake wing and rudders.

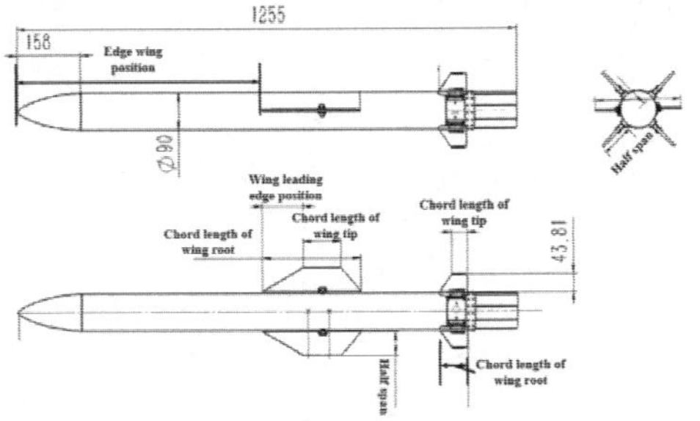


Figure 7 The design variables of the rocket aerodynamic discipline selected in this paper

In terms of the design variables of the rocket aerodynamic discipline selected in this paper, the rocket maintains zero angle at the booster section, and maintains the maximum lift-to-drag ratio gliding at the gliding section. The launch inclination angle and the slip time are selected as the design variables.

*2) Constraints and objective functions*

Referring to the rocket optimization design experience, the static stability margin, the maximum rise-to-drag ratio, etc are selected as the constraints and the wing size is limited. The constraint settings are shown in TABLE 3.

TABLE 3 CONSTRAINTS IN THE DESIGN OF ROCKET

| Constraints | Lower limit | Upper limit |
|---|---|---|
| **Position of strake wing(mm)** | 158 | 1255 |
| **Root-chord length of strake wing(mm)** | 220 | 270 |
| **Tip-chord length of strake wing(mm)** | 70 | 120 |
| **Position of the wing leading edge of strake wing(mm)** | 80 | 120 |
| **Half span of strake wing(mm)** | 40 | 60 |
| **Root-chord length of empennage(mm)** | 30 | 50 |
| **Tip-chord length of empennage(mm)** | 50 | 80 |
| **Full load longitudinal static stability margin** | 4% | – |
| **Full load lateral static stability margin** | 4% | – |
| **Absolute value of longitudinal stability ratio** | – | 1.0 |
| **Maximum Rise-to-drag ratio** | 3.0 | – |
| **Wing area of a single strake wing(m2)** | – | 0.014 |
| **Rudder surface area of a single tail rudder(m2)** | – | 0.004 |

The maximum range of the rocket is selected as the optimize the objective function.

## 5. Results and analysis of design optimization

### *5.1.Results of design optimization*

The smulation clculation is carried out using AstroLab. The engine performance calculation, mass characteristic calculation, aerodynamic characteristic calculation and ballistic calculation are carried out sequentially. The simulation process is plotted in Fig. 8.

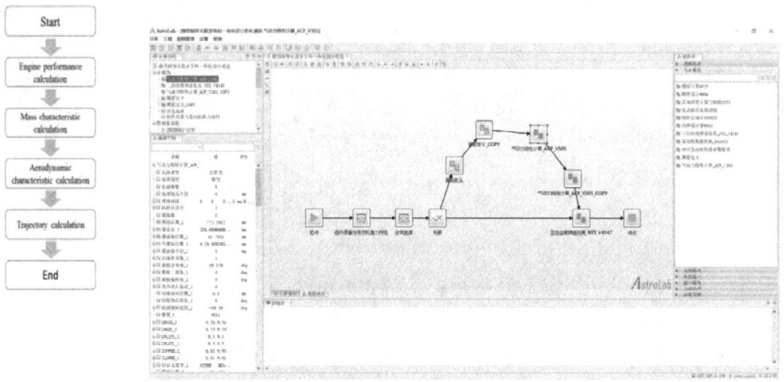


(a)Simulation process (b)Simulation process construction based on AstroLab

Figure 8 Simulation process

After iterating for 260 steps, the solution convergs to the global optimal solution, which takes 65 minutes, and the maximum range of the rocket is 1.29km. The convergence curve of the rocket objective function is plotted in Fig. 9, TABLE 4 shows the optimization results and the aerodynamic shape of the rocket design optimization is shown in Fig. 10.

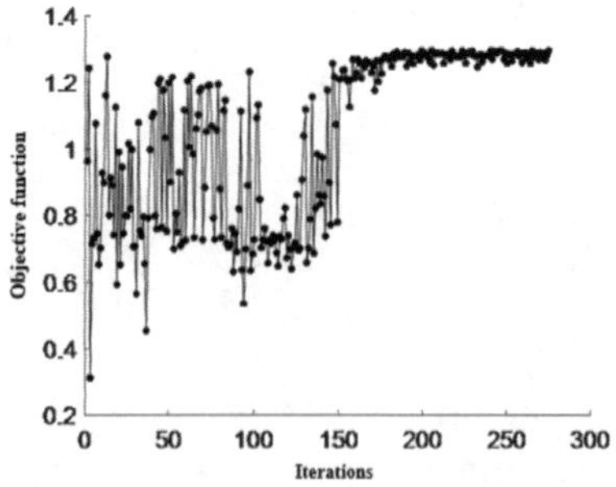


Figure 9 Convergence curve of the objective function

TABLE 4 The result of the optimization

| Target/Constraint | Constraint scope | Results of optimize |
|---|---|---|
| **Range(km)** | max (f) | 1.29 |
| **Full load longitudinal static stability margin** | ≮4% | 4.01% |
| **Full load lateral static stability margin** | ≮4% | 9.62% |
| **Absolute value of longitudinal stability ratio** | ≯1.0 | 0.4695 |

| | | |
|---|---|---|
| **Maximum Rise-to-drag ratio** | ≮3.0 | 3.19 |
| **Wing area of a single strake wing(m2)** | ≯0.014 | 0.0129 |
| **Rudder surface area of a single tail rudder(m2)** | ≯0.004 | 0.004 |

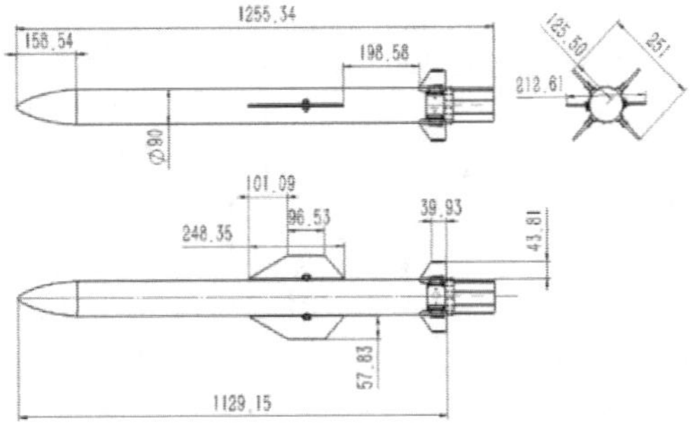


Figure 10 The shape design result of the booster gliding model guided rocket

*5.2.Performance analysis*

*1) Aerodynamic analysis*

The head vertex is chosen as the torque reference point, and the calculation result of the static aerodynamic characteristics of the pitch channel in the zero rudder deflection state is shown in Fig. 11. When the Mach number is 0.05 and the attack angle is 0°, the takeoff center of mass coefficient (relative head vertex) is 0.5938, and the rocket static stability margin is 4.01%.

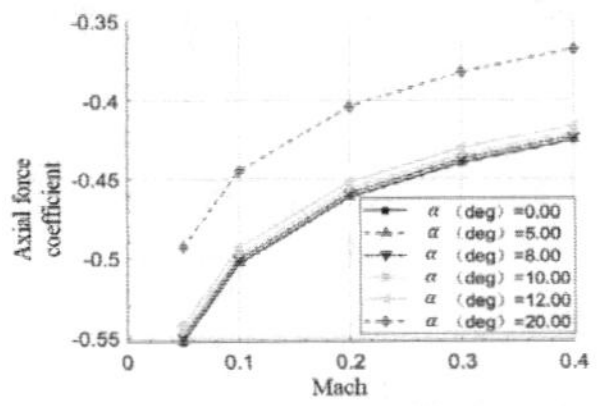


(a)Axial force coefficient in differet mach

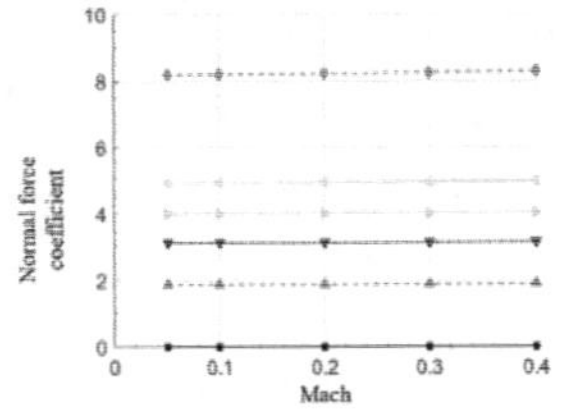


(b)Normal force coefficient in differet mach

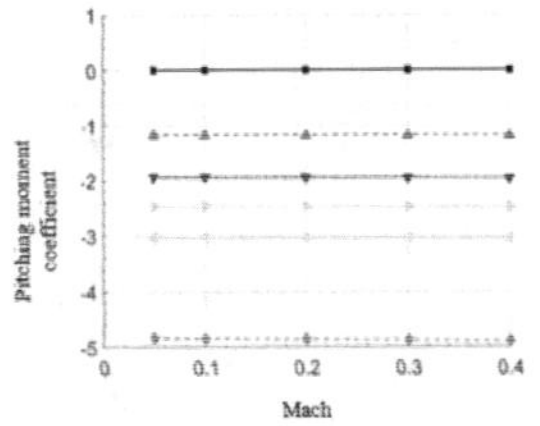


(c)Pitching moment coefficient in differet mach

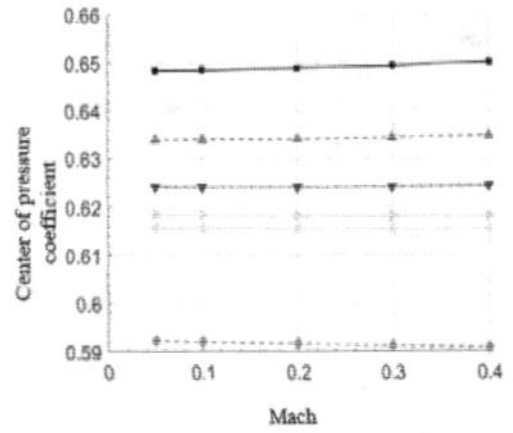


(d)Center of pressure coefficient in differet mach

Figure 11 Aerodynamic characteristics of the pitch channel in the zero rudder bias state

Calculation of the aerodynamic characteristics (depletion state) of the pitch channel leveling state is also completed(Fig. 12). The maximum lift-to-drag ratio of the rocket is 3.19.

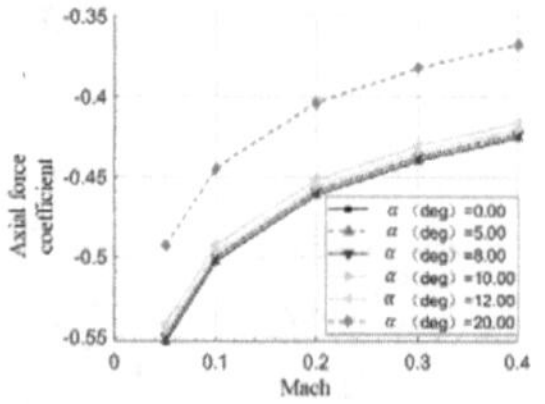


(a)Axial force coefficient in differet mach

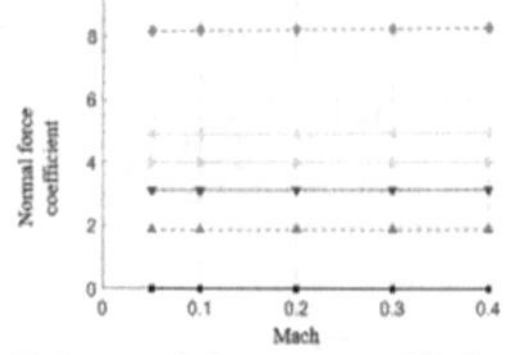


(b)Normal force coefficient in differet mach

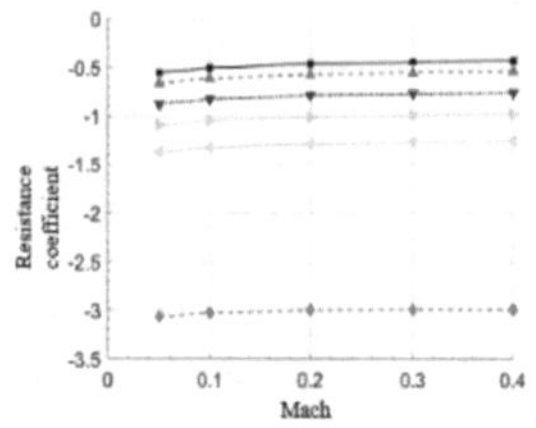


(c)Resistance coefficient in differet mach

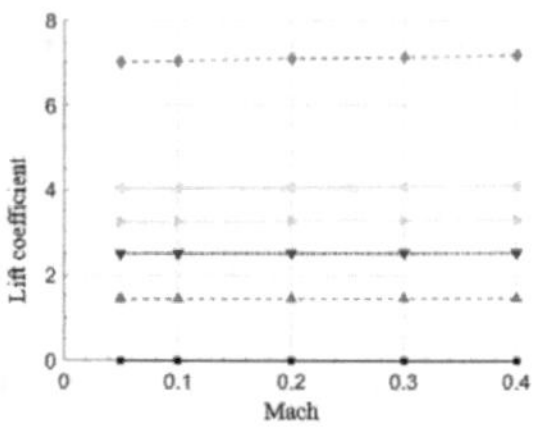


(d)Lift coefficient in differet mach

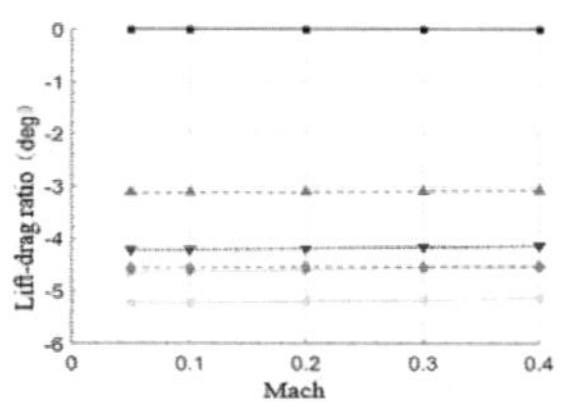


(e)Lift drag ratio in differet mach

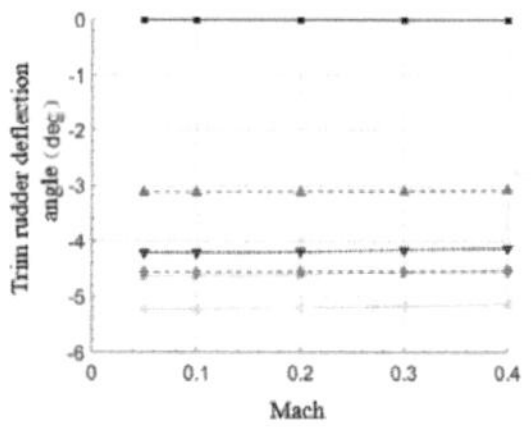


(f)Trim rudder deflection angle in differet mach

Figure 12 Aerodynamic characteristics of the pitch channel leveling state (Depleted state)

*2) Simulation analysis of boost glide ballistics*

The maximum range of the ballistic range is 1.29 km, the ballistic apex height is 240 m, and the total flight time is 28.45 seconds. The ballistic characteristic parameters are shown in TABLE 5.

TABLE 5 BALLISTIC FEATURE PARAMETERS

| Parameter name | Engine ignition | Out of the launcher | The engine shuts down | Ballistic vertices | Ballistic end |
|---|---|---|---|---|---|
| **Time(s)** | 0.01 | 0.24 | 1.51 | 7.60 | 28.45 |
| **Height (km)** | 0.00 | 0.00 | 0.06 | 0.24 | -0.01 |
| **Mach** | 0.00 | 0.04 | 0.28 | 0.15 | 0.13 |
| **Attack angle(°)** | 0.00 | 0.00 | 0.00 | 0.00 | 9.60 |
| **Range(km)** | 0.00 | 0.00 | 0.05 | 0.40 | 1.29 |
| **Velocity inclination(deg)** | 55.64 | 55.64 | 45.56 | -2.34 | -19.57 |

| Parameter name | Engine ignition | Out of the launcher | The engine shuts down | Ballistic vertices | Ballistic end |
|---|---|---|---|---|---|
| Dynamic pressure (kPa) | 0.00 | 0.10 | 5.34 | 1.50 | 1.27 |
| Axial overload factor (g) | 0.00 | 8.43 | -0.56 | -0.18 | -0.14 |
| Normal overload factor (g) | 0.00 | 0.00 | 0.00 | 0.00 | 1.00 |
| Velocity modulus (m/s) | 0.00 | 12.64 | 93.60 | 50.09 | 45.50 |
| Mass (kg) | 2.80 | 2.78 | 2.66 | 2.66 | 2.66 |
| Engine thrust (kN) | 0.00 | 0.23 | 0.00 | 0.00 | 0.00 |
| Pitch angle (deg) | 55.64 | 55.64 | 45.56 | -2.34 | -9.99 |

The maximum range ballistic curve is calculated using three-degree-of-freedom ballistic simulation software NTS(Fig. 13).

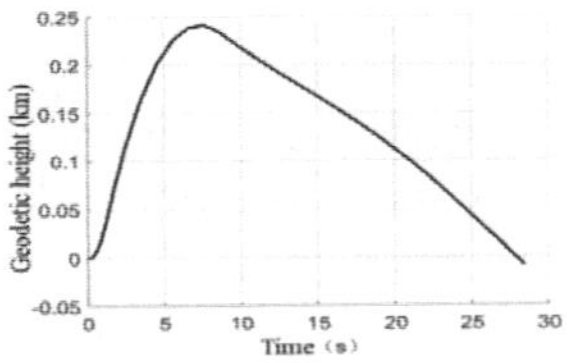

(a) Geodetic height in different time

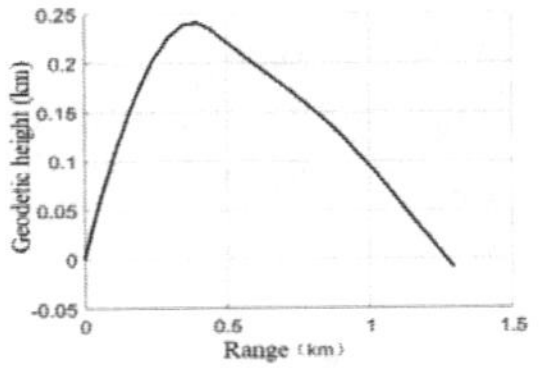

(b) Geodetic height in different range

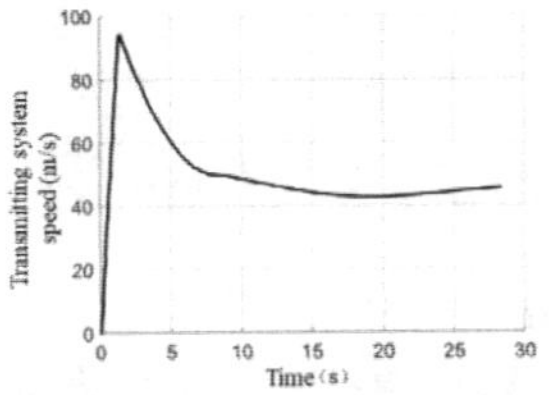

(c) Transmitting system speed in different time

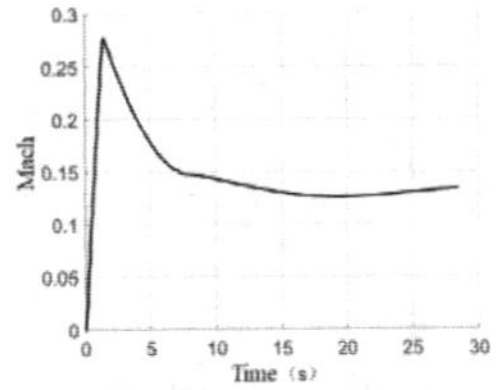

(d) Mach in different time

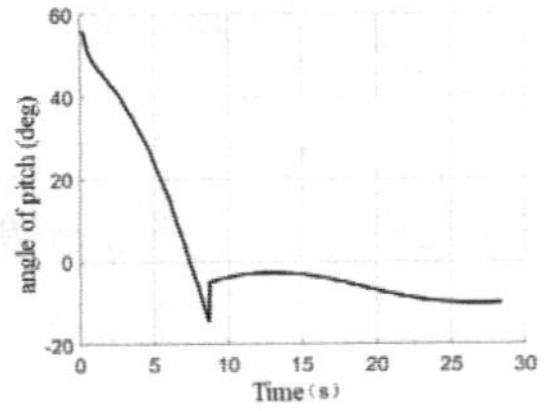

(e) Angle of pitch in different time

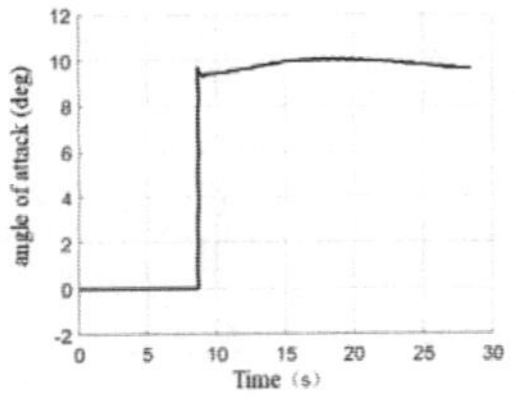

(f) Angle of attack in different time

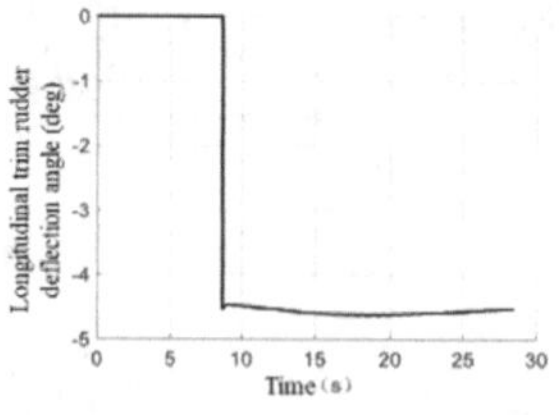


(g) Longitudinal trim rudder deflection angle in different time

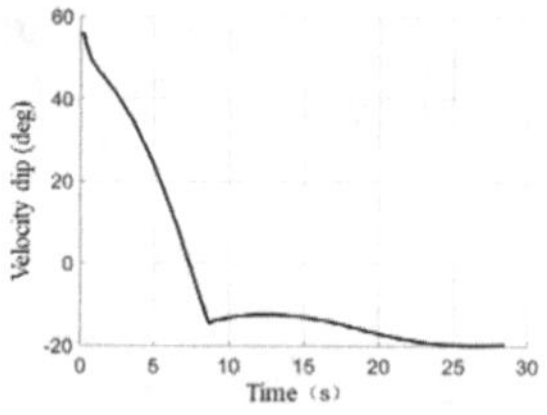


(h) Velocity dip in different time

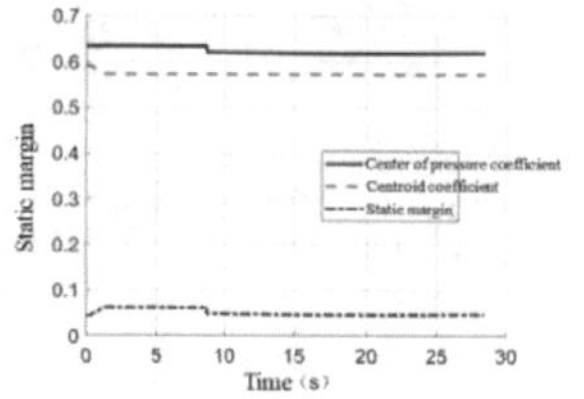


(i) Static margin in different time

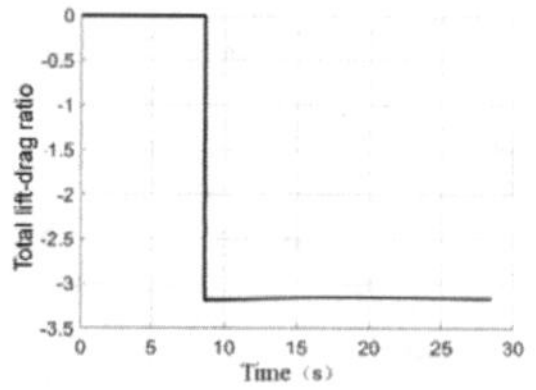


(k) Total lift-drag ratio in different time

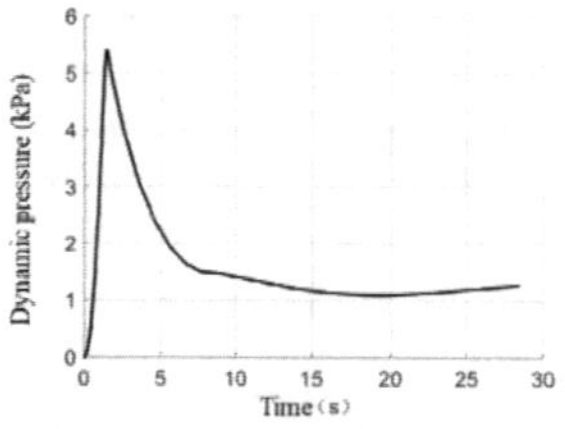


(l) Dynamic pressure in different time

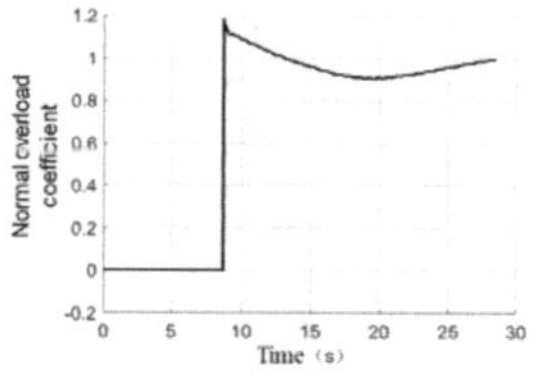


(m) Normal overload coefficient in different time

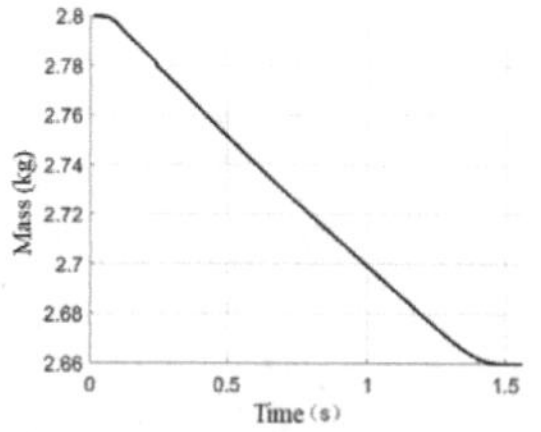


(n) Mass in different time

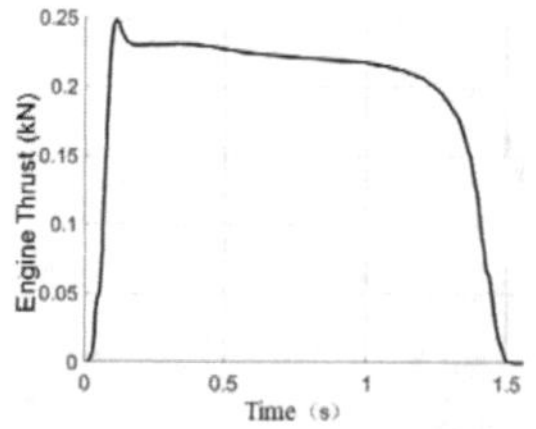


(o) Engine thrust in different time

Figure 13 Maximum range ballistic curve

The ballistic simulation with the launch inclination angle being in the range of 45°-75° is completed in the three-degree-of-freedom ballistic simulation program(NTS). When the rocket glides at the maximum lift-to-drag ratio at the apex of the trajectory. The scope of the range is 0.97km-1.29km, the maximum altitude change range is 0.16km-0.367km, the rocket flight time is greater than 24s, the maximum dynamic pressure is greater than 5kPa, and the maximum velocity modulus is greater than 90m/s. The ballistic characteristics of the different firing inclination angles are shown in TABLE 6.

TABLE 6 BALLISTIC CHARACTERISTICS OF THE ROCKET

| **Launch inclination (°)** | 45 | 50 | 55 | 60 | 65 | 70 | 75 |
|---|---|---|---|---|---|---|---|
| **Range (km)** | 1.26 | 1.28 | 1.29 | 1.206 | 1.146 | 1.087 | 0.970 |
| **Flight time** | 24.31 | 26.44 | 28.45 | 28.486 | 28.749 | 28.704 | 27.754 |

| Launch inclination (°) | 45 | 50 | 55 | 60 | 65 | 70 | 75 |
|---|---|---|---|---|---|---|---|
| (s) | | | | | | | |
| Maximum height (km) | 0.16 | 0.20 | 0.24 | 0.274 | 0.310 | 0.340 | 0.367 |
| Maximum dynamic pressure (kPa) | 5.63 | 5.52 | 5.34 | 5.32 | 5.24 | 5.17 | 5.12 |
| Maximum speed modulus (m/s) | 96.06 | 95.14 | 93.6 | 93.44 | 92.75 | 92.163 | 91.69 |
| Ballistic vertex speed (m/s) | 64.59 | 57.57 | 50.09 | 44.66 | 36.23 | 31.27 | 23.97 |

Setting the altitude of the open umbrella being 80m, maintainning gliding flight during the whole flight, the range can reach 1060m. After gliding for 10s, the negative angle of attack is pressed down, and the range is adjusted to about 950m. After gliding for 5s, the negative attack angle is pressed down, and the range is adjusted to about 800m. On the basis of gliding ballistics, the model guided rocket is pressed down at different times at negative attack angles. When the altitude of open-umbrella reaches 80m, the ballistic trajectory can be flexibly adjusted in the range of 590-1060m.

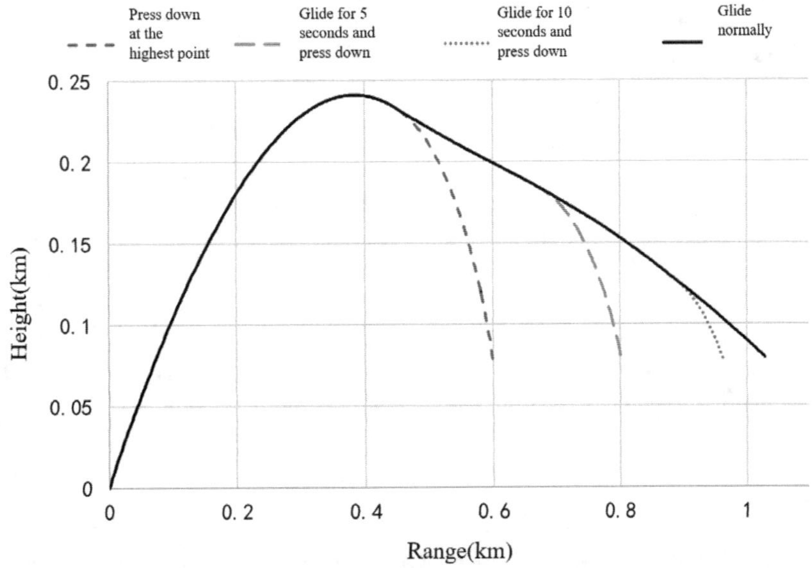


Figure 14 Negative attack angle down pressure rocket range-altitude curve at different times

## 6. Conclusion

This paper completes the design optimization of the boost-glide model guided rocket scheme, which can provide program support for the development of low-cost test rockets that meet the needs of the full-process and full-element high-level engineering practical training. The model guided rocket mission and modular layout design are carried out, and the rocket power scheme design is completed. When using two YSR-32 engines in parallel, the total engine thrust is 296.14N· S. A mathematical model of rocket aerodynamic/ballistic integrated design optimization problem is constructed, the aerodynamic layout and boost-gliding trajectory of the side strip wing + × tail rudder are used with the maximum range being the target function, and 8 design variables are used. Using the rocket overall rapid design optimization platform Astrolab, the rocket aerodynamic/ballistic performance integration simulation process is built, and the aerodynamic/ballistic integration design optimization of the kilometer-level range model guided rocket is completed. The maximum range of the optimization result is 1.29km, the take-off mass is 2.8kg, the shutdown point speed is 93.6m/s, and the maximum flight time is 28.4s.

## References

[1] Yu B.S.. (1996) Overall design of anti-aircraft missiles. Aerospace Publishing House,Beijing.
[2] Chen F.F.,Liu J.N. (2006) Design of guidance and control system for simple guided rocket. Journal of missiles and guidance.Beijing, pp 112-118.
[3] Nguyen C.M. (2016) Design of overall and attitude control system of spin-guided rocket. Dalian University of Technology.
[4] Peng Q.S.,Wang Y.P.,Chen H.B. (2013) Research on spin stability of sounding rocket flight attitude. Proceedings of the 26th National Symposium on Space Exploration of the Space Exploration Professional Committee of the Chinese Society of Space Science.pp 578-584.
[5] Liu X.J. (April 2017) Introduction to the overall design of missiles. National Defense Industry Press, Beijing.
[6] Zhang J.H. Design of low cost missile control system. Aeronautical computing technology.pp 172-177.
[7] Feng S.Z. (2012) Research on control strategy of pneumatic load simulator based on dynamic pressure feedback. Harbin Institute of technology.
[8] Qiang S.,Shen Z.,Yi D.Y.,etc.(2008) Ballistic Modeling and Simulation Based on Missile Design Principles. Tactical Missile Technology.pp 80-84,89.
[9] Wang S. (2017) Modular design of sounding rocket avionics equipment based on STM32.Science and education guide (Electronic Edition).pp 168-212.

# Research on optimization of wave glider wing plate

**Ganlin Wang [1,2], Liucun Zhu[1], Hong Wan[3], Jiyue Wang[1*], Haofeng Deng[1,2] and Congwei He[1,2]**

[1]Beibu Gulf University Advanced Science and Technology Research Institute, Qinzhou, Guangxi,China

[2]School of Mechanical and Marine Marine Engineering, Beibu Gulf University, Qinzhou, Guangxi,China

[3]Beibu Gulf University Advanced Science and Technology Research Institute, Qinzhou, Guangxi,China

[*]Corresponding author's e-mail: wangjiyue@bbgu.edu.cn

**Abstract.** In this study, a new wing plate was designed for the wave glider. The wing plate A and the wing plate C made of NACA0012 were extracted from the adult green sea turtle in the Pacific Ocean, and a new wing plate B was optimized by optimizing the design of the two, and set the control. Group wings D. In order to improve the utilization rate of wave energy of the wave glider, the hydrodynamic performance of the wave glider wing was studied. At the installation angle of 0-90°, the thrust FX in the forward direction of the four types of wings is simulated and calculated. The results show that at the angle of attack of 6° and 9°, the incoming flow velocity is 0.5m/s, 1m/s, 1.5 At m/s, the hydrodynamic performance of the wing plate increases with the increase of the installation angle. The hydrodynamic performance of the four kinds of wing plates is from good to bad: B-foil, A-foil, C-foil and D-foil . The research results help to understand the hydrodynamic performance of the wing plate, and provide a research basis for improving the wave energy utilization rate of the wave glider.

## 1. Introduction

The wave glider is a revolutionary unmanned sea surface mobile observation platform that can convert wave motion into forward propulsion. In theory, the endurance of the wave glider is unlimited. The wave glider has the advantages of long battery life, green environmental protection, position retention and satellite communication monitoring. Therefore, it has broad application prospects in marine surveys and observations [1-2]. In the design of hydraulic machinery, how to improve the hydrodynamic performance of hydraulic machinery is one of the most important issues [3]. As an important part of the power element in the wave glider, the structure of the wing plate has a huge impact on the overall hydrodynamic performance. Therefore, the research on the performance of the wing plate has an important impact on the development of the wave glider [4]. A lot of research has been done on wing panels over the decades. For example, Yuchen Shang and Juan J. Horillo used parametric methods and neural networks to study the performance of the NACA0012 hydrofoil tandem arrangement, and used a two-dimensional viscous numerical model to simulate the hydrofoil system [5]. Yang Yining carried out numerical simulation on the basis of the airfoil design of the hydromechanical blade. When Re=107, the hydrodynamics of the airfoil with different cambers were calculated [6]. Chen Jiawang et al. compared the different factors affecting the propulsion speed of the wing plate, and optimized the calculation results by comparing the simulation results with the test results [7]. Zhao Minsheng used the improved SST k-ω and Smagorinsky models to predict the unstable cavitation and periodic shedding of sheet cavitation, and analyzed the cavitation shape, shedding frequency, lift coefficient and drag coefficient of these three turbulence models [8]. D Raja Joesph, P Booma Devi & M Gopalsamy studied different angles of attack and found that pits of the same size could not be formed to provide the desired effect at higher Reynolds numbers [9].

The purpose of this study is to improve the hydrodynamic performance of the wave glider. Different from previous studies, this study regards improving the thrust $F_X$ in the forward direction of the wing as the best goal. The hydrodynamic performance of a wave glider wing is affected by many factors, such as the shape of the wing, structural dynamic characteristics, seawater velocity and operating depth. Considering the complexity of the actual sea conditions, it is difficult to monitor the structural dynamic characteristics of the wing plate and calculate its hydrodynamic performance on-site, and the use of software for simulation testing is widely used in this regard, so ANSYS is chosen for simulation. Based on the analysis of the NACA0012 wing and the forelimbs of adult green sea turtles in the Pacific, this study designed a new wing, and simulated the installation angle of the wing. The results show that the hydrodynamic performance of the wing increases with the increase of the installation angle. While enlarged, the newly designed wings have better hydrodynamic performance than other wings.

## 2. Establishment of the physical model

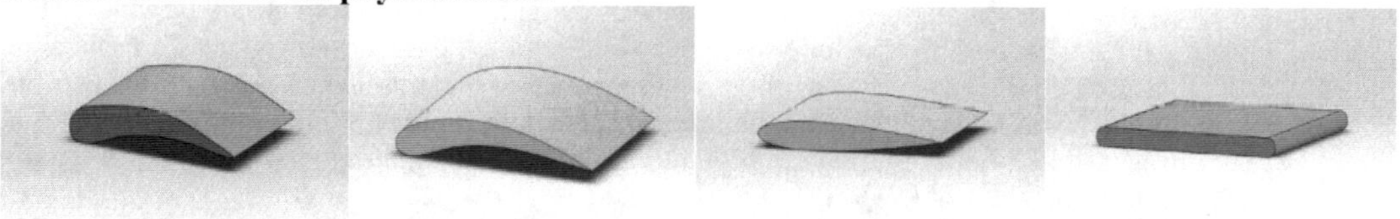

a. Bionic Wing plate A b. Design wing plate B c.NACA0012 wing plate C d. Refer wing plate D

Figure 1. Wing plate shape drawing

The wing plate is the key component of the wave glider to realize wave energy conversion, and the design of its shape and layout has a great influence on the propulsion effect [10]. Figure 1 shows the geometry of the four types of wings. Biomimicry is a science that studies the structure and properties of biological systems to provide new design ideas and working principles for engineering technology. Biomimetic technology studies the functional principles and action mechanisms of various biological systems as biological models, and finally realizes new technical design and manufactures better new instruments and machinery. In this study, the A-wing plate is based on the principle of bionics. By extracting the forelimbs of adult green sea turtles from the Pacific Ocean, it is concluded that the NACA0012 wing plate has good performance in the air and is widely used in wind tunnel experiments. Therefore, this study uses the NACA0012 wing plate as the C-wing plate, compare the advantages and disadvantages of its underwater performance, through the structural analysis and optimization design of the above two kinds of wing plates, we strive to design a wing plate with better hydrodynamic performance underwater, and obtain the B wing plate , In order to ensure the rigor of the research, a reference D-wing plate was set up. The four kinds of wings are now named as bionic A-wing plate, design B-wing plate, standard C-wing plate and regular D-wing plate. Figure 2 is the force analysis diagram of the wing, which mainly analyzes the influence of installation angle, attack angle and incoming flow speed on the hydrodynamic performance of the wing.

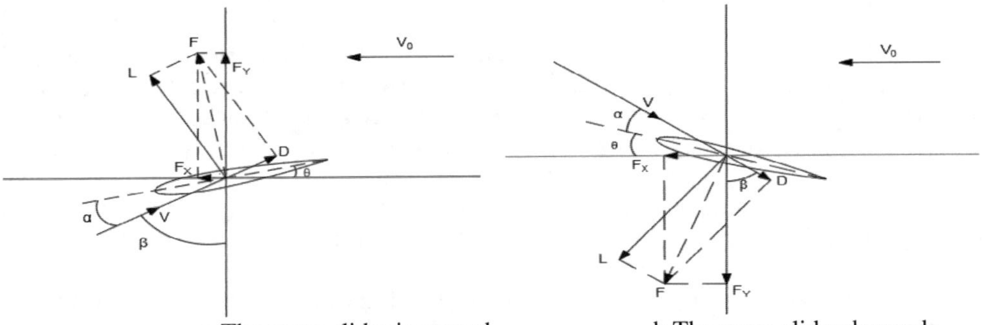


a.The wave glider is ascend b.The wave glider descends

Figure 2. Force analysis diagram of wing plate

There are some important dynamic parameters in the calculation and analysis of the flow field of the wing plate, and their definitions are as follows.

- V is the direction of the incoming flow velocity, and $V_0$ is the forward direction of the wing.
- Resistance D: The force that hinders the movement of an object. When an object moves in a liquid, the moving object is subjected to the force of the fluid to reduce its speed, and this force is also a resistance. The calculation formula is [11]:

$$D = 0.5\rho V^2 S_W C_{D(\alpha)} \tag{1}$$

- Lift L: A dimensionless quantity that refers to the ratio of the lift experienced by an object to the product of the airflow pressure and the reference area. The calculation formula is [11]:

$$L = 0.5\rho V^2 S_W C_{L(\alpha)} \tag{2}$$

- Angle of attack α: a fluid mechanics term. For the wing plate, the angle of attack defines the angle between the projection of the incoming flow velocity vector V on the longitudinal symmetry plane and the longitudinal axis of the wing plate.
- Pressure angle β: The pressure angle refers to the acute angle between the direction of force and the direction of motion without calculating friction.
- Installation angle θ: The installation angle refers to the angle between the chord line of the wing profile and the fuselage axis when the wing is installed on the fuselage. The installation angle is one of the parameters that affects the aerodynamic characteristics of the wing, and it affects the lift coefficient, zero lift angle, stall angle of attack and cruise resistance. According to the geometric relationship [11] :

$$\alpha + \beta + \theta = 90° \tag{3}$$

- In this study, the resultant force F is decomposed into the thrust force FX that pushes the wave glider forward in the X direction and the force FY that pushes the wave glider upward in the Y direction. The calculation formulas are [10]:

$$F_X = Lsin\beta - Dcos\beta \tag{4}$$

$$F_Y = Lcos\beta - Dsin\beta \tag{5}$$

## 3. Numerical computing

### *3.1. Generation of Computational Domain*

When establishing a computational domain, there are two computational domain division methods based on previous experience. One is to reduce the marine environment simulation to establish a rectangular computational domain; the other is to establish a three-dimensional rectangular computational domain with a hemispherical entrance. First choose to establish a rectangular computational domain, the computational domain of the wing plate is a rectangle of 1x6, The C-foil was used to verify the convergence of the computational domain, and the rectangular computational domain was used to calculate the lift-to-drag ratios of four 1-m-long wings. Analyzing the reason, it is found that the lower boundary of the large attack angle affects the calculation, and the simulation effect is not an open-ended result. The problem is found from the negative resistance when the attack angle is negative. The rectangular computing domain increases when the attack angle gradually increases, the boundary interferes with the incoming flow direction, changing the incoming flow angle of the wing plate, making it difficult to set the angle of attack, so it is necessary to change the computational domain to perform the correct simulation.

Considering the sufficient development of the flow and the convergence of the calculation, the computational domain is redesigned as a three-dimensional rectangular computational domain with a hemispherical inlet, which is the same as the wing plate, and the grids are 0.5m and 0.005m. Length X is 2.5e-002 m, Length Y is 2.5e-002 m, Length Z is 2.5e-002 m, Volume is 1.7878e-008 m³, and other parameters remain unchanged and continue the simulation calculation. Figure 3 shows the graph of the convergence verification of the new computational domain using the C-foil. It can be seen from the figure that the change of lift is basically linear, and the change of drag is nonlinear. When the angle of attack is positive, as the angle of attack increases, the lift and drag increase and the lift-to-drag ratio decreases. When the angle of attack is negative, as the angle of attack increases, the lift decreases to a negative value, the drag increases, and the lift-to-drag ratio decreases to a negative value. Negative values imply a change of direction, broken down to the forward direction, possibly from thrust to drag. The simulation effect is better than the rectangular computational domain, so this computational domain is selected for subsequent simulation calculations.

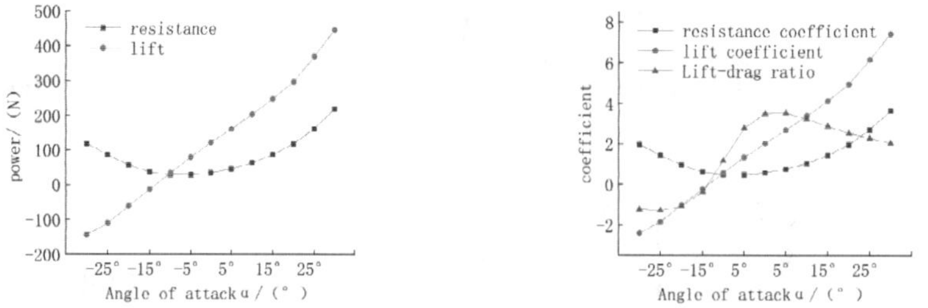


Figure 3. The new computational domain curve of the C-wing plate

*3.2. Setting the number of iteration steps*

Comparing the calculation results when the number of iteration steps is 1000, 1200, 1500, and 2000, we can see that the calculation result at 1200 can be regarded as convergence. Although the residual does not meet the setting, the value of the observation point has been stabilized, and there is no need to waste time to calculate to 2000. Since the flow difference between the inlet and the outlet is less than 1%, it can also be regarded as convergence. Selecting 1200 can calculate more data at the same time.

## 4. Simulation

*4.1. Seawater Viscosity Settings*

The angle of attack is -30°—30°, the interval is 5°, the temperature is 20°C, the water density is 998.2kg/m3, and the inflow velocity is 1m/s, and the viscosity and non-consideration of the four types of wings are considered respectively. Through the simulation calculation of viscosity, the lift and resistance of four types of wing plates are obtained, as shown in Figure 4.

a. No viscosity resistance curve b. Viscosity Resistance Curve

Figure 4. Viscosity curve

It can be seen from Figure 5 that when the attack angle is -30°—30°, the overall trends of a, b, c, and d are the same. After considering the viscosity of seawater, the resistance and lift are not much different. When the flow rate is very large, the viscous force and inertial force are in phase. Under the condition that the ratio is almost negligible, the use of the inviscid model can improve the calculation speed, and the calculation accuracy will not lose too much. Therefore, the existence of viscous force has no effect on the curve trend and comparison results, and the subsequent simulation will not consider seawater viscosity for calculation.

### *4.2. Influence of installation angle on hydrodynamics of wing plate*

In order to study the influence of the installation angle on the hydrodynamic performance of the four types of wings, four types of wings are selected, the temperature is 20°C, the water density is 998.2kg/m$^3$ in the Beibu Gulf at 20°C, and the installation angle is set to 0°—90°, interval of 5°, incoming velocity of 0.5m/s, 1m/s, 1.5m/s, angle of attack of 6° and 9°. Selecting as many installation angle ranges as possible is to fully guarantee the rigor of the research. The calculation domain is set as shown above. The above situations are simulated and calculated respectively, and the lift, drag, lift coefficient and drag coefficient of the wing plate are obtained. The focus of the research is to use $F_X$ as the standard to judge the hydrodynamic performance of the wing.

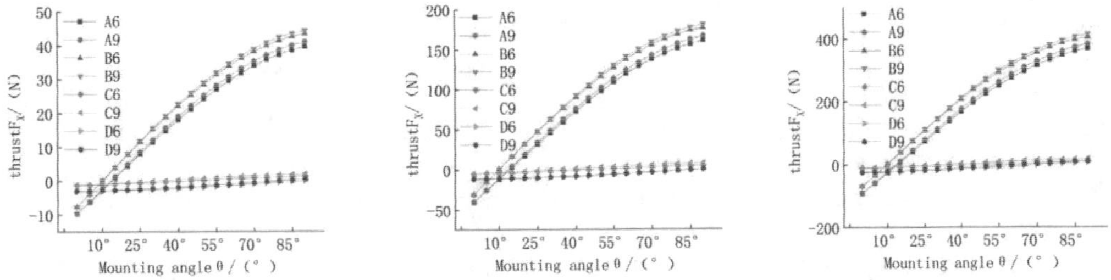


a. Incoming flow speed 0.5$F_X$ b. Incoming flow speed 1$F_X$ c. Incoming flow speed 1.5$F_X$

Figure 5. Incoming flow speed $F_X$ curve

It can be seen from Figure 5 that the thrust $F_X$ in the x-direction of the four blades changes linearly under the three incoming flow speeds, and the $F_X$ of the A and B blades increases with the increase of the installation angle and is in a parabolic shape. And the $F_X$ trends are basically the same under the three incoming velocities; while the $F_X$ of the C-wing and D-wing is relatively less affected by the installation angle. On the whole, under the three incoming flow speeds, the thrust in the x direction obtained by the B wing is the largest, followed by the A wing, and the thrust received by the D wing is the smallest. In addition, when the installation angle of the four types of wings is too small, the thrust is all negative, indicating that the installation angle should not be too small in practical applications. The four types of fenders basically have great changes with the increase of the installation angle. In general, the larger the installation angle is, the greater the incoming flow speed, and the better the hydrodynamic performance of the fender.

## 5. Conclusion

The purpose of this study is to improve the hydrodynamic and hydrodynamic performance of the wave glider. According to the analysis of the existing wing plate, the wing plate A is extracted according to the simulation idea, and because the NACA0012 wing plate has superior performance in the air, this study selects the NACA0012 wing plate. The plate is the wing plate C, and the two kinds of wing plates are optimized to obtain the airfoil B. In order to ensure the accuracy of the simulation results, the reference wing plate D is set. The influence of the installation angle on the hydrodynamic performance of the wing is studied through the simulation calculation of four kinds of wings. A comparative analysis of the thrust in the x-direction of the four types of wing plates is carried out, and the following conclusions are drawn:

- The influence of viscous force can be ignored during the hydrodynamic performance simulation calculation of the wave glider wing plate;
- When the incoming flow velocity is 0.5m/s, 1m/s, 1.5m/s, and the angle of attack is 6° and 9°, the thrust in the x direction of the wave glider wing plate increases with the increase of the installation angle ;
- The comprehensive comparison shows that the hydrodynamic performance of the four kinds of wing plates at different installation angles is from good to bad: B-wing plate, A-wing plate, C-wing plate and D-wing plate.

In this study, a design wing plate B with better hydrodynamic performance than the existing airfoil was obtained, which provided a research basis for improving the wave glider wing plate, thereby improving the wave energy utilization rate of the wave glider.

## Acknowledgments

This work was financially supported by Special Fund for Bagui Scholars of the Guangxi Zhuang Autonomous Region No.2019A08, Science&technology base and talent special project of Guangxi province GuiKeAD19245067 and Launching Project for High-level Talents of Beibu Gulf University 2018KYQD45.

## References

[1] Wang Peng,Tian Xinliang,Liang Xin,Guo Xiaoxian,Li Xin,Zhang Xiantao. Development of the control system for a wave driven glider[J]. Ocean Engineering,2021,229.
[2] Page Brian R.,Lambert Reeve,Mahmoudian Nina,Newby David H.,Foley Elizabeth L.,Kornack Thomas W.. Compact Quantum Magnetometer System on an Agile Underwater Glider[J]. Sensors,2021,21(4).
[3] Fuming Yang,Weichao Shi,Xu Zhou,Bin Guo,Dazheng Wang. Numerical investigation of a wave glider in head seas[J]. Ocean Engineering,2018,164.
[4] Cheong Snons,Kim YoungJun,Chun JongHwa,Kim JungKi,Huh Shin. Integrated Offshore Seismic Survey Using an Unmanned Wave Glider[J]. Energies,2021,14(2).
[5] Shang Yuchen,Horrillo Juan J.. Numerical Simulation and Hydrodynamic Performance Predicting of 2 Two-Dimensional Hydrofoils in Tandem Configuration[J]. Journal of Marine Science and Engineering,2021,9(5).
[6] Yang Yining. Analysis of the Influence of Camber on Hydrodynamic Characteristics of Airfoil Based on FLUENT[J]. Journal of Physics Conference Series,2020,1519(1).
[7] Chen Jiawang,Ge Yongqiang,Yao Chaoling,Zheng Binghuan. Dynamics Modeling of a Wave Glider With Optimal Wing Structure[J]. IEEE Access,2018,6.
[8] Zhao Minsheng,Wan Decheng,Gao Yangyang. Comparative Study of Different Turbulence Models for Cavitational Flows around NACA0012 Hydrofoil[J]. Journal of Marine Science and Engineering,2021,9(7).
[9] Joseph D. Raja,Devi P. Booma,Gopalsamy M.. Investigation on effect of square dimples on NACA0012 airfoil with various Reynolds numbers[J]. International Journal of Ambient Energy,2021,42(4).

[10] Yan Hao,Su Xiaozhen,Zhang Haozhou,Hang Jianwei,Zhou Ling,Liu Zhifeng,Wang Zhujiang. Design approach and hydrodynamic characteristics of a novel bionic airfoil[J]. Ocean Engineering,2020,216.
[11] Tom Daniel,Justin Manley,Neil Trenaman. The Wave Glider: enabling a new approach to persistent ocean observation and research[J]. Ocean Dynamics,2011,61(10).

# Research on global calibration technology of pressure sensitive paint based on airtight chamber

**Jinkui Shang, Ronghuan Zhao, Zhaonan Xu, Hongjie Zhong**

AVIC Aerodynamics Research Institute, Shenyang 110034, China

Jinkui Shang e-mail: 523664270@qq.com

**Abstract**—Pressure sensitive paint (PSP) technology has been widely applied in wind-tunnel test. Generally, PSP calibration data are obtained using a calibration sample in laboratory priori calibration device, which the sample is sprayed at the same time with the test model.In this study, taking the real test model as the calibration object, a global calibration technology of PSP coating on the surface of the test model is proposed, which is verified in the airtight chamber, and the effective PSP calibration data are obtained. A two-component fast response PSP coating was used in the study. In the validation test, a flying wing model is used, the flying wing model sprayed with PSP coating was placed in the airtight chamber. 11 groups of calibration tests were carried out in the environment of 0.06 MPa ~ 0.14 MPa and 30°C ~50°C, and the calibration curves obtained showed good consistency.

## 1. Introduction

Pressure sensitive paint (PSP) technology is a model surface pressure measurement technology using in wind-tunnel test, which is developed in the 1980s. It uses the fluorescence quenching effect of oxygen molecules between probe molecules (active ingredients in pressure sensitive paint), and reveals the local environmental pressure distribution by measuring the lifetime or intensity of the fluorescence emitted by the excited probe molecules[1]. PSP technology is a non-intrusive wind tunnel test technology, compared with conventional pressure measurement technology, it has many advantages, such as high spatial refinement, no flow field interference, short test cycle, etc., and has a wide range of extended applications, so it has been widely concerned.

Unsteady PSP technology can make a rapid response to the pressure change on the surface of the model, using fast response PSP. Compared with the steady-state PSP, the response time to the pressure change is usually 0.3 s ~ 0.5 s, the fast response PSP can respond within 1ms. Fast response PSP mainly improves the response speed to pressure changes by increasing the penetration rate of oxygen in the PSP coating, so fast response PSP coating is generally porous structure. Fast response PSP mainly includes two types, namely anodized aluminum PSP (AA-PSP) and polymer ceramic PSP (PC-PSP). Because PC-PSP are more convenient to use, they are more widely used in research[2~5]. But compared with 1 μs ~ 50 μs response time of AA-PSP, the response time of PC-PSP is usually between tens of microseconds and a few milliseconds, but PC-PSP with a response time of less than ten microseconds has been developed[6].

At present, the unsteady PSP technology based on PC-PSP has been increasingly used in engineering tests[7-10], but due to various reasons, the accuracy of fast response PSP test data need to be improved, which the error is usually greater than 1000 Pa[11]. There are many factors that affect the accuracy of PSP data. The well-known error factors include the fluorescent probe molecular distribution in the coating, coating thickness, illumination uniformity, temperature effect, coating photodegradation, alignment deviation between the reference image and the test image, etc., but in fact, the coating calibration error also has a critical impact on the accuracy of PSP test data.

The priori calibration based on laboratory samples is widely used at present. Through the calibration samples sprayed at the same time with the test model, the calibration coefficient of PSP coating is obtained in the laboratory by using the calibration system, and the PSP coating brightness ratio data is converted to pressure data based on the calibration data. The calibration process is simple and convenient, and the PSP coating calibration data can be obtained quickly, but this calibration method has an obvious defect. That is, it attempts to "generalize" with a small sample of data to describe the characteristics of the model's full surface PSP coating. Moreover, this calibration method does not consider the influence of excitation lighting / image acquisition environment, angle of attack and other factors on PSP test. In this paper, a PSP coating global calibration technology is proposed, and the calibration technology is studied and verified in the airtight chamber at AVIC Aerodynamics Research Institute.

## 2. PSP technology and convention priori laboratory calibration

Pressure sensitive paint is a kind of polymer material, which contains probe molecules and polymer adhesives. When the probe molecules are illuminated by light of special wavelength, the probe molecules (excited) will emit fluorescence (photoluminescence). When oxygen molecules exist in the environment, the oxygen molecules absorb the energy of the excited probe molecules by contacting with the probe molecules in the PSP coating, and the probe molecule loses the ability to emit fluorescence (known as oxygen fluorescence quenching). The oxygen fluorescence quenching rate depends on the local oxygen partial pressure, that is, the local air pressure, which establishes the relationship between the fluorescence intensity of the PSP coating and the local environmental pressure. The PSP technology uses this relationship to measure the pressure on the model surface.

The intensity of oxygen fluorescence quenching effect is related to the concentration of oxygen molecules in the environment, the intrinsic lifetime of the excited state of the probe molecule and the quenching rate of the excited state. Generally, Stern-Volmer formula can be used to express the relationship between the influence of oxygen fluorescence quenching effect of the PSP coating and the concentration of oxygen molecules [1]:

$$\frac{I_0}{I} = 1 + k_q \tau_0 [O_2] \tag{1}$$

In the formula, $I_0$ and $I$ are the fluorescence intensity of the PSP coating in the oxygen free environment and in the oxygen concentration $[O_2]$ environment respectively, and $k_q$ is the collision quenching rate constant of oxygen molecules, $\tau_0$ represents the excited state intrinsic lifetime of probe molecules, $[O_2]$ represents the concentration of oxygen molecule.

The following polynomial formula is usually used in wind tunnel tests[1]:

$$\frac{P}{P_{ref}} = A(T) + B(T)\frac{I_{ref}}{I} \tag{2}$$

Where, $I_{ref}$ and $P_{ref}$ represent the reference image intensity and reference pressure of the selected reference conditions, $A(T)$ and $B(T)$ are the calibration coefficients to be obtained, which is temperature dependent.

Generally, the calibration coefficient of PSP coating is obtained by applying the calibration sample sprayed at the same time with the test model in the laboratory calibration system, which includes a calibration cavity, an excitation light source, a scientific camera, and a temperature / pressure control system. Figure 1 shows the small PSP laboratory calibration system from AVIC Aerodynamics Research Institute.

Figure 1. the small PSP laboratory calibration system

## 3. Global calibration of PSP coating based on airtight chamber

PSP coating global calibration study is carried out in the airtight chamber, which is located at AVIC Aerodynamics Research Institute. the inner diameter of the airtight chamber is 2.6 meters and the length is 7 meters. Multiple optical windows are set on the chamber. The airtight chamber is connected with the medium pressure air source and vacuum air source, which can realize the pressure regulation of 0.02 MPa ~ 0.4 MPa. The working medium is air. The chamber is equipped with a heater, which can realize the temperature regulation by heating the air in the chamber. The test model used in the study is a small aspect ratio flying wing model designed and manufactured by the AVIC Aerodynamics Research Institute, which length is 1.3 meters, span is 1.1 meters Only the left wing of the flying wing model is used in the test. See Figure 2 for the PSP coating spraying area.

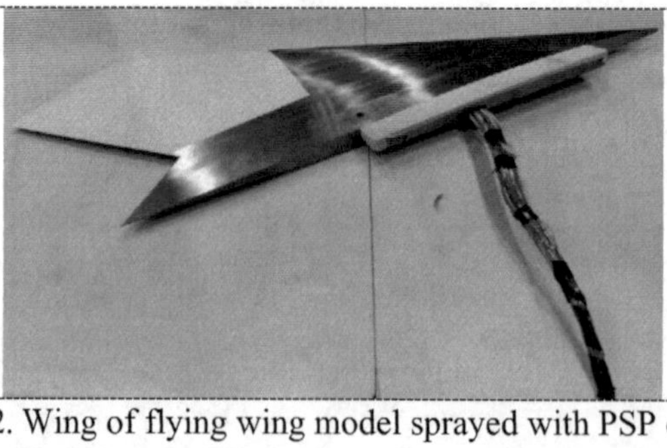

Figure 2. Wing of flying wing model sprayed with PSP coating.

The model is sprayed with self-made two-component fast response PC-PSP and placed inside the airtight chamber. The global calibration test simulates the real test environment by changing the ambient pressure and temperature, simulating the position and angle between test system and model. In this test, 11 groups tests were conducted, the environment air pressure is adjusted from high to low, and then from low to high, the adjust range is 0.06 MPa ~ 0.14 MPa, the air temperature is adjusted by the heater, 3 temperature points is carried out, namely, about 30°C, 40°C and 50°C. In the two-component PSP coating, green fluorescence is the reference channel and red fluorescence is the pressure channel.

Two PCO 2000 monochrome CCD cameras are used to photo PSP coating fluorescence images in this test. One of the two cameras is used as the green channel image acquisition equipment, the lens is equipped with BP532 filter, and the other takes the red channel image, the lens is equipped with BP650 filter. Two 10 W blue lasers manufactured by ZK Laser Company are used as the excitation light source to provide illumination for the PSP coating. The central wavelength of the excitation light source is 405nm. Figure 3 shows cameras and lasers at the test site, and the model and PSP coating picture in the test is shown in figure 4.

Before testing, turn on the excitation light source and warm up for minutes in order to ensure the stability of the laser. Then, adjust the temperature and pressure in the airtight chamber and maintain stability, capture 30 PSP images with two CCD cameras for each test condition, which takes about 10 seconds,until the data acquisition of all calibration States is completed. When the test is finished, turn off the excitation light source and capture dark images with two CCD cameras.

Figure 3. Cameras and light sources installed on the airtight chamber.

Figure 4. Model in calibration test.

The obtained PSP images, pressure and temperature data are processed by data process system from AVIC Aerodynamics Research Institute. The global calibration data processing is basically consistent with the convention priori laboratory calibration.Figure 5 shows the average intensity of red and green channel images from this calibration test, where the dot represents the red channel data, that is, the pressure signal intensity, and the plus sign represents the green channel data, that is, the reference signal intensity. Figure 6 shows the 11 calibration curves obtained in this test, and curves show good consistency.

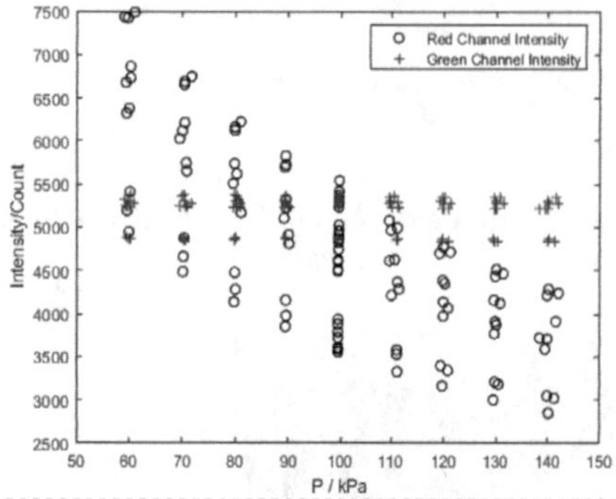


Figure 5. Brightness of calibration images.

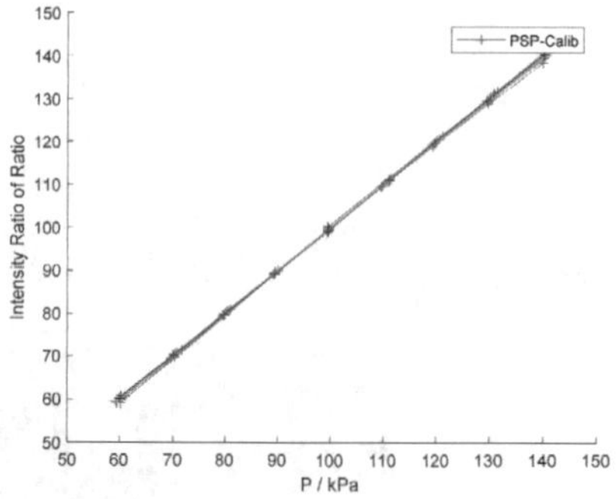


Figure 6. Calibration curves.

Due to the long illumination time to PSP coating in this calibration test, the photodegradation of PSP coating was corrected based on linear correction. And actually, the airtight chamber used in this test has the problem of poor sealing, and its air leakage rate is about 10pa/s, which may have a certain impact on the accuracy of calibration test data. However, in terms of this test, from the calibration curve, it has little impact on this test. Good consistency of 11 calibration curves, on the one hand, it shows that PSP coating has excellent pressure-sensitive and temperature-sensitive characteristics, on the other hand, good results have been achieved in this global calibration test.

## 4. Conclusion and Outlook

Based on the airtight chamber, using the real wind tunnel test model, under the condition of simulating the real test environment, the PSP coating global calibration technology is studied and verified, the calibration curve obtained in this test shows that the calibration technique can be used to get global calibration data of real test model PSP coating.

This technology can be popularized and applied to wind tunnels that can carry out similar calibration tests. Before or during the wind tunnel test, the PSP coating on the model surface can be globally calibrated, and reliable calibration data can be obtained under real test environmental conditions. In addition, this technology can also be used in small PSP coating areas, even get calibration coefficients of each pixel in calibration images, and then obtain proprietary calibration coefficients for each small area, so as to improve the accuracy of PSP data.

## References

[1] Liu T and Sullivan J (2005) *Pressure and temperature Sensitive Paint* (Springer-Verlag Berlin Heidelberg: Springer)

[2] Sugioka Y, Numata D, Asai K, Koike S, Nakakita K and Nakajima T (2018) Polymer/Ceramic Pressure-Sensitive Paint with Reduced Roughness for Unsteady Measurement in Transonic Flow *AIAA Journal* **56(6)** 2145–2156 doi:10.2514/1.j056304

[3] Merienne M, Le S, Lebrun F, Deleglise B and Sonnet D (2013) Transonic Buffeting Investigation using Unsteady Pressure-Sensitive Paint in a Large Wind Tunnel *51st AIAA Aerospace Sciences Meeting Including the New Horizons Forum and Aerospace Exposition*

[4] Woike M, Davis D, Clem M and Crafton J (2017) The Investigation of Shock Wave Boundary Layer Interactions Using Fast Pressure Sensitive Paint and Surface Stress Sensitive Film Measurement Techniques *33rd AIAA Aerodynamic Measurement Technology and Ground Testing Conference*

[5] Beberniss T and Ehrhardt D (2017) Dynamic Response of a Thin Panel Subjected to a Shock Wave Impingement and Thermal Buckling *AIAA Aviation*, Denver

[6] Egami Y, Sato Y and Konishi S (2019) Development of Sprayable Pressure-Sensitive Paint with a Response Time of Less Than 10μs *AIAA Journal* **1–6** doi:10.2514/1.j057434

[7] Sugioka Y, Nakakita K and Asai K (2016) Non-intrusive unsteady PSP technique for investigation of transonic buffetting ICAS-2016

[8] Schuster D, et al. (2016) Investigation of unsteady pressure-sensitive paint (uPSP) and a dynamic loads balance to predict launch vehicle buffet environments (No. NASA/TM-2016-219352)

[9] Walker M, Lee M and Mears L (2022) Cross-Validation of Computational and Experimental Distributed Surface Pressures on the Space Launch System *AIAA SCITECH 2022 Forum* (p. 0431)

[10] Ohmichi Y, Sugioka Y and Nakakita K (2022) Stable Dynamic Mode Decomposition Algorithm for Noisy Pressure-Sensitive-Paint Measurement Data. *AIAA Journal* **60**(**3**) 1965-1970

[11] Roozeboom N, *et al.*(2019) Development of Unsteady Pressure-Sensitive Paint Application on NASA Space Launch System *AIAA Aviation 2019 Forum* doi:10.2514/6.2019-3502

# Active vibration control of a piezoelectric composite laminate

**Hongxia Chen** [1,a], **Wenguang Liu** [2,b*]

[1]School of Aeronautical Manufacturing Engineering, Nanchang Hangkong University

[2]School of Aeronautical Manufacturing Engineering, Nanchang Hangkong University

[a]e-mail: 2357425741@qq.com

[*]Corresponding author: [b]liuwg14@nchu.edu.cn

**Abstract**—An active vibration control method of the piezoelectric composite laminated plate under the external disturbance is studied. Firstly, the dynamic model of the piezoelectric composite laminated plate is derived based on the classical laminated plate theory and the Hamilton's principle. And subsequently, a sliding mode observer is developed for the external disturbance, and the stability of the observer is verified by the Lyapunov equation, and also the LQR controller is designed based on the observed state. The particle swarm algorithm is adapted to optimize the weighting matrix. In the end, the impacts of the geometric parameters and the stacking sequences on the dynamic behavior of the piezoelectric composite material plate is analyzed. Results indicate that the sliding mode observer can accurately track the original state, and the vibration of the piezoelectric composite material plate can be suppressed effectively by the LQR control based on the particle swarm

## 1. Introduction

Composite materials are commonly used in the aerospace field for their advantages such as light weight, high strength and good heat resistance, and so on. Due to the long-term service in the vibration environment, fatigue damage of the composite structures is inevitably produced, which brings hidden dangers to the safety of the flying vehicle [1]. Therefore, the vibration control is significant to the composite structure design of aircrafts.

The vibration characteristic of composite materials has been studied by several researchers. Considering the damp-heat environmental factors, Zenkour et al. [2] discussed the effects of humidity and temperature on the vibration characteristics of laminated plate. Applying different cycle times and vibration stress, Longbiao et al. [3] analyzed the influences of composite properties and damage state on damping. Baharali et al. [4] predicted the frequency of delaminated multi-scale composite shells based on an approximate-analytical solution. Qi et al. [5] revealed increasing the fiber modulus can increase the frequency of the fiber-reinforced composite tube. Employing the equivalent spectrum method, Gao [6] predicted the fatigue life of composite laminated plate under random vibration loads by the vibration-based structural health monitoring method.

Piezoelectric materials are employed in active control of structural vibration due to their excellent power-to-electricity conversion properties. Kuriakose et al. [7] found that applying a voltage to the piezoelectric layer could restore the reduction of the flutter boundary caused by anisotropic damage in the composite sheet. Lu et al. [8] designed a robust controller and analyzed the effects of material damping, parameter uncertainty and aspect ratio on the control effect of piezoelectric composite cantilever plates in a hot and humid environment. Yao et al. [9] proposed a temperature feedback control method to realize active buckling control of piezoelectric composite laminate structures under thermal loading and uniaxial compression. Tian et al. [10] adopted the LQR control scheme and optimized the weighting matrix by a genetic algorithm to effectively suppress the vibration of piezoelectric composite beams. Quoc et al. [11] developed the constant velocity feedback strategy and discussed the influence of control gain on the active control of piezoelectric functionally graded carbon nanotubes. Xue et al. [12] adopted a delamination optimization method to make laminates with the highest critical buckling temperature. Wang et al. [13] designed an optimal control strategy to suppress the vibration, analyzed the effect of activation positions on the critical point of flutter in composite panels under the different piezoelectric actuators

Although researchers have made series of achievements in the research of vibration characteristics and piezoelectric active control of composite structures, few researchers have carried out active vibration control research on piezoelectric composite plates. In this paper, the dynamic model of a piezoelectric composite cantilever plate is established. A sliding mode observer is developed for external interference. Based on the observed state, the POS-LQR method is adapted to suppress the vibration of a piezoelectric composite plate, and the influence of stacking sequences on vibration control is discussed.

## 2. Motion equations

Consider a laminated plate with $n$ layers and dimension $a \times b \times h$ as shown in Fig. 1. The laminated plate is composed of an orthotropic composite material layer with piezoelectric sensor and actuator layer on the two surfaces. The thickness of the piezoelectric layers is $h_p$, the angle from the $x$-axis to the $x_1$-axis counterclockwise is assumed to be $\theta$.

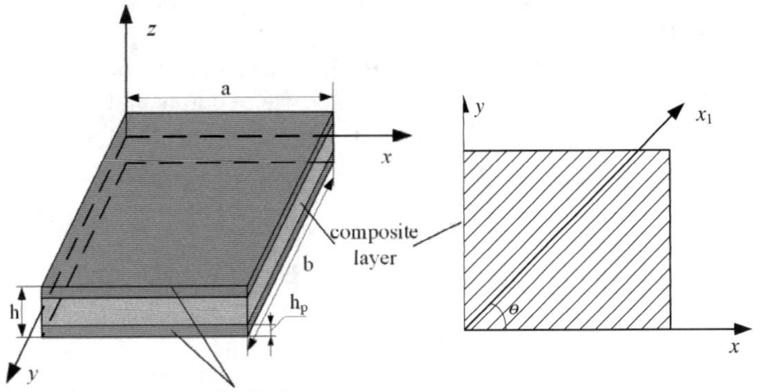


Fig. 1 The model diagram of piezoelectric composite laminated plate

Create a coordinate system $o$-$x$-$y$-$z$ with reference to the midplane of the plate. Based on the Kirchhoff hypothesis, the displacement field can be expressed by

$$\begin{aligned} u(x,y,z,t) &= u_0(x,y,t) - z\frac{\partial w_0(x,y,t)}{\partial x} \\ v(x,y,z,t) &= v_0(x,y,t) - z\frac{\partial w_0(x,y,t)}{\partial x} \\ w(x,y,z,t) &= w_0(x,y,t) \end{aligned} \tag{1}$$

Here, $u$, $v$ and $w$ are displacement components at any point, $u_0$, $v_0$ and $w_0$ are the displacement of the corresponding point on the system mid-surface, $t$ is time.

The linear constitutive equation of the $k$-th layer of the composite laminate can be written as [14]

$$\{\sigma\}^{(k)} = \{\overline{Q}_{ij}\}^{(k)} \{\varepsilon\}^{(k)} \tag{2}$$

Here, $\boldsymbol{\sigma}$ is the stress vector, $\boldsymbol{\varepsilon}$ is strain vector. $\overline{\boldsymbol{Q}}_{ij}$(i,j=1,2,6) is the transferred matrices by the stiffness matrix $\boldsymbol{Q}_{ij}$ of the composite laminate through the coordinate transformation matrix.

The constitutive equation of the piezoelectric layer can be presented as

$$\sigma_p = C_{pq}^D \varepsilon_q - h_{pk} D_k,\quad E_k = -h_{pk}\varepsilon_p + \beta_{kl}^\varepsilon D_l \tag{3}$$

where $\sigma_p$ is the stress vector, $\varepsilon_q$ is the strain vector, $C_{pq}^D$ is the piezoelectric material stiffness coefficient, $D_k$ is the electrical displacement vector, $h_{pk}$ is the piezoelectric stiffness constant, $\beta_{kl}^\varepsilon$ is the piezoelectric permittivity matrix, $E_k$ is the electric field vector, the subscript $p$, $q$=1,2,···,6， $k$, $l$=1,2,3.

According to the variational principle and Galerkin discretization, the kinetics motion of the piezoelectric composite laminated plate with cantilever boundary conditions can be expressed as

$$M\ddot{w} + Kw = F_a V_a + F_{ext} f \tag{4}$$

Here, $K$ and $M$ are the stiffness matrix and mass matrix of the laminated plate; $F_a$ and $F_{ext}$ are the external load matrix and the external force matrix; $w$ is the lateral displacement vector; $f$ is the external disturbance.

Substituting vector X=[$w$;$\dot{w}$] to Eq. (4), the state space model of the laminated plate can be written as

$$\dot{X} = AX + Bu + Lf,\quad y(t) = CX \tag{5}$$

where $A$, $B$ and $C$ are the input matrix, output matrix and transmission matrix. $L$ is the force matrix, $u$ is the input voltage, $y$(t) is the output voltage.

## 3. LQR controller based on particle swarm optimization

### 3.1. Sliding Mode Observer

The state estimation of the system with external disturbance is realized by using the sliding mode observer. The saturation function of the error between the system output and the observed output is used as the sliding mode control law, so as the sliding mode chattering can be suppressed. The sliding mode observer is given by [15]

$$\dot{\hat{X}} = A\hat{X} + Bu + H(y - \hat{y}) + L\upsilon,\quad \hat{y} = C\hat{X} \tag{6}$$

Here, $\hat{X}$ is the estimated state, $H$ is the observation gain, $\upsilon$ is the sliding mode control law, $\hat{y}$ is the observation output.

The state observation error is defined as $e = X - \hat{X}$, the sliding mode surface $s(t)$ and the control law $\upsilon$ is given by

$$s(t) = e_y = y - \hat{y} = Ce,\quad \upsilon = \varepsilon sat(s) \tag{7}$$

If the matrix pair ($A$, $C$) is detectable, there will exist a matrix $H$ as the $A-HC$ is stable, and for any symmetric positive definite matrix $Q_1$, there exists matrix $W_1$ and positive definite matrix $P_1$, so that Lyapunov equation (8) is satisfied

$$\begin{cases}(A - HC)^T P_1 + P_1(A - HC) = -Q_1 \\ L^T P_1 = W_1 C\end{cases} \tag{8}$$

The Lyapunov function $V$=0.5*$e^T P_1 e$ is considered. It follows that the time derivative of $V$ can be written as

$$\begin{aligned}\dot{V} &= \dot{e}^T P_1 e + e^T P_1 \dot{e} \\ &= e^T\left[(A - HC)^T P_1 + P_1(A - HC)\right]e + 2\left(L^T P_1 e f - L^T P_1 e\upsilon\right)^T \\ &= -e^T Q_1 e + 2W_1 e_y\left(f - \varepsilon sat(e_y)\right)\end{aligned} \tag{9}$$

where the$-e^T Q_1 e$ <0, if $e_y(f-\varepsilon sat(e_y))$<0, the $\dot{V}$ will be always negative, therefore

$$e_y > 0, \delta < \varepsilon; e_y < 0, \delta < \varepsilon \tag{10}$$

Here, $\delta$ is the boundary value for $f$, when the Eq. (10) is satisfied, then $\dot{V}<0$, the systematic observation error is asymptotically stable.

### *3.2. LQR Control Based on Particle Swarm Optimization*

According to the estimated state of the sliding mode observer, the LQR controller is adapted to gradually stabilize for the system state. The quadratic objective function is given by

$$J = \frac{1}{2}\int_0^\infty \left[ \hat{\boldsymbol{X}}^T \boldsymbol{Q}\hat{\boldsymbol{X}} + u^T \boldsymbol{R} u \right] \mathrm{d}t \tag{11}$$

where $\boldsymbol{Q}$ is the state weighting matrix, $\boldsymbol{R}$ is the control energy weighting matrix. The optimal control law that will minimize $J$ is described by

$$u(t) = -\boldsymbol{R}^{-1}\boldsymbol{B}^T\boldsymbol{P}\hat{\boldsymbol{X}} = -\boldsymbol{K}\hat{\boldsymbol{X}} \tag{12}$$

Here, $\boldsymbol{K}$ is the optimal control feedback gain matrix, $\boldsymbol{P}$ is the non-negative definite symmetric solution of the Riccati Equation

$$\boldsymbol{PA} + \boldsymbol{A}^T\boldsymbol{P} - \boldsymbol{PBR}^{-1}\boldsymbol{B}^T\boldsymbol{P} + \boldsymbol{Q} = 0 \tag{13}$$

The selection of weighting matrixes $\boldsymbol{Q}$ and $\boldsymbol{R}$ in LQR control is very important, which has a great impact on the control performance. Direct selection cannot obtain a good control effect. Therefore, the particle swarm optimization algorithm is adopted optimize weighting matrix $\boldsymbol{Q}$ and $\boldsymbol{R}$. The minimize quadratic objective function is selected as the fitness function, the update formula of particle velocity and position can be expressed as

$$\begin{aligned} V_i^{kg+1} &= \varpi(t) \times V_i^{kg} + c_1 r_1 \left(p_i^{kg} - X_i^{kg}\right) + c_2 r_2 \left(p_{ilocal}^{kg} - X_i^{kg}\right) \\ X_i^{kg+1} &= X_i^{kg} + V_i^{kg+1} \end{aligned} \tag{14}$$

Here, $\varpi(t)$ is inertia weight, $c_1$ and $c_2$ are local learning factor and global learning factor. $r_1$ and $r_2$ are random numbers in interval [0-1]. $V_{gi}^{k}$ and $X_{gi}^{k}$ are particle velocity and position at the current time. $V_i^{kg+1}$ and $X_i^{kg+1}$ are particle velocity and position at the next time. $p_{gi}^{k}$ is the optimal particle position in the current iteration, $p_{ilocal}^{kg}$ is the global optimal particle position.

The flow chart of particle swarm optimization LQR weight matrix is shown as

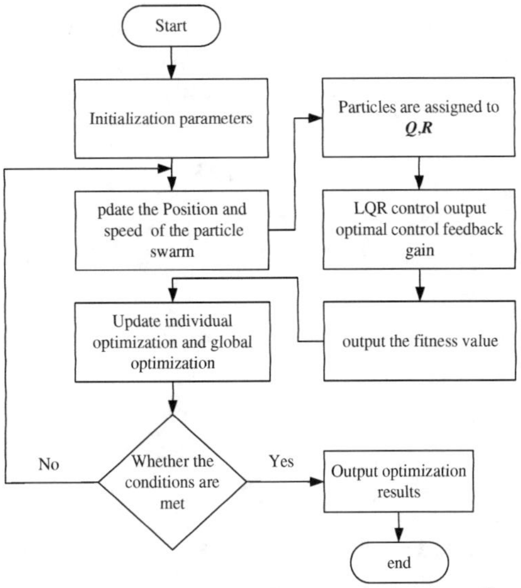


Fig. 2 Flow chart of particle swarm optimization algorithm

## 4. Results and Discussion

### 4.1. Vibration Characteristics

Considering the laminated plate structure shown in Figure 1, the laminated core is graphite/epoxy composite and the piezoelectric material is PZT-G_1195N, the material properties of laminated plate are shown in Table 1. The number of layers is assumed as 6, the natural frequencies of laminated plate are calculated theoretically. The results as shown in table 2 are basically consistent with COMSOL simulation, which verify the accuracy of the dynamic model. It can be seen from table 2 that the natural frequencies of laminated plate are disparity with the different stacking sequences and fiber angles, which are caused by different stiffness.

TABLE 1. MATERIAL PARAMETERS OF PIEZOELECTRIC COMPOSITE PLATE

| parameter | Graphite/Epoxy | PZT-G_1195N |
|---|---|---|
| Length(m) | $a_f$=1 | $a_p$=1 |
| Width(m) | $b_f$=1 | $b_p$=1 |
| Thickness(m) | $h_f$=0.002 | $h_p$=0.001 |
| Young's modulus (GPa) | $E_1$=181, $E_2$=10.3 | $E_p$=63 |
| Shear modulus (GPa) | $G_{12}$=7.17 | |
| Density(kg/m$^3$) | $\rho_f$=1580 | $\rho_p$=7600 |
| Poisson ratio | $v_{12}$=0.28 | $v_P$=0.3 |
| Piezoelectric stiffness constant(V/m) | -- | $h_{31}=h_{32}=-39.37\times10^8$ |
| Dielectric impermeability(m/F) | -- | $\beta_{33}=6.667\times10^7$ |

TABLE 2. NATURAL FREQUENCIES

| stacking sequences | frequency (Hz) | COMSOL | stacking sequences | frequency (Hz) | COMSOL |
|---|---|---|---|---|---|
| $[0^\circ,10^\circ]_s$ | 11.197 | 11.295 | $[0^\circ,10^\circ]_2$ | 11.111 | 11.099 |
| $[0^\circ,45^\circ]_s$ | 10.856 | 10.967 | $[0^\circ,45^\circ]_2$ | 9.6658 | 9.3546 |
| $[0^\circ,90^\circ]_s$ | 10.719 | 10.882 | $[0^\circ,90^\circ]_2$ | 9.0313 | 8.8774 |

The symmetrical laminates($p/0/\theta/\theta/0/p$) and antisym-metric laminates($p/0/\theta/0/\theta/p$) are investigated. Figure 3 illustrates the influence of aspect ratio and fiber angles on the frequency under different stacking sequences. From this figure, the natural frequency is smaller as the aspect ratio is larger. This is because as the aspect ratio increases that causes the mass of the laminated plate to increase and the stiffness to decrease, so that the natural frequency decreases. As the aspect ratio increases, the influence of fiber angle on frequency becomes smaller. Fig. 4 shows the frequency is different as thickness of piezoelectric layer under these stacking sequences. When the fiber angle of symmetrical laminates as same as antisymmetric laminates, as the thickness of the piezoelectric layer increases, the frequency first decreases sharply and then increases slowly.

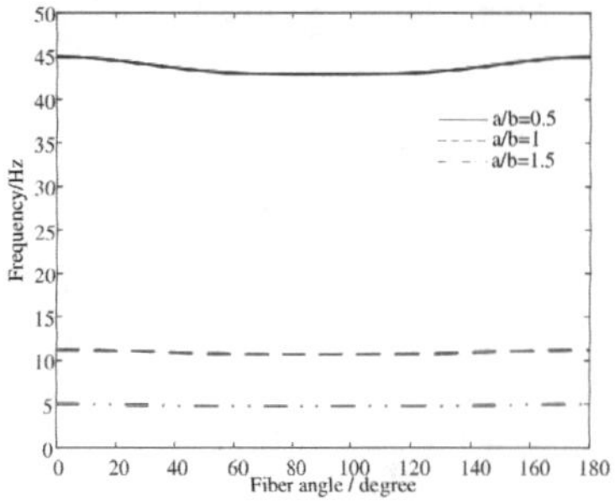


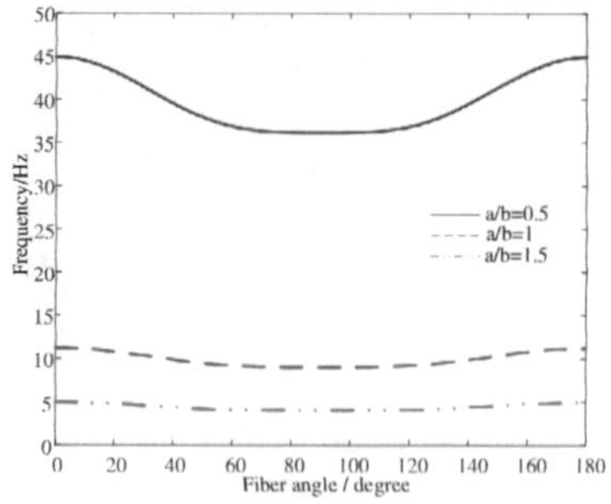


(a) symmetrical laminates (p/0/θ/θ/0/p) (b) antisymmetric laminates (p/0/θ/0/θ/p)

Fig. 3 Influence curve of aspect ratio on frequency

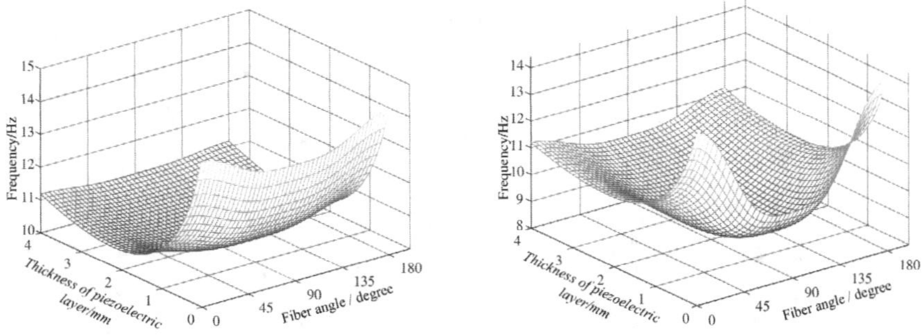


(a) symmetrical laminates (p/0/θ/θ/0/p) (b) antisymmetric laminates (p/0/θ/0/θ/p)
Fig. 4 Influence curve of piezoelectric layer thickness on frequency of cantilever plate

### *4.2. Vibration Control*

The sliding mode observer is used to estimate the state of the laminated plate with external interference. Take symmetric laminates (*p*/0/90/90/0/*p*) for example, set $f$=0.1sin(2π$\Omega t$), $\varepsilon$=3, the resonance frequency $\Omega$=10.719. Fig. 5-6 show that the amplitude of the laminated plate increases gradually during undamped resonance, and the observed displacement curve is basically consistent with the original system, which proves the sliding mode observer can effectively track the original system. Figure 8 shows that when there is external interference in the system, the sliding mode observation error can converge to a certain range, which is greatly different from the observed value, and the system is gradually stable.

Based on the sliding mode observer, the vibration of the piezoelectric composite laminated plate is suppressed by using POS-LOR control. The learning factors $c_1$=1.5, $c_2$=1.7, the number of particle swarm is 30, and the maximum number of iterations is Iter=100. In order to obtain better algorithm performance, the inertia weight decreases linearly from 0.9 to 0.1, so that it has good global search ability and local search ability before and after iteration. Follow the procedure shown in Fig. 2. The results in Fig 7 show that the appropriate weighting matrices $\boldsymbol{Q}$ and $\boldsymbol{R}$ are find quickly with the particle swarm optimization algorithm. Whether they are symmetric laminates or antisymmetric laminates, so as the optimal solution of the objective function and the best control effect is obtained. Figure 8-9 show that the amplitude of piezoelectric composite cantilever plate can be effectively suppressed with POS-LQR control.

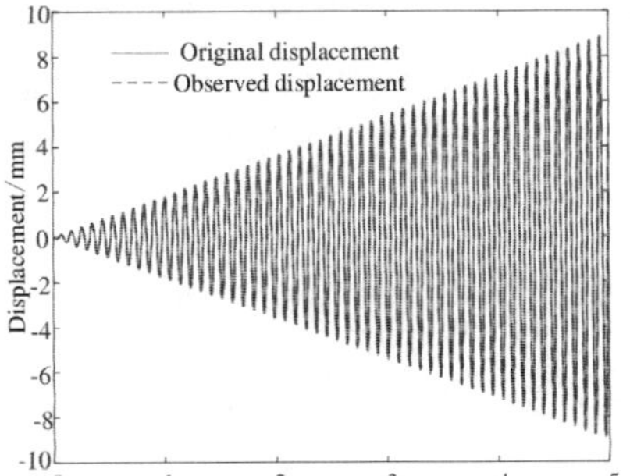


Fig. 5 Displacement curve of sliding mode observation

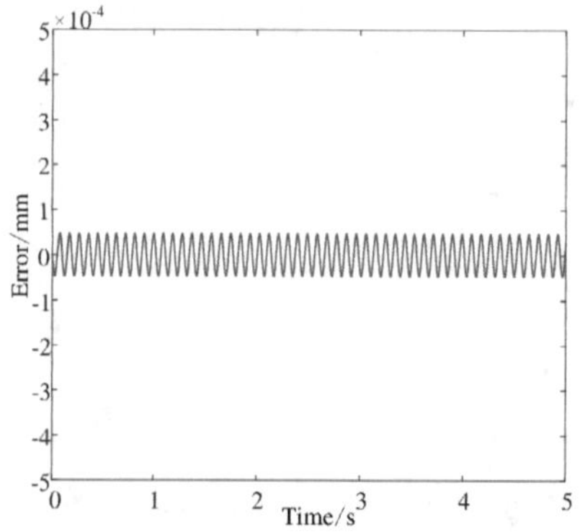

Fig. 6 Sliding mode observation output error

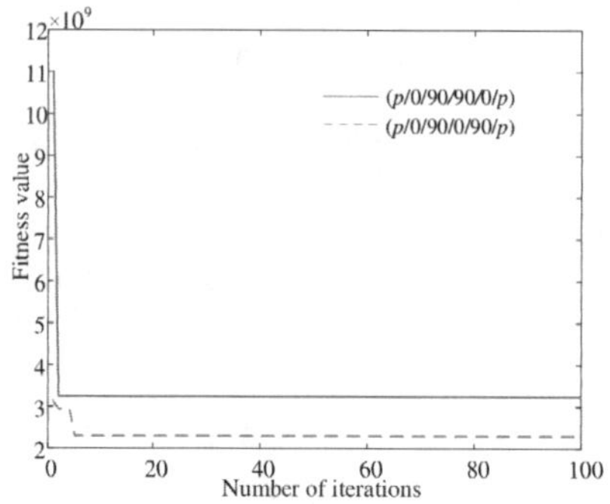

Fig. 7 Fitness function optimization process

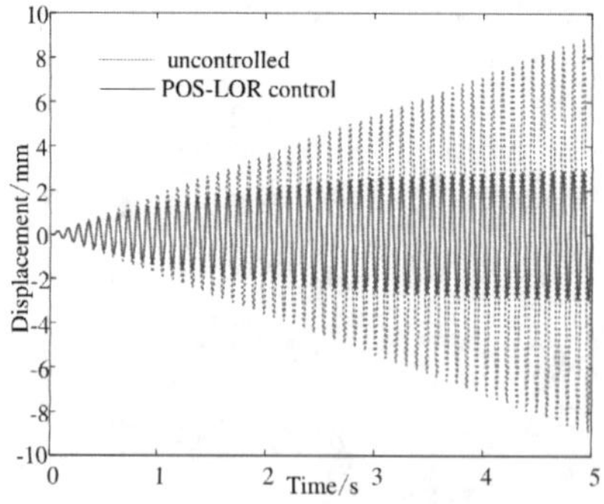

Fig. 8 POS-LQR control diagram of symmetrical laminates (p/0/90/90/0/p)

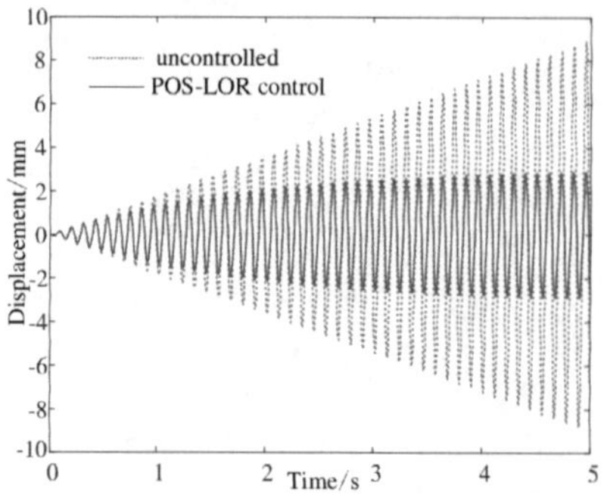

Fig. 9 POS-LQR control diagram of antisymmetric laminates (p/0/90/0/90/p)

## 5. Conclusion

The active vibration control of the piezoelectric composite laminated plate under external interference is investigated. The kinetic equation of the piezoelectric composite laminated plate is established. The sliding mode observer is used to track the original system state. The vibration of cantilever plate is controlled by POS-LQR method. Finally, the effects of geometric factors and stacking sequences on vibration characteristics are discussed, and the effectiveness of the controller is verified by SIMULINK simulation. The conclusions can be written as:

1) As the fiber angle increases, the natural frequency of the piezoelectric composite laminated plate decreases first and then increases. As the aspect ratio increases, the mass of the laminated plate becomes larger and the stiffness is smaller, so as the natural frequency decreases.

2) As the thickness of the piezoelectric layer increases, the frequency first decreases sharply and then increases slowly. Choosing the appropriate geometric size and stacking sequences can not only improve the stiffness of the piezoelectric composite plate, but also effectively avoid the influence of resonance.

3) The sliding mode observer is designed for external disturbance, which has good tracking performance and stability. The LQR controller is designed by using the observed state. The particle swarm optimization algorithm is used to optimize the weighting matrices $\boldsymbol{Q}$ and $\boldsymbol{R}$, which can improve the calculation speed and effectively control the vibration of the piezoelectric composite laminated plate.

## References

[1] Mahato, P.K., Maiti, D.K. (2010) Aeroelastic analysis of smart composite structures in hygro-thermal environment[J]. Composite Structures, 92(4): 1027-1038.
[2] Zenkour, A.M., El-Shahrany, H.D. (2021) Hygrothermal vibration of adaptive composite magnetostrictive laminates supported by elastic substrate medium[J]. European Journal of Mechanics - A/Solids, 85(19): 104140.
[3] Li, L. (2021) A cyclic-dependent vibration damping model of fiber-reinforced ceramic-matrix composites[J]. Proceedings of the Institution of Mechanical Engineers Part C Journal of Mechanical Engineering Science, 235(19): 4283-4295.
[4] Baharali, A.A., Yazdi, A.A. (2020) Analytical approach to study the vibration of delaminated multi-scale composite cylindrical shells[J]. Polymer Composites, 258: 153-172.
[5] Qi, L.L., Li, C.H., Yu, X.D., et al. (2021) Effect of reinforced fibers on the vibration characteristics of fibers reinforced composite shaft tubes with metal flanges[J]. Composite Structures, (1): 114460.
[6] Gao, D.Y., Yao, W.X., Wen, W.D., et al. (2021) Equivalent spectral method to estimate the fatigue life of composite laminates under random vibration loadings[J]. Mechanics of Composite Materials, 57(1): 101-114.
[7] Kuriakose, V.M., Sreehari, V. (2021) Study on passive flutter control of damaged composite laminates with piezoelectric patches employing finite element method[J]. Composite Structures, 269(399): 114021.
[8] Lu, S.F, Jiang, Y., Zhang, W., et al. (2020) Vibration suppression of cantilevered piezoelectric laminated composite rectangular plate subjected to aerodynamic force in hygrothermal environment[J]. European Journal of Mechanics - A/Solids, 83: 104002.
[9] Zhang, Y., Xue, Y., Yuan, W., et al. (2021) Active control of thermo-mechanical buckling of composite laminated plates using piezoelectric actuators[J]. Acta Mechanica Solida Sinica, 34(3): 1-12.
[10] Tian, J., Guo, Q., Shi, G. (2020) Laminated piezoelectric beam element for dynamic analysis of piezolaminated smart beams and GA-based LQR active vibration control[J]. Composite Structures, 112480.
[11] Quoc, T.H., Tham, V.V., Tu, T.M. (2021) Active vibration control of a piezoelectric functionally graded carbon nanotube-reinforced spherical shell panel[J]. Acta Mechanica, 232(3): 1005-1023.
[12] Xue, Y., Zhang, Y., Li J.Q. (2021) Optimization of thermal buckling control for composite laminates

with PFRC actuators using trigonometric shear deformation theory[J]. Journal of Mechanical Science and Technology, 35(1): 257-266.
[13] Wang, Z., Chen, H., Wang, G., et al. (2020) Prediction and active suppression of flutter in composite panel based on eigenvector orientation method - ScienceDirect[J]. Composite Structures, 262: 113422.
[14] Reddy, J.N. (1999) On laminated composite plates with integrated sensors and actuators[J]. Engineering Structures, 21(7): 568-593.
[15] Yan, X.G., Edwards, C. (2008) Robust sliding mode observer-based actuator fault detection and isolation for a class of nonlinear systems[J]. International Journal of Systems Science, 39(4): 349-359.

# Dynamic Response Analysis of the Aircraft Exposed to the Lateral Travelling Gust

**Zhiwei Jing[1], and Chu Tang[2*]**

[1] The First Aircraft Institute, Aviation Industry Corporation of China, Ltd., Xi'an, Shaanxi, 710089, China

[2]Unmanned System Research Institute, Northwestern Polytechnical University, Xi'an, Shaanxi, 710072, China

[*]Corresponding author's e-mail: tc@nwpu.edu.cn

**Abstract.** The travelling gust from detonation is an interesting and severe excitation for a flying aircraft, and it is characterized by high emanating speed and takes strong energy. Previous investigations on the dynamic response characteristics of the aircraft in the tail-on or head-on travelling gust have been carried out. This paper focuses on the lateral travelling gust and its influence on the dynamic response of the aircraft. Based on conventional flight dynamics equations, the dynamics modelling and simulation technique of the rigid aircraft exposed to the lateral travelling gust is established. By employing the static pressure distributions, one quasi-steady aerodynamics computation method is presented to determine the aerodynamic force and moment on the aircraft in the travelling gust. Simulation results show that the approach presented in this paper is effective and can deal with the engineering requirement. Results also indicate that the enveloping process of the travelling gust increases the aerodynamic moment of the $x$-axis significantly and subsequently influences the dynamic responses along the $x$-axis.

**Nomenclature**

| | | | |
|---|---|---|---|
| $V_A$ | : flight velocity, $m/s$ | $\boldsymbol{M}, \boldsymbol{m}$ | : moment vector, $N{\cdot}m$ |
| $V_S$ | : sound speed, $m/s$ | $\boldsymbol{r}$ | : location vector, $m$ |
| $V_\triangle$ | : enveloping velocity, $m/s$ | $\boldsymbol{n}$ | : normal vector |
| $U$ | : travelling gust velocity, $m/s$ | $s$ | : area, $m^2$ |
| $t$ | : time, $s$ | $p, q, r$ | : components of the angular velocity, (°)/$s$ |
| $C_p$ | : pressure coefficient | $u, v, w$ | : components of the velocity vector, $m/s$ |
| $\alpha$ | : angle of attack, (°) | $\psi, \theta, \phi$ | : Euler angles, (°) |
| $\beta$ | : sideslip angle, (°) | | |
| $\rho$ | : air density, $kg/m^3$ | | |
| $\bar{q}$ | : dynamic pressure, $N/m^2$ | | |
| $\boldsymbol{F}, \boldsymbol{f}$ | : force vector, $N$ | | |

**Subscripts**

| | | | |
|---|---|---|---|
| $i$ | : serial number of the grid | $x, y, z$ | : x-axis, y-axis, z-axis |

## 1. Introduction

The travelling gust caused by detonation is a particular excitation, which propagates at supersonic or sonic speed and has stronger disturbance energy than traditional gust. The in-flight aircraft encountering the travelling gust will demonstrate severe dynamic response to the incremental angle of attack and the airload. The travelling gust could significantly impact on the integrity and safety of the aircraft structure, and even endanger the flight safety. Therefore, it is necessary to carry out the study on the dynamic response prediction of the aircraft in the travelling gust.

Basically, the detonation field is a strong excitation. It is complicated and challenging to establish the coupling of the aircraft with a detonation. As regards the weak field, previous research has shown that the detonation wave can be represented as a travelling gust by neglecting the associated overpressure, and it emanates at sonic speed[1-2].

For the investigation of aircraft dynamic response to the travelling gust, one key challenge is the aircraft aerodynamics modelling method. Previous researchers often treated the travelling gust as the downwash effect upon the aircraft surface, [3] based on which the panel methods were established for calculating the aerodynamic load of the aircraft exposed to the travelling gust[4-6]. Some researchers used the moving vortex finite elementary solution to solve the initial boundary value problem of unsteady aerodynamic equations, while the others employed the circulation theory to directly determine the steady force and moment on the wing. However, these methods are linear aerodynamics calculation approaches, which could not deal with the nonlinear aerodynamic characteristics caused by stall or shock wave.

In terms of the aircraft dynamic response to the travelling gust, the studies on modelling and simulation technique have been conducted and some interesting results have been obtained, and they tend to focus on the tail-on travelling gust which comes from behind and below the aircraft[7]. However, the dynamic response analysis of the aircraft exposed to the lateral travelling gust, which comes from the lateral side of the wing, remains as a concern and it needs to be investigated in detail.

The primary objective of this paper is to establish the dynamics modelling and simulation technique for the rigid aircraft in the lateral travelling gust. Conventional flight dynamics equations are utilized to establish the dynamic response technique. By using the pressure distribution data from the wind tunnel or by CFD, one quasi-steady aerodynamics computation method, which can better deal with the enveloping process of the travelling gust, is derived and presented, and it is expected to show the nonlinear characteristics of the aerodynamic loads since the pressure distributions have nonlinear characteristics.

A large civil aircraft is chosen as the example aircraft to demonstrate the method presented in this paper. First, some special constraints are imposed on the example aircraft so as to verify the aerodynamics calculation method. Then, the effect of the travelling gust enveloping on the aircraft aerodynamics is investigated by designing two different aerodynamics calculation models, the first one considers the enveloping process of the travelling gust to the aircraft， the other one neglects the enveloping process. The comparison between the simulation results reveals that the enveloping process of the travelling gust has an important effect on the aerodynamic rolling moment of the aircraft, and this will increase the inertia load of the outer wing. This phenomenon should get much more attention in the engineering design.

## 2. Aerodynamics Modelling

Aerodynamics modelling technique is the key research in this paper. First, the relationship between the lateral travelling and the flying aircraft should be defined, such as enveloping velocity and angle of direction. Then, create triangle gridding on the whole aircraft. Since the travelling gust is an excitation field in the time domain, each grid will be influenced by different gust velocity, based on which the local variables of each grid can be updated. Finally, the pressure coefficient of each grid is interpolated by using the basic pressure distribution data and the total airload of the aircraft is obtained by summing all the grids.

### *2.1. Travelling gust*

When travelling gust approaches the flying aircraft from the starboard side or port side, shown in figure 1, it will envelop the aircraft from one wingtip to another. Such a disturbance process could produce a significant rolling moment on the aircraft. So, it is interesting to investigate the aircraft dynamic response caused by the lateral travelling gust.

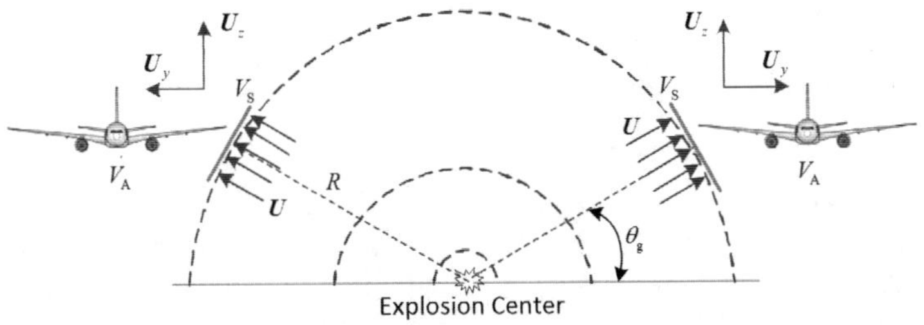


Figure 1. Illustration of the aircraft exposed to the lateral travelling gust.

Generally, travelling gust induced by detonation can be treated as a two-dimension gust field with horizontal gust and vertical gust. Since this paper focuses on the situation that the travelling gust acts on the left wing first, the travelling gust filed will be denoted by $U_y$ and $U_z$ which are demonstrated in figure 1.

The enveloping speed of the lateral travelling gust to the aircraft is defined as equation (1), where $\theta_g$ is the angle between the aircraft and the explosion centre.

$$V_\Delta = \frac{V_S}{\cos\theta_g} \quad (1)$$

*2.2. Determination of grid variables*

One important process is the gridding of the aircraft, which is used to calculate the aerodynamic loads on the aircraft. Figure 2 depicts some triangular grids on the wing. Each grid has five geometrical variables shown in figure 2, including the location vector $\boldsymbol{r}_i$, normal vector $\boldsymbol{n}_i$, area $s_i$ and its dimensionless coordinates ($\overline{y_i}$, $\overline{x_i}$).

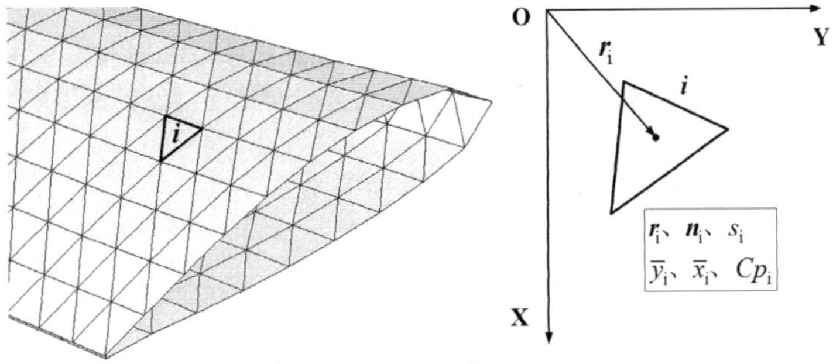


Figure 2. Schematic of grid variables.

The time axis of the travelling gust is chosen as the simulation time axis. Considering the time lag, the exact time of the $i^{th}$ grid in the travelling gust field can be determined as equation (2), where $t$ is the global simulation time, $\Delta y_i$ is the lateral distance between the $i^{th}$ grid and the left wingtip.

$$t_i = t - \frac{\Delta y_i}{V_\Delta} \quad (2)$$

Based on the original travelling gust field, the travelling gust velocity acting on the $i^{th}$ grid can be obtained through time interpolation by using $t_i$, and they are written as $U_y(t_i)$ and $U_z(t_i)$.

The body-axes components of the aircraft velocity vector are given as $u$, $v$ and $w$. Taking the travelling gust velocity into account, the corrected velocity vector of the $i^{th}$ grid at the global time $t$ takes the form equation (2), where $L_{bg}$ is transformation matrix from the ground-axes to the body-axes, which can be found in [8].

$$\begin{bmatrix} u_i(t) \\ v_i(t) \\ w_i(t) \end{bmatrix} = \begin{bmatrix} u(t) \\ v(t) \\ w(t) \end{bmatrix} + L_{\mathrm{bg}} \begin{bmatrix} 0 \\ -U_y(t_i) \\ +U_z(t_i) \end{bmatrix} \tag{3}$$

Given that the lateral travelling gust comes from the left wingtip, equation (2) and equation (3) indicate that the farther away from the left, the later the grid will be influenced by the gust. If the travelling gust has not acted on the aircraft or has left the aircraft, the travelling gust velocity $U_y(t_i)$ and $U_z(t_i)$ should be set to zero.

Then the local velocity of the $i^{th}$ grid is defined as

$$V_i(t) = \sqrt{u_i^2 + v_i^2 + w_i^2} \tag{4}$$

And the local dynamic pressure of the $i^{th}$ grid is given by equation (5), and, where $\rho$ is the air density.

$$\bar{q}_i(t) = \frac{1}{2}\rho V_i^2 \tag{5}$$

The angle of attack and the sideslip angle of the $i^{th}$ grid are determined by

$$\alpha_i(t) = atan\left[\frac{w_i(t)}{u_i(t)}\right] \quad , \quad \beta_i(t) = atan\left[\frac{v_i(t)}{V_i(t)}\right] \tag{6}$$

*2.3. Aerodynamics calculation*

To calculate the aerodynamic force and moment on the aircraft encountering the lateral travelling gust, it is necessary to obtain the pressure distribution data $C_p$ of the aircraft from the wind tunnel or by CFD. The pressure distributions are static data rather dynamic data.

Then, by using the flight velocity $M$, flight height $H$, trimmed deflection $\delta e$, $\alpha_i$, $\beta_i$ and the geometrical variables of the grid, the pressure coefficient at the centre of the $i^{th}$ grid is given by interpolation as follows

$$C_{pi}(t) = C_p(M, \beta_i, \alpha_i, \bar{y}_i, \bar{x}_i) + \Delta C_p(M, \delta e, \alpha_i, \bar{y}_i, \bar{x}_i) \tag{7}$$

The $i^{th}$ grid produces the aerodynamic force and moment for the aircraft, given by equation (8), where $\boldsymbol{f}_i$(t) and $\boldsymbol{m}_i$(t) should be calculated in the body-axes.

$$\begin{cases} \boldsymbol{f}_i(t) = \bar{q}_i(t) \cdot C_{pi}(t) \cdot s_i \cdot \boldsymbol{n}_i \\ \boldsymbol{m}_i(t) = \boldsymbol{r}_i \times \boldsymbol{f}_i(t) \end{cases} \tag{8}$$

Finally, the aerodynamic forces and moments acting on the complete aircraft can be obtained through integral of all grids, they are given by equation (9) and equation (10), respectively. Where $N$ is the number of grids of the complete aircraft.

$$\boldsymbol{F}_{\mathrm{A}}(t) = \begin{bmatrix} F_{x\mathrm{A}}(t) \\ F_{y\mathrm{A}}(t) \\ F_{z\mathrm{A}}(t) \end{bmatrix} = \sum_{i=1}^{N} \boldsymbol{f}_i(t) \tag{9}$$

$$\boldsymbol{M}_{\mathrm{A}}(t)=\begin{bmatrix} M_{x\mathrm{A}}(t) \\ M_{y\mathrm{A}}(t) \\ M_{z\mathrm{A}}(t) \end{bmatrix}=\sum_{i=1}^{N}\boldsymbol{m}_i(t) \tag{10}$$

It should be emphasized that the aerodynamics calculation technique presented in this paper has taking the enveloping process into account, and it aims to solve practical problems, and it is an engineering method.

## 3. Flight Dynamics

### *3.1. Kinetic equations*

The flight dynamics equations are established in the body-axes, and they come from [8]. The force equations of the aircraft are as equation (11), where $p$, $q$ and $r$ are the components of the angular velocity, $\psi$, $\theta$ and $\phi$ are the Euler angles, $F_{xf}$, $F_{yf}$ and $F_{zf}$ are the components of the external forces including the aerodynamic force vector in equation (9) and the engine thrust, $m$ is the aircraft mass, $g$ is the gravity acceleration.

$$\begin{cases} \dfrac{du}{dt}=-g\sin\theta+rv-qw+\dfrac{F_{xf}}{m} \\ \dfrac{dv}{dt}=g\sin\phi\cos\theta-ur+pw+\dfrac{F_{yf}}{m} \\ \dfrac{dw}{dt}=g\cos\phi\cos\theta+uq-pv+\dfrac{F_{zf}}{m} \end{cases} \tag{11}$$

The moment equations of the aircraft take the form equation (12), where $I_{xx}$, $I_{yy}$, $I_{zz}$ and $I_{xz}$ are the components of the inertia matrix of the aircraft, $L$, $M$ and $N$ are the components of the external moments including the aerodynamic moment vector in equation (10), the moment caused by the engine thrust and the aerodynamic damping moment.

$$\begin{cases} \dfrac{dp}{dt}=\dfrac{\left(I_{xx}+I_{zz}-I_{yy}\right)I_{xz}}{I_{xx}I_{zz}-I_{xz}^2}pq+\dfrac{I_{yy}I_{zz}-I_{zz}^2-I_{xz}^2}{I_{xx}I_{zz}-I_{xz}^2}qr+\dfrac{I_{zz}}{I_{xx}I_{zz}-I_{xz}^2}L+\dfrac{I_{xz}}{I_{xx}I_{zz}-I_{xz}^2}N \\ \dfrac{dq}{dt}=\dfrac{I_{xz}}{I_{yy}}\left(r^2-p^2\right)+\dfrac{I_{zz}-I_{xx}}{I_{yy}}pr+\dfrac{M}{I_{yy}} \\ \dfrac{dr}{dt}=\dfrac{I_{xx}^2+I_{xz}^2-I_{xx}I_{yy}}{I_{xx}I_{zz}-I_{xz}^2}pq+\dfrac{\left(I_{yy}-I_{zz}-I_{xx}\right)I_{xz}}{I_{xx}I_{zz}-I_{xz}^2}qr+\dfrac{I_{xz}}{I_{xx}I_{zz}-I_{xz}^2}L+\dfrac{I_{xx}}{I_{xx}I_{zz}-I_{xz}^2}N \end{cases} \tag{12}$$

To determine the Euler angles, the kinematic equations of the aircraft are required and given as follows

$$\begin{cases} \dfrac{d\psi}{dt}=q\left(\sin\phi/\cos\theta\right)+r\left(\cos\phi/\cos\theta\right) \\ \dfrac{d\theta}{dt}=q\cos\phi-r\sin\phi \\ \dfrac{d\phi}{dt}=p+q\sin\phi\tan\theta+r\cos\phi\tan\theta \end{cases} \tag{13}$$

*3.2. Simulation flowchart*

Figure 3 presents the dynamic response simulation process of the aircraft undergoing the travelling gust. Generally, this process is similar to that of a common flight dynamics simulation. The Fourth-Order Runge-Kutta method is utilized to solve the nonlinear ordinary differential equations of the flight dynamics. The key calculation in the whole simulation process is to determine the aerodynamic forces due to the lateral travelling gust, which is shown in the dashed box of figure 3. It should be pointed out that the calculation method produces quasi-steady aerodynamic forces rather than unsteady aerodynamic forces.

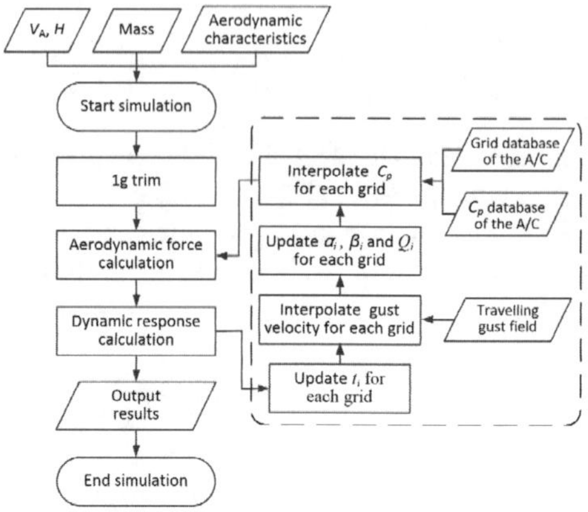


Figure 3. Flowchart of dynamic response simulation.

When considering the enveloping process, the cost of the aerodynamics computation will be considerable as a result of repeating interpolation and sum operations of the all the grids. So, the dynamic response simulation often takes relatively long time.

## 4. Example and Results

*4.1. Simulation parameters*

A large twin-engine civil transport aircraft, as shown in figure 4, is chosen as an example to demonstrate the approach described above. The aircraft has a 61-meter wingspan.

In the following simulation, the flight velocity $M$ is 0.6, flight height $H$ is 3000 m, $\theta g$ is 13° as illustrated in figure 1.

To have a better comparison and illustrate the problem, two calculation methods for the aircraft aerodynamics have been conducted:

1) The enveloping process of the lateral travelling gust to the aircraft is considered, which is presented in this paper and denoted with 'with enveloping'.

2) The enveloping process is not considered, which is the same as the traditional gust and denoted with 'without enveloping', actually the whole aircraft is supposed to experience the same gust velocity at the same time.

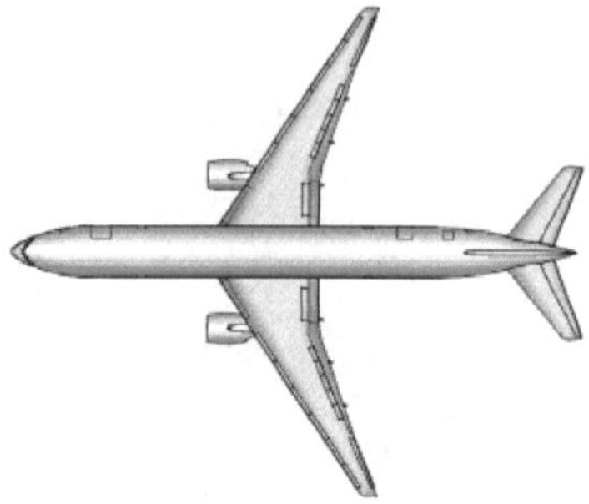

Figure 4. Schematic of the example aircraft.

### *4.2. Verification of aerodynamics calculation*

In order to verify the aerodynamics calculation technique of a rigid aircraft exposed to the lateral travelling gust, some special treatments are needed for the example aircraft: 1) the freedoms of rigid body motion are constrained so as to eliminate the contribution of the aircraft's own motions to the aerodynamic forces, 2) the simulation parameters are from chapter 4.1, 3) a lateral travelling gust filed with the step profile, as shown in figure 5, is imposed on the aircraft.

The lift force coefficient and the rolling moment coefficient are illustrated in figure 6 and 7, respectively. The figures indicate that the aerodynamic results denoted with 'without enveloping' reach the steady values immediately, while the results denoted with 'with enveloping' arrive at the steady values after a dynamic process.

As shown in figure 6, the lift force coefficient reaches its peak when the step-profile travelling gust surrounds the aircraft completely.

Unlike the lift force coefficient, the rolling moment coefficient shown in figure 7 reaches its peak as the travelling gust just surrounds the left wing and does not arrive at the right wing yet, in which case the step-profile travelling gust can generate the maximum aerodynamic rolling moment. It should be noticed that the overshoot of the aerodynamic rolling moment is remarkably great.

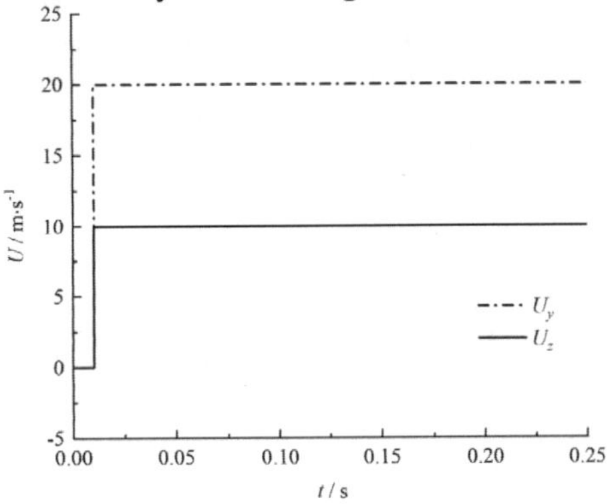


Figure 5. The lateral travelling gust with step-profile.

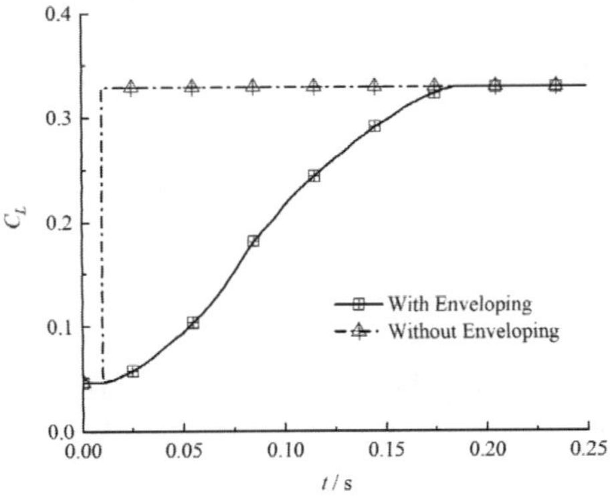


Figure 6. Time history of aircraft lift force coefficient.

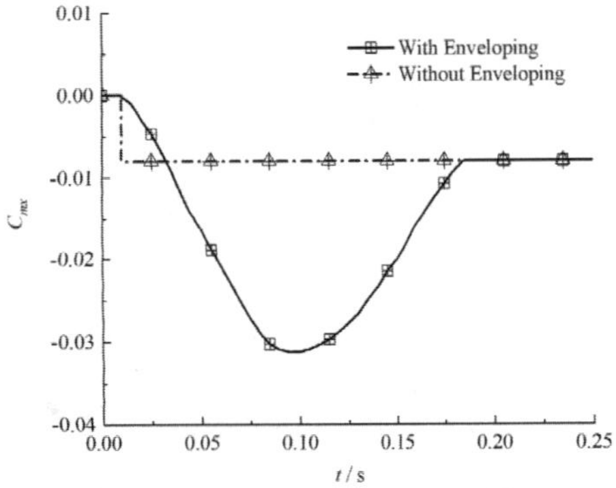


Figure 7. Time history of aircraft rolling moment coefficient.

### *4.3. Simulation results*

Some main simulation results are shown in figure 8~13. These results demonstrate reasonable dynamic trends and amplitudes.

Figure 8 shows the dimensional velocity profiles of the lateral travelling gust, which has horizontal and vertical disturbance energy on the aircraft. The gust velocity profiles along the two directions are similar in the shape.

Generally, it can be seen from Figure 9~12 that the parameters denoted with ‘with enveloping’ are very close to those denoted with ‘without enveloping’, and they share the same change trend. Furthermore, the amplitude of the parameter denoted with ‘with enveloping’ is slightly greater, and its peak occurs a little later.

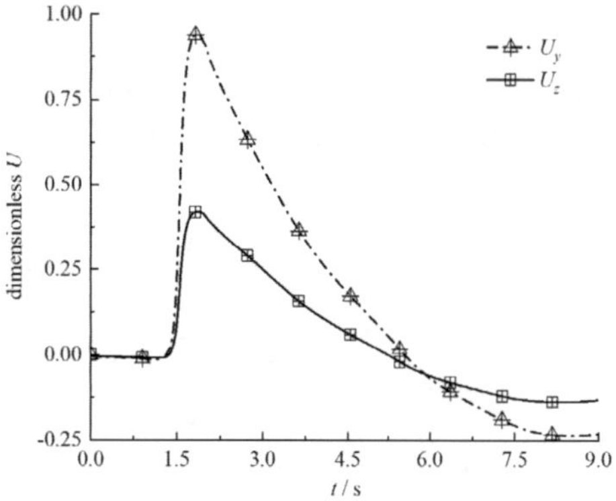


Figure 8. The velocity profiles of the lateral travelling gust.

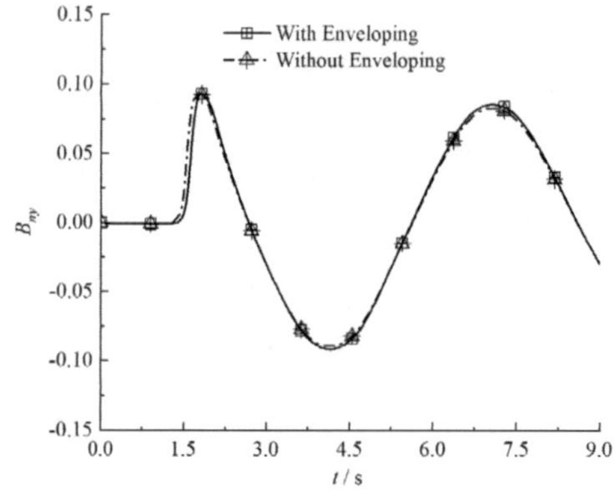


Figure 9. Time history of the lateral load factor.

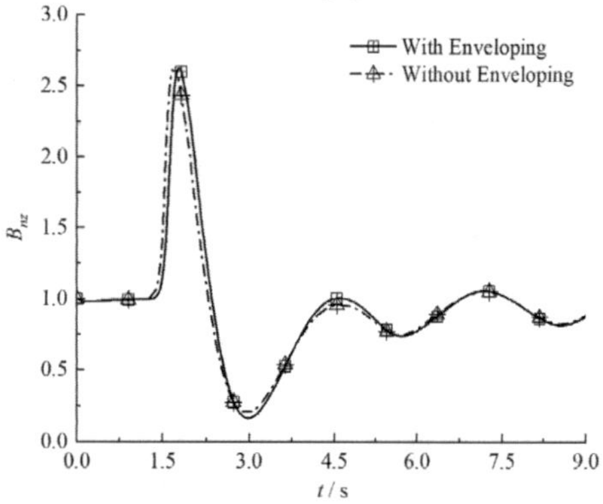


Figure 10. Time history of the normal load factor.

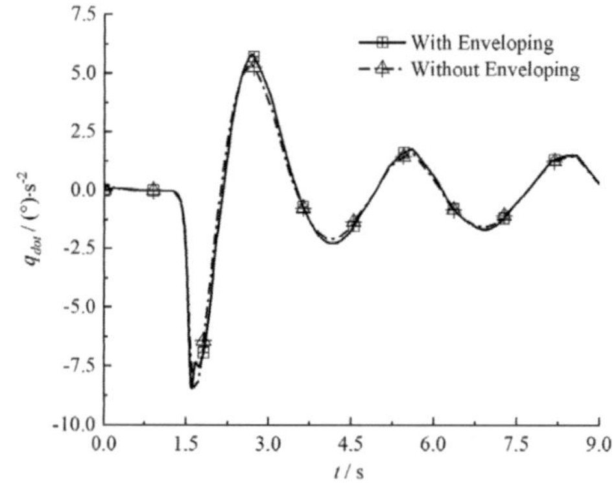


Figure 11. Time history of the pitching angular acceleration.

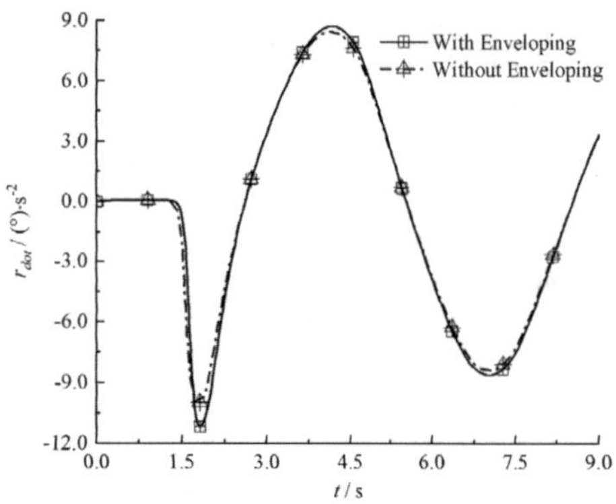


Figure 12. Time history of the yawing angular acceleration.

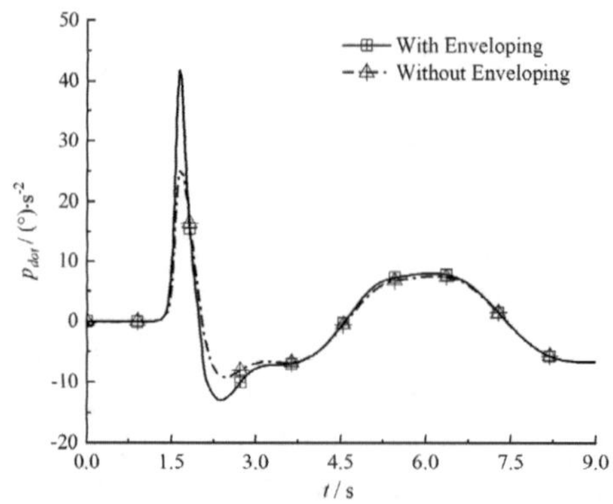


Figure 13. Time history of the rolling angular acceleration.

It can be seen from figure 9 and 12 that both the lateral load factor and the yawing angular acceleration decay slowly, the main reason is that the example aircraft has a small aerodynamic yawing damping.

Figure 13 shows that the lateral travelling gust can induce a remarkable increase in the normal load factor, while the tail-on travelling gust produces a relatively small rise, as shown in Ref. 7). The essential reason is that the tail-on travelling gust reduces the airspeed of the aircraft considerably, which will result in a significant reduction in the dynamic pressure.

## 5. Conclusions

To study the dynamic characteristics of the rigid aircraft in the lateral travelling gust, the dynamics modelling and simulation technique is established.

One quasi-steady aerodynamics computation method is presented by employing the static pressure distribution data from the wind tunnel or by CFD, which shows good effect in dealing with the enveloping process of the lateral travelling gust to the aircraft.

An example aircraft with large wingspan is analyzed. The simulation results reveal that the approach presented in this paper is effective and can satisfy the requirement of the engineering design. The enveloping process of the lateral travelling gust results in a significant increase in the aerodynamic rolling moment and mainly impacts on the response of the x-axis, while the enveloping process has little influence on the responses of the y-axis and z-axis. It should be noticed in the engineering practice.

## References

[1] Goodman, T. R., Sargent, T. P., Some Evidence Justifying the Representation of Blasts as Traveling Gusts, Journal of the Aero/Space Sciences, 9 (1959): 608-608.

[2] Ruetenik, J. R., Traveling-Gust Method for Airfoils in Strong Blast Waves, Journal of the Aerospace Sciences, 10(1960): 795-796.

[3] Drischler, J. A., Diederich, F. W., Lift and Moment Response to Penetration of Sharp-Edged Travelling Gust with Application to Penetration of Weak Blast Waves, NACA TN 3956, 1957.

[4] Wen, G. B., Sun, Z. S., Numerical Computation of Unsteady Subsonic Aerodynamic Forces on Wing Exposed to Travelling Gust, Acta Aerodynamical Sinical, 1(1981): 44-52 (in Chinese).

[5] Wen, G. B., Sun, Z. S., Numerical Computation of Unsteady Subsonic Aerodynamic Forces on Wing-Body-Tail Exposed to Travelling Gust, Acta Aeronautical et Astronautical Sinical, 3(1981): 23-30 (in Chinese).

[6] Jing, Z., Tang, C., Modelling Technique of the Aircraft Unsteady Aerodynamics Due to the Travelling Gust, International Journal of Plant Engineering and Management, 2(2022): 86-102.

[7] Jing, Z., Tang, C.: Dynamic Response Modelling and Simulation of Aircraft in the Travelling Gust, Flight Dynamics., 3 (2021): 8-13 (in Chinese).

[8] Stevens, B. L. and Lewis, F. L.: Aircraft Control and Simulation (2nd Edition), John Wiley & Sons, Inc., New Jersey, 2003, pp. 107-115.

# Design and Analysis of the Wing Functioning as a Ramp for a Spitfire Flight Simulator

**Yichong Zhang**

Department of Aeronautics, Imperial College London, London, SW7 5QT, UK

E-mail: yz621@ic.ac.uk

**Abstract.** This paper designed and analysed the wing for a Spitfire Mk9 Flight Simulator using three-dimensional modeling, finite element analysis and fatigue analysis methodologies with 3D EXPERIENCE, ABAQUS/FE-SAFE software. The results demonstrated that by using the wing structure as the access ramp to the cockpit simulator, the accessibility, safety and structural integrity could be improved vastly in comparison with the existing Spitfire flight simulators.

## 1. Introduction

With the development of transportation, aviation industry has gained high popularity in recent years. Flight simulator, a professional equipment used for training pilots in the past, now becomes approachable for the public. The Spitfire fighter is one of the most famous aircrafts in World War II. Thus, Spitfire simulators have high popularity due to its classic design and historical significance. Figure 1 shows a Spitfire Mk9 cockpit flight simulator in the University of Nottingham workshop.

Figure 1. Spitfire Mk9 simulator in University of Nottingham
(Permission is granted)

For a large proportion of the Spitfire simulators in use, steps are used to assist people in entering the simulator. However, the distinct drawback exists for use of steps. The structure of these device is oversimple. Without a wing structure, people could not acquire a better understanding of aircraft structure when using simulator. In addition, the lack of safety protection equipment could lead to potential hazards and a deficiency in the aspect of accessibility. Thus, to solve these problems, it is worth investigating a wing structure which can function as a ramp and a study on designing a wing structure for the Spitfire flight simulator is needed. In this paper, the 3D model built by 3D EXPERIENCE of the wing structure was demonstrated, and structural integrity of the model was tested through finite element analysis and fatigue analysis methods by using ABAQUS. Considering the integrity, the wing was designed based on the original structure of the Spitfire Mk9 fighter. Ideally, by using the wing structure which consists of wing body, pedestals and handrail as the access ramp to the cockpit, the Spitfire flight simulator will have better accessibility, integrity, and safety performance.

## 2. Key methodologies

The process of this study is shown in Figure 2.

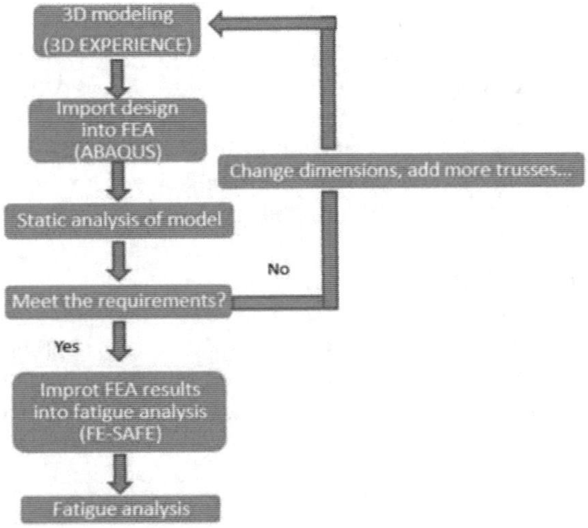


Figure 2. Flow chart of design and analysis process

### 2.1. *Basic principles of finite element method*

The first step in any finite element simulation is to use a set of finite elements to discretize the actual structure. In the stress analysis, the displacement of each node is the fundamental variable for ABAQUS calculation. After evaluating the strain-displacement and stress-displacement relationships, the global stiffness matrix and load vector are obtained. The equilibrium equation can be rewritten as:

$$[K]_{Assembly}[u]_{Assembly} = [P]_{Assembly} \tag{1}$$

where $[K]_{Assembly}$ is the global stiffness matrix, $[u]_{Assembly}$ is the global displacement matrix and $[P]_{Assembly}$ is the global load matrix.

In ABAQUS/Standard, from above equation system, the unknown displacement variables can be solved and then, element stress can be obtained by using the displacement results.

### 2.2. *Brown-Miller algorithm of fatigue analysis*

In this study, the Brown-Miller algorithm is selected for fatigue analysis. This is a critical-plane, multi-axial fatigue algorithm using planes both normal to the surface and at 45 degrees to the surface. The complete Brown-Miller strain-life equation is demonstrated below.

$$\frac{\gamma_{max}}{2} + \frac{\Delta\varepsilon_n}{2} = 1.65\frac{\sigma_f'}{E}\left(2N_f\right)^b + 1.75\varepsilon_f'\left(2N_f\right)^c \tag{2}$$

The Brown-Miller equation suggests that the maximum fatigue damage appears on the plane which sustains the maximum shear strain amplitude, and the damage is a function of both this shear strain and strain perpendicular to this plane. This equation is a recommended algorithm for most ductile metals at room temperature and is the default algorithm for most materials in the FE-SAFE materials database since it gives the most realistic life estimates [1].

## 3. Material and loadcase

### *3.1. Material*
The material used for wing body is aluminium alloy 2024 due to its high strength and fatigue resistance properties and the material used for handrail is S355 structural steel due to its superior strength, machinability, and weldability.

### *3.2. Loadcase for finite element analysis*
As an access ramp, the wing body needs to withstand the weight of a person. Based on the foot size survey done in 2019[2], the length of one foot with shoes is assumed to be 320 mm and the width is assumed to be 120 mm. From the data published on Medical News Today [3], the average weight in adults (man and woman combined) is 71 kg in Europe. In this paper, the weight for one person is set to 80 kg. Therefore, the load applied on the wing body when a person stands with one foot approximately equals 0.02 MPa. For handrail, according to the health and safety regulation of handrail design [4], the strength requirement of the handrail is to withstand an 890 N force applied in a downward or outward direction on the top edge of the handrail. Thus, the load applied to the handrail is an 890 N point load.

### *3.3. Loadcase for fatigue analysis*
In FE-SAFE, the loading used for the chain test is the constant amplitude loading because each cycle has the same load amplitude and the load ratio R is set to -1. When R is equal to -1, the wing body experiences both tension and compression load during the fatigue analysis. Ideally, the wing body will not experience tension load when acting as an access ramp. The only type of load the structure will sustain is the compression. Nevertheless, to further examine the strength of the wing, the performance of the structure under alternating load is studied. The performance of wing body under 0.02 MPa, 0.04 MPa and 0.06 MPa pressure load is studied.

## 4. Results

### *4.1. Three-dimensional model*
Considering the integrity, the wing body is designed based on the structure of the Spitfire Mk9 fighter. The outline of the wing is made based on the aerofoil STN0 [5] and the slope of aerofoil STN0 meets the health and safety requirements [6]. In the purpose of maintaining strength and reducing weight, based on the skin thickness layout [5], the skin thickness is set to 2 mm. The internal structure of Spitfire wing is modelled based on the drawings from Warbirds Restoration [7]. The complete wing structure is shown in Figure 3. Considering the popular science education purpose as well as the weight control, half of the wing is designed not covered by skin. For that section, it is designed to be simply exposed to the air.

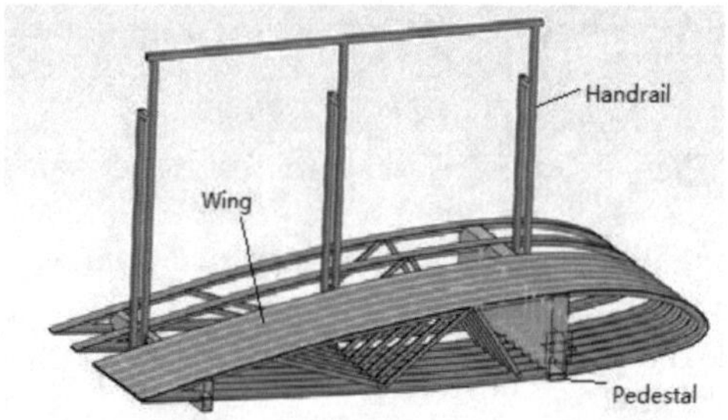


Figure 3. Wing with pedestals and handrail

### *4.2. Finite element analysis of wing body*
For the section functioning as a ramp, the inner structure consists of repetitive ribs. Taking advantage of symmetry to reduce problem size is a common skill in engineering practice. In addition, when a

person steps on the wing with one foot, normally three or more ribs support the person due to the rib spacing. Therefore, for this model, the performance of the model containing three ribs can reflect the strength of the wing body. The wing structure is imported into ABAQUS as a part model to avoid the convergence issues. Since the wing structure is covered by skin, when people stand on it, the stress under their feet will be distributed through the skin. Thus, the load is set to be pressure whose magnitude is corresponding to the weight of a person. The downward displacement of the wing structure is restricted as the boundary condition considering the wing is supported at the front and rear spar in real life. For spars, the rectangular prism element is chosen for mesh and the rest part of model is meshed with tetrahedral element. Simulation shows that the maximum displacement is 2.59 mm and the maximum stress is 97 MPa (shown in Figure 4 and Figure 5). This value is much smaller than the yield strength (325 MPa) of aluminium alloy 2024.

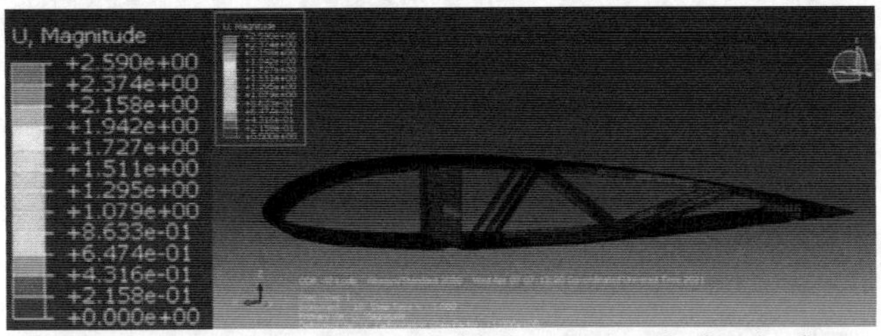

Figure 4. Contour plots of displacement of wing body

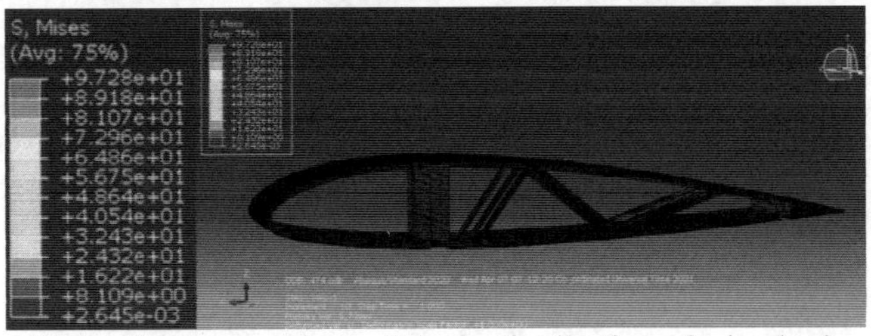

Figure 5. Contour plots of stress distribution of wing body

These results are based on the assumption that a person stands on the wing with single foot. In reality, when people stand on the wing with two feet, the stress applied to the wing under one foot will be halved and there should be no doubt that the maximum stress and displacement in real life will be much smaller in comparison with the simulation results. Thus, this wing structure is safe and reliable. The analysis of the model with three ribs can reasonably simulate the real-life working condition of the wing and the results illustrates that this structure could achieve the role as a ramp for the simulator.

*4.3. Finite element analysis of handrail*

According to the regulation of handrail design [4], the strength requirement of the handrail is to withstand an 890 N force applied in a downward or outward direction on the top edge of the handrail. Therefore, the load is point force whose magnitude is 890 N and the downward displacement of handrail is restricted as boundary condition. The contour plots of stress distribution are shown in Figure 6 and Figure 7. From the results, the maximum stress under both working condition is below the yield strength of the material showing that the structure is safe and reliable.

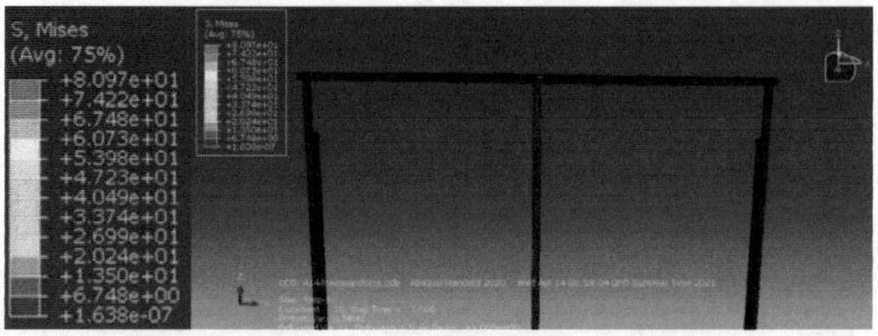


Figure 6. Contour plots of stress distribution with downward force

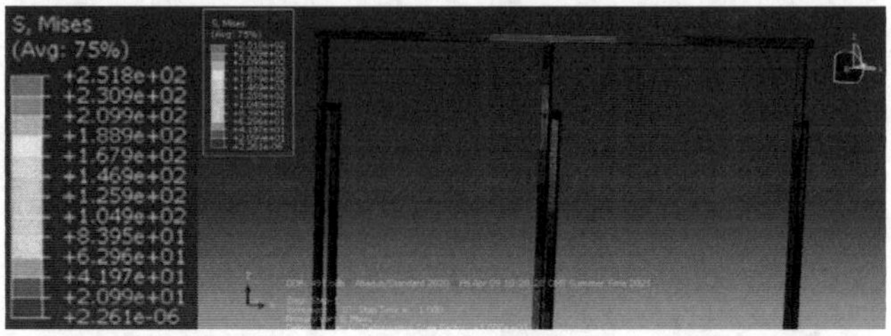


Figure 7. Contour plots of stress distribution with outward force

### *4.4. Fatigue analysis of the wing structure*

FE-SAFE uses the finite element analysis results from ABAQUS to simulate the fatigue phenomenon. The import model is assigned with the aluminium alloy 2024 using the material database in FE-SAFE. The Brown-Miller algorithm is selected, and the load ratio R is set to -1. After the fatigue analysis of the wing body in general loading condition where the load is equivalent to 80 kg weight, the results show that there is no damage occurred in the structure within $10^5$ cycles. By doing the analysis under 120 kg weight and 160 kg weight, the fatigue feature first appears with the 160 kg weight. The fatigue life contour plot of the wing structure is shown in Figure 8. The minimum life repeat is 3.934 in the contour plot which represents $10^{3.934}$ and is approximately equal to 8590. This result show that the minimum fatigue life of the wing structure is 8590 cycles, that is, after the completion of these cycles, the fatigue crack begins to generate. The minimum fatigue life appears at the rear truss and this result is identical with the results of statics analysis.

Figure 8. Fatigue analysis results

Considering the real-life loadcase, the fatigue phenomenon due to the continuous loading is not likely to occur for this wing structure. In general, after one visitor goes into the simulator, there will be at least a few minutes time before the next loading. However, the fatigue analysis does proof the strength of the wing structure in another perspective.

## 5. Conclusion

The Spitfire Mk9 flight simulator wing structure designed in this paper gives a proper solution for increasing the accessibility, safety, integrity, and science education significance of the existing Spitfire flight simulators. The 3D model is built using 3D EXPERIENCE and the strength of the wing and the handrail are studied using finite element analysis software ABAQUS. The simulation results show that the wing structure based on the original Spitfire fighter could maintain the strength without excessive deformation when sustaining the weight of a person. Under the loadcase when 80 kg load is applied, the maximum displacement of the wing structure is 2.59 mm, and the maximum stress is 97 MPa which is far below the yield strength of the aluminium alloy 2024. The stress concentration area is on the truss near the rear spar. The fatigue analysis conducted by the software FE-SAFE shows the fatigue crack will not occur in the wing structure within $10^5$ cycles when 80 kg load is applied. The minimum fatigue life of the wing structure is 8590 cycles when 160 kg load with load ratio $R = -1$ is applied and the minimum fatigue life appears at the rear truss and this result is identical with the results of statics analysis.

## 6. Further work

(1) The detailed finite element analysis and fatigue analysis of the welded joint are needed.

(2) The plasticity effects need to be considered.

(3) Additional physical experiments are necessary for further validating of the strength of the wing structure and handrail.

**Acknowledgments**

The author would like to thank Dr. Mark Jabbal for providing the support through the process of researching and writing this paper. The author also wants to give special thanks to Prof. Xuechang Zhang for providing continuous encouragement throughout the year of study.

## Nomenclature

$b$ – Fatigue strength exponent
$c$ – Fatigue ductility exponent
$E$ – Young's modulus
$\gamma_{max}$ – Maximum shear strain and the
$\varepsilon_n$ – Strain perpendicular to the maximum shear strain
$2N_f$ – Endurance in reversals
$\sigma_f'$ – Fatigue strength coefficient
$\varepsilon_f'$ – Fatigue ductility coefficient

## References

[1] Dassault System. (2020) fe-safe User Guide. https://help.3ds.com/2020/English/DSSIMULIA_Established/FesafeUserPdf/FesafeUser.pdf?contextscope=all&id=9fd74efa02b14f4eb8b6c646be6123b3.
[2] Jurca, A., Žabkar, J. and Džeroski, S. (2019) Analysis of 1.2 million foot scans from North America, Europe and Asia. Scientific reports., 9(1): 1-10.
[3] Medical News Today. (2018) Is there an average weight for man. https://www.medicalnewstoday.com/articles/320917.
[4] United States Department of Labor. (2003) Walking and Working Surfaces; Personal Protective Equipment (Fall Protection Systems). https://www.osha.gov/laws-regs/federalregister/2003-05-02.
[5] Monforton, P. H. (2008) Spitfire Mk. IX & XVI Engineered. Monforton Press, Ottawa.
[6] Ministry of Housing, Communities & Local Government. (2013) Protection from falling, collision and impact: Approved Document K. https://www.gov.uk/government/publications/protection-from-falling-collision-and-impact-approved-document-k.
[7] Warbirds Restoration. (2015) Spitfire Mk IX & XVI. https://warbirdsrestoration.com/shop/spitfire/mk-ix/spitfire-mk-ix/.

# Physical and Chemical Analysis of Contact Points of relay on EMU

**Ping Sun[1,2], Junying Zhang[1,2], Anji Dong[1,2], Lei Zhang[1,2], Wenjun Wang[1,2], Nan Wu[1,2], Abo Dou[3*]**

[1]CRRC Changchun Railway Vehicles Co., Ltd., Changchun, Jilin, 130000, China
[2]National Engineering Research Center of Railway Vehicles, Changchun, Jilin, 130000, China

[3]Centre Testing International Group Co., Ltd., Shenzhen, Guangdong, 51800, China

[*]Corresponding author's e-mail: douabo@cti-cert.com

**Abstract:** Electromagnetic relay is an essential part in electric control system on EMU. Because of frequent operation of on-off, external vibration, temperature rising and corrosive gas, etc when the EMU is in operation may cause wear and oxidation of the surface of contact points. Therefore, the electrical contact resistance increased, which cause contact electrical performance failure. The scanning electron microscopy (SEM), Fourier transform infrared spectroscopy (FT-IR) was used to observe and measure the surface and section of contact points. The experimental result shows that the surface layer wear and contained phosphate and the section contained O, F, S, P, Ag, Ni elements. Thereafter the mechanism of efficacy losing of electromagnetic relay was discussed.

## 1. Introduction

The electromagnetic relay is responsible for the power distribution and logic control functions of the EMU. The failure of the relay affects the reliability of the system, and its failure often leads to the failure of important equipment or loops on the EMU, causing deceleration and even parking, affecting the driving order and safety.

The contact point is one of the core components of the relay, which is directly responsible for the connection and disconnection of the circuit. According to relevant statistical data, most relay failures occur in the contact part, and its failure rate accounts for more than 90% of the whole relay failure rate[1]. Systematically carry out product characteristics and fault analysis and research, accumulate sufficient fault handling experience to provide support for effective life evaluation and design selection.

## 2. Pretreatment Experimental Method

The sample tested and analyzed is from the fault relay, from which the contact piece with elastic pieces is removed and the contact head part is cut off. The Carl Zeiss AXIO Imager.A1m metallurgical microscope was used as a guide to make slices. The PerkinElmer Spotlight 200i Fourier transform infrared spectroscopy(FT-IR) was used to analyzed the composition of surface oxide layer. The Zeiss SIGMA 500 scanning electron microscope(SEM) was used to observe the morphology, observe at different magnification and take SEM images of different samples. The Oxford 51-XMX1121 energy dispersive spectroscopy(EDS) was used to analyzed the composition.

In order to understand the failure of the relay more accurately, the relay contact including stationary contact and moving contact which are uneven with black spots visible to the naked eye only drying treatment which acetone, absolute ethanol and other solvents are not applicable.

## 3. The Surface Analysis of Contact Points of Relay on EMU

The surface is uneven and partially black pockmarked with the naked eye, it can be comparative analysis by different area.

### *3.1. Metallurgical Microscope Morphology*

The morphology test results are shown as figure 1, the contact areas are focus on the centre of the whole area[2], and shown as many of bumps and holes, and the research shows that the uneven wear can cause the increase of contraction resistance[3].

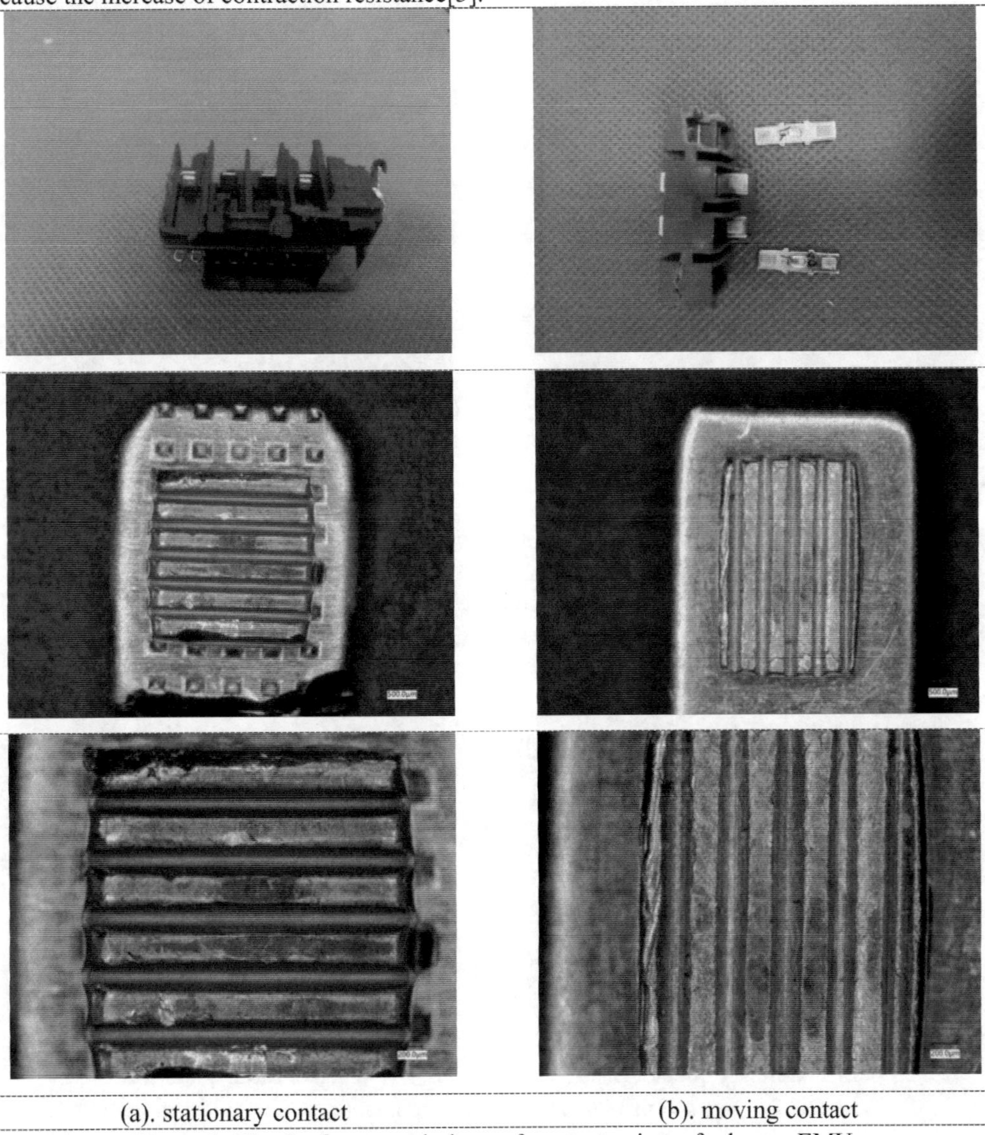

(a). stationary contact (b). moving contact

Figure 1. The Surface morphology of contact points of relay on EMU.

### *3.2. FT-IR Analysis*

The FT-IR test results are shown as figure 2(a): the stationary contact surface contain phosphate, while the figure 2(b): moving contact does not. The source of phosphate is protective agent, oxidation or foreign matter, as the residual protective agent is in conflict with the oxide layer, so it cannot be a protective agent.

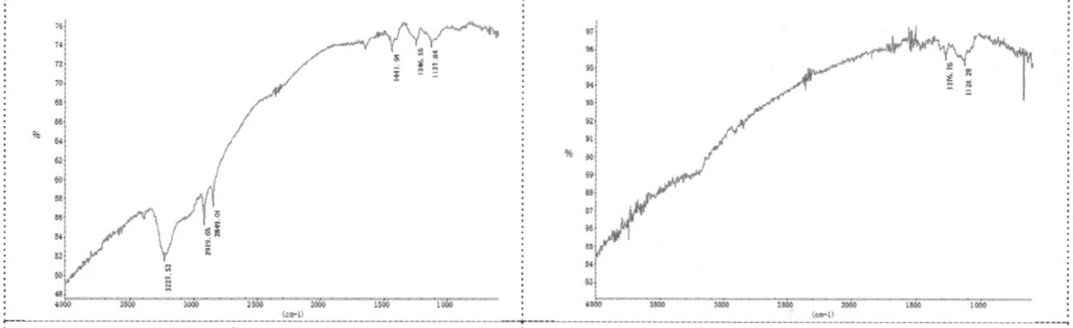

(a). stationary contact (b). moving contact

Figure 2. The Surface FT-IR analysis of contact points of relay on EMU.

### *3.3. SEM-EDS Analysis*

The SEM-EDS analysis results are shown as figure 3, the morphology result shows the contact points wear down in operation on EMU, the research shows that the wear can cause the increase of electrical clearance and reduce the opportunities of contact[3].

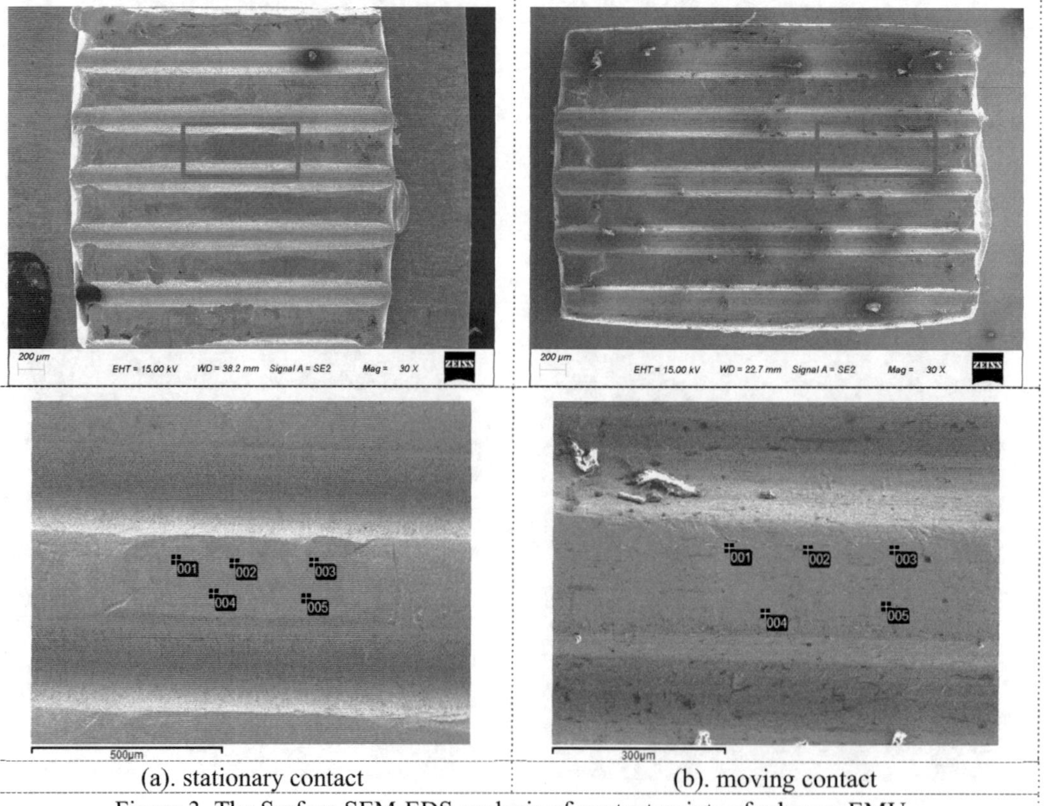


(a). stationary contact (b). moving contact

Figure 3. The Surface SEM-EDS analysis of contact points of relay on EMU.

The SEM-EDS test results are shown as table 1 and table 2, the stationary contact surface contain more P, S, O element than moving contact, it shows more oxidized in stationary contact, and the same foreign matter amount(F element) in both contacts.

Table 1. The Surface SEM-EDS analysis of the stationary contact points of relay on EMU.

| No. | The Element Composition(W/W,%) of The Stationary Contact | | | | | |
|---|---|---|---|---|---|---|
| | O | F | P | S | Ni | Ag |
| 001 | 11.8 | 4.5 | 2.0 | 1.1 | 13.2 | 67.2 |
| 002 | 11.9 | 4.0 | 1.3 | 1.0 | 1.7 | 80.2 |
| 003 | 12.6 | 2.4 | 1.9 | 1.4 | 0.7 | 80.9 |
| 004 | 7.1 | 1.0 | 0.8 | 0.8 | 0.3 | 90.0 |
| 005 | 7.7 | 2.8 | 1.1 | 1.3 | 0.5 | 86.6 |

The No. corresponds to figure 1 (a)

Table 2. The Surface SEM-EDS analysis of the moving contact points of relay on EMU.

| No. | The Element Composition(W/W,%) of The Moving Contact | | | | | |
|---|---|---|---|---|---|---|
| | O | F | P | S | Ni | Ag |
| 001 | 6.5 | 3.6 | 0.8 | 0.6 | 0.7 | 87.7 |
| 002 | 13.5 | 1.2 | 0.8 | 0.4 | / | 81.6 |
| 003 | 1.6 | 1.0 | 0.2 | / | / | 97. |
| 004 | 3.5 | 1.1 | 0.3 | 0.5 | 1.4 | 93.2 |
| 005 | 6.9 | 0.7 | 0.4 | / | 1.5 | 90.4 |

The No. corresponds to figure 1 (b)

## 4. The Section of contact points of relay on EMU

On the basis of surface analysis, Slice the surface analysis sampling area, inlay samples, polish with 600 mesh sandpaper and observe with SEM.

### *4.1. SEM morphology*

The morphology test results are shown as figure 4, in the process, electric welding is first followed by stamping, and the direction of extrusion between stationary contact and moving contact is vertical.

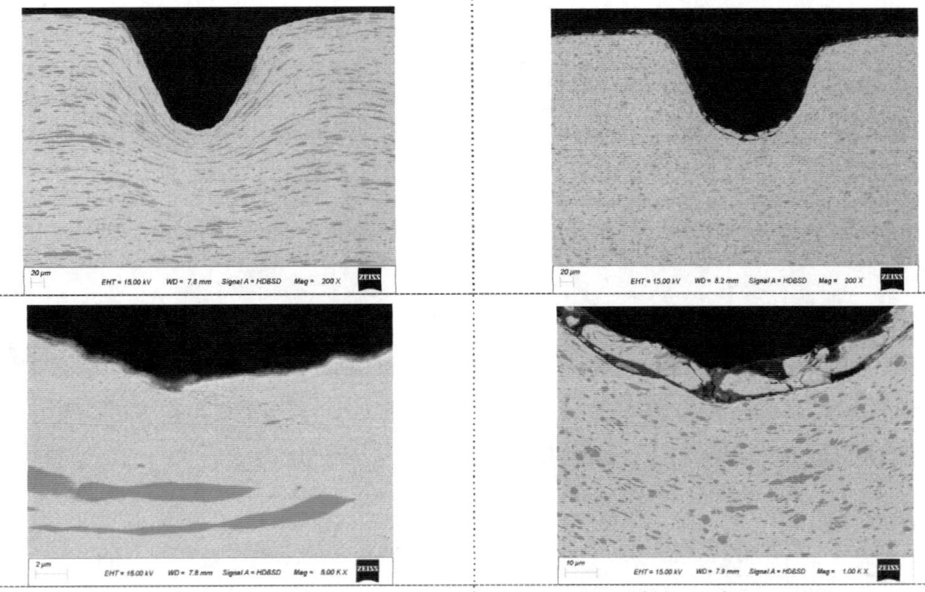

(a). stationary contact (b). moving contact

Figure 4. The Section SEM-EDS analysis of contact points of relay on EMU.

### *4.2. SEM-EDS Analysis*

The morphology and the composition distribution test result of the line scan by SEM and EDS are shown as figure 5, figure 6 and figure 7. According to the analyses, the polluted layer is composed of oxides, impurities and contact alloy elements, and uneven distribution the uniformity of Ag-Ni alloys.

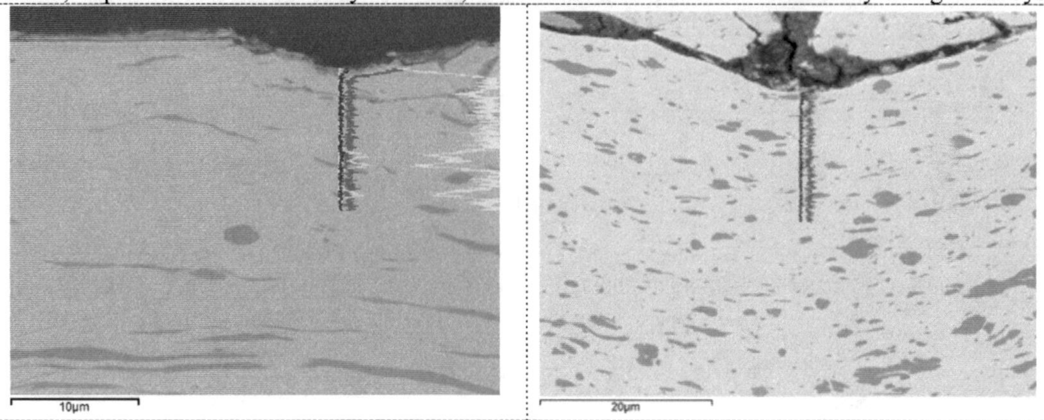


(a). stationary contact (b). moving contact

Figure 5. The Section SEM-EDS analysis of contact points of relay on EMU.

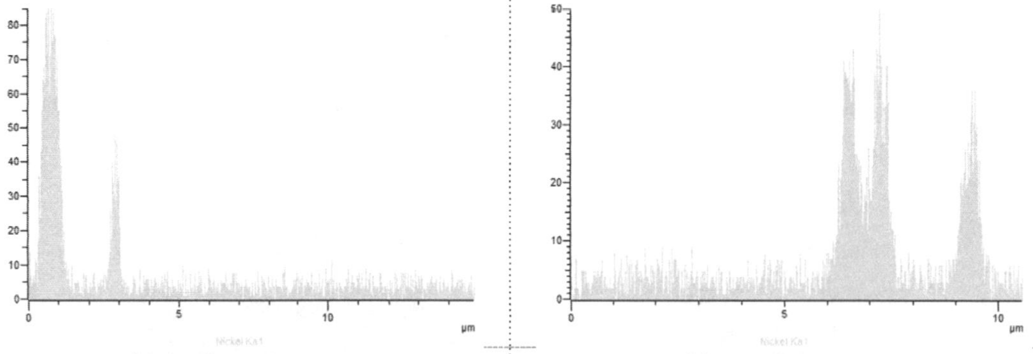


(a). stationary contact (b). moving contact

Figure 6. The Section SEM-EDS analysis of contact points of relay on EMU.

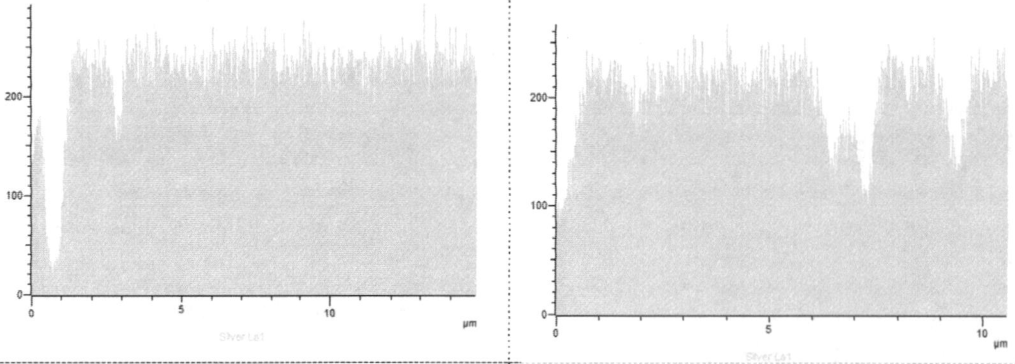


(a). stationary contact (b). moving contact

Figure 7. The Section SEM-EDS analysis of contact points of relay on EMU.

## 5. Conclusion

The physical and chemical analysis result shows the electrical performance failure because of the abnormal of contact resistance which mainly cause by increase of contraction resistance and sheet resistance, and when the clearance caused by wear also cause abnormal contact.

### *5.1. Contraction Resistance*

According to formula (1), contraction resistance is mainly related to resistivity, the number and size of contact surfaces on the micro level, etc. The larger the number of contact surfaces per unit area is, the larger the contact area is, and the resistance will decrease. On the contrary, the smaller it will increase[4].

$$R_E = \frac{\rho}{2\sum_i a_i} + \frac{3\pi\rho}{32}\left(\sum_{i\neq j}\frac{a_i a_j}{S_{ij}}\right)/\left(\sum_i a_i\right)^2 \tag{1}$$

Whereas: ρ is the resistivity of the contact, a is the micro peak radius of the contact, and Sij is the spacing of micro planes.

### *5.2. Sheet Resistance*

According to the polluted layer contained O, S element, it can increase the sheet resistance because AgS is not conductive(AgO is conductive) and impurities may not be conductive.

### *5.3. Electrical Clearance*

According to the research[5] , when the clearance caused by wear is greater than the distance between the overtravel distance and the electrical clearance , it can cause abnormal conduction.

## References

[1] Li Zh.B, Xu J.J., Huang L., Xu J. (2007) Mechanism Analysis on Dynamic Welding of Electromagnetic Relay Contacts, Electric Switchgear, 5: 1-3, 28.
[2] Liu, J.A., Wang W.J., Li B., Gong L. Yu Y. (2022) Study on electric-deoxidation technology of contactor in rail transit. In: 2022 International Conference on Electronic Information Technology. Chen Du. pp. 12254 0P.
[3] Li W.H., Wang L.G., Zhou L.L., Guan X., (2017) The 3D surface topography analysis of railway relay contact, Journal of Railway Science and Engineering, 14(8): 1752-1757.
[4] Zhao ZH.Y. (2020) Factors Affecting Contact Resistance Change of Railway Signal Relay, Railway Signalling & Communication Engineering, 17(8): 13-18, 23.
[5] Liang H.M., Zhai G. F. (2014) Three Stages Deign Of Aerospace Electromagnetic Relay. Harbin Industrial University Press, Harbin.

# Application Research of Civil Aviation High Frequency Communication System

***Kaikai Si, Qing Lu[a], Xiaoyan Guo[b], Bin Gao[c]**

Shanghai Aircraft Design and Research Institute, Pudong New Area Shanghai China

Email: *1571901582@qq.com; [a]luqing@comac.cc; [b]guoxiaoyan@comac.cc; [c]gaobin@comac.cc

**Abstract:** High frequency communication system is an important part of civil aviation radio communication system. This paper introduces the typical structure and function of civil aircraft high frequency communication system. Discussing the high frequency system architecture mainly and introducing the components of aircraft high frequency transceiver, high frequency coupler and antenna in detail. This paper discusses its principle and function, so as to provide some reference for the design of civil aviation high frequency communication system.

## 1. INTRODUCTION

Up to now, high-frequency communication system is an important means of communication in civil aviation communication. Although satellite communication solves the problem of aircraft long-distance communication to a certain extent.Considering economy, safety and redundancy, high frequency communication is still an important part of aircraft radio communication[1].

This paper mainly studies the application and design of aircraft high frequency communication system, and introduces in detail the composition of civil aircraft high-frequency communication system, the functional requirements of civil aircraft high-frequency communication system, the composition unit of high-frequency transceiver, the function of each unit, the composition unit and function of coupler and common antenna, etc[1]. It summarizes the design experience of previous high-frequency communication system, so as to provide reference for the design and optimization of high-frequency communication system in the future.

## 2. INTRODUCTION TO HF COMMUNICATION SYSTEM

There are three main modes of modern civil aviation radio communication: very high frequency communication, satellite communication and high frequency communication. Among them, high frequency communication mainly refers to the signal with frequency of 3MHz-30MHz and wavelength of 10M-100M[2]. It has become an indispensable part of civil aviation radio communication because of its strong anti-interference ability, low interception, low cost and long communication distance[1]. High frequency communication system is mainly used for remote areas, transoceanic and off shore operations and general airport aircraft to provide emergency and blind ground to air communication services[3]. With the development of communication technology, the shortcomings of civil aviation high-frequency communication are gradually overcome, and the advantages are gradually highlighted, so it has attracted more and more attention.

## 3. SYSTEM ARCHITECTURE

The high-frequency communication system mainly consists of high-frequency transceiver, high-frequency coupler, high-frequency feeder and high-frequency antenna. The typical architecture of high-frequency communication system is shown in Figure 1[3]. The tuning control board is mainly used for frequency tuning and sends commands to the high-frequency transceiver through the radio interface unit. The high-frequency transceiver is mainly used to receive and send voice or data signals. Its work needs to rely on the tuning command of the tuning control board. For the transceiver of data signal: the radio interface unit mainly sends the data signal to the high-frequency transceiver through ARINC 429 bus, and then generates the RF signal through the modulation of the high-frequency coupler, which is sent by the antenna. The data receiving process is opposite to the data transmitting process[2]. For voice transceiver: the audio control board converts the analog voice signal into digital signal, converts the audio signal into analog signal through ARINC 429 bus through the radio interface unit, and transmits it to the high-frequency transceiver through the radio interface unit. The high-frequency transceiver modulates the signal through the coupler to generate RF signal, which is sent through the antenna. The voice receiving process is opposite to the sending process.

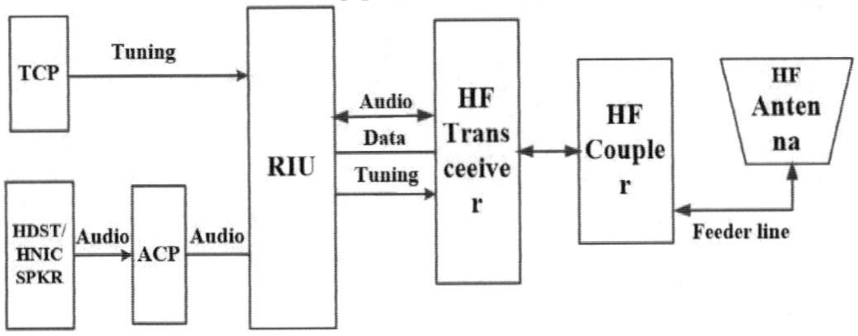


Figure 1 Typical architecture of HF communication system

## 4. FUNCTION

### *4.1. Functions of civil aviation HF communication system*

The main function of civil aviation high frequency communication system is to provide long-distance communication between aircraft or between aircraft and ground. In addition, civil aircraft high frequency communication system shall also support the following functions[5]:

- It has the function of two-way voice communication, which can voice transceiver in the upper sideband (USB), lower sideband (LSB) and equal amplitude modulation (AME) mode;
- Selecting call signal output (when not in high frequency data link mode);
- The working frequency is between 2MHz and 30MHz, and the frequency interval is 100Hz;
- High frequency system detection function (including manual self detection function of high frequency transceiver and continuous BIT Self detection function);
- High frequency system operation data recording function and flight data recording function;
- CAS alarm function of high frequency communication system;
- High frequency system maintenance information broadcasting function and configuration data broadcasting function;

### *4.2. Functional operation description*

The voice communication function of HF system is mainly controlled by tuning control panel (TCP), audio control panel (ACP) and pressing the call switch (PTT). When the aircraft is powered on, the high-frequency system will be in standby mode, and the high-frequency system will be tuned and mode controlled through the tuning control board.

When TCP selects HF data mode, the transceiver enters the data mode and does not receive or transmit voice signals; When the HF voice mode is selected by TCP, the HF radio station can be tuned on TCP. When the frequency tuned by TCP matches the frequency of the target radio station, the HF system can receive and send voice signals. HF only supports analog voice and cannot select bypass

ACP and RIU for direct communication through AJP. If the aircraft is refueling on the ground, HF will be inhibited automatically.

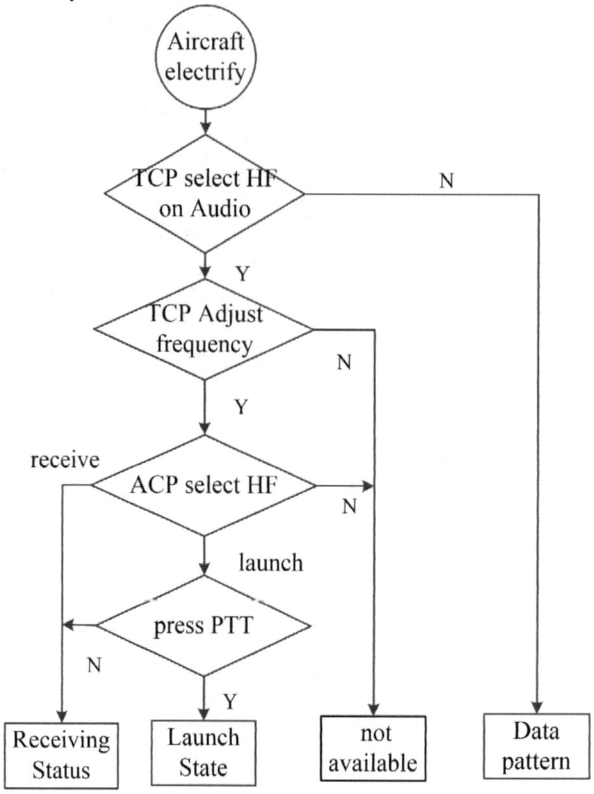


Figure 2 Operation flow of HF system

## 5. HF TRANSCEIVER

The transmission power of civil aviation high-frequency transmitter is generally 125w, 400W, 1000W, etc. Considering the construction cost and communication distance, the high-frequency transmitter with transmission power below 400W is generally selected. The high-frequency transceiver is composed of power supply unit, business processing unit, RF processing unit, power amplification unit, RF transceiver control unit, rear panel and front panel. The principle block diagram is shown in Figure 3.

The power supply unit mainly provides power for the whole high-frequency transceiver. It rectifies and filters the single-phase 115V and 400Hz power supplies sent by the avionics system, integrates them into the rated voltage required by each unit, and then outputs them to each unit[6]. The service processing unit allocates corresponding resources for all external interfaces of the high-frequency transceiver, provides the underlying physical channel, receives the avionics system control instructions, completes the audio signal processing with the system and realizes the zero if digital processing of the received signal and excitation signal. The RF processing unit performs up conversion, D / A, excitation amplification and other processing on the transmitted digital modulation signal sent by the service processing unit. Filter, amplify, A / D and digitize the received RF signal, detect and control the working state of the power amplification unit at the same time, and protect the power amplifier. The power amplification unit includes two parts: a power amplifier and a harmonic filter. The power amplifier amplifies the power of the RF excitation signal sent by the RF processing unit, amplifies the RF excitation signal to the rated RF power, and the harmonic filter attenuates each harmonic of the RF power signal output by the power amplifier. The RF transceiver control unit mainly completes the sampling of the output power signal of the power amplification unit and the switching of receiving and transmitting RF channels. The rear panel mainly completes signal filtering, transient voltage suppression and RF signal transfer with the airborne equipment of avionics system. The front panel mainly displays the working status through the indicator light, including LRU status, control fault and no launch status[4].

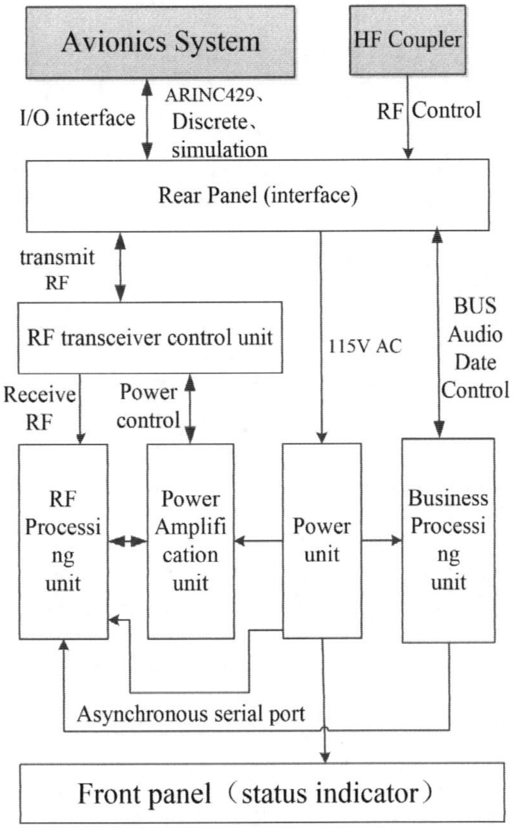


Figure 3 Schematic block diagram of HF transceiver

## 6. HF COUPLER

The high frequency coupler is composed of power supply unit, control unit and matching network. Its principle block diagram is shown in Figure 4. During signal receiving, the high-frequency coupler transmits the RF signal received from the high-frequency antenna to the high-frequency transceiver. When transmitting the signal, the high-frequency coupler receives the tuning command and frequency information, the control unit samples and processes the RF signal, calculates the more accurate impedance and phase value, controls the relay on and off according to the obtained impedance and phase information, and adjusts the parameters of the matching network unit until the high-frequency coupler completes the impedance matching.

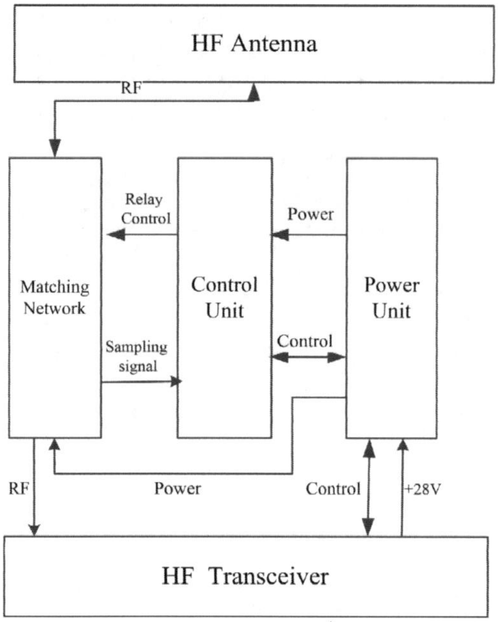


Figure 4 Schematic block diagram of HF coupler

## 7. HF ANTENNA

High frequency antenna is an energy conversion device. It can convert the energy in the form of high-frequency current into radio wave energy of the same frequency and transmit it outward. It can also accept the radio wave energy within the working range and convert it into high-frequency current energy of the same frequency. At present, the antenna of commercial aircraft high frequency communication system mainly includes two parts, airborne antenna and base station antenna.

### *7.1. Airborne antenna*

The high-frequency airborne antenna is generally installed at the front edge of the vertical stabilizer, and the antenna coupler is inside the vertical stabilizer. As shown in Figure5[2].

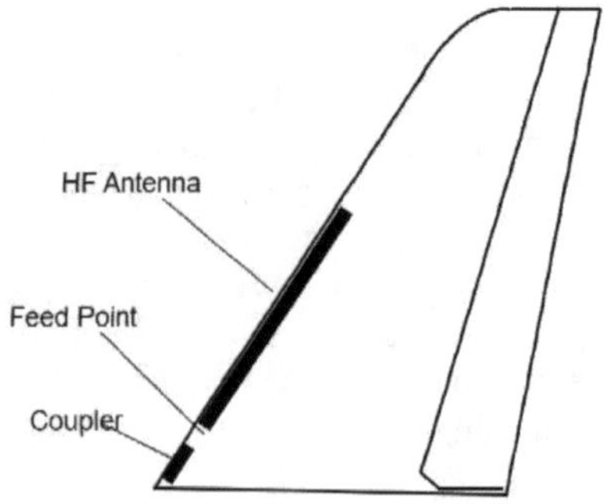


Figure 5 Installation position of HF antenna

Concealed antennas are generally used for civil aviation airborne high-frequency antennas, including cable antenna, notch antenna, loop antenna, sleeve antenna, etc. The cable antenna is simple and efficient, but it is easy to freeze and affect the aerodynamic performance of the body. Notched antennas are difficult to layout and cannot be used in composites. The real part of the low-frequency impedance of the loop antenna is very small, which leads to low efficiency. The antenna efficiency of sleeve antenna is also very low. In order to achieve good communication effect, the airborne antenna must meet the requirement that the horizontal pattern is omnidirectional in the whole working frequency band, and its gain should reach more than - 20dbi. The input impedance of the antenna needs to meet the requirement that the real part is greater than 0.1Ω. Considering the influence of the overall layout of the aircraft, the smaller the size of the concealed airborne antenna is, the better without affecting the antenna performance.

### *7.2. Base station antenna*
The base station antenna plays an important role in the communication quality of the high-frequency communication system. The working state of the base station antenna can be divided into receiving and transmitting. For the receiving working state, high gain omnidirectivity is the best characteristic. For the transmitting working state, no blind area, high elevation, long distance and high radiation are the best characteristics. At present, the high-frequency base station antennas widely built in major high-frequency communication centers are roughly divided into three wire antennas, cage antennas, log periodic antennas, etc. Three wire antenna and log periodic antenna are briefly introduced[1].

#### *7.2.1. Three wire antenna*
Three wire antenna is a kind of low gain horizontal polarization antenna, which is widely used. Its power capacity is 1kw, up to 2kw. The broadband antenna can be erected horizontally(Figure6) or obliquely in an inverted V-shape (Figure7). The horizontal three wire antenna equipped with 125w high-frequency radio can realize long-distance communication up to 3000km. However, considering the direction of the horizontal erection of the antenna, its radiation intensity in each direction is different. If the direction is north-south, the east-west direction is the wide side, and the radiation is stronger than the narrow side in the north-south direction, This results in a communication blind area for the horizontally erected three wire antenna, and the best communication direction is perpendicular to the erection direction. Inverted V-shaped erection will erect the vibrator at an included angle of 55° to the ground, which eliminates the influence of directionality on the radiation intensity. It can take into account the horizontal polarization wave and vertical polarization wave. Equipped with 125w high-frequency radio station, it can meet the communication without blind area within 1500km[1]. Three wire antenna is widely used in China because of its simple structure, convenient erection and low cost.

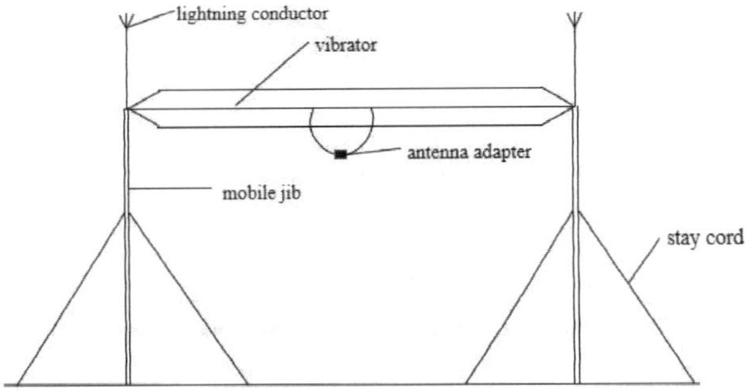


Figure 6 Schematic diagram of horizontal erection of three wire antenna

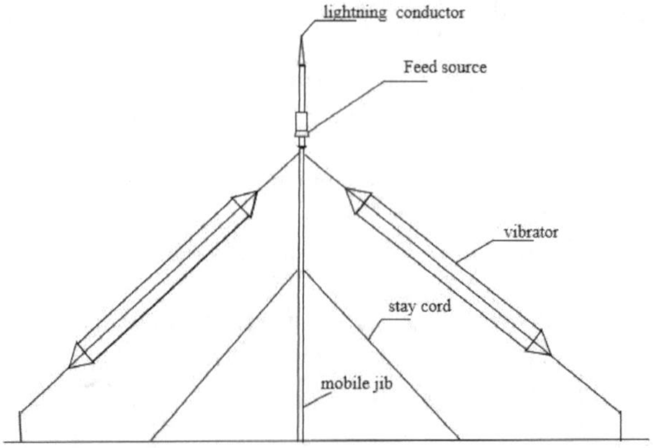


Figure 7 Schematic diagram of inverted V erection of three wire antenna

### *7.2.2. Log periodic antenna*

Log periodic antenna is a high gain directional antenna, which has the following characteristics: 1. Wide working frequency band and high gain. The working frequency band is 4-30MHz, which can meet the requirements of aircraft high-frequency communication. The main lobe gain can reach up to 12dBi and high gain[6]. 2. Good directivity and long communication distance; Log periodic antenna is a high gain directional antenna. Although the gain is very high, it has the problem of communication blind area. 3. It works stably and has a long service life.

The high frequency log periodic antenna is a non frequency varying antenna with wide working frequency band, stable input impedance and good voltage standing wave ratio in band. The horizontal placement of the antenna body can easily change the communication azimuth, which is conducive to one-to-many point and remote mobile target communication. The antenna has compact overall structure, small floor area and convenient erection, but it has high cost and low cost performance.

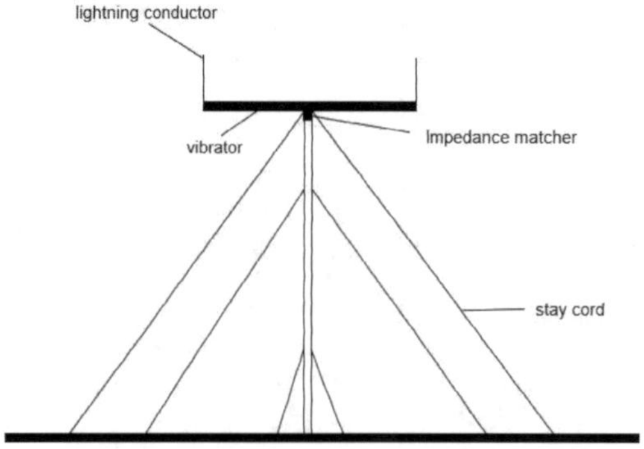


Figure 8 Architecture of log periodic antenna

## 8. CONCLUSIONS

As an important part of civil aviation radio communication system, high frequency communication system will play a more and more important role in civil aviation because of its low cost, simplicity and efficiency. This paper mainly introduces the components of high frequency system, and discusses the function and operation description of civil aviation high frequency communication. Exploring the design principles of transceiver, coupler and antenna has certain guiding significance for the environmental construction of civil aviation high-frequency communication laboratory and the planning and construction of civil aviation high-frequency station.

**References**

[1] Liu,H.W.(2018) Application of high frequency communication in civil aviation ground to air communication service.J. Wireless Internet technology,12,16-18

[2] Wu,H.M(2017) Analysis on the test flight method of civil aircraft high frequency communication system.J.Electronics World.10:141-142.

[3] Li,W.B,Jiang,X.H,(2009) Research on HF blind zone free communication and its application,Communication system and network technology,35:14-15

[4] Cui,Y.M.Wei,Y.(2014) Shortwave blind zone free communication technology and its application,Communications Technology,47:1135-1138

[5] Chen,J.L.Han,X.Fan,B.(2017) Research on high frequency communication technology.J.Post and telecommunications design techonlogy,12:15-19

[6] Wu,C.H(2012) Smoothness indicator of weno scheme for resolving short wave.J.Transactions of Nanjing University of Aeronautics & Astronautics,42:43-46

# Advances in Resilience Engineering for Air Traffic Management Applications

**Jianmin Yi[a], Cunbao Ma[b*], Jin Zhao**

School of Civil Aviation. Northwestern Polytechnical University, Xi'an, China

[a]yijianmin@mail.nwpu.edu.cn; [b]caecmacb@nwpu.edu.cn

[*] Corresponding author: caecmacb@nwpu.edu.cn

**Abstract**—As an emerging safety perspective, resilience engineering is attracting increasingly attentions in aerospace industry. Numerous researches have been carried out to explore ATM resilience properties from different aspects. Nevertheless, there appear to be significant variations on the research objects, issues concerned, methods employed and indexes used. As far as we know, there seems to be an absence of a dedicated literature review to explore the application status of resilience in ATM area. To this end, our work is concentrated on the research situation and progress of resilience in ATM. A wide range of databases are retrieved to obtain an extensive coverage of the literature. Subsequently, the research works are reviewed on the basis of different classifications followed by a brief introduction of related findings and implications. Attempting to provide some reference for further studies, promising directions are pointed out in view of the challenges and requirements. It is desired that this work may throw some light on further explorations of ATM resilience.

## 1. Introduction

In recent decades, unprecedented challenges to aviation safety have been brought about by the rapid development in science and technology. Safety issues that consist of complex mechanisms, new failure modes, complicated interactions and many unknown events are emerging and delaying the development process. Traditional safety concepts, for example SAE ARP4761, have shown deficiencies in recent years with the rapidly development of techniques. Confronting the challenges, deficiencies of existing methods and theories gradually appear. In 2006, Hollnagel, et al. published a book to introduce the concept of resilience engineering distinct from traditional thinking on safety [1]. In 2007, EUROCONTORL launched a project to understand the emerging area of resilience. Then in 2009, a white paper describing the concepts and realization principles of resilience engineering in ATM domain was released [2]. The white paper regards resilience engineering as an important direction of ATM safety. In 2013, EUROCONTROL published another white paper as a guideline for the application of resilience engineering in ATM [3].

As is obviously shown in the above demonstration, resilience has attracted extensive attentions and become a research front especially in the ATM safety. Aviation is an industry with extreme safety requirements, while ATM plays an important role in assuring aviation safety. When traditional perspectives and methods for safety assurance seem to enter the bottleneck, this new concept has shown great development potentials. It is of great significance to explore advanced methods for complex ATM system resilience which is an innovative perspective that breaks conventional thinking. Numerous articles concerning ATM resilience have been published with an increasing trend year by year. Nevertheless, to the best of the author's knowledge, there seems to be a lack for dedicated reviews on this topic. For the purpose of exploring the application status as well as developing potentials of this new safety concept in ATM, also in an attempt to shed some light on the research status, gaps, together with further directions, this work widely explores advances on the topic.

## 2. METHOD

This work follows the research method named PRISMA (preferred reporting items for systematic reviews and meta-analyses) proposed by Moher, et al. [4]. Five databases are selected for the retrieval for a wide inclusion, namely Web of science, Engineering Village, Science Direct, AIAA ARC and IEEE Xplore in January 2022. 50 papers are finally included for the study. The collected literature are divided into particular groups from different perspectives according to their characteristics. Taking account of the comprehensive information of the papers, this work focuses on the following five classifications: (1) Research objects; (2) Disturbance scenarios; (3) Research methods; (4) Evaluation metrics; (5) Interactions between different factors.

## 3. RESULTS

### *3.1. Objects studied in ATM resilience*

Research objects studied in the collected literature are divided into 8 categories, namely ATM system, ATC (Air Traffic Control), ATFM (Air Traffic Flow Management), Air traffic network, Airspace system, AAM (Advanced Air Mobility), TBO (Trajectory Based Operations) and Airport. In the first category, ATM system concerns the overall consideration talking about system design, definition and measurement, safety/risk assessment, etc. The category named ATC include ATC, ATC sector or ATC sector network related studies. Researches that focus on technical or human resources during ATC tasks are also classified into this category. Likewise, there are other categories labeled as ATFM, Air traffic network, Airspace system, AAM, TBO and Airport respectively.

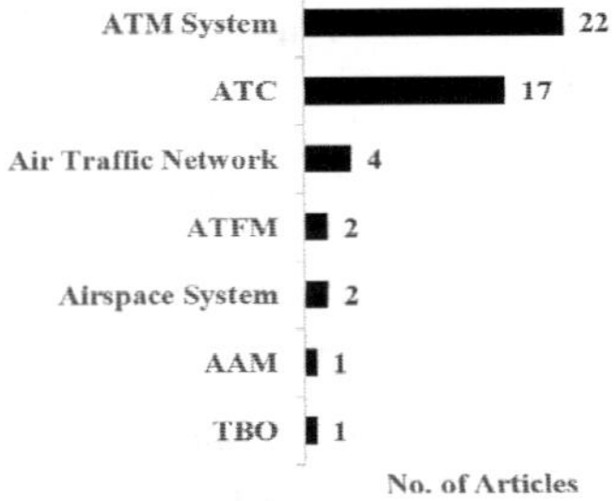


Figure 1. Distribution of research objects

As can be seen from Fig. 1, 44% (22) of the literature focus on the whole ATM system. 34% (17) of the literature are related to ATC, Air traffic network accounts for 8% (4) of the totality. ATFM and Airspace system makes up 4% (2) separately. For the last three ones, there is 1 paper for each group classified as AAM, TBO and Airport with a proportion of 2% (1). In the category of ATM, 3 papers contribute to ATM system design and development [5-7]. 15 papers are related to the analysis, assessment and management of ATM resilience [8-22]. 4 papers concerned about resilience definitions and measurements [23-26].

Among the 17 papers in the category of ATC, 8 of them payed attention to human factors [27-34]. 7 papers focused on technical problems such as data transmission resilience and communication network reliability in ATC system [35-41]. The left 2 papers studied the resilience of ATC sectors [42,43].

2 researches are classified into the category of ATFM [44,45]. Four papers are included into the category of Air traffic network [46-49].Two studies are with the perspective of Airspace system [50,51]. Classifications of AAM [52], TBO [53] and Airport [54] includes 1 paper respectively.

### *3.2. Disturbances considered in ATM resilience*

Disturbances taken into account in the literature can be organized into 2 groups, namely general and specific situations. Studies have not mentioned disturbances are not included. 7 papers considered general disturbances related to the following aspects and the interactions thereof: (1) human operators (human behavior and their communications); (2) technical systems (such as airborne equipment and ATC tools); (3) external environments (such as weather affects and flock of birds). 39 papers (a large majority of the literature) defined specific disturbance in their works, which are divided into detailed groups as shown in Fig. 2.

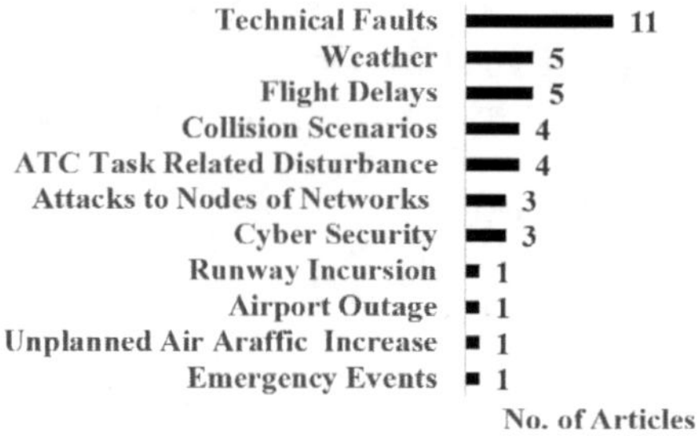


Figure 2. Detailed categories of specific disturbance

11 papers (28%) studied the resilience properties under failures, error or degradation of technical systems [7,10-12,21,24,35,36,41,42,53]. Edwards, et al. additionally take human operator breakdown into account [53]. 5 researches (13%) took bad weather conditions as a disturbance scenario [8,17,28,29,54]. 5 papers (13%) focused on flight delays to measure the resilience of air traffic systems [16,44,45,48,49]. 4 out of the literature (10%) payed attention to collision scenarios [13,14,26,32]. 4 papers (10%) focused on ATC task related disturbance [30,31,33,34]. 3 studies (8%) explored the resilience of the whole networks subject to different attacks to their nodes [43,47,50]. Three papers (8%) considered cyber security threats to study the system resilience under different technical attacks [38-40]. Patriarca, et al. discussed runway incursion associated disturbances [20]. Ryerson talked about the disturbance of flight divert due to the outage of a target airport [46]. Felici, et al. took unplanned increase of air traffic that exceeds the sector capacity as a disturbance [19]. Foster, et al explored the ATC resilience subject to emergency events (the closure of US airspace following the terrorist attack of 911) [32].

### *3.3. Methods employed in ATM resilience*

Various methods were employed to study the resilience performance of ATM. Such as resilience principles and methods, complex network theory, machine learnings, etc., which are grouped into different categories as depicted in Fig. 3 with the amounts. It should be noted that one paper may use one or more methods, hence one paper can be classified into different groups. Studies without specific methods are not included.

TABLE 1. REPRESENTATIVE METRICS OF COMPLEX NETWORK THEORY

| Metric | Expression | Note |
|---|---|---|
| Degree | $K_i = \sum_{j=1}^{N} e_{ij}$ | For the ATC sector network with *N* nodes, the degree $K_i$ of node *i* represents the number of sectors which have flight flow relationships with node *i*. |
| Betweenness | $B_i = \sum_{j \neq k} \frac{n_{ik}(i)}{n_{jk}}$ | Where $n_{jk}$ is the total number of shortest paths from sector *j* to sector *k*, and $n_{jk}(i)$ is the number of shortest paths through sector *i* from sector *j* to sector *k*. |

As per the categorization shown in Fig. 3, the most used method is test and simulations with 11 papers [13,17,18,20,21,23,30,31,39,40,54,]. A process of agent-based modelling and simulation (ABMS) for resilience engineering was shown in Fig. 4 adopted from [17]. 7 of the literature used resilience principles and methods [6,8,9,14,20,26,52]. Fig. 5 shows a sketch map of a function as a part of a whole FRAM model. 6 researches employed complex network theory [6,43,47-50]. Representative metrics used are shown in Table 1 [43]. 5 researches took the advantage of optimization methods [10-12,24,49]. Resilience management was taken as an optimization problem presented by (1) in the work of [10].

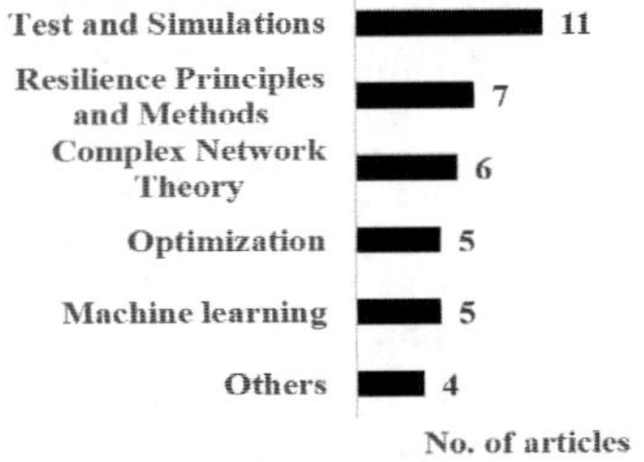


Figure 3. Methods employed in the studies

5 papers made use of machine learning approaches [16,22,33,44,50]. For instance, SVM, k-NN and RVFL were explored in the research of Sarlija, et al. [33]. 4 papers employed other methods namely statistical analysis [46], meta-population epidemic process model [49], discrete events model [38] and model of adaptation [34].

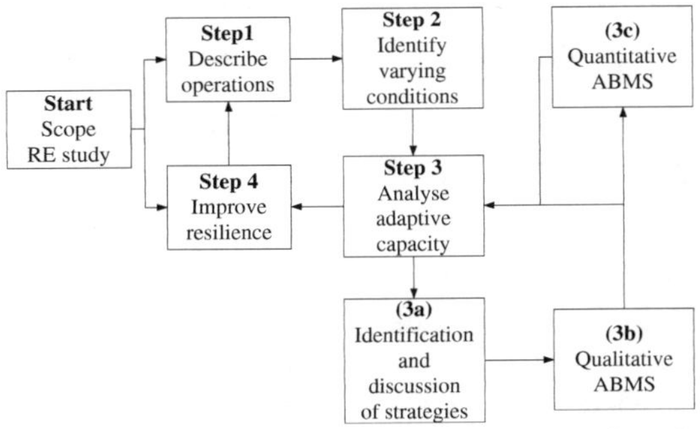


Figure 4. Process of ABMS for resilience engineering

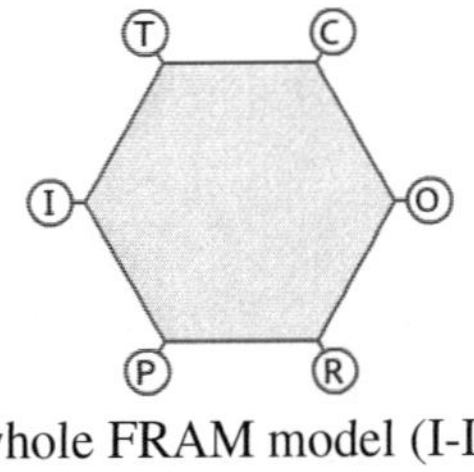


Figure 5. A function as a part of a whole FRAM model (I-Input; O-Output; P-Precondition; R-Resources or Execution Conditions; C-Control; T-Time)

$$S_{\mathrm{opt}} = \arg\min_{S_1 \in \Gamma^{(\mathbb{S})}} \mathrm{RL}_{\mathbb{S}}(S_l) = \arg\min_{S_1 \in \Gamma^{(\mathbb{S})}} d(S_0, S_l) \tag{1}$$

Where $S_{opt}$ is the best alternative flow in the scenario $\mathbb{S}$, $d(\cdot)$ is a flow distance function, $\Gamma^{(\mathbb{S})}$ is the set of alternative flows, $S_l \in \Gamma^{(\mathbb{S})}$ is an alternative route, and $S_0$ is to the nominal flow.

### *3.4.Metrics of ATM Resilience*

A variety of metrics have been applied to study the characteristics of ATM resilience. Among the included literature, 8 researches defined special indicators, namely cost resilience metric [25], Resilience Impact Factor (RIF) [13], Resilience Index (RI) [43], Task work scores of controllers [31], Service Quality Level(SL) and Operating System Affordance (OSA)RAGS levels [9], function of the four abilities [23], resilience score calculated by resilience factors [51], recovery rate and maximum eigenvalue[49]. Specifically defined mathematical descriptions for resilience are partially exhibited from (2) to (4) referring to the work of [25], [43] and [51] respectively. 35 papers used existing properties to measure resilience of different objects, for instance, the predictability and efficiency of available resources [5], ATM key performance areas (KPAs) [10-12,24,27], complex network theoretic measures [47,48,50,]. The left 7 papers did not define resilience metrics, which put more emphasize on other problems, e.g. exploring resilience at interconnected system levels [15], modelling and the analysis approach [8], key characteristics of ATM system resilience [14], etc.

$$R_C = \frac{\sum_u^d C_u(t) - \sum_u^d \sum_m C_u^m(t) - \sum_m C_m(t)}{\sum_u^d C_u(t)} \tag{2}$$

Where $R_c$ is the cost resilience metric, $\sum_m$ is the summation of events causally affected by the mechanism, and $\sum_u^d$ for the disturbance, $C_u(t)$ denotes the cost associated with a disrupted flight or passenger at time t in the absence of a mechanism, and $C_u^m(t)$ in the presence of a mechanism, $C_m(t)$ denotes any tactical costs associated with each investment mechanism.

$$R = \frac{\int_{t_d}^{t_{ns}} P(t)dt}{\int_{t_d}^{t_{ns}} P(t_0)dt} \tag{3}$$

Where $t_d$ is the time that system performance starts to decline and then recovers to the original phase at time $t_{ns}$, $P(t_0)$ is the target value of system performance and $P(t)$ is the function over time.

$$Resilience = 1 - \Pi[1 - R(component_i)] \tag{4}$$

Where $R(component_1)$ is resilience factor attributed to system availability, $R(component_2)$ is resilience factor attributed to system diversity, $R(component_3)$ is resilience factor attributed to system monitor and control.

### *3.5. Interactions between different factors*

To shed light on the interactions between different factors of socio-technical systems, the numbers of studies on this issue is shown in Fig. 6. Among the 35 papers (excluding 15 papers without clearly identified factors), 23 papers considered the interactions of two or more factors, 12 papers considered single factor including human, technique or management. Specifically, 9 papers considered the human, technique and environment interactions [5,6,8,9,12,17,23,37,42,]. 8 papers took into account human, technique and organization [7,14,15,18,20,26,32,41]. Six papers focused on the single factor of human [19,28-31,33]. 5 papers paid attention to technique issues [35,36,38,39,51].

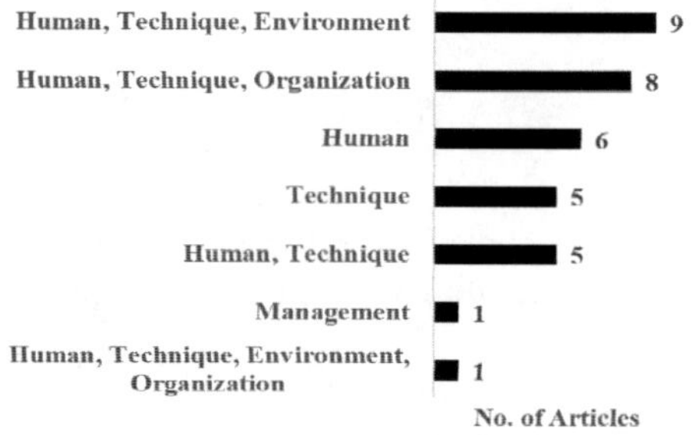


Figure 6. Interactions between different factors in the studies

5 papers concerned about human and technique interactions [11,21,24,52,53]. There is 1 paper considering management factor [34] and 1 paper combining the interactions of all the 4 factors namely human, technique, environment and organization [27].

## 4. DISCUSSIONS

### *4.1. Findings and implications*

According to the results presented in section III, A variety of objects in the field of air traffic management have been studied and most focuses were centered on ATM system and ATC. There are a wide range of research objects in the literature reviewed, varying from ATM system (top level) to components (basic level). Nevertheless, efforts should be made to study more specific topics, so as to explore the resilience properties under disturbances in depth. More particular situations are recommended to be taken into consideration to perform more detailed investigations. A large proportion of the papers considered specific scenarios as disturbance origins, while it is necessary to further explain the rationalities and necessities of the selected scenarios during disturbance identification and analysis. There seem to be insufficient systematic analysis on the origins of variations although a large variety of disturbances have been taken into account.

Considering the mainstream methods of current studies, simulations are able to provide a large amount of data as a useful analysis method, while inevitably dependences on the correctness, accuracy and assumptions in the simulation process affect the results significantly. Resilience principles and methods are specifically developed to deal with resilience issues, while some of them seemingly provides with evaluation forms rather than methods (e.g. RAG). Additionally, FRAM is a qualitative analysis method and may require further improvements to be a more effective tool for resilience analysis. Complex network theory offers theoretical supports for analyzing the impact of attacked nodes on the whole network, while few internal characteristics can be explored except the macroscopic properties of the whole network. Optimization methods employed in the studies enable researchers to explore the reallocation strategies of resources after disturbances. However, optimizations put more emphasis on the reallocation of tasks, which suggests the recovery performance of the system mainly and may overlook other resilience properties. Machine learning methods appear to perform well on the occurrence prediction of disturbances, which presents the ability of foresight. Nevertheless, similar insufficiencies as optimizations may exist in the method application process.

Most of the researches use existing properties to measure resilience performance, while part of them have defined special indicators referring to their study targets. No matter existing or created metrics, more safety related characters are suggested to be taken into account. As for the consideration of the interactions between different factors, more than half of the researches have taken into account two or more factors of the socio-technical system consists of human, technique, environment and organization. While systematic models dealing with the interactions are expected to be developed in future works.

### *4.2. Challenges and requirements*

As a complex socio-technical system, ATM is affected largely by the interactions and variations of its components. The challenge that resilience is faced with is to express the variability which decides the performance of the system. As Patriarca, et al. concluded in their research [55], attentions are increasingly payed to resilience modeling instead of defining. Modeling issue of resilience need to be further investigated. Besides, models developed to characterize resilience properties should help to understand the interactions between technology, human, environment and management. Moreover, what should be paid special attention to is the dynamic property of resilience that varies with time, system states, configurations, types or intensities of disturbance and so on. Additionally, there is necessity to develop methodologies and techniques to identify as well as to assess the internal or external threats originated from the interactions of the elements, so as to reasonably determine the variabilities and disturbances.

According to the accident analysis report from Airbus [56], over the last 20 years, a majority of commercial aviation accidents occurred during approach and landing phases. Among the 5 hull loss accidents recorded in 2021, 3 occurred in the approach and landing phases. The leading causes of fatal accidents are LOC-I (Loss of Control In-flight) (36%), CFIT (Controlled Flight into Terrain) (21%), RE (Runway Excursion) (17%). Simply SCF (System/Component Failure or Malfunction) related fatal accidents accounts for merely 7% of the total. As suggested in the Airbus report, more attentions in ATM resilience researches should be attracted to the approach and landing phases. Disturbances concerning LOC-I, CFIT and RE deserve further considerations when determining variation origins. Moreover, disturbances considered ought to be the interactions of the elements constituting the socio-technical system rather than single technical failures.

*4.3. Further directions*

The perspective of resilience engineering emphasize that either successes or failures are emergent phenomena which originates from the combinations or interactions between system components rather than the status of single elements. To realize a resilient system, concepts distinct from traditional modes of thinking should be based on. The start point ought to be targeted at the interacting mechanisms between events and functions. In order to take a step forward in resilience engineering investigations, the following methodologies may be attempted to throw light on promising research areas:

*1) System engineering*

System engineering provides an idea to interact and coordinate the relationships between different components and the system to achieve better overall performance. System engineering takes its components as a whole and considers problem from a systematic perspective. Above mentioned features of system engineering fit well into resilience engineering researches offering theoretical basis and effective tools. For instance, model based system engineering (MBSE) is especially suitable for the analysis of resilience properties owing to its flexibility and convenience in modeling and analyzing.

*2) Complexity theory*

Complex systems are composed of a large quantity of components with complex interactions, and the overall behavior of the system cannot be obtained by the behavior of its components. Individuals in complex systems have a strong coupling relationship with each other. The behavior of elements in a complex system follows certain rules, the states and behaviors can be adjusted according to external environment and information they received. Complex system theory regards complex system as a nonlinear system with high uncertainty, and studies its randomness under uncertainties. Resilience researches can draw from complexity theories to understand the overall performance under disturbances and to organize the internal structures.

*3) Control theory*

The same as safety, resilience can be regarded as a control problem. By imposing certain actions on the objects under control, controlling activities enable the controlled system to operate as expected. The goal of a control action is to eliminate deviations and errors. Consequently, resilience properties can be interpreted as the abilities of a system to maintain its expected state in control without excessively deviating from the predetermined control objectives in the presence of disturbances. Control model may be built to study the resilience of a system with its components, external disturbance and the interactions of different parts in a whole control loop. Many control theories and methods support resilience studies, such as robust control theory that can assist with the stability problem under uncertainties.

*4) Chaos engineering*

Chaos is a seemingly random motion in a deterministic dynamic system. Its essence is the extreme sensitivity of the long-term behavior of the system to the initial conditions. Chaos theory belongs to nonlinear science and has become a research hotspot of nonlinear science. Chaos improves our ability to describe, understand and even predict, so as to help us in dealing with instabilities. In resilience engineering, success as well as failure originate from the nonlinear combination of system components' behaviors. So chaos engineering may bring some inspirations to understand the generation principles and performance forecasting of system resilience.

To summarize, it seems to be necessary to characterize and model the property of emergence as a basic for further researches of resilience engineering. A key point worth noting is that efforts should be concentrated on the designing and developing stage to realize intrinsic resilient systems. It is undoubtedly much more significant to develop systems with considerable resilience properties than to analyze and assess the resilience of existing systems.

## 5. Conclusions

To explore the application status and developing potentials of resilience in ATM, this paper performed a comprehensive review supported by PRISMA method. Literature collected were summarized and analyzed based on different classifications from 5 aspects. Brief implications and comments were proposed with regard to each aspects respectively. The results reveal some problems remain to be solved in further studies. Based on the challenges ATM resilience is confronted with and actual requirements from accident analysis reports, promising directions are recommended as an attempt to throw some light on subsequent researches.

## References

[1] Hollnagel, E., Woods, D.D., Leveson, N. "Resilience Engineering: Concepts and Precepts " Ashgate Publishing, Aldershot, UK. 2006.
[2] EUROCONTROL . "A White Paper on Resilience Engineering for ATM ". 2009.
[3] EUROCONTROL. "From Safety-I to Safety-II: A White Paper." 2013.
[4] Moher, D., et al. "Preferred reporting items for systematic reviews and meta-analyses: the PRISMA statement." Int J Surg 8(5): 336-341, 2010.
[5] Woltjer, R., et al. "Towards understanding work-as-done in air traffic management safety assessment and design." Reliability Engineering and System Safety 141: 115-130, 2015.
[6] Weiland, L. V., et al. "Ensuring sustainable and resilient air traffic management systems for south africa with complexity and whole-of-society theory approaches." South African Journal of Industrial Engineering 31(3): 83-96, 2020.
[7] Yenson, S. K., et al. Exploring human-system resiliency in air traffic management technologies. 34th Digital Avionics Systems Conference, DASC 2015, September 13, 2015 - September 17, 2015.
[8] Ragosta, M., et al. Concept maps for integrating modeling techniques for the analysis and re-design of partly-autonomous interactive systems. 5th International Conference on Application and Theory of Automation in Command and Control Systems, ATACCS 2015, September 30, 2015 - October 2, 2015.
[9] Patriarca, R., et al. "Resilience engineering to assess risks for the air traffic management system: A new systemic method." International Journal of Reliability and Safety 10(4): 323-345, 2016
[10] Pascarella, D., et al. "An Analytical Approach for Optimal Resilience Management in Future ATM Systems." Intelligent Distributed Computing Ix, Idc'2015 616: 415-425, 2016.
[11] Palumbo, R., et al. Modeling Approach for Resilience Engineering of the Future ATM System. 15th AIAA Aviation Technology, Integration, and Operations Conference, 2015.
[12] Errico, A., et al. Simulation Approach to the Resilience Engineering Assessment of the ATM System in Crisis Scenarios. AIAA Modeling and Simulation Technologies Conference, 2016.
[13] Marshall, C. J., et al. "Context-Driven Autonomy for Enhanced System Resilience in Emergent Operating Environments." IEEE Systems Journal 13(3): 2130-2141, 2019.
[14] De Carvalho, P. V. R. "The use of Functional Resonance Analysis Method (FRAM) in a mid-air collision to understand some characteristics of the air traffic management system resilience." Reliability Engineering and System Safety 96(11): 1482-1498, 2011.
[15] Woltjer, R. Exploring Resilience at Interconnected System Levels in Air Traffic Management. Exploring Resilience: A Scientific Journey from Practice to Theory. S. Wiig and B. Fahlbruch: 105-112, 2019.
[16] Sanaei, R., et al. "A study of capacity regulations to define European air traffic management network states." Transportation Planning and Technology 44(4): 337-355, 2021.
[17] Stroeve, S. H. and M. H. C. Everdij. "Agent-based modelling and mental simulation for resilience engineering in air transport." Safety Science 93: 29-49, 2017.
[18] Forster, P., et al. An Approach for Attribute- and Performance-Based Evaluation of Interdependent Critical Infrastructures. 5th ENRI International Workshop on ATM/CNS, EIWAC 2017, November 14, 2017 - November 16, 2017.
[19] Felici, M., et al. Supporting resilience in Air Traffic Management. 2008 RISE/EFTS Joint International Workshop on Software Engineering for Resilient Systems, SERENE'08, November, 2008.

[20] Patriarca, R., et al. "A Monte Carlo evolution of the Functional Resonance Analysis Method (FRAM) to assess performance variability in complex systems." Safety Science 91: 49-60, 2017
[21] Brooker, P. "SESAR safety decision-making: Lessons from environmental, nuclear and defence modelling." Safety Science 48(7): 831-844, 2010.
[22] Sanaei, R., et al. "Toward ATM Resiliency: A Deep CNN to Predict Number of Delayed Flights and ATFM Delay." Aerospace 8(2):1-22, 2021.
[23] Pasquini, A., et al. Towards a measure of Resilience. 5th International Conference on Application and Theory of Automation in Command and Control Systems, ATACCS 2015, September 30, 2015 - October 2, 2015.
[24] Filippone, E., et al. "Resilience management problem in ATM systems as a shortest path problem." Journal of Air Transport Management 56: 57-65, 2016.
[25] Cook, A., et al. "Measuring the cost of resilience." Journal of Air Transport Management 56: 38-47, 2016.
[26] de Carvalho, P. V. R. and B. Ferreira. "Modeling activities in air traffic control systems: antecedents and consequences of a mid-air collision." Work-a Journal of Prevention Assessment & Rehabilitation 41: 232-239,2012.
[27] Stroeve, S. H., et al. "Analysis of the roles of pilots and controllers in the resilience of air traffic management." Safety Science 76: 215-227, 2015.
[28] Joyekurun, R.. Weather hazards in ATM: Designing for resilient operations. 25th Anniversary Conference of the European Association for Cognitive Ergonomics, EACE, August 28, 2007 - August 31, 2007.
[29] Joyekurun, R., et al. A functional collaborative work unit for analysing hazardous flight approaches. 51st Annual Meeting of the Human Factors and Ergonomics Society, HFES 2007, October 1, 2007 - October 5, 2007.
[30] Ćosić, K., et al. "Stress Resilience Assessment Based on Physiological Features in Selection of Air Traffic Controllers." IEEE Access 7: 41989-42005, 2019.
[31] Malakis, S., et al. "Managing emergencies and abnormal situations in air traffic control (part I): Taskwork strategies." Applied Ergonomics 41(4): 620-627, 2010.
[32] Foster, C. J., et al. "Adaptation as a source of safety in complex socio-technical systems: A literature review and model development." Safety Science 118: 617-631, 2019.
[33] Sarlija, M., et al. "Prediction of Task Performance from Physiological Features of Stress Resilience." IEEE Journal of Biomedical and Health Informatics 25(6): 2150-2161, 2021.
[34] Foster, C. J., et al. "A very temporary operating instruction: Uncovering emergence and adaptation in air traffic control." RELIABILITY ENGINEERING & SYSTEM SAFETY 208:1-14, 2021.
[35] Delicado, F., et al. Error resilient in MPEG-2 video transmission over wireless ATM networks. 2000 IEEE Conference on High Performance Switching and Routing, ATM 2000 - Joint IEEE ATM Workshop 2000 and 3rd International Conference on ATM, ICATM 2000, June, 2000.
[36] Kabashkin, I. Resilient communication network of Air Traffic Management system. 2016 Advances in Wireless and Optical Communications, RTUWO 2016, November 3, 2016 - November 4, 2016.
[37] Edwards, T. and P. Lee. Designing graceful degradation into complex systems: The interaction between causes of degradation and the association with degradation prevention and recovery. 18th AIAA Aviation Technology, Integration, and Operations Conference, 2018, June, 2018.
[38] Kamaruzzaman, M. R. B., et al. Analyzing Cascading Effects of Spoofing Attacks on ADS-B Using a Discrete Model of Air Traffic Control Responses and AGMOD Dynamics. 2021 IEEE International Conference on Cyber Security and Resilience (CSR).
[39] Leonardi, M., et al. "On Jamming Attacks in Crowdsourced Air Traffic Surveillance." IEEE Aerospace and Electronic Systems Magazine 36(6): 44-54, 2021.

[40] Vernotte, A., et al. "A domain-specific language to design false data injection tests for air traffic control systems." INTERNATIONAL JOURNAL ON SOFTWARE TOOLS FOR TECHNOLOGY TRANSFER,24 (2) , pp.127-158, 2022.
[41] Hollnagel, E., et al. "A day when (Almost) nothing happened." Safety Science 147: 1-6,2022.
[42] Jaksic, Z. and M. Janic. "Modeling resilience of the ATC (Air Traffic Control) sectors." Journal of Air Transport Management 89:1-20, 2020.
[43] Wang, X. L., et al. "Vulnerability and Resilience Analysis of the Air Traffic Control Sector Network in China." Sustainability 12(9):1-18, 2020.
[44] Sanaei, R., et al. Machine Learning Application in Network Resiliency based on Capacity Regulations. 38th IEEE/AIAA Digital Avionics Systems Conference, DASC 2019, September, 2019.
[45] Belkoura, S., et al. "Generation and recovery of airborne delays in air transport." Transportation Research Part C: Emerging Technologies 69: 436-450, 2016.
[46] Ryerson, M. S. "Diversion ahead: Modeling the factors driving diversion airport choice." Journal of Infrastructure Systems 24(1):1-11, 2018.
[47] Yassien, Y., et al. "Air Transportation Infrastructure Robustness Assessment for Proactive Systemic Risk Management." Journal of Management in Engineering 36(4):1-13, 2020.
[48] Cook, A., et al. "Applying complexity science to air traffic management." Journal of Air Transport Management 42: 149-158, 2015.
[49] Baspinar, B., et al. "Event-Based Air Transport Network Resiliency Management with Meta-Population Epidemic Model." Journal of Aerospace Information Systems 18(9): 632-644, 2021.
[50] Marzuoli, A., et al. Resilience of the national airspace system structure: A data-driven network approach. 33rd Digital Avionics Systems Conference, DASC 2014, October, 2014.
[51] Uhing, J. and T. S. Guyah. Measuring Resiliency for Withstanding and Rapid Recovery of NAS Events. 2020 Integrated Communications Navigation and Surveillance Conference, ICNS 2020, September 8, 2020 - September 10, 2020, Herndon, VA, United states, Institute of Electrical and Electronics Engineers Inc.
[52] Wing, D. J., et al. Achieving resilient in-flight performance for advanced air mobility through simplified vehicle operations. AIAA AVIATION 2020 FORUM, June 15, 2020 - June 19, 2020, Virtual, Online, American Institute of Aeronautics and Astronautics Inc, AIAA.
[53] Edwards, T. and P. Lee. Towards designing graceful degradation into trajectory based operations: A human-machine system integration approach. 17th AIAA Aviation Technology, Integration, and Operations Conference, 2017, June, 2017.
[54] Fines, K., et al. "Agent-Based Distributed Planning and Coordination for Resilient Airport Surface Movement Operations." Aerospace 7,48:1-25, 2020.
[55] Patriarca, R., et al. "Resilience engineering: Current status of the research and future challenges." Safety Science 102: 79-100, 2018.
[56] Airbus. "Statistical Analysis of Comercial Aviation Accidents 1958-2021.", 2022.

# Study on bottom thermal environment of launch vehicle during high altitude flight

**Chen-geng ZHAO[1,a], Zhong-yi SUN[1,b], Yi-fei SU[1,c], Yi-chen WANG[1,d],Gui-gao LE[1,*]**

[1]School of mechanical engineering, Nanjing University of technology, Nanjing, Jiangsu 210094

[a]e-mail: 2846822647@qq.com; [b]e-mail: 982658200@qq.com;

[c]e-mail: yifeisue@163.com; [d]e-mail: 496234810@qq.com

[*] Corresponding author: leguigao@ njust.edu.cn

**Abstract**—Aiming at the problem of the thermal environment at the bottom of the launch vehicle in the high-altitude flight stage, taking the launch vehicle with core stage four-way machine as the model, the coupled convection / radiation heat transfer problem at the high-altitude of the launch vehicle is deeply studied through the numerical simulation method, the gas jet flow model of the launch vehicle in the high-altitude flight stage is established, and the flow field information of the rocket under different working conditions and the heat flow at the bottom of the rocket are obtained. The numerical results are compared with the experimental data to verify the effectiveness of the selected model and calculation method. The research shows that the external flow field of the rocket body will have a great impact on the thermal environment at the bottom. With the continuous increase of flight altitude, the convective heat flow at the bottom of the rocket body first rises and then decreases. In the high-altitude stage, the total heat flow at the bottom of the rocket body is dominated by radiant heat flow, the convective heat flow outside the main nozzle is dominant, and the radiant heat flow outside the traveling jet nozzle is dominant.

## 1. INTRODUCTION

When the launch vehicle engine works at high altitude, the gas flow expands rapidly after entering the external environment because of the low external environmental pressure. After the rocket takes off, as the flight altitude gradually increases, the external atmospheric pressure gradually decreases, and the engine gas expansion angle gradually increases, forming a plume phenomenon. During the second stage flight, the engine plume forms a recirculation heating zone at the bottom, and the high-temperature gas has a strong thermal radiation effect on the structure around the bottom of the second stage[1].

The two-stage rocket spends most of its time flying at high altitude. After the jet in the thrust chamber of the two-stage traveling block leaves the nozzle, it will expand at a large angle. Part of the jet will spray onto the outer surface of the nozzle of the main engine, affecting the heat transfer and structure of the nozzle of the main engine[2-3]. According to the flight sequence, in the later stage of the second stage operation, the main engine is shut down, and the traveling block continues to work for hundreds of seconds. The jet of the traveling block heats the nozzle of the main engine more seriously, and may even burn through the stainless steel nozzle of the main engine, causing adverse lateral interference to the flight attitude of the rocket. In order to obtain the thermal environment at the bottom of the second stage flight section of the rocket, and analyze the impact of the jet of the traveling block and the exhaust of the main engine and the turbine of the traveling block on the thermal environment at the bottom of the main engine and the bottom, it is necessary to establish a gas jet thermal simulation model to analyze the thermal environment at the bottom of the second stage.

Compared with foreign research on the thermal environment at the bottom of launch vehicles, China has started relatively late at present. In particular, there are few flight test data of liquid launch vehicles in China, and most domestic and foreign researchers' research on the thermal environment at the bottom of launch vehicles is the research results of numerical simulation. Due to the limitation of computing resources, most researchers have simplified the numerical simulation to various degrees, for example, only one quarter of the The grid has no boundary layer and thermal radiation, which greatly affect the correctness of the calculation results[4]. In this paper, CFD software is mainly used to numerically calculate the jet flow field of a two-stage liquid launch vehicle engine, and through large-scale in-depth research, the changes of heat flow at the bottom of the launch vehicle under various flight altitude conditions are analyzed.

## 2. Calculation Model and Method

### *2.1. Computational model*

The calculation model is shown in Figure 1. The core stage section is bundled with 1 main engine and 4 traveling block engines, plus the main engine turbine exhaust pipe, traveling block turbine exhaust pipe, support and arrow body. The center of the main engine nozzle outlet section is defined as the coordinate origin o (0, 0, 0), along the arrow head direction as the Z axis, and the coordinate system is fixed on the arrow body.

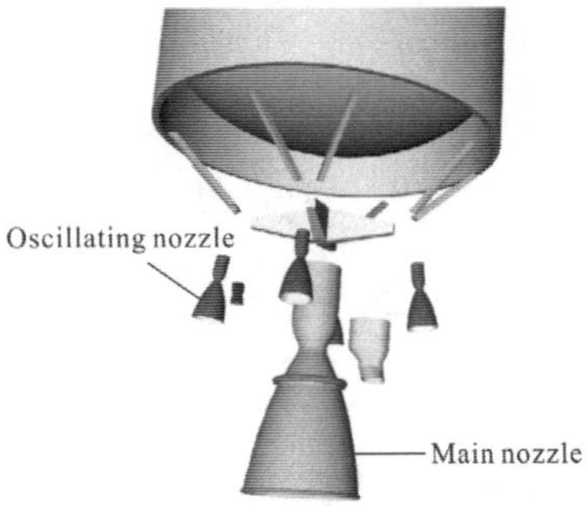


Fig.1 computational model

### *2.2. Computational grid*

For the high-altitude flight stage model of the launch vehicle, the block grid division method is adopted, and the grids of the main engine nozzle, the traveling block nozzle and the important jet region are densified. At the same time, in order to improve the smoothness and orthogonality of the computational grid, the structural grid is used in the computational domain. The grid along the far end of the arrow body gradually becomes sparse, and the total number of grids reaches 12.23 million, as shown in Figure 2.

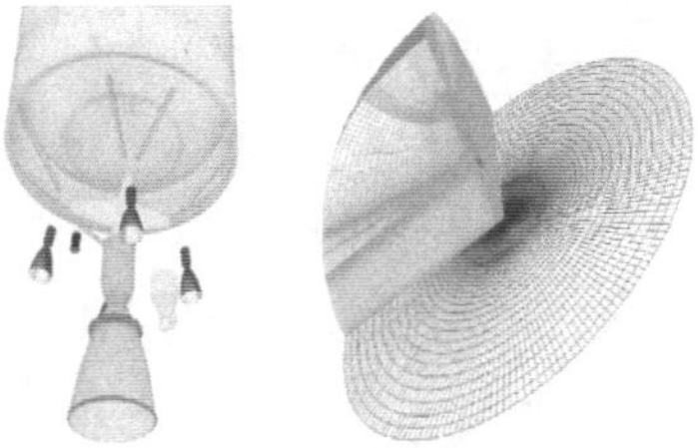

Fig.2 Computational grid

*2.3.governing equation*

Multicomponent N-S equation

It is proposed to adopt the flow model of real gas and air multicomponent transport. Before establishing the gas jet control equation, the following basic assumptions are made for the gas jet: (1) the gas jet meets the continuum assumption; (2) The gas jet is a compressible pure gas phase medium; (3) There is no chemical reaction inside the gas jet; (4) The multi-component mixed flow model of gas and air is adopted, and various components meet the ideal gas equation of state. The multicomponent transport equation is

$$\frac{\partial}{\partial t}(\rho Y_i)+\nabla\cdot\vec{J}_i+R_i+S_i=0 \qquad i=1,\cdots N \tag{1}$$

Component diffusion flux

$$\nabla\cdot\vec{J}_i=-\rho D_{i,m}\nabla Y_i \tag{2}$$

In rectangular coordinate system, the conservative form of compressible N-S equation of single component is:

$$\frac{\partial}{\partial_t}\iiint_V QdV+\oint\!\!\oint_S[(E-E_v)\vec{i}+(F-F_v)\vec{j}+(G-G_v)\vec{k})]\cdot d\vec{S}=0 \tag{3}$$

Of which:

$$\begin{cases} Q=(\rho,\rho u,\rho \upsilon,\rho w,\rho E)^T \\ E=(\rho u,\rho u^2+p,\rho u\upsilon,\rho uw,\rho u(E+P/\rho))^T \\ F=(\rho\upsilon,\rho u\upsilon,\rho\upsilon^2+p,\rho\upsilon w,\rho\upsilon(E+P/\rho))^T \\ G=(\rho w,\rho uw,\rho\upsilon w,\rho w^2+p,\rho w(E+P/\rho))^T \end{cases} \tag{4}$$

$$\begin{cases} E_v=(0,\tau_{xx},\tau_{xy},\theta_x)^T \\ F_v=(0,\tau_{yx},\tau_{yy},\theta_y)^T \\ G_v=(0,\tau_{zx},\tau_{zy},\theta_z)^T \end{cases} \tag{5}$$

$$\begin{cases} \tau_{xx}=2\mu(2u_x-\upsilon_y-w_z)/3 \\ \tau_{yy}=2\mu(2\upsilon_y-u_x-w_z)/3 \\ \tau_{zz}=2\mu(2w_z-u_x-\upsilon_y)/3 \\ \tau_{xy}=\mu(2u_y+\upsilon_x) \\ \tau_{xz}=\mu(2u_z+w_x) \\ \tau_{yz}=\mu(2\upsilon_z+w_y) \end{cases} \tag{6}$$

$$\begin{cases} \theta_x=u\tau_{xx}+\upsilon\tau_{xy}+w\tau_{xz}+K\partial T/\partial x \\ \theta_y=u\tau_{xy}+\upsilon\tau_{yy}+w\tau_{yz}+K\partial T/\partial y \\ \theta_z=u\tau_{xz}+\upsilon\tau_{xz}+w\tau_{zz}+K\partial T/\partial z \end{cases} \tag{7}$$

$$d\vec{S} = (n_x\vec{i} + n_y\vec{j} + n_z\vec{k})dS \tag{8}$$

The total energy per unit volume of gas is

$$P = (\gamma - 1)[\rho E - \rho(u^2 + v^2 + w^2)/2] \tag{9}$$

The turbulence model uses K- ε Model.

k- ε Model equation

$$\frac{\partial}{\partial t}(\rho k)+\frac{\partial}{\partial x_i}(\rho k u_i)=\frac{\partial}{\partial x_j}(\alpha_k \mu_{eff}\frac{\partial k}{\partial x_j})+G_K+G_b-\rho\varepsilon-Y_M+S_K \tag{10}$$

$$\frac{\partial}{\partial t}(\rho\varepsilon)+\frac{\partial}{\partial x_i}(\rho\varepsilon u_i)=\frac{\partial}{\partial x_j}(\alpha_\varepsilon \mu_{eff}\frac{\partial \varepsilon}{\partial x_j})+C_{1\varepsilon}\frac{\varepsilon}{K}(G_K+C_{3\varepsilon}G_b)-C_{2\varepsilon}\rho\frac{\varepsilon^2}{K}-R_\varepsilon+S_\varepsilon \tag{11}$$

Thermal radiation equation

In the radiation heat transfer model, the integral differential equation of radiation heat transfer is the basic starting point. The equation is

$$S \cdot \frac{\partial I}{\partial x_j} = k_\alpha I_b - (k_\alpha + k_s)I + \frac{K_S}{4\pi}\int H(K,\bar{K})I_\alpha \bar{K} \tag{12}$$

*2.4.numerical method*

N-S equation

As the high-altitude rocket is a typical hypersonic flow, under the continuous thrust acceleration of the engine, the rocket speed is as high as 2000m/s, and the plume Mach number is also very large. It is necessary to use the AUSM numerical scheme which can capture the strong shock wave to effectively suppress the non physical oscillation after the strong shock wave.

AUSM series format system has strong working and learning ability to identify physical discontinuities such as shock wave and simple contact. At the same time, it has excellent positive stability. It does not form the special phenomenon of ruby in the calculation of hypersonic flow. Therefore, it is easy to be popularized to other hyperbolic systems. It can better solve the problem of shock wave transmission at low speed and has the advantages of high calculation efficiency. This requires that it can not only solve the shock wave in the inviscid internal flow field with high resolution, but also accurately solve the shear flow in the viscous internal flow field.

Due to the advantages of AUSM series format, this calculation will use the structural finite volume method designed by this format to realize the spatial structure discretization of the model, and this method can also be used for the hybrid grid composed of structural grid and unstructured grid. AUSM format is the most basic format in this series, which was first proposed by Liou. Other formats after that were improved on its basis.

Thermal radiation equation

The discrete coordinate method (do) was first proposed by Chandrasekhar and developed by Carlson B G and Lathrop K D. Recently, fiveland and Jamaluddin studied its application in radiation heat transfer simulation. It discretizes the radiation heat transfer into m discrete directions, each direction corresponds to a certain solid angle, within which the radiation intensity is a fixed value.

## 3. Resule and Discussion

After the launch of the launch vehicle, due to the continuous rise of the flight altitude and the change of the angle of attack, and according to whether the main engine nozzle is started or not, the main parameters that change include the ambient pressure, temperature and flight Mach number. In order to study the changes of the thermal environment at the bottom of the launch vehicle at different altitudes, the following four typical operating conditions will be selected for numerical simulation. Table 1 shows the free flow condition parameters of these four operating conditions. The temperature at the inlet of the main engine is 660k and the pressure is 0.38mpa. The temperature at the inlet of the traveling block is 760K and the pressure is 0.2MPa. The gas components include COOH, H2O, CO2, Co, H2, oh and air to simulate a more realistic flight environment.

TABLE 1. FREE FLOW CONDITION PARAMETERS

| Number | h/km | P/MPa | T/k | Main nozzle |
|---|---|---|---|---|
| 1 | 75 | 2.3880 | 208 | on |
| 2 | 100 | 0.0320 | 195 | on |
| 3 | 100 | 0.0071 | 240 | on |
| 4 | 110 | 0.0071 | 240 | off |

*3.1.Mach number distribution at different altitudes*

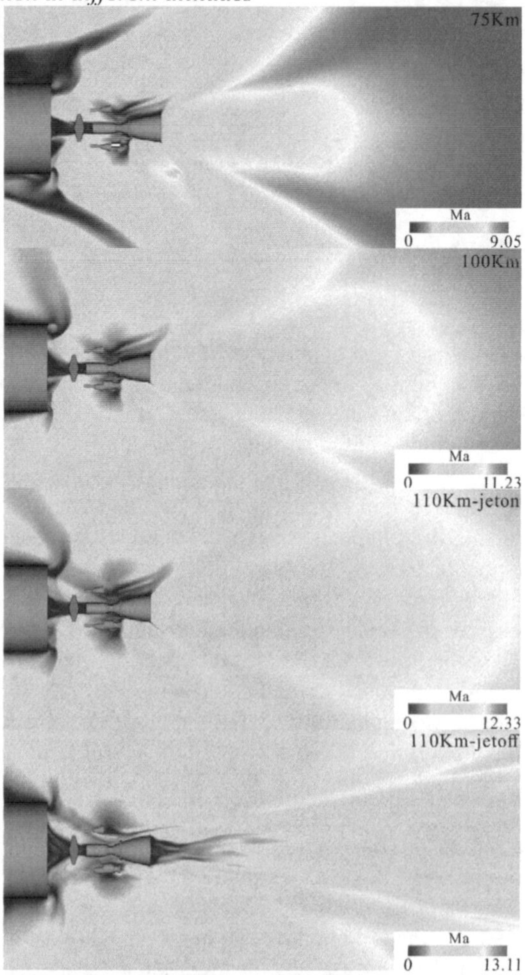


Fig.3 Mach cloud diagram at different altitudes

Because the total pressure of the rocket engine is much greater than the external ambient pressure, the gas flow will produce barrel pressure fluctuations around the nozzle outlet. However, because the rocket flying height is too high, the jet wave nodes in the gas flow field are not prominent.

It can be seen that at 75KM, due to the great difference between the total pressure in the nozzle and the external atmospheric pressure, the rocket gas flow expands strongly during the injection process, and is compressed by the upstream gas flow, forming a large plume area downstream of the rocket body.

At 100km, the compression degree of the upstream flow on the tail plume of the arrow body is weakened, which is caused by the decrease of atmospheric pressure, density and temperature. The maximum Mach number of the plume further increases, and the jet can fully expand in the farther downstream area of the arrow body. When the rocket altitude rises to 110km, due to the continuous decrease of external pressure and temperature

Compression continues to weaken and jet expansion continues to strengthen. This is because the gas discharged from the nozzle of the traveling crane is above the main engine, and the gas flow discharged from the turbine exhaust pipe of the main engine and the turbine exhaust pipe of the traveling crane also expands. When the nozzle of the main engine is shut down under condition 4, the barrel compression wave disappears.

### *3.2. Temperature distribution at different heights*

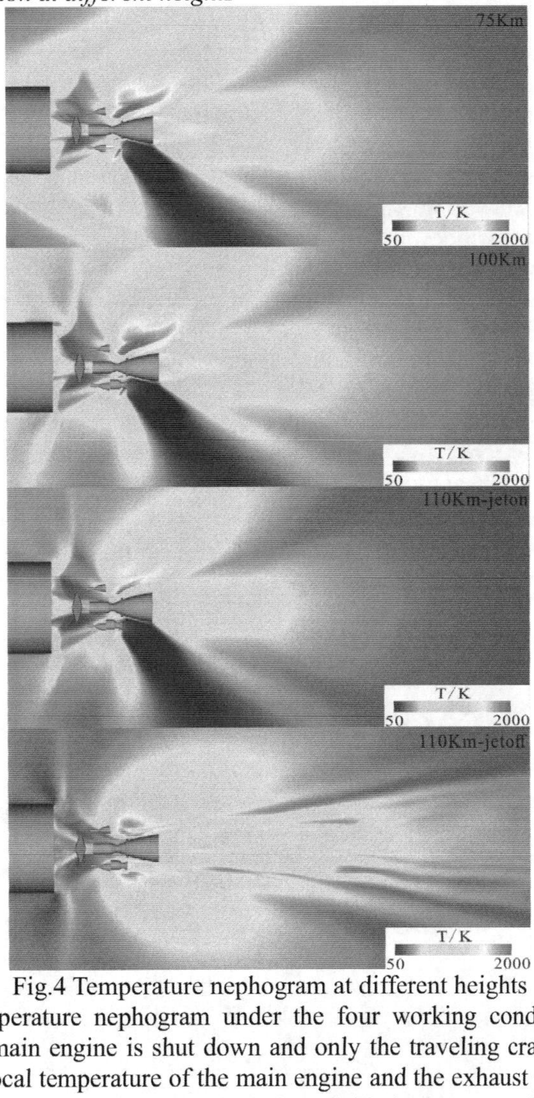


Fig.4 Temperature nephogram at different heights

Figure 4 shows the temperature nephogram under the four working conditions of 75KM, 100km, 110km and 110km (the main engine is shut down and only the traveling crane works). It can be seen from the figure that the local temperature of the main engine and the exhaust pipe of the main engine is high, and the temperature of other parts is relatively low. With the increase of altitude, the jet expansion at the nozzle tail increases. The expanded jet flow boundary and the supersonic autonomous incoming flow generate a shear layer. The high-temperature gas flow from the nozzle quickly returns to the lower part of the launch vehicle under the influence of the shear layer flow, resulting in convective heat flow.

The free incoming flow, engine jet and shock wave form a closed region between the lower part of the rocket and the nozzle side wall, The convective heat flow generates a recirculation system in the closed shear layer to continue convective heat transfer to the lower part of the rocket, resulting in the gradual increase of the convective heat flow value at the lower part due to the increase of altitude in the early stage. However, with the increasing altitude, the tail jet flow and shock wave expand more violently, which makes the closed area gradually narrow.

*3.3.Heat flow distribution at different heights*

Figure 5 shows the change of total heat flow distribution at the bottom of the launch vehicle with flight altitude. The size of the peak heat flow depends on the impact strength of high-temperature gas.

As the flight altitude increases, the external pressure decreases and the jet expansion continues to strengthen, resulting in more backflow and increased convective heat transfer at the bottom of the arrow. On the other hand, as the altitude increases, the Mach number of the incoming flow in front increases, which hinders the backflow at the bottom of the arrow. Therefore, the convective heat at the bottom of the arrow decreases. When the altitude is low, the impact of external pressure on the convective heat flow at the bottom of the rocket is greater than the incoming flow velocity, so the convective heat flow increases with the increase of altitude. When the altitude is high, the influence of external pressure on the convective heat flow at the bottom of the rocket body is smaller than the incoming flow velocity, so the convective heat flow shows a downward trend. Therefore, in the high altitude stage, the total heat flow at the bottom of the arrow body is dominated by the radiant heat flow.

Figure 6 shows the distribution of radiant heat flow at the bottom of the arrow at different heights. It can be seen that the radiant heat flow at the bottom of the arrow accounts for a large proportion of the total heat flow at high altitude, which is greater than the impact of convective heat flow. When the altitude is 75KM, the radiant heat flow reaches the maximum value. When the altitude is 110km, the main engine is shut down, and the radiant heat flow decreases significantly.

At the same time, the jet from the exhaust pipe of the downstream turbine and the main turbine at different heights causes the convective heat flow on the outer wall of the main engine nozzle. Affected by the gas flow from the upstream turbine and the main engine nozzle, the local convective heat flow on the bottom cooling ring near the main engine pipe is large. The total heat flow distribution on the surface of the main engine and the traveling block at different heights is shown in Figure 7 and figure 8.

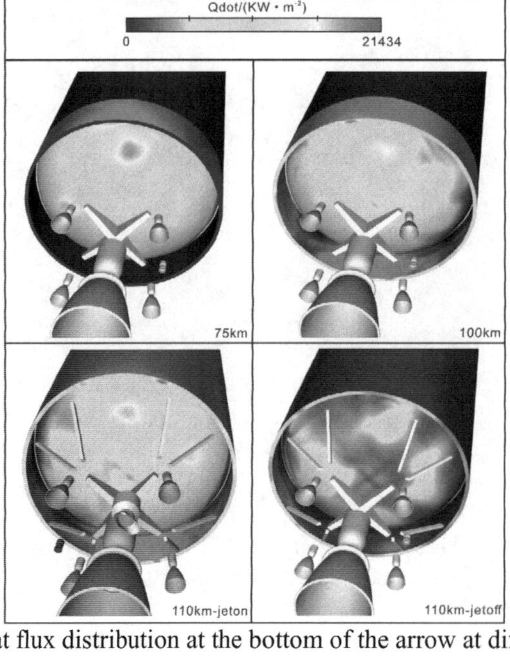


Fig.5 Total heat flux distribution at the bottom of the arrow at different heights

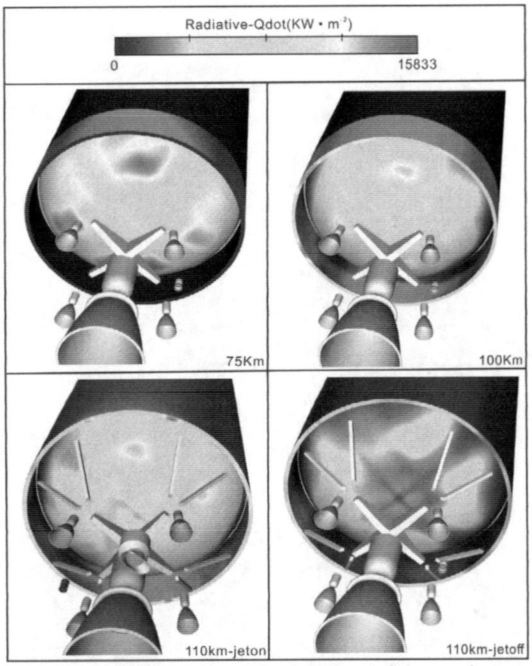


Fig.6 Radiation heat flux distribution at the bottom of the rocket at different altitudes

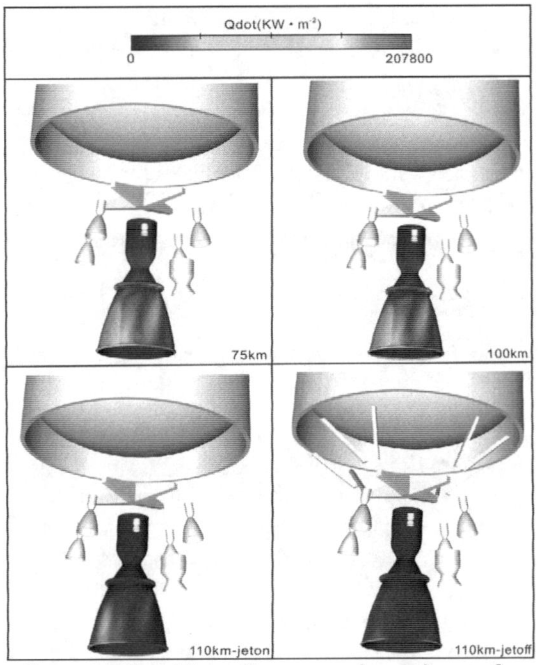


Fig.7 Distribution diagram of total heat flow on main engine surface at different heights

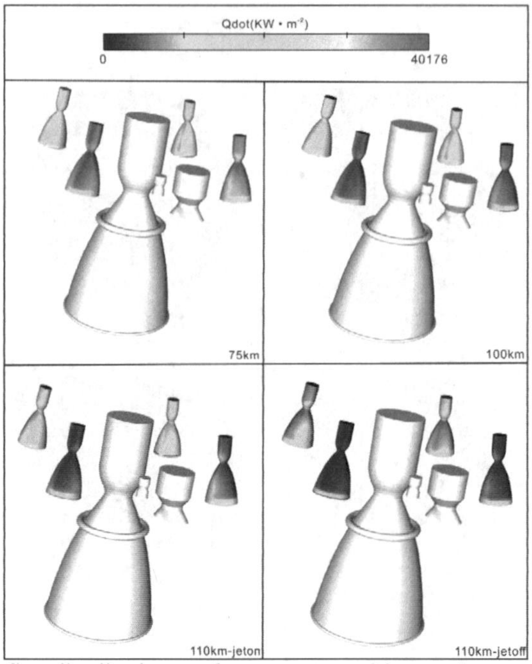

Fig.8 Total heat flux distribution on downstream turbine surface at different heights

*3.4.Comparison between numerical simulation and experiment*

A two-stage launch vehicle is selected as the research object of flight experiment. Compared with the traditional LOX kerosene engine, LOX methane engine has the advantages of low carbon deposition and high specific impulse, so LOX methane is selected as the propellant of launch vehicle. Two heat flow sensors are placed at the bottom of the experimental rocket. Figure 9 shows the installation position of the sensors. The launch vehicle is tested in the actual trajectory flight to obtain the value measured by the sensor at the bottom of the rocket.

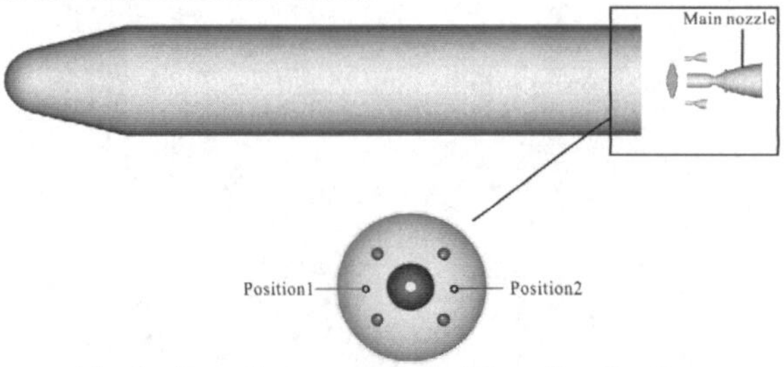

Fig.9 Heat flow sensor position distribution

At the same time, in the process of numerical calculation, two monitoring points 1 and 2 are placed at the same position as the heat flow sensor, and the heat flow parameters of the monitoring points are obtained. By comparing the feedback data from the sensor at the bottom of the rocket with the simulation data, the accuracy of the numerical calculation method selected in this paper is verified. Figure 10 is the comparison curve between the total heat flow measured by the flight experiment of the two sensors and the total heat flow calculated by the two monitoring points. It can be seen from the figure that the degree of coincidence between the two is relatively high or close. 130s after the launch of the launch vehicle, the total heat flow appeared the first peak, which corresponded to the vicinity of the 75KM working condition. With the further increase of the flight altitude, the total heat flow showed a slight downward trend. 180s after the launch, the total heat flow reached the peak for the second time, which corresponded to the vicinity of the 110km working condition.

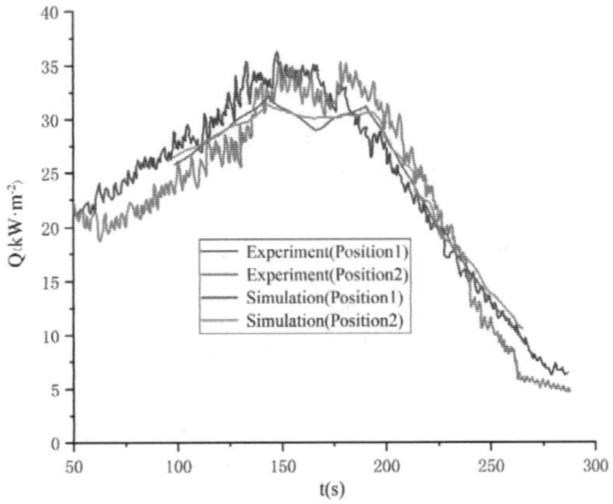


Fig.10 Total heat flow comparison

Figure 11 shows the comparison between the Mach cloud and isosurface diagrams of the ground test run experiment and numerical calculation of the launch vehicle engine. From the figure, it can be seen that the barrel compression wave and Mach node in the nozzle gas jet basically perfectly correspond to the ground experiment, which further verifies the reliability and effectiveness of the model selected in this paper.

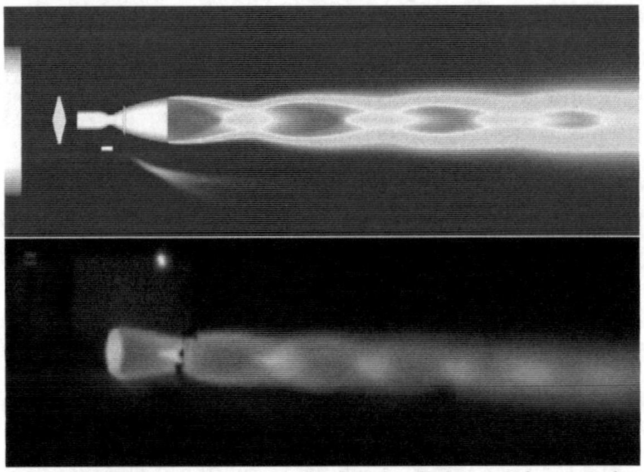

Fig.11 Comparison between numerical calculation and ground experiment

## 4. Conclusions

(1) The numerical simulation results are in good agreement with the experimental results, which verifies the accuracy and effectiveness of the numerical method used in this paper

(2) In the high altitude stage, the total heat flow at the bottom of the rocket body is mainly radiant heat flow, and the four states account for more than 78.79% of the total heat flow. Turn off the main engine. When only the traveling block works, the heat flow on the side wall of the arrow body is minimum.

(3) The exhaust pipe jet of the turbine of the traveling block and the main engine causes the convective heat flow on the outer wall of the main engine nozzle. Affected by the gas flow of the traveling block and the main engine nozzle, the local convective heat flow on the bottom cooling ring near the main engine nozzle is large. This is because the flow is stagnant on the slit of the cooling ring, and the convective heat values at the other three heights are relatively similar.

(4) After the nozzle of the main engine is shut down, the impact of the exhaust pipe of the traveling block and the turbine on the convective heat flow of the nozzle of the main engine increases, because the main engine has no gas outflow expansion obstruction, and the exhaust of the traveling block is easier to expand and flow to the outer wall of the main nozzle. After the main engine is shut down, the

heat flow distribution at the bottom of the arrow body decreases. Since only the gas of the traveling crane expands to the bottom of the arrow body, it is far less than the large amount of under expanded gas jet flowing to the bottom of the arrow body when the main engine is working.

(5) The influence factors of the thermal environment at the bottom of the launch vehicle in the high altitude flight phase studied in this paper provide a certain basis and reference value for the safety design of the thermal protection device at the bottom of the rocket.

## References

[1] Yan Zhijiang, Shen Dan, Wu Yansen, Pu Pengyu, Gong Yukun Study on thermal environment at the bottom of multi nozzle launch vehicle [J]. Missile and space delivery technology, 2021 (01): 105-109+114.

[2] Su Hong, Xu Shanshu, He Wei, Shen Dan Study on the thermal environment of the bottom of a new generation of large rocket [J]. Missile and space delivery technology, 2021 (05): 20-24 + 49

[3] Zhou zhitan, Ding Yifu, Le guigao, Liang Xiaoyang, sun Peijie, Sheng Yinghua Study on thermal environment of liquid launch vehicle bottom in high altitude flight environment [J] .Journal of Astronautics, 2019,40 (05): 577-584

[4] Shi Yanan Study on the effect of nozzle plume on the thermal environment at the bottom of the bundled launch vehicle [d].Beijing University of technology, 2016

[5] Zhou Yipeng, Zhu Dingqiang. Numerical calculation of radiant heat flow on nozzle wall of liquid rocket engine [J]. Rocket propulsion, 2015, 41 (3): 27-32

[6] Editorial board of the world space vehicle collection Encyclopedia of world space vehicles [M].Beijing: Aerospace publishing house, 1996: 76-77

[7] Wang Xun Study on the influence of engine jet interference on bottom thermal environment [J].General aerospace technology, 2021,5 (02): 24-30

[8] Hideyo N, et al. Numerical analysis of plume heating environment for H-II a launch vehicle during powered ascent[R].AIAA 2007-5505, 2007.

[9] Mehta M, el at. Numerical base heating sensitivity study for a four-rocket engine core configuration[J].Journal of Spacecraft and Rockets, 2013, 50(3): 509-526.

[10] Tang Z G, Liu G, Mou B, et al. Effects of turbulence models on the numerical simulation of nozzle jet[J].Acta Areodynamica Sinica 2010 28(2):188 — 196

[11] Cai Guo-biao, Wang Hui-yu, Zu Guo-jun, et al. Mont carlo direct simulation of small nozzle plume in vacuum [J].Journal of Propulsion Technology.1997, 18(2): 44 — 49.

[12] Reijasse P, Delery J. Investigation of the flow past the ARIANE 5 launcher afterbody [J].Journal of Spacecraft & Rockets, 1994, 31( 2) : 208 — 214.

[13] Dai Xinyi Study on internal process and bottom convective thermal environment of solid bound rocket [D].Shanghai: Shanghai Jiaotong University, 2014.

[14] Yan Zhijiang, Wu Yansen, Zhang Yunlong, Shen Dan, Tang Wei Review of the research on the thermal environment of the bottom jet of the first stage of launch vehicle at home and abroad [J].Manned spaceflight, 2021,27 (02): 262-268.

# Research on modular scheme of aviation emergency rescue for natural disasters

**Yueyue Li[1, a], Zhenlei Liu[1, b*], Keyu Chen[2, c], Ji Qi[1, d], Chuang Li[1, e]**

[1] College of aerospace, Shenyang Aerospace University, Shenyang, Liaoning, China

[2]Southwest Automation Research Institute, Mianyang, Sichuan, China

[a]e-mail: yueyuel08@163.com, [b*]email: alex@sau.edu.cn,

[c]e-mail: 775817820@qq.com, [d]e-mail: 215340275@qq.com,

[e]e-mail: 1158263278@qq.com

**Abstract.** Aviation emergency rescue is the core part of China's emergency rescue system and the main rescue force, which plays an important role in disaster relief, rescue materials transfer, casualty transportation, air rescue, major health and epidemic prevention. Based on H145 helicopter, the modular design concept is adopted to redesign the aviation rescue equipment, and the modular structure layout of the cabin is completed for major natural disasters. Aiming at the path planning problem of rescue helicopter, based on the improved ant colony algorithm, the optimal path planning of helicopter is completed by comprehensively considering the constraints of helicopter flight performance and flight environment. Through the use of medical data interaction technology, a standardized transport information service platform is established, which solves the problem of medical information island between the transfer unit, the transport unit and the evacuation hospital when the helicopter performs the aviation emergency rescue task, realizes the interconnection of medical data, effectively shortens the response time of aviation emergency rescue and improves the rescue efficiency.

## 1. Introduction

Aviation emergency rescue refers to the response to major accidents disasters and accidents, by using aviation technology and technical equipment, aircraft as the main carrier, the implementation of emergency patients rapid transport, air emergency rescue, medical supplies delivered, search and rescue flight activities [1-3]. In order to promote the construction of China's aviation emergency rescue system and form a unified command, both special and comprehensive, responsive and up-down linkage emergency management system with Chinese characteristics. The State Council issued the 14th Five-Year National Emergency System Planning, which clearly points out that it is necessary to accelerate the construction of aviation emergency rescue forces and improve the comprehensive aviation rescue capacity [4]. Aviation emergency rescue is the fastest, safest and most efficient rescue method to deal with all kinds of major natural disasters and emergencies. It has the characteristics of fast response, motor ability and wide rescue range. It shows the characteristics of fast policy focus, fast social demand appeal, fast emergency disposal and fast capacity-building guarantee [5-6].

## 2. Research on aviation emergency rescue mission and equipment

### *2.1. Analysis of aviation emergency rescue task types*

In 2021, a total of 107million people were affected by various natural disasters throughout the year, 867 people died and disappeared due to the disaster, 5.738 million people were urgently relocated, 616 forest fires occurred, and the affected forest area was about 4292 hectares, direct economic loss 334.02 billion yuan [7], the randomness and immeasurability of sudden major natural disasters have caused huge casualties and resource losses. Due to the characteristics of high rescue efficiency and fast response speed, aviation emergency rescue has played an enormous role in the transportation of the wounded, air first aid, emergency rescue and disaster relief, and rescue materials transfer.

According to the frequency of major natural disasters and safety accidents in China, aviation emergency rescue is divided into eight task types, which are air reconnaissance and detection, air resource scheduling, air fire and rescue, air emergency transportation, air search and rescue, air special payload, air emergency communication and international rescue.

Table 1. Types and contents of aviation emergency rescue tasks

| Task type | Task content |
|---|---|
| Air reconnaissance and detection | Survey, mapping and monitoring of affected areas and accident sites; Transmit relevant information to the command system efficiently. |
| Air resource scheduling | The rescue forces such as air, ground, helicopter and UAV in the disaster area are unified dispatched to provide air guidance services. |
| Air fire and rescue | Airborne hanging equipment such as buckets, engine-bellied water tanks and fire-extinguishing water cannons are used to carry out fire extinguishing operations for special fires such as forests and high-rise buildings. |
| Air emergency transportation | By means of air transportation, airdrop, and aircraft landing, rescue forces, equipment, and disaster relief materials are rapidly delivered to the affected areas and the scene of the accident to transfer and evacuate people in distress and casualties. |
| Air search and rescue | Use graphic transmission equipment, tweeters, search lights, combined with hand-held GPS and other auxiliary equipment to carry out air search and positioning of people in distress and trapped. |
| Air special payload | When the ground traffic is blocked or the traffic is inconvenient, the rescue equipment and materials will be accurately delivered by hanging within 1 hour. |
| Air emergency communication | When the communication signal is not covered or the communication facilities are damaged, the communication needs of rescue operations are guaranteed by establishing air base stations and relay platforms. |
| International rescue | Carrying out foreign-related emergency rescue tasks such as overseas evacuation, emergency rescue and emergency medical treatment; assume the responsibilities and obligations for international humanitarian assistance. |

### *2.2. Analysis of aviation emergency rescue equipment system*

The aviation emergency rescue equipment system is mainly composed of helicopters, supplemented by fixed-wing plane and other aircraft (UAVs, airships.). The helicopter can realize the functions of free take-off and landing, and vertical hover. When carrying out aviation emergency rescue tasks, it is not limited by terrain and has strong adaptability to the rescue environment. It has become the most important aviation emergency rescue equipment. Unmanned aerial vehicle (UAV) based on remote sensing technology has incomparable advantages in in-depth field search and reconnaissance in disaster areas. It can also carry out disaster exploration and topographic mapping under poor flight conditions, poor rescue environment and unable to ensure the safety of rescue personnel. It is an auxiliary aviation rescue equipment. Fixed-wing plane has stronger rescue materials transfer and people transportation capacity than helicopters, and can absorb a large amount of water for air forest fire fighting after modification.

## 3. Aviation emergency rescue scheme design

### *3.1. Demand analysis of modified rescue helicopter*

Aviation emergency rescue can be divided into natural disaster rescue, accident disaster rescue, public health emergency rescue and public safety emergency rescue according to the types of emergencies [9]. The traditional helicopter with transport function has been unable to meet the task requirements of the helicopter in terms of infrastructure, airborne precision medical instruments, medical personnel seats, and fixed devices. The purchase of medical rescue helicopters alone has high cost and low efficiency. In order to ensure the life safety of rescued personnel, the emergency rescue service system has a strong demand for helicopters to transform medical equipment [10-12]. The refitting design of helicopters and loading of medical equipment have become an important basis for the realization of aviation emergency rescue, which is of great significance for the implementation of tasks such as batch transfer of patients in major natural disasters and on-board rescue [13].

H145 helicopter has superior performance and is an ideal model for performing aviation emergency rescue tasks. The cabin combines the concept of Electro Magnetic Susceptibility (EMS), and the loading height and fixed guide shaft installation method provide excellent ergonomics design for rescue doctors and injured patients. The cabin can be quickly and easily modified for different emergency rescue tasks. The main rotor and fenestron tail rotor of H145 helicopter can easily load and unload patients from the side or back of the helicopter even in working condition. The tail adopts the mussel cabin door structure design. The medical stretcher can quickly enter and exit the cabin without excessive adjustment through the tail door, which reduces the burden of medical staff when getting on and off the plane. The side sliding door on both sides also provides convenience for the use of rescue winch. It is an ideal model for emergency rescue tasks. This paper will focus on major natural disasters and redesign the H145 helicopter.

### *3.2. Design of multi - casualty batch transportation process*

The batch transportation function performed by helicopter is to timely rescue and maintain the stable vital signs of patients, prevent the development of the disease, reduce the pain of patients, stabilize the injury, prevent secondary injury, and reduce the disability and death rate. It provides rapid and efficient first aid monitoring and rapid and safe air transportation, which is mostly used for the transfer treatment of patients within 400 km. The refitted rescue helicopter is mainly used in the transportation process of the wounded. The rescue tasks such as fracture location, bandage hemostasis, on-board monitoring and rescue are carried out to maintain the vital signs of the wounded. Based on the characteristics of helicopter transportation, the emergency, unconventional and sudden of major natural disasters, this paper puts forward the process design of batch transportation of multiple casualties in major natural disasters, which is the premise and guarantee for the successful completion of aviation emergency rescue tasks.

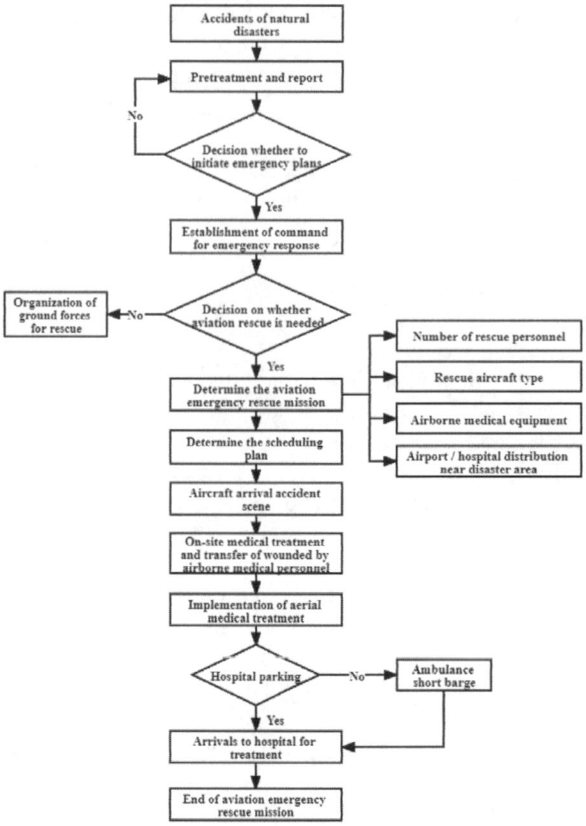


Figure 1. Multi - casualty batch transportation process for major natural disasters

## 4. Modular design of aviation emergency rescue equipment

Based on the H145 helicopter, the aviation emergency rescue equipment is decomposed into relatively independent standard modules for the batch transportation of casualties in major natural disasters. By adopting standardized design criteria, the unified electrical interface, power interface, oxygen interface, compressed air interface and general fixed interface are standardized to ensure that each module of the helicopter works independently. The functional test and repair / maintenance of a single module are convenient. The combination of different modules can meet the personalized needs of various rescue tasks, effectively save rescue time and improve rescue efficiency [14-15].

### *4.1. Modular design of airborne medical equipment*

The modular design of aviation rescue equipment for the transportation of the wounded in major natural disasters takes full account of the actual use and ergonomics. The power module, emergency goods storage module, oxygen output module and intravenous infusion module are carried out. The functional modules are clearly divided and coordinated.

The overall layout of the rescue helicopter cabin and the modular modification of airborne medical equipment are realized. The power supply module consists of two 28V DC power three-pole interface, two 220V 50 / 60Hz AC power three-pole interface and inverter control panel, which solves the problem of medical equipment interruption due to power failure, and further ensures the life safety of the wounded in the transport process. Emergency goods storage module can realize the classification and placement function of medical consumables, drugs, equipment, and tools. In order to ensure the convenience of goods access and improve the efficiency of rescue under unexpected conditions. The oxygen output module consists of two oxygen interfaces and a compressed air interface, and the surface reads the oxygen pressure value through the pointer. Intravenous infusion module can meet the trauma bandage treatment and life energy cycle of the mild wounded. Based on the modular design concept, according to the different types of rescue tasks, airborne rescue equipment can be installed and disassembled in a short time, and task attribute conversion can be implemented.

### *4.2. Modular layout of batch transportation of the mild wounded*

According to the condition of the injured patients, the injured patients with different conditions are classified according to the medical evacuation, and the severity of the disease is distinguished. First, the severe cases are rescued and the mild cases are stable. The overall layout of the rescue helicopter cabin for the mild and severe injured patients is designed, which can effectively ensure the orderly rescue work and improve the efficiency of aviation emergency rescue. The modular design of batch transportation of the mild wounded takes into account the light condition and no fracture of the injured, and adopts a multi-sit layout structure, which is shown in Figure 2.

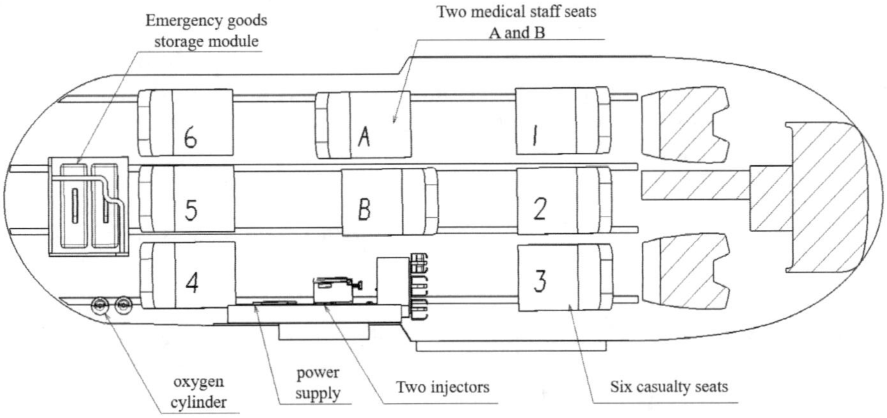


Figure 2. Modular layout of batch transportation of the mild wounded

The rescue helicopter can carry two medical staff and six injured patients. Before the injured patients get on board, the first-line emergency medical team carries out preliminary injury classification, recovery, fixation and rescue according to the injury situation. After entering the cabin, the medical staff faces the front and rear rows of the wounded respectively, so that the wounded are in the field of vision. According to the patients' injury, intravenous infusion and microinjection are carried out to facilitate the observation of their vital signs. The modular loading diagram of batch transportation for mild injuries is shown in Figure 3.

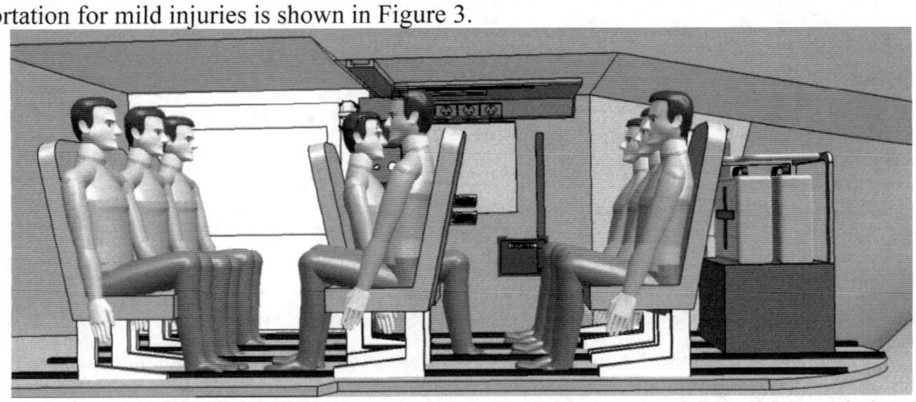

Figure 3. Modular loading diagram of batch transportation of the mild wounded

Helicopter transport of the wounded need to comprehensively consider constraints such as weather changes, terrain obstacles, transport requirements of the wounded, and rescue aircraft performance (flight altitude, flight time and longest flight distance), so as to reasonably and scientifically plan transport routes. With the help of 5G technology, remote sensing detection technology and intelligent obstacle avoidance technology, based on the improved ant colony algorithm, the disaster location is regarded as a nest, and the evacuation hospital is regarded as a food source. The simulation model of H145 helicopter ant colony algorithm is established by setting the initial pheromone concentration of each node differently. The H145 helicopter track is the feasible solution of the transportation route. The shortest path in the feasible solution is the optimal path between the disaster occurrence point and the evacuation hospital, which effectively solves the problems existing in the process of batch transportation of mild injuries, improves the efficiency and speed of path planning, and verifies the adaptability and effectiveness of ant colony algorithm.

### *4.3. Modular layout of batch transportation of the severe wounded*

The modular design of batch transportation of severe injuries according to the condition of the wounded, the multi-position layout structure is used to leave sufficient head care space, lighting space and the required space for the wounded to lean, which improves the helicopter cabin's utilization rate of the internal space. At the same time, it meets the emergency medical rescue of the upper and lower layers of the wounded, and has the emergency disposal ability of medical oxygen supply, ECG monitoring, fracture location, shock prevention and infection. The modular layout of batch transportation of mild injuries is shown in Figure 4 and Figure 5.

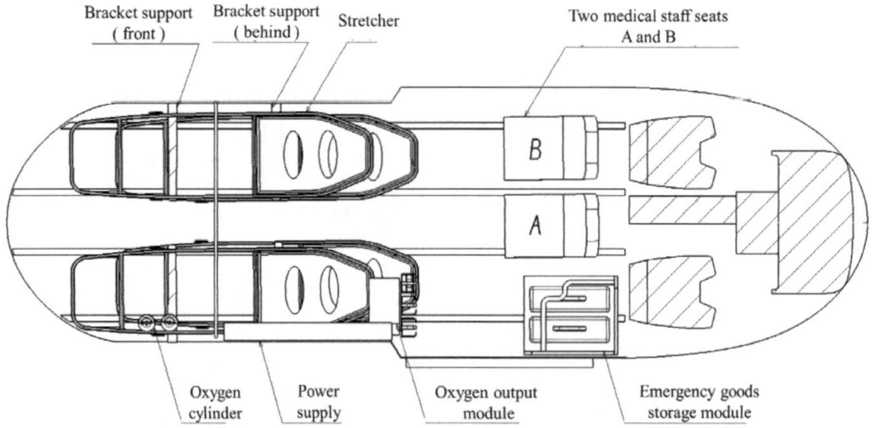


Figure 4. Modular layout of batch transportation of the severe wounded (1)

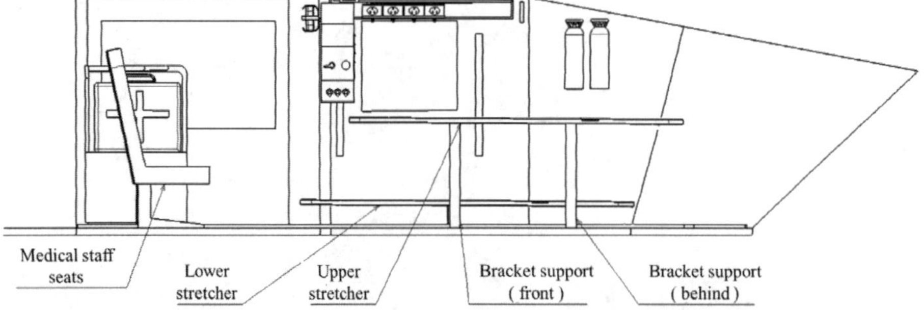


Figure 5. Modular layout of batch transportation of the severe wounded (2)

The medical cabin is equipped with two medical staff seats and four stretchers. There is a 350 mm wide passage between the two stretchers, which provides enough space for the movement of medical staff and care for the injured patients. The modular layout of batch transportation of severe injured patients and all personnel loading forms are shown in Figure 6.

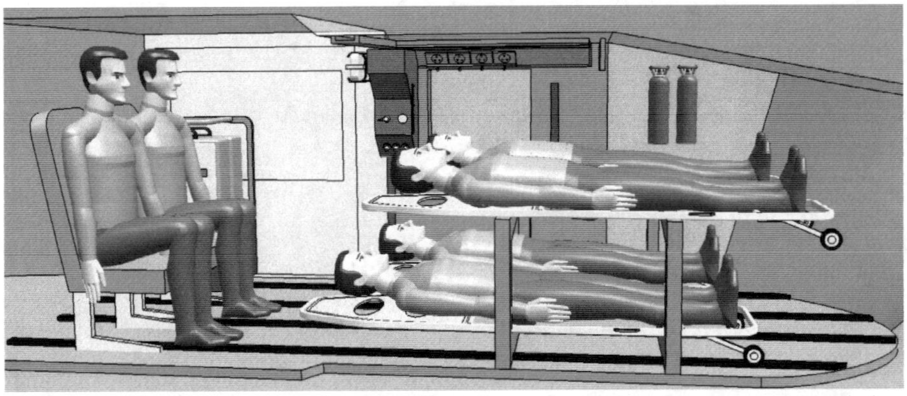

Figure 6. Modular loading diagram of batch transportation of the severe wounded

Due to the critical condition of the critically injured, the helicopter needs to be treated in time after arriving at the hospital. In order to realize the interconnection of the injured information, it is necessary to establish the information sharing database between the transfer unit, the transport unit and the receiving hospital to ensure the continuity of the information transfer in the transfer process. By using medical data interaction technology, an API (Application Programming Interface) interface for transhipment units and receiving hospitals is designed. Based on tornado, a Web service is built, and a unified standard, unified format and convenient sharing information service platform for casualty transhipment is constructed, which effectively solves the information island between units. The receiving hospital can directly obtain data by accessing the API interface address, and realize the real-time continuous transmission of the number and basic situation of severe casualties, medical record information, and disease change information in the process of transport. The landing point of rescue helicopter and the expected arrival time of transport are predicted in advance, and the treatment plan is planned in advance according to the condition of severe casualties.

## 5. Conclusion

(1) Helicopter is the most common and effective emergency rescue equipment to perform aviation emergency rescue tasks. Common rescue equipment includes helicopters, fixed wing aircraft and unmanned aerial vehicles; Helicopter is the core equipment of aviation emergency rescue because of its strong mobility, short rescue time, strong environmental adaptability and high safety. It plays an important role in the implementation of material transportation, batch transportation of multiple casualties, and rescue and disaster relief.

(2) Based on the needs of society, the development of national aviation emergency rescue system and the construction of aviation emergency rescue team, H145 helicopter is selected as the ideal model to perform emergency rescue tasks. For major natural disasters, the batch transfer process of multiple casualties is designed, which shortens the response time of aviation emergency rescue, ensures the orderly rescue process, and greatly improves the efficiency of aviation emergency rescue.

(3) Applying the modular design concept to the helicopter refitting design is conducive to realizing the independent work of a single module, and the combination of multiple modules can meet the personalized needs of varieties rescue tasks. Based on ant colony algorithm, reasonably and scientifically plan the helicopter transfer route; The unified and standardized API interface realizes the interconnection of patients' medical data, and solves the problems of data not sharing and information transmission not in time between units when performing aviation emergency rescue tasks.

## References

[1] Lou Y F. Research on Strengthening the Aviation Emergency Rescue System: Take Henan Province as an Example [J]. China Emergency Rescue, 2021(06):64-68.

[2] Wang R D, Pan W J, Ren J. Research on Multi-objective Model of Resource Scheduling Based on General Aviation Emergency Rescue[C]//.Proceedings of 2018 3rd

International Conference on Computational Modeling, Simulation and Applied Mathematics (CMSAM 2018).Proceedings of 2018 3rd International Conference on Computational Modeling,2018:127-136.
[3] Bruhn J D, Williams K A, Aghababian R. True costs of air medical vs. ground ambulance systems.[J]. Air medical journal,1993,12(8).
[4] The State Council issued the 14th Five-Year National Emergency System Planning[J]. China Emergency Management, 2022(02):4.
[5] Wang L F, Zhang Z H, Shen Z N, et al. Aviation medical rescue research and equipment development[J]. Science and Technology & Innovation, 2022(07):49-51+58.
[6] Guo A B, Liu B, Fu L, et al. Research on key technologies of air-ground collaboration in aviation emergency rescue[J]. Journal of Natural Disasters,2022,31(01):157-167.
[7] The Ministry of Emergency Management issued the National Basic Situation of Natural Disasters in 2021.https://www.mem.gov.cn/xw/yjglbgzdt/202201/t20220123_407204.shtml.
[8] Development Analysis of China Aviation Emergency Rescue, 2014[J]. M2 Presswire,2014.
[9] Air Rescue Operation Mode Research of China. [D]. Civil Aviation University of China, 2018.
[10] Davis, A. R., Bush, C., Harvey, J. C. and Foley, M. F., "Fresnel lenses in rear projection displays," SID Int. Symp. Digest Tech. Papers 32(1), 934-937 (2001).
[11] Christoph Kessler. Helicopter emergency medical service: motivation for focused research [J]. CEAS Aeronaut J6, 2015(3):337-394.
[12] Yu D F, Li J J, Guan H S. Research on Airworthiness Verification Method of Emergency Medical Equipment Modification Test for a Certain Type of Light Helicopter[J]. Medical Journal of National Defending Forces in Southwest China, 2020,30(09):850-851.
[13] Zhang X K, Li Y J, Cao Y Y, et al. Refitting optimization design and airworthiness compliance verification of a certain type of helicopter medical floor[J]. Science Technology and Engineering, 2021,21(29):12765-12771.
[14] Yang X X. Characteristics of modular design and its application in Engineering[J]. Chemical Engineering & Equipment,2022(02):177-178.
[15] Gökdağ İstemihan,Acar Erdem. Application of a modular topology optimization framework to an aerospace bracket design[J]. Materials Testing,2022,64(7).

# Acceleration and performance analysis of a compressible Euler solver with CUDA

**Yuqiao Chen[1,a], Jianhan Liang[1*], Meng Ding[1], Lin Zhang[1*], Qingdi Guan[1], and Xinxin Wang[1]**

[1]College of Aerospace Science and Engineering, National University of Defense Technology, Changsha, Hunan, 410073, China

[a]email: chenyuqiao77@163.com

*Corresponding author's e-mail: jhleon@vip.sina.com, zhlin151@163.com

**Abstract.** To develop high performance computing methods for compressible flow calculation, a GPU-accelerated compressible flow solver is developed with Compute Unified Device Architecture (CUDA). The WENO5 scheme is adopted for spatial discretization, and the third-order Runge-Kutta scheme is used for time discretization. According to the algorithm and programming model, the heterogeneous computing method of the solver is designed. Different kernels are designed to implement different computing functions, and shared memory is used for time-advanced computations. The solver is verified by the one-dimensional shock tube case, and a good acceleration effect is obtained with the increase of the grid size. And the impact of execution configuration on kernel performance was investigated. When the block size is reduced under different grid sizes, the speedup changes in the same way, but the performance parameters change differently.

## 1. Introduction

Compared to CPUs, most of the transistors in GPUs are designed as arithmetic logic units(ALU). Therefore, GPU is more suitable for large-scale data parallel computing. At first, researchers needed to call graphics API to use GPUs for computing, which was challenging to program. After NVIDIA released CUDA[1] (Compute Unified Device Architecture), using GPU computing became simple and easy, and General-Purpose Graphics Processing Unit (GPGPU) has been applied in many fields[2-4].

In 2007, Brandvik et al. [5] first implemented the acceleration of CFD solvers using the CUDA programming model on the GPU. In 2011, Phillips et al. [6] extended a single GPU to multiple GPUs. Dawes et al. [7] constructed an unstructured grid Navier-Stokes solver using CUDA with a speedup of 10. Wei et al. [8] compute the 2D unsteady compressible Euler equation discretized with the finite-volume method on a nested Cartesian grid, and speedup ratios of 9.98-14.04 are achieved.

In this study, a GPU-based CFD solver using shared memory is built and verified by a one-dimensional shock tube case. The influence of execution configuration on computational performance is studied in-depth, which provides a valuable reference for CFD acceleration with GPU parallel computing.

## 2. Governing equations and calculation modal

### *2.1. Governing equations*

Without considering the influence of heat source, heat conduction, and body force, the one-dimensional unsteady Euler equation:

$$u_t + f_x = 0 \tag{1}$$

Of which:

$$u = \begin{pmatrix} \rho \\ \rho u \\ E \end{pmatrix}, \quad f = \begin{pmatrix} \rho u \\ \rho u^2 + p \\ (E + p)u \end{pmatrix} \tag{2}$$

Where ρ is the fluid density, p is the pressure, u is the velocity component in the x-direction, e is the internal energy per unit mass, and E is the total energy per unit volume. Ideal gas state equation:

$$p = (\gamma - 1)\rho e = (\gamma - 1)\left(E - \frac{1}{2}\rho u^2\right) \tag{3}$$

### *2.2. 1D shock tube problem*

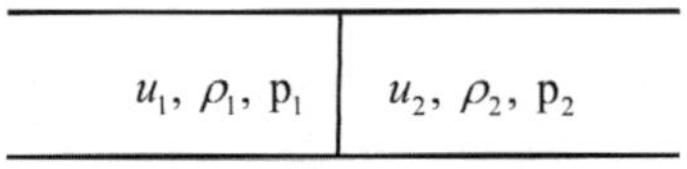


Figure 1. The SOD shock tube problem

In this study, the classical case of a one-dimensional shock tube is used for calculation and verification, and the initial conditions are set as follows:

$$\begin{cases} (u_1, \rho_1, p_1) = (0, 1, 1) & 0 \le x < 0.5 \\ (u_2, \rho_2, p_2) = (0, 0.125, 0.1) & 0.5 \le x \le 1 \end{cases} \tag{4}$$

## 3. GPU implementation and numerical experiment

### *3.1. Parallel calculation flow*

Rational design of GPU parallel program flow is essential for computing performance. As shown in Figure 1, preprocessing, such as initialization, and postprocessing, such as output results, are performed on the CPU. The core solving part takes the most time and is suitable for parallel computing with GPU. The data transfer between CPU and GPU is time-consuming. Data transfer is performed only once at the beginning and end of the calculation to get the optimal computing performance.

Appropriate settings of kernels and allocation of memory resources are also crucial for computing speed. The core solving part of this solver is divided into five parts. The first kernel function is to solve the global time step. The second is to split the flux vector using the Steger-Warming scheme, and the third is the flux reconstruction kernel. After calculating the three kernels, we can get the right-hand side term and time step, and transfer them to the Runge-Kutta kernel to get the flow field data next time. The update of boundary conditions for one-dimensional problems is simple, so boundary condition handling is included in the Runge-Kutta kernel. In the Runge-Kutta time advancement kernel, threads do not need to read the memory of threads in other blocks, which is suitable for using shared memory to calculate. So before the calculation, first transfer the data in the global memory to the shared memory. Note that thread synchronization must be done to ensure that all threads in the block complete the data transfer.

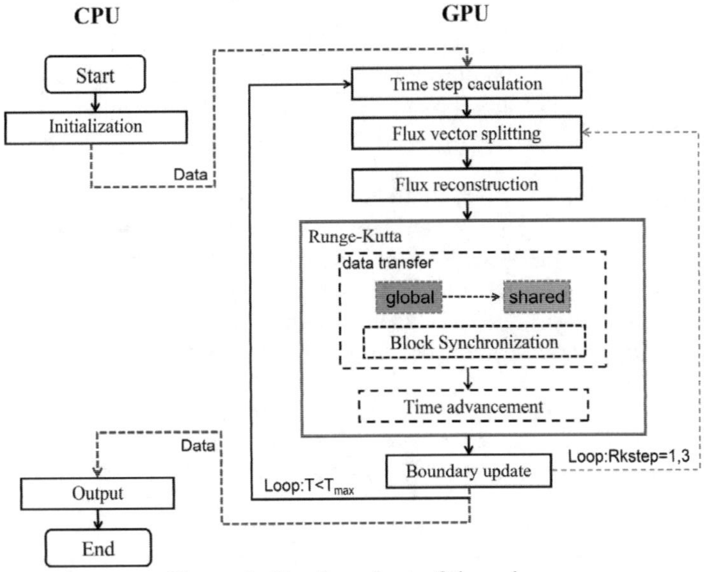


Figure 2. The flowchart of the solver

### *3.2. Accelerated effect of the slover*

The shock tube case was selected to verify the accuracy of the solver, and the initial conditions are as above. The calculation results at time T=0.2 is shown in Figure 2. The WENO5 scheme has high calculation accuracy, the wave structure can be clearly seen. The expansion wave is formed between x=0.2 and x=0.5. Near x=0.68, the velocity and pressure remain unchanged, and the density jumps, forming contact discontinuity. Compared with the exact solution, it can be seen that the calculation results of the solver are reasonable and accurate.

Test the speedup under different grids, and the block size is (1024,1,1). The result is shown in Figure 3. As the scale of grids increases, the speedup increases. When the quantity of computation is small, the computation speed of the solver based on the CPU/GPU heterogeneous system is slower than the CPU. The number of meshes increased from 1000 to 100000, and the speedup also increased by two orders of magnitude to 36.37. Therefore, the solver has high computational performance and can achieve a good speedup ratio. GPU parallel computing is more suitable for large-scale data parallel computing. Since data transmission is time-consuming, it cannot show its advantages when the amount of calculation is small.

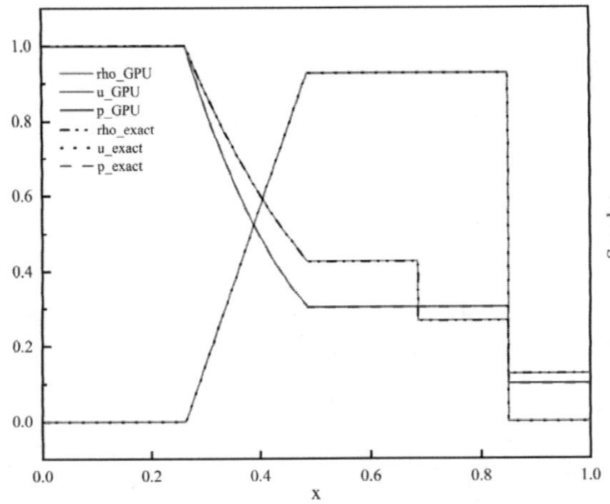


Figure3. The results at with grid size 10000

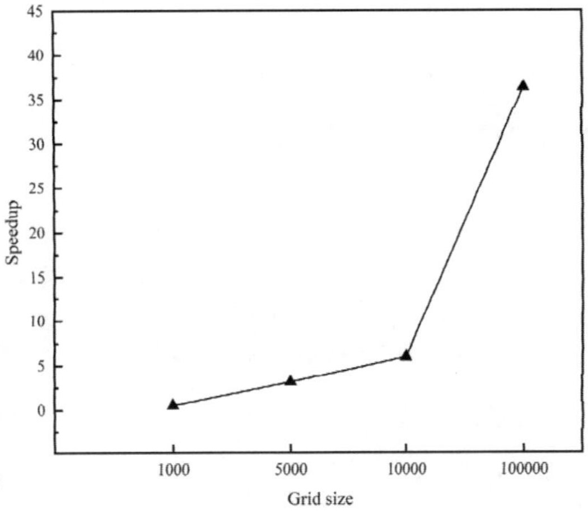


Figure 4. The speedup at different grid sizes

### *3.3. Execution configuration*

Thread block execution configuration can have a significant impact on computing performance. Whether the computing grid is 10000 or 100000, the speedup increases as the block scale decreases. When the grid size is 100000, the speedup reaches 46.54 when the thread block size is 128, which is 27.05% faster than the thread block size is 1024. When the grid size is 10000, the speedup is 3 times higher when the thread block is 128 than 1024.

Whether the computing grid is 10000 or 100000, the memory throughput of the two main kernels increases as the block scale decreases. When the grid size is 1000, throughput increases sharply, but the occupancy decreases when the speedup increases as the block size decrease. On the contrary, when the grid size is 100000, the occupancy rate increases as the speedup increases. When the grid size is small, and the block size is large, the number of blocks is less than the multiprocessors, so the computing resources of the GPU cannot be fully utilized. Reduce the block size and increase the number of blocks to ensure that each multiprocessor has at least one block, which can fully utilize the available hardware resources. Because the mesh size is small, the number of threads working on each multiprocessor decreases as the block size decreases, resulting in a decrease in the occupancy rate. But the number of multiprocessors that can work simultaneously increases, so the computing speed will still increase.

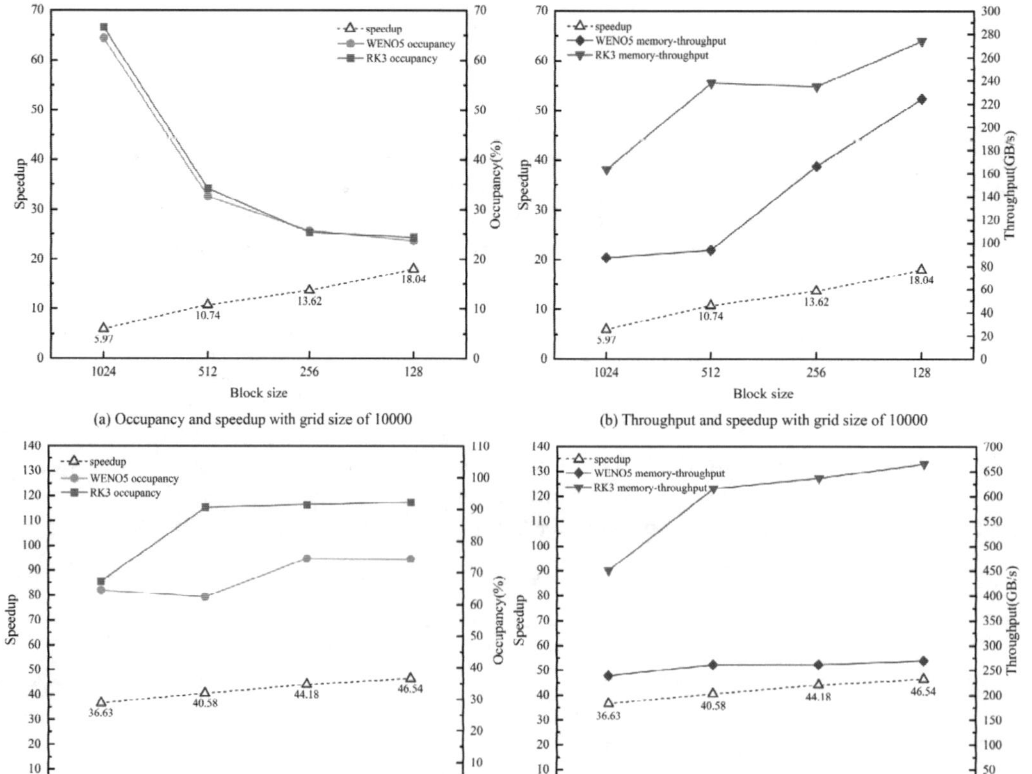


(a) Occupancy and speedup with grid size of 10000 (b) Throughput and speedup with grid size of 10000

(c) Occupancy and speedup with grid size of 100000 (d) Throughput and speedup with grid size of 100000

Figure 5. Impact of Execution Configuration on Compute Performance at different grid sizes

## 4. Conclusions

This study builds a CFD solver on a heterogeneous system consisting of an Intel i9 CPU and an RTX3090 GPU using shared memory. The classic case of a one-dimensional shock tube is tested to verify the rationality and accuracy of the solver, and the acceleration performance is studied. The conclusions are as follows:

(1) The solver based on a CPU/GPU heterogeneous system is suitable for computing large-scale grids. When the quantity of computation is small, the computing speed is very slow or even lower than calculated based on the CPU. A higher speedup ratio can be achieved with the increased quantity of computation.

(2) When the grid size is small, the calculation speed can be improved by reducing the block size. When the block size is reduced, the memory throughput increases, and the utilization rate of the multiprocessor increases. So the computing resources can be more fully utilized to improve the computing performance. When the grid size is large, the occupancy of kernels will increase when the block size is reduced, and the calculation speed will increase.

**Acknowledgments**
This paper was supported by the Key Laboratory Foundation in China (Grant No. 6142703200310).

**References**

[1] NVIDIA. (2021) NVIDIA CUDA C++ Programming Guide. NVIDIA Corporation.
[2] Bonelli, F., Tuttafesta, M., Colonna, G., Cutrone, L., Pascazio, G. (2017) An mpi-cuda approach for hypersonic flows with detailed state-to-state air kinetics using a gpu cluster. Computer Physics Communications, 219:178-195.
[3] Neic, A., Liebmann, M., Hoetzl, E., Mitchell, L., Vigmond, E. J., Haase, G., Plank, G. (2012) Accelerating cardiac bidomain simulations using graphics processing units. IEEE Transactions on Biomedical Engineering, 59:2281-2290.
[4] Song, J. P., Ross, J. A., Shires, D. R., Richie, D. A., Henz, B. J., Nguyen, L. H. (2010) Hybrid core acceleration of UWB SIRE radar signal processing. IEEE Transactions on Parallel & Distributed Systems, 23:46-57.
[5] Brandvik, T., Pullan, G. (2007). Acceleration of a two-dimensional Euler flow solver using commodity graphics hardware. Proceedings of the Institution of Mechanical Engineers, Part C: Journal of Mechanical Engineering Science, 221: 1745-1748.
[6] Phillips, E. H., Zhang, Y., Davis, R. L., Owens, J. D. (2011) Acceleration of 2-D compressible flow solvers with graphics processing unit clusters. Journal of Aerospace Computing Information & Communication, 8:237-249.
[7] Dawes, W. N., Dhanasekaran, C., Kellar, W. (2011) Acceleration of an Unstructured Hybrid Mesh RANS Solver by Porting to GPU Architectures. 49th AIAA Aerospace Sciences Meeting including the New Horizons Forum and Aerospace Exposition.
[8] Wei, F., Jin, L., Liu, J., Ding, F., Zheng, X. (2019) Acceleration of a 2D unsteady Euler solver with GPU on nested Cartesian grid. Acta Astronautica.

# Aerodynamic Analysis of Large Aspect Ratio Parafoil

**Chongxian Li [1,2], Xinfeng Li [1], Hongen Zhong[1]**

[1]Technology and Engineering Center for Space Utilization, Chinese Academy of Sciences.Beijing,100094, China

[2]University of Chinese Academy of Sciences.Beijing,100049, China

Corresponding author's e-mail: lichongxian20@mails.ucas.ac.cn

**Abstract.** Large aspect ratio parafoils are more attractive compared with moderate ones less than 3.5, since they either improves load-carrying capacity with the same aero-dynamical area or decreases taxing speed with the same load. Based on its original airfoil, one open-source elliptical parafoil with 4.7 AR is modified while scaling the canopy geometric parameters, then CFD simulations, solving the incompressible Reynolds time-averaged Navier-Stokes equation with the (SST) K-ω turbulence model in the finite volume method, are adopted to analyze its aerodynamics in forward taxing mode as well as steering mode with lateral wind. With the numerical simulation results of this canopy, lift and drag are analyzed according to their attack angle profiles for taxing. The steering mode with the deflection of one-side flap increases the drag coefficient significantly, even it raises its lift simultaneously, reduces the lift-drag ratio, and generate rolling and yaw moments; The rolling moment due to lateral wind during steering mode has two phases of rising linked by a short straight line when sideslip angle increases, while yaw moment increases to a peak value then reduces, finally followed by a relatively stable trend. Compared with the above two moments, the pitching moment changes moderately with the increase of the sideslip angle, although it has an overall decreasing trend; The simulation of the modified canopy provides a calculation basis for its following manufacture and dynamical modelling.

## 1. Introduction

A ram-air parafoil is a flexible fabric wing that maintains its designed shape with gliding speed. It consists of airfoil holed ribs sandwiched by upper surface and lower one, which are inflated into cells during flying, as shown in Figure 1. Large aspect ratio parafoils usually refer to ram-air parafoils more than 3.5, which are widely used in high-mobility powered parafoils. For earth reentry mission[1-3] it would be advisable to exploit a parafoil in the last stage of the aerodynamic decelerator system instead of a near reticular splash-down with ballistic parachute. A large aspect ratio parafoil has a lower gliding angle, its landing vertical speed is small, the impact is low, and it has a higher carrying capacity. The higher lift-to-drag ratio allows length-extended parafoil to have the same vertical velocity with a lower canopy area, or improves horizontal component with the same aerodynamical area to enhance its capability of precise landing by gliding compensation of horizontal error or lateral wind. With lower total velocity despite of larger horizontal component, "wave riding" flight can be realized in low-power assistance or even unpowered scenario, an intermittent-powered parafoil is proposed to winder its air-route in dense-atmosphere planetary mission[4-5] like Europa exploration,etc.

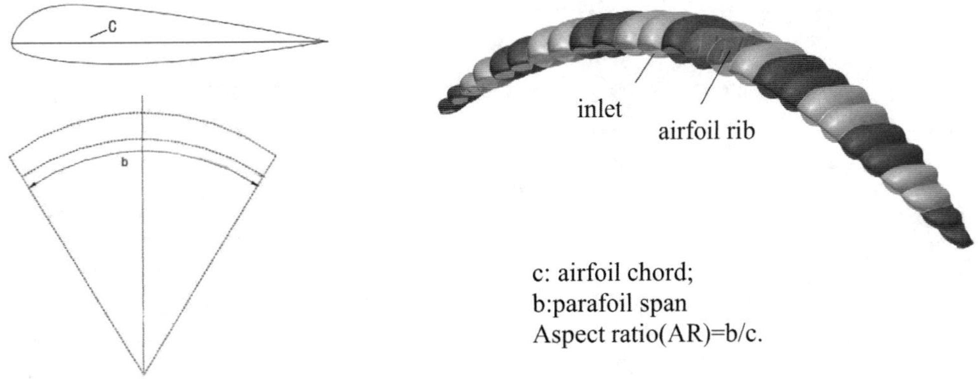


Figure 1. Parafoil model schematic

The National Aeronautics and Space Administration (NASA) and Notre Dame University first conducted the classic parafoil (small and medium aspect ratio) wind tunnel test in 1967-1968[6]. Subsequent researches have further developed on this basis. Mohammad A. Mohammadi [7]and Hamid Johari adopted the finite volume method to solve the incompressible Navier-Stokes equations, analyzed the flow fields and forces about the high-performance parafoil,and concluded that the leading edge opening had a great influence on the drag before the parafoil stalled , it was proposed that improving the geometry of the leading edge opening was beneficial to enhance the lift-drag ratio and the overall performance of the parafoil. Furfaro[8] and Uddin[9] analyzed the influence of the existence of leading edge opening on the aerodynamic performance of parafoils by wind tunnel tests; Li Yang[10]applied the preprocessing method of unsteady flow to reveal the aerodynamic characteristics of two-dimensional airfoil and three-dimensional parafoil under the condition of trailing edge flap deflection,then analyzed that the regular fluctuation of lift-drag coefficient was due to the shedding and generation of vortices appear on the surface of the parafoil; Zhu Xu [10]and Cao Yihua established three-dimensional parafoil models of different shapes and adopted cfd simulation calculation to analyze the influence of the plane shape on the aerodynamic performance of the parafoil.

Considering the characteristics of simple geometry and easy deployment, the research on parafoils in aerodynamic analysis mostly focuses on aerodynamic calculations with aspect ratios within 3.5, but there are not many researches on parafoils with large aspect ratios. Combining the lift line theory[12] and the relevant aerodynamic research on different shapes of parafoils, it can be proved that the elliptical wing is a plane shape with the best lift and drag characteristics. Therefore, based on a certain elliptical parafoil that has been scaled down, this paper analyzes the aerodynamic characteristics of the large aspect ratio parafoil of the specific wind field.

## 2. Large AR parafoil model design

The equivalent scaled elliptical parafoil of the GNULAB airfoil is established from the open-source parafoil parameters (Table 1).According to the relevant experience of parafoil design, the cut angle of the leading edge of the parafoil is set to be 50°(between the notch line and the airfoil chord), and the cut height is 45%(the airfoil maximum thickness *0.45).The turning is achieved by pulling down the trailing edge flap on the side of the parafoil,shown in Figure 2.

Figure 2. Airfoil leading edge cut and trailing edge flap deflection definition

Table 1. Parafoil model design parameters

| Indicator item | Flat area | Wingspan (straight state) | Aspect ratio | Curvature Projected Area | Projected wingspan | Nominal maximum load |
|---|---|---|---|---|---|---|
| | 6.26m$^2$ | 5.42m | 4.7 | 5.48 m$^2$ | 3.62 | 25Kg |

## 3. Simulation and Analysis

### *3.1 Method validation*

According to the parafoil configuration modeling used by NASA Langley Research Center, numerical simulation methods are used to calculate aerodynamic parameters and compare with the experimental data in the literature to ensure that the settings of the calculation domain, mesh quantity and quality, and boundary conditions are applicable.

The full flow field unstructured mesh is used and local refinement is performed within the range of one chord length near the parafoil. The number of grids is between 2 million and 3 million. The inlet is the velocity inlet, which is 10m/s. The outlet is the free outflow, and the other boundaries are free-slip wall conditions. The finite volume method is used to solve the incompressible Reynolds time-averaged Navier-Stokes (RANS) equation in a three-dimensional coordinate system .

Reynolds averaging is to decompose the instantaneous variables in the Navier-Stokes equation into two parts: average and pulsating:

$$\boldsymbol{u_i = \overline{u_i} + u_i'} \tag{1}$$

Substitute into the continuity and momentum equations to get:

$$\frac{\partial P}{\partial t} + \frac{\partial(\rho u_i)}{\partial x_i} = 0 \tag{2}$$

$$\rho \frac{du_i}{dt} = \frac{\partial P}{\partial x_i} + \frac{\partial}{\partial x_j}\left[u\left(\frac{\partial u_i}{\partial x_j} + \frac{\partial u_j}{\partial x_i} - \frac{2}{3}\delta_{ij}\frac{\partial u_t}{\partial x_t}\right)\right] + \frac{\partial}{\partial x_j}\left(-\rho\overline{u_i' u_j'}\right) \tag{3}$$

The shear stress transport (SST) k-ω is used in the solution process. A two-equation turbulence model is applied to simulate turbulent flow, and the SIMPLE method and the second-order upwind style is utilized to solve.

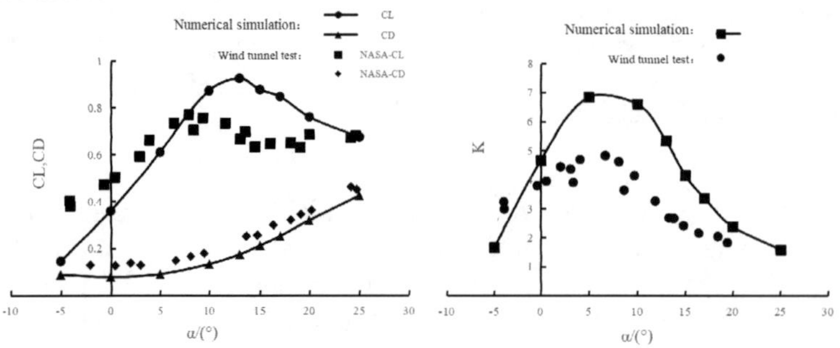


Figure 3. Comparison of numerical simulation and wind tunnel test

The comparison of pneumatic data is shown in Figure 3.(1) The numerical simulation drag coefficient is slightly lower than the experimental data, and the lift coefficient trend is basically consistent; (2) The change trend of lift-drag ratio is consistent and the maximum lift-drag ratio corresponds to the same angle of attack. The reason for the deviation is that the parafoil model is regarded as a rigid body in the numerical calculation and the numerical simulation bias and other reasons.

In summary, the parafoil aerodynamic coefficient calculated by the numerical simulation method has a certain accuracy, and the aerodynamic variation trend with the angle of attack is well fitted, so it is reliable.

*3.2 Analysis with forward winds*

As a finite span structure, the parafoil is quite different from the two-dimensional airfoil, and its spanwise flow also affects the aerodynamic performance of the parafoil.

Figure 4 shows that in the taxiing state, the pressure on the bottom of parafoil increases gradually and a separation vortex is generated at the upper trailing edge with the increase of the angle of attack.In the nominal attack angle range from 0° to about 8°, there is no detachment vortex on the upper surface of the entire parafoil, so the aerodynamic performance of the parafoil within this range should be improved with the increase of the attack angle. Due to the existence of the dihedral angle,two symmetrical vortices are gradually generated above the middle of the upper surface of the parafoil when the angle of attack exceeds 8°.

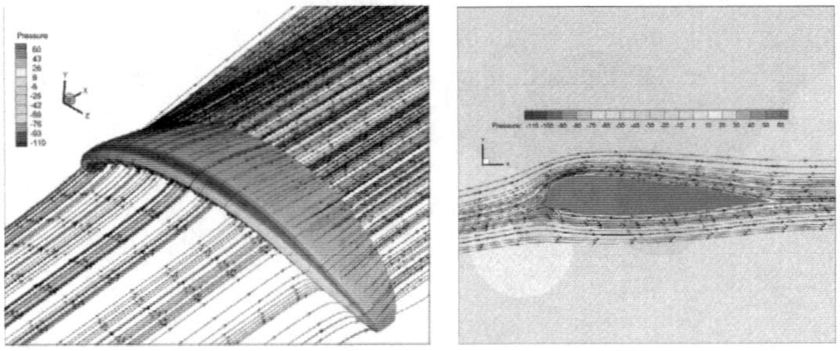

(a) $\alpha$ =7.5°, flow field and pressure diagram of the parafoil

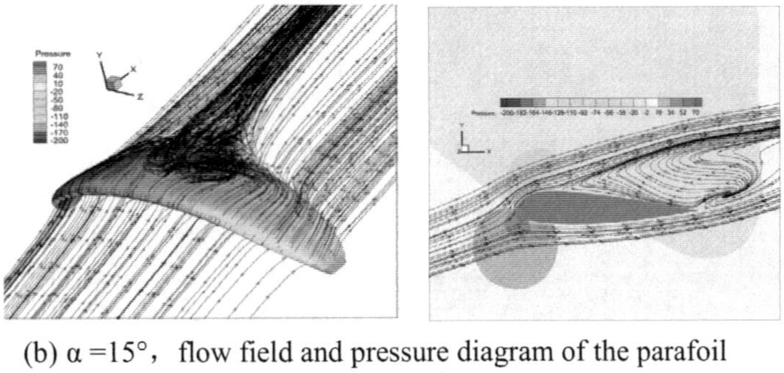

(b) $\alpha$ =15°, flow field and pressure diagram of the parafoil

Figure 4. The three-dimensional flow field of the undeflected parafoil and the pressure diagram of the section at the middle of the parafoil

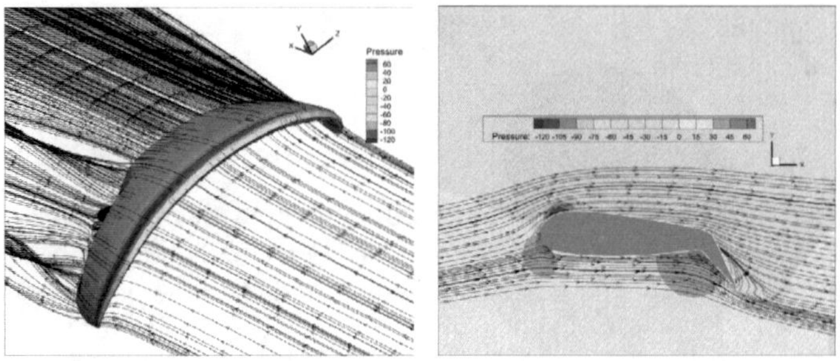

(a) α =7.5° (b) α =7.5°

Figure 5. The three-dimensional flow field of the deflected parafoil and the pressure diagram of the section at the deflected position of the parafoil

The plot in figure 4 and figure 5 shows that when α=7.5°,the pressure of the bottom surface of deflected parafoil is higher than that of the undeflected parafoil and the negative pressure area of the top surface expands along the flap deflection position .The combined effect of the two result in the increase of lift. The flap deflection of the trailing edge will cause a vortex behind the deflection position,which will have adverse effects on the aerodynamic performance of parafoils, such as loss of lift and increase in drag.

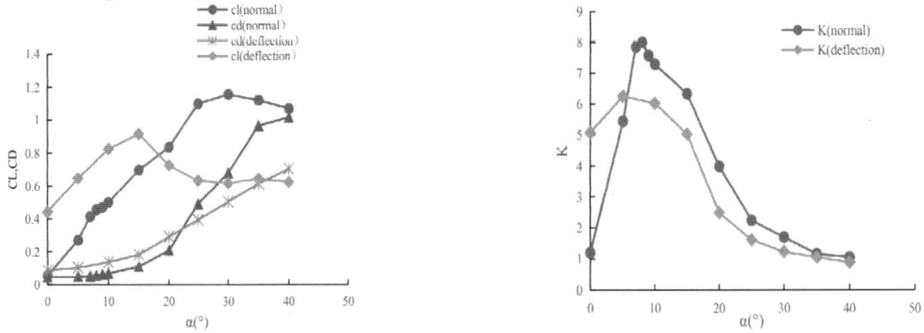


(a) Comparison of aerodynamic coefficients of flap deflected parafoil and undeflected parafoil

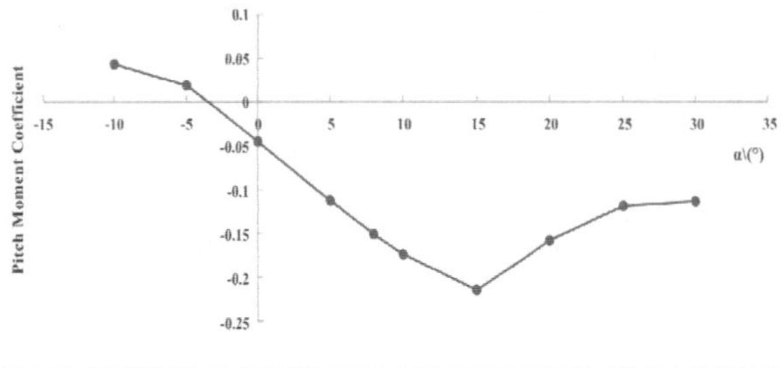


(b) Pitching moment coefficient curve

Figure 6. Aerodynamic coefficient comparison and pitching moment coefficient curve

In Figure 6, the following conclusions can be drawn:

(1) In the range from -10° to 15°, the derivative of the pitch moment coefficient to the angle of attack is negative, which indicates that the parafoil is in a longitudinal static equilibrium state at this stage.

(2) The deflection of the trailing edge increases the lift coefficient and the drag coefficient;

(3) The maximum lift-to-drag ratio of the parafoil with deflected flaps is less than the undeflected parafoil,and the position of the angle of attack at which the maximum lift-to-drag ratio occurs is relatively forward.

*3.3 Steering analysis with lateral winds*

By setting the rigging angle, the parafoil fly at the angle of attack corresponding to the better lift-to-drag ratio.We realized the turning of parafoils by pulling down the trailing edge of one side. Therefore,lateral wind is introduced on the basis of the above forward wind conditions, on the premise that the positive angle of attack is 8°(close to the maximum lift-to-drag ratio),the lateral wind is set to 4° to 28° (4° increments)to simulate the parafoil.

The following figures are the moment cofficient diagram and the pressure cloud maps of the upper and lower surfaces of the parafoil when the sideslip angle is 0°,12°, and 16°:

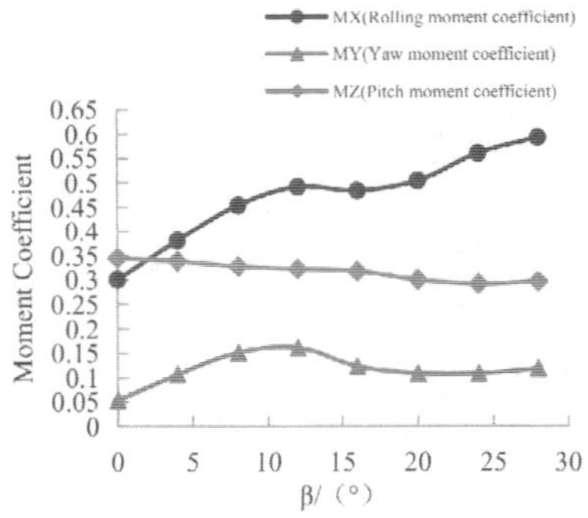


Figure 7. Curve of parafoil moment coefficient with sideslip angle

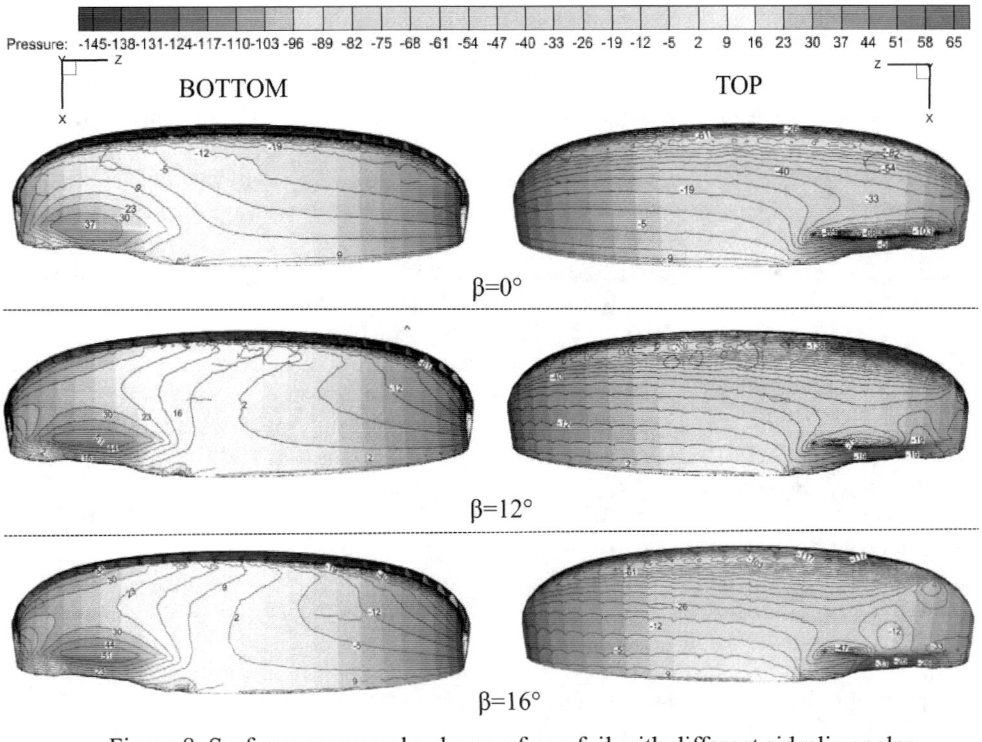


Figure 8. Surface pressure cloud map of parafoil with different sideslip angles

(1) The pitch moment coefficient decreases slowly with the increase of the sideslip angle; Rolling moment coefficient is relatively gentle between 10° and 20°. Yaw moment has a trend of increasing first and then decreasing, with a peak value;

(2) The existence of the sideslip angle causes a large negative pressure in the leading edge area corresponding to the deflected side of the parafoil. With the increase of the sideslip angle, the pressure distribution on the upper surface is basically similar, and the pressure at the deflection position of the lower surface gradually increases, and the pressure distribution expands to the undeflected area on the other side. (3) When the trailing edge of the parafoil is pulled down, the windward area is increased, which means an increase in one-sided drag and an increase in the lift. In response to the trailing edge pull-down, the parafoil will be coupled with the yaw and roll. Considering the moment of inertia and the hindered deflection of the payload, the centrifugal force on the payload will make the parafoil roll in the correct direction and keep turning . The yaw effect is more obvious than the roll.

## 4. Conclusion

By analyzing the elliptical parafoil , the parafoil with undeflected flaps maintains longitudinal static balance in a wide range of angle of attack.The deflection of the unilateral trailing edge flap of the parafoil increase the lift coefficient and the drag coefficient, but the lift-drag ratio decreases, and the angle of attack corresponding to the maximum lift-drag ratio appears earlier. The deflection of the trailing edge flap makes the symmetrical vortex that should be generated in the middle of the undeflected parafoil shift to the position of the deflection flap, and the vortex intensifies with the increase of the angle of attack. Under the simultaneous action of forward wind and lateral wind, the moment coefficient of the elliptical parafoil varies with the slip angle and the change of the moment coefficient is relatively stable. Combined with the pressure changes on the upper and lower surfaces of the parafoil, it can be seen that the deflection of the trailing edge flap will produce a relatively obvious coupling between the yaw moment and the rolling moment,which must be considered in the safety design of the parafoil.

## References

[1] Wegereef, W., Doejaaren, F., Benolol, S,et al. (2011). FASTWING CL flight tests with a high-glide ram-air parachute for 6000kg payloads. In 21st AIAA Aerodynamic Decelerator Systems Technology Conference and Seminar. Dublin, Ireland (p. 2596).

[2] Velasquez, C., & Benedict, Z. (2015, June). Parafoil Recovery of a High Altitude Balloon Payload. In Academic High Altitude Conference (Vol. 2015, No. 1). Iowa State University Digital Press. Ames, Iowa.

[3] Avanzini, G., Guglieri, G., & Torasso, A. (2012). Multibody analysis of terminal phase for a reentry vehicle: a comparative study. *Journal of aircraft*, *49*(6), 1940-1952.

[4] Shaw, D. O. (2016). The Mars Parafoil Airship (MPA). In AIAA SPACE 2016 (p. 5599).

[5] Balossino, A., Battocchio, L., Giacci, M., Giudotti, G., Rufolo, G., Denaro, A., & Paletta, N. (2017). Conceptual design of the descent subsystem for the safe atmospheric reentry flight of space rider. In 7th European Conference for Aeronautics and Space Sciences (EUCASS).

[6] Nicolaides, J. D. (1971). Parafoil wind tunnel tests. Notre Dame Univ in Dept of Aerospace and Mechanical Engineering.

[7] Mohammadi, M. A., & Johari, H. (2010). Computation of flow over a high-performance parafoil canopy. Journal of Aircraft, 47(4), 1338-1345.

[8] Furfaro, J., Hailey, C., Behr, V,et al. (1997). An experimental study of parafoil sections. In 14th Aerodynamic Decelerator Systems Technology Conference. San Francisco, CA, U.S.A. (p. 1527).

[9] Uddin, M. N., & Mashud, M. (2010). Wind tunnel test of a paraglider (flexible) wing canopy. International Journal of Mechatronics Engineering IJMME-IJENS, 10(3), 7-13.

[10] LI Yang. (2004) .Numerical Simulation of Ram-air Parachute with Flap Deflection[D]. Changsha: National University of Defense Technology. (in Chinese)

[11] ZHU Xu, CAO Yihua. (2011). Numerical Simulation of Planform Geometry Effect on Parafoil Aerodynamic Performance[J]. Acta Aeronautica et Astronautica Sinica, 32(11): 1998-2007. (in Chinese)

[12] Lingards, J. S. (1995). Precision Aerial Delivery Seminar, Ram-air Parachute Design. In 13th AIAA Aerodynamic Decelerator System Technology Conference. Clearwater Beach, FL, U.S.A.

# Simulation of Combustion Performance of High Salinity Organic Waste Liquid Incinerator

**Wen-kai Li[1,a], Yong-kang Qian[2,b], Xin-yu Qian[3,c]**

[1]Beijing Aerospace Propulsion Institute, Beijing, The China

[2]Beijing Aerospace Propulsion Institute, Beijing, The China

[3]Changzheng Engineering Co. ,Limited. , Beijing, The China

[a]374298106@qq.com; [b] qianyk2006@126.com; [c]xinyuqian2000@outlook.com

**ABSTRACT:** High-salinity organic waste liquid is one of the important industrial pollution sources and the main by-product of petrochemical enterprises in China. Incineration is the mainstream technology for treating this kind of waste liquid. A three-dimensional numerical simulation of a practical 18t/h waste liquid incinerator was carried out using Fluent. The velocity, temperature, component concentration and particle trajectory distribution in the incinerator were obtained, proposing a corresponding incinerator performance optimization scheme according to the simulation results. The results demonstrate that the numerical combustion model established in this paper can accurately simulate the distribution of flow, combustion and heat transfer in the incinerator, verifying the reliability of the numerical model; The overall temperature and flow field organization of this type of waste liquid incinerator is relatively reasonable, with heat exchange reported only in the flue corner and some heat exchange surfaces. The flow field and temperature field distribution in the incinerator can be effectively modified so as to improve the boiler operation and maintenance cycle by optimizing the corner flow line and furnace structure.

## 1. INTRODUCTION

The high salinity organic waste liquid was characterized as high concentration of organic matter, high toxicity, complex composition, and involvement of a large amount of alkali (sodium hydroxide) or alkali metal salts (sodium sulfate, sodium nitrate, etc.). Given the strong inhibitory effect of high concentration of alkali and alkali metal salts on microorganisms, incineration is currently the mainstream technology approach, which has absorbed a number of in-depth studies by many researchers at home and abroad. For example, the combustion tests have been conducted on an industrial scale waste liquid incinerator and a pilot scale waste liquid incinerator respectively by researchers (Kaul et al. 1998, Ehrhardt et al. 1998). It is concluded that the combustion temperature of 850~950 °C can induce the removal rate of toxic and harmful substances in the concentrated waste liquid to reach more than 99%, and the emission values of various pollutants reduced to lower than the relevant emission standards; In a work (Li et al. 2007) organic waste liquid fuel has been found with the characteristics of high ignition point and low calorific value. Therefore, the addition of auxiliary fuel is an effective method to elevate the waste liquid treatment rate when treating waste liquid by incineration method.

In addition, due to the factors of large volume, high temperature and pressure, long operation time, high experimental cost, and the difficulty in obtaining some significant parameters with experimental means during the operation of the high salinity organic waste liquid incinerator, for predicting the flow, combustion and heat transfer process in the incinerator, so as to contributing to exploring the causes and solutions of the operation issues of the operation boiler actually put into, and even further optimizing the design of the new boiler, numerical simulation is required. A number of researchers at home and abroad have also completed a lot of work in this area. The research (Cardoso et al. 2012) using WinGEMS to conduct numerical simulation of alkali recovery boiler shows that WinGEMS can accurately simulate the temperature field in the furnace, while fails to obtain the ideal distribution of combustion gas in the furnace. At the same time, it is demonstrated that the expansion index and combustion time of black liquid droplets can be correctly simulated with PSL model; PSL company (Yuan 2009) calculated the commercial cooperation of alkali recovery boilers around the world using the self-built combustion model, reporting that the results were precise in comparison with the actual parameters; In a work (Zhang 2015) numerical simulation on an actual alkali recovery boiler were conducted by programming the black liquor numerical model (Frederick 1991) in cooperation with Wuhan Boiler Energy Engineering Co., Ltd. to, and the simulation results show robust capability in prediction of the velocity field, temperature field, component field and particle motion trajectory in the furnace, and qualitative prediction pf the mechanical carrying and slagging conditions in the furnace.

However, the numerical simulation of high salinity organic waste liquid incinerator still lacks conduction at home and abroad. Based on the combustion mechanism of high salinity organic waste liquid, a numerical model suitable for related simulation calculation is proposed in this paper, simulating a practical 18t/h caprolactam waste liquid incinerator with Fluent. The flow field, temperature field, concentration of each component and particle trajectory distribution in the furnace are obtained. Considering the characteristics of flow and heat transfer, the corresponding furnace performance optimization scheme is proposed based on the simulation results, providing a possible theoretical basis for the performance optimization design of this type of boiler.

## 2. Mathematical model

The liquid droplets of waste liquid is assumed to remain spherical during the combustion process. The combustion process can be categorized into three stages: drying, pyrolysis and surface reaction. In the drying stage, the moisture inside the particles is evenly distributed, with the drying rate only controlled by the external heat transfer, and only the moisture decreased during the drying process, the content of other components remaining unchanged, without the effect of solids inside the particles on the evaporation process of moisture; In the pyrolysis stage, the particles expand nonlinearly, with the effect of temperature gradient inside the droplet ignored, and the pyrolysis rate of droplet particles is only affected by itself; In the surface reaction stage, the specific surface area of the droplet is constant, the surface reaction rate is determined by the chemical kinetics and diffusion process (Connolly 2006), and the droplet particle diameter only depends on the internal solid content. After completing the solid reaction, the particle diameter remains unchanged (Smith 1982). The flow, energy, momentum, gas phase combustion, particle motion and combustion equations in the waste liquid incinerator are as follows:

### *2.1 General model*

The waste liquid in the furnace is mainly composed of gas phase after atomization and evaporation, so the flow of gas can be described by the traditional continuity equation, energy equation and momentum equation:

$$\frac{\partial \rho}{\partial t} + \nabla \cdot (\rho \vec{v}) = S_m \tag{1}$$

$$\frac{\partial}{\partial t}(\rho E) + \nabla \cdot (\vec{v}(\rho E + p)) = \nabla \cdot \left[k_{eff}\nabla T - \sum h_j \vec{J_j} + \left(\overline{\overline{\tau_{eff}}} \cdot \vec{v}\right)\right] + S_h \tag{2}$$

$$\frac{\partial}{\partial t}(\rho\vec{v}) + \nabla\cdot(\rho\vec{v}\vec{v}) = -\nabla p + \nabla\cdot(\bar{\bar{\tau}}) + \rho\vec{g} + \vec{F} \tag{3}$$

Where $S_m$, $S_h$: Source item; $k_{eff}$: Effective thermal conductivity of gas under turbulence; $\overline{\overline{\tau_{eff}}}$: Viscous dissipation of fluid in turbulent flow; $\vec{F}$: External volume force; $\bar{\bar{\tau}}$: Newtonian stress tensor, $\bar{\bar{\tau}} = \mu\left[(\nabla\vec{v} + \nabla\vec{v}^T) - \frac{2}{3}\nabla\cdot\vec{v}I\right]$.

DPM model is as follows:

$$\frac{du_p}{dt} = F_D(\vec{u} - \vec{u}_p) + \frac{\vec{g}(\rho_p - \rho)}{\rho_p} \tag{4}$$

Where $F_D(\vec{u} - \vec{u}_p)$: Drag force on waste liquid particles, $F_D = \frac{18\mu}{\rho_p d_p^2}\frac{C_D\mathrm{Re}}{24}$;

$C_D$: Drag coefficient, $C_D = a_1 + \frac{a_2}{\mathrm{Re}} + \frac{a_3}{\mathrm{Re}^2}$, $a_1$、$a_2$、$a_3$: Constants related to Re.

After the velocity distribution of each particle in the furnace is calculated by equation 4, the corresponding particle trajectory can be obtained by integrating the velocity distribution of each particle twice.

$$\frac{dx}{dt} = u_p \tag{5}$$

Calculation of gas phase combustion of waste liquid using component transport model:

$$\frac{\partial\rho f_k}{\partial t} + \nabla\cdot(\rho u_i f_k) = \nabla\cdot(\Gamma_k\nabla f_k) + R_k + S_k \tag{6}$$

Where $R_k$: Net rate of product formation in the reaction;

$S_k$: Additional generation rate of chemical reaction of discrete phase or custom item.

The interaction between turbulence and chemical reaction in gas phase combustion is calculated using the EDC (finite rate / vortex dissipation) model in Fluent:

$$R_{k,r} = (v''_{k,r} - v'_{k,r})\left[k_{f,r}\prod_{j=1}^{N}(c_{j,r})^{\eta_{j,r}}\right] \tag{7}$$

$$R_{k,r} = min\left\{v'_{k,r}M_{w,k}A\rho\frac{\varepsilon}{k}\min_R\left(\frac{Y_R}{v'_{R,r}M_{w,R}}\right), v'_{k,r}M_{w,k}AB\rho\frac{\varepsilon}{k}\frac{\sum Y_p}{\sum_{j=1}^{N}v'_{j,r}M_{w,j}}\right\} \tag{8}$$

In the EDC model, the chemical kinetic rate and vortex dissipation rate are simultaneously calculated, and the net reaction rate is generally the smaller of the two rates.

Where $v'_{k,r}$: Stoichiometric coefficient of the $k_{th}$ reactant in the $r_{th}$ reaction; $c_{j,r}$: Molar concentration of component j in the $r_{th}$ reaction; $\eta_{j,r}$: Reaction rate index of component j in the $r_{th}$ reaction; $k_{f,r}$: Reaction rate constant of the $r_{th}$ reaction; $A_r$、$\beta_r$、$E_r$、$R$: The $r_{th}$ pre reaction factor coefficient, temperature index, reaction activation energy and gas constant; $Y_p$, $Y_R$: Mass fraction of product P and reactant R; A=4.0, B=0.5; $\frac{\varepsilon}{k}$: Reciprocal of vortex mixing time scale.

### *2.2 Waste liquid particle combustion model*

The internal mass changes of waste liquid particles during drying are as follows:

$$\frac{dm_p}{dt}h_{fg} = hA_p(T_\infty - T_p) + \varepsilon_p A_p\sigma(\theta_R^4 - T_p^4) - m_p c_p\frac{dT_p}{dt} \tag{9}$$

Where $h_{fg}$: Latent heat of water evaporation; $\theta_R$: Radiation temperature; $h$: Heat transfer coefficient.

The change of waste liquid particle diameter and boiling point temperature during drying is as follows:

$$d_p = 1.5d_0\frac{dT_p}{dt} \tag{10}$$

$$T_{bp} = 373 + 50 \cdot y_s^{2.84} \tag{11}$$

Where $d_0$: Initial droplet diameter; $y_s$: Solid content in droplet.

In each time step of CFD calculation, the decrease of particle mass is equal to the increase of continuous phase moisture:

$$S_{H_2O} = \Delta m_p \tag{12}$$

Where $S_{H2O}$: Continuous phase component source term.

The calculation formula of pyrolysis rate in pyrolysis stage is as follows:

$$R_1 = A_1 e^{-(E_1/RT_p)} \quad R_2 = A_2 e^{-(E_2/RT_p)} \tag{13}$$

Where $R_1$, $R_2$: Low and high temperature pyrolysis rates.

The volatilization rate and the change formula of waste liquid quality are shown in Eqs. 14 and 15:

$$\frac{m_v(t)}{(1-f_{w,0})m_{p,0} - m_a} = \int_0^t (\alpha_1 R_1 + \alpha_2 R_2)\exp\left(-\int_0^t (R1 + R2)dt\right)dt \tag{14}$$

$$m_p = (1 - f_{w,0})m_{p,0} - m_v(\mathrm{t}) \tag{15}$$

Where $\alpha_1$, $\alpha_2$: yield factor; $m_a$: The mass of the particles when the volatiles of the waste liquid particles are completely precipitated; $m_v(t)$: The mass of volatile matter that has been precipitated at time t.

The change rule of particle diameter and surface temperature of waste liquid in pyrolysis stage is as follows:

$$d_p = d_{dry} + (d_{max} - d_{dry}) \cdot \left(\frac{m_{dry} - m_p}{m_{dry} - m_a}\right)^{0.8} \tag{16}$$

$$m_p c_p \frac{dT_p}{dt} = hA_p(T_\infty - T_p) - \frac{dm_p}{dt}h_{fg} + \varepsilon_p A_p \sigma(\theta_R^4 - T_p^4) \tag{17}$$

Where $d_{dry}, d_{max}$: The particle diameter at the time of drying completion and the maximum diameter of particle pyrolysis completion; $m_{dry}$: Particle quality at the time of drying.

Similarly, the increase in the source term of the continuous phase during the pyrolysis process corresponds to the decrease in the particle mass:

$$S_{\mathrm{i}} = \Delta m_p \tag{18}$$

The law of mass change and diameter change of waste liquid particles in the surface reaction stage is shown in Eqs. 19 and 20:

$$\frac{dm_p}{dt} = \sum_{i=1}^{N} R_i \tag{19}$$

$$d_p = \varphi d_0 + (d_{max} - \varphi d_0) \cdot \left(\frac{f_c}{f_{c0}}\right)^{1/3} \tag{20}$$

Where $\varphi$: waste fluid contraction factor; $f_c, f_{c0}$: Carbon content and initial carbon content.

The variation law of particle temperature in the surface reaction stage is as follows:

$$m_p c_p \frac{dT_p}{dt} = hA_p(T_g - T_p) + f_h \sum_{i=1}^{N} R_i Q_i + \varepsilon_p A_p \sigma(\theta_R^4 - T_p^4) \tag{21}$$

Where $f_h$: The proportion of heat absorbed by the waste liquid particles during the surface reaction process; $Q_i$: The $i_{th}$ surface reaction releases heat.

The variation law of the source term of the continuous phase components in the surface reaction process is shown in Eq. 22:

$$S_k = \sum_{i=1}^{N} n \cdot R_i \cdot \Delta t \cdot \frac{v''_{ik} \cdot M_{gik}}{v'_{ij} \cdot M_{sij}} \tag{22}$$

Where for reactant *k*, *n* = -1; for product k, n = 1; $v'_{ij}$: Stoichiometric coefficient of surface component j in reaction i; $v''_{ik}$: Stoichiometric coefficient of gas phase component k in reaction i; $M_{sij}$, $M_{gik}$: Relative molecular mass of components j, k.

During the surface reaction process, various alkali metals are involved mainly in the chemical reaction in the salt-containing waste liquid, and the kinetic parameters of some of the chemical reactions involved are listed in Table 1:

Table 1. Some kinetic parameters of surface reaction.

| Reaction | Reaction Rate Expression | *A* | *Ea*/(J/mol) |
|---|---|---|---|
| $Na_2SO_4+2C \rightarrow Na_2S+2CO_2$ | $k \cdot [x_{Na_2SO}]^{1.4} \cdot [C] \cdot V_p$ | $3.79\times10^3$ | $7.81\times10^4$ |
| $C + 0.5O_2 \rightarrow CO_2$ | $k \cdot A_s \cdot [C] \cdot p_{O_2}$ | $0.75\times10^3$ | $1.80\times10^5$ |
| $C + O_2 \rightarrow 2CO$ | $k \cdot A_s \cdot [C] \cdot c_{CO_2}/(c_{CO_2} + 3.4c_{CO})$ | $3.94\times10^8$ | $2.50\times10^5$ |
| $C + H_2O \rightarrow CO + H_2$ | $k \cdot A_s \cdot [C] \cdot c_{H_2O}/(c_{H_2O} + 1.42c_{CH_2})$ | $1.60\times10^7$ | $2.10\times10^5$ |
| $Na_2S + 2O_2 \rightarrow Na_2SO_4$ | $k \cdot A_s \cdot [Na_2S] \cdot p_{O_2}$ | $0.75\times10^3$ | $1.80\times10^5$ |

## 3. Geometric modeling and boundary conditions

The geometric model of the incinerator simulated in this paper is shown in Figures 1:

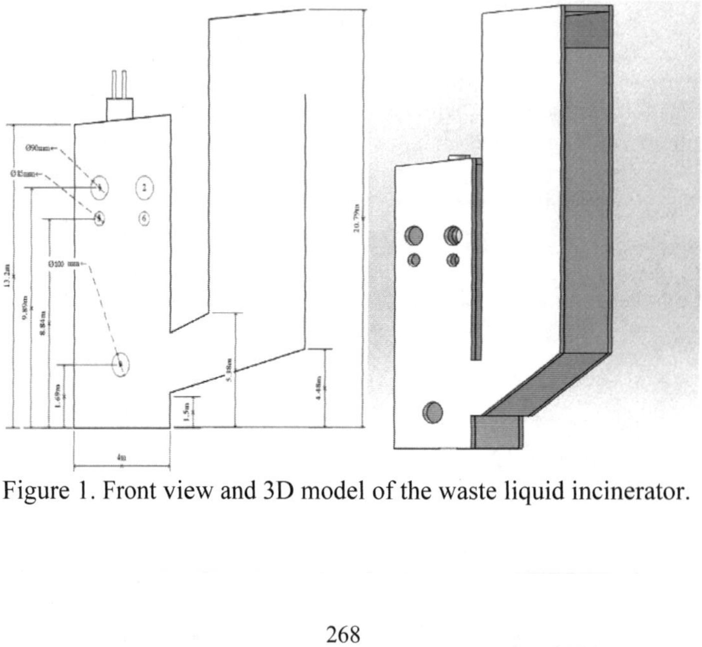

Figure 1. Front view and 3D model of the waste liquid incinerator.

The calculation model is 100 times smaller in proportion to the actual size of the boiler.

The total treatment capacity of the caprolactam waste liquid incineration boiler is 18t/h, composed of waste alkali liquor with 12t/h, and X-oil waste liquor with 6t/h, and the components of waste alkali liquor, X-oil and fuel are listed in Table 2:

Table 2. Incinerator waste liquid and gas components.

| Waste lye solids | $Na_2CO_3$ | $Na_2SO_4$ | C |
|---|---|---|---|
| Quality fraction (%) | 89.7 | 1.2 | 9.1 |
| X-oil | $CH_3OH$ | $C_2H_6O$ | $H_2O$ |
| Quality fraction (%) | 53 | 33 | 14 |
| Gas | $CH_4$ | $N_2$ | |
| Mole fraction (%) | 99 | 1 | |
| Air | $O_2$ | $N_2$ | $H_2O$ |
| Mole fraction (%) | 22.82 | 76.4 | 0.78 |

The velocity boundary condition is adopted in calculation process, and the flow setting of is listed in Table 3:

Table 3. Inlet flow of caprolactam waste liquid incinerator.

| Gas flow(kg/s) | Air flow(kg/s) | Waste lye flow(kg/s) | X-oil flow(kg/s) |
|---|---|---|---|
| 0.8 | 11.5 | 3.3 | 1.67 |

The waste lye and heavy oil are injected into the furnace after atomized by the top spray gun, adopting R-R distribution for particle size of the atomized particles of the two waste liquids. The minimum/maximum diameter of waste lye droplets is 100/400 μm, and SMD is 200 μm; the minimum/maximum diameter of heavy oil droplets is 80/200μ m, and SMD is 120 μm.

## 4. Result and discussion

### *4.1 Analysis of Numerical Simulation Results*

The section of the incinerator with X=2 m is selected, with the velocity field, temperature field and oxygen concentration field depicted in Figure 2, where the front and rear wall airflows collide in the center of the furnace, the high temperature area in the furnace is mainly concentrated in the middle and lower parts of the furnace, with the core combustion temperature around 1450 °C. The velocity of the primary air nozzle area is evenly mixed, providing favorable conditions for the combustion in the central area of the furnace; the airflow in the secondary air area is partially turned, which effectively strengthens the heat exchange of the water wall and avoids the risk of slagging on the wall; the air flow in the tertiary air area is fully better, the unburned waste liquid is fully supplemented, with the rise of the flue gas particles at the bottom punctually blocked; the heat load in the corner area of the flue is large, with high risk to wear and slagging; the right flue reports uneven heat exchange, heat transfer design requires to be modified.

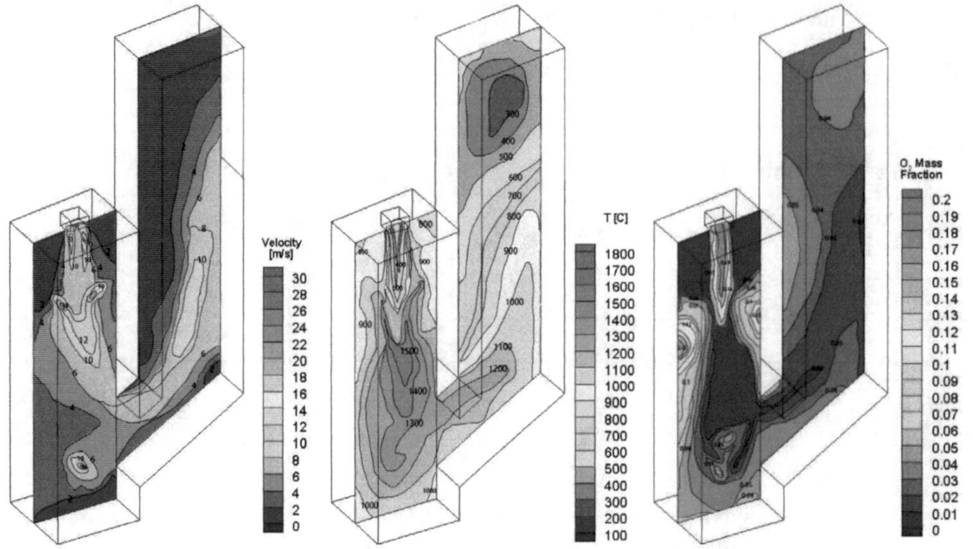


Figure 2. Furnace velocity, temperature and oxygen concentration distribution.

In Figure 3 exhibits the combustibles distribution and particle trajectories for X=2 m. An obvious enrichment area of combustibles is formed in the middle of the furnace from top to bottom, in which the fuel reacts with oxygen and releases a lot of heat; some combustibles are still carried with the flue gas particles, this insufficiently complete reaction may be resulted from insufficient oxygen supply or excess fuel. Therefore, the oxygen-fuel ratio and air distribution can be optimized to promote the combustibles to burn as completely as possible, thereby further optimizing the combustion efficiency of the boiler.

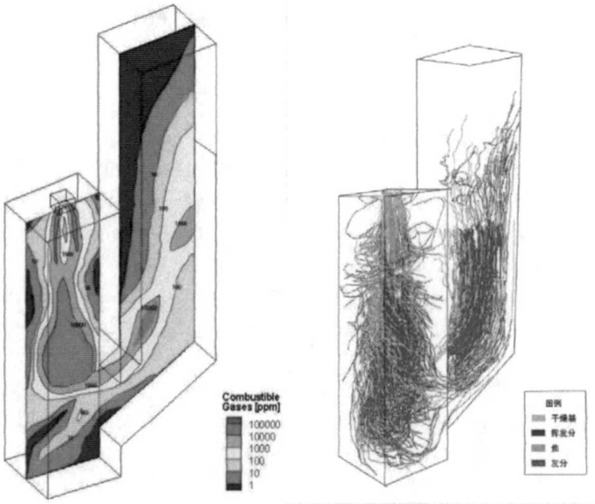


Figure 3. Combustibles and particle trajectory distribution.

In Table 4 lists the calculated temperature, theoretical temperature and actual temperature of some parameter points in the furnace:

Table 4. Comparison of calculation results with actual operating parameters and theoretical parameters.

| Type | Calculated | Operating | Theoretical |
|---|---|---|---|
| core combustion temperature | 1453℃ | 1350℃ | 1495℃ |
| The temperature at point A | 1165℃ | 1063℃ | 1125℃ |

| | | | |
|---|---|---|---|
| The temperature at point B | 1086℃ | 986℃ | 1033℃ |
| The temperature at point C | 896℃ | 842℃ | 864℃ |
| The temperature at point D | 473℃ | 464℃ | 456℃ |

As Table 4 shows, point A represents the average temperature before the turning angle of the flue, point B for the average temperature after the turning angle of the flue, point C for the average temperature of the Z=10 m section, and point D for the average temperature at the outlet of the flue. It can be concluded that the error of the calculation results is smaller than the actual operating parameters and theoretical parameters, which is indicated to basically reflect the combustion and flow conditions in the incinerator.

### *4.2 Combustion performance optimization*

The streamline optimization design is adopted for the corner area of the furnace, enable the gas from the furnace to enter the heat exchange screen and the flue for a smooth transition avoiding the occurrence of high heat load points; the excess volume at the bottom of the furnace is removed, and the bottom of the furnace is designed as an inclined surface, which contributes to the ash on the bottom of the furnace accumulating and sliding out naturally under the action of gravity, reduces the cost of manual cleaning and materials, and also saves the installation of boiler materials. The verification result of the scheme reports that the material and boundary conditions are consistent with the calculations above, and the velocity field and temperature field distribution in the furnace obtained are shown in Figure 4:

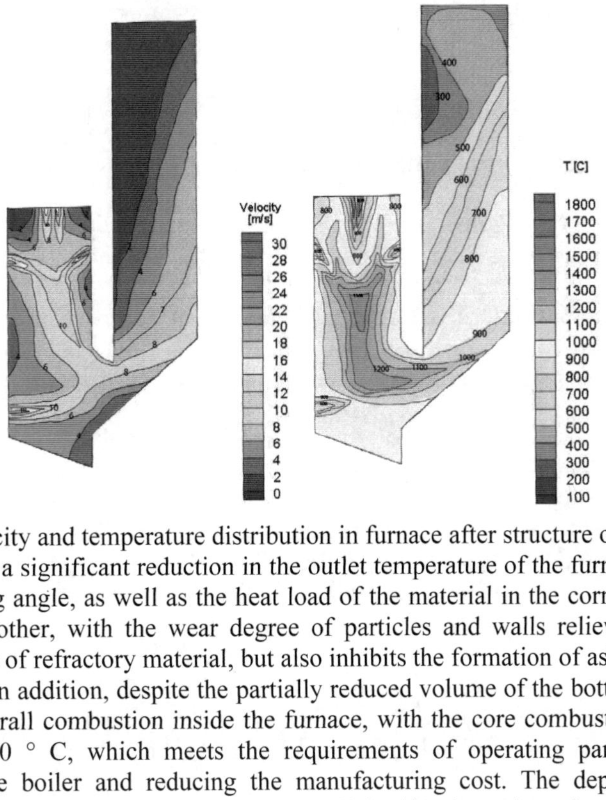


Figure 4. Velocity and temperature distribution in furnace after structure optimization.

The Figure 4 shows a significant reduction in the outlet temperature of the furnace after the smooth transition of the turning angle, as well as the heat load of the material in the corner area, and the flue gas circulation is smoother, with the wear degree of particles and walls relieved, which not only elevates the service life of refractory material, but also inhibits the formation of ash and slagging in the wall area of this area. In addition, despite the partially reduced volume of the bottom of the furnace, it does not affect the overall combustion inside the furnace, with the core combustion temperature still remaining around 1370 ° C, which meets the requirements of operating parameters, saving the installation area of the boiler and reducing the manufacturing cost. The deposited ash can also naturally slide to the lower right corner of the furnace elicited by the action of gravity, so as to realize the continuous cleaning of the ash deposits and elevate the operation and maintenance cycle of the boiler.

Table 5 compares some temperature parameters before and after optimization:

Table 5. Comparison of some parameters before and after optimization.

| Type | Before optimization | After optimization | Operating |
|---|---|---|---|
| core combustion temperature | 1453℃ | 1368℃ | 1350℃ |
| The temperature at point A | 1165℃ | 1121℃ | 1063℃ |
| The temperature at point B | 1086℃ | 937℃ | 986℃ |
| The temperature at point C | 896℃ | 886℃ | 842℃ |
| The temperature at point D | 473℃ | 442℃ | 464℃ |

In Table 5, points A-D reports the same as in Table 4. It can be obviously seen that the flue gas temperature is greatly lowered after the flue is bent after structural optimization, reducing the heat load of the furnace wall and the risk of ash and slagging in this area.

## 5. Conclusions

In this paper, with the combustion characteristics of high-salt organic waste liquid combined, a corresponding numerical model suitable is proposed, and a three-dimensional numerical simulation of its flow and combustion process is conducted using Fluent with an actual operating waste liquid incinerator as a model. The results demonstrate that the proposed waste liquid combustion model can more accurately predict the flow, combustion, heat transfer and particle trajectory movement in the furnace; the design of this waste liquid incinerator is reasonable in general, with the relatively uniform combustion structure, only a certain amount is reported in the local area. However, considering the numerical simulation results, the optimization of bending angle and structural design of the boiler can be carried out to reduce the ash and slagging and particle wear in the furnace, improving the combustion performance of the boiler.

## References

[1] Cardoso M, Costa G A A, Oliveira E D D, et al. (2012). Simulation of eucalyptus kraft black liquor combustion in industrial Recovery Boilers. J. Latin American Applied Research. 42, 197.
[2] Connolly T S. (2006). Carbon dioxide pyrolysis and gasification of kraft black liquor char. D. New Hampshire: The University of Maine. 32-40.
[3] Ehrhardt K, Ehret A. & Leuckel W. (1998). Experimental study on the dependence of burnout on the operation conditions and physical properties in wastewater incineration. J. Symposium (International) on Combustion. 27, 1293-1299.
[4] Frederick W J. (1991). Combustion process in black liquor recovery: Analysis and Interpretation of combustion rate data and an engineering model. C. USA.
[5] Kaul S N, Nandy T, Deshpande C V, et al. (1998). Application of full-scale evaporation-incineration technology for hazardous wastewater. J. Water Science and Technology. 38, 363-372.
[6] Lü Hong-jun & Li Xiao-dong. (2007). Investment on combustion characteristic and dynamics of high-concentration organic waste water. J. Boiler Technology. 38, 74-78. (in Chinese)
[7] Smith I W. (1982). The combustion rates of coal chars: A review. J. Symposium (International) on Combustion. 19, 1045-1065.
[8] Yuan Jian-wei. (2009). Computer Simulation of Alkali Recovery Boiler and Its Application in Combustion System Design. J. China Paper. 28, 39. (in Chinese)
[9] Zhang Jian-xin. (2015). Modeling and Numerical Simulation of Black Liquor Combustion Process. D. Harbin: Harbin Institute of Technology. (in Chinese)

# Numerical Study on the Water Takeoff Performance of Unmanned Seaplane

**Chengyi Wu*, Yichong Wang, Aihua Wang, Hang Chen and Fei Zhang**

Shanghai Electro-Mechanical Engineering Institute, Shanghai, 201109, China

*Corresponding author's e-mail: wuchengyibuaa@163.com

**Abstract.** The research on the water takeoff performance of unmanned seaplane is of great significance to its overall optimization design and taxiing stability control system. This paper proposes a new numerical method for solving the water takeoff performance of unmanned seaplane. The aerodynamics and hydrodynamics are solved separately by decoupling solution principle. The influence of aerodynamic force and moment, engine thrust and moment on hydrodynamic force and taxiing attitude is considered synthetically by the vertical approximate balance hypothesis and instantaneous pitch balance hypothesis. The discrete velocity piecewise solution is used to solve the takeoff process. The results show that the taxi resistance peak appeared in the range of 0.3~0.35 times of takeoff velocity, and the greater is the trim angle, the smaller are the hydrodynamic resistance and hydrodynamic moment when the unmanned seaplane taxiing at medium or high speed. What's more, compared with the fixed trim angle hypothesis, the combined trim angle method is more consistent with the actual takeoff process.

## 1. Introduction

In recent years, unmanned seaplane has gradually become a new research hotspot in the field of UAVs. The unmanned seaplane not only has the good accessibility in water areas, but also has the characteristics of flexibility and low risk, which has broad application prospects in water areas such as oceans and lakes [1-3]. Unmanned seaplane generally requires the ability to take off and land at short distances autonomously on the water surface. Therefore, the research on the water takeoff performance plays an important role in its overall optimization design and taxiing stability control system design.

During the water takeoff process, the unmanned seaplane is subjected to various forces, such as gravity, engine thrust, aerodynamic force, and hydrodynamic force [4]. These forces of different properties make the takeoff process become a complex motion. There are two main methods to study the water separation characteristics of traditional seaplanes. The first method is to carry out model test. Qiao Wuqi. [5] and Shi Shengzhe. [6] realized the measurement of water resistance and the estimation of takeoff distance by towing tank test. The second method is to use a series of resistance maps, theoretical or empirical formulas to predict the characteristics of water resistance. Referencing to the research of high speed planning hull, Wu Qingwei et al. [7], Sun Jianjun et al. [8] propose a semi-theoretical and semi-empirical formula for estimating the water resistance characteristics of seaplanes in hydrostatic water. Among the above research methods, the model test method is the most accurate, but its test cost is high and the period is long. At the same time, the second methods above also have shortcomings such as a narrow scope of application.

With the development of computational fluid dynamics technology, the current CFD software has the ability to accurately solve unsteady multiphase flow problems, which provides the possibility to numerically solve the water takeoff characteristics of seaplanes. The application of numerical calculation method can not only reduce the design cost and shorten the design cycle, but also study the flow field information from a microscopic perspective. In the existing numerical research on the water takeoff characteristics of seaplanes, Qiu Liangjun et al. [9] discretized this process in the time domain, and solved the aerodynamic force and hydrodynamic force by the decoupling method. Li Zhi et al. [10] discretized the process in the velocity domain, and solved the water takeoff characteristics according to the velocity sections. In the above studies, it is assumed that the seaplanes take off from the water with a constant pitch angle, ignoring the pitch change. Actually, in the initial taxiing stage, the speed is low which cannot make the seaplanes maintain a fixed pitch angle. What's more, taxiing with a fixed pitch angle also ignores the influence of various forces on the attitude of seaplanes during the taxiing process.

Therefore, on the basis of previous research, this paper proposes a new idea to numerically solve the water takeoff characteristics of the unmanned seaplane. The decoupling method is used to solve the aerodynamic and hydrodynamic forces separately. The pitch instantaneous balance hypothesis and the vertical approximate equilibrium state hypothesis are proposed to consider the influence of aerodynamic force and moment, engine thrust and moment on hydrodynamic force and taxi attitude comprehensively. And then the strategy of combined pitch angle is adopted to complete the solution of the water takeoff process, which is in line with the actual takeoff process as much as possible.

## 2. Solution Method

### *2.1. Decoupling Method*

In CFD simulation, aerodynamic calculation and hydrodynamic calculation have different requirements in terms of calculation grid, solution method, calculation efficiency, etc. [11]. In order to obtain good calculation accuracy, aerodynamic calculation needs to divide the whole shape into grids, which will make the grids number to be millions or even tens of millions. However, the requirement for hydrodynamic calculation is only about 1 million. What's more, the height of the first layer of the boundary layer grid is generally 0.01 millimetre for aerodynamic calculation, and 1 millimetre for hydrodynamic calculation, which is very different. In terms of the solution method, the steady solution method can be selected for the aerodynamic calculation, while the unsteady method needs to be used for the hydrodynamic calculation to solve the two-phase flow problem. Last but not least, hydrodynamic calculation is a transient process, which requires a shorter time step and more internal iterations. As a result, directly constructing the water-air mixed flow field for coupled solution has low computational efficiency and is not suitable for the preliminary design stage.

In order to take the efficiency and accuracy into consideration, the decoupling method of aerodynamic and hydrodynamic calculation will be adopted in this paper. Firstly, the whole seaplane model is used to solve the aerodynamic characteristics in a single air flow field. Then, the VOF (volume of fluid) model will be used to construct a two-phase flow field based on the results of aerodynamic calculation. And the hydrodynamic and motion characteristics of the seaplane hull in the water-air mixed flow field are solved using the overlapping dynamic mesh and the DFBI (dynamic fluid body interaction) six-degree-of-freedom model.

### *2.2. Instantaneous Pitch Balance Hypothesis*

In the water takeoff process, the motion equation contains 3 mass centre kinematic equations and 3 rotational kinematic equations. However, the lateral movement, lateral rotation, and heading rotation rarely occur actually, which have little impact on the hydrodynamic characteristics. Generally, the above motion and rotation will be ignored when solving the water takeoff process.

The hydrodynamic force of seaplane can be divided into three types: static buoyancy, hydrodynamic lift and hydrodynamic resistance. Static buoyancy is a function of waterline height and trim angle, while hydrodynamic resistance and hydrodynamic lift are related to a function of speed, waterline height and trim angle. When the seaplane taxing at a fixed speed steadily, there will be a unique waterline height and trim angle corresponding to the speed. This paper will use the overlapping dynamic grid and the DFBI model to simulate the free heave and pitch motion, and then solve the waterline height, pitch angle at a given taxiing speed. What's more, in order to consider the influence of aerodynamic force and engine thrust on hydrodynamic characteristics comprehensively, this paper proposes an instantaneous pitch balance hypothesis: at any speed instant, the unmanned seaplane is in a state of torque balance, that is, the seaplane goes from a speed equilibrium state to the other speed equilibrium state is completed instantaneously, which is ignoring the process of state transition.

### *2.3. Vertical Approximate Balance Hypothesis*

Based on the above conditions, the six-degree-of-freedom equation of the unmanned seaplane can be transformed into the accelerated linear motion in the heading and vertical direction. At the same time, imagine such an ideal situation: during the takeoff process, the hydrodynamic resistance is ignored. Then the above each motion can be regarded as a uniformly accelerated linear motion. Therefore, the dynamics and kinematics equations of seaplane under ideal conditions can be obtained as follows.

$$\begin{cases} F_T - R_a - R_h = m\dfrac{dV}{dt} \\ F_v = L_a + L_h + D_w - mg = ma_v \\ V_{TO} = \dfrac{F_T}{m}t \\ h_{st} = \dfrac{1}{2}a_v t^2 \end{cases} \tag{1}$$

For the unmanned seaplane in this paper, the take-off weight is 104kg, the engine thrust is 400N, the initial waterline height is 0.216m, and the take-off speed is 21m/s. Substituting the above conditions into the above formula, it can be solved that the required time for accelerating to the take-off speed is 5.46s, the vertical acceleration is 0.0145m/s$^2$, and the vertical resultant force is just 1.5N under the ideal condition. From the above result, it can be found that the vertical resultant force is about one thousandth of the seaplane take-off gravity. Therefore, the vertical approximate equilibrium hypothesis can be put forward, which means that the vertical resultant force of seaplane is approximately 0, and the vertical acceleration motion won't be taken into consideration during the takeoff process. Finally, the kinematic equations of unmanned seaplane in the takeoff process can be transformed into the following form.

$$\begin{cases} F_T - R_a - R_h = m\dfrac{dV}{dt} \\ L_a + L_h + D_w - mg \approx 0 \end{cases} \tag{2}$$

### *2.4. Velocity Piecewise Solution*

Based on the assumption of instantaneous pitch balance and vertical approximate balance, the takeoff process can be discretized to several speed segments. In each speed segment, the seaplane satisfies the kinematic equations (2). When the speed segment is sufficiently finely divided, the heading motion in each speed segment can be approximated as a uniform acceleration linear motion. Therefore, as long as the starting speed and ending speed of the speed segment are given, the information such as the forces, heading acceleration and waterline height under the speed segment can be solved by simulation calculation. And then the kinematic equations can be solved to obtain the motion state parameters such as taxiing time and taxiing distance of each segment. Finally, these parameters will be summed in segments to complete the solution of the entire water takeoff process. For the state solution of other velocity points in each discrete segment, it can be obtained by linear interpolation.

Figure 1 shows the whole solution flow of water takeoff process. By means of discrete segmentation, each sub-stage is independent of each other during the simulation calculation, and the two sub-stages before and after only transmit the speed information that has been determined during the segmentation. Therefore, after the discrete segmentation is completed, the simulation calculation of each sub-stage can be performed in parallel, which can improve the computational efficiency.

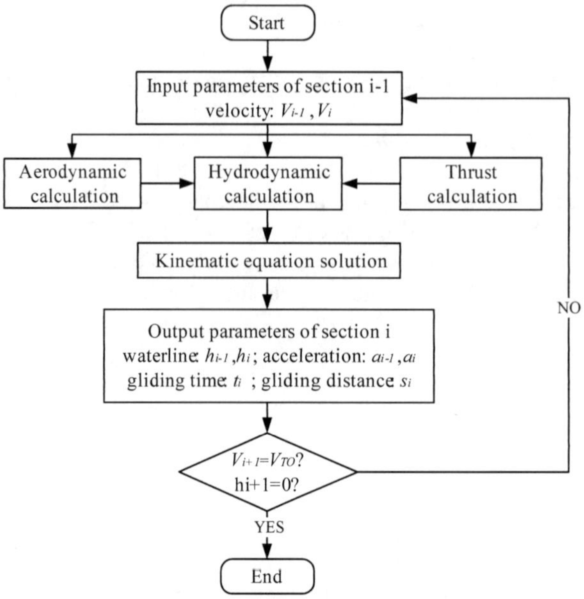


Figure 1. The takeoff solution process.

## 3. Validation Case

Based on the principle of decoupling solution, the aerodynamic force and hydrodynamic force can be solved separately when studying the water takeoff characteristics. And the research method can refer to the ship hydrodynamics with more experimental data, especially the numerical research methods of high-speed planning hull. In the study of simulating the hydrodynamic characteristics of high-speed planning hull on the water surface, the Fridsma model is a commonly used standard verification model [12]. The Fridsma model has a relatively simple shape and has a large amount of water tank test data that can be used for validation [13]. In this paper, one of the configurations of the Fridsma model is selected to carry out CFD simulation calculation in STAR CCM+ software. The calculation results will be compared with the experiment data to verify the effectiveness of the CFD method adopted in this paper. The geometric shape of the Fridsma model is shown in Figure 2.

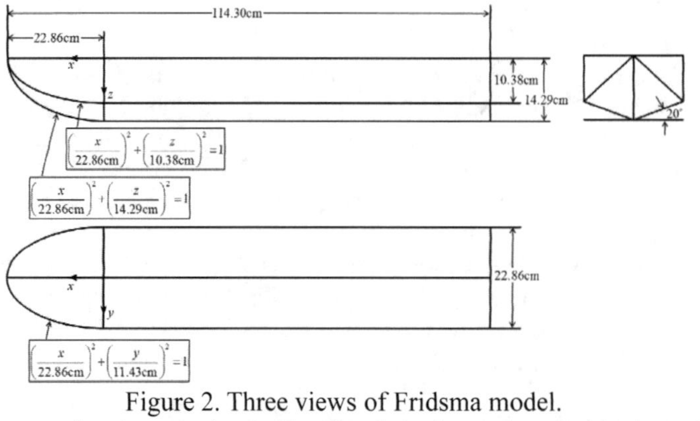


Figure 2. Three views of Fridsma model.

In this paper, the overlapping moving grid technology is used to divide the computing grid. The far-field background grid adopts the octree method to generate the hexahedral grid. What's more, the cut volume grid is used in the overlapping area, and the interface adopts the linear interpolation format to convey the flow field information between overlapping and background regions [14]. The meshing of the Fridsma model is shown in Figure 3.

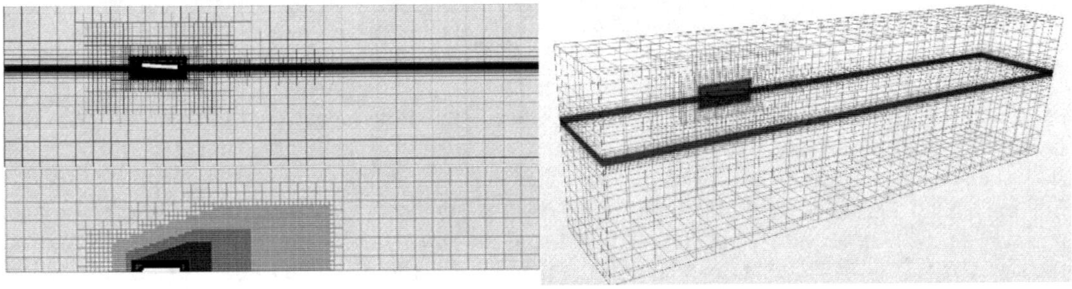

Figure 3. Meshing of Fridsma model.

In terms of calculation settings, the turbulence model and the VOF model are used to solve the unsteady Reynolds average equation. The DFBI model built in STAR CCM+ is selected to simulate the six-degree-of-freedom motion of the hull. Since only the pitch and heave motions of the Fridsma model were allowed in the tank test, only the Z-direction movement and the Y-direction rotation were activated in the six-degree-of-freedom solver settings of the DFBI model.

According to the test data of the Frisma model, the motion characteristics of the Fridsma model under five working conditions are calculated and compared with the experimental data. The comparison results are shown in Figure 4. It can be seen that the CFD calculation results of water resistance are in good agreement with the experimental data, which shows that the positive and negative errors of each data point are within 6%, and the average absolute error is 3.43%. The CFD calculation results of the pitch angle are basically consistent with the experimental data, and the errors of most data points are within 6%. Comparing the actual difference value of the pitch angle, it can be found that the errors are all within 0.7°. In general, for the pitch angle prediction of the unmanned seaplane, this error value meets the accuracy requirements.

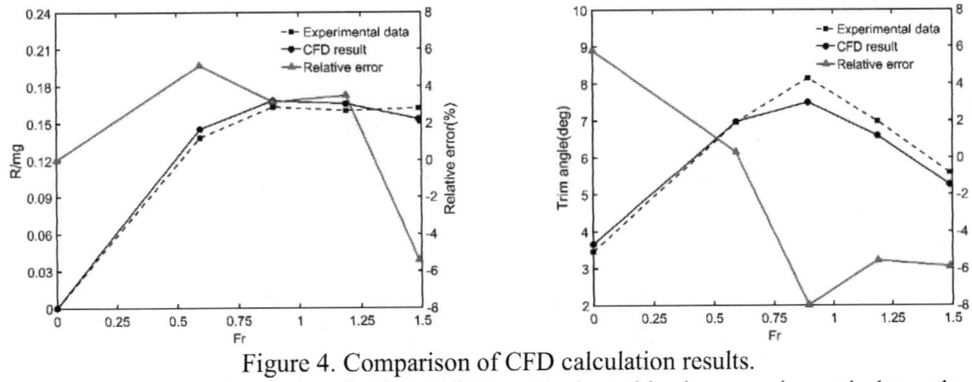


Figure 4. Comparison of CFD calculation results.

Therefore, by comparing the CFD calculation results with the experimental data, the CFD calculation method adopted in this paper meets certain accuracy requirements and is feasible. The same meshing technology and calculation settings will be used for the subsequent simulation calculations of hydrodynamic characteristics of the unmanned seaplane.

## 4. Calculation Model

In this paper, the main overall parameters of the unmanned seaplane are shown in Table 1. In the aerodynamic and hydrodynamic simulation calculations, the semi-model configuration is used, so the subsequent calculation input and calculation results are half of the actual value.

Table 1. The main parameters of unmanned seaplane.

| Parameter | Value | Parameter | Value |
|---|---|---|---|
| Takeoff weight | 104kg | Wing area | 3.0m$^2$ |
| Hull length | 4.0m | Full thrust | 360N |
| Hull width | 0.4m | CG position | (1.65m,0.0m,0.15m) |
| Hull height | 0.57m | Thrust arm | 0.49m |

Considering the complexity and computational efficiency of the model comprehensively, the aerodynamic calculation adopts the unstructured grid of the whole layout. The grid height of the first layer of the boundary layer is 0.01mm, and the y+ value is close to 1. Regions with drastic changes in flow such as the planning step of fuselage are refined, and the number of aerodynamic grids is about 7 million. Based on the principle of decoupling solution, the wings as well as tails and buoy were removed in the hydrodynamic calculation model, and the hull was meshed. The meshing method was the same as that of the Fridsma model, and the number of hydrodynamic grids is about 1.5 million. The meshing of calculation models are shown in Figure 5.

Figure 5. The surface mesh of aerodynamic model (left) and hydrodynamic model (right).

## 5. Result Analysis

In order to better study the water takeoff characteristics of the unmanned seaplane, this paper not only analyses the hydrodynamic characteristics under the fixed trim angle by referring to the research methods of the existing literature, but also compares the water resistance characteristics and moment characteristics between four different working conditions, which are the fixed trim angle of 3.7° (trim angle at mooring), the fixed trim angle of 5.0°, the fixed trim angle of 7.0° and the free trim.

### *5.1. Analysis of Fixed Trim Angle Condition*

This paper takes a fixed trim angle of 5.0° as an example to carry out the analysis. Figure 6 shows the water surface waveforms under different taxiing speeds. With the increase of the taxiing speed, the wave height of the wave system at the head of the hull increases, and the angle of the wave system at the stern gradually decreases as the speed increases. These changing laws are in line with the waveform change trend of ship hydrodynamics [15].

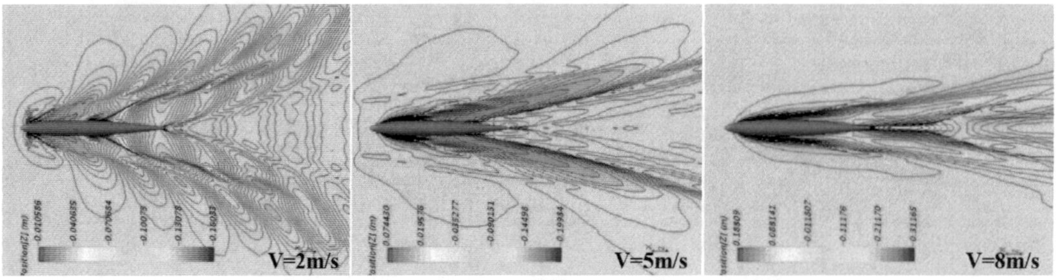


Figure 6. Waveforms at different taxiing speeds.

Figure 7 and Figure 8 show the hydrodynamic characteristics of fixed trim angle condition. Generally, the takeoff process of a seaplane can be divided into four stages: sailing stage (0~0.25$V_{TO}$), transition stage (0.25$V_{TO}$~0.5$V_{TO}$), taxiing stage (0.5$V_{TO}$~0.8$V_{TO}$) and takeoff stage (0.8$V_{TO}$~$V_{TO}$).

In the sailing stage, there is a significant feature that the hydrodynamic lift is negative. Because of the low speed, the aerodynamic lift is small, which means that the static buoyancy needs to be increased to balance the gravity. At this stage, the waterline height will first increase and the decrease as shown in Figure 7(d).

In the transition stage, there is an obvious resistance peak, and its corresponding speed is called "uplift speed", which is 7m/s in this paper as shown in Figure 7(a). At this stage, the air cushion phenomenon appears at the planning step of hull. As shown in the Figure 8, the sliding area of the rear body begins to decrease and the hydraulic force and bowing moment of the rear body decrease accordingly. In order to maintain balance, the hydraulic force of front body continuously increases, which makes the total hydrodynamic moment of the seaplane increase sharply until it reaches the maximum value.

In the taxiing stage, the height of the waterline decreases continuously, and the rear body gradually separates from the water surface, which means that the seaplane gets into a single fault-step taxiing state. As shown in Figure 8(c), the immersed area of hull decreases and is concentrated on the front sliding surface, and the hydrodynamic resistance continues to decrease. What's more, the distance between the action point and the centre of gravity is also reduced, so that the hydrodynamic moment decreases.

In the takeoff stage, the speed is continuously increased to the takeoff speed, the hydrodynamic resistance and the waterline height are rapidly reduced to zero, and then the unmanned seaplane completes the water takeoff process.

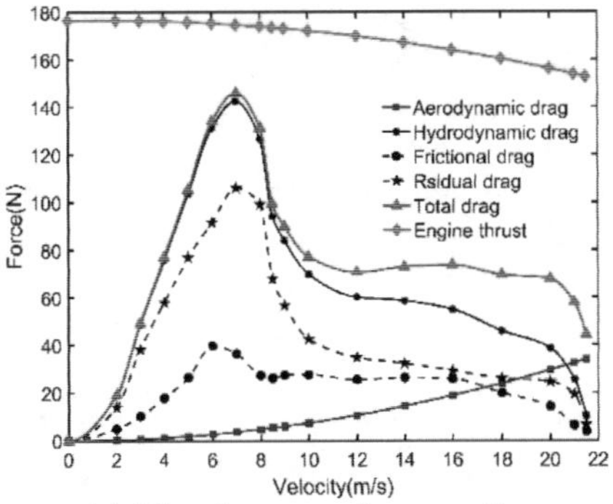


(a) The changes of course force

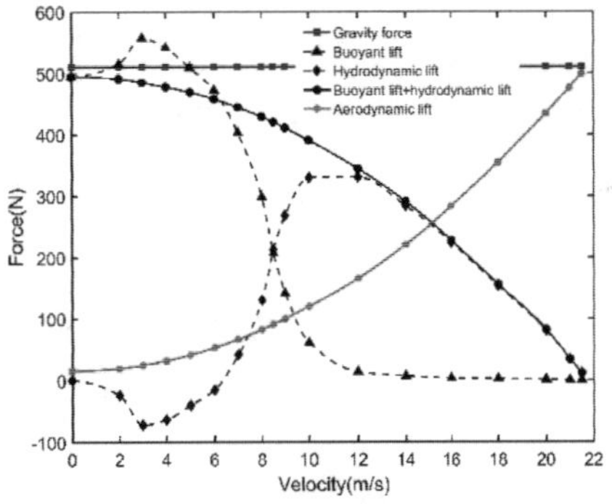


(b) The changes of vertical force

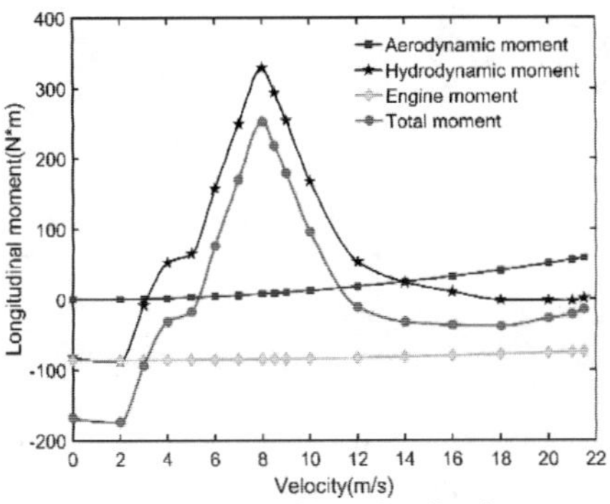


(c) The changes of longitudinal moment

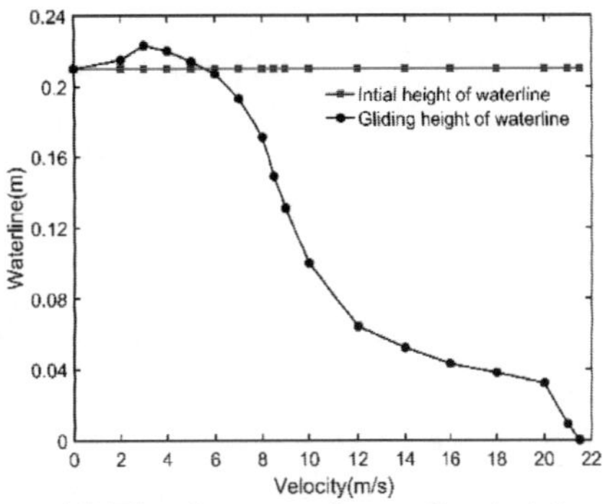


(d) The changes of waterline height

Figure 7. Volume fraction of water phase of hull at different speeds.

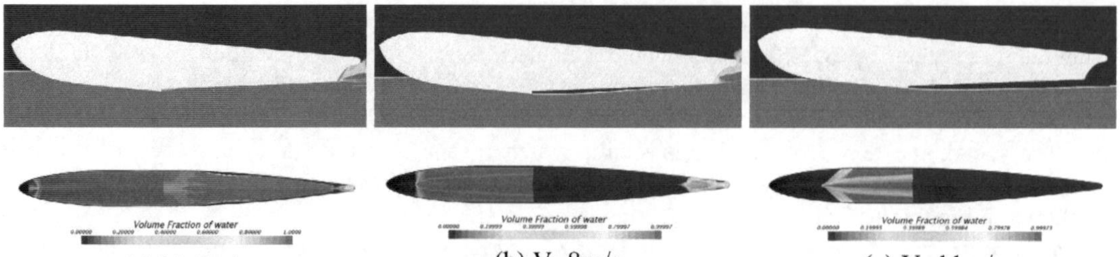


(a) V=5m/s (b) V=8m/s (c) V=11m/s

Figure 8. Volume fraction of water phase of hull at different speeds.

*5.2. Compared Analysis of Different Conditions*

The comparison of hydrodynamic characteristics under four different working conditions is shown in Figure 9. Among them, only the calculation data in the range of 0~8.5m/s is given in the free trim condition. The reason is that without manipulation, the unmanned seaplane in this paper occurs dolphin motion after the taxiing speed is greater than 9m/s, which make it impossible to achieve pitch balance and vertical force balance under constant taxiing speed. However, in the actual seaplane taxiing out of the water, the tail rudder can be manipulated to provide a pitching moment to help recover from the unstable state.

In the low-speed taxiing section (V≤5m/s), the change trend and value of the hydrodynamic resistance of the four working conditions are basically the same. However, there are obvious differences in the performance of the hydrodynamic moment, which indicates that the trim angle is mainly affecting the total hydrodynamic moment of seaplane. The larger the trim angle is, the greater the hydraulic force of the rear body is, and the greater the bow moment is. From Figure 9, it can also find that the hydrodynamic moment under the free trim condition is always kept in balance with the aerodynamic moment and the engine moment. What's more, the smaller hydrodynamic moment received by the unmanned seaplane, the easier it is to taxi with a constant trim angle, and the smaller control capability of the horizontal tail is required. When taxiing at medium or high speed, the larger the fixed trim angle is, the smaller the hydrodynamic resistance and the hydrodynamic moment are, while the hydrodynamic resistance under the free trim condition is the smallest. The resistance peaks under the four conditions all appear at the speed range of 0.3 to 0.35 times the takeoff speed.

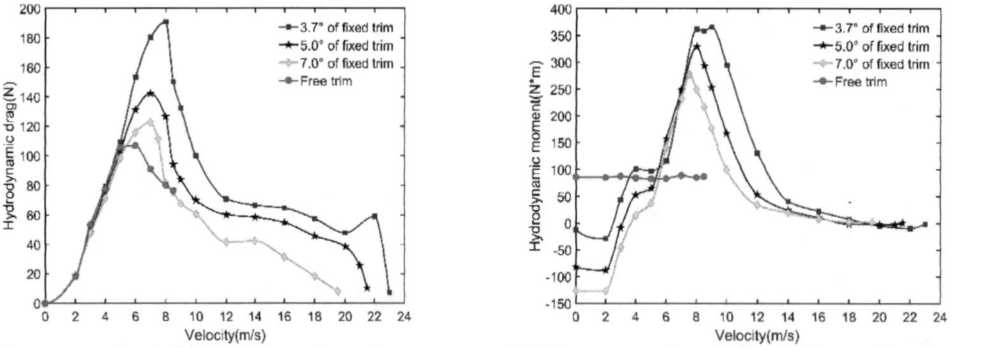


(a) The comparison of hydrodynamic resistance (b) The comparison of hydrodynamic moment

Figure 9. Hydrodynamics of different conditions.

*5.3. Calculation of Takeoff Process*

For the unmanned seaplane in this paper, its free-trim taxiing will show unstable motion when the taxiing speed is greater than 9m/s, and a control torque must be applied to maintain the stability of the taxiing motion attitude in actual situation. Therefore, this paper proposes a takeoff strategy with a combined trim angle: in the low-speed taxiing section (0~8.5m/s), it is free-trim taxiing, and after

8.5m/s, a horizontal tail control moment is applied to maintain the attitude at 7° trim angle until seaplane reaches its takeoff speed. Figure 10(a) shows the balance torque required in the taxiing process under the four working conditions. Combining with the aforementioned resistance characteristic curves, it can be seen that this takeoff strategy not only reduces the resistance during high-speed taxiing, but also makes the hydrodynamic torque as well as aerodynamic torque and engine torque to maintain a balance with each other when taxiing at low speed rang, which means there is no need to increase the horizontal tail control. On the contrary, each fixed pitch angle condition requires a balance torque of at least 100N·m during taxiing at the speed of 0~2m/s, which is seriously inconsistent with the actual situation. Therefore, compared with the assumption of takeoff with a fixed trim angle all the time, the take-off strategy of combined trim angle is more suitable for the real taxiing process and is feasible.

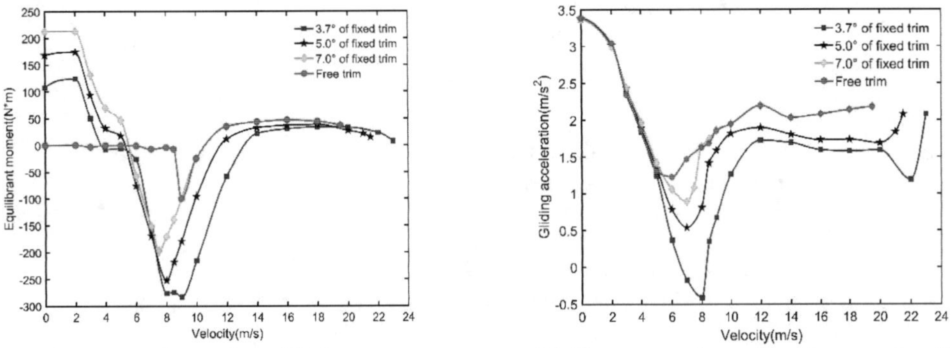


(a) The comparison of equilibrium torque (b) The comparison of taxiing acceleration

Figure 10. The comparison of equilibrium torque and taxiing acceleration.

According to the course force of the whole aircraft under the aforementioned working conditions, the acceleration of the whole aircraft under different taxiing speeds is obtained as shown in Figure 10(b). According to the principle of discrete interpolation calculation, the takeoff process in this paper is discretely segmented by speed, and the kinematic equation is solved for each speed segment based on the assumption of uniform acceleration linear motion. Then the taxiing time and taxiing distance in a single speed segment are obtained. By means of segmental summation, the total taxiing time and taxiing distance of the whole aircraft are calculated. Since the peak resistance exceeds the maximum thrust when taxiing with a fixed pitch angle of 3.7°, this taxiing condition cannot be successfully taken off from the water. The calculation results of the take-off performance for the other working conditions are as follows.

$$\begin{cases} t_{5.0^\circ} = 13.8\text{s} & s_{5.0^\circ} = 146.2\text{m} \\ t_{7.0^\circ} = 10.4\text{s} & s_{7.0^\circ} = 100.0\text{m} \\ t_{combine} = 9.9\text{s} & s_{combine} = 96.8\text{m} \end{cases} \tag{3}$$

Through the above calculation and analysis, the unmanned seaplane in this paper adopts the water takeoff strategy with combined trim angle, and its take-off taxiing time is 9.9s and the take-off taxiing distance is 96.8m.

**6. Conclusion**

(1) During the taxiing process of unmanned seaplane on the water surface, the resistance peak appears in the range of 0.3~0.35 times the take-off speed in the transition stage. When taxiing at medium or high speed, the larger the trim angle is, the smaller the hydrodynamic resistance and hydrodynamic moment are.

(2) Based on the assumption of instantaneous pitch balance and vertical approximate balance, the water takeoff characteristics of the unmanned seaplane under the free trim condition is being simulated

and calculated, which can comprehensively take the influence of aerodynamic force and aerodynamic moment as well as the thrust and engine moment into consideration.

The analysis method and the above conclusions of this paper have certain reference significance for the overall design, water takeoff performance research and flight control system design of the unmanned seaplane. In the next step, a detailed numerical simulation analysis of the unstable movement phenomenon of "dolphin motion" can be carried out to further improve the water takeoff performance of the unmanned seaplane in this paper.

**References**

[1] Yang X, Wang T, Liang J, et al. Survey on the novel hybrid aquatic–aerial amphibious aircraft: Aquatic unmanned aerial vehicle[J]. Progress in Aerospace Sciences, 2015, 74: 131-151.
[2] Johnson E R. American flying boats and amphibious aircraft: An illustrated history[M]. Jefferson, North Carolina: McFarland & Company Inc. Publishers, 2009: 213-228.
[3] WANG Cheng, NI Yi, LI Liang, et al. Development status and thinking of early warning technology for small commercial unmanned aerial vehicles[J]. Air & Space Defense, 2021, 3(1): 83-90.
[4] QIU Liangjun. Integrated aerodynamic and hydrodynamic optimization of amphibious aircraft. Shanghai: Shanghai Jiao Tong University, 2013, 27-28.
[5] Qiao Wuqi, YE Yashui, DAI Sizong, et al. Test determination of water resistance and approximate calculation of take-off distance for hull seaplane[J]. Journal of Nanjing University of Aeronautics and Astronautics, 1964, 6: 68-89.
[6] SHI Shengzhe, ZHENG Yaxiong, WU Bin, et al. Hydrodynamic regression analysis of seaplane single hull tests[J]. Aeronautical Computing Technology, 2015, 45(4): 97-100.
[7] WU Qingwei, GAO Xiaopeng, WU Bin. A method to evaluate the resistance of seaplane sliding in still water[J]. Ship & Ocean Engineering, 2013, 42(3): 154-157.
[8] SUN Jianjun, MA Dongli. Estimation of gliding resistance in still water of hull seaplane at medium and high speed[J]. Journal of Beijing University of Aeronautics and Astronautics, 2015, 41(5): 925-929.
[9] Qiu L, Song W. Efficient decoupled hydrodynamic and aerodynamic analysis of amphibious aircraft water takeoff process[J]. Journal of aircraft, 2013, 50(5): 1369-1379.
[10] LI Zhi, MA Dongli, YANG Muqing, et al. A method to solve the out-of-water characteristics of water UAV based on piecewise interpolation[J]. Acta Aeronautica et Astronautica Sinica, 2017, 12.
[11] Burg C.O.E., Screenivas K. Vassalos D., et al. Computational of Free Surface Turbulent Flow Solutions: Validation and Verification[A]. 32nd AIAA Fluid Dynamics Conference and Exhibit[C]. St. Louis, Missouri, 2002: 2977-3002.
[12] Savitsky D. Hydrodynamic design of planing hulls[J]. Marine Technology and SNAME News, 1964, 1(04): 71-95.
[13] Fridsma G., “A systematic study of the rough water performance of planing boats,” Stevens Institute of Technology, Report 1275, 1969.
[14] LI Ming, LI Minggao, GAO Feng, et al. STAR-CCM+ and flow field calculation[M]. Beijing: China Machine Press,2011.
[15] LU Xiaoping. Principle of Ship[M]. Beijing: National Defense Industry Press, 2009: 103-130.

# Comparative study on the performance of turbulence models in flow separation of large expansion ratio nozzle

**Le Wang[1,a], Zhengyu Tian[1,b*], Hang Yu[1,c], Ye Zhang[1,d], Hua Li[1,e]**

College of Aerospace Science and Engineering National University of Defense Technology, Changsha, China

[a]E-mail: 1246102019@qq.com; [b*]E-mail: tianzhengyu_kd@163.com;

[c]yuhang18@nudt.edu.cn; [d]zhangye17a@nudt.edu.cn; [e]lihuakd@tom.com

**Abstract**—The asymmetry of flow separation inside the nozzle with large expansion ratio is the main factor for the generation of side-load in the engine, so the accurate prediction of the flow separation is of great significance. The turbulence model is the key to the prediction, but in different scenarios, the performance of the turbulence model can introduce huge uncertainties. In this paper, a typical thrust-optimized nozzle (VOLVO-S1) is simulated and analyzed for three widely used turbulence models: **SST** $k$ - $\omega$ 、 **RNG** $k$ - $\varepsilon$ and Spalart-Allmaras model. The results show that all these models can effectively predict the mode of shock separation phenomenon. The comprehensive performance of **SST** $k$ - $\omega$ is the best; The Spalart-Allmaras model has better prediction performance under low pressure ratio, and the **RNG** $k$ - $\varepsilon$ model has better prediction effect under high pressure ratio.

## 1. Introduction

The nozzle is an important component for thrust generation in the rocket engine. For the main engine of a launch vehicle, in order to obtain better high-altitude thrust performance, a nozzle with a large expansion ratio (the ratio of nozzle exit area to throat area) is often used. For example, the compressed truncated perfect nozzle (CTP) is used in the Russian RD-0120 and Japanese LE-7 engines; The thrust-optimized parabolic (TOP) nozzles, also known as thrust-optimized profile (TOC) nozzles are used in American J-2S rocket engine, Space Shuttle Main Engine (SSME) and the European Vulcain engine. Although the high expansion ratio nozzle provides excellent high-altitude performance, it also jeopardizes the sea-level operation. When the nozzle operates at low altitude, the high backpressure makes the flow to be in an over-expanded state, and separate from the wall in the diffuser section of the nozzle. The asymmetry of flow separation inside the nozzle is the main factor for the generation of side-load in the engine. There are two distinctive flow separation phenomena inside the nozzle, named free shock separation (later referred to as FSS) and restricted shock separation (later referred to as RSS).

FSS is widely present in all types of convergent–divergent nozzles, as shown in Fig.1, When the nozzle exit pressure is lower than the ambient pressure, the gas generates an oblique shock wave to match the external environmental pressure, at the same time, the gas detaches from the wall and continues to flow in the form of supersonic free jets. The supersonic jet maintains a certain distance from the wall, and the ambient gas outside the nozzle flows into the nozzle along the wall, forming a recirculation zone.

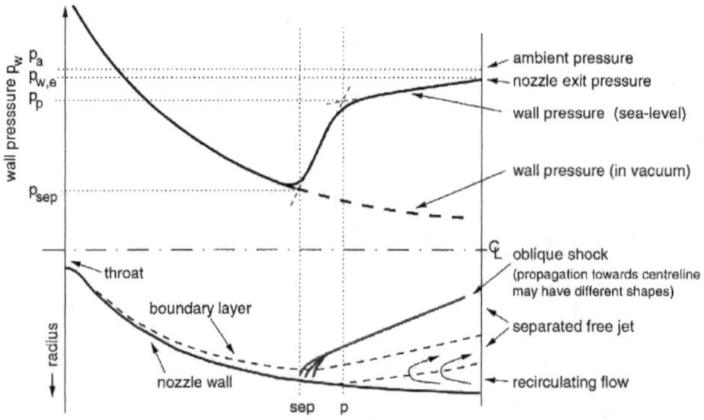

Figure 1. Phenomenologica sketch of free shock separation [1]

RSS usually occurs when the TOP, TOC or CTP type nozzle is operating within a specific pressure ratio range. Its basic structure is shown in Fig.2. In the RSS phenomenon, the jet is pushed back to the wall after leaving the wall for a short time, so it is called restricted shock separation. Behind the separation point, a small separation vortex generates. The reflection of the shock wave on the wall leads to a large increase in the wall pressure. The detail of the flow structure can be found in these references[1-3].

The significant side-load caused by the asymmetry of flow separation can cause structural failure of thrust system. For this reason, the phenomenon of flow separation is widely studied.

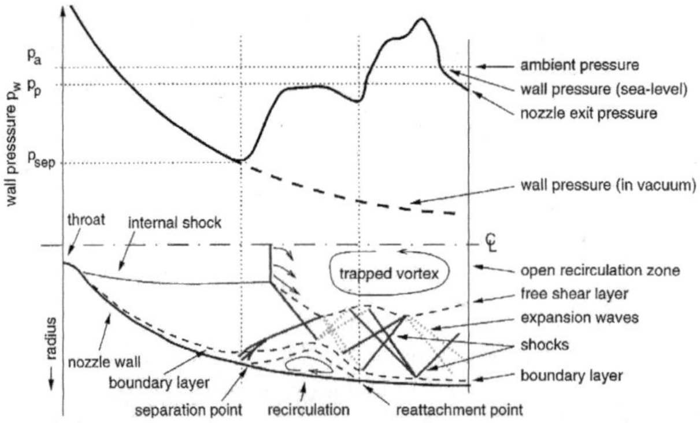

Figure 2. Phenomenological sketch of restricted shock separation [1]

It can be seen that the accurate prediction of the flow separation inside the nozzle is extremely important. Since the flow separation of a large expansion nozzle involves many complex flow phenomena such as: separation vortex, shock-shock interaction and shock-boundary layer interaction, it is a challenge to accurately simulate the flow separation phenomenon inside the nozzle. The turbulence model is a key element in the simulation of these flow problems. In the past numerical research of flow separation inside the nozzle, various RANS models and their correction models have been widely used and studied. As early as the last century, researchers have noticed the considerable differences between different turbulence models in dealing with flow separation inside the nozzle. In 1999, Östlund[4] tested four turbulence models that based on $k$ - $\omega$ two equations and found that only the Menter's SST model was able to capture the correct type of separation during startup and shutdown, but the wall pressure downstream of the separation was significantly different from the experimental data. Baloni et al[5] used the SST $k$ - $\omega$ turbulence model to conduct a two-dimensional steady state simulation of a bell-shaped nozzle. The numerical simulation results and analysis results are in terms of shock structure, shock position, normal shock and asymmetric lambda shock size have good consistency. R. Stark et al[6] conduct a comparative study on the prediction performance of multiple two-equation turbulence models that based on $k$-$\omega$ equation and Spalart-Allmaras turbulence models. The nozzle used in the experiment is ideal truncated profile nozzles (TIC nozzles). Calculation results of multiple turbulence models are compared with the experimental results. The results of this study show that the vast majority of prediction results tend to underestimate the location of the position of separation point (closer to the upstream). The S-A and SST $k$-$\omega$ models show more reasonable prediction results than others. Nebbache et al[7] used the SST $k$-$\omega$ turbulence model to simulate a sub-size TOC nozzle. The results show that when the pressure ratio is equal to 10.2 and 15.4 respectively, the predicted wall pressure is completely consistent with the experimental value. However, as the PR increases to 20.0, there is a slight difference between the numerical and experimental results, and when the PR is further increased to 23.9, the difference becomes very significant. Under such flow condition, the flow still maintains FSS in the experiment, but the numerical result is RSS. As PR continues to rise to 37.9 and 41.6, the numerical prediction results are in good agreement with the experimental results. Martelli E et al[8] used the Spalart-Allmaras one-equation model simulated a TOP nozzle in 2D axisymmetric. The simulation results capture the flow characteristics well under different shock separation phenomenon. At the same time, the hysteresis phenomenon in the transition of shock separation mode is effectively predicted. Huang et al [9] carried out a numerical simulation study on the overexpansion flow inside the nozzle, and compared the prediction performance difference between the RNG $k$-$\varepsilon$ turbulence model combined with the standard wall function and the enhanced wall function. Their article shows that the prediction result of the RNG $k$-$\varepsilon$ model combined with the enhanced wall function is in agree with the experimental values, and the performance is better than the standard wall function.

To sum up, the applicability of turbulence models has great uncertainty under different nozzles and different pressure ratios, which makes it extremely necessary to evaluate the applicability of different turbulence models. Therefore, this paper compares and evaluates the performance of three turbulence models widely used in previous numerical simulation studies, including two two-equation models: SST $k$-$\omega$、 RNG k-ε and a one-equation Spalart-Allmaras model under different pressure ratios. The differences in predictions from three turbulence models will be explained in combination with turbulence model properties. Based on the above results, the selection strategy of turbulence models for simulating different conditions will be given.

## 2. Physical Models and Numerical Methods

### *2.1. Physical Models*

The nozzle used in this paper is the S1 nozzle (VOLVO-S1) developed by VOLVO Airlines. The S1 nozzle is a typical TOP nozzle, which can produce optimum thrust at specified values of nozzle length and ambient pressure.

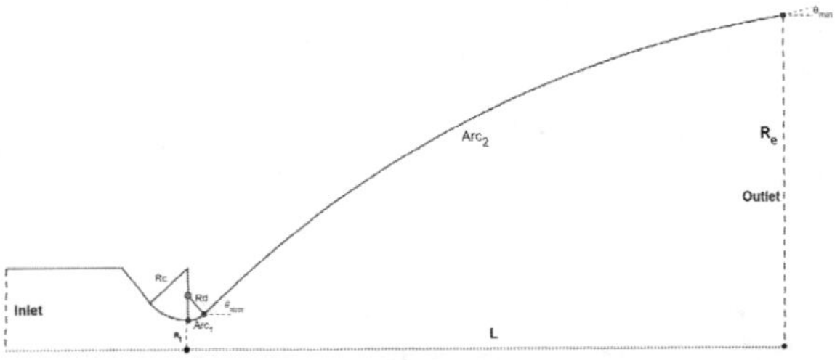


Figure 3. VOLVO-S1 nozzle contour

Since the configuration of the convergent section has little effect on the flow, here the contour of the diffusion section is only briefly described. The configuration of the S1 nozzle is shown in Fig.3. The throat radius of the nozzle is $R_t$ =33.54mm. The diffusion section is composed of two parts. First part is the expansion part (Arc1), which is composed of an arc with a radius of 0.5 $R_t$ , and starting position of Arc1 is tangent to the nozzle throat, the end point is tangent to the start point of the compression section of the second part to ensure that the first derivative of the diffusion section is continuous. The compression section (Arc2) is controlled by a parabolic curve, and its control curve equation is shown in (1).

$$\left(\frac{y}{Rt}+b\frac{x}{Rt}\right)^2+c\frac{x}{Rt}+d\frac{y}{Rt}+e=0 \quad (1)$$

The four parameters (b, c, d, e) of the control curve are controlled by the length of the expansion section L, the nozzle outlet radius $R_e$, the maximum inclination angle of the expansion section $\theta_{max}$ and the minimum inclination angle $\theta_{min}$ . In the S1 nozzle: $R_e$=350mm 、 L=150mm 、 $\theta_{max}$=35° 、 $\theta_{min}$=4°

*2.2.Boundary Conditions and Numerical Methods*

In order to impose an atmospheric environment close to the actual state at the nozzle outlet, this paper combines previous studies[10], and conservatively considers the expansion of the external flow field of the nozzle to exclude the calculation difference due to insufficient expansion of the computational domain. As shown in Fig.4, the computational domain is extended 100 $R_t$ axially back from the nozzle exit, radially extend 50 $R_t$ from the outer wall at the nozzle exit, and extend 30 $R_t$ axially forward from the nozzle outlet. The wall thickness of the nozzle exit is 0.33 $R_t$ .

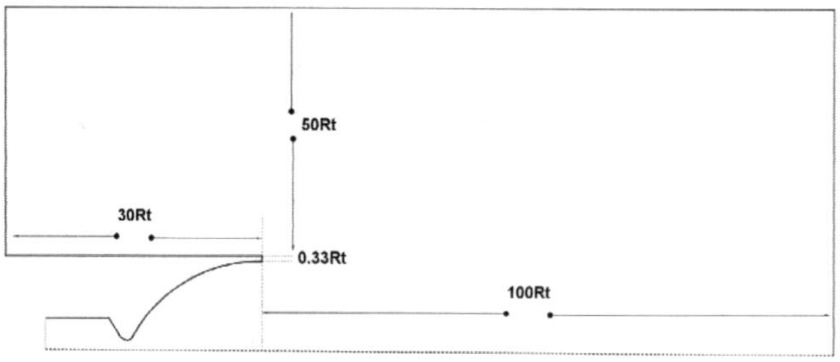


Figure 4. Computational domain and boundary conditions

According to the experiments carried out by Mattsson J[11] et al, the boundary conditions of this research are set as shown in the Fig.4. The solid blue and solid green lines are both pressure inlet conditions, denoted as Pressure-Inlet1 and Pressure-Inlet2, respectively. The red solid line is set as the pressure outlet and recorded as Pressure-Outlet. The bottom black dotted line is the symmetry axis. Wall is set to adiabatic and no-slip condition, as shown by the black solid line in the Fig.4. Using turbulence intensity and hydraulic diameter to specify the flow turbulence characteristics. For specific parameters, see TABLE 1.

The equations solved here are the steady, 2-D axisymmetric Reynolds-averaged compressible Navier-Stokes equations. The perfect gas relation and a constant ratio of specific heats are used in the computation. The relationship between viscosity and temperature is established by Sutherland's formula. Using the second-order upwind style for spatial discretization. The governing equations are closed using the eddy-viscosity model (EVM), which linearly relates the Reynolds stress to the mean strain rate via eddy-current viscosity. It should be noted that the applies to locally balanced flows, where the time scale of turbulence is on the same order of magnitude as that of mean strain. However, in highly non-equilibrium flows, the time scale of mean strain is significantly smaller than that of turbulent flow. Therefore, the linear eddy-viscous assumption produces unrealistically high Reynolds stress values, resulting in very high turbulent kinetic energy *k*, so Production Limiter are used in this paper to correct the turbulent kinetic energy of the two-equation models. The three turbulence models used in this paper will be briefly introduced later.

TABLE 1. Boundary condition

| Boundary | Condition | | | |
|---|---|---|---|---|
| | Total Temperature [K] | Total Pressure [Pa] | Turbulent Intensity [%] | Hydraulic Diameter [m] |
| Pressure-Inlet1 | 300 | 101325 | 0.1 | 1 |
| Pressure-Inlet2 | 450 | —— | 1.9 | 0.2 |
| Pressure-Outlet | 300 | 101325 | 1.9 | 1.8 |

*1) Spalart-Allmaras*

Spalart-Allmaras[12] (hereafter referred to as S-A) is a one-equation model that designed specifically for aerospace applications involving wall-bounded flows and has been shown to give good results for boundary layers subjected to adverse pressure gradients. In its original form, the S-A model is effectively a low-Reynolds number model, requiring the viscosity-affected region of the boundary layer to be properly resolved (providing a mesh of $y^+ \sim 1$ at the boundary layer).

*2)* RNG $k-\varepsilon$

The RNG $k-\varepsilon$ [13](hereinafter referred to as RNG) model was derived using a statistical technique called renormalization group theory. It is similar in form to the standard $k-\varepsilon$ model, but with a certain degree of improvement: The RNG model has an additional term in its $\varepsilon$ equation that improves the accuracy for rapidly flows; And RNG theory provides an analytically derived differential formulation for effective viscosity that accounts for low Reynolds number effects, but its effectiveness depends on appropriate treatment of the near-wall region.

*3)* SST $k$ - $\omega$

The SST $k-\omega$[14](hereinafter referred to as SST) combines the advantage of the $k$ - $\omega$ model and $k-\varepsilon$ model, and in addition accounts for the transport of the turbulence shear stress in the definition of the turbulent viscosity to obtain appropriate transport behavior. The modified formula is shown in (2). These features make the SST model more accurate and reliable for a wider class of flows (for example, adverse pressure gradient flows, airfoils, transonic shock waves) than the standard $k-\varepsilon$ and the BSL $k-\varepsilon$ models.

$$\mu_t = \frac{\rho k}{\omega} \frac{1}{\max\left[\frac{1}{a^*}, \frac{SF_2}{a_1 \omega}\right]} \tag{2}$$

## 3. Grid independence verification

*3.1 Grid Generation*

In this paper, the mesh type is quadrilateral structure mesh. In order to refine the key areas, the grid is divided into four blocks including the nozzle constriction section A, the expansion section B, the jet section C and the external flow field.

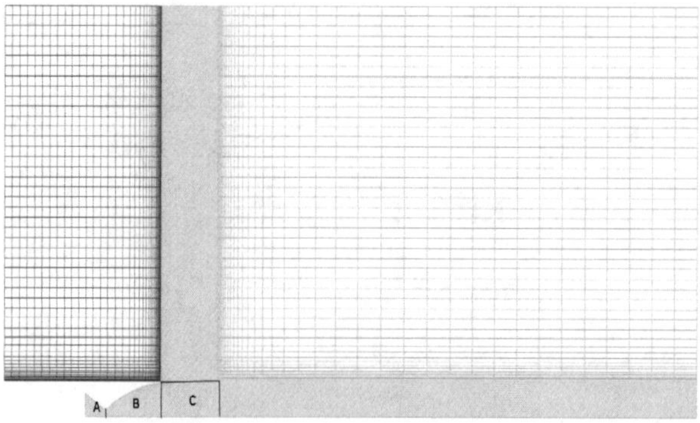

Figure 5. Schematic of grid structure

Four sets of grids are generated in this paper, and the grid independence verification is carried out for two typical PRs (pressure ratio: the ratio of the total pressure of the combustion chamber $P_c$ to the ambient pressure $P_a$ ). The height of the first layer of the four sets of grids near the wall is set to 1μm to ensure $y^+ \sim 1$, the normal growth rate of the wall grid is 1.2, the number of grids in the boundary layer is about 20. The aspect ratio $\Delta y / \Delta x$ of the main grid inside of the nozzle is kept as 2. The main difference between the four sets of grids is the axial and radial grid spacing of the nozzle diffusion section and the jet section. TABLE 2 lists the mesh size of the four sets of mesh diffusion sections and the total element of meshes.

TABLE 2. grid parameters

| Grid Number | Grid Properties | |
|---|---|---|
| | Size of diffusion section | Total Element |
| mesh1 | 1.5mm*3mm | 55546 |
| mesh 2 | 1mm*2mm | 89939 |
| mesh 3 | 0.75mm*1.5mm | 140168 |
| mesh 4 | 0.5mm*1mm | 243329 |

*3.2.Independence Test*

The TOP nozzle has two completely different separation phenomena (FSS and RSS) at different pressure ratios, so the grid independence is carried out for the two shock separation modes. On this basis, the grid used in this paper for follow-up research are selected. The SST turbulence model is used for calculation and verification in the process of mesh independence research.

1)*Free Shock Separation*

In this section, four meshes are calculated under the condition of PR=14. The calculation results are shown in the Fig.6. The axial distance is normalized with the throat radius $R_t$, and the y-direction is the pressure distribution along the wall, which is normalized using the ambient pressure $P_a$. The same normalization method will be used in the following text and will not be repeated. "exp" is the experimental data[11].

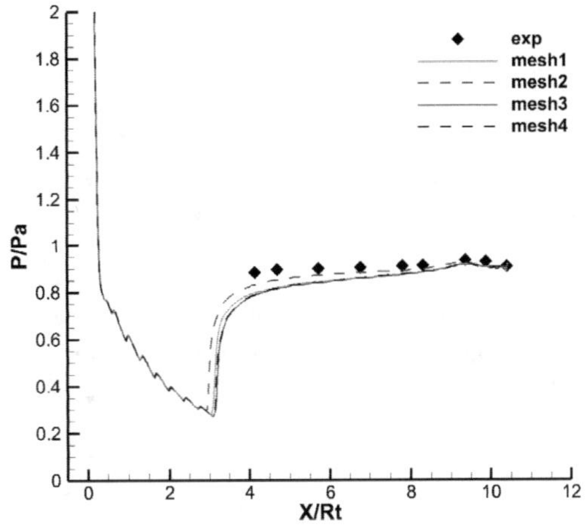


Figure 6. Results of wall pressure for different meshes at PR=14

As shown in the Fig.6, the four sets of meshes are in agreement with the experimental results. When the number of mesh element increases gradually, the wall pressure curve is gradually unified to the result that calculated by the set of meshes with the largest number of total element (mesh4). The calculation results using mesh3 is almost identical to the results using mesh4.

2） *Restricted Shock Separation*

For research on grid independence in RSS mode, the PR=16 case is calculated in this section. Except for PR, the calculation conditions are exactly same as the FSS mode. The calculation results are shown in the Fig.7.

Fig.7 shows that the numerical prediction results are in good agreement with the experimental data. It can be seen that only the green curve which is the calculation result of mesh1 is relatively deviated from the results of other grids, and the result of mesh2 basically coincides with the result of mesh4. The results of mesh3 and mesh4 are almost indistinguishable, and are consistent with the experimental data.

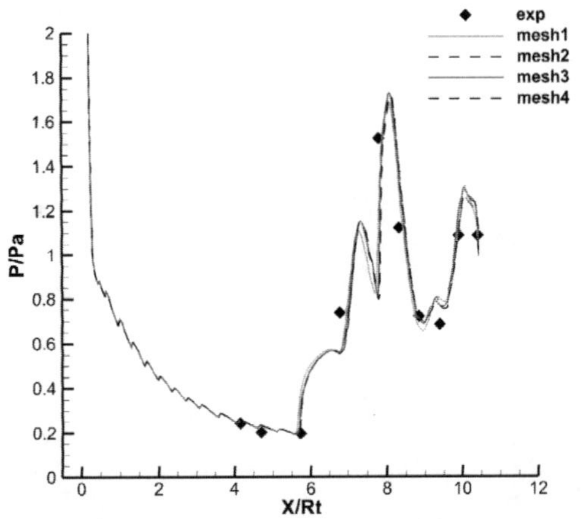


Figure 7. Results of wall pressure for different meshes at PR=16

Combining the calculation results under the two conditions, it can be seen that when mesh2 is used, the grid almost no longer affects the calculation results in the case of PR=16, but for the condition of

PR=14, the calculation result of mesh2 shows that the mesh still interferes with the calculation results. Therefore, mesh3 is adopted for the subsequent research.

## 4. Results & discussion

In this section, the prediction performance of each turbulence models under different pressure ratios is discussed by comparing the calculation results under different turbulence models with the experimental data. The content of this section is based on three turbulence models and five calculation conditions, including two FSS modes, two RSS modes, and a critical mode.

### *4.1.Free Shock Separation*

In this section, numerical calculations are carried out for the two PRs that form the FSS. The prediction results of these three turbulence models are discussed from the aspects of shock separation mode, separation position, wall pressure, and Mach stem position. It should be noted that since the pressure monitoring points in the experiment is not a continuous distribution, the position of the separation point measured in the experiment cannot be completely and accurately pointed out.

1) PR=*12.9*

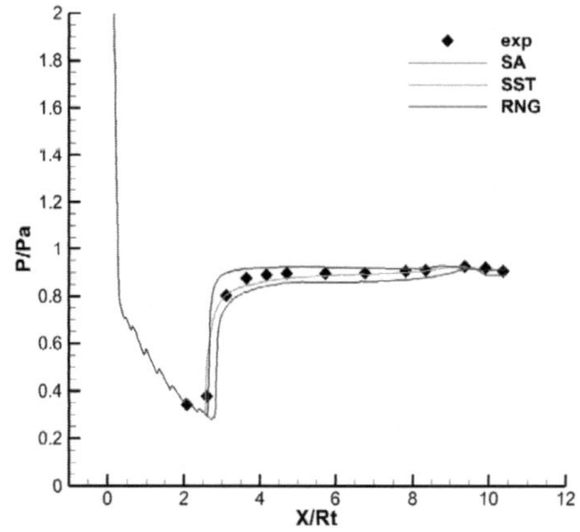


Figure 8. Results of wall pressure for different models at PR=12.9

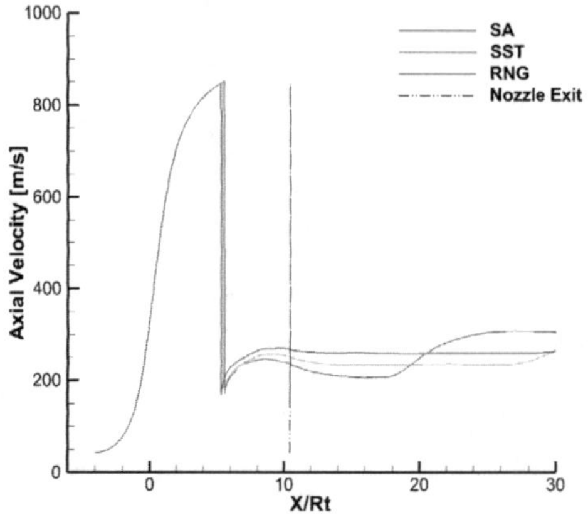


Figure 9. Results of Axial Velocity for different models at PR=12.9

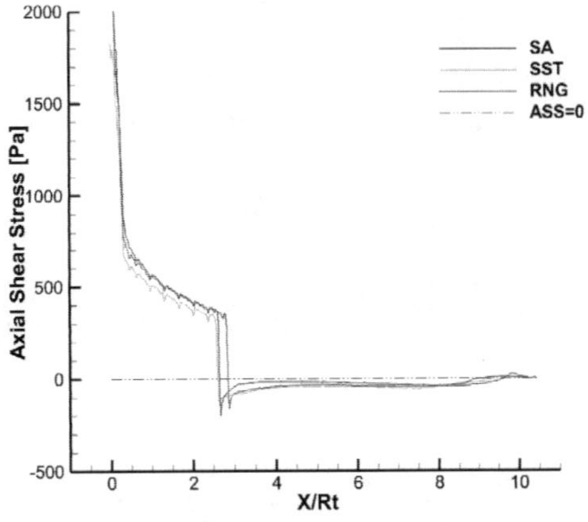


Figure 10. Results of Axial Shear Stress for different models at PR=12.9

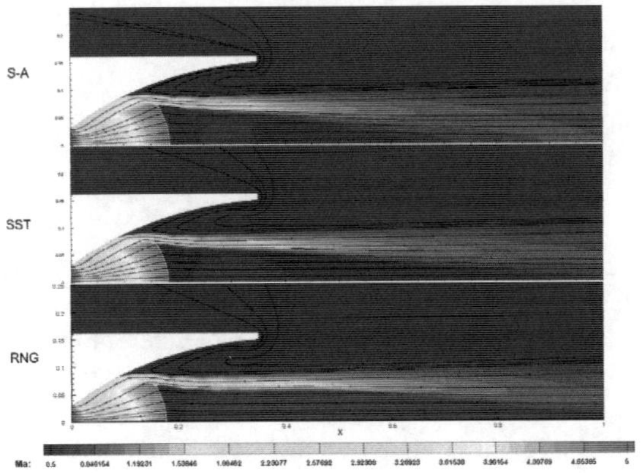


Figure 11. Mach number contour and streamlines for different models at PR=12.9

From the wall pressure distribution curve and the contour of Mach number, it can be observed that when PR is 12.9, all three turbulence models accurately predicted the FSS mode, but the position of the separation point were overpredicted (the predicted position was downstream of the experimental separation position). Among these three models, the SST and RNG models perform relatively well in the prediction of separation points. In terms of wall pressure, the SST model is most consistent with the experimental value; The pressure calculation result of the SA model is lower than the experimental value and the prediction results of others; The wall pressure predicted by the RNG model rapidly recover to the level close to ambient pressure behind separation position and remains at this value until reaching the nozzle outlet. This characteristic is quite different from the SA and SST models (pressure recovery trend is relatively slow) for the predicted trend of wall pressure. It is also worth noting that the position of the separation point predicted by the RNG model is always closer to the upstream than the result with other models, and this difference can also be observed in the following multiple cases. Regarding this difference, some qualitative conclusions can be obtained by analyzing the prediction results of the turbulent kinetic energy of the SST model and the RNG model.

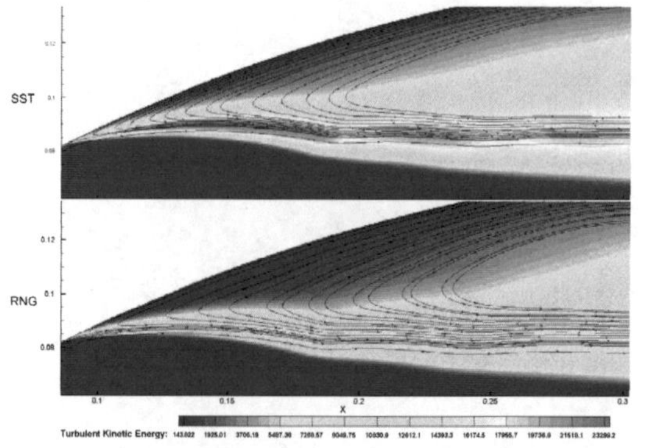


Figure 12. Turbulent kinetic energy contour and streamlines for SST and RNG models at PR=12.9

Fig.12 shows the contour of turbulent kinetic energy of the SST model and the RNG model behind the separation position. It can be seen that the turbulent kinetic energy level predicted by SST is much higher than that of RNG. Greater turbulent kinetic energy means greater jet mixing, which also means that the jet will spread faster, with the resulting effect of reducing the effective angle between the boundary of the separating jet and the nozzle wall. The cross-sectional area of flow entrainment at the nozzle exit is also reduced. Hence the increased entrainment speed reduces static pressure on the wall. The separation location is also pushed downstream leading to its over-prediction. The RNG's relatively low prediction of turbulent kinetic energy leads to a rapid recovery of wall pressure behind the separation position and a higher prediction of wall pressure and an underprediction of the separation point (closer upstream).

It can be seen from the contour of turbulent kinetic energy and the streamlines that the global flow structure predicted by the three models are consistent, but in the RNG model, the distance that the gas sustaining jet form travels downstream is farther than the other models. This also shows that the prediction of jet spread by the RNG model is conservative compared to the other models. This is consistent with the previous analysis that lower levels of turbulence predicted under the RNG model.

Fig.9 is the axial velocity curve on the axis of symmetry. The black dotted line in the figure is the exit position of the nozzle. When the flow crosses the normal shock wave, the velocity will drop off a cliff, so the position where the axial velocity drops rapidly in the figure is the shock position. It can be seen that the predictions of these three turbulence models for the position of the Mach stem are relatively consistent, and the difference in the axial positions between these models is less than 5 mm through measurement. At the same time, it can be observed that the magnitude of the axial velocity after passing through the shock wave is consistent in the prediction results of these three models, which indicates that these predictions of the shock wave intensity are reasonable and consistent. However, in the subsonic region downstream of the Mach stem, the velocity distribution on the axis shows a large discrepancy, indicating that these three models are inaccurate for the prediction of the flow structure in the subsonic region. Fig.10 shows the variation of the axial shear stress (ASS) of the wall with the axial position. The ASS is equal to zero means that the recirculation vortex appears in this area. It can be seen that the position of ASS=0 in the SA model is biased downstream relative to the other models. It can be noticed that these three shear stress curves intersect with ASS=0 when the X/Rt equal to 9, 9.3 and 9.6 respectively, manifesting that a small recirculation vortex is captured near the outlet. This phenomenon is consistent with the previous research results.

2） *PR=14.4*

Similarly, from the wall pressure curve, the qualitative conclusion shown by these three models is consistent with the result when PR=12.9. The position of the separation point predicted by RNG is relatively close to the upstream, and the pressure recover rapidly to the level close to ambient pressure behind separation position. The reason for this distinction has been discussed in the previous section:

the difference in the prediction of turbulent kinetic energy levels by different models. This difference can also be observed in Fig.17. The previous view is further supported by this result.

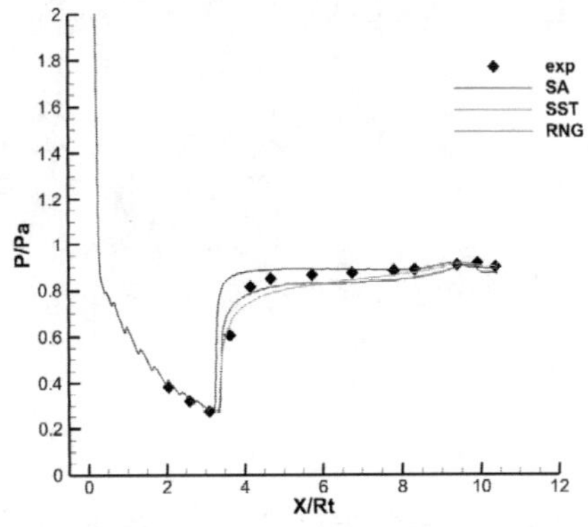


Figure 13. Results of wall pressure for different models at PR=14.4

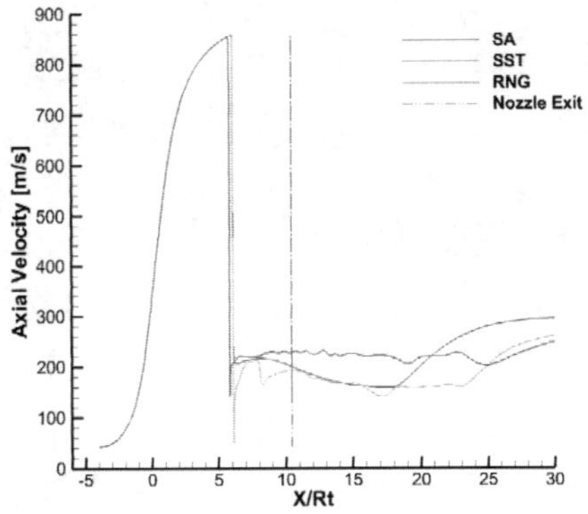


Figure 14. Results of axial velocity for different models at PR=14.4

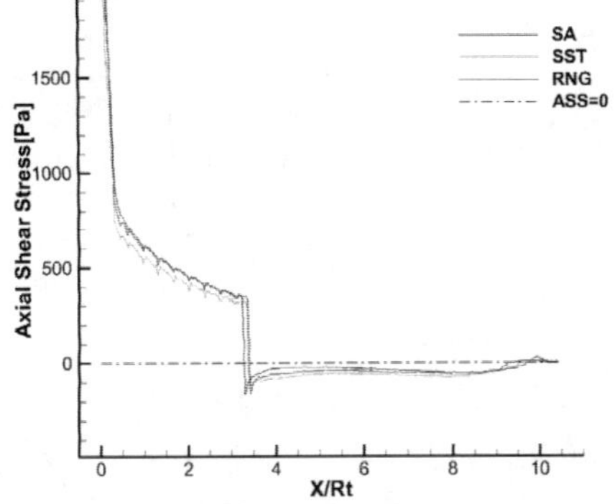


Figure 15. Results of axial shear stress for different models at PR=14.4

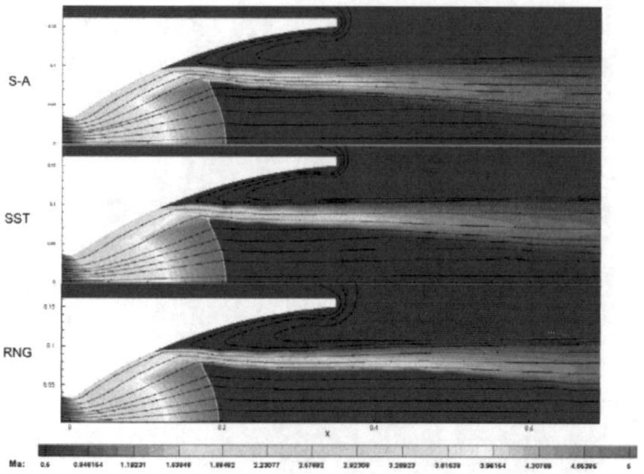


Figure 16. Mach number contour and streamlines for different models at PR=14.4

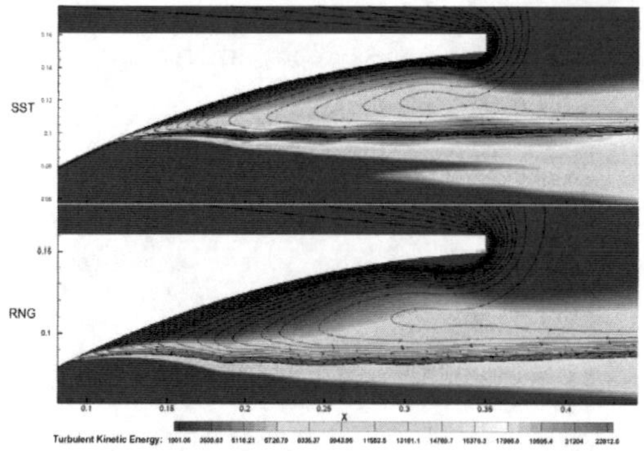


Figure 17. Turbulent kinetic energy contour and streamlines for SST and RNG models at PR=14.4

Judging from the comparison between the predicted results and the experimental results, these three models have very little distinction in predicting the position of the separation point and compare well with the experimental results. The results of RNG are slightly closer to the upstream, and the results of SST and SA are consistent and slightly biased to the downstream. In the recovery stage, the results of SA and SST are both smaller than the experimental values, while the predicted results of the RNG model are higher. This feature is consistent with the result when PR=12.9.

Comprehensively considered, the prediction results of the three models are close to these experimental values, and the prediction results of SA are relatively more accurate. The prediction results of all models for the axial shear stress in the recirculation zone are basically same. From the axial velocity curve, the SST and SA models are completely coincident and biased downstream for the Mach stem position prediction. The shock intensity predicted by the SST model is much stronger than the other. The contour of Mach number and streamline map show that the global flow structure predicted by these three models is consistent but the results of RNG show that the jet propagation distance is longer. This result is consistent with the previous conclusion that the prediction of jet diffusion by RNG is smaller than other models.

*4.2. Critical Pressure Ratio*

In the experiment, the shock separation phenomenon conversion pressure ratio of the engine start-up stage is not clearly given. The only thing that can be determined is that the mode conversion PR is between 14.4 and 16.2[11]. In this paper, PR=15.1 is selected to analyze the prediction performance of each turbulence model under the critical pressure ratio of mode conversion. The relevant calculation results are as follows.

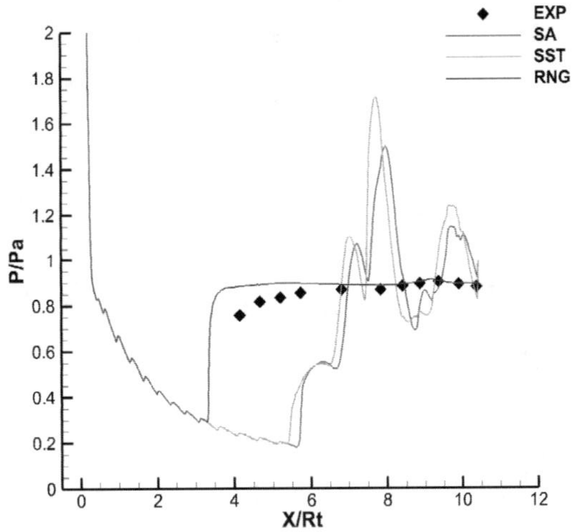


Figure 18. Results of wall pressure for different models at PR=15.1

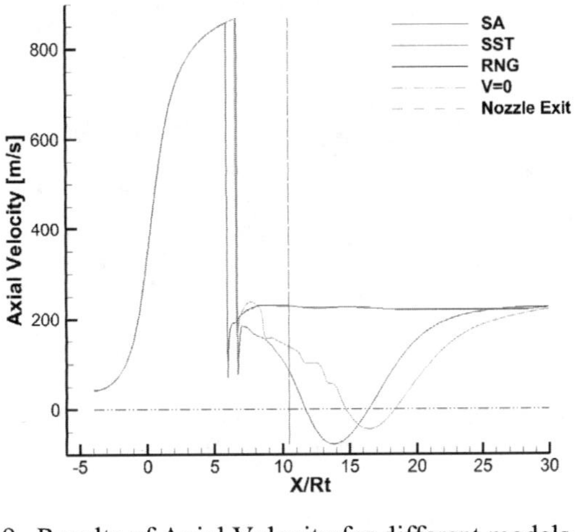


Figure 19. Results of Axial Velocity for different models at PR=15.1

From the pressure curve and the counter of Mach number, it can be clearly seen that the prediction results of the three models are very different when the pressure ratio is 15.1. The separation mode displayed by the SST and SA models is restricted shock separation, which is completely different from the FSS mode shown in the experimental results, and the RNG model accurately predicts the separation mode under this condition. It should also be noted that the prediction of wall pressure in the front section of the recirculation zone (X/Rt<7) is higher than the experimental value. As the axial position gets closer to the nozzle outlet, the calculated results gradually approach the experimental values. From the wall shear stress curves, it also observes the variations in the prediction results of different models for the separation mode. The difference of the separation model makes the detail of the flow under each turbulence models are completely different, so this section will not describe it too much.

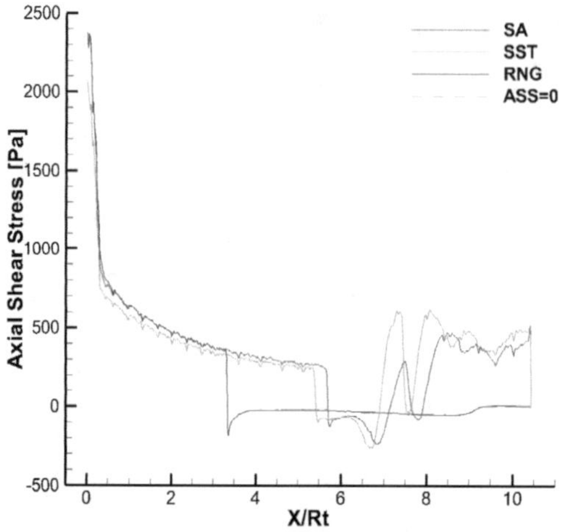


Figure 20. Results of Axial Shear Stress for different models at PR=15.1

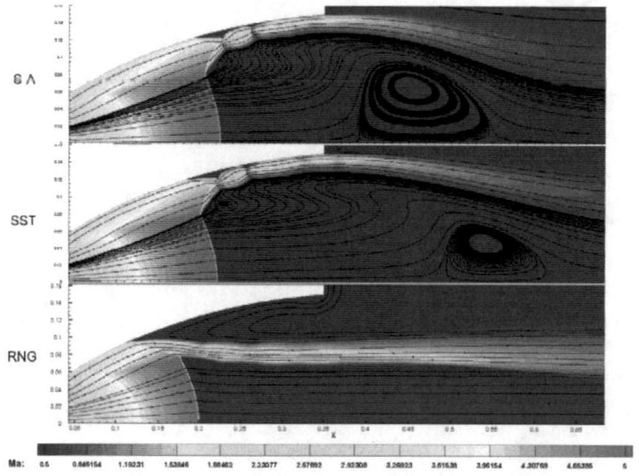


Figure 21. Mach number contour and streamlines for different models at PR=15.1

*4.3.Restricted Shock Separation*

In this section, the calculation and analysis of the two PRs that form the RSS model are made, the results are as follows.

1） PR=*16.2*

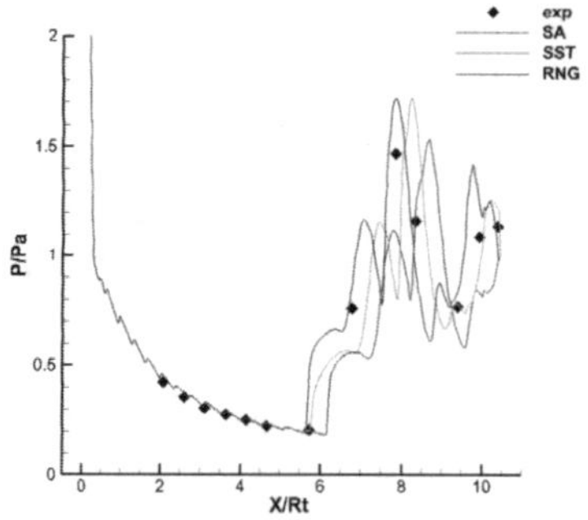


Figure 22. Results of wall pressure for different models at PR=16.2

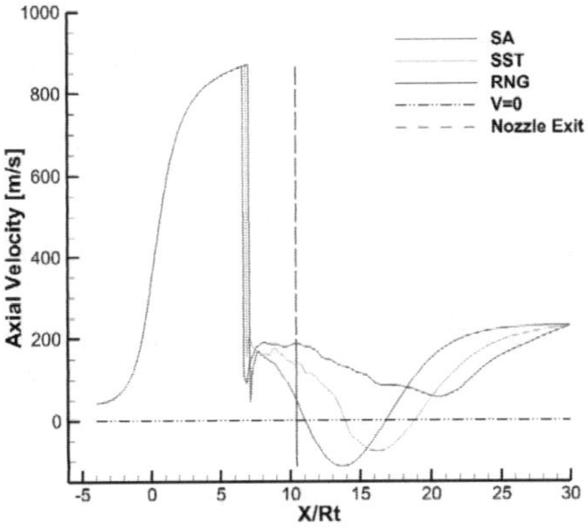


Figure 23. Results of Axial Velocity for different models at PR=16.2

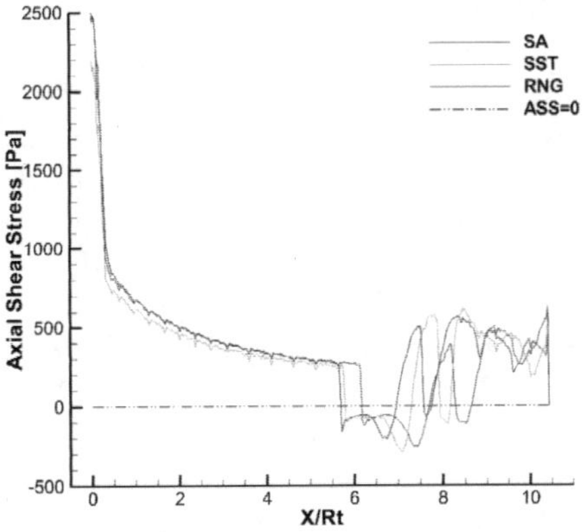


Figure 24. Results of Axial Shear Stress for different models at PR=16.2

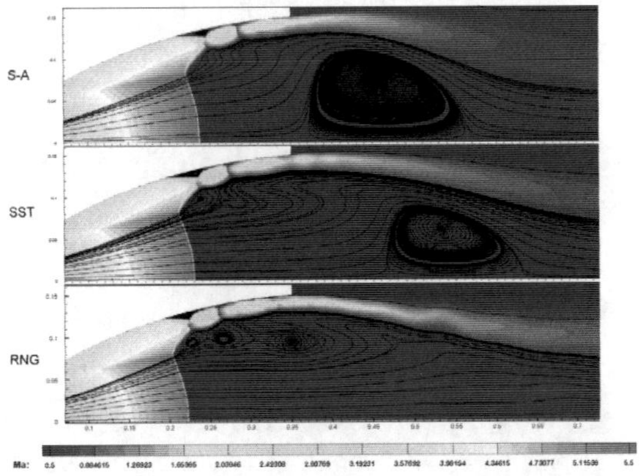


Figure 25. Mach contours and streamlines for different models at PR=16.2

The pressure distribution curves show that all these three turbulence models accurately predict the shock separation mode (RSS), which is consistent with the experimental results. The SST and RNG turbulence models predict the separation location in agreement, while the SA model predicts relatively downstream. The results of the pressure distribution after separation showed that the SST and RNG models provide better prediction performance. It can be noticed that the prediction results of the RNG model only compare well with the experimental results when X/Rt is less than 8.5, and the prediction performance is poor when X/Rt>8.5. On the contrary, SST predicts the results with large deviation when X/Rt is less than 8.5, and agree well with experimental data when X/Rt>8.5.

In the axial shear stress curve, the prediction results of these three models all show that there are two separation vortices near the wall, and the scale of the first separation vortex is larger than the second separation vortex. However, these three predictions on the specific details of the separation vortex do not show consistency. From the axial velocity curve, there are differences in the position of the shock between the results of these three models, and the axial distance difference predicted by different models is about 6 mm. The shock intensity predicted by the SA model is greater than the RNG model, and the shock intensity predicted by the SST model is the weakest. Near the exit of the nozzle (X=0.5m), the axial velocity drops below zero in the prediction results of the SST and SA models, which indicates that a trapped vortex is generated here. This structure can be observed more intuitively in the streamline diagram about this structure. It can be seen that there is only one large-scale trapped vortex in the flow field in the SA model, while in the results of the SST model, in addition to a large-scale trapped vortex at the axis, there is also a small-scale trapped vortex below the jet. Meanwhile, the size of the trapped vortex at the axis position is smaller than the predicted result of SA. In the RNG results, no large-scale trapped vortex was generated near the axis, and multiple small-scale vortices could be observed below the jet. Previous numerical studies have shown that when the flow structure is RSS, a large-scale trapped vortex always appear at the downstream axis of the Mach stem[15]. The results of SST and SA confirm this view, while the RNG simulation results contradict this view. It is worth noting that the RNG simulation results predict the wall pressure distribution better than the SA model.

2） PR=*19.5*

From the wall pressure curve, all these models accurately predict the shock separation mode (RSS) under this PR. The position of the separation point predicted by the SST model and the RNG model is coincident and close to the experimental data, while the predicted result of the SA model is located downstream of the actual separation point. It can be noticed that the prediction of the wall pressure downstream of the separation position by the SST model and the SA model is consistent, but there is a large deviation compared with the experimental data. It is worth noting that the RNG model is in excellent agreement with the experimental results at this pressure ratio. The wall shear stress curve also reflects this phenomenon. As predicted at PR=16.2, all three models capture the existence of two separation vortices at the wall, but they do not give consistent results for the detail of flow structure.

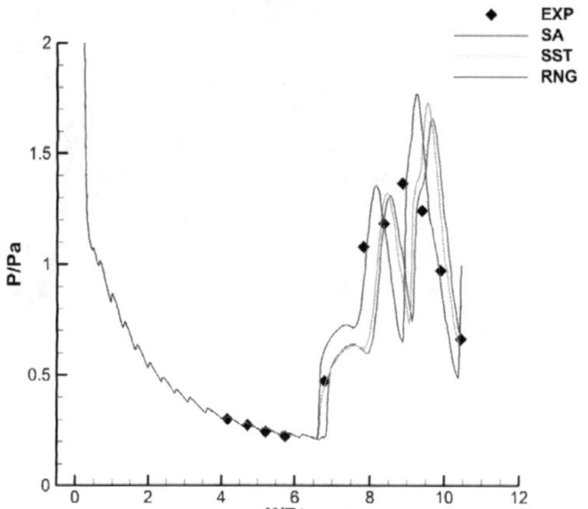


Figure 26. Results of wall pressure for different models at PR=19.5

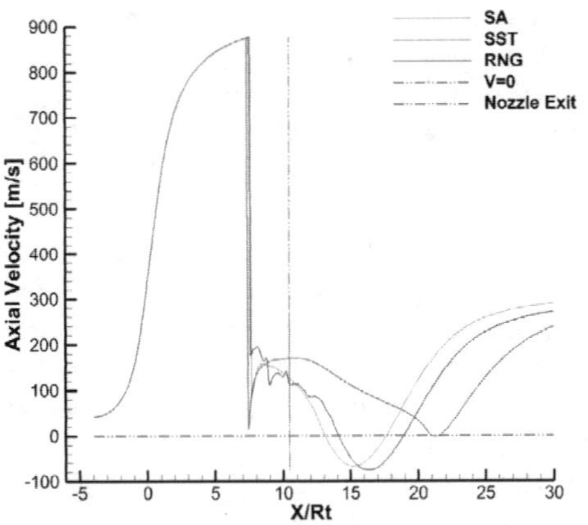


Figure 27. Results of Axial Velocity for different models at PR=19.5

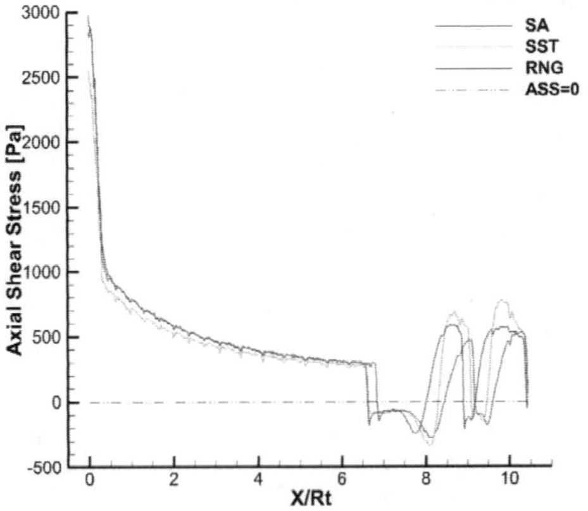


Figure 28. Results of Axial Shear Stress for different models at PR=19.5

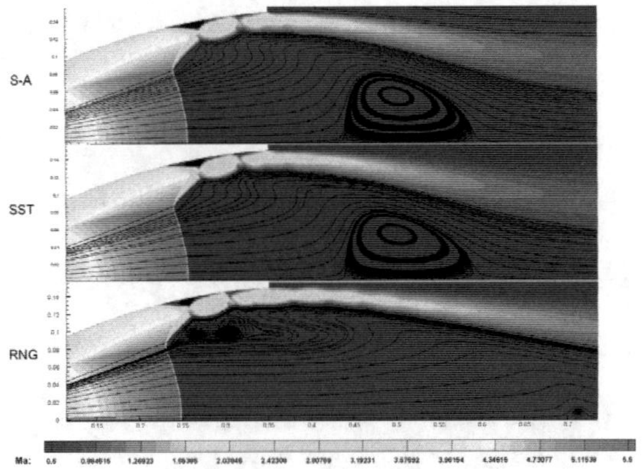


Figure 29. Mach contours and streamlines for different models at PR=19.5

The axial velocity curve shows that the prediction of the shock wave position by SA and SST is consistent, but the Mach stem position predicted by RNG is biased upstream, and the axial distance from the others is about 8 mm. The shock wave intensity predicted by RNG is the strongest, followed by SA, and SST is the weakest. It can be seen from the counter of Mach number that a trapped vortex is captured near the axis in all models, but the size of the trapped vortex that predict by RNG model is much smaller than others. There are still multiple small-scale trapped vortices below the jet in the prediction results of the RNG model, while other models do not predict this structure.

## 5. Conclusion

In this paper, the performance of three RANS turbulence models which are widely used in large expansion ratio nozzle separation flows is properly evaluated. The evaluation work is carried out under five PRs of the three separation modes, and the grid interference is excluded. The performance analysis of the turbulence model was carried out and the following conclusions were obtained:

*1) All these three models can accurately predict the separation phenomenon under the vast majority of PRs, but under the critical PR, both the SST and SA models predict the separation mode unreasonably, and only the RNG model is consistent with the experimental results.*

*2) The SA model has lower computational cost, and has better prediction performance under the condition of low pressure ratio. The prediction effect under high pressure ratio is relatively weak, and there is a certain deviation. It could be used in the case that the calculation accuracy is low and the simulation with a limit on computational cost .*

*3) Although the SST model shows poor simulation performance at critical pressure ratios, it exhibits the best comprehensive performance for most conditions. Therefore, the SST model is recommended for general-accuracy simulations, but should be modified for more accurate simulations, especially of critical mode.*

*4) The prediction performance of the RNG model at low pressure ratio is weak, but it accurately predicts the separation mode under the critical PR. The results of the two PRs that form the RSS mode are in good agreement with the experiments. It should be noted that the simulation of the form and position of the trapped vortex that generated downstream of the Mach stem is inconsistent with the previous research results, and the specific reasons need to be further explored.*

*5) The difference between the results of the separation position and wall pressure recovery of the RNG model and the prediction results of the SST model is due to the difference in the prediction of the turbulent kinetic energy. The higher turbulence kinetic energy level means that the supersonic jet spreading inside the nozzle will be faster, the resulting effect is that the effective angle between the separating jet boundary and the nozzle wall is reduced. The entrainment cross-sectional area at the nozzle outlet is also reduced, which results in an increase in entrainment velocity and a decrease in static pressure on the wall. This result will cause the position of the separation point to move downstream. In the RNG model, the lower turbulent kinetic energy level causes the separation position to move upstream, and the wall pressure rises faster and closer to the atmospheric pressure behind the separation point.*

## References

[1] Frey, M., Hagemann, G. (1999). Flow separation and side-loads in rocket nozzles. In 35th Joint Propulsion Conference and Exhibit ,p. 2815.
[2] Hagemann, G., Frey, M., Koschel, W. (2002). Appearance of restricted shock separation in rocket nozzles. Journal of Propulsion and Power, 18(3), p.577-584.
[3] Nasuti, F., Onofri, M. (2009). Shock structure in separated nozzle flows. Shock Waves, 19(3), p.229-237.
[4] Ostlund, J., Jaran, M. (1999). Assessment of turbulence models in overexpanded rocket nozzle flow simulations. In 35th Joint Propulsion Conference and Exhibit, p. 2583.
[5] Baloni, B. D., Kumar, S. P., Channiwala, S. A. (2017). Computational analysis of bell nozzles. In 4th International Conference of Fluid Flow, Heat and Mass Transfer, Toronto, Canada, Vol. 110.

[6] Stark, R. , Hagemann, G. (2007). Current status of numerical flow prediction for separated nozzle flows. In 2th European Conference For Aerospace Sciences (EUCASS). DLR.
[7] Nebbache, A. , Pilinski, C. (2006). Pulsatory phenomenon in a thrust optimized contour nozzle. Aerospace science and technology, 10(4), p.295-308.
[8] Martelli E, Nasuti F, Onofri M. (2010) Numerical calculation of FSS/RSS transition in highly overexpanded rocket nozzle flows[J]. Shock Waves, 20(2), p.139-146.
[9] Hung H Y,Wang Q . (2007) Numerical investigation on internal and external flows for axisymmetric convergent-divergent nozzles on over-expansion state. Journal of Aerospace Power, 22(7):p.1069-1073.
[10] Yaravintelimath, A., Raghunandan, B. N., Moríñigo, J. A. (2016). Numerical prediction of nozzle flow separation: Issue of turbulence modeling. Aerospace Science and Technology, 50, p.31-43.
[11] Mattsson, J., Högmann, U., Torngren, L. (1999). A Sub Scale Test Programme on Investigation of Flow Separation & Side Loads in Rocket Nozzles. In Aerothermodynamics for space vehicles , Vol. 426, p. 373.
[12] Spalart, P., Allmaras, S. (1992). A one-equation turbulence model for aerodynamic flows. In 30th aerospace sciences meeting and exhibit. p. 439.
[13] Orszag, S. A. , Yakhot, V. , Flannery, W. S. , Boysan, F. , Patel, B. (1993). Renormalization group modeling and turbulence simulations.
[14] Menter, F. R. (1994). Two-equation eddy-viscosity turbulence models for engineering applications. AIAA Journal, 32(8), p.1598-1605.
[15] Chen, C. L., Chakravarthy, S. R., & Hung, C. M. (1994). Numerical investigation of separated nozzle flows. AIAA Journal, 32(9), p.1836-1843.

# Design of mini airborne emergency power supply based on super capacitor

**Jian Li[1*], Guoxing Li[2] and Yonglin Guo[3]**

[1]Chinese Flight Test Establishment,Xi'an ,Shanxi, China

[2]Chinese Flight Test Establishment,Xi'an ,Shanxi, China

[3]Chinese Flight Test Establishment,Xi'an ,Shanxi, China

*Corresponding author's e-mail: lijian19890219@163.com

**Abstract:** In order to solve the problem that it is difficult to install emergency battery with limited space on some flight planes, a micro emergency power supply based on super capacitor was designed in combination with the good energy storage characteristics of super capacitor. The selection calculation of capacitor and the design process of main control module and charge and discharge module are discussed in detail. In the laboratory environment, a small airborne test system was built to carry out verification experiments. The results show that the device meets the basic requirements of the system for backup power supply, and has the advantage of taking up less space and avoiding maintenance, which provides a new choice for the backup power supply selection of the future in-flight test system.

## 1. Introduction

During the flight test, the power supply quality of the airborne test system is directly related to the stability and reliability of the test system. The short-term fluctuation or power failure of the airborne power supply system may cause abnormal operation or even physical damage of some equipment in the airborne test system. In order to cope with the impact of the fluctuation of the power supply system on the machine or the short-term power failure on the test system, it is usually necessary to provide power supply protection to the key equipment on the machine. The power fluctuation on the machine mainly depends on the adjustment of the test equipment itself, and the short-term power failure protection mainly uses emergency batteries. to fulfill. Although the test equipment itself has a certain ability to adapt to power fluctuations, because the test equipment needs to weigh many factors such as equipment size and power consumption in the design, the power supply system of the airborne test equipment itself has a limited adaptability to power supply fluctuations.

The power distribution system on the aircraft can be divided into two parts: AC power distribution and DC power distribution according to the nature of power distribution properties. The airborne test and acquisition system supplies power to the entire system by drawing the electrical energy on the DC bus bar. When the bus bar voltage fluctuates or the system power supply is interrupted, the short-term battery life of the test system itself is particularly important. At present, during the flight test of military aircraft at home and abroad, silver-zinc storage batteries [1] are selected to ensure the uninterrupted power supply of the aircraft. When the aircraft power supply system fails, the contactor is switched to the storage battery for uninterrupted testing. The system is powered by common power. However, due to the long switching time of the contactor, a brief power-off phenomenon occurs during the conversion process, causing the test equipment to be powered on again, which may also cause data interruption. At the same time, due to the self-discharge of the battery, it needs to be removed to the ground for charging and maintenance after the battery is exhausted. At the same time, the on-board emergency battery itself has the problem of continuous voltage drop during use. When the voltage drops to the critical threshold voltage of the test equipment, it is easy to cause the instability of the test equipment itself.

Considering the good charging and discharging characteristics of super capacitors and the advantages of taking up less space, this scheme intends to use wide-voltage power regulation technology to ensure the stable and reliable operation of the airborne test system. And when the power supply system on the machine is short-term power off, it can realize short-term, stable and reliable online power supply to the key test equipment on the machine, and improve the stability and reliability of the on-board test system. At the same time, the super capacitor is used as the energy storage unit of the emergency power supply device to realize the on-board fast charging of the energy storage unit. The use of program control realizes the charging management of the super capacitor energy storage unit, eliminating the need for manual maintenance and improving the automation of the equipment work.

## 2. The specific design of the super capacitor battery

*2.1 Super capacitor energy storage technology*

The super capacitor has a fast charging speed, which can reach more than 95% of the capacity after charging for 10 seconds to 10 minutes. At the same time, the cycle life is long, and the number of deep charging and discharging can reach 500,000 times. High energy conversion efficiency, small process loss, high current energy cycle efficiency ≥ 90%; and super capacitor charge and discharge safety factor is high, long-term maintenance-free, good ultra-low temperature performance, temperature range -40 ~ 70 °C; There is no negative impact on its lifespan [2].

The internal resistance of the super capacitor hinders its rapid discharge. The time constant of the super capacitor is 1~2s, and it takes about 5 to 0s to completely discharge the RC circuit [3]. Super capacitors can be quickly charged and discharged, and the peak current is limited by its internal resistance. Super capacitors have ultra-low series equivalent resistance, and the discharge power density is more than ten times that of lithium-ion batteries, and can be discharged at large currents; It can be charged to any potential within the rated voltage range and can be completely discharged, while the battery is limited by its own chemical reaction to work in a narrower voltage range.

*2.2 Selection of capacitors and calculation of parameters*

In this scheme, the super capacitor is selected as the energy storage unit of the emergency power supply. Considering that the airborne test and acquisition system works at DC 28V, the emergency power supply time is set to be no less than 30s and the working power is 140W. After comparing the devices of many companies, this scheme uses the super capacitor produced by Boaiger Electronic Technology Co. Ltd. Since a single capacitor has a withstand voltage of 3V and a capacity of 70F, a capacitor module with a withstand voltage of 30V and a capacity of 5F is obtained by connecting 10 cell capacitors in series [4]. Usually,

$$C = Q / U \tag{1}$$

$$Q = I * T \tag{2}$$

In formulas (1) and (2): C - rated capacity of capacitor; U - voltage across capacitors; I - capacitor charging current;

According to formulas (1) and (2), the charging time of 30V/5F super capacitor (calculated by the average charging current of 1A) is:

$$T = C\text{*U/I=30*5/1=150S} \tag{3}$$

In formula (3): T-capacitor charging time.

2.3 *Design principle and composition of emergency power supply*

In the normal working state of the test system, in order to avoid the charging and discharging of the emergency power supply affecting the original machine, this scheme considers the charging current to be controlled at about 1A [5], and a bypass control is designed at the same time, only when the power supply fluctuation exceeds a certain threshold, the The discharge operation is performed only by the emergency power supply. The system functional block diagram of the airborne emergency device based on super capacitor is shown in Figure 1. It is mainly composed of a voltage regulation circuit, an energy storage power supply charging circuit, a discharge circuit and a bypass control circuit to realize the voltage regulation of the external input power supply. When the external input power fluctuates, the regulated voltage is guaranteed to be stable, and it has overvoltage and overcurrent protection functions.

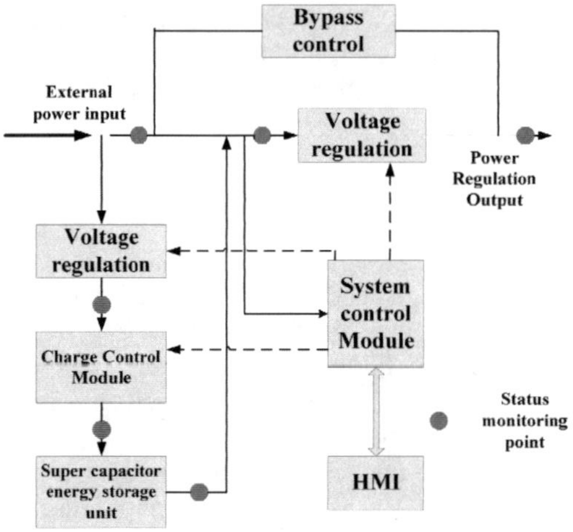


Figure 1. Functional block diagram of emergency power supply

The energy storage circuit is mainly composed of three parts: voltage regulation, charging control, and super capacitor energy storage unit. When the external power supply is normal, the power is obtained from the external power supply to charge the super capacitor energy storage unit. When the external power supply is abnormal, the system power supply changes. Provided by the energy storage unit, and ensure that the power supply continues during switching, and there is no power interruption.

The system control unit completes the power supply control of the entire system through the ARM controller, including the control of the charging process of the super capacitor energy storage unit, the control of human-computer interaction, the acquisition of the detection status of multiple status monitoring points, and the realization of the self-detection function of the working status, and The detection information can be displayed through the man-machine interface (control bus/status indication).

*2.4 Charging circuit scheme and software design*

*2.4.1 Schematic design of charging circuit*

If the charging circuit directly charges the capacitor [6], the charging current of the circuit drops sharply with the charging process due to the existence of the load resistance, and the whole charging process lasts for about an hour, without taking advantage of the high current charging of the super capacitor [7]. In this design scheme, the switching of the charging circuit is controlled by the power tube, and the corresponding load resistance is selected according to the change of the voltage across the capacitor, so as to control the entire charging circuit to be controlled within a certain current range. To sum up, this scheme designs a constant voltage current limiting charging circuit based on ARM controller. The circuit can not only control the charging voltage, but also realize the charging current in the range of 0.56~1.17A without affecting the normal power supply of the original test system [8].

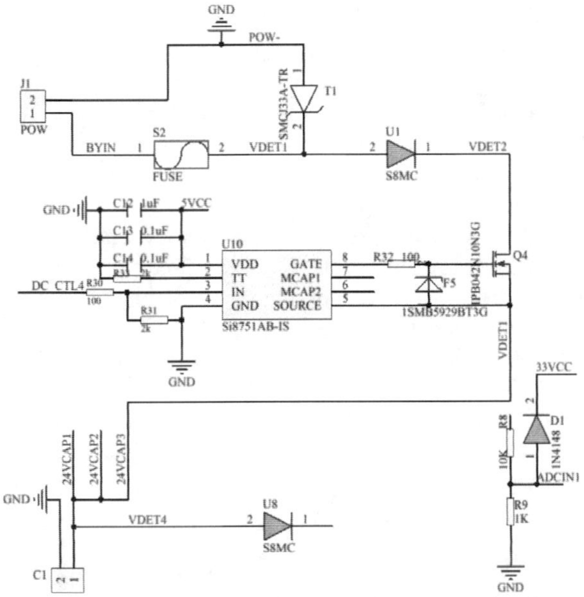

Figure 2. Schematic diagram of charging circuit

*2.4.2 Design of charging circuit switching process*

In order to achieve constant voltage and constant current charging, as the voltage across the capacitor continues to rise, the current-limiting resistor must be continuously switched for shunt charging. As shown in Figure 3, the IN pin of the driver chip is controlled by the ARM chip, so that the start of the triode is controlled through the gate level of the field effect transistor, thereby realizing the charging of the super capacitor. During the whole charging process, with the increase of the capacitor voltage, the current will become smaller under the premise that the charging voltage remains unchanged. Therefore, it is considered to control the charging current by adding a resistor. The charging voltage is 24V, and the resistors are 2 ohms, 12 ohms, and 25 ohms respectively. Through calculation, the current can be maintained at 0.6A to 1.2A. The entire charging process can be fully charged in about 3 minutes.

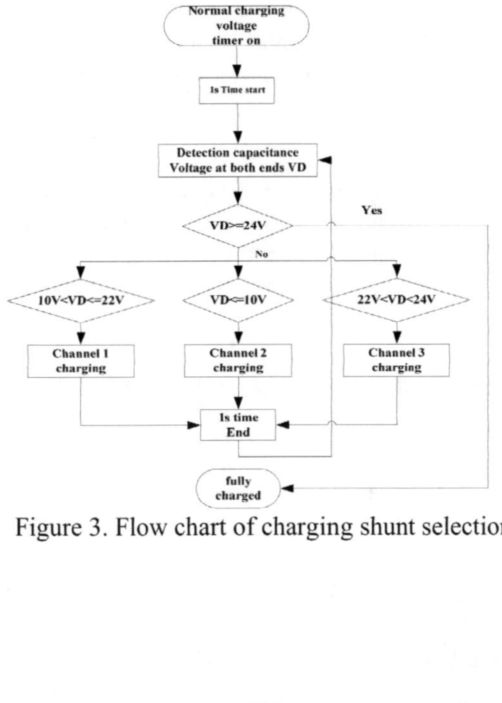

Figure 3. Flow chart of charging shunt selection

*2.5 Design of main control circuit*

The main voltage regulation circuit is responsible for regulating and regulating the externally input power supply and then outputting it, and realizing the overvoltage and overcurrent protection functions of the output voltage; the main control circuit is used to control various functions within the entire device, including charging management functions, Human-computer interaction function, status monitoring function. In order to realize the self-diagnosis of the internal working state of the device, as shown by the red dot in Figure 1, the monitoring of the input state of the external power supply, the monitoring of the output state of the power supply of the device, and the output state of the charging voltage regulation circuit are realized through the state monitoring point. At the same time, the human-computer interaction interface circuit completes the display of the internal working status of the equipment through the status display light.

As shown in Figure 4, the device adjusts the voltage of the four state points to between 0V and 3.3V through the voltage divider adjustment circuit, and completes the collection and analysis of the four-channel voltage signals through the ARM microcontroller, and then performs the display circuit through the display circuit. display, and at the same time realize data sharing with other devices in the test system through the control bus.

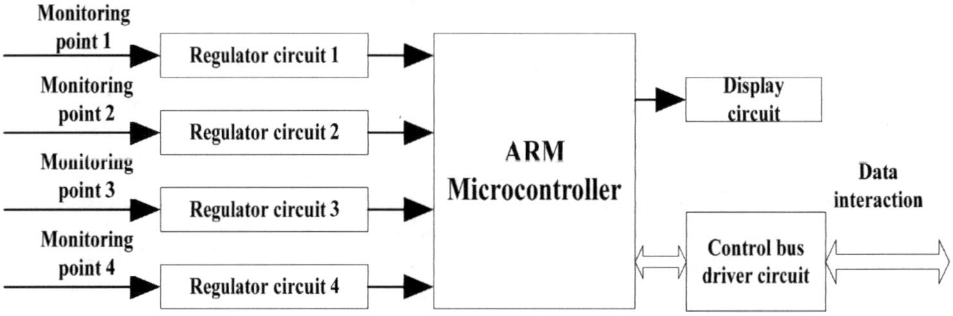


**Figure 4.** Block diagram of working state monitoring

*2.6 Design of power supply circuit scheme*

Since the airborne environment requires a continuous and reliable DC 28V power supply [8], and the voltage across the capacitor is bound to decrease with the continuous increase of power consumption during the use of the capacitor, so this scheme uses a voltage stabilizer chip to realize the voltage stabilizer function. , the chip can continuously output 28V DC voltage under the premise that the input voltage is not less than 9V, which meets the design requirements. At the same time, in order to actively turn off the capacitor power supply circuit if necessary, this scheme designs a MOS transistor driver chip to realize the cutoff function of the output circuit. At the same time, in this scheme, AD samples the voltage of key points to judge and perform corresponding control operations [9-10], such as ADCIN5 and ADCIN6, in order to protect the ARM chip from being damaged, it is necessary to control the AD sampling below 4V, such as As shown in the figure, a bypass diode and a 3.3V power supply are designed near the voltage divider sampling point. The clamping function of the sampling point is realized by the bypass diode, and the ARM control chip will not be damaged when the voltage has an excessive shock [11-12].

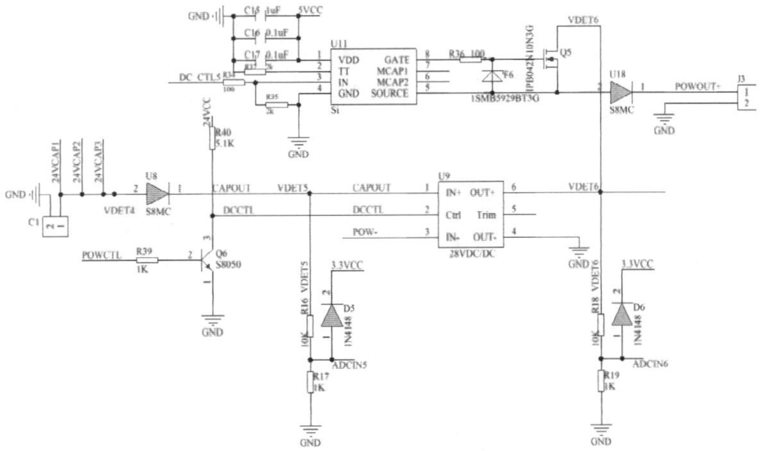

Figure 5. Schematic diagram of discharge circuit

**3.Joint test experiment and analysis**

After completing the production of the PCB board, the principle prototype was made. It was verified by the laboratory joint test that it takes about 120 seconds to fully charge the equipment, which is almost the same as the theoretical calculation. At the same time, the cockpit integrated display, a commonly used key equipment in the airborne test system, is selected as the load equipment, and the current consumption of this equipment is about 0.9 amperes. As shown in Figure 6, the equipment connection diagram (a) and the actual output voltage and input voltage waveform of the laboratory (b) are respectively given. In Figure 6.(b), the input voltage and output of the emergency power supply are monitored by an oscilloscope. The waveform of the voltage, as shown in Figure 6.(a), the output voltage is stable at about 28V when the input voltage fluctuates widely between 10V and 30V. In addition, after disconnecting the external equipment, the on-board emergency power supply can also be used. Stable output of 28V DC voltage ensures the normal operation of the load. Through experiments, it is found that under the premise that the load working current is 1 ampere, when the external power supply is cut off, the onboard emergency battery can continue to supply power for no less than 170 seconds, which meets the initial design requirements.

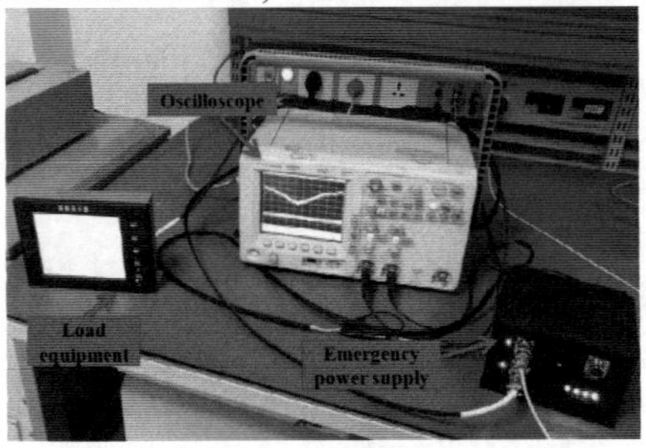


(a). Joint test diagram

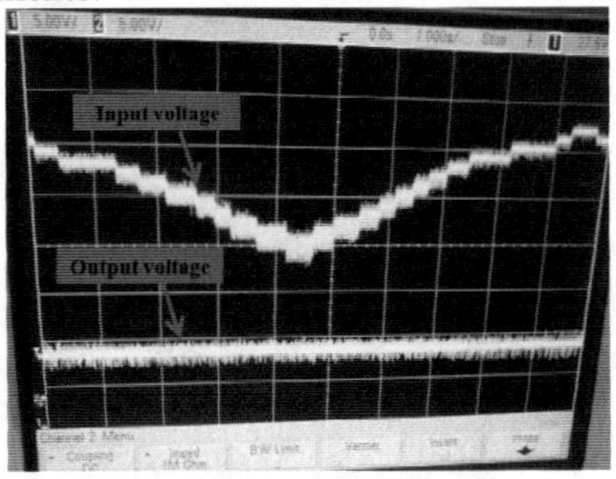


(b). Waveform monitoring diagram

Figure 6. Alignment and waveform

**4. Conclusion**

Aiming at the problem of limited installation space on some test aircrafts, and considering the good charging and discharging characteristics of super capacitors, a micro emergency power supply design scheme based on super capacitors is proposed. In order to avoid the impact on the original machine test system, the constant voltage and constant current charging method of about 1 ampere is selected after full evaluation and calculation, and the output voltage is regulated by voltage regulation to ensure the quality of the output power supply. The joint test experiment of the principle prototype shows that the device meets the basic requirements of the backup power supply in the flight test, and the size of the device is less than one-fifth of the silver-zinc battery mentioned, which further verifies the reliability and effectiveness of the design. With the further development of capacitive energy storage technology, the device will play a very important role in the field of flight test.

**References**

[1] Zhang Haiying, Hao Linzhao. Research on power supply and distribution technology for aircraft modification [J]. Heilongjiang: Science and Technology Innovation and Application, 2020, 10(9): 167-168.
[2] Li Yue. Research on super capacitor energy storage system [D]. Beijing: Beijing Jiaotong University, 2015.
[3] Liao Chuanping.Super capacitor battery[J].Chemical Bulletin,2014,77(9):866-867.
[4] Xi Yong. Research and design of low-power super capacitor independent power supply [D]. Kunming: Kunming University of Science and Technology, 2014.
[5] Liu Guannan. Research on bidirectional DC-DC conversion technology based on super capacitor [D]. Harbin: Harbin Institute of Technology, 2013.
[6] Meng Yanjing, Zhang Shangzhou, Chen Jingwen, etc. Influence of super capacitors on the energy efficiency of super capacitors [J]. Electronic Devices, 2014, 37(1): 14-16.
[7] Shan Jinsheng, Wu Lifeng, Guan Yong, etc. Current status and prospect of super capacitor modeling [J]. Electronic Components and Materials, 2013, 32(8): 6-8.
[8] Hao Guoliang, Research on the calculation method of super capacitor state of charge [D]. Beijing: North China Electric Power University, 2012.
[9] Yuan Jinlei. Research on super capacitor constant power charge and discharge control system [D]. Dalian: Dalian University of Technology, 2015.
[10] Zhang Yuan. Research on super capacitor voltage equalization technology [D]. Taiyuan: Taiyuan University of Technology, 2015.
[11] Dai Juan. Development of a super capacitor intelligent charger [D]. Chongqing: Southwest University, 2013.
[12] Chen Fang. Research on bidirectional DC/DC converter based on super capacitor [D]. Hangzhou: Zhejiang University, 2012.

# Overall design and sensitivity analysis of solar aircraft based on feasible region planning method

**Liang Zhang[1, a], Dongli Ma[1, b], Muqing Yang[1]*, Yuan Yao[1, c]**

[1]School of Aeronautic Science and Engineering, Beihang University, Beijing, China, 100191

[a]email: microbeast@126.com; [b]email: madongli@buaa.edu.cn;

*email: qingfengrumu@163.com; [c]email: fanfysuziyy@163.com

**Abstract**—Solar aircraft has the ability of 24-hour uninterrupted cruise, so it has become a research hotspot all over the world. The existing overall design method of solar aircraft generally obtains the design scheme through the iterative operation of the balance equations. However, the overall parameters satisfying the boundary conditions are not unique, but distributed in a specific interval. This paper presents a method to construct the radial feasible region of the overall parameters based on the balance equations, and studies the sensitivity of the feasible region to the task indicators and performance indicators respectively, which provides suggestions for the optimization direction of the subsystem in the overall design stage. Finally, based on the above methods and research conclusions, the design scheme of a 15m wingspan solar aircraft is introduced. This paper provides an intuitive overall design method for solar aircraft, and offers scientific basis for multidisciplinary collaborative optimization.

## 1. INTRODUCTION

Solar aircraft uses photovoltaic (PV) cells to convert light energy into electricity. Part of these energy is used to maintain daytime cruise, and the other is stored in secondary battery for night cruise. Theoretically, it can complete weeks or even months of uninterrupted flight [1, 2]. Compared with conventional aircraft, solar aircraft shows the coupling between multiple subsystems in the overall design stage. For example, the wing area is not only related to the aerodynamic performance, but also directly affects the energy acquisition ability of the aircraft [3]. Therefore, the overall design method of solar aircraft is special.

The earliest overall design method of solar aircraft was proposed by Youngblood et al. [4] in 1984. The core of this method is: the required value of the cruise factor $C_L^{1.5}/C_D$ is obtained from the energy balance equation, and the calculated value of $C_L^{1.5}/C_D$ is obtained from the aerodynamic formula. The rationality of the overall design parameters is judged based on whether the two factors match the aircraft scale. Noth et al. [5] improved the method of Youngblood, further refined the calculation formula of energy and mass, and built a more perfect overall design method of solar aircraft, which is widely adopted by scholars today. Romeo et al. [6] took the HELIPLAT aircraft as the research object, elaborated the aerodynamic characteristic and structural weight estimation method, and introduced in detail the influence of wing root tip ratio, wing tip configuration, inflection point position and other factors on the aerodynamic characteristics. In recent years, due to the development of computing technology, the overall design method of solar aircraft has been further improved. Maleki et al. [7]

introduced the "technical level" parameter representing the ratio of load weight to total weight, which can assist in the estimation of takeoff mass and wing parameters. Morrisey et al. [8] proposed a multidisciplinary optimization method for solar aircraft, calibrated the analysis module according to the existing data, and verified the final integrated optimization design program.

Although the design methods introduced above are diverse, the core of the methods is consistent, that is to obtain the optimal parameters by solving the energy and mass balance equations. In fact, the balance equations may not take the equal sign, and the overall parameter feasible region can be constructed by geometric method. This paper introduces the feasible region construction method, and studies the sensitivity of radial feasible region to task indicators and performance indicators respectively. On this basis, the design scheme of a 15m wingspan solar aircraft are proposed. The research conclusion of this paper has reference significance for the overall design of solar aircraft.

## 2. Feasible region planning method

### *2.1. Energy balance*

As shown in Figure 1, during the day, the solar aircraft uses the PV cells to absorb light energy and convert it into electric energy. Among them, part of the energy is used for daytime flight and equipment power consumption, and the other is stored as gravitational potential energy and electric energy for night cruise of the aircraft.

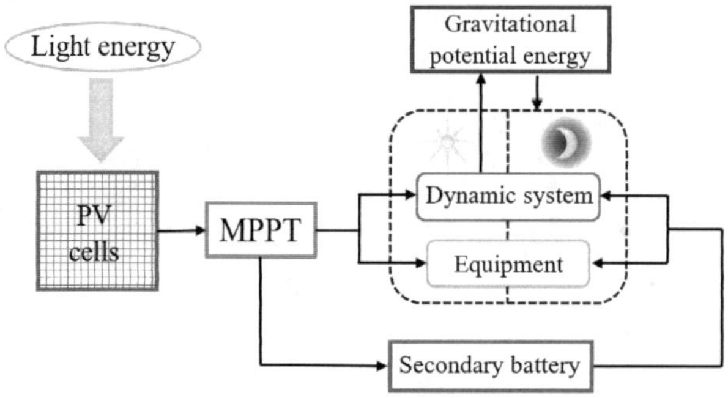


Figure 1 Solar aircraft energy cycle framework

To realize 24-hour uninterrupted flight, the light energy absorbed during the day should be sufficient to supply the aircraft for daytime flight, secondary battery charging and climbing energy storage. The energy balance equation is expressed as

$$E_{\mathrm{avi}} \geq E_{\mathrm{req,d}} + E_{\mathrm{bat}} + E_{\mathrm{climb}} \tag{1}$$

where $E_{\mathrm{avi}}$ is the energy obtained during the day, $E_{\mathrm{req,d}}$ is the energy for daytime flight and equipment operation, $E_{\mathrm{bat}}$ is energy stored in secondary battery, and $E_{\mathrm{climb}}$ is the energy for potential energy storage.

$E_{\mathrm{avi}}$ is related to the solar radiation intensity and the absorption capacity of the photovoltaic cell, which is expressed as [9, 10]

$$E_{\mathrm{avi}} = E_{\mathrm{sc}} S \eta_{\mathrm{ar}} \eta_{\mathrm{sc}} \eta_{\mathrm{MPPT}} \tag{2}$$

where $E_{\mathrm{sc}}$ is the total amount of solar radiation in a day, $S$ is the wing area, $\eta_{\mathrm{ar}}$ is the PV cell lamination rate, $\eta_{\mathrm{sc}}$ is the conversion efficiency of PV cells, $\eta_{\mathrm{MPPT}}$ is the conversion efficiency of the MPPT module, and $\eta_{\mathrm{MPPT}}$ is usually taken as 0.95.

$E_{\mathrm{req,d}}$ includes the energy supplied to the motor and equipment, which is expressed as

$$E_{\mathrm{req,d}} = \left( \frac{P_{\mathrm{cruise}}}{\eta_{\mathrm{prop}} \eta_{\mathrm{motor}} \eta_{\mathrm{s\text{-}m}}} + \frac{P_{\mathrm{equ}}}{\eta_{\mathrm{DC}} \eta_{\mathrm{s\text{-}e}}} \right) T_{\mathrm{d}} \tag{3}$$

where $P_{\mathrm{cruise}}$ is the cruise power of the solar aircraft, $P_{\mathrm{equ}}$ is the average power of the equipment,

$\eta_{prop}$ is the propeller efficiency in cruise, $\eta_{motor}$ is the motor efficiency in cruise, $T_d$ is the daytime duration, $\eta_{DC}$ is the conversion efficiency of step-down module, and $\eta_{DC}$ is usually taken as 0.9.

$E_{climb}$ provides a new energy storage method for solar aircraft, which is expressed as

$$E_{climb} = \frac{mg\Delta H}{\eta_{climb}\eta_{prop}\eta_{motor}\eta_{s\text{-}m}} \tag{4}$$

where $m$ is the mass of the solar aircraft, $g$ is the acceleration of gravity, $g$=9.8m/s$^2$, $\Delta H$ is the climbing height in gravity energy storage stage, $\eta_{climb}$ is the climb efficiency coefficient, and $\eta_{glide}$ is the sliding efficiency coefficient.

$E_{bat}$ accounts for nearly half of the total energy obtained by the aircraft, which is expressed as

$$E_{bat} = \left( \frac{P_{cruise}T_n - mg\Delta H\eta_{glide}}{\eta_{prop}\eta_{motor}\eta_{b\text{-}m}} + \frac{P_{equ}T_n}{\eta_{DC}\eta_{b\text{-}e}} \right) \frac{1}{\eta_{bat}\eta_{s\text{-}b}} \tag{5}$$

where $T_n$ is the night time length.

The definition and value of cable efficiency in Equation (1) ~ (5) is shown in Table 1.

Table 1 Definition of efficiency coefficient

| Symbol | Definition | Value |
|---|---|---|
| $\eta_{s\text{-}m}$ | PV cells to motor | 90% |
| $\eta_{s\text{-}e}$ | PV cells to equipment | 85% |
| $\eta_{s\text{-}b}$ | PV cells to battery | 85% |
| $\eta_{b\text{-}m}$ | battery to motor | 95% |
| $\eta_{b\text{-}e}$ | battery to equipment | 85% |

*2.2. Mass balance*

Different from conventional fuel aircraft, the total mass of solar aircraft remains unchanged during flight, and the mass balance equation is expressed as

$$m \geq m_{bat} + m_{sc} + m_{MPPT} + m_{dis} + m_{dyn} + m_{str} + m_{cab} + m_{equ} \tag{6}$$

$m_{bat}$ is the mass of the secondary battery, which is expressed as

$$m_{bat} = \frac{E_{bat}\eta_{s\text{-}b}}{\rho_{bat}} \tag{7}$$

where $\rho_{bat}$ is the energy density of the secondary battery.

$m_{sc}$ is the mass of the PV cells, which is expressed as

$$m_{sc} = S\eta_{ar}\rho_{sc} \tag{8}$$

where $\rho_{sc}$ is the area density of the PV cells.

$m_{MPPT}$ is the mass of the MPPT, and is proportional to the maximum power of the module, which is expressed as

$$m_{MPPT} = \frac{P_{sc,max}S\eta_{ar}\eta_{sc}}{\rho_{MPPT}} \tag{9}$$

where $P_{sc,max}$ is the maximum optical radiation power, $\rho_{MPPT}$ is the power to weight ratio of the MPPT, and $\rho_{MPPT}$ is usually taken as 0.5kg/kw.

$m_{dis}$ is the mass of the DC/DC module, which is expressed as

$$m_{dis} = \frac{P_{equ}}{\rho_{DC}} \tag{10}$$

where $\rho_{DC}$ is the power to weight ratio of the DC/DC module.

$m_{dyn}$ is the mass of the propulsion system, including the motor and propeller, which is expressed as

$$m_{dyn} = \frac{k_P P_{cruise}}{\eta_{prop}\eta_{motor}\rho_{dyn}} \tag{11}$$

where $k_P$ is the ratio of maximum output power to cruise output power of the motor, and $\rho_{dyn}$ is the power to weight ratio of the power system.

$m_{str}$ is the aircraft structural mass, which is expressed as

$$m_{str} = k_{str} S^{x_1} \mathrm{AR}^{x_2} \tag{12}$$

where $k_{str}$ is the process level factor, and the smaller the value, the higher the processing level. AR is the aspect ratio of the wing; $x_1$ is the area factor, taken as 1.1; $x_2$ is the wingspan factor, taken as 1.3.

$m_{cab}$ is the mass of the cable, and $m_{equ}$ is the mass of the equipment, including flight control system, data link system, payload, etc.

### 2.3. Feasible region

Equations (1) and (6) contain two unknown quantities: mass $m$ and wing area $S$, and other parameters are known. Therefore, the feasible region can be constructed through the equations:

$$\begin{cases} f_E(m,S) = E_{req,d} + E_{bat} + E_{climb} - E_{avi} \le 0 \\ f_M(m,S) = m_{bat} + m_{sc} + m_{MPPT} + m_{dis} + \\ \qquad m_{dyn} + m_{str} + m_{cab} + m_{equ} - m \le 0 \end{cases} \tag{13}$$

As shown in Figure 2, the feasible region formed by the balance equations is radial and widens with the increase of wing area.

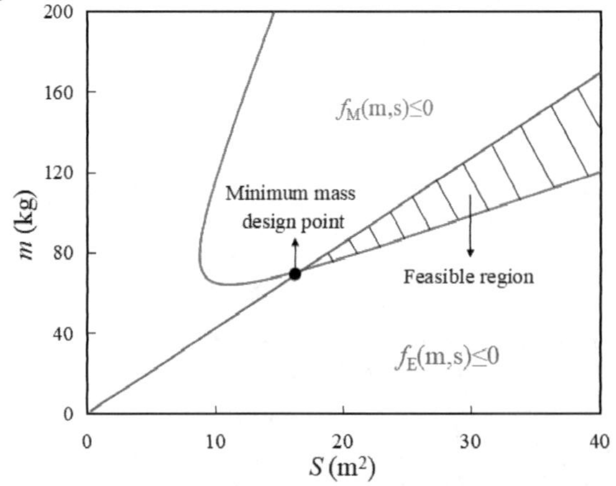


Figure 2 Feasible region of overall parameters

When both equations take the equal sign, the mass and wing area of the aircraft reach the minimum value at the same time. As the wing area continues to increase, the value range of the mass changes from point to interval, and the interval range gradually increases, which means that when the wing area is taken as a constant value, the mass is constrained by both the upper and lower boundaries. Similarly, when the takeoff mass is taken as a constant value, the wing area is constrained by both the left and right boundaries.

### 2.4. Minimum mass solution

The Newton iteration method of nonlinear equations is used to solve the minimum mass design point of solar aircraft. Take $\boldsymbol{x} = [m, S]^{T}$, $\boldsymbol{F}(\boldsymbol{x}) = [f_E, f_M]^{T}$, and the iteration formula is

$$x^{(k+1)} = x^{(k)} - F'(x^{(k)})^{-1} F'(x^{(k)}) \quad (k = 0,1,\cdots,) \tag{14}$$

where $F'(x)^{-1} = \begin{bmatrix} \frac{\partial f_E(x)}{\partial m} & \frac{\partial f_E(x)}{\partial S} \\ \frac{\partial f_M(x)}{\partial m} & \frac{\partial f_M(x)}{\partial S} \end{bmatrix}^{-1}$, and note that $x^{(k+1)} - x^{(k)} = \Delta x^{(k)}$. The linear equations are expressed as

$$F'(x^{(k)})\Delta x^{(k)} = -F(x^{(k)}) \tag{15}$$

The minimum mass solution is solved by the iterative process shown in Figure 3. If the number of iterations exceeds the set value $n$, it is considered that the equations have no solution. If the difference between $x^{(k+1)}$ and $x^{(k)}$ is less than the set value $\varepsilon$, the solution is convergent.

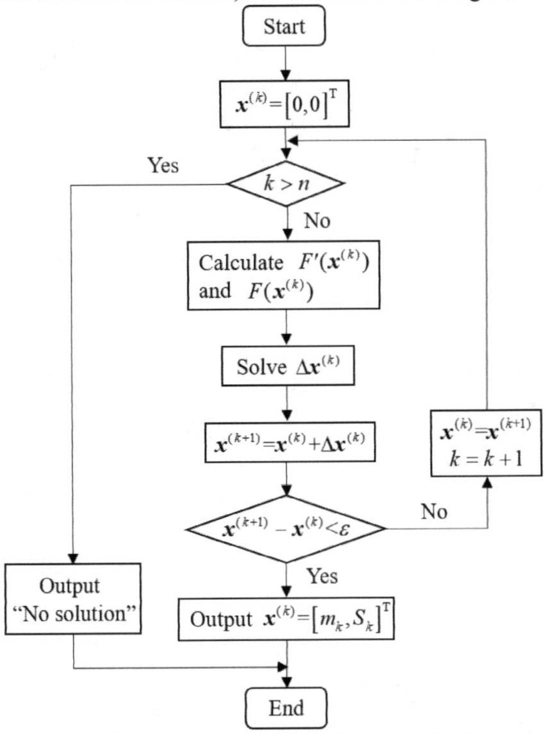


Figure 3 Minimum mass design point solving process

## 3. Sensitivity analysis

In this section, the overall design input parameters are divided into two types of indicators: task indicators and performance indicators. The task indicators determine the approximate scale of the solar aircraft, while the performance indicators represent the technical progressiveness of the solar aircraft.

### *3.1. Task indicators*

**Cruise height $H$:** Fix the other three task indicators as: $m_{load}$=10kg, $\phi$=40°, $n_d$=265. As shown in Figure 4, with the increase of $H$, the boundary curves shrinks in the direction of the feasible region, resulting in the gradual reduction of the included angle of the feasible region. As $H$ increases, the energy balance curve moves down evenly, while the mass balance curve moves up sharply, which also leads to the sharp contraction of the feasible region, until it does not exist. As shown in Figure 5, with the increase of $H$, the minimum mass and minimum wing area gradually increases. When $H$ exceeds 18km, although the parameter feasible region still exists, limited by processing technology and engineering expenses, the mass and wing area of the design scheme have become infeasible, and this region should be avoided in the scheme design. When $H$ reaches 19.3km, there are no feasible solution.

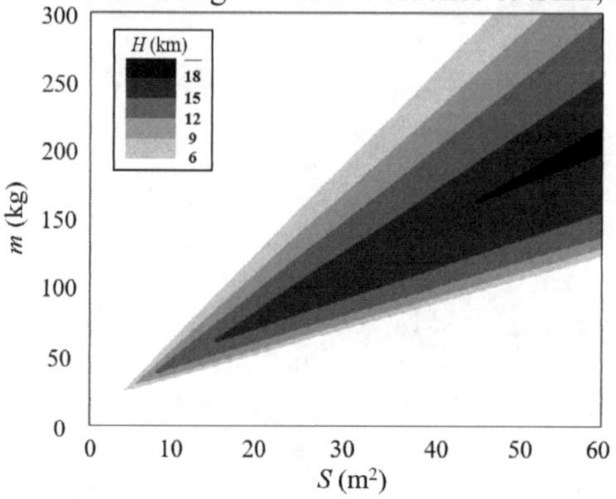


Figure 4 Impact of $H$ on feasible area

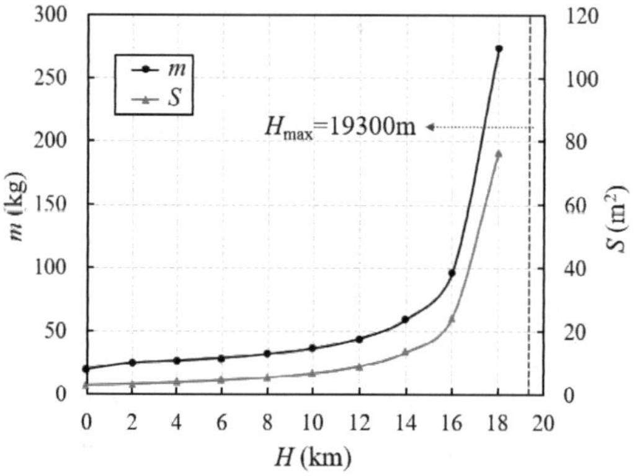


Figure 5 Impact of $H$ on design parameters

**Mission load weight $m_{load}$:** Fix the other three task indicators as: $H$=15km, $\Phi$=40°, $n_d$=265. As shown in Figure 6, as the load mass increases, the mass balance equation curve moves upward in a uniform amplitude, and the slope decreases slightly. Due that the energy consumption brought by the load increment is very small compared with the total energy, the energy balance curve hardly changes. Therefore, with the change of load mass, the feasible region only shifts, but not shrinks. As shown in Figure 7, the minimum mass and wing area of the solar aircraft increase linearly as the load mass increasing. Theoretically, there is no upper limit. However, affected by the processing technology and airport scale, there is still an upper limit on the load mass that can be carried by the solar aircraft.

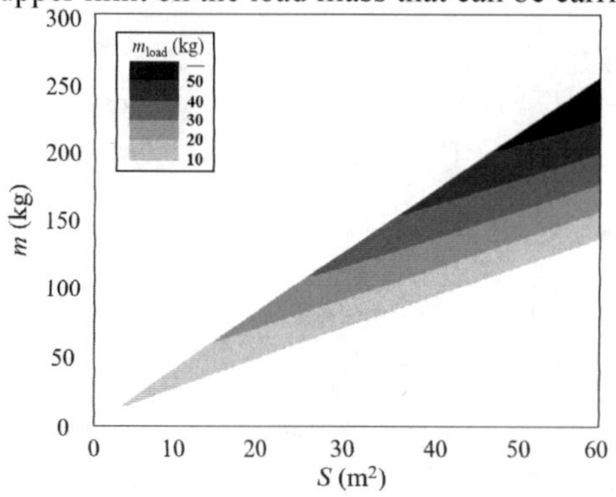


Figure 6 Impact of $H$ on feasible area

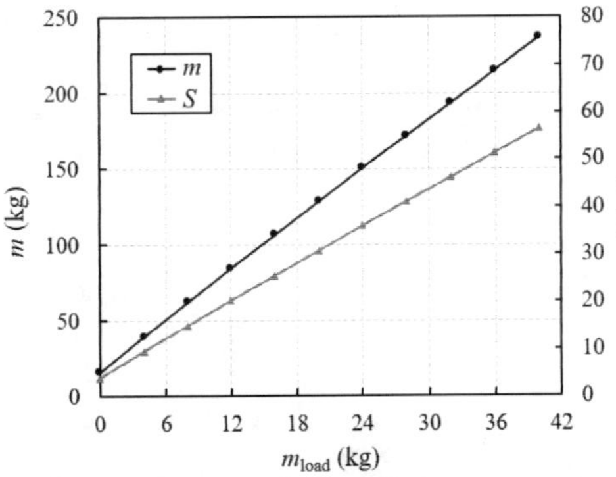


Figure 7 Impact of $H$ on design parameters

**Flight latitude $\Phi$:** Fix the other three task indicators as: $H$=15km, $m_{load}$=10kg, $n_d$=265. As shown in Figure 8, with the increase of latitude, the energy balance curve moves downward, resulting in the contraction of the feasible region. The change of latitude will also cause the mass balance curve to move down, making the feasible region expand, but the range is small. Since the contraction amplitude is much larger than the expansion amplitude, the feasible region will not exist at high latitude. As shown in Figure 9, when the latitude exceeds 55°, the minimum mass and wing area increase significantly, and the feasibility of the scheme also decreases. When the latitude reaches 72°, the light intensity is not enough to support the aircraft to complete 24-hour uninterrupted flight, and the overall parameters no longer have a feasible solution.

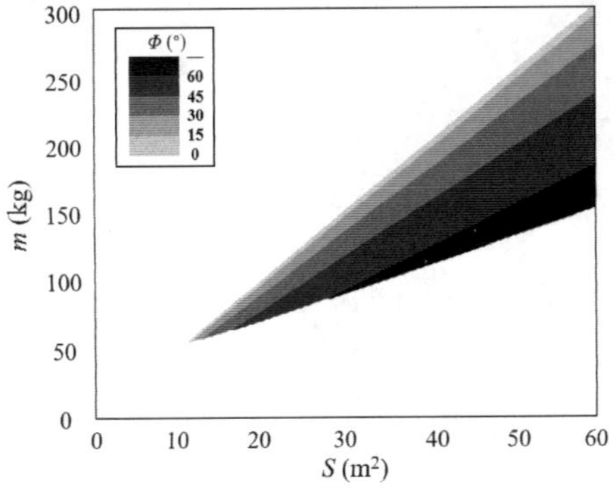


Figure 8 Impact of $H$ on feasible area

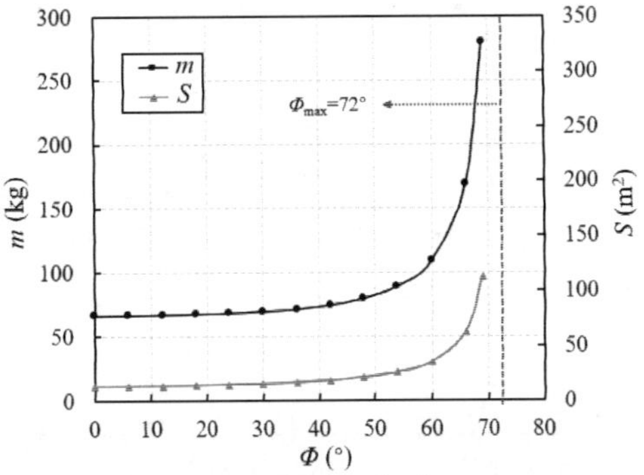


Figure 9 Impact of $H$ on design parameters

**Flight date serial number $n_d$:** Fix the other three task indicators as: $H$=15km, $m_{load}$=10kg, $\Phi$=40°. As shown in Figure 10, with the number of flight date away from the summer solstice, the energy balance boundary moves downward, and the moving amplitude is more and more obvious. The mass balance boundary moves upward slightly, and the amplitude is approximately uniform. As a result, the feasible area of the scheme gradually shrinks with the increase of the flight date serial number, until it does not exist. As shown in Figure 11, When the flight date is after the autumnal equinox ($n_d$=265), the aircraft mass and wing area increase significantly, and the feasibility of the scheme becomes lower. When the flight date approaches the winter solstice (nd=351), there is no feasible solution.

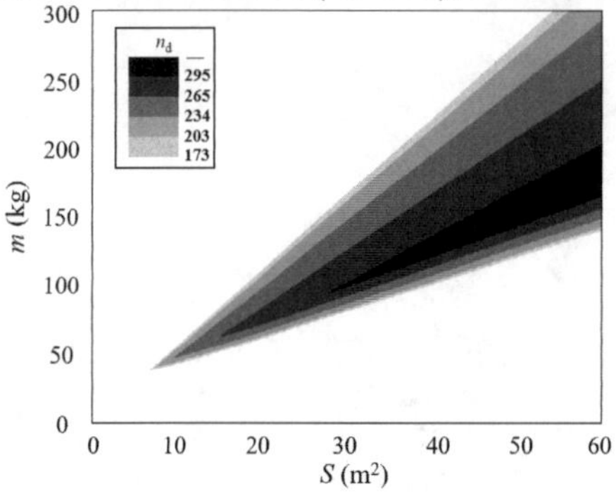


Figure 10 Impact of *H* on feasible area

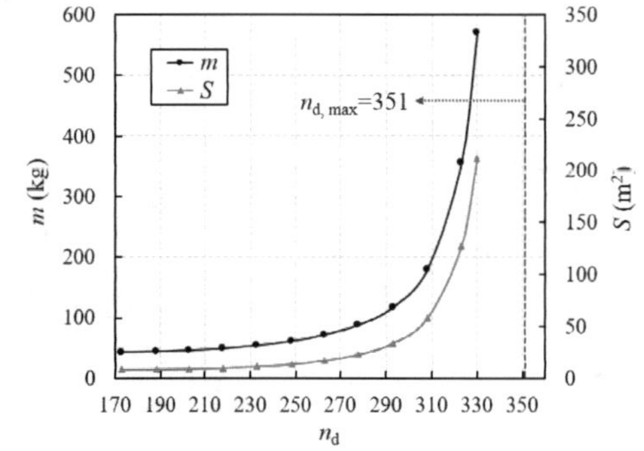


Figure 11 Impact of *H* on design parameters

### 3.2. *Performance indicators*

At the current technical level, when the parameter of a certain performance indicator is increased by a fixed proportion through technical update, the reduction of the minimum takeoff mass reflects the sensitivity.

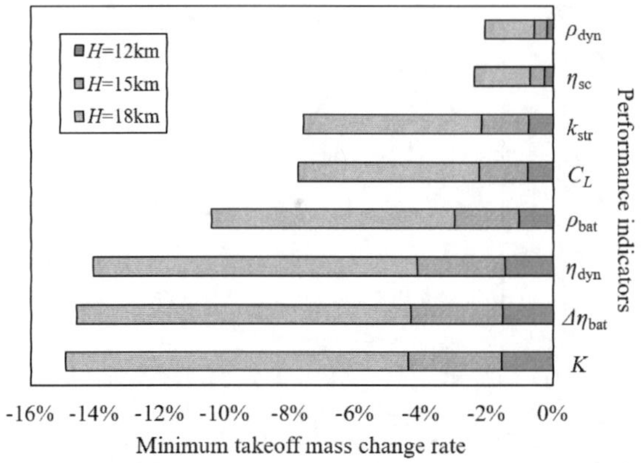


Figure 12 Sensitivity of minimum takeoff mass to technical indicators

As shown in Figure 12, increase each performance indicator by 1% respectively, and the reduction of minimum takeoff mass becomes larger with the increase of cruise altitude. However, the order of sensitivity does not change with the increase of cruise altitude. The technical innovation of lift drag ratio $K$, cable transmission efficiency $\eta_{cab}$ and power system efficiency $\eta_{dyn}$ has the greatest contribution to the feasibility of the scheme, and should be the preferred optimization indicators in overall design stage. The improvement of secondary battery energy density also has a great impact on the minimum takeoff mass, which is a key parameter to be considered. The power density of the propulsion system and the conversion efficiency of PV cells have little impact on the minimum takeoff mass, and there is no need to put forward too high requirements for the progressiveness of these two indicators.

## 4. Design results

The design objectives of the solar aircraft include: payload mass≥1kg, flight latitude of 40°N, night flight altitude of 15000m, and continuous cruise time≥7 days.

Through the overall design method introduced above, the minimum takeoff mass and wing area meeting the mission requirements are obtained. On this basis, combined with the technical implementation difficulty of each subsystem and the risk caused by index fluctuation, the design scheme with high feasibility is found in the feasible region. The overall design parameters of the scheme are shown in Table 2.

Table 2 Overall design parameters

| **Parameters** | **Value** |
|---|---|
| Takeoff mass | 35.0 kg |
| Wing area | 13.3 $m^2$ |
| Wing load | 2.6 $kg/m^2$ |
| Night flight altitude | 15000 m |
| Cruise speed | 16.5 m/s |
| Flight latitude | 40°N |

Calculated by the mass balance equation, the weight distribution of the aircraft is shown in Table 3 and Figure 13. Among them, the weight of structure and secondary battery accounts for more than one-third of the total weight of the whole aircraft respectively, and the weight of other parts is less than one-third of the total weight, which means the weight reduction of solar aircraft should focus on the secondary battery and structure quality.

Table 3 Mass distribution of the aircraft

| **Components** | **Mass/kg** | **Weight ratio** |
|---|---|---|
| Secondary battery | 13.08 | 37.38% |
| Structure | 12.24 | 34.96% |
| Photovoltaic cells | 3.59 | 10.26% |
| MPPT | 1.13 | 3.22% |
| DC/DC module | 0.06 | 0.17% |
| Airborne cable | 1.22 | 3.48% |
| Propulsion system | 1.94 | 5.55% |
| Equipment | 1.74 | 4.97% |
| Total | 35.00 | 100% |

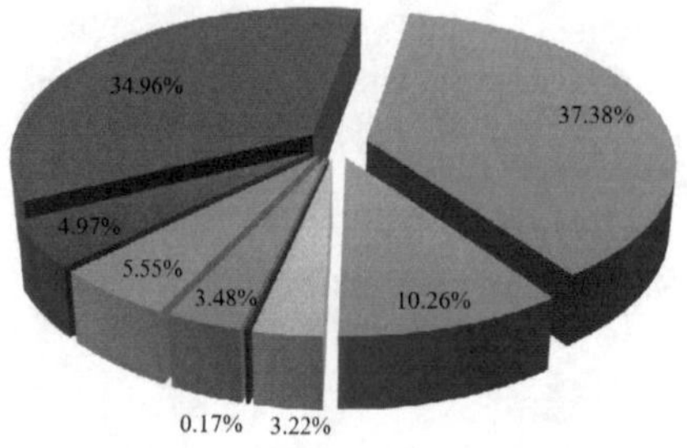


Figure 13 Component weight ratio

Calculated by the energy balance equation, the relevant parameters of the energy and propulsion system are obtained, as shown in Table 4. The PV cell laying rate is only about 60%, which indicates that the current PV cell efficiency is surplus to support the solar aircraft to complete uninterrupted flight.

Table 4 Energy and propulsion parameters

| Parameters | Value |
|---|---|
| PV cells area | 7.98 $m^2$ |
| Maximum PV absorbed power | 2.25 kW |
| Secondary battery capacity | 4.58 kWh |
| Maximum discharge power | 1.41 kW |
| Maximum input power of motor | 1.30 kW |
| Required power for level flight | 0.34 kW |

Compared with flying wing layout, tandem wing layout and canard layout, the conventional layout has high technical maturity and good stability, which is the most mainstream pneumatic layout scheme of solar aircraft. This paper also adopts the conventional layout, as shown in Figure 14.

Figure 14 Appearance of solar powered aircraft

## 5. Conclusion

In this paper, a feasible region construction method for overall parameters of solar aircraft is proposed, and based on this method, parameter sensitivity analysis and aircraft design are carried out. The main conclusions include:

(1) The feasible region of the overall parameters of the solar aircraft is constructed by the equilibrium equations. In the feasible region, when the wing area is taken as a fixed value, the takeoff mass is limited by the upper and lower boundaries at the same time.

(2) For task indicators, the load mass mainly affects the energy balance, the flight latitude mainly affects the mass balance, and the flight altitude and the number of flight days affect both the energy balance and the mass balance.

(3) For performance indicators, lift drag ratio, cable transmission efficiency and propulsion system efficiency have the greatest impact on the overall parameters, while the power density of the propulsion system and the conversion efficiency of photovoltaic cells have little impact.

**REFERENCES**

[1] P. Rajendran, and H. Smith. "Future trend analysis on the design and performance of solar-powered electric unmanned aerial vehicles," *Adv. Mater. Res.*, vol. 1125, pp. 635-640, 2015.

[2] X. F. Zhu, G, Zheng, and Z. X. Hou. "Solar-powered airplanes: A historical perspective and future challenges," *Prog. Aerosp. Sci.* vol. 71, pp. 36-53, 2014.

[3] D. L. Ma, L. Zhang, M. Q. Yang, X. L. Xia, and S. Q. Wang. "Review of ultra-long-endurance solar-powered unmanned aerial vehicle in Key Technologies," *Acta. Aeronaut. Astronaut. Sin.* vol. 41, pp. 623418, 2020.

[4] J. W. Youngblood, T. A. Talay, and R. J. Pegg. "Design of long-endurance unmanned airplanes incorporating solar and fuel cell propulsion," in *AIAA/SAE/ASME 20th Joint Propulsion Conference*, 1984, p. 1-11.

[5] A. Noth. "Design of solar powered airplanes for continuous flight," Ph.D. Thesis, Ecole Polytechnique Fédérale de Lausanne, Zurich, Suisse, 2007.

[6] G. Romeo, G. Frulla, and E. Cestino, "Design of solar high altitude long endurance aircraft for multi payload & operations," *Aerosp. Sci. Technol.*, vol. 10, no. 6, pp. 541-550, Sep. 2006.

[7] M. H. Maleki. "Conceptual Design Method for Solar Powered Aircrafts," in *49th AIAA Aerospace Sciences Meeting including the New Horizons Forum and Aerospace Exposition*, Jan, 2011.

[8] B. Morrisey, and R. Mcdonald. "Multidisciplinary design optimization of an extreme aspect ratio HALE UAV," in *9th AIAA Aviation Technology, Integration, and Operations Conference (ATIO)*, Sep, 2009.

[9] C. A. Gueymard. "Revised composite extraterrestrial spectrum based on recent solar irradiance observations," *Sol. Energy.* vol. 169, pp. 434-440, May. 2018.

[10] I. Reda, and A. Andreas. "Solar position algorithm for solar radiation applications," *Sol. Energy.* vol. 76, pp. 577-589, May. 2004.

# Bearing capacity analysis of large deflection thin plate for civil aircraft

**Kaiwen Zhang[1], Zhouli He[1] and Longhui Huang[1*]**

[1]Shanghai Aircraft Design And Research Institute, Shanghai 201210, China

[*]Corresponding author's e-mail: huanglonghui@comac.cc

**Abstract.** At present, the analysis of thin plate structure with small deflection and linear deformation is very mature, but the thin plate structure with large deflection and nonlinear deformation is often used in practical engineering design. Aiming at the thin plate structure with large deflection under out of plane load, this paper studies the bearing capacity of thin plate structure with large deflection under out of plane load by combining theoretical analysis, finite element simulation analysis and experimental analysis. According to the research, it is found that the process of theoretical analysis is complex and the workload and difficulty are too large. In practical engineering, some analysis results can be obtained quickly by using empirical formula and looking up the chart, but the application scope of these analysis results is small, and only the local stress results can be obtained. At present, designers usually use finite element simulation analysis to get the stress distribution of the whole structure. It should be noted that the geometric nonlinearity must be considered in the finite element analysis. The simulation results can provide a certain reference for engineering research and can be applied in practical engineering.

## 1. Introduction

At present, thin plate structure is widely used in the main load-bearing structures of civil aircraft. The excellent bearing capacity of thin plate structure is reflected in that as long as the thickness is thin, it has a certain bearing capacity, low manufacturing cost and high material utilization rate. It is precisely because of the remarkable advantages of thin plate structure that researchers have taken a fancy to that, thin plate structure has been widely used in the actual civil aircraft structure. With the extensive use and promotion of thin plate structure in civil aircraft, researchers found that the thin plate structure is prone to large deflection deformation when bearing large out of plane loads at some positions, and the occurrence of large deflection deformation does not mean that the structure is damaged. In the case of large deflection deformation, the structure can still have a certain bearing capacity [1]. With the progress of the times and the expansion of demand, airlines have increasingly stringent requirements for cost, and making full use of the bearing capacity of the structure after large deflection deformation to improve the structural utilization is an effective means to reduce costs. Therefore, the research on the large deflection deformation bearing capacity of thin plate structure has very important research and development significance in practical engineering applications. The research on the structural stability of thin plates has achieved a lot in recent 100 years, but engineers found that there is still a certain bearing capacity after large deflection deformation, and it is still a short time to start to study the bearing capacity in the case of geometric nonlinearity after large deflection. Especially after finding that using the bearing capacity after large deflection deformation can bring better benefits in engineering, the research on large deflection began to develop rapidly [2-3]. The problem of structural stability was first proposed by Euler in the 18th century, and then after hundreds of years of research in theory, experiment and practical engineering application, the theoretical analysis method has been constantly improved [4]. Researchers design and analyze thin plate structures based on the theories of elasticity, material mechanics and structural mechanics. However, the derivation formula of large deflection theory is often applicable to the ideal situation under certain assumptions, and the process is very complicated [5-6]. And the theoretical analysis needs a variety of assumptions, and there are many constraints for the actual structural analysis. The analytical solution obtained through theoretical analysis is too difficult, and solving the analytical solution when encountering engineering problems seriously affects the work efficiency. The ultimate bearing capacity of civil aircraft thin plate structure under out of plane load needs to consider geometric nonlinearity and material nonlinearity, and the classical elastic stability theory cannot cover such problems [7]. It is difficult to obtain an analytical solution for the bearing capacity of large deflection thin plate structures. In a large number of experiments and engineering applications, researchers summarized and sorted out a simplified analysis method combining theory and experience, which can quickly solve and get some required results. It has important reference significance for the application of finite element method to analyze elastoplasticity of thin plate structure with large deflection bearing capacity [8]. With the popularization of computers, large-scale finite element software generally has the analysis function of analyzing large deflection deformation structures, but it is worth noting that the finite element software analysis is usually simplified, and the boundary conditions and parameter settings need to be consistent with the actual structure as far as possible, so that the analysis results can reflect the real situation. In this paper, through engineering methods, finite element simulation and combined with the test data in engineering practice, the bearing capacity of large deflection thin plates under out of plane load is analyzed and studied, and a verification method that can effectively solve practical engineering problems is summarized and sorted out, which can provide a certain reference for the study of the bearing capacity of thin plates in practical engineering.

## 2. Theoretical analysis

In civil aircraft structures, the boundary constraints of thin plates are generally between simply supported and fixed supported. What kind of boundary constraint conditions are used in the theoretical solution has a great impact on the analysis results. At present, it is widely used that the average value of the analysis results of the two support cases is closer to the actual engineering problems [9].An example of solving the equation for the analytical solution of large deflection is as follows [10]:Suppose a rectangular thin plate with four simply supported edges has a length of $a$, a width of $b$ and a thickness of $t$. The applied load is uniform compression load along the length direction. The differential equation of large deflection is as follows:

$$D\nabla^4\overline{W} = \tilde{L}(\overline{W}+\overline{W}^*,\overline{F}) \tag{1}$$

$$\nabla^4\overline{F} = \frac{1}{2}Et\tilde{L}(\overline{W}+2\overline{W}^*,\overline{W}) \tag{2}$$

Where: $\overline{W}^*(X,Y)$ represents the initial deflection of the rectangular thin plate; $\overline{W}(X,Y)$ represents the additional deflection under load, and $\overline{F}(X,Y)$ represents the stress function. From the above formula and the simply supported boundary conditions on four sides, it is assumed that the dimensionless parameters are as follows:

$$x = \pi X / a \tag{3}$$

$$y = \pi Y / b \tag{4}$$

$$\beta = a / b \tag{5}$$

The large deflection balance equation is derived as follows:

$$\lambda_x = \lambda_x^{(0)} + \lambda_x^{(2)} W_m^2 + \lambda_x^{(4)} W_m^4 + \cdots \tag{6}$$

$$\delta_x = \delta_x^{(0)} + \delta_x^{(2)} W_m^2 + \delta_x^{(4)} W_m^4 + \cdots \tag{7}$$

Where:

$$(\lambda_x^{(0)},\ \lambda_x^{(2)},\ \lambda_x^{(4)}) = (S_0,\ S_2,\ S_4) / 4\beta^2 \tag{8}$$

$$S_0 = \frac{(m^2 + n^2\beta^2)^2}{(1+\mu)\ m^2} \tag{9}$$

$$S_2 = \frac{1}{16}\frac{m^4 + n^4\beta^4}{m^2}(1+2\mu) \tag{10}$$

$$S_4 = \frac{1}{256m^2}\{2(1+\mu)^2(1+2\mu)^2(m^4+n^4\beta^4)\ \ (\frac{m^4}{J_{13}} + \frac{n^4\beta^4}{J_{31}})\ \text{-}(1+\mu)\ \ (1+2\mu)[2(1+\mu)^2 + (1+2\mu)](\frac{m^8}{J_{13}} + \frac{n^8\beta^8}{J_{31}})\} \tag{11}$$

Where: *m*. *n*, *J* are dimensionless parameters, $\mu$ is Poisson's ratio

The above analytical solution is more complex, and where the accuracy is not so high in engineering applications, we usually need to combine the test data and use a relatively simple engineering method to solve it.

Example: take the fuselage skin of a civil aircraft as an example. The skin material is 2024 aluminum alloy. According to the relevant data system, the elastic modulus and Poisson's ratio of 2024 aluminum alloy are shown in Table 1. The structure is shown in Figure 1. According to the actual flight simulation analysis, when the maximum velocity and pressure occurs and the corrected airspeed is 387 knots, the load is the aerodynamic pressure perpendicular to the skin w=6915pa. According to the skin materials (see Table 2) and structural parameters (see Table 2), combined with figure 2 and figure 3, the composite stress of the calculated structure under the boundary conditions of simple support and fixed support is solved respectively, and the calculation results (see Table 3).

Table 1. Material properties.

| Modulus of elasticity($E$/GPa) | Poisson's ratio $\mu$ |
|---|---|
| 72 | 0.3 |

Table 2. The dimensions of the structure.

| | |
|---|---|
| Unit number | 1 |
| Length $a$/mm | 730 |
| Width $b$/mm | 424 |
| Thickness $t$/mm | 0.8 |

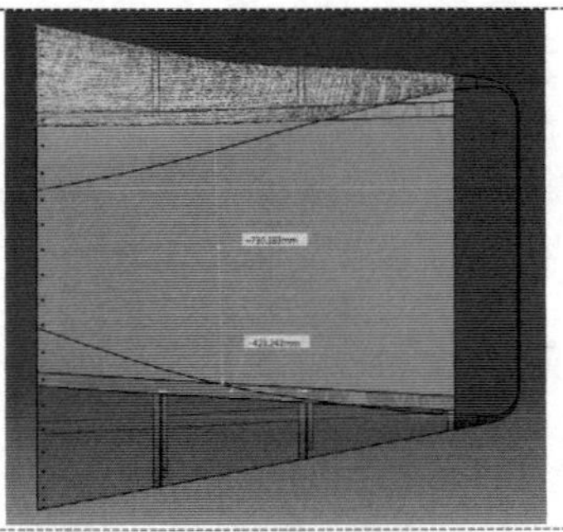

Figure 1. Structure diagram.

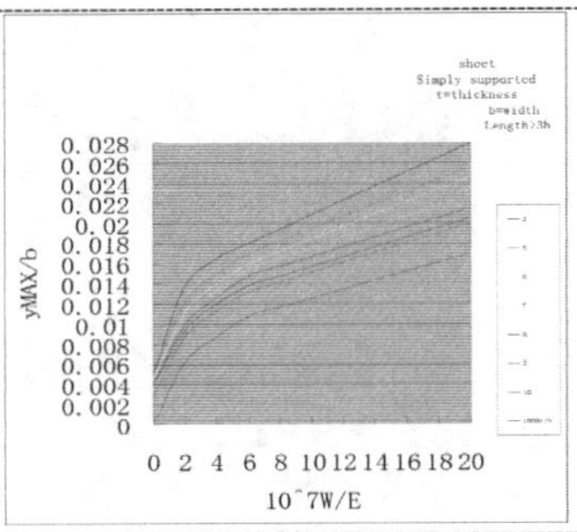


Figure 2. Composite stress of simply supported thin plate.

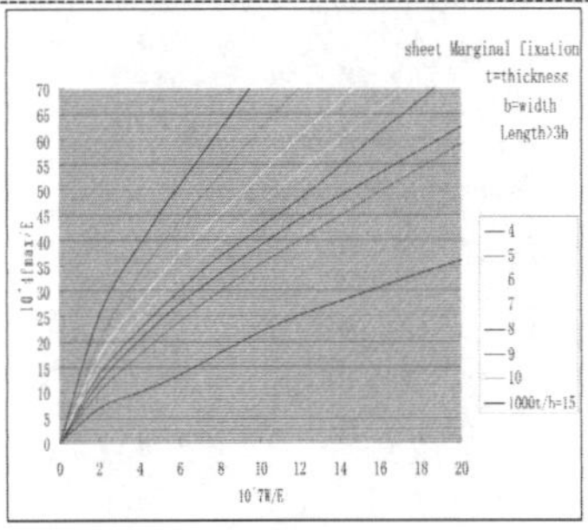


Figure 3. Composite stress of fixed supported thin plate.

Table 3. Simply supported and fixed supported composite.

| Unit number | Simply supported ($\sigma_1$/MPa) | Fixed support ($\sigma_2$/MPa) | Average value ($\sigma_3$/MPa) |
|---|---|---|---|
| 1 | 43 | 145 | 94 |

## 3. Simulation analysis

Because the calculation of large deflection bearing capacity of thin plate structure is cumbersome, engineering researchers usually solve it by solving differential equations and finite element method. In practical engineering, the amount of work to solve by numerical method is huge, and the efficiency is too low in engineering practice, which affects the whole analysis progress. Modern engineering researchers mainly use large-scale commercial software for analysis.

According to the example structure in the previous section, use finite element software to create a model, as shown in Figure 4. The element type is quadrilateral plate element, the element size is 10mm, and the material property is 2024 aluminum alloy. Set the same boundary conditions and loads as the example structure, and simulate and solve the structural stress distribution without considering geometric nonlinearity and considering geometric nonlinearity by NASTRAN solver. Figure 5 shows the Mises stress of the structure obtained by finite element analysis without considering geometric nonlinearity. The maximum stress of the skin occurs in the middle of the long side, and the maximum stress value is 236mpa. Figure 6 shows the Mises stress considering geometric nonlinearity. The maximum stress of the skin also occurs in the middle of the long side, and the maximum stress value is 119mpa.

According to the example structure in the previous section, use finite element software to create a model, as shown in Figure 4. The element type is quadrilateral plate element, the element size is 10mm, and the material property is 2024 aluminum alloy. Set the same boundary conditions and loads as the example structure, and simulate and solve the structural stress distribution without considering geometric nonlinearity and considering geometric nonlinearity by NASTRAN solver. Figure 5 shows the Mises stress of the structure obtained by finite element analysis without considering geometric nonlinearity. The maximum stress of the skin occurs in the middle of the long side, and the maximum stress value is 236mpa. Figure 6 shows the Mises stress considering geometric nonlinearity. The maximum stress of the skin also occurs in the middle of the long side, and the maximum stress value is 119mpa.

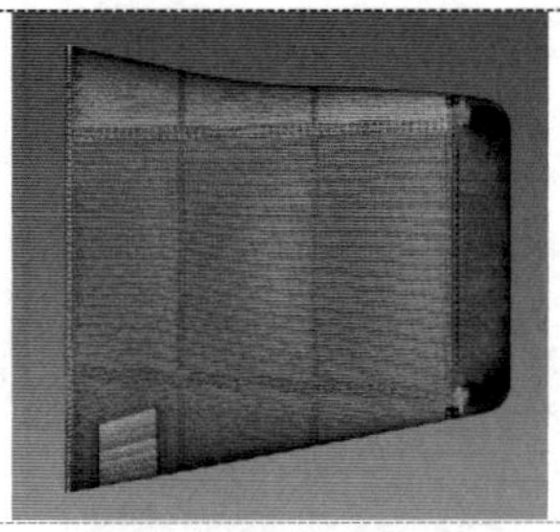

Figure 4. Finite element modeling diagram.

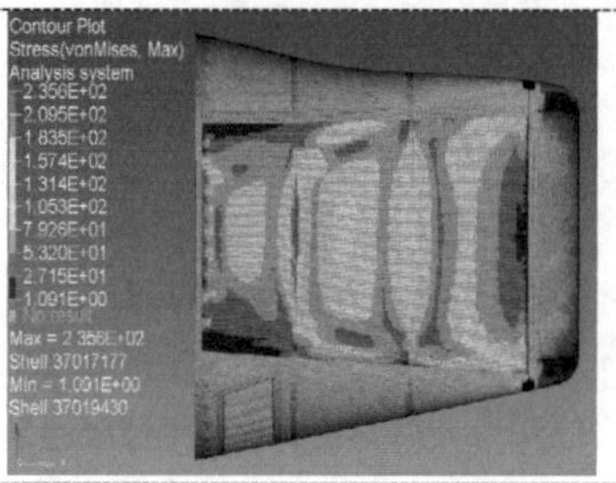


Figure 5. Stress nephogram without considering geometric nonlinearity.

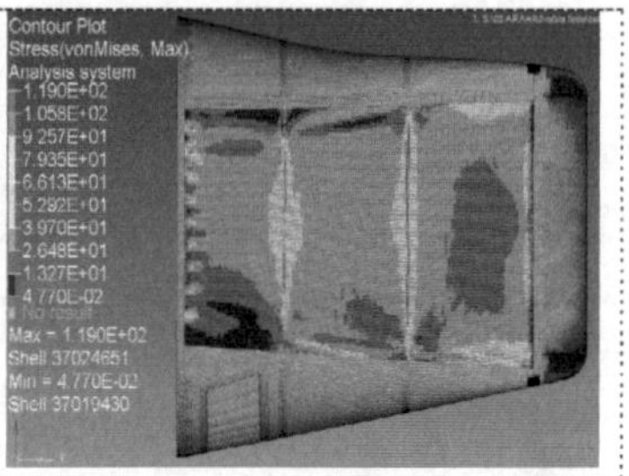


Figure 6. Stress nephogram considering geometric nonlinearity.

## 4. Test analysis

The physical object of the fuselage skin structure of a civil aircraft and the placement of patches on the structure are shown in Figure 7.

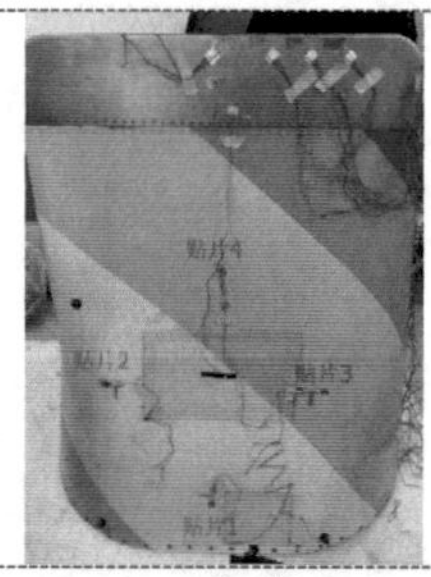

Figure 7. Strain gage patch diagram of test structure.

Use the monitoring equipment to measure the real-time strain data of the skin in the actual flight process. The comparison results between the simulation analysis values and the test measured values are shown in Table 4. According to the results in Table 4, the minimum strain measured by the test patch is at the position of patch 3. The measured working strain at the position of patch 3 is -650, and the converted stress is 46.8mpa. The simulation analysis shows that the working stress at the position of patch 3 is -47.9mpa. The absolute value of the working stress obtained from the simulation analysis is larger than the absolute value of the working stress obtained from the test. The maximum error of the simulation analysis stress value than the test measured stress value is no more than 10%, which shows that the simulation analysis result has a certain conservative amount, and the error between the simulation result and the actual test is small. The modeling and analysis method and results in this paper can be used as a reference in practical engineering applications.

Table 4. Simulation and measurement results.

| Strain gauge number | Simulated stress value（MPa） | Measured stress value （MPa） | Average error |
|---|---|---|---|
| 1 | 16 | -14.4 | 10% |
| 2 | 24.7 | 22.3 | 10% |
| 3 | -47.9 | -46.8 | 2% |

## 5. Conclusion

This paper takes thin plate structure as the research object, and mainly analyzes the bearing capacity of thin plate structure under large deflection deformation. In this paper, the bearing capacity of thin plate structure under large deflection deformation is analyzed through three aspects: large deflection theory, finite element simulation and experimental test. According to the analysis results in this paper, the following three conclusions can be drawn:

(1) When the thickness of thin plate structure is small and subjected to a certain amount of out of plane load, the stress results calculated by linear elastic analysis method are obviously too conservative, which will pay a large weight cost and reduce the economic benefits of the structure. The analysis results without considering geometric nonlinearity are obviously unreasonable. Geometric nonlinearity must be considered in finite element modeling and analysis of large deflection thin plate structures;

(2) When the simplified theoretical method is used to calculate the thin plate on the actual aircraft structure, it is sometimes not conservative to use the average value of the simply supported and fixed supported composite stress, and some corrections should be considered in engineering application;

(3) In this paper, the boundary of thin plate structure is rivet connection or bolt connection. It can be found that the analysis results of rivet connection or bolt connection are closer to the analysis results of fixed support. In practical engineering, when analyzing the strength of some non important thin plate structures that do not need accurate analysis, the analysis method of fixed support theory can be used conservatively.

## References

[1] Du Hongzeng，Tian Xiuyun，Zhang Xing. Analysis of thin-walled structures considering post buckling of shear plates[J]. Journal of Aeronautics (Engineering Edition)，1986:43-47.

[2] Shi Xudong, Wang Xinwei. Buckling analysis of rectangular plates subjected to in-plane non uniformly distributed loads [J]. AJournal of Aeronautics, 2006(11):1113-1116.

[3] Guo Yanlin, Chen Guodong, LAN Yong. Research on ultimate bearing capacity of three tube shuttle steel lattice columns [J]. Journal of building structure, 2002 (05): 25-30.

[4] Chen Shaofan. Guide to stability design of steel structures [M]. Beijing: China Construction Industry Press, 1996.6:175-179.

[5] Chen Tieyun, Shen Huizhong. Structural buckling [M]. Shanghai: Shanghai Science and Technology Literature Press, 1993:97-110.

[6] Wongkaew, Chen. Consideration of out-of-plane buckling in advanced analysis for planar steel frame design[J]. Journal of Constructional Steel Research, 2002,58(5):943-965.

[7] Editor in chief of aircraft design manual. Aircraft Design Manual Volume 9: load, strength and stiffness [S]. Beijing: Aviation Industry Press, 2001.12:317-324.

[8] Cui Degang. Structural stability design manual [M]. Beijing: Aviation Industry Press, 2006.5:13-16.

[9] Niu Chunyun. Structural stress analysis and dimensional design of practical aircraft [M]. Beijing: Aviation Industry Press, 2009.12:404-410.

[10] Shen Huishen. Post buckling behavior of plates and shells [M]. Shanghai: Shanghai Science and technology, 2002:27-29.

# A New Design of the 3D Deployable Solar Array in the Aerospace Field and Kinematic Analysis during Deployment

**Jiayi Sun***

School of Aerospace, Mechanical and Mechatronic Engineering, The University of Sydney, Sydney, Australia

*Corresponding author e-mail: jsun8177@uni.sydney.edu.au

**Abstract**—The paper focuses on a new design of the 3D deployable solar array for spacecraft and landers. Compared with existing solar array designs, the main advantage of this new design is the further increase in the area of the solar array. The principles of spatial rings and domes are used for this new design. This design consists of a large dome and four smaller rings. The model is constructed using Solidworks, and different configurations during the deployment are shown. The kinematic analysis shows that the length ratios of elements are independent of the input angle θ. It proves that this deployable structure has only one mobility. The configuration can be determined by a specific input angle. When the input angle θ increases, the top-view area of the structure increases, while the maximum height of the structure decreases. The volume of the increases and then decreases with the increasing θ. The maximum volume occurs at θ = 110°. In addition, the shape of the top-view geometry for a particular design is determined by two design parameters. It is found that the length ratios of elements increase with the increasing value of two design parameters.

## 1. Introduction

Since the 1950s, the deployable structure has experienced rapid development and has been widely used in several fields, such as aerospace, architecture and medical devices [1]. The most significant advantage of deployable structures is that their configurations can be transformed to meet the specific requirements. In aerospace, the 3D deployable structures are widely used for solar arrays and antennas on the satellites, spacecraft and landers. Due to the size limitation of the launch vehicles, deployable structures are applied to reduce the size of solar arrays and antennas so that they can fit well in the launch vehicles [2].

The solar array technology has evolved in the past decades in order to accommodate the increasing power requirements [3]. In 1958, the U.S Vanguard I spacecraft was launched. It was the first time for solar arrays to be used in aerospace, with a simple spacecraft body-mounted configuration [4]. Then, the solar array technology evolved to single-panel flip-out configuration. Nowadays, the kinematically complex multi-panel deployable structures become the main solar array technology [3].

There are several common types of deployable mechanisms for aerospace applications, such as zig-zag mechanism, z-folding mechanism, fan mechanism and hub-wrapping origami mechanism. The zig-zag mechanism is mainly used for solar arrays on satellites. Fig. 1 shows an example of the zig-zag mechanism, which is used on the NOAA GOES-R satellite [5]. Zig-zag solar arrays always consist of several separated solar panels, which can be assembled, folded and deployed.

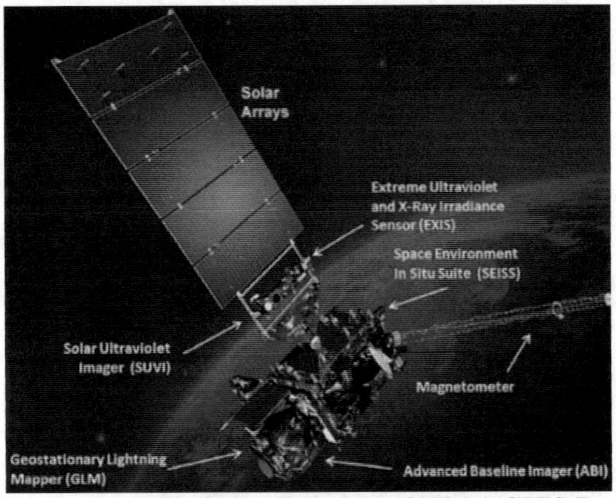


Figure 1. Zig-zag mechanism for solar arrays of NOAA GPES-R satellite [5]
(This image has been authorized to use)

Fig. 2 shows a typical z-folding mechanism for solar arrays on the International Space Station (ISS) [6]. Such kind of solar arrays with large area can provide high power and has increasing importance for spacecraft to travel deeper into space.

The main applications of the fan mechanism are on the solar arrays of landers and spacecraft, such as the NASA's Insight Mars Lander and American Lucy Spacecraft [7-8]. Fig. 3 and Fig. 4 show NASA's Insight Mars Lander and American Lucy Spacecraft, respectively. Both of them consist of several separated solar panels. When the structure is folded, all the solar panels overlap with each other. During the deployment, each piece of solar panel rotates about the center point to its specific position.

Figure 2. Z-folding mechanism for solar arrays of ISS [6]
(This image has been authorized to use)

Figure 3. Fan mechanism for the solar array of NASA's Insight Mars Lander [7]
(This image has been authorized to use)

Figure 4. Fan mechanism for the solar array of American Lucy Spacecraft [8]
(This image has been authorized to use)

The hub-wrapping origami mechanism can be used for solar arrays and antennas. Origami is the art of paper folding, which has provided a lot of inspiration for structural engineers to create new deployable structure designs [9]. Brigham Young University (BYU) engineers have worked with NASA through the Jet Propulsion Lab to create a new solar array design using the hub-wrapping origami mechanism. Fig. 5 is the computer-generated model of this solar array design [10]. An origami technique called Miura folding can be used for origami-inspired solar array design. Miura folding is a kind of Japanese tradition origami and has inspired the solar array design for aerospace engineering [11]. However, since the thickness of origami is negligibly small, it is a challenge to accommodate the thickness of solar panels in the aerospace applications [12].

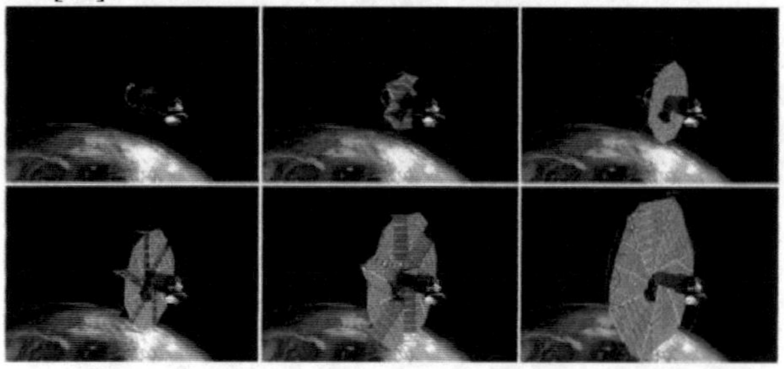

Figure 5. Deployment configurations of the hub-wrapping origami mechanism for solar arrays [10]
(This image has been authorized to use)

The deployment of the subsystem (i.e. antennas or solar arrays) with hinge mechanisms can be controlled by motors. The hinge mechanisms can be divided into two categories, which are the rigid hinge and the flexible hinge [13]. Another approach to achieving the deployment of structures is using the shape memory composite, which has the self-deployable property [14].

According to the evolution of the solar array technology, scientists have always been looking for a more powerful and efficient solar array design [4]. However, from Fig. 3 and Fig. 4, it can be found that the size of the fully deployed solar array is restricted by the main body of the spacecraft or lander. The limited size may result in the limitation of higher power. A possible method to provide higher power is to increase the area of the solar array. Therefore, the main purpose of this new design is to increase the size of solar arrays using the fundamental principle of spatial rings and domes [15].

This paper focuses on a new design of the 3D deployable solar array in the aerospace field, as well as some kinematic analysis during its deployment. In Section 2, the model design method will be introduced, including the top-view geometry design and the inner, intermediate and outer elements design for each ring. The analysis method of the mobility and the kinematic relationships will be introduced in this section as well. In Section 3, the design configurations during deployment will be shown. The results of the mobility analysis and the kinematic relationships in terms of the top-view area, the maximum height and the volume of the whole structure will be calculated, plotted and discussed. In addition, the effect of the design parameters will be analyzed as well.

## 2. Methods

### *2.1. Model design methods*

The conceptual design needs to be considered from two aspects. One aspect is the top-view geometry design of this deployable structure. According to Ref. [15], three concepts of ring can be chosen for this design. The dome can be designed by adding more outer rings to the inner ring structure, regardless of which concepts the rings are. Another aspect to be considered is the design of the inner, intermediate and outer elements of each ring. There are several patterns of the intermediate elements design, depending on whether all the intersections of a pair of elements are at the same level [15].

Combining the top-view geometry and elements design, a 3D conceptual model can be constructed using Solidworks. When constructing the model, the zero size assumption is made, which means all the elements are simplified as lines with zero thickness and width. Therefore, the revolute joints connecting linkages can be simplified as a point.

*1) Top-view geometry design*

Fig. 6 shows the top view of the conceptual design, which consists of a large dome and four small rings. The dome is a combination of two rings designed using different concepts, and four rings are completely the same as each other. Since each ring shares a pair of outer elements with the dome, four small rings can also be driven together with the dome.

In Fig. 6, some nodes, angles and distances are labelled, which will be referred to in the following. According to the design of the inner ring of the dome (in yellow), the angle $\alpha_1$ is equal to 60°. The angle $\alpha$ of the outer ring of the dome (in green) is half of the angle $\alpha_1$, which equals 30°. The design of four small rings (in blue) results in an angle $\alpha_2$ equal to 60° as well.

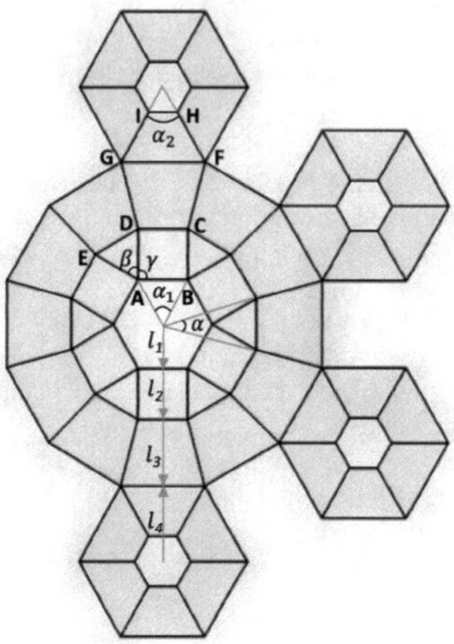


Figure 6. Top view of the design

*2) Inner, intermediate and outer elements design*

Figs. 7, 8 and 9 show the designs of the inner, intermediate and outer elements for three different kinds of rings, respectively. Elements with different lengths are labelled in *a, b, c* and *d*. There is only one input angle $\theta$ in the entire design, which is the angle between two elements in each inner, outer or intermediate plane. In addition, all the intersections of two elements in inner, outer and intermediate planes are at the same level.

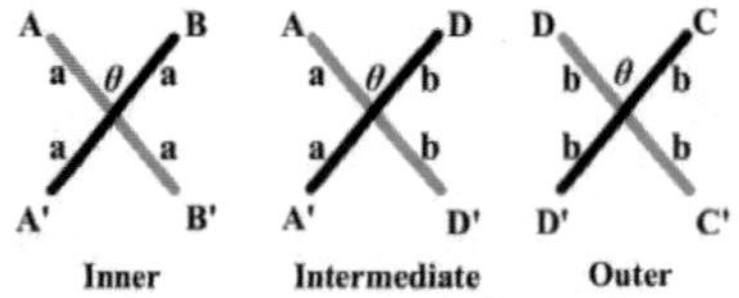


Figure 7. Inner, intermediate and outer elements design for the inner ring of the dome

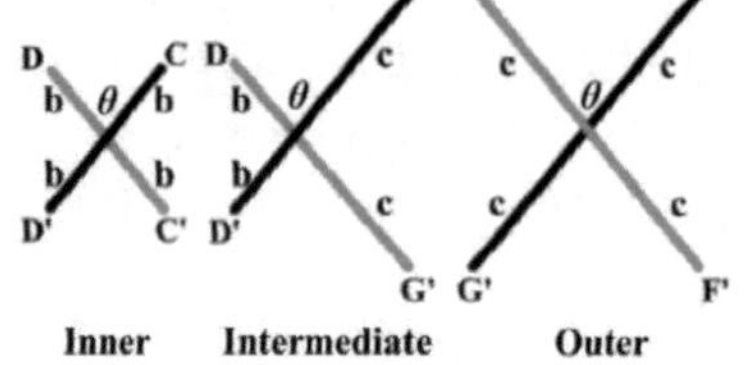


Figure 8. Inner, intermediate and outer elements design for the outer ring of the dome

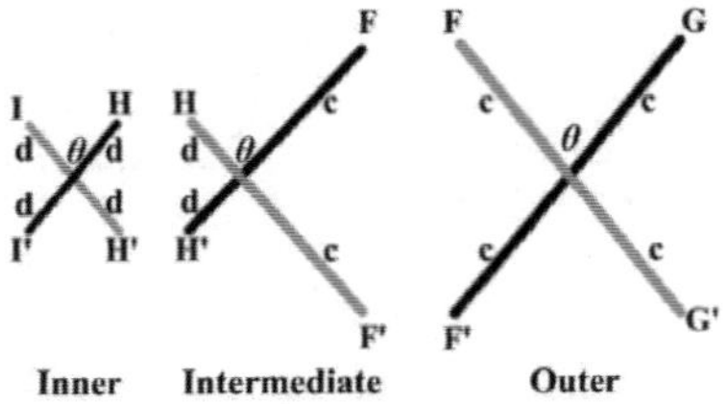


Figure 9. Inner, intermediate and outer elements design for four small rings

*2.2.Kinematic analysis method*

*1) Mobility analysis method*

The mobility (i.e. degree of freedom) is an important aspect to be analyzed for a deployable structure. The value of mobility can indicate the number of inputs required to determine a specific configuration of the structure. In order to analyze the mobility, three length ratios *b/a*, *c/b* and *c/d* need to be calculated first. These ratios can be calculated based on the geometry equations of top-view geometry, as well as the inner, intermediate and outer elements during motion.

For the inner ring of the dome, Eq.1 and Eq.2 can be obtained based on the top-view geometry in Fig. 6.

$$\frac{1}{2}DE = AD\sin\frac{\beta}{2} \tag{1}$$

$$AB = CD + 2AD\cos\gamma \ \left(\text{where } \gamma = \frac{\pi}{2} - \frac{\beta - \alpha_1}{2}\right) \tag{2}$$

According to the inner, outer and intermediate plane during motion in Fig. 7, *AB*, *CD* and *AD* in Eq.2 can be expressed as Eq.3, Eq.4 and Eq.5.

$$AB = 2a\sin\frac{\theta}{2} \tag{3}$$

$$CD = 2b\sin\frac{\theta}{2} \tag{4}$$

$$AD = (a+b)\sin\frac{\theta}{2} \tag{5}$$

Based on Eq.1 to Eq.5, the ratio *b/a* can be calculated using Eq.6.

$$\frac{b}{a} = \frac{\left(1 + \sqrt{2}\sin\frac{\alpha_1}{4}\right)^2}{2\cos^2\frac{\alpha_1}{4}} \tag{6}$$

For the outer ring of the dome, Eq.7 can be obtained based on the top-view geometry shown in Fig. 6.

$$2DG\sin\frac{\alpha}{2} = FG - CD \tag{7}$$

According to the inner, outer and intermediate plane during motion shown in Fig. 8, *CD*, *FG* and *DG* in Eq.7 can be expressed as Eq.8, Eq.9 and Eq.10.

$$CD = 2b\sin\frac{\theta}{2} \tag{8}$$

$$GF = 2c\sin\frac{\theta}{2} \tag{9}$$

$$DG = (b+c)\sin\frac{\theta}{2} \tag{10}$$

Based on Eq.7 to Eq.10, the ratio *c/b* can be computed by Eq.11.

$$\frac{c}{b} = \frac{1 + \sin\frac{\alpha}{2}}{1 - \sin\frac{\alpha}{2}} \tag{11}$$

For the four smaller rings, Eq.12 can be obtained based on the top-view geometry in Fig. 6.

$$2FH\sin\frac{\alpha_2}{2} = FG - HI \tag{12}$$

According to the inner, outer and intermediate plane during motion in Fig. 9, *HI*, *FG* and *FH* in Eq.12 can be expressed as Eq.13, Eq.14 and Eq.15.

$$HI = 2d\sin\frac{\theta}{2} \tag{13}$$

$$FG = 2c\sin\frac{\theta}{2} \tag{14}$$

$$FH = (d + c)\sin\frac{\theta}{2} \tag{15}$$

Based on the Eq.12 to Eq.15, the ratio *c/d* can be calculated by Eq.16.

$$\frac{c}{d} = \frac{1 + \sin\frac{\alpha_2}{2}}{1 - \sin\frac{\alpha_2}{2}} \tag{16}$$

The mobility of the deployable structure can be then evaluated by analyzing expressions of three length ratios.

2) *Analysis methods kinematic relationships*

Since this 3D deployable solar array design is supposed to be applied in the spacecraft and lander, it is also meaningful to analyze the kinematic relationships of the top-view area, the height and the volume related to the only input $\theta$.

The top-view area of the whole structure can be calculated using Eq.17. The top-view area of the whole structure, the dome and each ring are defined as $S$, $S_{dome}$ and $S_{ring}$, respectively.

$$\begin{aligned}
S &= S_{dome} + 4 \times S_{ring} \\
&= 12 \times \frac{1}{2} \times GF \times (l_1 + l_2 + l_3) + 4 \times 6 \times \frac{1}{2} \times FG \times l_4 \\
&= 12c\sin\frac{\theta}{2}\left(\frac{a\sin\frac{\theta}{2}}{\tan\frac{\alpha_1}{2}} + (a+b)\sin\frac{\theta}{2} + (b+c)\sin\frac{\theta}{2}\right)\cos\frac{\alpha_1}{4} \\
&\quad + 24\frac{c^2\sin^2\frac{\theta}{2}}{\tan\frac{\alpha_2}{2}} \\
&= \left[\left(\frac{12ac}{\tan\frac{\alpha_1}{2}} + 12ac + 24bc + 12c^2\right)\cos\frac{\alpha_1}{4} + \frac{24c^2}{\tan\frac{\alpha_2}{2}}\right]\sin^2\frac{\theta}{2}
\end{aligned} \tag{17}$$

where $l_1$, $l_2$, $l_3$ and $l_4$ are labelled in Fig. 6.

The height to be analyzed is the maximum height of the whole structure, which occurs at the outer planes of the dome. According to the outer elements design shown in Fig. 8, the maximum height of the deployable structure can be calculated using Eq.18. The height is defined as $H$.

$$H = 2c\cos\frac{\theta}{2} \tag{18}$$

Specifically, this study emphasizes the volume of the whole structure instead of the volume of the linkages. In other words, it can reflect the space the structure will take during the deployment. When calculating the volume, the structure is simplified as a combination of a twelve-prism (for the dome) and four hexagonal prisms (for smaller rings). The volume of the whole structure can be defined as *V* and calculated by Eq.19.

$$V = S \times H$$

$$= \left[\left(\frac{24ac^2}{\tan\frac{\alpha_1}{2}} + 24ac^2 + 48bc^2 + 24c^3\right)\cos\frac{\alpha_1}{4} + \frac{48c^3}{\tan\frac{\alpha_2}{2}}\right]\sin^2\frac{\theta}{2}\cos\frac{\theta}{2} \tag{19}$$

According to Fig. 6, $\alpha_1$ and $\alpha_2$ are design parameters, which are constants for a particular design. Therefore, these two parameters are not inputs for the structure. However, it is still meaningful to analyze the relationship between these two angles and the length ratios of elements (i.e. *b/a, c/a, d/a*) in that they can influence the shape of the structure. In this analysis part, $\alpha_1$ and $\alpha_2$ are set to change simultaneously and have the same value. Two new notations are introduced that $n_{z1} = n_{z2} = 2\pi/\alpha_1 = 2\pi/\alpha_2$. The expressions of length ratios of elements are obtained in the mobility analysis part. When $n_{z1}$ and $n_{z2}$ increase, the top-view shape of the dome and four rings are closer to circle. The larger area can be then achieved when the solar array is fully deployed and in use.

## 3. Results and Discussions

### 3.1. Configurations during deployment

Figs. 10, 11, 12, 13 and 14 show the five typical configurations during the deployment of the structure with input angles $\theta$ equal to 0°, 60°, 90°, 120° and 180° respectively. These Figs. are obtained from the same viewpoint and on the same scale. Due to the zero size assumption, all the elements are coincident and the structure becomes a line when it is fully folded (Fig. 10), while the structure becomes a plane when fully deployed (Fig. 14).

Figure 10. Solidworks model with the input θ = 0° (fully folded) (scale up three times)

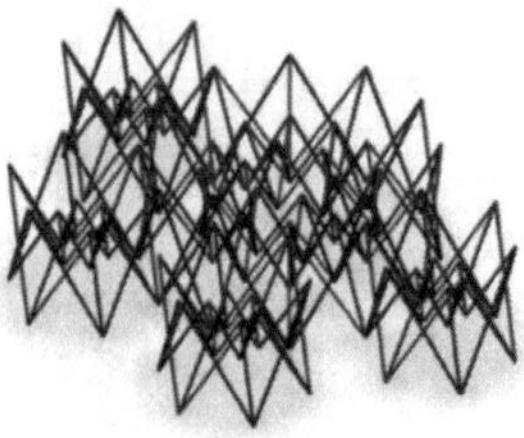

Figure 11. Solidworks model with the input θ = 60°

Figure 12. Solidworks model with the input θ = 90°

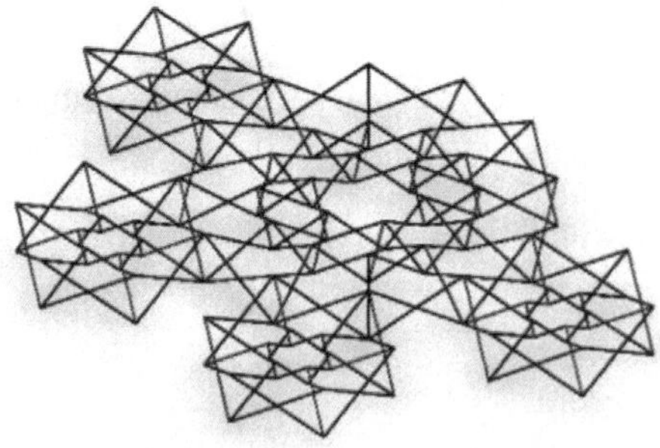

Figure 13. Solidworks model with the input θ = 120°

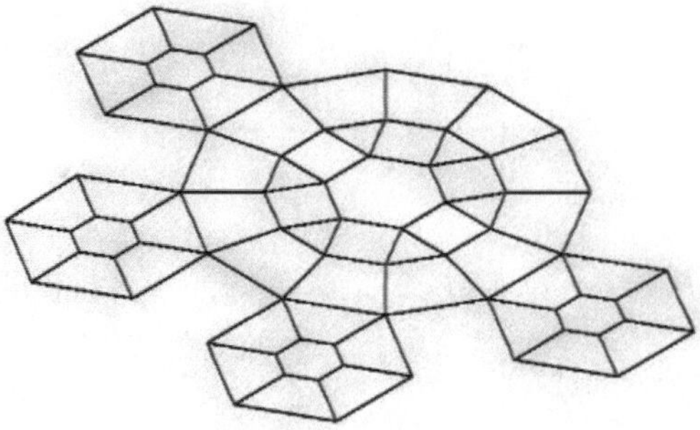

Figure 14. Solidworks model with the input θ = 180° (fully deployed)

Journal of Physics: Conference Series **2364** (2022) 012039 doi:10.1088/1742-6596/2364/1/012039

*3.2.Analysis of mobility*

According to Fig. 6, $\alpha_1 = 60°$, $\alpha = 30°$ and $\alpha_2 = 60°$ for this particular design. Substituting these values into Eq.6, 11 and 16, three length ratios are $b/a = 1$, $c/b = 1.698$ and $c/d = 3$. Here, $a$ is set to a unit length. Then, $b = a$, $c = 1.698a$ and $d = 0.566a$.

In general, $\alpha_1$, $\alpha$, $\alpha_2$, $a$, $b$, $c$ and $d$ are all design parameters, which are constants for a particular design. The only input is the angle $\theta$, which is the angle between a pair of elements in any inner, outer and intermediate planes. When the angle $\theta$ is fixed at a specific value, the corresponding structure configuration can be determined. Therefore, this 3D deployable structure design for solar arrays has one mobility.

*3.3.Kinematic relationships between top-view area, height, volume and input θ*

Since $a$, $b$, $c$, $\alpha_1$ and $\alpha_2$ in Eq.17 are all constants, the top-view area of the whole structure is proportional to $\sin^2\frac{\theta}{2}$. Fig. 15 shows the relationship between the top-view area and the input $\theta$. It is known that the input θ increases when the structure deploys. From Fig. 15, the top-view area of the structure increases with the increasing input angle $\theta$ at an increasing rate when $\theta$ ranges from 0° to 90°, while the top-view area increases with increasing input angle $\theta$ at a decreasing rate when $\theta$ ranges from 90° to 180°.

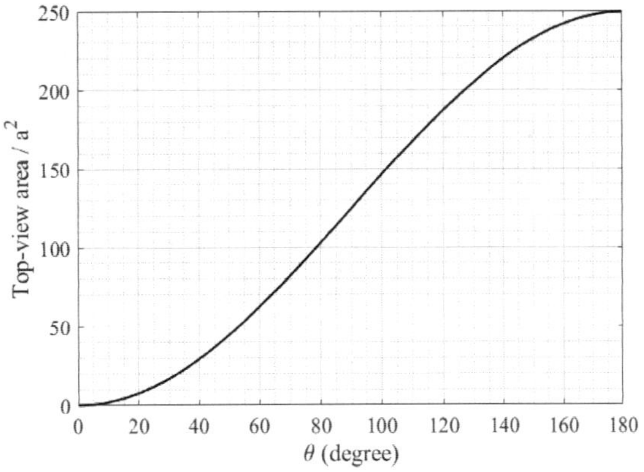


Figure 15. Kinematic relationship between the top-view area and the input angle θ

Since $c$ in Eq.18 is a constant, the maximum height of the structure is proportional to $\cos\frac{\theta}{2}$. Fig. 16 indicates the relationship between the maximum height and the input angle $\theta$. From this Fig., the maximum height of the structure decreases with the increasing input angle $\theta$ during deployment.

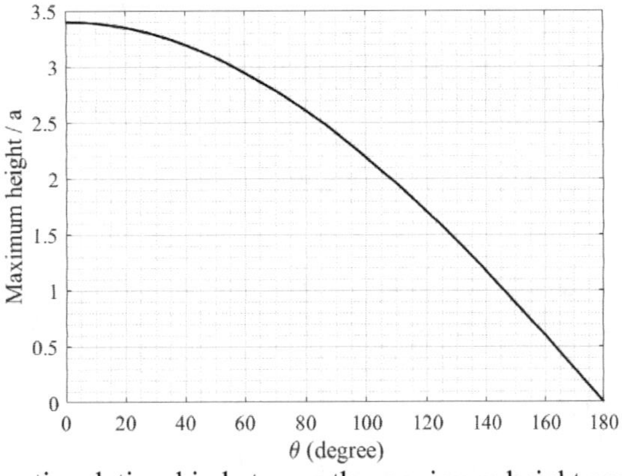


Figure 16. Kinematic relationship between the maximum height and the input angle θ

Since $a$, $b$, $c$, $\alpha_1$ and $\alpha_2$ in Eq.19 are all constants, the volume of the whole structure is proportional to $\sin^2\frac{\theta}{2}\cos\frac{\theta}{2}$. Fig. 17 reflects the relationship between the volume of the structure and the input angle $\theta$. When the input angle $\theta$ increases during deployment, the volume of the structure increases firstly and then decreases. The maximum volume is 325.653 $a^3$, which occurs at $\theta = 110°$.

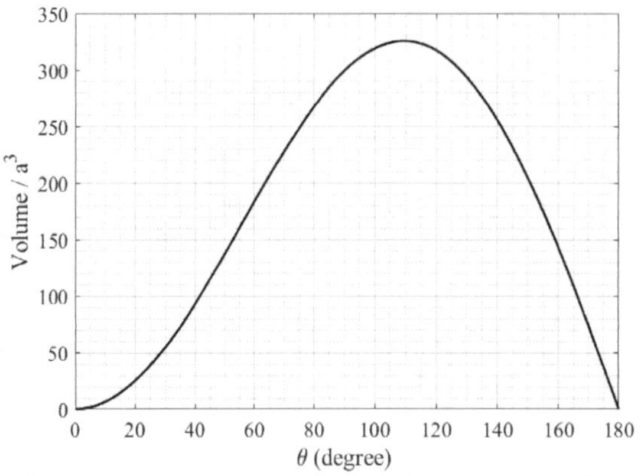


Figure 17. Kinematic relationship between the volume and the input angle θ

### *3.4. Kinematic relationship between length ratios of elements and design parameters $\alpha_1$ and $\alpha_2$*

In Fig. 18, length ratios *b/a*, *c/a* and *d/a* vs. $n_{z1}$ and $n_{z2}$ are plotted, where a is designed as a unit length as mentioned in previous sections. According to this Fig., the length ratios of elements *b/a*, *c/a* and *d/a* decrease with increasing $n_{z1}$ and $n_{z2}$, and the decreasing rates slow down. Moreover, some values cannot be chosen for $n_{z1}$ and $n_{z2}$ due to the geometry design of the dome and ring, which needs to be considered carefully when determining the design parameters.

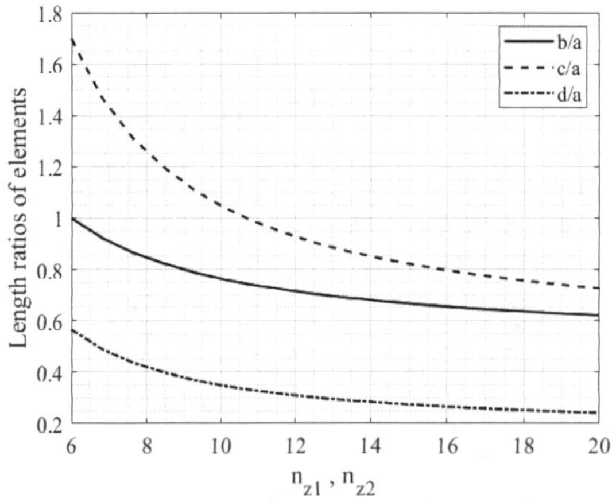


Figure 18. Kinematic relationship between the length of ratios and the shape of the top-view geometry

## 4. Conclusion

In this paper, a new design of the 3D deployable solar array for spacecraft and landers has been created. A model of this design has been constructed using Solidworks. Several configurations during the deployment have been shown. Based on the kinematic analysis, it has been found that the design has one mobility, which is the only input angle $\theta$. Some kinematic relationships have been analyzed as well. When the input angle $\theta$ increases, the top-view area increases, the maximum height decreases, and the volume of the whole structure increases firstly and then decreases. In addition, the effect of two design parameters $\alpha_1$ and $\alpha_2$ on length ratios of elements has been discussed. This design consists of a large dome and four small rings. Using this design method, the area of the deployable solar array can be further increased. Thus, more power can be provided by solar arrays, which improves the performance of the spacecraft or lander.

## References

[1] G. E. Fenci and N. G. Currie, "Deployable structures classification: a review," International Journal of Space Structures, vol. 32, no. 2, pp. 112-130, 2017.

[2] S. D. Guest, "Deployable structures: concepts and analysis," Ph.D. dissertation, University of Cambridge, Cambridge, 1994.

[3] P. A. Jones and B. R. Spence, "Spacecraft solar array technology trend," IEEE Aerospace and Electronic Systems Magazine, vol. 26, no. 8, pp. 17-28, 2011.

[4] H. S. Rauschenbach, Solar cell array design handbook: the principles and technology of photovoltaic energy conversion, New York: Springer, 1976.

[5] R. Garner, "NOAA GOES-R satellite black wing ready for flight," NASA, 24 June 2014. [Online]. Available: https://www.nasa.gov/content/goddard/noaa-goes-r-satellite-black-wing-ready-for-flight/. [Accessed: 09 May 2022].

[6] "Cost-saving method yields solar cells for exploration, gadgets," NASA, 2016. [Online]. Available: https://spinoff.nasa.gov/Spinoff2016/ee_5.html. [Accessed: 09 May 2022].

[7] T. Greicius, "InSight Mars Lander," NASA, 13 December 2019. [Online]. Available: https://www.nasa.gov/mission_pages/insight/main/index.html. [Accessed: 09 May 2022].

[8] R. Garner, "NASA's Lucy stretches its wings in successful solar panel deployment test," NASA, 07 April 2021. [Online]. Available: https://www.nasa.gov/feature/goddard/2021/nasa-lucy-stretches-its-wings-in-successful-solar-panel-deployment-test. [Accessed: 09 May 2022].

[9] P. Wang-Iverson, R. J. Lang and M. YIM, Origami 5: Fifth international meeting of origami science, mathematics, and education, CRC Press, 2016.

[10] "Origami solves a space problem," Kurzweil, 3 December 2013. [Online]. Available: https://www.kurzweilai.net/origami-solves-a-space-problem#!prettyPhoto. [Accessed: 07 May 2022].

[11] Y. Nishiyama, "Miura folding: applying origami to space exploration," International Journal of Pure and Applied Mathematics, vol. 79, no. 2, pp. 269-279, 2012.

[12] E. Landau, "Solar power, origami-style," NASA, 07 August 2017. [Online]. Available: https://www.nasa.gov/jpl/news/origami-style-solar-power-20140814. [Accessed: 07 May 2022].

[13] J. W. Jeong, Y. I. Yoo, J. J. Lee, J. H. Lim and K. W. Kim, "Development of a tape spring hinge with a SMA latch for a satellite solar array deployment using the independence axiom," IERI Procedia, vol. 1, pp. 225-231, 2012.

[14] L. Santo, F. Quadrini, A. Accettura and W. Villadei, "Shape memory composites for self-deployable structures in aerospace applications," Procedia Engineering, vol. 88, pp. 42-47, 2014.

[15] Z. You and Y. Chen, Motion structures: deployable structural assemblies of mechanisms, London: CRC Press, 2014.

# Design of a Foldable Solar Panel with Radial and Circumferential Folding Motion

**Jiaxuan He[1,a†], Zeqi He[2,b†], Cexiao Hu[3,*,†]**

[1]Institute of International Education, Hunan Institute of Engineering, XiangTan, China

[2]School of Mechanical and Electrical Engineering and Automation, Harbin Institute of Technology Shenzhen, China

[3]School of Aerospace Engineering, University of Bristol, Bristol, UK

[a]e-mail: 2317257194@qq.com; [b]e-mail: 190310204@stu.hit.edu.cn

* Corresponding author e-mail: pw20122@bristol.ac.uk

[†]These authors contributed equally to this work

**Abstract**—This paper focuses on designing a foldable solar panel that can be folded both circumferentially and radially simultaneously. Most of the existing foldable solar panels have only one movement mode when folded. In contrast, the solar panel designed in this paper has two different movement modes when folded, which has the advantage of further reducing the area of the solar panel when it is fully folded. In this paper, the solar panel can achieve circumferential motion based on the motion principle of the folding fan, and the solar panel can achieve radial motion based on the principle of the slider mechanism. Then the two separate motions are unified by improving the scissors-like element structure. In addition, this paper adopts SolidWorks modeling, CAXA drawing, and calculation to prove the normal operation of the structure. Finally, the control variable method is used to analyze the structural skeleton's radial and circumferential folding capacity. The critical parameter $\alpha_2$ (the central angle when the structure is fully unfolded) is obtained, affecting the mechanism's contracting capacity. When $\alpha 2$ increases, the radial contraction ability of the structure first increases and then decreases, and the circumferential folding ability of the structure continues to increase. At the same time, the radial and circumferential shrinkage capabilities of the structure and the layout of the solar panel are comprehensively considered, and the value of $\alpha 2$ is obtained when the folding ratio is as small as possible.

## 1. Introduction

With the development and combination of solar panel technology and the aerospace field, many functional branches of scientific research have been born. Among them, the foldable solar panel, as the energy supply system on the spacecraft, plays a vital role in developing the spacecraft. The energy supply of spacecraft is mainly divided into nuclear power generation and solar power generation. When solar energy can provide more power, it can provide more energy for spacecraft on long-duration or long-distance missions, enabling a more stable working environment, such as those on Mars or more distant planets. In the future, setting up a solar power generation system in space can be used as an alternative energy source for the earth. For spacecraft in extraterrestrial orbit, solar power generation can significantly reduce the quality of chemical catalysts carried by spacecraft; thereby, it has a very favorable effect on the life span and working environment of the spacecraft [1-3]. For the foldable solar panel structure on the spacecraft, the main structure is a suspended structure; one end of the solar panel is connected to the spacecraft, while the other end can be freely extended [4]. However, the spacecraft's space is limited, and all facilities need to be folded into the spacecraft's fairing before launch. The fairing and solar folding panels are opened only when launched to the designated location. Then, the solar panels usually need to start working. [5]. Therefore, under the condition that the solar folding panel does not affect other facilities of the spacecraft and its normal operation, having a more significant folding ratio will provide a more stable and durable working environment for the entire spacecraft.

Nuclear power generation is also one of the essential choices for the spacecraft energy supply system. The advantage of using nuclear power to generate electricity is that it can provide enough energy for very long-term tasks, and it can supply energy very stably during use. At the same time, nuclear energy is a country's strategic energy. The main advantages of metal sealing reflect small size, high efficiency, and long life. Therefore, nuclear energy is very attractive for people to carry out strategic tasks such as exploring survivable planets [6]. However, the disadvantages of nuclear energy are also obvious. The nuclear energy reaction is irreversible at the current level of human technology. Moreover, it cannot be stopped once the reaction starts while the reaction is in progress. This is very restrictive for the maintenance of the spacecraft. At the same time, because the nuclear power generation system is fixed on the spacecraft and the shape is not easy to change, it may lead to the overall design of the spacecraft not matching the atomic power generation system, and the expected effect cannot be achieved. Until now, most spacecraft did not need to perform a specific strategic mission, and as more and more civilian spacecraft are launched. Using the solar power generation system as the power source of the spacecraft gradually shows its advantages. The solar power generation system has the advantages of low cost, good maintainability, and adjustable solar panel structure to suit the spacecraft. Therefore, compared with nuclear power generation systems, solar power generation is more in line with people's needs on most spacecraft.

Compared with the high power of nuclear power generation, the low power of solar power generation is one of the main weaknesses of solar power generation. One of the key metrics for increasing the power of a folded solar panel is the folding rate. Under the same storage volume, a larger folded solar panel can unfold a larger effective area and provide more energy. At present, the folding forms of solar folding panels can be divided into two states: flat-folded and superimposed (as shown in Figure 1[7] and Figure 2[8]), respectively. Flat folding can be divided into axial and parallel folding, and overlapping forms can be divided into telescopic and coiling. Considering the lifetime and fold times of the unique solar panel material, the folded conditions are usually overlapping, and the axial and parallel folded forms are mainly used [9]. Compared with the existing solar folding panels on the market, there are two kinds of space-expandable skeleton design and rigid folding through origami technology. At the same time, more attention is now being paid to the emerging technology of rigidly folded origami to design foldable solar panels, such as solar panels designed using the Miura folding method [10] (as shown in Figure 3[11]). Although, origami techniques can maximize the folded material. However, this requires very high performance in the solar panel itself, and often requires materials with good design flexibility. On the contrary, creating the space deployment skeleton has relatively low-performance requirements on the solar panel itself. And through the structure's plan, the effect of increasing the expansion ratio can also be achieved.

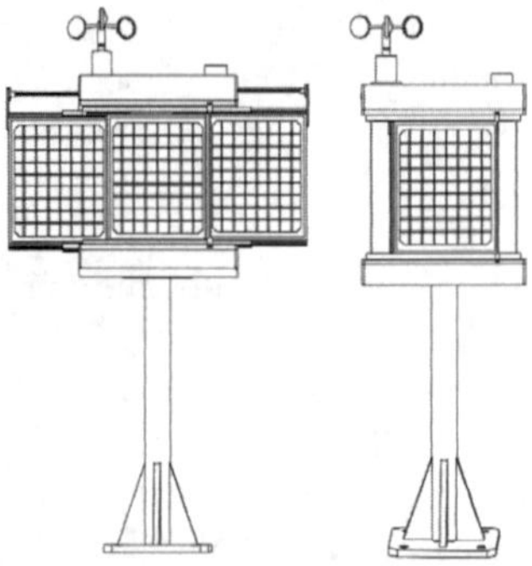

Figure 1. The schematic diagram of the "superimposed" form model in the folded form is selected from [7]

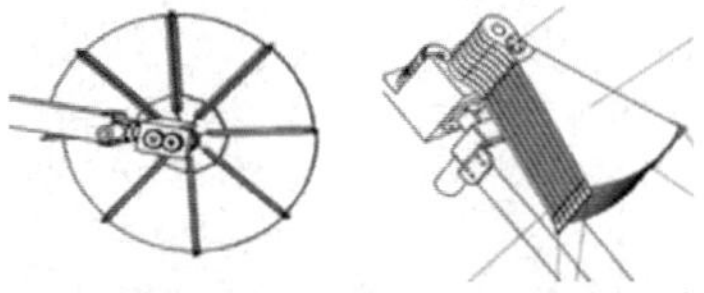

Figure 2. The schematic diagram of the "flat-fold" form model in the folded form is selected from [8]

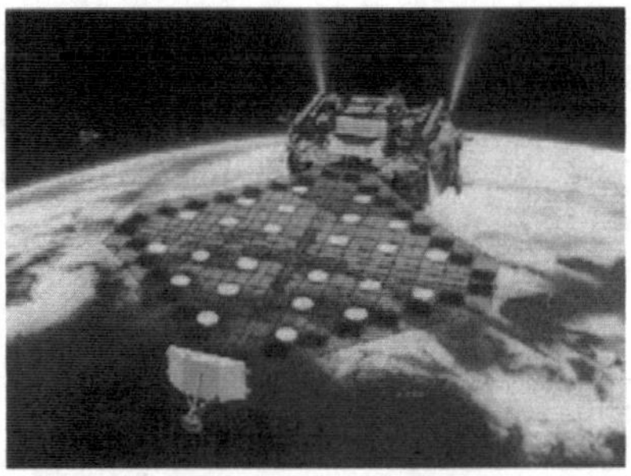

Figure 3. Miura's foldable solar panel is selected from [11]

The "scissor" structure is known as one of the joint structures that can be unfolded or closed. Due to the geometric distribution of the scissor structure, the scissor structure can be extended from the motion of a single structural unit to the complex simultaneous movement of multiple structural units [12-13]. (As shown in Figure 4), it is one of the most basic structures in the scissor structure, and the application of this structure can be found in many practical designs. For example, the scissor structure is used to create the eXtendable Solar Array System (XSAS) [14]. The scissor-like structure is a type of structure formed by extraordinary changes in the scissor structure. While inheriting the advantages of the basic scissors structure, the scissors-like structure can also absorb the advantages of other structures, such as increasing the folding ratio of the structure. Therefore, this paper proposes a structural design method for the skeleton of a solar foldable windsurfing panel based on a scissor-like structure. The structure provides a "motion source," and the structure can be contracted in both the circumferential and radial directions simultaneously. The purpose of this is to maximize the expansion rate of the structure, thereby increasing the area where the solar panels are loaded and the efficiency of solar power generation from the side.

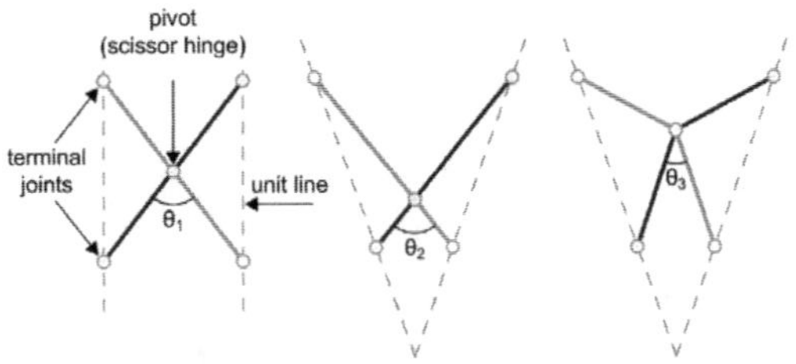


Figure 4. The basic structural unit of the scissor structure is selected from [14]

This article focuses on designing a new foldable solar panel design that can be used in aerospace and discusses the optimal folding ratio for the structure. Section II introduces the design method of the model, including the formal design idea of the structural skeleton and the composition of the two basic parts of the solar panel shape design, and discusses the preconditions for the stable movement of the structure through kinematics analysis. The radial and circumferential directions of the structure. Section III focuses on the geometrical and kinematic analysis to calculate the influence of the opening and closing angle of the model on the expansion ratio of the element in the circumferential and radial directions and discusses the influence of the parameter changes of the structural components on the structural motion. The conclusion is presented in Section 4.

## 2. Design method

### *2.1Design of circumferential folding of the structure*

The foldable solar panels studied in this paper can be folded in the circumferential and radial directions. The principle of circumferential folding of this solar panel is similar to the unfolding principle of traditional folding fans. The principle of traditional folding fan expansion is that the fan bones take the fan nails as the central axis, and each fan bone rotates at a certain Angle around the fan nails. When it is expanded, the fan bones are arranged in an equidistant order to form a fan [15]. The realization of the circumferential motion of this structure is to replace the fan bone with a jointed arm and replace the fan nail with a rotating shaft. The largest central angle between the two jointed arms (when the structure is fully unfolded) is denoted by $\alpha_1$, and the smallest central angle between the two jointed arms (when the structure is fully folded) is denoted by $\alpha_2$. The joint arm's length (80cm) and the Angle design of $\alpha_1$ and $\alpha_2$ can be determined according to needs. The design concept of the structure is to modify and adjust based on the skeleton structure of the folding fan so that the solar panel can be folded in the circumferential direction while shrinking radially.

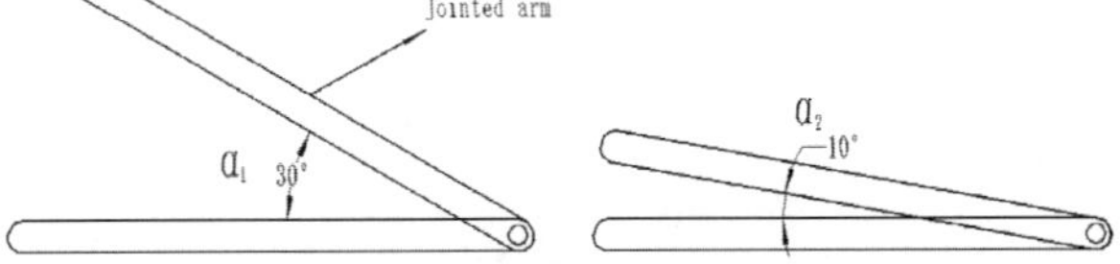


Figure 5. Schematic diagram of the circumferential folding of the structure

### *2.2.Design of radial shrinkage of structure*

In order to radial shrink the solar panel, a slider mechanism can be used to achieve this. Figure 6 and figure 7 show a schematic diagram of the structure of the solar panel shrinking in the radial direction. The blue outline in Figures 6 and 7 is the solar panel. First, a slot with a certain length is opened in the joint arm so that a circular slide block can move in the slot. Next, tie a slide bar under the solar panel, and open a slot with a smaller

length on the joint at the left end of the joint arm. Finally, the slide block is connected with the slide bar to contract the solar panel radially.

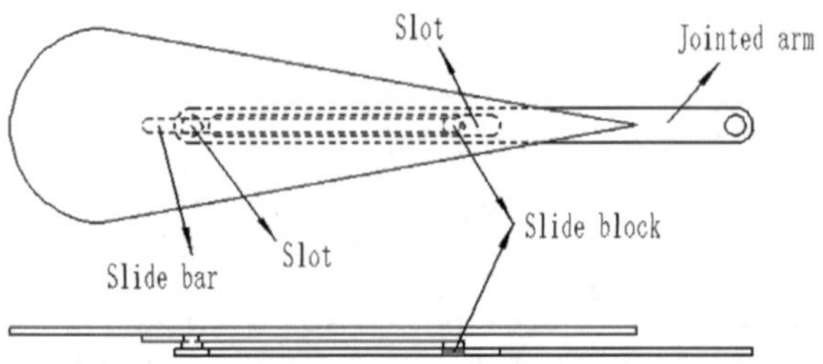


Figure 6. Schematic diagram of the radial shrinkage of the solar panel (before stretching)

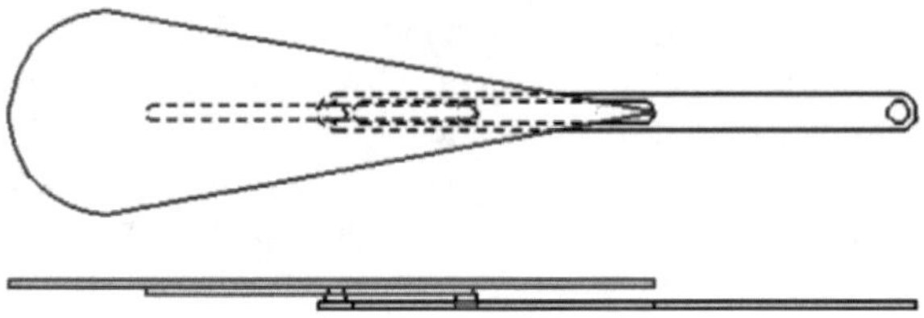

Figure 7. Schematic diagram of the radial shrinkage of the solar panel (after stretching)

*2.3.Structural design that associates circumferential folds and radial contractions*

*1) The design of connecting part*

So far, solar panels can shrink both circumferentially and radially, but the two movements are independent. Next, a mechanism is needed to associate the two movements, and this function can be achieved by incorporating a scissor-like element into this structure. The so-called scissor-like element structure refers to a stretchable and foldable structure that a plurality of scissor-like elements can form through hinged endpoints. Because each rod length and hinge position are different, different forms of scissor-like element structures can be formed [16]. The image below shows the arched structure:

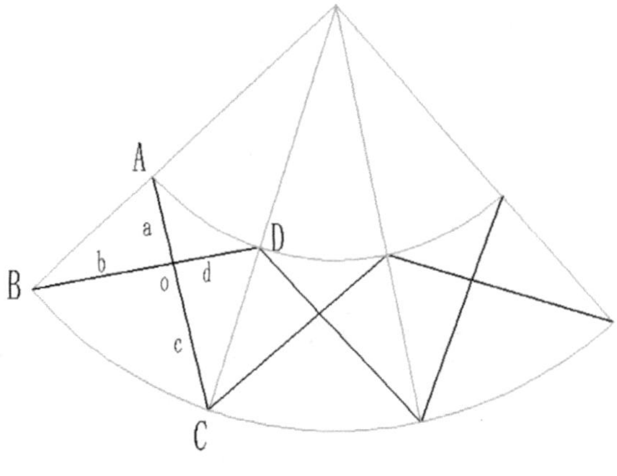


Figure 8. Arched structure, a=c, b=d and a=b [16]

The mechanism designed in this paper is based on improving the arch structure. For scissor-like elements, the two rods' point of intersection (O) is hinged. However, for the structure designed in this paper, the intersection of the two rods is not hinged, and the two ends of one rod are respectively connected with the endpoint of one jointed arm and the slider on the other adjacent jointed arm. The rod is located above the two joint arms. The other rod is connected in the same way as the previous rod, and the only difference is that the rod is located below the two joint arms. The purpose of this design can effectively improve the size of the solar panel, and the specific reason will be explained in the solar panel size design section. The rod connecting the two joint arms will be referred to as connecting rod.

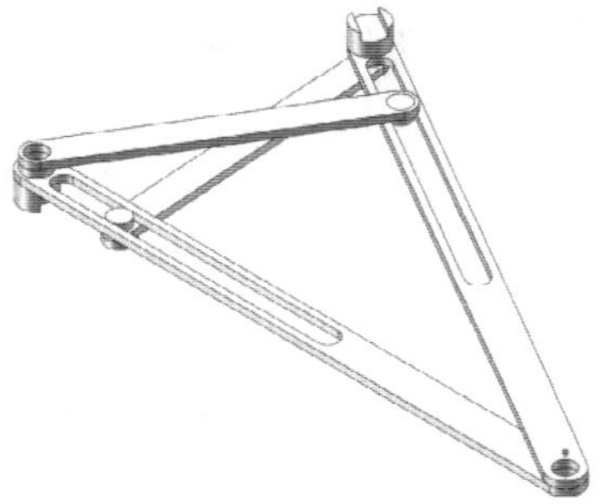

Figure 9. Connecting rod structure

*2) Analysis of the conditions required for the smooth operation of the structural skeleton*

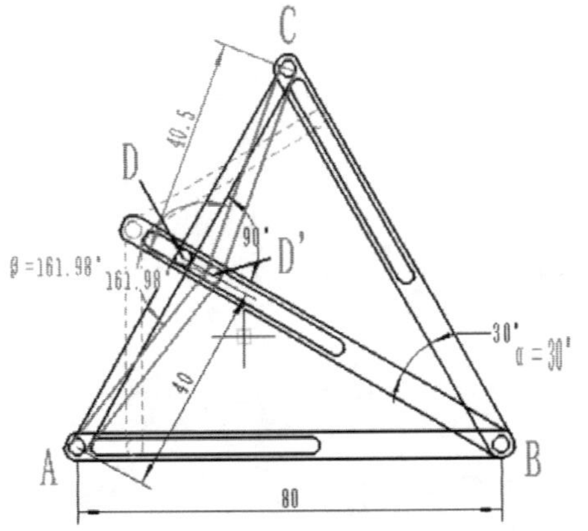


Figure 10. Schematic diagram of the structural skeleton

Since the values of the jointed arm and $\alpha$ are defined, the first thing to consider is the length of the connecting bar. In ΔBCD, the shortest rod length of the connecting bar (CD) can be easily determined by the equation (1)

$$CD = BC\sin\alpha_1 \tag{1}$$

The structure can move when the connecting bar meets the minimum bar length, but the movement will be stuck or disconnected. When there are three or more jointed arms in the structure, since the connecting bar is the minimum rod length, there will be one or more equilateral triangle structures (the triangle shown in figure 10 is ΔABC). The triangular structure is stable, which creates significant resistance when the entire structure contracts. Therefore, when setting the connecting bar, its length should be greater than the shortest bar length. In other words, ∠ADC should be less than 180°. The angle is defined as β. It can be seen that β decreases from 180° to 161.98° as the length of the connecting rod in FIG. 10 increases from 40 cm to 40.5 cm. Therefore, the formula (2) and (3) that satisfies the normal working of the structural skeleton is

$$CD > BC\sin\alpha_1 \tag{2}$$

Or

$$\beta < 180°\left(\beta = \pi - 2\arccos\frac{CD}{CD'}\right) \tag{3}$$

*2.4.Solar panel design*

*1) Determination of the installation location of solar panels*

For a joint arm, two solar panels need to be installed above it. A slide bar is fastened below the first solar panel and fixed on the slide block, and the second panel is connected with the first panel through a cylindrical hinge. Considering that the first solar panel needs to carry the weight of the second solar panel when the structure is fully unfolded, the first solar panel is made of monocrystalline silicon with higher hardness, and the second panel is made of lighter CIGS film solar cells as materials.

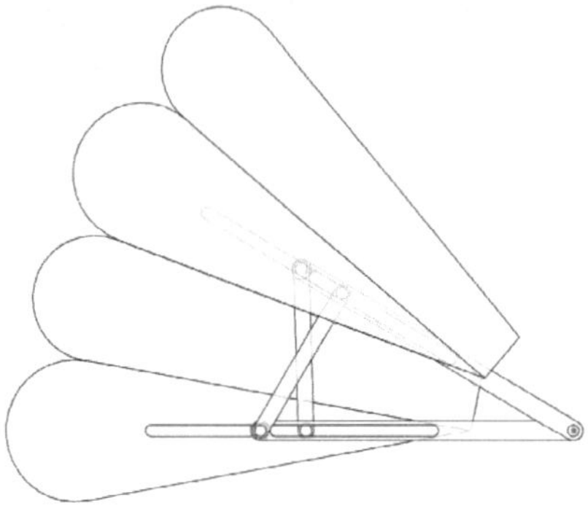

Figure 11. Solar panel installation diagram

*2) Dimensional design and area calculation of solar panels*

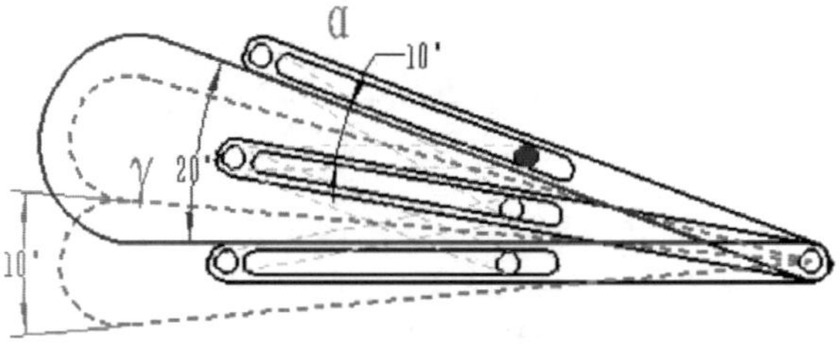


Figure 12. Schematic diagram of the relevant parameters of the first solar panel

After ensuring that the structural skeleton can function normally, the next step is to determine the conditions the solar panels need to meet for the entire structure to function correctly. When the structure is fully folded, the maximum central angle of the first solar panel is defined as $\gamma$. The calculation of $\gamma$ can be obtained by Equation (4)

$$\gamma = 2\alpha_2 \tag{4}$$

If the angle is greater than the value of $\gamma$, the layout of the solar panel will collide with the slider at the same level. As can see in figure 12, the blue slider is in the same plane as the solar panel. The side length of the solar panel is designed according to the size of the structural frame, and the side length of the solar panel in this paper is 80 cm. As mentioned above, placing two connecting bars above and below the two jointed arms can optimize the size of the solar panel. If two connecting bars are placed on top of two joined arms, in order to prevent collision between parts, the maximum central angle $\gamma$ of the solar panel will be reduced by one-half ($\gamma=\alpha_2$), as shown by the red dotted outline in figure 12. In this way, the area of the first solar panel will be greatly reduced, which will affect the installation of the second solar panel.

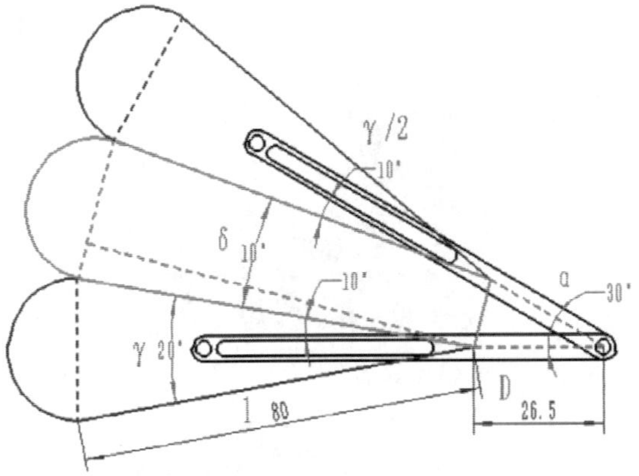


Figure 13. Parameters related to solar panel area calculation

The size of the second solar panel is not strictly limited. When the structure is fully unfolded, the second solar panel only needs to fill the vacant area between the first solar panel.

In order to be able to calculate the ratio between the fully folded and fully unfolded footprint of the entire mechanism, the area of the solar panel needs to be calculated. In order to simplify the area calculation of the solar panel, the area calculation of the blue solar panel is simplified to the area calculation of the triangle. The known side length is 80 cm (denoted by l), the maximum central angle γ is 20°, and the distance from the end of the first solar panel to the axis of rotation is 26.5 cm. This distance is also the travel of the slider (denoted by D), then the area formula (5) is ($cm^2$):

$$S_{triangle} = l^2 \sin\frac{\gamma}{2}\cos\frac{\gamma}{2} \tag{5}$$

Simplify the area calculation of the purple solar panel to the area calculation of the trapezoid. It is necessary to determine the angle between the two waists (defined as δ). The calculation of δ can be obtained by Equation (6)

$$\delta = \alpha_1 - \gamma \tag{6}$$

Calculation formula (7) of Trapezoid top：

$$L_{top} = 2D\sin\frac{\alpha_1}{2} \tag{7}$$

Calculation formula (8) of Trapezoid bottom:

$$L_{bottom} = 2l\sin\frac{\delta}{2} + 2D\sin\frac{\alpha_1}{2} \tag{8}$$

Calculation formula (9) of Trapezoid high:

$$H = l\cos\frac{\delta}{2} \tag{9}$$

Calculation formula (10) of Trapezoid area:

$$S_{trapezoid} = \frac{l\cos\frac{\delta}{2}\left(2l\sin\frac{\delta}{2} + 4D\sin\frac{\alpha_1}{2}\right)}{2} \tag{10}$$

*2.5. Movement of the structure*

The model shown below is only a partial model of the structure. The unfolding and folding of this model are done in two steps. The first step is for the blue solar panel to carry the purple solar panel for circumferential and radial expansion.

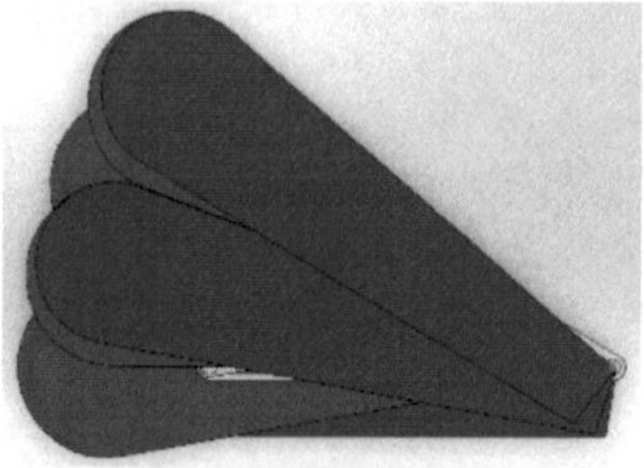

(a) The condition when the total center angle is 30 degrees

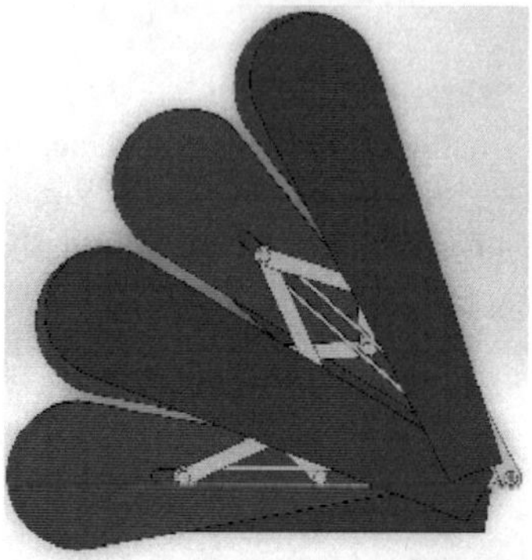

(b) The condition when the total center angle is 60 degrees

(c) The condition when the total center angle is 90 degrees

Figure 14. The first step to expand

The second step is that all purple solar panels based on blue panels simultaneously perform a 180° circumferential unfolding motion.

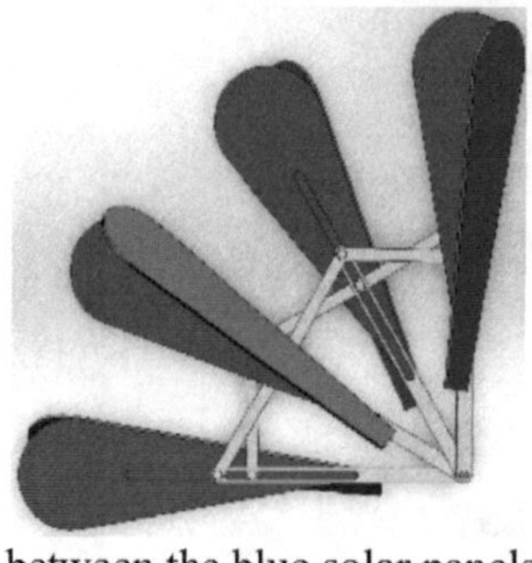

(a) The condition when the angles between the blue solar panels and the purple solar panels are 60 degrees

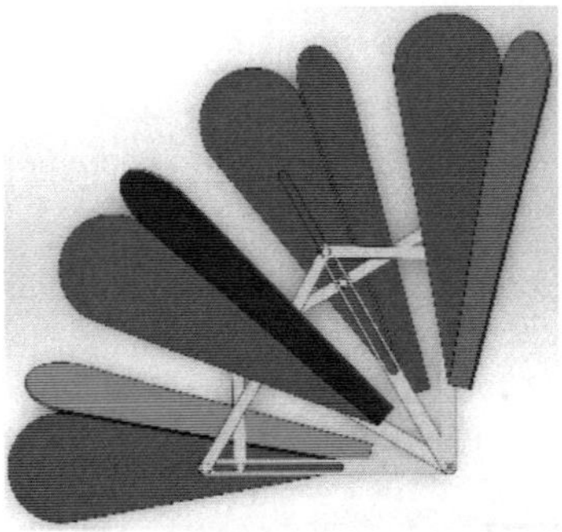

(b) The condition when the angles between the blue solar panels and the purple solar panels are 120 degrees

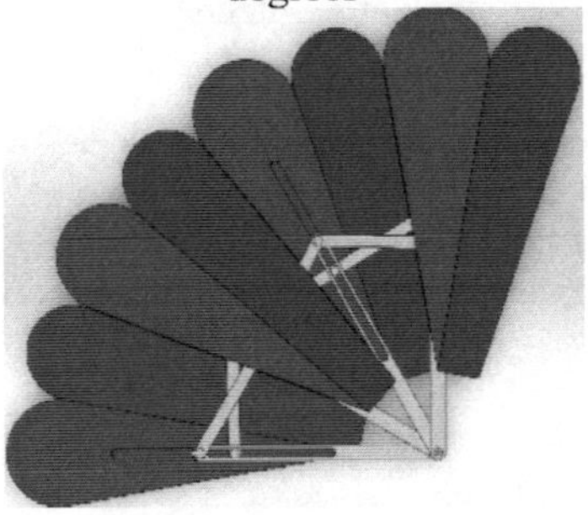

(c) The condition when the angles between the blue solar panels and the purple solar panels are 180 degrees

Figure 15. The second step is to expand.

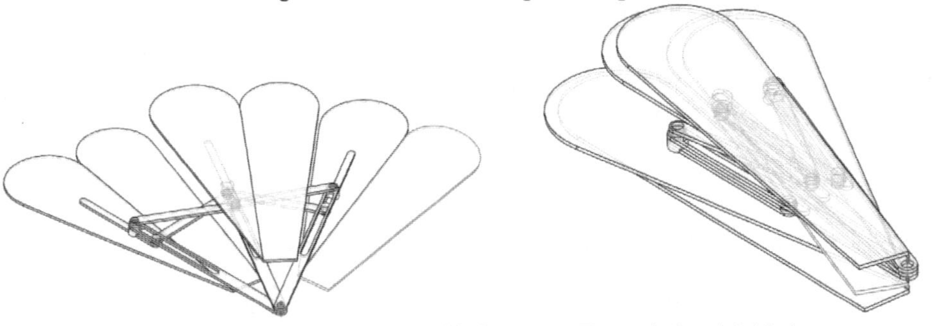

Figure 16. Part Perspective with Structure Expanded and folded

The structural components are stacked on top of each other, and the installation method and normal operating conditions of this structure have been described in detail above. Therefore, in an ideal state, the components of the entire structure do not collide during operation, and the structure can be smoothly folded and unfolded.

### *2.6.Analysis method for folding ratio when structure unfolds and folds*

The folding ratio of a structure is the ratio between the footprint of the structure when it is fully folded and the structure's footprint when it is fully unfolded. Taking into account the mechanism's circumferential and radial shrinkage capabilities will directly affect the folding ratio. In order to obtain the minimum folding rate of the mechanism, this paper will analyze the parameters affecting the circumferential and radial shrinkage capacity of the mechanism. Furthermore, the control variable method is used to analyze the parameters and the most favorable parameter value for the contractility of the mechanism is obtained. Finally, calculate the floor area when the mechanism is fully unfolded and folded, and then obtain the minimum folding ratio.

## 3. Result and discussion

*3.1.Parametric Analysis of Structural Skeletons*

*1) Influence of connecting bar on radial shrinkage capacity*

The stroke of the slider and the center Angle of the mechanism plays a decisive role in the mechanism's area reduction rate when it is fully folded. The stroke of the slider reflects the radial shrinkage capacity of the mechanism. The larger the stroke, the stronger the radial contraction capacity of the mechanism, and vice versa. There are three parameters affecting the stroke of the slider, the length of connecting bar, and the size of the center angle when the mechanism is fully folded and fully expanded. The values of the two parameters are set to constant, and the value of one parameter is changed to explore the variation of the slider stroke.

First, when the center Angle of the mechanism is unchanged (10° when fully folded; 30° when fully expanded), test the influence of connecting bar length changes on the stroke of the slider. By the cosine law equation (11):

$$CD^2 = CB^2 + BD^2 - 2 \cdot CB \cdot BD \cdot \cos\alpha_1 \tag{11}$$

When the value of CD is changed, the value of BD when α is equal to 30° and 10° is decomposed and calculated. The stroke of the slider can be obtained by subtracting the value of BD (at 30°) and BD (at 10°).

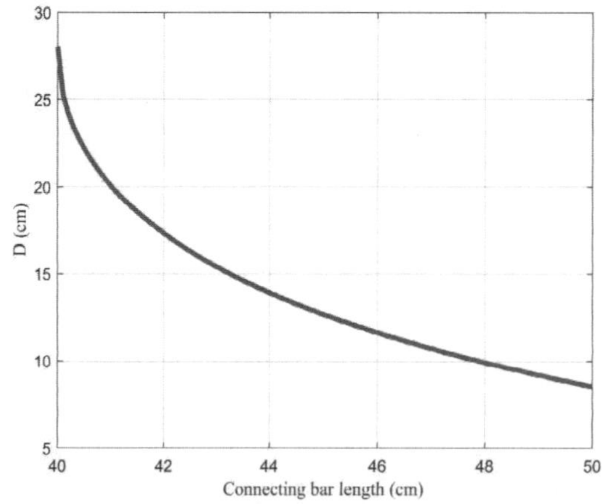


Figure 17. The stroke of the slider changes with the connecting bar length

It can be seen from Figure 17 that when the length of the connecting rod increases, the stroke of the slider is shortened, which harms the area reduction rate of the mechanism after folding. Because the shorter the stroke of the slider, the means that the radial shrinkage of the mechanism is worse. Selecting a short connecting bar can enhance the contraction ability of the mechanism, but the force carried by the mechanism will be larger. The parameter to evaluate the force on the mechanism is β, and the greater the value of β, the greater the force on the mechanism.

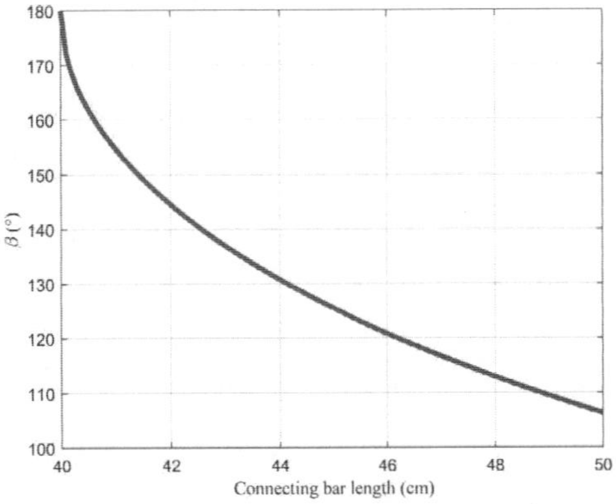


Figure 18. The β changes with the connecting bar length

When the β angle is larger, the connecting bar will produce a plastic deformation or fracture to prevent excessive force. The skeleton of the mechanism can be made of materials with better strength and stiffness, which ensures that the stroke of the slider will not have too much loss and ensures that the strength of the structure is enough to withstand the force brought by the motor. In general, it is appropriate to take the minimum length of the connecting bar plus 0.5 cm to 1cm as the actual length of the connecting bar under the condition of ensuring the normal operation of the structure and sufficient strength of the components in the structural skeleton.

*2) Influence of α1 on radial shrinkage capacity*

Secondly, when the length of the connecting bar is the minimum value, set the center Angle (when the mechanism is fully folded) to 10°, change the size of the center Angle (when the mechanism is fully expanded), and test the influence of the change of the center Angle on the slider stroke. According to the law of cosines, Figure 19 shows how the stroke varies with the $\alpha_1$.

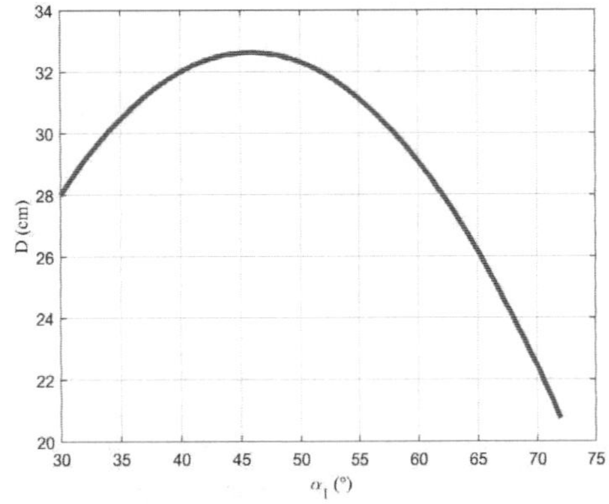


Figure 19. The stroke of the slider varies with the center Angle(when the mechanism is fully unfolded)

As can be seen from figure 19, when the center Angle is from 30° to 72°, the stroke of the slider increases first and then decreases. When the center Angle is 46°, the maximum stroke of the slider is 32.62 cm. Since the total angle of the mechanism when fully expanded is 360°, it is not appropriate to select a 46° center Angle. Suitable center Angle degrees are 30°, 36°,40°, 45°, 60°, and 72°. The 45° central Angle corresponds to the slider with the greatest travel between the five angles. Therefore, 45° is used as the center Angle when the mechanism is fully expanded.

*3) Influence of α2 on radial shrinkage capacity*

Lastly, when the length of the connecting bar is the minimum value, set the center Angle (when the mechanism is fully expanded) to 45°, change the size of the center Angle (when the mechanism is fully folded), and test the influence of the change of the center Angle on the slider stroke. According to the law of cosines, figure 20 shows how the stroke varies with the $\alpha_2$.

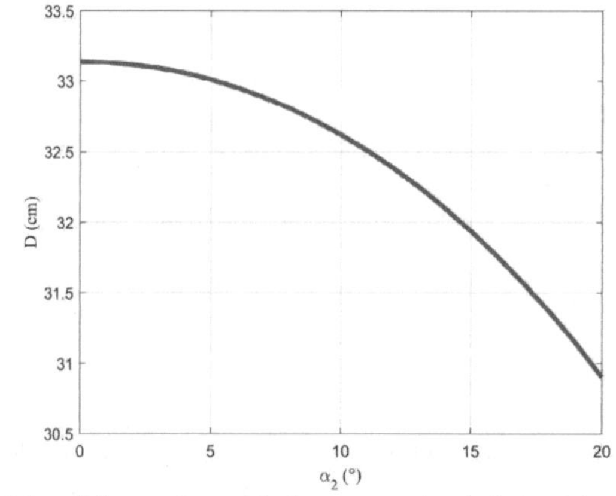


Figure 20. The stroke of the slider varies with the center Angle(when the mechanism is fully folded)

As can be seen from figure 20, when the center Angle increases from 0° to 10°, the stroke of the slider is not significantly reduced, from 33.13 to 32.62. When the center Angle increases from 10° to 20°, the stroke of the slider decreases significantly, from 32.62 to 30.90. However, in general, the reduction of stroke during the whole process does not influence the increase of area shrinkage rate after the mechanism is fully folded. In theory, the smaller the center Angle is, the stronger the radial shrinkage capability of the mechanism is. However, considering the fit and thickness between the bars and the reasonable size of solar panels, it is more appropriate to take 10° as the center Angle when the mechanism is fully folded.

*4) Influence of α1 and α2 on the circumferential shrinkage capacity*

The parameters that directly affect the circumferential folding ability of the structure are $\alpha_1$ and $\alpha_2$. It is more appropriate to take $\alpha_2$ as 10° (the reason has been explained above). The circumferential folding ability of the structure is analyzed by changing the size of α2. Regardless of the values of $\alpha_1$ and $\alpha_2$, the total central angle when the structure is fully unfolded is always 360°. It is now only necessary to obtain the total central angle when the structure is fully folded to determine the circumferential shrinkage capacity of the structure. When the structure is fully shrunk, the smaller the total central angle, the stronger the circumferential shrinkage of the structure, and vice versa. The values of $\alpha_1$ are 30°, 36°, 40°, 45°, 60°, and 72°, which are taken to form an open circle when the structure is fully unfolded.

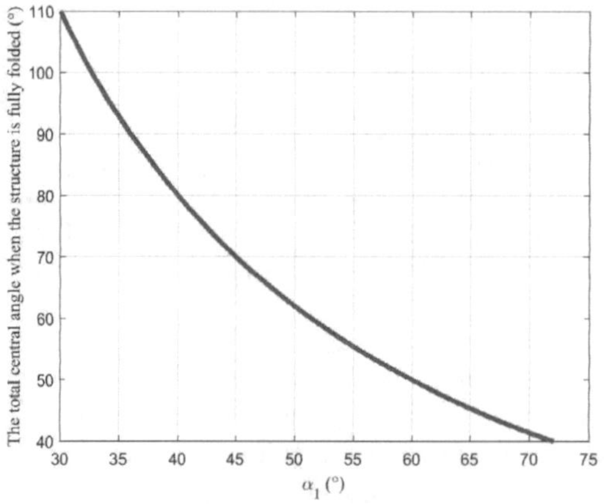


Figure 21. The total central angle when the structure is fully folded varies with α1

It can be seen from figure 21 that with the increase of $\alpha_1$, the total central angle of the structure becomes smaller when the structure is fully folded, and the circumferential shrinkage ability of the structure becomes stronger.

*3.2.Explore the folding ratio of structures*

It can be concluded from the previous results that when $\alpha_1$ is 45°, the radial shrinkage ability of the structure is the strongest; when $\alpha_1$ is 72°, the radial shrinkage ability of the structure is the worst. However, when $\alpha_1$ is 72°, the circumferential folding ability of the structure is the strongest; when $\alpha_1$ is 45°, the circumferential shrinking ability of the structure is average. It can be seen that for $\alpha_1$, the determination of the value of $\alpha_1$ is difficult to benefit both the radial shrinkage and the circumferential folding ability of the structure. The same is true for other parameters. Most of them cannot have a favorable effect on the structure simultaneously. The relationship between them is mutually restrictive, such as the relationship between the length of the connecting bar and β mentioned in the previous results. The structure's radial shrinkage and circumferential folding ability were discussed separately in the past. Now, to make the folding ratio of the structure as small as possible and the structure as reasonable as possible, it is necessary to combine the radial shrinkage ability and the circumferential folding ability of the structure and analyze them together.

The specific operation is to calculate the folding ratio of the structure using the parameters derived from the previous results that are beneficial to the radial shrinkage and circumferential capability of the structure. The area calculation method uses CAXA to draw a plane sketch when the structure is fully unfolded and folded, and CAXA can be used to calculate the area of the outline of the sketch. For the drawing of the sketch, the area of the first solar panel and $\alpha_2$ is a fixed value; the area of the second solar panel, the stroke of the slider, and $\alpha_1$ are variables. The parameters for calculating the solar panel area can be determined according to the structure, and a simplified algorithm is used to calculate the area. The calculation formula is given in the method section. $\alpha_2$ is set as 10°, and $\alpha_1$ is taken as 30°, 36°, 40°, 45°, 60°, and 72°. The stroke of the slider is obtained by $\alpha_1$, $\alpha_2$, and connecting bar length (the shortest). Figure 22 is a sketch of the structure when it is fully unfolded and folded for different $\alpha_1$ conditions.

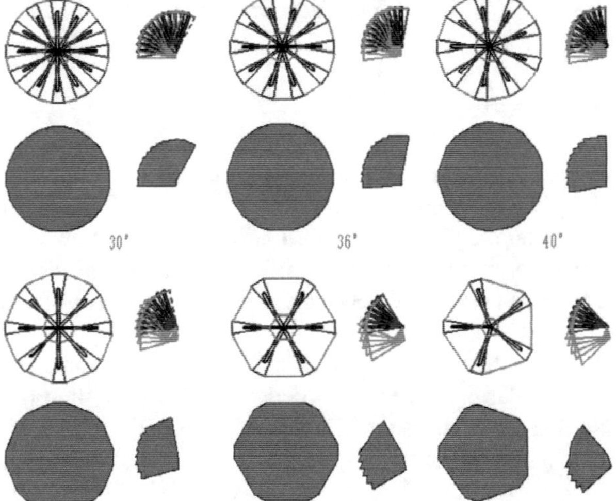


Figure 22. Structural sketches correspond to different α1. The color image shows a top view of the structure when fully unfolded and folded. The gray figure is the projection of the structure based on the top view, and the structure's footprint can be calculated based on the projection using CAXA.

From the appearance of the structure alone, when $\alpha_1$ is 30°, 36°, 45°, and 60°, the figure 22 presents when the structure is fully unfolded is similar to a circle, and the structure is more beautiful. When $\alpha_1$ is 60° and 72°, the projection of structure profile when the structure is fully unfolded presents a hexagon and a pentagon, respectively, and the structure also maintains a beautiful appearance. However, the unfolded area has a particular loss compared to the first four. The area of the second solar panel is much larger than the area of the first solar panel, which means that when the structure is unfolded, the first solar panel may not be able to support the weight of the second solar panel. Next, the folding ratios of the structures corresponding to different $\alpha_1$ are discussed, and it is most reasonable to discuss which structure in depth.

It is shown in figure 23 below that when $\alpha_1$ is equal to 60° and 72°, the folding ratio of the structure is large, and the form of the structure is unreasonable. Therefore, the structures corresponding to these two angles are not reasonable. When $\alpha_1$ is equal to 30°, the form and folding ratio of the structure are good. However, considering that there are most parts in the structure, and it is relatively troublesome in production and processing, it is not considered. When $\alpha_1$ is equal to 36°, 40°, and 45°, the folding ratio of the structure is about 0.2, and when $\alpha_1$ is equal to 40°, the folding ratio of the structure is the smallest, which means that the structure has the strongest shrinkage ability. Among these three angles, the structure corresponding to 36° has the most parts, but the first solar panel bears the lightest mass of the second solar panel. The structure corresponding to 45° has the fewest components, but the first solar panel bears the heaviest mass of the second solar panel. The structure corresponding to 40° is a balance between the structure corresponding to 36° and the structure corresponding to 45°, and the folding ratio is the smallest. In general, the most suitable structure is when $\alpha_1$ is 40°.

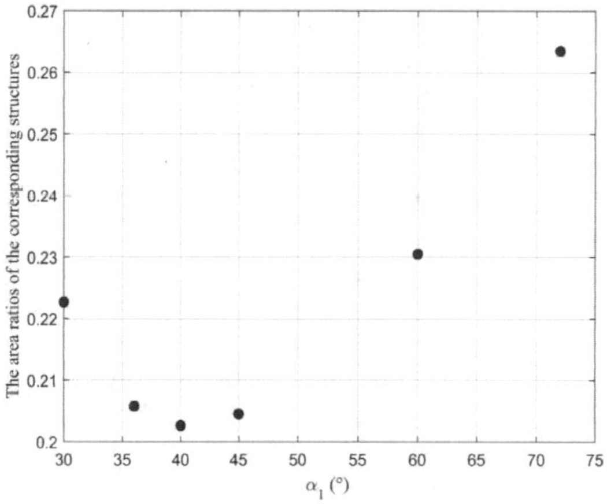


Figure 23. The folding ratios of the corresponding structures under different conditions of α1

When the folding ratio of the structure is analyzed, it can be seen that the radial shrinkage ability of the structural skeleton is the main factor that makes the folding ratio smaller. Discussing the shrinkage ability of the skeleton separately, reducing the circumferential folding ability of the structural skeleton to the structural folding ratio is theoretically not weaker than the radial shrinkage ability of the structural skeleton. The main reason is that the area occupied by the second solar panel is too large when the structure is folded, which limits the ability to shrink in the circumferential direction. When $\alpha_1$ is equal to 60° and 72°, if the second solar panel can be folded multiple times, then when the structure is fully folded, the footprint of the second solar panel will be similar to the footprint of the first solar panel. So the folding ratio of the structures corresponding to 60° and 70° will be smaller than 0.2. The critical problem is that there is no way to effectively drive the multiple unfolding and folding of the second solar panel.

## 4. Conclusion

In this paper, the structural design of a solar panel is carried out. The structural design can be divided into a skeleton part and a solar panel part. In the design of the skeleton part, the circumferential shrinkage and radial shrinkage are first considered separately. Then joint shrinkage is considered and designed so that the entire structure can be folded and unfolded smoothly. With the help of 2D and 3D models, analysis can be performed intuitively. The geometric conditions that the structure needs to meet to work smoothly are obtained through some calculations. In the design of solar panels, by designing the size, placement, and folding methods of the panels, it is ensured that the structure can be folded and unfolded smoothly. At the same time, it is closely arranged when unfolded. The overall movement process of the structure is explained in detail in the text, and 3D models are provided to make the movement more intuitive.

The article also analyzes the kinematics analysis, and the influence of different geometric parameters on the shrinkage performance of the whole structure is obtained quantitatively. The capacity of radial shrinkage can be reflected by the stroke of the slider. By studying the three influencing factors of connecting rod length, fully undeployed central angle, and fully deployed central angle, the influence of each on the radial shrinkage capacity is analyzed, and a comprehensive analysis is carried out to arrive at a solution that maximizes radial shrinkage capacity. Next, the circumferential shrinkage ability analysis is carried out, and the trend of the degree of circumferential shrinkage changing with the change of the angle when fully expanded is studied. Finally, the folding ratio of the overall shrinkage and expansion of the structure is studied, and the optimal ratio can be achieved from an overall perspective. The values of each parameter under the optimal ratio are obtained through comprehensive analysis.

## References

[1] K.O. Brauer, "Expandable structures for spacecraft." In A New Era in Space Transportation (pp. 89-97). Pergamon. (1977).

[2] M. Nazario, "Advancement of a 30kW Solar Electric Propulsion System Capability for NASA Human and Robotic Exploration Missions." (No. E-664449). (2013).

[3] K. Reed, and H. J. Willenberg, "Early commercial demonstration of space solar power using ultra-lightweight arrays." Acta Astronautica, 65(9-10), 1250-1260. (2009).

[4] D.X. Li, Flexible Spacecraft Structural Dynamics, Science Press. 2010.

[5] S. LIU, "Development and Application of Origami in Space Structure." Spacecraft Recovery & Remote Sensing, 41(6), 114-128. (2020).

[6] P. G. Bailey, and V. L. Teofilo, "Space nuclear power integration studies for a surveillance satellite mission." In Proceedings of the 24th Intersociety Energy Conversion Engineering Conference (pp. 1269-1274). IEEE. (1989, August)

[7] Y.G. Miao, Y.H. He, and Y. Li, "Design of Automatic Folding Solar Panels." China Academic Journal Electronic Publishing House.2021-681(2021)

[8] J. Wang, S. Zhu, Y. Yang, M.Y. Qi, J. Wu, and J.J. Cao, "A Design for Folding Solar Panels Automatically". Journal of Changshu Institute of Technology（Natural Sciences）.2022-36-2(2022)

[9] B. Meng, "The Research and Design of Folding Solar Panel and Tracking System for Vehicle". Engineering Technology ii Album,2018-01. (2018)

[10] P. Sareh, and S.D. Guest, "Tessellating variations on the Miura fold pattern." In IASS-APCS Symposium, Seoul, South Korea (Vol. 21, pp. 1070-1078). (2012).

[11] K. Miura, Map fold a la Miura style, its physical characteristics and application to the space science. Research of Pattern Formation, 77-90. (1994).

[12] T. Langbecker, "Kinematic analysis of deployable scissor structures." International Journal of Space Structures, 14(1), 1-15. (1999).

[13] F. Maden, Y. Akgün , G. Klper , Ş. Gür , M. Yar, and K. Korkmaz, "A Critical Review on Classification and Terminology of Scissor Structures." Journal of the International Association for Shell and Spatial Structures, 60(1), 47–64. (2019).

[14] N. McKay, A. Chou, D. Becker, C. Hoffman, and J. Walters, "Research and development of the eXtendable solar array system." In 50th AIAA Aerospace Sciences Meeting including the New Horizons Forum and Aerospace Exposition (p. 143). (2012).

[15] L. FuGe. The Breeze Comes Slowly——A Study on the Design Art of Folding Fans. Engineering Science and Technology Vol.1. 2018-03,2017.

[16] Q. Juan. Dynamics Analysis of Damped Scissor Deployable Flexible Truss. Basic Science. 2016-03, (2016).

# Bidirectional coupling method for solving the electromagnetic-joule thermal properties of electromagnetic launcher

**Pengxiang Zhang*, Tao Shu**

Air and Missile Defense College, Air Force Engineering University, Xi'an Shaanxi 710051, China

*Corresponding author's e-mail: qianxing979@163.com

**ABSTRACT:** The actual working process of the electromagnetic launcher is very complex, and using simulation software to solve it is the main method of research at this stage. The one-way coupling method will ignore the temperature-dependent characteristics of the material. In order to better simulate the actual working process, this paper uses the bidirectional coupling simulation method to solve the electromagnetic force and Joule heat of the four-rail electromagnetic launcher, and compares it with the unidirectional coupling. The simulation results show that the bidirectional coupling is more accurate than the unidirectional coupling, and the increase of the number of iterations will improve the solution accuracy, but affect the solution speed.

## 1. Introduction

The working process of electromagnetic launcher is a complex multi-physical field coupling process, in which the Joule thermal characteristics of electromagnetic launcher are calculated as the coupling of electromagnetic field and temperature field. The Joule heat and friction heat generated during the operation of the electromagnetic launcher will lead to the increase of the temperature of the armature and the rail, and the temperature change of the armature and the rail will lead to the change of its conductivity and thermal conductivity, and the change of these factors not only affects the calculation of the Joule heat characteristics of the electromagnetic launcher, but also affects the analysis of the electromagnetic characteristics of the electromagnetic launcher under the consideration of temperature. The analysis of the electromagnetic characteristics under the temperature conditions is considered. In this paper, the influence of two coupling methods, unidirectional coupling and bidirectional coupling, on the simulation of Joule heat and electromagnetic force is compared and analyzed, and the influence of the number of coupling iterations in bidirectional coupling on the simulation results is analyzed, which provides a more accurate calculation method for the simulation of Joule heat of electromagnetic launcher.

## 2. Physical model and simulation conditions

The four-rail electromagnetic launcher model used in this study is shown in Figure 1. The rail and armature are designed with a hyperbolic structure, and the four rails are distributed in a uniform circular array with the armature as the center. The trapezoidal pulse current flows in from the two opposite rails, flows through the armature, and then flows out from the remaining two rails. The armature adopts the hollow design shown in the figure, which can provide a good electromagnetic shielding environment for the smart munitions carried in it. The cost and material properties of the electromagnetic launcher for the rail and armature are fully considered, using aluminum as the armature material and copper as the rail material, and the material properties are shown in Table 1.

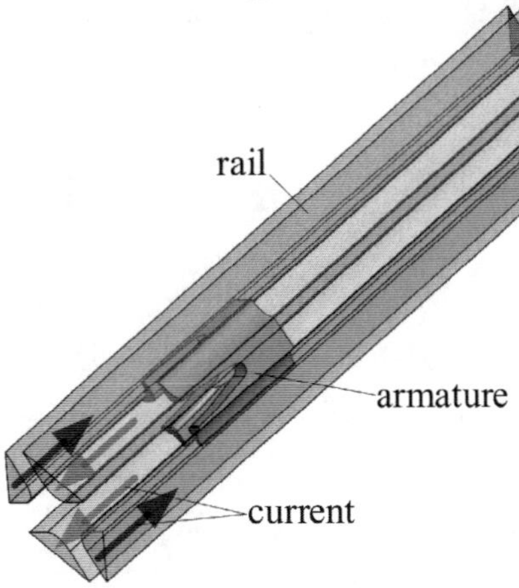


(a) Three-dimensional model

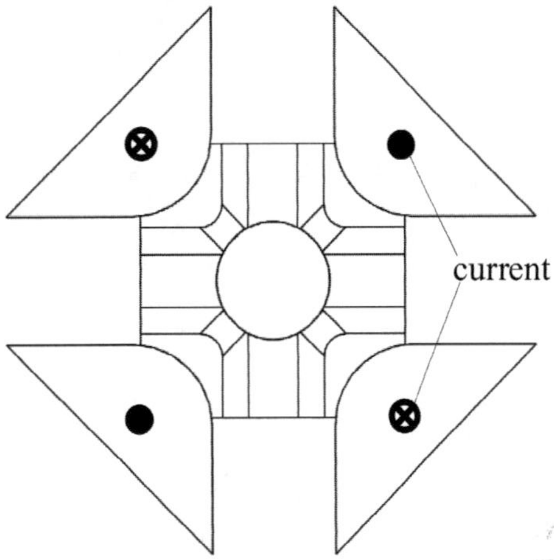


(b) Top view of the model

Figure 1 Four-rail electromagnetic launcher model

Table.1 Material parameters of electromagnetic rail gun

| | Material | density / ($kg \cdot m^{-3}$) | Conductivity/ ($S \cdot m^{-1}$) | relative permeability /- | specific heat capacity/ J/(kg · K) |
|---|---|---|---|---|---|
| rail | copper | $8.66\times10^{3}$ | $5.80\times10^{7}$ | 0.999991 | 385 |
| armature | aluminum | $2.70\times10^{3}$ | $3.76\times10^{7}$ | 1.000021 | 900 |

The trapezoidal current excitation used in this simulation is shown in Figure 2, with a current rising edge phase of 0-1ms and a current amplitude of 600kA.

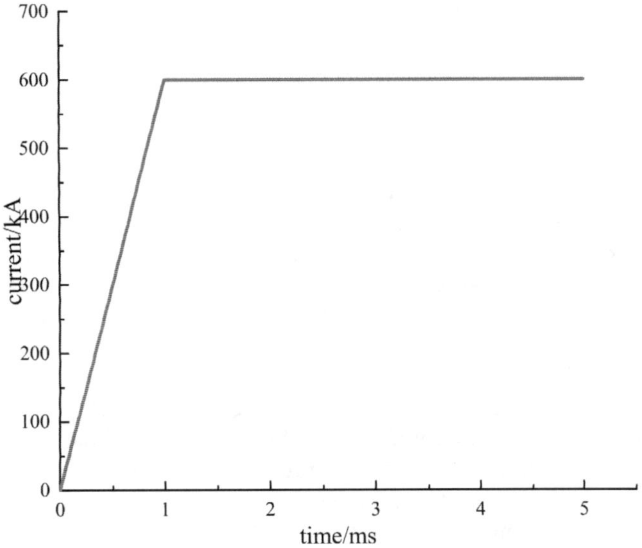


Figure 2 Simulated current waveform

## 3. Simulation analysis

*3.1 Effect of coupling method on electromagnetic force and Joule heat*

The workbench platform is used to build the unidirectional coupling and bidirectional coupling electromagnetic launcher electro-thermal coupling calculation project shown in Figure 3 respectively, setting the number of iterations of bidirectional coupling simulation to 3 and the ambient temperature to 22°C.

(a) Unidirectional coupling method (b) Bidirectional coupling method

Figure 3 Two coupling methods to calculate Joule heat

The temperature change curve of the electromagnetic launcher calculated by the two methods is shown in Figure 4.

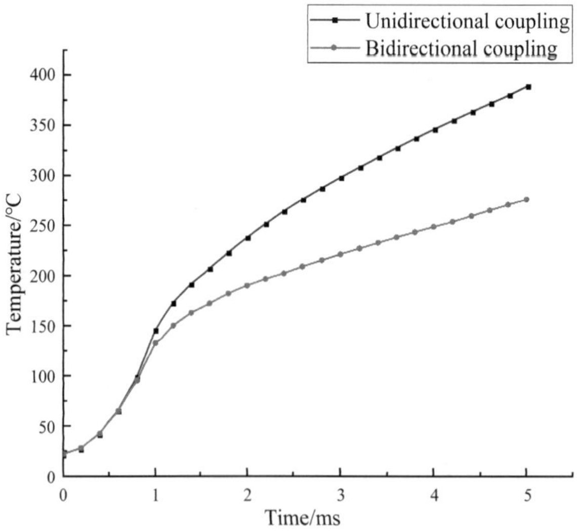


Figure 4 Temperature rise curve of electromagnetic launcher calculated by two coupling methods

As shown in the figure, the overall maximum temperature of the electromagnetic launcher reached 388.52°C in the unidirectional coupling case, while the overall maximum temperature of the electromagnetic launcher was 276.14°C in the bidirectional coupling condition. In the current rising edge stage, i.e., before t=1ms, the temperature difference between the two calculation conditions is not significant, which is because the current in the rail is still relatively small at the early stage of energization, and the temperature rise of the electromagnetic launcher is small, which has less impact on the conductivity of the material, so the temperature in the bidirectional coupling calculation condition is not much different from the unidirectional coupling. With time, the temperature difference between the two calculation conditions becomes larger and larger, which is because the conductivity of the armature and the orbit keeps decreasing as the temperature rises, so the temperature difference between the bidirectional coupling calculation conditions and the unidirectional coupling is larger and more in line with the actual temperature rise trend of the electromagnetic launcher. It can be seen that after considering the temperature dependence of the electromagnetic launcher material, the temperature variation of the electromagnetic launcher is much different than the ideal case.

The variation curves of the electromagnetic thrust on the armature for the unidirectional and bidirectional coupling cases in Maxwell are shown in Figure5.

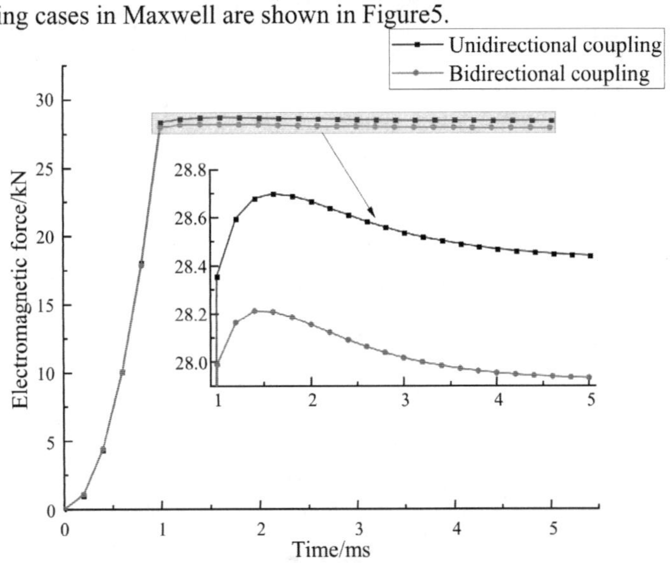


Figure 5 Variation curve of electromagnetic thrust on the armature calculated by two coupling methods

As shown in Figure 5, the peak electromagnetic force is 28.7kN under unidirectional coupling and 28.21kN under bidirectional coupling, and the difference between the electromagnetic force obtained by the two calculation methods is not significant in the rising current phase, i.e., before t=1ms. This is because the conductivity of the copper rail and aluminum armature decreases when the temperature rises, which causes the current density in the rail and armature to decrease and the magnetic field generated to become smaller, and the electromagnetic thrust on the armature also decreases. It can also be seen from the figure that the electromagnetic thrust on the armature increases during the rising current phase and decreases after t=1ms. During the rising phase, the current increases and the magnetic field increases, so the electromagnetic thrust on the armature increases, and the changing magnetic field also generates eddy currents in the armature, and after the current reaches the maximum value, the current stops changing, so the magnetic field also stops changing, so the eddy currents in the armature gradually decrease, and the skin effect of the current gradually weakens, so the electromagnetic force gradually decreases.

*3.2 Effect of the number of bi-directional coupling iterations on simulation results*

In the process of electromagnetic-temperature bi-directional coupling, the setting of the number of iterations affects the accuracy of the solution and thus the overall temperature variation. Now set the number of iterations to 3, 5 and 10 to analyze the effect of iteration number on the simulation results. Figure 6 shows the temperature rise curve of electromagnetic launcher with different number of iterations.

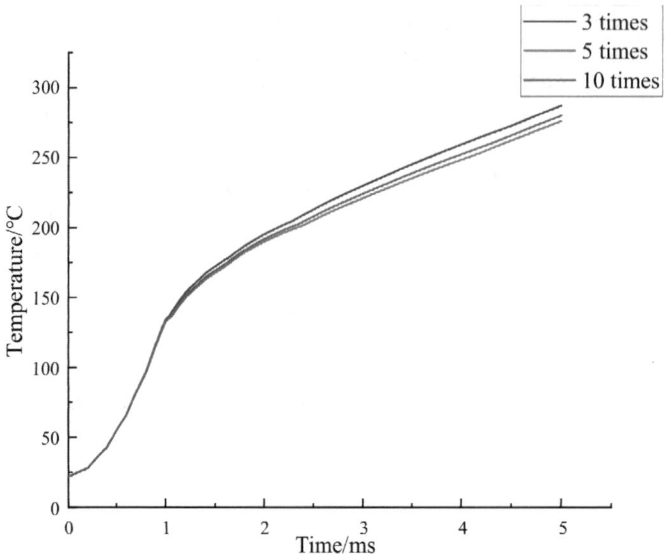


Figure 6 Temperature rise curve of electromagnetic launcher under different iterations

As shown in Figure 6, the maximum temperature of the electromagnetic launcher is 287.5°C for the simulation condition with iteration number 3, 276.14°C for the simulation condition with iteration number 5, and 280.15°C for the simulation condition with iteration number 10. From the figure, it can be seen that the temperature profiles of the three are not very different before t=1ms, while after t=1ms, the temperature profile of iteration number 10 is closer to that of iteration number 5, while the temperature profile of iteration number 3 is significantly larger, which indicates that the temperature profile is solved more accurately as the number of iterations increases. Figure 7 shows the rising curves of the electromagnetic force on the armature for different iterations.

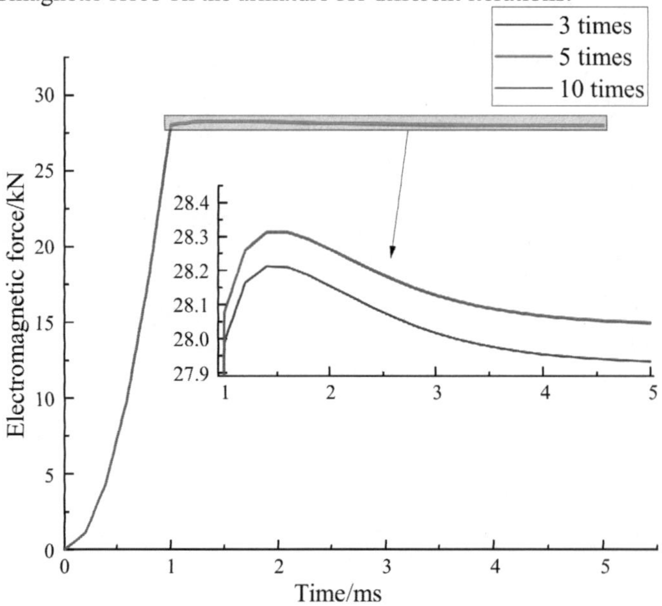


Figure 7 Curve of electromagnetic force on the armature under different iterations

As shown in the figure, the electromagnetic force curve with iteration number 3 deviates more than the other two, while the electromagnetic force curves with iteration number 5 and 10 almost overlap, which indicates that the accuracy of the electromagnetic force solution improves significantly with the increase of iteration number, and the simulation condition with iteration number 5 is accurate enough. Therefore, in the subsequent simulation of Joule heat and electromagnetic force of the electromagnetic launcher, the number of iterations of 5 can meet the computational requirements under the premise of ensuring the accuracy.

## 4. Conclusion

This paper analyzed and compared two coupling methods for solving Joule heat and electromagnetic force of electromagnetic launcher using ANSYS, and came to the following conclusions.

(1) the results of electromagnetic force and Joule heat calculated by bidirectional coupling are more accurate than those of unidirectional coupling and are more consistent with the actual working process of electromagnetic launchers.

(2) The higher the number of iterations of the bidirectional coupling, the higher the accuracy of the solution, and in the model of this paper, the number of iterations is set to 5 to meet the calculation requirements.

## References

[1] Lu J Y, Feng J H, Li K et al. Review of research on ultra-high speed missile pellet [J / OL]. *Journal of Harbin Engineering University*, 2021, {4} (10): 1-12.

[2] Ma X P, Liao X, Chen B. Review of research status of electromagnetic launched hypervelocity guided projectile at home and abroad [J]. *Air and space defense*, 2021,4 (02): 87-92.

[3] KERRISK J F. Electrical and thermal modeling of railguns[J]. IEEE Transactions on Magnetics, 1984, 20(2): 399-402.

[4] Lu T G. Analysis of pivot rail contact of electromagnetic railgun under temperature field [D]. Yanshan University, 2020.

[5] Ma W M, Lu J Y, Li X P. Electromagnetic launch ultra-high speed integrated projectile [J]. *Journal of National University of Defense Science and technology*, 2019,41(4): 1-10.

[6] Li X J, Wan M, Wang Z H et al. Numerical simulation of rail temperature field and thermal stress of electromagnetic railgun [J]. Fire and command and control, 2017,42 (02): 154-158.

[7] Gong F, Wei W, Guo H H. Analysis of electrothermal properties of commonly used materials for electromagnetic rail guns[J]. Journal of Ballistics,2018,30(03):78-81+87.

[8] Fan W, Su Z Z, Zhang T et al. Temporal and spatial distribution of transient temperature rise of enhanced electromagnetic railgun [J / OL]. *High Voltage technology*, {3}, {4} {5}: 1-10.

[9] Praneeth S R N, Singh B, Khatait J P. Study on effect of rail filet radius in electromagnetic railgun[J]. IEEE Transactions on Plasma Science, 2021, 49(9): 2997-3002.

[10] Zhang H H, Li S, Gao X, et al. Distribution Characteristics of Electromagnetic Field and Temperature Field of Different Caliber Electromagnetic Railguns[J]. IEEE Transactions on Plasma Science, 2020, PP(99): 1-8.

# The Bricard Linkage Network Constructed by Bennett Linkage: Design Method and Kinematic Analysis

**Ziyi Huang[1,a†], Wentao Jiang[2,b†], Xuan Pan[3,c†], Zekai Tao[4*,†]**

[1]South China University of the Technology, School of Civil Engineering and Transportation, Guangzhou, China

[2]Chongqing University, College of Mechanical and Vehicle Engineering, Chongqing, China

[3]Nanjing Agricultural University, Mechanical Design manufacture and Automation Major, Nanjing, China

[4]University of Edinburgh, School of Engineering,Edinburgh, UK

[a]e-mail: 1127664113@qq.com; [b]e-mail: 20192129@cqu.edu.cn;

[c]e-mail: px_880728@163.com

*Corresponding author e-mail: s2269547@ed.ac.uk

[†]These authors contributed equally to this work.

**Abstract**—The paper focuses on a Bricard Linkage network constructed by Bennett Linkage and discusses the possibility of connecting Bricard linkage using Bennett linkage as the connection. To simplify the problem, accessible geometric parameters are set without much loss of generality. This new design can also evolve into some new mechanisms by connecting the linkages in different ways. Based on the designed connection idea, the corresponding three-dimensional model is created by Onshape. Using Onshape, the movement and deployment of each mechanism could be observed. Then the kinematic analysis is carried out with Adams software, to find out how the kinematic angles vary as time increases. A physical model is made to observe the deployment process more clearly, indicating a potential application of a retractable table. Since the problem has been simplified by setting accessible parameters, the mechanism may show some further performance worthy of being explored if the parameters are more complicated.

## 1. Introduction

In 1903, Bennett [1] described a new mechanism, an innovative spatial connection system comprised of four revolute joints and structured uniquely. The linkage consists of four links connected by four hinges. In general, the degree of freedom of a spatial association comprised of four rotating pairs is zero, but this linkage acquires one degree of freedom following Bennett's particular arrangement.

Unlike Bennett's four-bar linkage, Bricard [2] described three movable octahedrons and, afterwards, three types of movable associations: linear symmetry, plane symmetry, and trihedral six-bar link [3]. It is a unique overconstrained mechanism with three symmetrical planes at any point. In 1980, Baker [4] presented the closed-loop equations of the six Bricard linkages and did a thorough analysis.

In the modern age, it's found that two Bennett linkages may be connected by joining two corners to produce a new Bennett linkage [5]. For Bennett Linkages Network, it shows that the connections are at the rectangle's corners. Based on its symmetric property, the links at the two pairs of opposite corners play the role of extension in two directions, respectively.

The idea of the Bennett linkage network carried out by Chen, and You [5] has brought many applications to real life. For example, people now could use this knowledge of the Bennett linkage network to build a deployable frame for tents based on the deployable arch concept. This is broadly applied to the military field to build an army camp or shelter [6] quickly.

The Bricard linkage has six corners divided into two groups with a different symmetric property. In one group, all the kinematic angles are equivalent. Chen and You [5] give the inspiration of connecting Bricard linkages by joining corners into a Bennett linkage as well. However, there was no study to explore the feasibility of linking several Bricard linkages in such a way. The properties and practicability of its network structure remain unclear, which is the problem this paper wishes to solve. But much research on a single Bricard linkage has been conducted, so some performance of the Bricard linkage network might be extrapolated by the behaviour and property of just one Bricard linkage.

Guo et al. [7] also similarly give an idea of the Bricard linkage network. Different from the conception of Bennett linkage connection, this network combines each Bricard linkage at just one point. It's more likely to extend the bars of one Bricard linkage to form many other Bricard linkages. This simplifies since the connection is just a hinge, so a more complicated network could be achieved. However, this also causes the singularity of possibility. This paper introduces a practical mobility analysis method to determine the mobility of mechanical networks, and in this paper, the method will be used with Adams simulation and graph analysis.

Thus, this paper aims to carry out a new conception of the Bricard linkage network, including the possible ways to connect each Bricard linkage, some potential applications and the kinematic analysis of the mechanism. The layout of this paper is as follows. The geometric and kinematic features of Bennett and Bricard linkages are briefly introduced in Section II, and the construction of the connection between the two Bricard linkages. Some types of network structures connected by Bricard links are explored and assessed in Section III. Finally, the conclusion is drawn in Section IV.

## 2. METHOD

Bricard linkage is a significant type of spatial single-ring connection that Bricard proposed. It is a unique six-link overconstrained mechanism with three symmetrical planes at any point. Bennett devised the Bennett linkage [1]. The linkage comprises four rigid links joined by four hinges, a process known as rotation in the mechanical engineering literature, thus the name "moving 4R ring."

The Bricard Linkage and Bennett Linkage figures are shown in Figures 1 and 2, respectively.

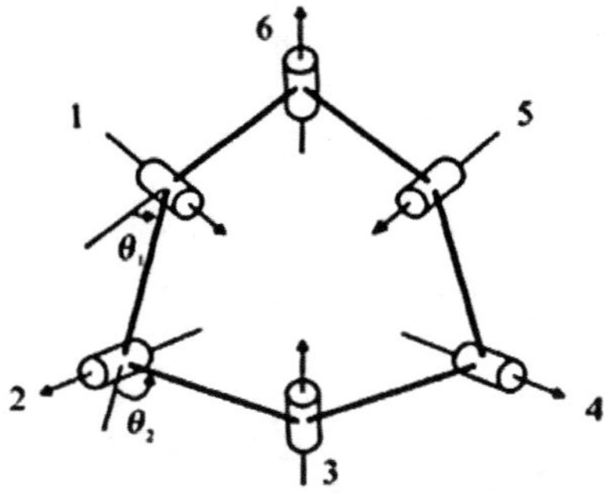


Figure 1. Bricard Linkage

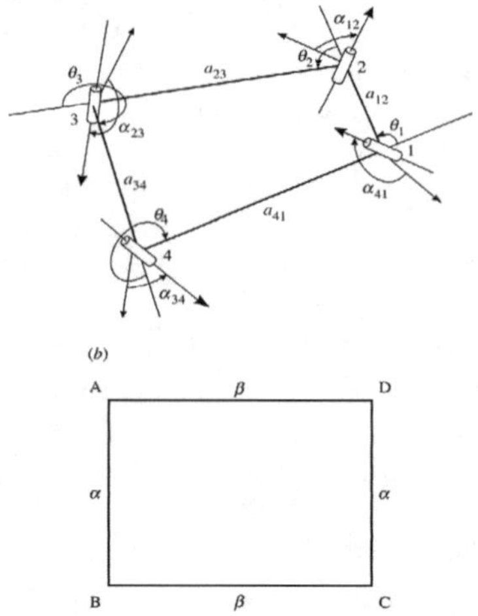


Figure 2. (a) A Bennett Linkage. (b) A schematic of linkage [5]

According to Chen and You's research, two Bennett linkages may be connected by joining two corners to produce a new Bennett linkage [5]. Using a similar method to make two corners meet and form a connection, Bricard Linkages can also be linked together. The two connecting mechanisms are depicted in Figure 3.

Figure 3. The extrapolation from Bennett Linkage connection to Bricard Linkage connection

## 2.1. Geometric parameters setting

### 2.1.1. For Bricard Linkage

As seen in Figure1, the Bricard Linkage's offset $R$ between any two adjacent linkages equals zero. Yan Chen, Zhong You, and Tibor Tarnai [8] reveal the geometric characteristics of the planar symmetric Bricard connection. In addition to $R = 0$, the following criteria (1) and (2) must be met.

$$a_{12} = a_{23} = a_{34} = a_{45} = a_{56} = a_{61} = l \quad (1)$$

$$\alpha_{12} = \alpha_{34} = \alpha_{56} = \alpha, \alpha_{23} = \alpha_{45} = \alpha_{61} = 360^{\circ} - \alpha \quad (2)$$

The linkage that satisfies the preceding requirements has triple rotational symmetry and three planes of symmetry.

### 2.1.2. For Bennett Linkage

Bennett 4R Linkage's one-dimensional motion has the following geometric conditions (3) and (4).

(a)The two opposite links have the same length and corner, i.e.

$$a_{12} = a_{34} = a, a_{23} = a_{41} = b \quad (3)$$

$$\alpha_{12} = \alpha_{34} = \alpha, \alpha_{23} = \alpha_{41} = \beta \quad (4)$$

(b)The lengths and corners should satisfy (5) and (6).

$$\frac{\sin\alpha}{a} = \frac{\sin\beta}{b} \quad (5)$$

$$R_i = 0 (i = 1, 2, 3, 4) \quad (6)$$

### 2.2. Kinematic parameters setting

In Figure 4, we define link length $a$ as the shortest distance between the axes of two adjacent joints, offset R as the distance between two link lengths, $\alpha_{i(i+1)}$ as the twist angle of rotation from $z_i$ to $z_{i(i+1)}$ positively about $x_{i(i+1)}$, $\theta_{i(i+1)}$ as the angle of rotation from $x_i$ to $x_{i(i+1)}$ positively about $z_i$ (kinematic variable). All offsets R are zero in the mechanism we use.

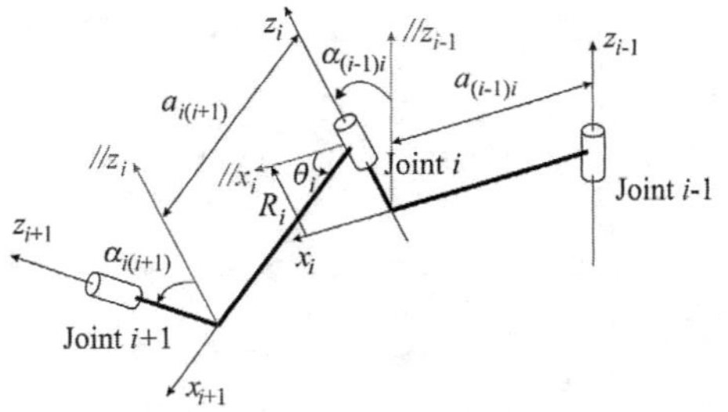


Figure 4. Coordinate systems for two links connected by a revolute joint

The transformation matrix is transforming the $i^{th}$ coordinate system into the $i+1^{th}$ coordinate system. Denavit and Hartenberg [9] introduced a notation to express the transformation matrix $T_{i(i+1)}$ (9) in a single four-dimensional matrix described by using the homogeneous coordinates containing the coordinates of point P in the $O_i$ frame.

$$p'_{(i+1)} = T_{i(i+1)} p'_i \tag{7}$$

$$p'_i = \begin{bmatrix} 1 \\ x_i \\ y_i \\ z_i \end{bmatrix} \tag{8}$$

$$T_{i(i+1)} = \begin{bmatrix} 1 & 0 & 0 & 0 \\ -a_{i(i+1)} & \cos\theta_i & \sin\theta_i & 0 \\ -R_i \sin a_{i(i+1)} & -\cos a_{i(i+1)} \sin\theta_i & \cos a_{i(i+1)} \cos\theta_i & \sin a_{i(i+1)} \\ -R_i \cos a_{i(i+1)} & \sin a_{i(i+1)} \sin\theta_i & -\sin a_{i(i+1)} \cos\theta_i & \cos a_{i(i+1)} \end{bmatrix} \tag{9}$$

#### 2.2.1. For Bennett Linkage

This linkage-type has the following mathematical relations (10), (11).

$$\begin{aligned} p'_2 &= T_{12} p'_1 \\ p'_3 &= T_{23} p'_2 \\ p'_4 &= T_{34} p'_3 \\ p'_1 &= T_{41} p'_4 \end{aligned} \tag{10}$$

$$\begin{aligned} p'_1 &= T_{41} p'_4 = T_{41} T_{34} p'_3 = T_{41} T_{34} T_{23} p'_2 = T_{41} T_{34} T_{23} T_{12} p'_1 \\ T_{41} T_{12} T_{23} T_{34} &= I \end{aligned} \tag{11}$$

In a four-dimensional matrix-vector expression of such a Bennett linkage, we have the transform matrixes (12), (13), (14) and (15).

$$T_{41} = \begin{bmatrix} 1 & 0 & 0 & 0 \\ -b & \cos\theta_4 & \sin\theta_4 & 0 \\ 0 & -\cos\beta\sin\theta_4 & \cos\beta\cos\theta_4 & \sin\beta \\ 0 & \sin\beta\sin\theta_4 & -\sin\beta\cos\theta_4 & \cos\beta \end{bmatrix} \tag{12}$$

$$T_{12} = \begin{bmatrix} 1 & 0 & 0 & 0 \\ -a & \cos\theta_1 & \sin\theta_1 & 0 \\ 0 & -\cos\alpha\sin\theta_1 & \cos\alpha\cos\theta_1 & \sin\alpha \\ 0 & \sin\alpha\sin\theta_1 & -\sin\alpha\cos\theta_1 & \cos\alpha \end{bmatrix} \quad (13)$$

$$T_{23} = \begin{bmatrix} 1 & 0 & 0 & 0 \\ -b & \cos\theta_2 & \sin\theta_2 & 0 \\ 0 & -\cos\beta\sin\theta_2 & \cos\beta\cos\theta_2 & \sin\beta \\ 0 & \sin\beta\sin\theta_2 & -\sin\beta\cos\theta_2 & \cos\beta \end{bmatrix} \quad (14)$$

$$T_{34} = \begin{bmatrix} 1 & 0 & 0 & 0 \\ -a & \cos\theta_3 & \sin\theta_3 & 0 \\ 0 & -\cos\alpha\sin\theta_3 & \cos\alpha\cos\theta_3 & \sin\alpha \\ 0 & \sin\alpha\sin\theta_3 & -\sin\alpha\cos\theta_3 & \cos\alpha \end{bmatrix} \quad (15)$$

Then it's figured out from equation (11) that the kinematic relation of Bennett linkage is (16).

$$\theta_1 + \theta_3 = 2\pi, \theta_2 + \theta_4 = 2\pi, \tan\frac{\theta_1}{2}\tan\frac{\theta_2}{2} = \frac{\sin\frac{1}{2}(\alpha+\beta)}{\sin\frac{1}{2}(\alpha-\beta)} \quad (16)$$

### 2.2.2. *For Bricard Linkage*

Similarly, the kinematic result of Bricard linkage could be derived from the closure equation (17).

$$T_{61}T_{12}T_{23}T_{34}T_{45}T_{56} = I \quad (17)$$

Because of threefold symmetry, the six kinematic variables must satisfy the following conditions (18).

$$\theta_1 = \theta_3 = \theta_5 = \theta, \theta_2 = \theta_4 = \theta_6 = \varphi \quad (18)$$

In a four-dimensional matrix-vector expression of such a Bricard linkage, the transform matrixes are (19) and (20).

$$T_{23} = T_{45} = T_{61} = \begin{bmatrix} 1 & 0 & 0 & 0 \\ -a & \cos\varphi & \sin\varphi & 0 \\ 0 & -\cos\alpha\sin\varphi & \cos\alpha\cos\varphi & -\sin\alpha \\ 0 & -\sin\alpha\sin\varphi & \sin\alpha\cos\varphi & \cos\alpha \end{bmatrix} \quad (19)$$

$$T_{12} = T_{34} = T_{56} = \begin{bmatrix} 1 & 0 & 0 & 0 \\ -a & \cos\theta & \sin\theta & 0 \\ 0 & -\cos\alpha\sin\theta & \cos\alpha\cos\theta & -\sin\alpha \\ 0 & \sin\alpha\sin\theta & -\sin\alpha\cos\theta & \cos\alpha \end{bmatrix} \quad (20)$$

Calculate with equation (17), the kinematic relation of Bricard linkage is thus (21).

$$\begin{aligned} &\theta\cos^2 a + \sin^2 \alpha(\cos\theta + \cos\varphi) + (1+\cos^2 a)\cos\theta\cos\varphi - 2\cos\alpha\sin\theta\sin\varphi = 0, \\ &0^\circ \leq \alpha \leq 180 \end{aligned} \quad (21)$$

This equation (21) shows that the kinematic relations can be determined for any given twist angle.

*2.3. Combine two Bricard linkages by Bennett linkage*

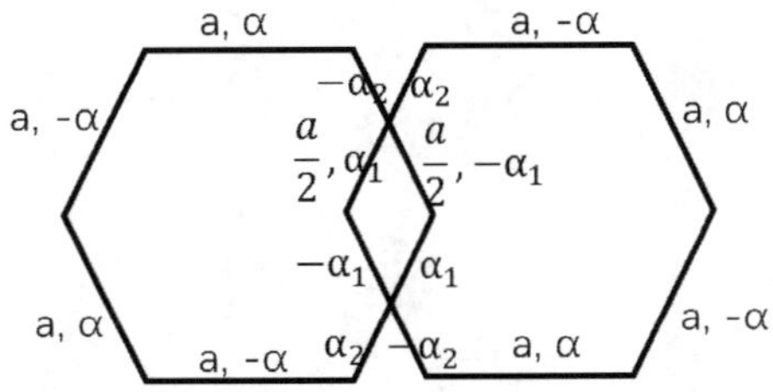


Figure 5. Geometric parameters of the connection element

Two identical Bricard linkages are connected by making the corners of two Bricard linkages meet to form Bennett linkages as a connection. To retain one degree of freedom, every parameter of two Bricard linkages and one Bennett linkage must first satisfy their geometric relations. To make the problem easier to analyze, assume the length of Bricard linkage is a, and all the Bennett linkage lengths are $\frac{a}{2}$.

For a Bennett linkage, put a=b in the equation (3), it turns out that $\alpha = \beta$ or $\alpha + \beta = \pi$. However, according to Equation (16), $\alpha = \beta$ is not allowed as the denominator becomes 0. If $\alpha = \beta$, this will not be a Bennett linkage but a 2R open kinematic chain because there are two overlap linkages, as a result that $\theta$ must be a constant as $\pi$ to satisfy the equation. When making the model, a revolute joint will remain unchanged after a twist angle of $180^\circ$ or $0^\circ$. Therefore, the equation $\alpha + \beta = \pi$ can be replaced with $\alpha + \beta = 0$ so that $\beta = -\alpha$, which will make no difference except for analyzing the kinematic relation.

With the above description, the geometric parameters of the connection element can be determined, shown in figure 5, where $a$ and $\pm\alpha$ mean the length and twist angle respectively of Bricard linkage, $\pm\alpha_1$ means the twist angle of Bennett linkage, and $\alpha_1 + \alpha_2 = \alpha$

## 3. RESULTS

With the results of section II, many possibilities of connecting the elements can be done. Notice that, in Kutzbach criterion [10], $f_i$ is 1 in this case, which means the more joints the structure has, the smaller the DoFs calculation result will be. Since the Bricard linkage and Bennett linkage have 0 and even negative DoFs, the connection will add more joints, so this criterion can't determine whether the construction is movable.

Because the network has a simple construction with only bars and joints, to observe the network's performance more virtually, The Onshape modelling tool is used to build up the network simulation. A hinge is made at each connection between every two bars, acting like a revolute joint.

### *3.1. Form a ring*

Considering the three-fold symmetric property of Bricard linkage, we can connect once for every two corners and finally form a ring.

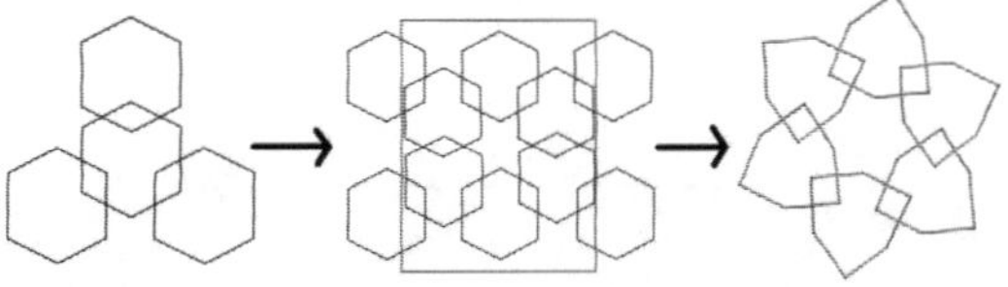

Figure 6. The ring conception is extrapolated from the most left figure, which comprises three connection elements (or four Bricard linkages network)

The figure above shows that the ring comes from the left 4R (nR means n Bricard linkages are contained in the network) network. This 4R connection has some interesting behaviors shown in Figure

7. Similar to the element of Bricard linkage, this connection could be folded into a compact volume (a) but also extended into a large surface (b). This indicates a potential application of a retractable table.

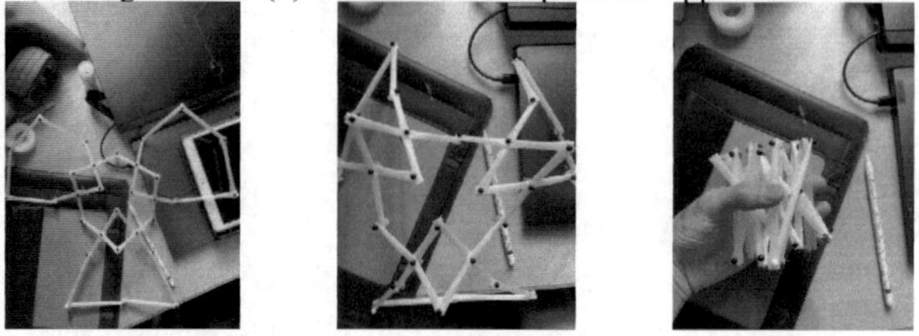

(a) (b) (c)

Figure 7 Pictures of prototype: (a) Folded in a compact volume (b) Half deployed (c) Deployed completely into a plane look

However, for this 6R Bricard ring it won't show this behavior. This connection has transferred to a structure rather than a mechanism because of its closure. Imagine one of six Bricard linkages is moving, the other 5 linkages should move towards this moving one in the ring's radial direction.

However, if one Bennett connection between two Bricard linkages is removed, it can have a 6R to 5R behavior shown below in Figure 8.

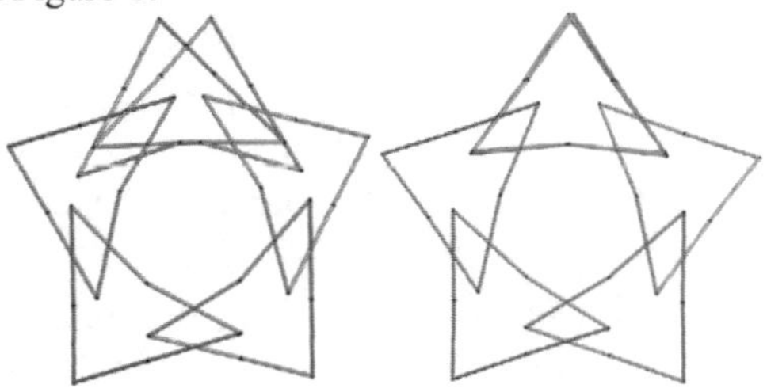

Figure 8. The mechanism is folded from 6 elements to 5 elements

Actually, when one Bennett connection is removed, the 6R ring could fold to make some Bricard elements overlap, so it looks like a less R ring. And here is a figure of when it's overlap to 5R and 4R.

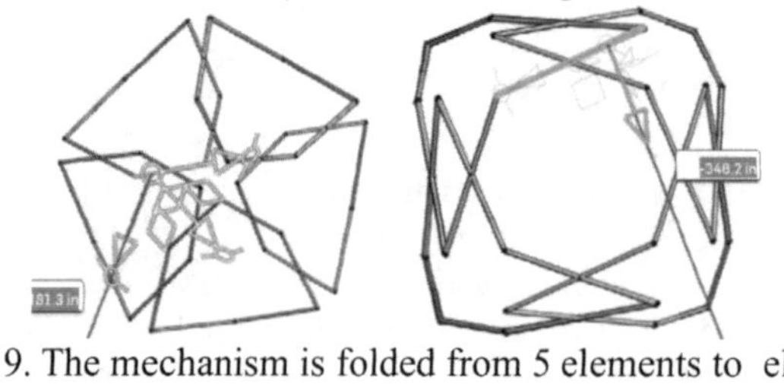

Figure 9. The mechanism is folded from 5 elements to  elements

### *3.2. Adjacent connection*

This is shown in Figure 10 below.

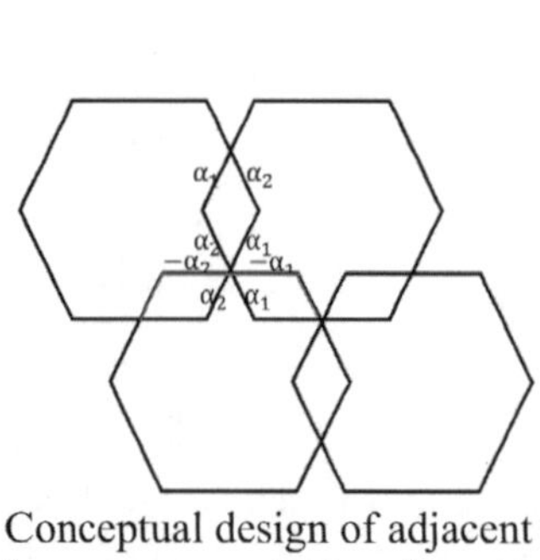

Figure 10. Conceptual design of adjacent connection

In the figure, we set the geometric parameters starting from the red parts and keep in mind that the twist angle α of Bricard linkage equals to $\alpha_1 + \alpha_2$ explained in section II. As a result, the Bennett linkage at the upside can be determined. However, as demonstrated before, in the connection Bennett linkage the two twist angles should be opposite numbers, which means that in this connection $\alpha_1 + \alpha_2 = \alpha = 0$.

Therefore, if we want to achieve this mechanism, the twist angles of Bricard linkage will be all zero, and each Bricard element will become a planar linkage which has 3 DoFs according to 2D Kutzbach criterion [10].

Therefore, this mechanism will be very unstable and not ideal.

### 3.3. Parallel connection

This one is a bit different. A common link will be contained in two Bricard elements and used twice, which is shown in Figure 11 below.

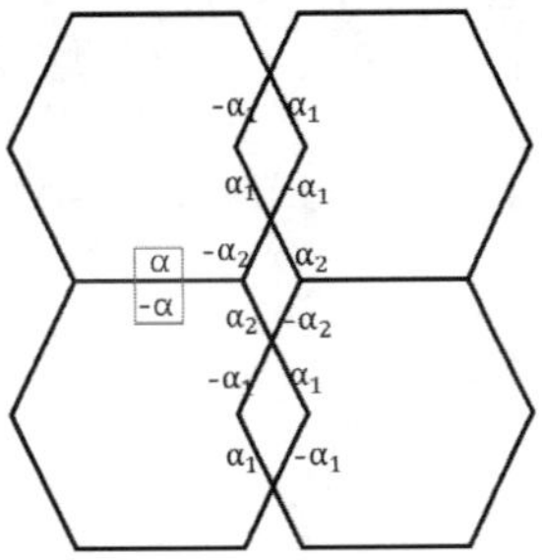


Figure 11. Conceptual design of adjacent connection

First, we still derive from the Bennett linkage at the connections at the middle. However, a conflict occurs at the red part marked in the figure. From the upper Bricard element, we know that for this link the twist angle should be $\alpha$, but it's opposite as $-\alpha$ if we derive from the Bricard element at the other side. Since that we can't let $\alpha$ be 0 because it will make the Bricard element a 2D planar linkage again, we try to find a value for $\alpha$ that can solve this conflict.

This value will be $90^\circ$. As is mentioned before, a revolute joint will remain the same position if the twist angle is within a period of $180^\circ$. Therefore, the same revolute joint at the same position rotates after a twist angle of $90^\circ$ or $-90^\circ$ will get the same position, because the angle difference is: $90^\circ - (-90^\circ) = 180^\circ$. Here is the model build of this idea:

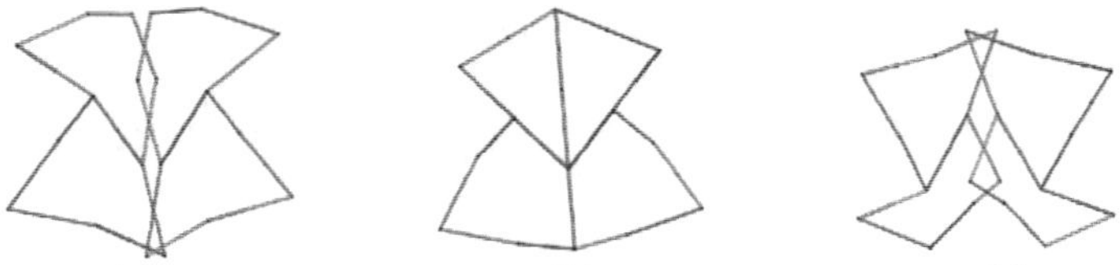

(a) (b) (c)

Figure 12. Pictures of prototype. Subfigures (a),(b),(c)are three stages when deploying.

### 3.4. Kinematic analysis

All three ideas come from the connection element shown in Figure 5. Here the analysis of the physical model shown in Figure 7 is carried out. This model comprises 3 connection elements. The three-dimensional simulated model made by Onshape is imported into Adams system for kinematic analysis. An angular velocity of 30 degrees per second driving force was applied to one link in this model as input. Adams analysis system can be used to find the relationship between the kinematic parameters and time. Here we want to find out how the kinematic angles of the Bricard Linkage and the Bennett in the connection element vary with time. Linkage. Thus, the kinematic parameters are chosen as $\theta_1$ and $\theta_2$ explained in Figure 2. This is because one angle $\theta_1$ is the common angle of both the Bricard Linkage and the Bennett linkage. The angles analyzed are shown in Figure 13, And the curves of both angles varying with time from Figure 13 to (a) in Figure 7 are shown in Figure 14 and Figure 15.

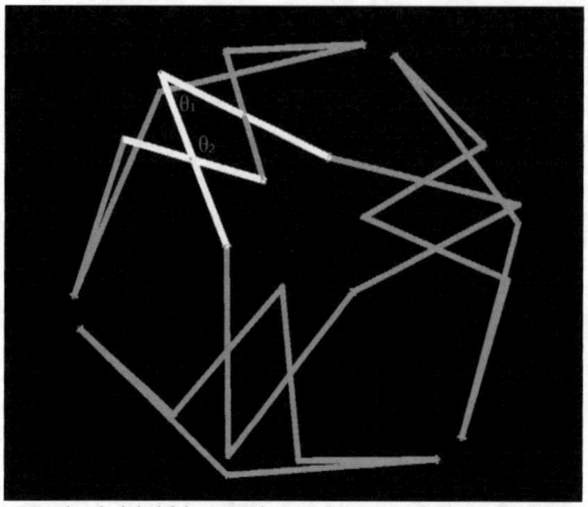


Figure 13. the initial kinematic angles set of the Adams analysis

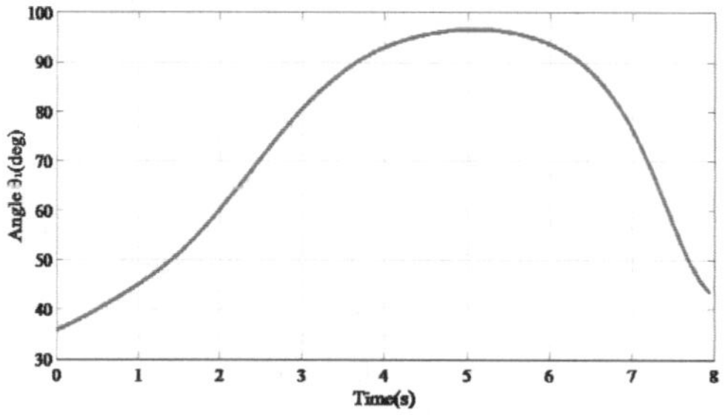


Figure 14. The kinematic angles ($\theta_1$) of the Bricard Linkage changing with time

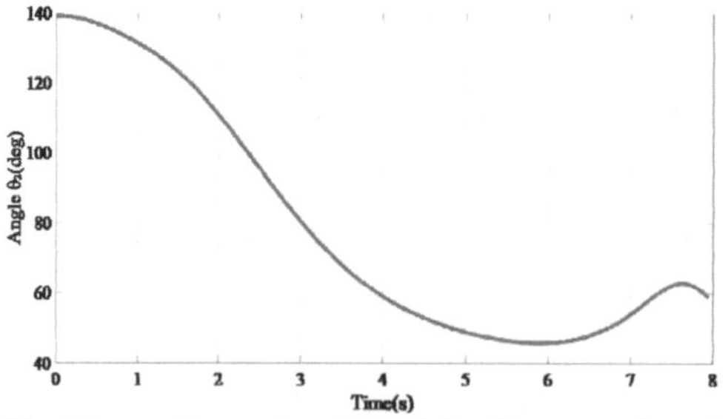


Figure 15. The kinematic angles ($\theta_2$) of the Bennett changing with time

From the above two curve diagrams, it could be concluded that during the whole movement of the mechanism, the angle $\theta_1$ first increases to a peak of 96 degrees and then decreases, showing a variation range of 36 degrees to 96 degrees. The angle $\theta_2$ first decreases to a minimum of 45 degrees and then increases in the motion process, The variation range of $\theta$ angle is 45 degrees to 140 degrees.

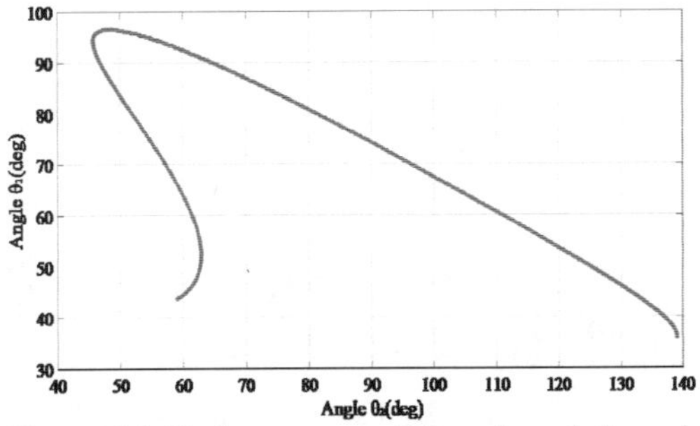


Figure 16. Fitting curves of $\theta_1$ angle and $\theta_2$ angle

The change of two angles with time was fitted into a curve, as shown in Figure.16.

## 4. CONCLUSION

This paper discusses the possibility of connecting Bricard linkage using Bennett linkage. During the process, to make the ideas easier to be understood, the geometric parameters are set and simplified with the accordance to the definition of the linkage's connection. Then, based on the connection element comprised of 2 Bricard linkages and 1 Bennett linkage at the joint after the geometric parameters setting, 3 ideas of extension are proposed the feasibility of each is explored, and the first one and third one are thought to be ideal and good. Since 3 ideas are based on the connection element, at the end, Adams kinematic analysis is carried out to check how the element moves, so that the movements of more constructions could be extrapolated. However, there could be more possibilities without the limitation and simplification of the geometric parameter, that means, if these parameters are set to be more complex, some further constructions could be explored.

## REFERENCES

[1] G. T. Bennett, "A new mechanism," Engineering, vol. 76, no. 777, 1903.
[2] R. Bricard, "Mémoire sur la théorie de l'octaèdre articulé," Journal de mathématiques pures et appliquées, Liouville, vol. 3, no. 5, pp. 113–148, 1897.
[3] R. Bricard, "Leçons de cinématique," Gauthier-Villars, Paris, vol. 2, pp. 7–12, 1927.
[4] J. E. Baker, "An analysis of Bricard linkages," Mechanism and Machine Theory, vol. 15, no. 4, pp. 267–286, 1980.
[5] Y. Chen and Z. You, "Mobile assemblies based on the Bennett linkage," Proc. R. Soc. A.4611229– 1245, 2005.
[6] T. C. Tran and J.Y.R. Liew, Department of Civil Engineering, National University of Singapore, BLK E1A, #02-18, 1 Engineering Drive 2, Singapore 117576.
[7] OX., H. Guo, R. Liu, F. Meng, Q. Chen, Y. Xu and R. Liu, "Mobility Analysis of the Threefold-Symmetric Bricard Linkage and Its Network," ASME. J. Mechanisms Robotics, February 2020, vol. 12, no. 1, pp. 1.
[8] Y. Chen, Z. You and Tibor Tarnai, "Threefold-symmetric Bricard linkages for deployable structures," International Journal of Solids and Structures, vol. 42, issue 8, pp. 2287- 2301, 2005.
[9] J. Denavit and R. S. Hartenberg, "A Kinematic Notation for Lower-Pair Mechanisms Based on Matrices," ASME. J. Appl. Mech, June 1955, vol. 22, no. 2, pp. 215–221.
[10] Kutzbach criterion, Jorge Angeles Clifford Truesdell. Rational Kinematics. Springer. p. Chapter 6, p. 78ff.

# THE BELLOWS DESIGNING METHOD OF DIVIDED WORKING CONDITIONS FOR LAUNCH VEHICLE

**Yi Zhang[1,2,*], Xu Li[3], CongFei Wang[4], Wei Huang[1,2], Gang Shi [1,2] and Yi Wang[1,2]**

[1] Beijing Institute of Astronautical Systems Engineering, Beijing, 100076, China

[2] Beijing Key Laboratory of Cryogenic Technology Research, Beijing 100076, China

[3] Logistics Center, China Academy of Launch Vehicle Technology, Beijing 100076, China

[4] China Academy of Launch Vehicle Technology, Beijing 100076, China

13811906267@163.com

**Abstract.** The working environment of the pipeline system of the launch vehicle is very complex. Bellows in piping system should endure high temperature, corrosive medium, vibration and other conditions, which require for high reliability. However, due to the restrictions of fixed location, weight reduce requirement, the number and location of bellows are strictly required. There is a great difference for the bellows design method between the launch vehicle and the ground pipeline. In this paper, a bellows designing method for launch vehicle is introduced in detail. The method calculates the compensation amount of bellows under divided working conditions, and checks the fatigue damage coefficient as the main failure judgement basis. It has been proved by practice that the designing method of divided working conditions can solve the design problems of the bellows of launch vehicles effectively.

## 1. Introduction

Bellows are important compensation element in pipe system, and they are revolving thin-walled shell with corrugated generatrix. The bellows in the pipeline could not only compensate the working displacement and installation deviation, but also reduce vibration and noise, and optimize the working environment of the system. Bellows can compensate for radial, axial and angular displacement deviation when working, and can also partially absorb and isolate vibration and water hammer energy. At present, bellows are widely used in pipeline system for thermal power, nuclear power, chemical engineering and launch vehicle.

The main function of the pressurization feed systems of launch vehicle is to deliver propellant to rocket engine and pressurize for the tank [1], which involves the design of long-distance delivery pipeline, and a large number of bellows need to be arranged in the pipeline system for installation and work compensation. Because the pipe support must be placed on the load-bearing structure, which is located at both ends of the tank, higher requirements are put forward for the compensation of bellows. At the same time, the pipeline of the launch vehicle is often arranged in the space inside the rocket, and the bellows is constrained by the space of other nearby instruments and equipment, so there are few installable positions. These reasons lead to higher compensation requirements of a single bellows for launch vehicles.

The launch vehicle pipeline is used to transport low-temperature propellant and high-temperature propellant gas vaporized by the engine. The working conditions of the bellows on the pipeline in each flight state of the rocket are significantly different. Therefore, targeted design should be carried out according to the characteristics of bellows of launch vehicle under various working conditions.

American Expansion Joint Association Standard [2] puts forward the calibration index of bellows under multi-working conditions-fatigue damage coefficient. This paper will use this index to briefly introduce the design of bellows for launch vehicle.

## 2. Design characteristics of bellows for launch vehicle

### *2.1. Complex using environment*

Bellows used in oil pipelines and heating pipelines on the ground have a single working condition, and the working temperature, working pressure and displacement compensation will not change significantly. Under different working conditions of launch vehicle bellows, the working pressure, temperature and displacement compensation will change significantly.

Several main working conditions of the launch vehicle include: assembly, transportation, filling, pressurization and flight. The assembly process is similar to that of ground products. The bellows are connected to the pipeline system by flange connection or welding. In this process, the bellows compensate the assembly error and the manufacturing error of other products of the pipeline system. The transportation process is the condition that the launch vehicle transported to the launch site in the assembly plant. In the transportation process, in order to ensure the integrity of the rocket pipeline system, it is necessary to fill a protective pressure of no more than 0.1MPa in the pipeline system, so that the pipeline, tank and other structural parts have sufficient rigidity. The filling process is the condition that the rocket fills propellant at the launch site. In this working condition, the increase of propellant causes the pipeline to be displaced by gravity. At the same time, in the process of filling low-temperature propellant, the tank and long-distance pipeline will also expand with heat and contract with cold. The pressurization condition is that the launch vehicle fills the tank with pressure after filling to ensure the starting pressure of the rocket launcher. Flight condition is the process from the ignition of launch vehicle engine to the separation of load. During the flight, the rocket is affected by factors such as airflow impact, engine noise and so on. At the same time, the medium inside the pipeline will flow violently, which will cause the bellows to bear large internal pressure and temperature changes.

### *2.2. Material special*

Propellant or propellant vapor is often in the pipeline where the bellows of launch vehicle is located, so the bellows material must be resistant to propellant or propellant vapor. In addition, the medium in the propellant conveying pipeline of the launch vehicle is low-temperature propellant such as liquid hydrogen and liquid oxygen, and the medium in the pressurization pipeline is high-temperature steam, so the bellows material also have special properties of mechanics and physics. For example, the new generation launch vehicle uses liquid hydrogen liquid oxygen propellant, and the pipeline where the bellows is located often bears high-temperature and high-speed gas-oxygen impact. If the bellows material is not properly selected, under the impact of high-temperature and high-speed gas and oxygen, the material will easily be destroyed and the pipeline leakage will have a fatal impact on the rocket safety. According to engineering experience, 1Cr18Ni9Ti, 0Cr18Ni9, 022Cr17Ni12Mo2, high-temperature alloy and other materials are often selected as bellows materials for launch vehicles.

### *2.3. High reliability requirements*

The launch vehicle is different from the ground products. If the bellows in the pipeline system leaks, the launch will be delayed in some cases, and the property loss or launch failure will be caused in some cases. The impact is immeasurable. Therefore, the reliability of the bellows of the launch vehicle is extremely high. When designing bellows, it is necessary not only to strictly control raw materials and dimensional accuracy, but also to carry out sufficient ground verification tests on finished products.

### *2.4. Application of design method*

The design of pipeline system and related pipeline components is carried out by combining empirical formula and finite element analysis in engineering. EJMA also introduces the engineering empirical formula in detail. For systems with uncomplicated system structure and relatively simple working conditions, geometric formulas are often used for manual calculation and preliminary design to meet the requirements.

The bellows working condition of launch vehicle is complex. If the previous extreme envelope method is used to extract the working index, the practical experience proves that it is difficult to check it successfully. During the design, the checking method of different working conditions is carried out, and whether the superimposed value of fatigue damage coefficient in each working condition is greater than 1 is judged as the checking basis.

## 3. design flow of bellows for launch vehicle

Launch vehicle is a complex systematic product, and requires constant data exchange and close coordination among related subsystems. Bellows are the most basic single component in ground equipment, but in the design process of bellows for launch vehicle, it is necessary to obtain input conditions from other subsystems.

The bellows design of launch vehicle is included in the pipeline design. On the basis of determining the bellows type, interface, limit size, compensation amount, temperature and other special working conditions, a detailed stand-alone design is carried out. The main process is as follows:

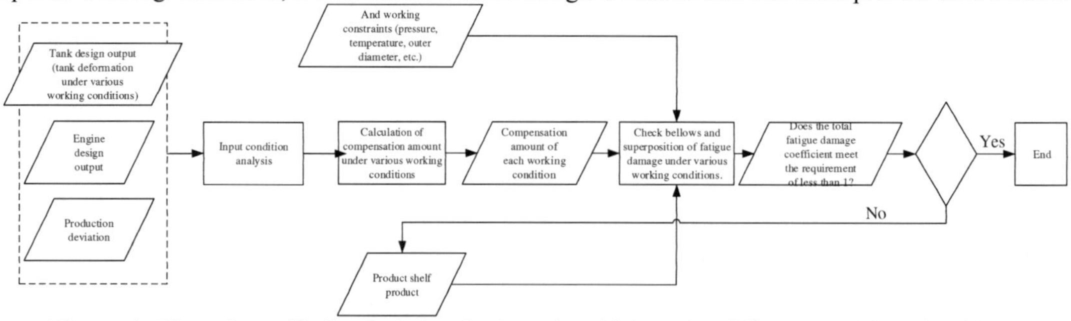


**Figure 1.** Flow chart of bellows design for launch vehicle under different working conditions

### *3.1. Input condition analysis*

Carry out comprehensive analysis on various input conditions, including tank design output, engine design output, production and processing deviation, etc.

Tank design output: At the same time of pipeline design, tank designer calculates the deformation of tank front bottom, rear bottom and column section under various working conditions, which can be used by bellows design .

Engine design output: During the working process of the engine, due to thermal expansion, cold contraction, vibration and other reasons, there will be axial displacement in the engine frame, which will affect the calculation of bellows compensation.

Manufacturing deviation: the tolerance of tank, flange and other structural parts in the production process will affect the compensation amount of bellows.

### *3.2. Calculation of compensation amount under various working conditions*

The bellows of launch vehicle experience five working conditions, so it is necessary to determine the compensation amount in each working condition for performance check. The whole pipeline compensation distribution is divided according to the layout of rocket supports, and the pipeline compensation between adjacent fixed supports is calculated separately.

The compensation amount under each working condition consists of two parts: compensation caused by manufacturing tolerance of pipeline and its components and work compensation caused by temperature, internal pressure and overload. The manufacturing tolerance mainly includes the manufacturing tolerance of single products such as tanks, short wall-board, flanges, pipelines, etc., which can be enveloped according to the limit tolerance. Transportation compensation takes into account pipeline deformation caused by tank deformation under transportation protection pressure. The decisive factors of the compensation amount in filling conditions are the pipeline deformation caused by the tank expansion and contraction in the process of filling low-temperature propellant and the pipeline deformation caused by the tank sinking under the influence of gravity after propellant filling. The pressurization process of launch vehicle is carried out after the tank is filled, so the pressurization condition is the result of two factors: filling deformation and pressurization deformation, and the two compensation amounts need to be superimposed in calculation. Compared with other ground conditions, the flight conditions of launch vehicle are significantly different. Bellows bear not only static load, but also dynamic load. At present, in the design of single bellows, only the check under static load can be considered, while the dynamic load is verified by system-level numerical simulation or ground test.

### *3.3. Check bellows and superposition of fatigue damage under various working conditions.*

After the calculation of compensation amount under each working condition is completed, considering the working and constraint conditions, the bellows model meeting the external conditions such as interface type and limit size is selected in the product shelf for checking. This work has a repeated iterative process. When the strength, stability and life of the selected model parameters do not meet the requirements, it is necessary to modify the model parameters and check them again until a satisfactory check result is obtained.

Calculate the fatigue life of bellows according to each working condition. If the fatigue life design requirements of each working condition are as follows $N_1$、$N_2$、$N_3 \cdots\cdots N_K$ .The calculated values of fatigue life under various working conditions are as follows $n_1$、$n_2$、$n_3 \cdots\cdots n_K$ .The total fatigue damage coefficient is:

$$D = \frac{n_1}{N_1} + \frac{n_2}{N_2} + \frac{n_3}{N_3} + \ldots + \frac{n_K}{N_K} \tag{1}$$

If D is less than 1, the bellows design meets the requirements; otherwise, adjust the model parameters until a satisfactory check result is obtained.

## 4. An example of bellows design for launch vehicle

### *4.1. Example introduction*

In this paper, the design of bellows in the oxygen tank pipeline of a launch vehicle is selected as an example. The function of the oxygen tank conveying pipe is to convey the propellant in the oxygen tank to the engine combustion chamber. The delivery pipeline is long, the starting point is the bottom of the oxygen tank on the left in the figure, and the ending point is the engine area on the right.

In the pipeline design, it is proposed to set a U-shaped bellows on the left side and the right side of the thrust bracket respectively for compensation. Among them, the left pipeline mainly compensates for the deformation of the bottom of the oxygen tank, the self-deformation of the front section of the conveying pipeline at low temperature and the deformation of the column section of the combustion tank. The right pipeline mainly compensates the deformation of the engine frame under the influence of inertia load and self-thrust, as well as the self-deformation of the conveying pipeline at low temperature.

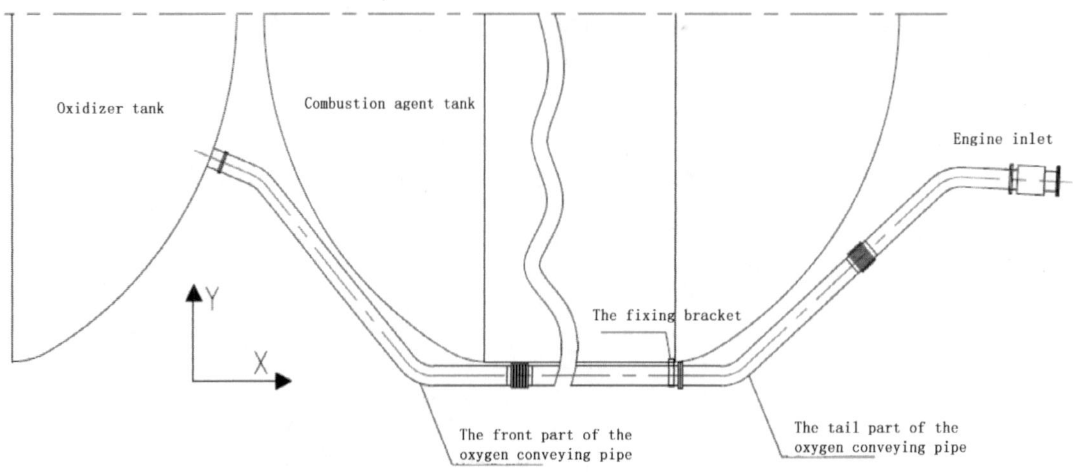


**Figure 2.**Schematic diagram of oxygen pipeline installation

For a launch vehicle, the X axis is the axial direction of the rocket body, and the Y axis is the radial direction. Propellant flows out from the oxygen tank through the delivery pipe, and goes along the side wall of the combustion tank to the engine at the back of the combustion tank. In the process of pipeline design, it has been determined that a U-shaped corrugated pipe is set at the front section of the oxygen delivery pipe and a U-shaped corrugated pipe is set at the tail section of the oxygen delivery pipe.

### *4.2. Input condition analysis*
As can be seen from Figure 2, the bellows of the oxygen delivery pipe is located on the oxygen delivery pipe and connects the tank with the engine. It is necessary to analyze the input conditions.

#### *4.2.1. Tank analysis*
At the same time, the deformation of the tank under various working conditions is calculated in the pipeline design, and the results are shown in Table 1:

**Table 1.** Deformation of tank under various working conditions

| Work condition | Oxidizer tank | | | Combustion agent tank | | |
|---|---|---|---|---|---|---|
| | front bottom | column section | rear bottom | front bottom | column section | rear bottom |
| Transportation | +3 | +2.5 | +3 | +3 | +2.5 | +3 |
| Filling（Influence of propellant weight） | 0 | 0 | +1 | 0 | +0.10 | +0.60 |
| Filling（Low temperature propellant effect） | -4 | -35 | -4 | 0 | 0 | 0 |

| | | | | | | |
|---|---|---|---|---|---|---|
| Pressurization | +4 | +5 | +5 | +5 | +1 | +4 |
| Flight | +6 | -40 | +2 | +4 | +1 | +4 |

Where "+" is expansion, "-" is compression, and all deformations are along the X axis.

*4.2.2. Engine analysis*
During the rocket flight, the deformation of the engine frame is +5mm, and the manufacturing tolerance of the engine support is ±2mm.

*4.2.3. Engine analysis*
The relevant production units provided production tolerances, all of which were in the X-axis direction: the manufacturing tolerance of the back bottom flange of the oxygen tank is (1~3)mm, the manufacturing tolerance of the section between the tanks is ±2mm, the manufacturing tolerance of the combustion tank body is±5mm, and the manufacturing tolerance of the rear short shell of the combustion tank is±2mm.

*4.3. Calculation of compensation amount under various working conditions*

*4.3.1. Compensation amount distribution*
The fixing bracket divides the oxygen box conveying pipe into two parts: the front part of the oxygen conveying pipe and the tail part of the oxygen conveying pipe. Therefore, the compensation distribution of the whole piping system is also considered separately with the fixed bracket as the node, as shown in Figure 3.

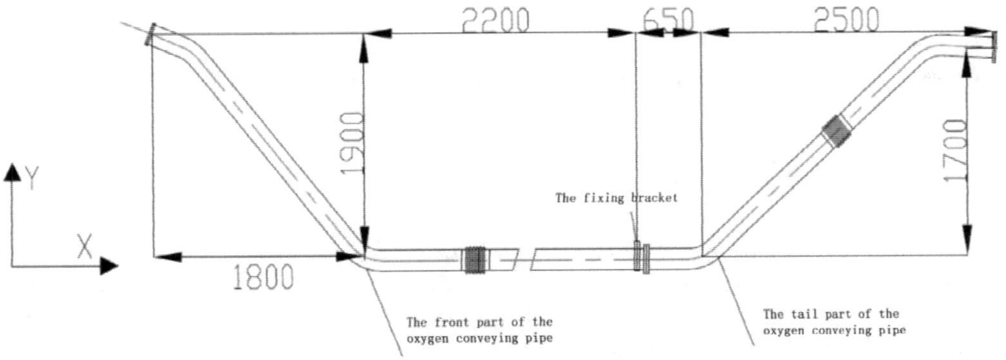


**Figure 3.**Dimension drawing of oxygen delivery pipe (unit: mm)

The whole oxygen delivery pipe is fixed on the short shell behind the combustion tank by a fixing bracket. The compensation amount of bellows in the front section of the oxygen delivery pipe is mainly affected by the machining error of structural parts such as the tank, the deformation of the back bottom of the oxygen box and the deformation of the column section of the combustion box. The compensation amount of bellows in the tail section of oxygen delivery pipe is mainly affected by the machining errors of structural parts such as tank and engine frame, and the deformation of engine frame when working.

In the subsequent calculation process, considering the different directions, the oxygen delivery pipe can be divided into four parts according to the dimensions along the X axis, from left to right, they are L1 (1800mm long in the X direction), L2 (2200mm long in the X direction), L3 (650mm long in the X direction) and L4 (2500mm long in the X direction).

*4.3.2. Calculation of compensation amount in front section of conveying pipe*
The front section of the conveying pipe includes L1 and L2, and the bellows is located in L2.

Set $\delta_X^1$ as the axial compensation amount of bellows, and $\delta_Y^1$ is the radial compensation amount of bellows.

Assembly condition: For the bellows in the front section of the conveying pipe, the assembly errors to be compensated include the manufacturing tolerance of the rear bottom flange of the oxygen tank, the manufacturing tolerance of the section between tanks and the manufacturing tolerance of the combustion tank. The input data introduced in Section 4.2.3 are superimposed.

$$\delta_{X\ (\text{assembly})}^1 = (-6 \sim +10)\ mm \tag{2}$$

Transport condition: Under transport condition, the bellows should not only compensate the deformation of the box body under transport condition, but also consider the production and processing errors. The deformation of the box body under the transportation condition should consider the deformation of the back bottom of the oxygen box and the deformation of the column section of the fuel box. See Table 2 for the data.

$$\delta_{X\ (\text{sum of assembly})}^1 = \delta_{X\ (\text{assembly})}^1 + \delta_{X\ (\text{transport})}^1 = (-6 \sim +10) + (+3) + (+2.5) = (-0.5 \sim +15.5)mm \tag{3}$$

Filling condition: The filling condition is complicated. Bellows are compensated for production and processing errors, box deformation caused by propellant weight, box deformation caused by low-temperature propellant and pipeline deformation caused by low-temperature propellant. Because the L1 pipeline is not perpendicular to the coordinate axis, the deformation of pipeline affected by low-temperature propellant can be divided into X direction and Y direction. The pipeline material is 1Cr18Ni9Ti, and its linear expansion coefficient is $\alpha = 13.5\times10^{-6}(mm/K)$

$$\begin{aligned}\delta_{X\ (\text{sum of filling})}^1 &= \delta_{X\ (\text{assembly})}^1 + \delta_{X\ (\text{propellant weight})}^1 + \delta_{X\ (\text{low temperature of tank})}^1 + \delta_{X\ (\text{low temperature of pipe})}^1 \\ &= (-6 \sim +10) + (-1+0.10) + (+4) + (1800\times203\times13.5\times10^{-6}) \\ &= (+2 \sim +18)\ mm\end{aligned} \tag{4}$$

$$\delta_{Y\ (\text{filling})}^1 = \delta_{Y\ (\text{low tempeture of pipe})}^1 = -1900\times203\times13.5\times10^{-6} = -5.2mm \tag{5}$$

Pressurized working condition: Pressurized working condition is compensated for the tank deformation caused by production and processing errors and pressurization.

$$\delta_{X\ (\text{sum of pressurization})}^1 = \delta_{X\ (\text{assembly})}^1 + \delta_{X\ (\text{pressurization})}^1 = (-6 \sim +10) + (-5) + (+1) = (-10 \sim +6)mm \tag{6}$$

$$\delta_{Y\ (\text{pressurization})}^1 = \delta_{Y\ (\text{sum of filling})}^1 = -5.2mm \tag{7}$$

*4.3.3. Calculation of compensation amount at the tail section of conveying pipe*
The rear section of the conveying pipe includes L3 and L4, and the bellows is located in L4. It can be seen from fig. 3 that the L4 part is not perpendicular to the x and y axes, so the compensation amount of the rocket and the compensation amount of the bellows at the rear section of the oxygen delivery pipe need to be converted.

Set $\delta_X^2$ as the axial compensation amount of bellows, and $\delta_Y^2$ is the radial compensation amount of bellows.

Assembly condition: For the bellows at the rear section of the conveying pipe, the assembly errors to be compensated include the manufacturing tolerance of the short shell behind the combustion tank (±2mm) and the manufacturing tolerance of the engine support (±2mm). The input data introduced in Section 3.2.4 are superimposed.

$$\delta_{X\ (\text{assembly})}^2 \times \frac{2500}{\sqrt{2500^2+1700^2}} = \pm4 \Rightarrow \delta_{X\ (\text{assembly})}^2 = \pm4.8mm \tag{8}$$

$$\delta_{Y\ (\text{assembly})}^2 = 0mm \tag{9}$$

Transportation condition: Besides the assembly error, the factors that affect the compensation amount of the bellows at the tail section of the oxygen delivery pipe include the deformation of the short shell behind the burning box and the deformation of the engine frame. The latter two are zero under the transportation condition, so the compensation amount of bellows under the transportation condition is the same as that under the assembly condition.

$$\delta^2_{\mathrm{X\ (assemblly)}} = \pm 4.8mm \tag{10}$$

$$\delta^2_{\mathrm{Y\ (assembly)}} = 0mm \tag{11}$$

Filling condition: Bellows compensate for production and processing errors, tank structure deformation caused by propellant weight, tank structure deformation caused by low-temperature propellant and pipeline deformation caused by low-temperature propellant. For the bellows at the tail section of the oxygen delivery pipe, it is mainly affected by the structural deformation of the short shell behind the combustion tank and the engine frame, and these two factors are zero in the processing condition.

Because the L4 pipeline is not perpendicular to the coordinate axis, the deformation of pipeline affected by low-temperature propellant can be divided into X direction and Y direction. The pipeline material is 1Cr18Ni9Ti, and its linear expansion coefficient is $\alpha = 13.5\times10^{-6}(mm/K)$

$$\begin{aligned}&\delta^2_{\mathrm{X\ (filling)}} \times \frac{2500}{\sqrt{2500^2+1700^2}} = \delta^2_{\mathrm{X\ (assembly)}} + \delta^2_{\mathrm{X\ (low\ temperature\ of\ pipe)}}\\&=(-4.8\sim+4.8)+[-(650+2500)\times203\times13.5\times10^{-6}]\\&\Rightarrow \delta^2_{\mathrm{X\ (sum\ of\ filling)}} = (-16.2\sim-4.6)\ mm\end{aligned} \tag{12}$$

$$\begin{aligned}&\delta^2_{\mathrm{Y\ (filling)}} \times \frac{1700}{\sqrt{2500^2+1700^2}} = \delta^2_{\mathrm{Y\ (low\ temperature\ of\ pipe)}} = -1700\times203\times13.5\times10^{-6}\\&\Rightarrow \delta^2_{\mathrm{Y\ (filling)}} = -8.3mm\end{aligned} \tag{13}$$

Pressurized working condition: Under Pressurized working condition, the deformation of the short shell behind the combustion tank and the deformation of the engine frame are all zero, and the compensation of bellows is only the production and processing error.

$$\delta^2_{\mathrm{X\ (pressurization)}} = \pm 4.8mm \tag{14}$$

$$\delta^2_{\mathrm{Y\ (pressurization)}} = \delta^2_{\mathrm{Y\ (filling)}} = -8.3mm \tag{15}$$

Flight condition: The compensation for flight condition is the manufacturing error and the deformation of engine frame caused by flight.

$$\begin{aligned}&\delta^2_{\mathrm{X\ (flight)}} \times \frac{2500}{\sqrt{2500^2+1700^2}} = \delta^2_{\mathrm{X\ (assembly)}} + \delta^2_{\mathrm{X\ (frame)}} = (-4.8\sim+4.8)+(-5\sim+5)\\&\Rightarrow \delta^2_{\mathrm{X\ (sum\ of\ flight)}} = \pm 11.85mm\end{aligned} \tag{16}$$

$$\delta^2_{\mathrm{Y\ (flight)}} = 0mm \tag{17}$$

Through calculation, the compensation requirements of corrugated pipe in transmission pipeline under various working conditions are shown in Table 2:

**Table 2.** Compensation index of bellows under various working conditions (mm)

| | Bellows in front section of conveying pipe | | Bellows at rear section of conveying pipe | |
|---|---|---|---|---|
| | axial | radial | axial | radial |
| Assembly | +10 | 0 | +4.8 | 0 |
| Transportation | +15.5 | 0 | +4.8 | 0 |
| Filling | +18 | -5.2 | -16.2 | +8.3 |
| Pressurization | -10 | -5.2 | +4.8 | -8.3 |
| Flight | +9 | 0 | 11.85 | 0 |

### *4.4. Check bellows and superposition of fatigue damage under various working conditions.*

#### *4.4.1. Pipeline design output*

The working and constraint conditions of the pipeline where the bellows is located are shown in Table 3.

**Table 3.** Working and constraint conditions of bellows

| | Assembly | Transportation | Filling | Pressurization | Flight |
|---|---|---|---|---|---|
| Working temperature（K） | 293 | 293 | 90 | 90 | 90 |
| Working Pressure（MPa） | 0 | 0.06 | 0.20 | 0.35 | 0.35 |
| Working life（frequency） | 5 | 10 | 5 | 5 | 1 |
| Bellows form | U | | | | |
| Inner diameter(mm) | Φ46 | | | | |
| Maximum outside diameter(mm) | Φ80 | | | | |
| Thickness(mm) | 1.0 | | | | |

#### *4.4.2. Product shelf product*

Before checking the parameter, firstly, according to the bellows type spectrum library, select the bellows model that meets the dimensions. The main parameters of the model selected in this paper are:

- Outside diameter is 80mm
- Outside diameter of straight section is 60mm.
- Wave distance is 9mm
- Wave number is 10
- Layers number is 2
- Thickness of single layer is 0.3mm

In this example, the medium in the bellows is low-temperature liquid oxygen, and 022Cr17Ni12Mo2 is selected as the bellows material. The following table shows the mechanical properties of 022Cr17Ni12Mo2 under various working conditions.

**Table 4.** Mechanical performance indicators of corrugated pipe materials under various working conditions

| | Working temperature（K） | $R_{P0.2}$ MPa | $R_m$ MPa | E GPa |
|---|---|---|---|---|
| Assembly | 293 | 205 | 520 | 198 |
| Transportation | 293 | 205 | 520 | 198 |
| Filling | 90 | 380 | 1625 | 214 |
| Pressurization | 90 | 380 | 1625 | 214 |
| Flight | 90 | 380 | 1625 | 214 |

*4.4.3. Check the fatigue damage superposition under various working conditions.*
Using the input conditions obtained above, the bellow check is carried out by using the check formula [3][4]. On the premise of meeting the strength, stability and other indicators, the fatigue life check results of the bellows in each working condition (considering 10 times safety factor) are as follows:

**Table 5.** Check result of fatigue life of conveying pipe

| | Bellows in front section of conveying pipe | Bellows at rear section of conveying pipe |
|---|---|---|
| Assembly | 409004 | Enormous |
| Transportation | 9717 | Enormous |
| Filling | 298 | 159 |
| Pressurization | 1063 | 668 |
| Flight | Enormous | Enormous |

The fatigue damage coefficients under various working conditions are superimposed as follows:
Bellows in front section of conveying pipe:

$$\begin{aligned} D &= \frac{n_1}{N_1} + \frac{n_2}{N_2} + \frac{n_3}{N_3} + \frac{n_4}{N_3} + \frac{n_5}{N_5} \\ &= \frac{5}{409004} + \frac{10}{9717} + \frac{5}{298} + \frac{5}{1063} + \frac{1}{enormous} \\ &= 0.0225 < 1 \end{aligned} \tag{18}$$

Bellows at rear section of conveying pipe:

$$\begin{aligned} D &= \frac{n_1}{N_1} + \frac{n_2}{N_2} + \frac{n_3}{N_3} + \frac{n_4}{N_3} + \frac{n_5}{N_5} \\ &= \frac{5}{enormous} + \frac{10}{enormous} + \frac{5}{159} + \frac{5}{668} + \frac{1}{enormous} \\ &= 0.0389 < 1 \end{aligned} \tag{19}$$

It can be seen that the bellows parameters selected in this example can meet the design requirements.

## 5. Conclusion

This paper systematically introduces the design method of bellows of launch vehicle under different working conditions from the aspects of material selection, index input, compensation calculation and qualification judgment. This method focuses on the analysis of the working conditions of the system under various working conditions, and successfully solves the difficult problem of bellows design of launch vehicle. In the design process using this method, the following two points should be paid attention to at the same time:

- 1. Because the existing checking formula is mainly for the design of ground pipeline, and the estimation of the flight vibration reliability of launch vehicle is insufficient, the vibration fatigue analysis and test of pipeline where bellows is located should be carried out separately, and its analysis methods include finite element analysis method and bellows random vibration equivalent method.
- 2. In the process of compensation check, the position of bellows clamp bracket should be fully considered.

## References

[1] Liao, S.Y. (2007) Rocket propulsion pressurized conveying system. National Defense Industry Press, BeiJing.
[2] EJMA-2011,Standards of the Expansion Joint Manufactures Association Inc.[S]

[3] Li,Y.S. (2003) Practical technology of corrugated expansion joint: design, manufacture and application.Chemical industry press,BeiJing.
[4] GB/T12777-2008,General specification for metal bellows expansion joints.[S]

# Numerical simulation of jet atomization process of swirl nozzle

**Na Gao[1,2*], Xiaobin Zhang[3]**

[1]AVIC Aerodynamics Research Institute, Harbin, 150001, China,

[2]Low Speed High Reynolds Number Aeronautical Key Laboratory, Harbin, 150001 China,

[3]College of Aerospace and Civil Engineering, Harbin Engineering University, Harbin, 150001, China,

*Corresponding author's e-mail: gn0427@163.com

**Abstract.** As the initial conditions of evaporative combustion, the jet atomization formed by the swirl nozzle has the characteristics of large distribution space and small atomization particle size, which is very important for the study of the jet atomization of the swirl nozzle. In this paper, the cross-scale VTD model is used to comprehensively consider the actual situation of multi-scale and multi-phase flow coupling in the atomization process. The evolution process and the breaking shape of the rotating cone liquid film during the atomization and crushing process are studied. The results show that the complete broken shape formed by the rotating cone liquid film can be divided into three areas along the axial direction: continuous liquid film, primary crushing, and secondary crushing. Among them, the appearance of irregularly distributed small holes is a significant feature of primary crushing. The sign of secondary crushing is that the liquid filaments are broken into small-scale droplet particles; after the primary and secondary crushing, the fully developed cone-shaped liquid film will appear to move up the position of the primary crushing cavity and reduce the scope. At the same time, the initial position of the secondary crushing will move up.

## 1. Introduction

Nozzle structure is widely used in various fields [1-2], according to the nozzle function can be divided into two types of combustion equipment and non-combustion equipment. Applications in non-combustion equipment such as mechanical cooling in industry, industrial boilers, and 3D printing, etc., but the importance of nozzles is more reflected in a variety of combustion devices, such as in the aerospace field, from high-performance aviation engines, liquid rocket engines to scramjets, etc., all have extremely high requirements for the nozzle structure design. As one of the most important components in aerospace engines, the nozzle is a key component of the thrust chamber of the liquid rocket engine-the component unit of the injector. Its function is to atomize the liquid fuel to form small-scale droplets and make the liquid fuel which is mixed with oxygen increase the gas-liquid contact area and achieve the goal of full combustion of the liquid fuel. Therefore, the nozzle atomization performance design is directly related to the overall combustion effect of the engine, and the atomization process is both a transient process and a variable flow resistance process. The study found that the gas-liquid two-phase atomization can be divided into one phase according to the different crushing mechanism. Crushing and secondary crushing [3], the process of jet crushing to form larger size droplets in the initial stage of atomization is called primary crushing, and the size of the crushing is generally above the millimeter level; while the secondary crushing is the droplets or liquid formed by the primary crushing. Under the influence of air resistance, the process of further crushing and atomization of the silk mainly occurs in the gas-liquid two-phase mixing area, and the droplet particle size is in the range of a few microns to a few hundred microns. The research on the atomization and crushing process of swirl nozzles has become a hot and difficult problem at present, so a large number of scholars have carried out experiments and numerical simulations.

After a lot of research, scholars have found that although multiphase flow methods such as the VOF model can solve the single-phase momentum equation while ensuring the conservation of mass, the detailed characteristics of the jet atomization process can be obtained[4]. The secondary crushing simulation for atomization is limited. Only when the grid size is smaller than the secondary crushing droplet size, it can be used to study the secondary crushing phenomenon. This makes the grid requirement the primary problem that restricts the simulation; the DPM model is a This is a commonly used solution method for discrete particles[5], but in the same way, the DPM model does not have the continuous phase solution ability for the overall jet problem, so it is not suitable for the simulation before the secondary crushing process, so it is completely unsuitable It is used for the simulation of the liquid from the ejection section to the crushing section. The atomization process is a typical cross-scale process. In the simulation, both the continuous phase and the discrete phase participate in the flow. However, there are few studies on the multi-scale algorithm of the atomization process, and it is mainly realized by the following methods. Herrmann[6] proposed a multi-scale simulation method based on Level Set; Kim[7] used an improved Level Set method to directly solve the two-phase fluid. When the closed interface volume is less than twice the cell, the droplet is converted into LPT droplets, so as to achieve the conversion of Euler phase to discrete items; Tomar[8] also used a similar method and added grid adaptation, based on the open-source software Gerris, taking into account the liquid film breakage and droplet fusion process, using The two-way coupling method of VOF model and LPT carried out a multi-scale simulation of atomization breaking.

Based on the VOF-to-DPM model (VTD model), the author has carried out a numerical simulation study on the development process of the atomized liquid film formed by the rotating nozzle. The VOF model is used for the primary crushing, and the DPM model is used for the secondary crushing. The VOF model captures the phase interface. Combining the advantages of DPM with the advantages of DPM for the dispersion of discrete particles, it can better solve the problem of cross-scale, in order to accurately obtain the crushing characteristics of the conical liquid film under the action of the swirl nozzle. On this basis, the evolution process and crushing morphology of the rotating cone liquid film under rotating conditions are analyzed, and the influence of different jet parameters on the characteristics of the cone liquid film during the crushing process is explored.

## 2. Numerical calculation method

### *2.1. Physical model*

This paper mainly studies the breaking phenomenon of the rotating cone liquid film formed after the annular liquid film is ejected from the nozzle expansion section. Considering the calculation conditions and the periodic characteristics of the rotating cone liquid film, the cone liquid film atomization interval Simplified to a 1/4 cone, so that while reducing the amount of calculation, it can accurately reflect the actual situation of the breaking of the rotating cone liquid film. The calculation domain is mainly composed of 6 boundaries, which are the gas boundary corresponding to the gas core, the liquid boundary corresponding to the annular liquid film, and the periodic boundary. The specific model is shown in Figure 1, and the hydraulic diameter of the liquid inlet is 0.85mm. In order to better capture the two-phase interface, this paper uses adaptive grid technology to encrypt the two-phase interface with 4 steps, and the minimum encryption unit volume threshold is $2e^{-12}m^3$.

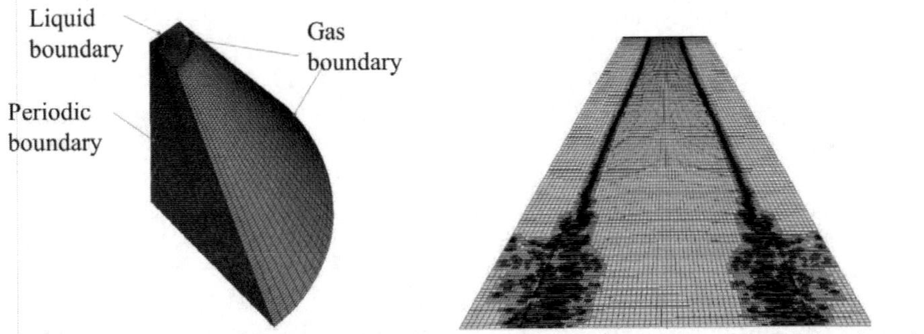

(a)Calculation model (b)Adaptive grid when rotating cone liquid film is fully developed
Fig.1 Mesh division

### *2.2. Boundary conditions and solutions*
Set the liquid boundary as the velocity inlet, and set the same side and bottom surface of the gas boundary as the pressure outlet, and the pressure value is one atmosphere. In this paper, the LES model is used to simulate the turbulence effect in the atomization process, and the VTD model is used to study the atomization breaking process.

### *2.3. Model validation*
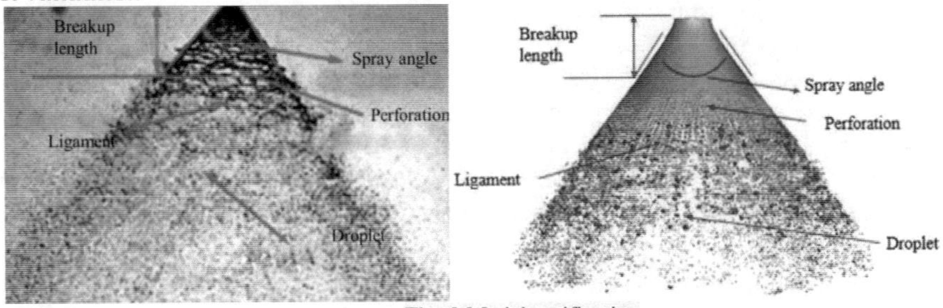

Fig. 2 Model verification

Figure 2 are numerical calculations and images of the cone-shaped liquid film breaking process taken by Wang et al. It can be seen from the figure that the calculation results in this paper can meticulously capture various detailed structures such as liquid filaments and droplets during the atomization process, which is basically consistent with the cone-shaped liquid film atomization and the crushing process taken by high-speed photography. It can be seen that the calculation method based on the cross-scale VTD model can well capture the details of the atomization and crushing process of the cone liquid film.

## 3. Numerical calculation method

### *3.1. The evolution of the overall shape of the rotating cone liquid film*
Swirling jet is a common atomization method, and scholars continue to put forward new understandings on the broken distribution morphology and formation mechanism of the liquid film. The initial shape of the rotating cone-shaped liquid film ejected from the annular nozzle into the calculation domain will be affected by the gas-phase interaction in the flow field and the centrifugal force of the liquid film, both of which will cause the flow field to change, and then the liquid film will be broken and destroyed. The secondary crushing has an impact. Therefore, it is necessary to conduct an in-depth comparative analysis of the breaking mechanism of the rotating cone liquid film. The following will analyze in detail the overall evolution of the primary and secondary crushing involved in the injection of the rotating cone-shaped liquid film from the annular inlet to the complete atomization of the liquid film.

Figure 3 shows the overall morphological evolution of the rotating cone liquid film from the annular inlet to the complete development at different times. This is to analyze the liquid jet entering the static gas environment to the complete development to form the rotating cone liquid film and the process of atomization and fragmentation.

(1) At t=0.5ms, in the initial stage of the liquid ejecting from the annular inlet, the tangential velocity of the liquid film will develop freely and gradually convert to radial velocity due to the loss of the centripetal force constraint provided by the wall, which makes the ring The cone-shaped rotating jet still develops in a conical posture and causes its initial section to have an outward arc. The cone angle of the liquid film will gradually increase. Due to the resistance of the gas to the jet, the head of the rotating cone-shaped liquid film There will be a lagging eversion circle.

(2) When t=1.0ms, the cone-shaped liquid film continues to develop, the head of the liquid film still has an outward turning circle, and the liquid film between the eversion circle and the cone-shaped liquid film has undergone a second break, the outer ring There is still a complete circle of liquid ribbon, and there are a small number of discrete particle clusters between the liquid filament and the tapered liquid film on the periphery of the lagging turning circle.

(3) When t=1.5ms, a large number of cavities and broken liquid filaments appear in the liquid film and form a cross-network distribution. The main body of the cone-shaped liquid film has completed a crushing process, and the lagging eversion ring is completely broken into discrete phase droplets. There are newly formed discrete particle clumps at the head of the liquid film, which indicates that the cone-shaped liquid film body has begun to enter the secondary crushing area.

(4) When t=2.0ms, the broken liquid filaments continue to be broken into smaller size liquid filaments, the liquid filaments at the tip of the cone nozzle become sparse, almost all the liquid filaments have undergone secondary crushing, and the front end is full of discrete phases. Agglomerates of particles, continuous droplets become sparse along the axis direction, at this moment the cone-shaped liquid film has been roughly formed.

(5) When t=2.5ms, the rotating cone-shaped liquid film has been fully developed. During the process from the last moment to full development, the disturbance wave downstream of the cone-shaped liquid film will affect the movement and development of the upstream liquid, thus promoting the rotation Compared with the crushing form at t=2.0ms, the primary crushing of the cone-shaped liquid film will cause the position of the primary crushing cavity to move up, the range will decrease, and the starting position of the secondary crushing will move up.

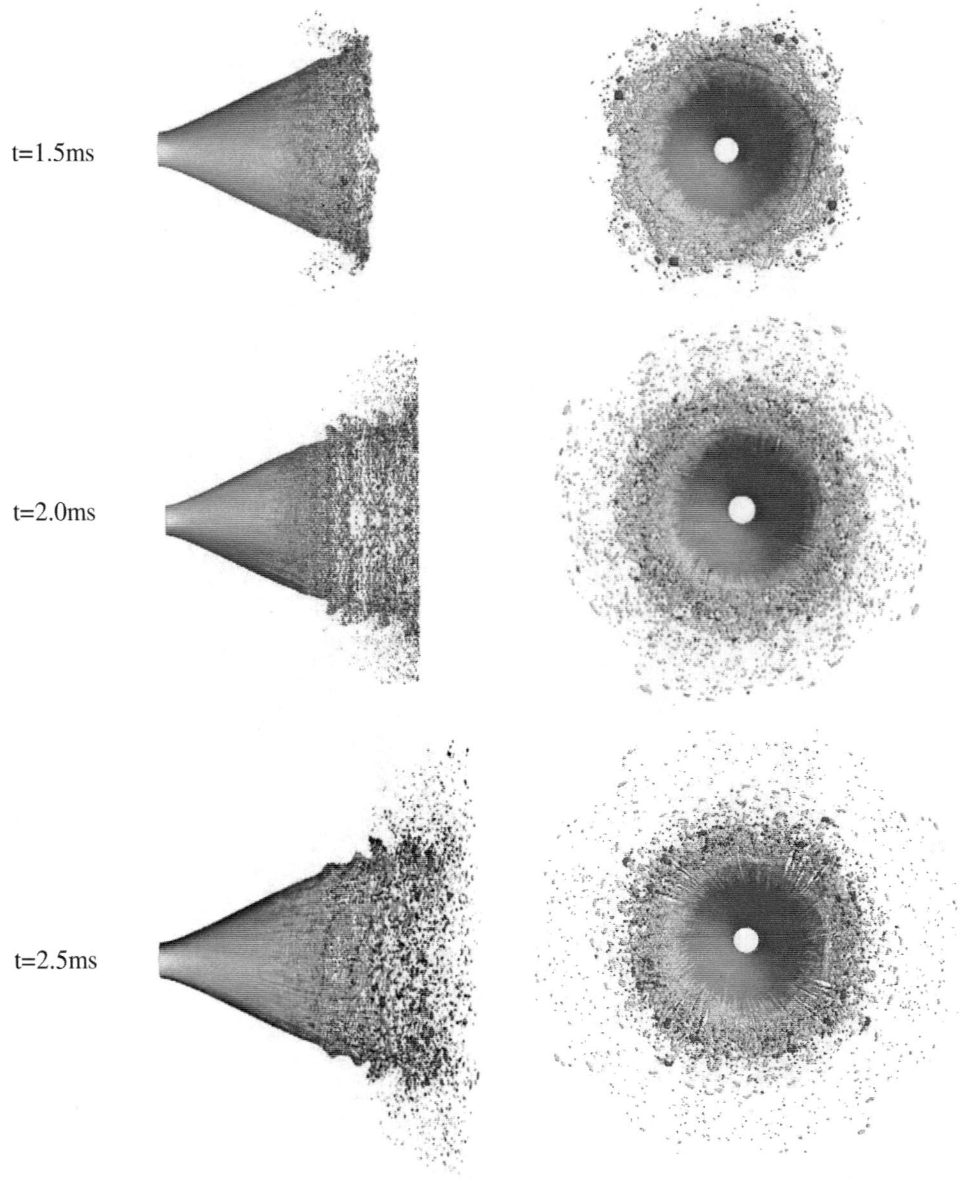


Fig.3 The atomized and broken shape of the rotating cone liquid film at different times

### *3.2. Breaking process of rotating cone liquid film*

The complete liquid film breaking form formed by the rotating cone liquid film after fully developed can be divided into three areas. Along the axial direction, it is the continuous liquid film area first, and then it is divided into the primary crushing area and the secondary crushing area in turn, as shown in Figure 4. As shown in the left picture, the different areas have been divided by solid lines in the picture. For the breaking of the rotating cone liquid film, in the continuous liquid film stage, the jet mainly presents a uniform and smooth liquid film distribution after ejection, and the formed cone angle gradually increases until it is stable. The right picture of Figure 4 shows a partial enlargement of the distribution of the liquid film once broken. It can be found that the appearance of irregularly distributed small holes is the most significant feature when a break occurs. The pulling force generated by the fluctuation of the liquid film surface overcomes the liquid film. The surface tension of the film itself causes the liquid film to appear small holes at the troughs of the surface waves. The small holes expand with time. Under the influence of the axial surface wave, the small holes will merge with adjacent holes on the same axis to form a band-shaped hole. The liquid film is torn to a certain extent to form a ribbon-like liquid filament, which causes the liquid film to break axially. The morphology of the liquid film transitions from a smooth and uniform continuous liquid film to a dense networked liquid film structure. At the end of a crush, a large number of band-shaped liquid filaments will be formed. As can be seen from the right figure of Figure 4, these fluid filaments are in cross-section. The liquid phase distribution is shown as discontinuous point-like distributions with similar scales.

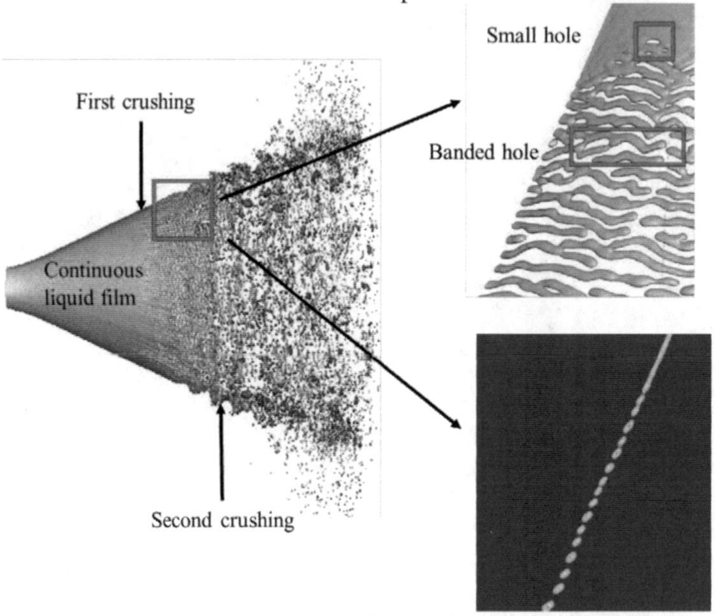


Fig.4 The shape distribution of the broken cone liquid film

The sign of the secondary breakage of the liquid film atomization is that the liquid filaments are broken to form small-scale droplet particles. Figure 5 describes the secondary breakage of the rotating cone-shaped liquid film. In Figure 5, droplets begin to appear between the liquid filaments. Particles (red), which means that the liquid film has entered the secondary crushing stage, and it can be seen from Figure 5 that the distribution of the liquid filaments captured by the VOF changes from dense to sparse in the axial direction. This is due to the continuous conversion of the liquid filaments in the secondary crushing area DPM droplet particles, with the downward movement of the unfolding liquid film, the size of the liquid filaments becomes smaller, and the degree of conversion continues to increase.

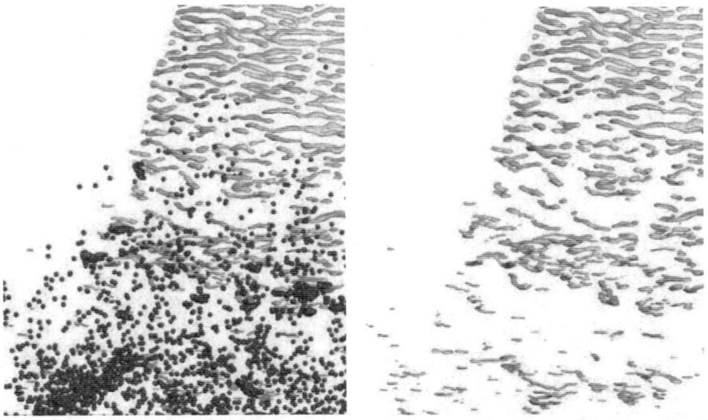

（a）VTD liquid mist distribution form （b）Continuous phase liquid mist distribution form

Fig. 5 Local distribution of secondary crushing of cone liquid film

The coupling of the VTD model used in this article during the atomization process is mainly for the secondary crushing process. Figure 6 strips the large-scale continuous droplets from the small-scale discrete phase droplets in the rotating cone liquid film. It is found that the discrete phase droplets are distributed in an annular shape, and the inner circle is relatively dense and discrete. The droplet particles are in the form of larger particle clusters. These large droplets instantly undergo a large number of secondary crushing. It can be seen from the above that the asphericity is 0.5, and the volume equivalent sphere diameter ranges from 0.1 to 0.5. There is still a small amount of larger droplets, Their asphericity and block size do not meet the transformation requirements and continue to spread. The area where the secondary crushing occurs is much larger than the continuous liquid film area and the primary crushing area.

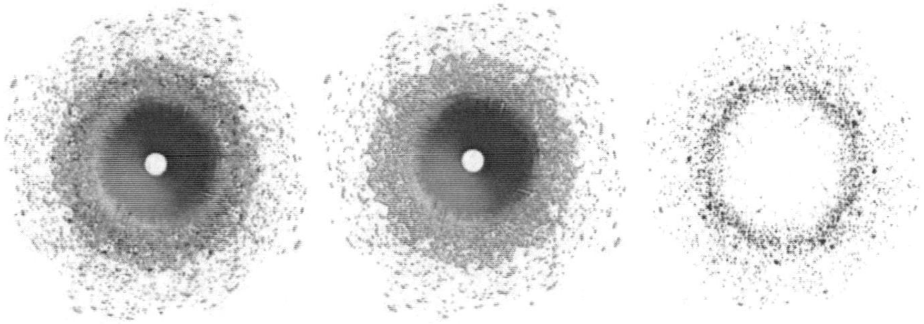

（a）Coupling phase （b）Continuous phase （c）Discrete phase

Fig.6 Coupling schematic diagram of the rotating cone liquid film VTD model

**4. Conclusion**

Based on the adaptive grid technology, this paper uses the cross-scale VTD model to comprehensively consider the actual situation of multi-scale and multi-phase flow coupling in the atomization process. The atomization and fragmentation process of the swirl nozzle jet is studied, and the conclusions are as follows :

1). Due to the resistance of the gas to the jet, the rotating cone liquid film will produce a lagging eversion ring on the head. At the same time, after the primary crushing and secondary crushing, the fully developed cone-shaped liquid film will move up the position of the primary crushing cavity, reduce the scope, and move up the starting position of the secondary crushing compared with the previous stage.

2). The complete crushing form formed by the rotating cone-shaped liquid film can be divided into three regions along the axial direction: continuous liquid film, primary crushing, and secondary crushing, and the appearance of irregularly distributed small holes is a significant feature of primary crushing. The sign of secondary breakage is that the liquid filament breaks into small-scale droplet particles.

References

[1] Gan, X.H. Jet Fuel Nozzle Technology for Aviation Gas Turbine[M]. National Defense Industry Press: Beijing, China, 2006.
[2] Pinero, L.R.; Bowers, G.E. Power electronics for a miniaturized arcjet[R]. AIAA. 1996, 96-2961.
[3] Smith, T.F.; Shen, Z.F.; Friedman, J.N. Evaluation of coefficients for the weighted sum of gray gases model[J]. Journal of Heat Transfer. 1982, 104(4), 602-608.
[4] Wan, W.Y.; Raza, A.; Chen, X.Y. Effect of Height and Geometry of Stepped Spillway on Inception Point Location[J]. Applied Sciences. 2019, 9(10), 2091.
[5] Zhang, X.B.; Zhu, W.B.; Qian, L.; Li, Miao. Research on Performance Prediction Model of Impeller-Type Breather[J]. Applied Sciences. 2019, 9(17), 3504.
[6] Herrmann, M. Refined Level Set Grid Method for Rracking Interfaces[R]. Center for Turbulence Reseach Annual Research Briefs, 2005.
[7] Kim, D.; Herrmann, M.; Moin, P. The breakup of a round liquid jet by a coaxial flow of gas using the refined level set grid method[R]. APS division of fluid dynamics meeting abstracts, 2006.
[8] Tomar, G.; Fuster, D.; Zaleski, S.; Popinetc, S. Multiscale Simulations of Primary Atomization[J]. Computers&Fluids. 2010, 39(10), 1864-1874.

# Formation of $H_2CO$ and $CH_3OH$ in the Protoplanetary Disks

**Lin Xiao[1,a], Xia Liu [2,*] , Li Zhang[3,b]**

[1]School of Science, Changchun Institute of Technology, Changchun, China

[2]School of Civil Engineering, Changchun Institute of Technology, Changchun, China

[3]School of Civil Engineering, Changchun Institute of Technology, Changchun, China

[a]xiaolin1702@126.com; *e-mail: liuxia@ccit.edu.cn; [b]dbcy9999@163.com

**Abstract**—It is known that the complex organic molecules in planetary systems may be formed by the organic chemistry of gases and particles in protoplanetary disks. Among these molecules, $H_2CO$ and $CH_3OH$ are potential precursors of essential organic compounds like glycolaldehyde. We calculate the physical and chemical structures of the protoplanetary disks around LkCa 15 and MWC 480 stars, compare the obtained column densities with the observed values. Our results mostly show reasonable agreements with observations. We identify the formation of $H_2CO$ to be through the both gas-phase and dust surface reactions, while $CH_3OH$ is dominantly formed by surface chemistry in the original astrolabe and then released into the gas-phase by surface exothermic chemical reactions in the LkCa 15 disk. We studied the synthesis of complex organic molecules ($NH_2CHO$ and $CH_3OCHO$) in the protoplanetary disk to predict the observable species and Molecular abundance. In the disks around LkCa 15 and MWC 480 stars, we suggest that the strong spectral line observation of $NH_2CHO$ belonging to band 4 may obtain good results with ALMA, the observation of $CH_3OCHO$ may be challenging.

## 1. Introduction

Planets are formed among the protoplanetary disks (PPDs) [1] while protoplanetary disks are formed from the molecular cloud core collapse. Protoplanetary disks are composed of gas and dust particles [2]. Organic molecules in planetary systems can promote prebiotic chemistry leading to formation of the different building blocks of RNA and proteins [3]. There is well known active chemistry going on in protoplanetary disks since molecules have been observed toward these sources [4, 5, 6].

Among the molecules that were detected toward protoplanetary disks, $H_2CO$ and $CH_3OH$ caused much interest due to two reasons: Firstly, these molecules play an important role in organic chemistry. $H_2CO$ is the precursor of some organic molecules. $H_2CO$ is the precursor of some organic molecules (e.g., $NH_2CHO$ or $CH_3OCHO$), that are considered potential precursors of many organic compounds essential to life. Secondly, these two molecules are believed to be abundant in these regions [6]. In general, in gas-phase chemistry, $H_2CO$ can form through neutral-neutral reactions, but these reactions require a higher temperature environment in the inner disk or surface layers. In grain-surface chemistry, $H_2CO$ can form via CO ice hydrogenation, and can generate $CH_3OH$. In contrast, it is a common belief that $CH_3OH$ only forms through ice chemistry. The reaction related to the formation of $H_2CO$ and $CH_3OH$ is shown in Fig.1.

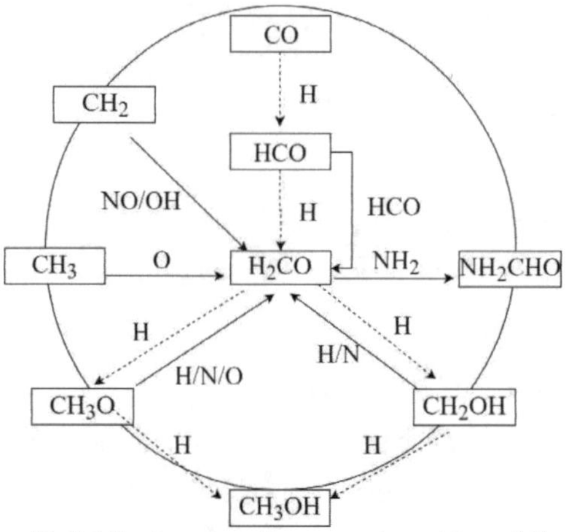


Figure 1. The Gas-phase (Solid line) and surface reactions (dotted line) of $H_2CO$ and $CH_3OH$.

The Atacama Large Millimeter/ submillimeter Array (ALMA) has been used to search for $H_2CO$ toward various disks. $H_2CO$ is one of the most common species detected in the protoplanetary disks. We found that the detection rate is higher in T Tauri disks than Herbig Ae/Be disks. However, both column densities are a few times $10^{12}cm^{-2}$.

So far, $CH_3OH$ has been observed in TW Hya disk, and the column density is $(3\text{-}6)\times10^{12}$ $cm^{-2}$, that is one order of magnitude lower than that predicted by the model [6]. Upper limits on $CH_3OH$ in the HD 163296 disk is $5\times10^{12}$ $cm^{-2}$ and the results of [6] calculation show that the methanol abundant of HD 163296 disk is relatively less than that of TW Hya disk. HD 163296 is a Herbig Ae star, while TW Hya is a T Tauri star. Reference finded a $3\sigma$ upper limit column density of $8\times10^{12}$ $cm^{-2}$ in MWC 480 and LkCa 15.

In this study, we investigated the distribution and chemistry of $H_2CO$ and $CH_3OH$ in the disks around LkCa 15 and MWC 480. LkCa 15 is a young low-mass star (T Tauri), while MWC 480 is an intermediate-mass (Herbig Ae/Be) star. Their mass and luminosity are quite different, which could introduce different effects on the chemical composition of the disk. Some people studied the sulfur chemistry in protoplanetary disks, and reproduced the observations of MWC 480 and LkCa 15 by model prediction. We compared modelling results with the observations, and predicted the column density of $CH_3OH$ in these two disks.

## 2. The disk model

Although MWC 480 and LkCa 15 are both in Taurus scluster. However, their age, luminosity, and mass are different. MWC 480 is more than 20 times brighter than LkCa 15, and its stellar mass is about twice that of LkCa 15, Table 1 shows some parameters of these two disks that are determined by observations. We consider models of axisymmetric disk in steady state around MWC 480 or LkCa 15 star. In the models, the disk specifically along the radius r and height z, using 1+1D disk modeling. The physical structure parameters used for calculation are shown in Table 2.

TABLE 1. Stellar and disks Characteristics of the MWC 480 and LkCa 15

| **Disk** | **Stellar Type** | **Age*** **(Myr)** | $L_*$ **($L_\odot$)** | $M_*$ **($M_\odot$)** | **Disk Mass** **($M_\odot$)** | $T_{eff}$ **(K)** |
|---|---|---|---|---|---|---|
| LkCa 15 | K3-K5 | 0.9-4.3 | 1 | 1 | 0.05-0.1 | 4350 |
| MWC 480 | A4.5±2 | 6 | 19-25 | 1.7-2.3 | 0.11 | 8330 |

TABLE 2. PHYSICAL PARAMETERS

| **Parameters** | **MWC 480** | **LkCa 15** |
|---|---|---|
| Stellar mass: $M_*$ ($M_\odot$) | 1.8 | 1 |
| Disk mass: $M_d$ ($M_\odot$) | 0.18 | 0.03 |
| Stellar luminosity: $L_*$ ($L_\odot$) | 25 | 1 |
| Characteristic radius: $R_0$ (AU) | 100 | 100 |
| Outer cut-off radius: $r$ (AU) | 500 | 550 |
| Midplane temperature at $R_0$ (K) | 30 | 15 |
| Atmosphere temperature at $R_0$ (K) | 48 | 20 |
| Surface density at $R_0$: | 5.7 | 0.9 |
| Temperature power-law index: $q$ | 0.5 | 0.5 |
| Vertical temperature gradient index: $\delta$ | 2 | 2 |
| UV Flux of reference: $f_{uv, R0}$ | 8500 | 2550 |

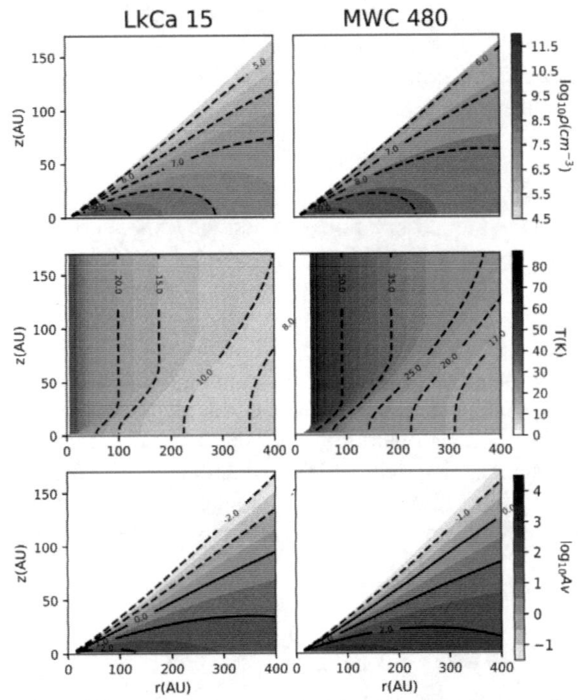


Figure 2. Number density of H, temperature, and extinction distributions in the LkCa 15 (left) and MWC 480 (right) disks.

To reveal the chemical processes in the disk, we input the outcome of MWC 480 and LkCa 15 physical models (see Fig.2) into Nautilus v1.1. To understand the chemistry of the disk at the time of planetary formation, we started from the molecular cloud stage. We ran the dense cloud chemistry models for $10^6$ yr at the first stage. The physical conditions we used are as the following: the grain and gas temperature of 10 K, the gas density of $n_H = 2\times10^4$ cm$^{-3}$, the visual extinction of 15 mag and the cosmic ionization rate of $1.3\times10^{-17}$ s$^{-1}$. The element abundances are shown in Table 3. The output composition of the molecular cloud was used as the initial condition for the disk.

TABLE 3. INITIAL ABUNDANCES

| **Species** | $n_i / n_H$ | **Reference** |
|---|---|---|
| $H_2$ | 0.5 | |
| He | $9.0\times10^{-2}$ | 1 |
| $C^+$ | $1.7\times10^{-4}$ | 2 |
| N | $6.2\times10^{-5}$ | 2 |
| O | $2.4\times10^{-4}$ | 3 |
| $S^+$ | $8.0\times10^{-9}$ | 4 |
| $Si^+$ | $8.0\times10^{-8}$ | 4 |
| $Fe^+$ | $3.0\times10^{-9}$ | 4 |
| $Na^+$ | $2.0\times10^{-9}$ | 4 |
| $Me^+$ | $7.0\times10^{-9}$ | 4 |
| $P^+$ | $2.0\times10^{-10}$ | 4 |
| $Cl^+$ | $1.0\times10^{-9}$ | 4 |
| $F^+$ | $6.7\times10^{-9}$ | 5 |

In the second stage, we simulated the chemical composition of the disk for a million years. Although this time range is a little short compared with the age of the disk, some studies found that if the growth of dust particles is neglected, the chemical evolution will become slow enough after one million years, thus this is an reasonable approximate.

## 3. Simulation Results

We present the results of column density distributions of various molecules in the LkCa 15 and MWC 480 and compare them with the observational results in this section,. Then we discuss on the distribution and formation paths of $H_2CO$ and $CH_3OH$ in these two disks.

### *3.1.CO*

In Figure 3, we show the calculated distribution of CO ice in the LkCa 15 and MWC 480 disks. The icy CO abundance in the MWC 480 disk is higher than that in the LkCa 15 disk, especially at the middle height. In the vertical range of the outer disk, the MWC 480 and LkCa 15 disks both have temperatures below 30 K, which is not high enough here to trigger the thermal desorption of CO ice, controlled by photodesorption form interstellar and stellar FUV radiation. At the middle height, the extinction of MWC 480 disk is one order of magnitude higher than that of LkCa 15. Therefore, for MWC 480, FUV is more shielded. The photodesorption of CO ice is ineffective for MWC 480, comparing with LkCa 15.

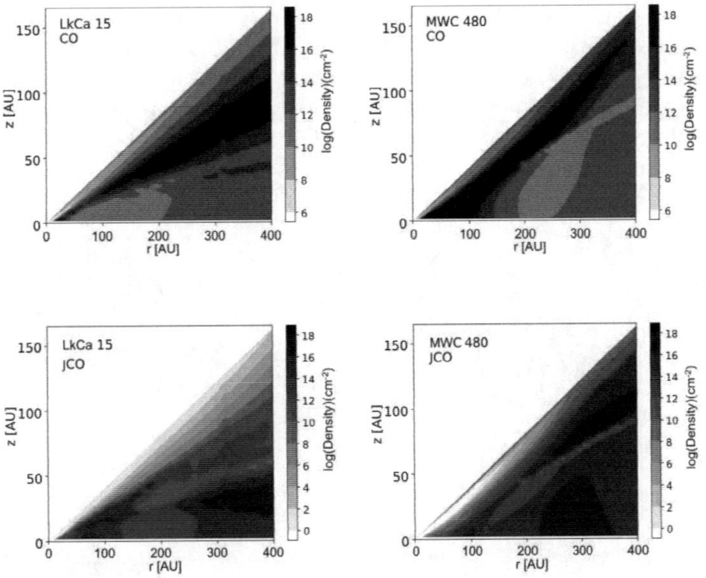


Figure 3. CO and CO ices distributions in the LkCa 15 (left) and MWC 480 (right) disks.

*3.2.$N_2H^+$*

Reference [5] investigated the spatial distribution of $N_2H^+$ in TW Hya disks. $N_2H^+$ has been detected in many disks. Fig.4 shows $N_2H^+$ distributions in our disk model. $N_2H^+$ is confined to a layer closer to the midplane in LkCa 15 disk, only in the region of R < 50 AU, where the column density of CO reaches $10^{12}$ $cm^{-2}$. The predicted column density is one or two orders of magnitude lower than the observed value of $N_2H^+$ in T Tauri disks similar to LkCa 15. The column density of $N_2H^+$ in MWC 480 is higher between 100-300 AU, reaching $10^{11}$-$10^{12}$ $cm^{-2}$. This is consistent with the observation range of HD 163296 disk. In our model results, the column density of $N_2H^+$ in LkCa 15 is lower than in MWC 480 at the outer disk region (R > 50 AU). The low column density of $N_2H^+$ in LkCa 15 disk is due to the use of the three-phase model to prevent the evaporation of $N_2H^+$'s precursor $N_2$, making the production reaction in effective. In these two disks, there is a negative correlation between CO and $N_2H^+$ in the gas phase. This is not because $N_2H^+$ is destroyed by CO, but the formation paths of CO and $N_2H^+$ are the same, there is a competitive relationship between them.

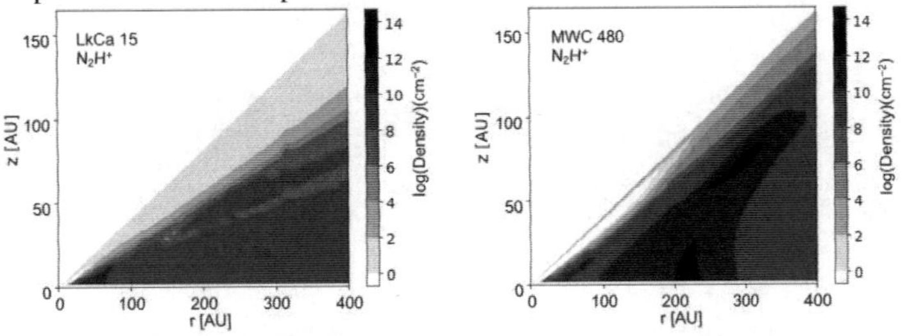


Figure 4. $N_2H^+$ distribution in the LkCa 15 (left) and MWC 480 (right) disks.

*3.3.$H_2CO$*

In the LkCa 15 disk model (see Fig. 5), $H_2CO$ column density at intermediate heights is consistent with the observations ($10^{12}$ $cm^{-2}$). It is reduced by two orders of magnitude in the region near the star (< 50 AU). For MWC 480, the column density of $H_2CO$ is between $10^{10}$-$10^{12}$ $cm^{-2}$ in most regions, and reaches the maximum in the region near the middle plane of 250-350 AU. There is more $H_2CO$ at the middle height in LkCa 15 than in MWC 480, which is consistent with the survey of $H_2CO$ in both disks. Formaldehyde ice showed the same variation pattern as CO ice in both disks, and the abundance of $H_2CO$ ice in MWC 480 below the surface of the disc was greater than that in LkCa 15. In the experiment, the freezing temperature of $H_2CO$ is 70-90 K, which is much higher than temperature of our two disks, therefore the thermal desorption of $H_2CO$ from the ice sheet is inhibited. Due to the high extinction and high density of the two disks, the optical resolution of $H_2CO$ can be neglected.

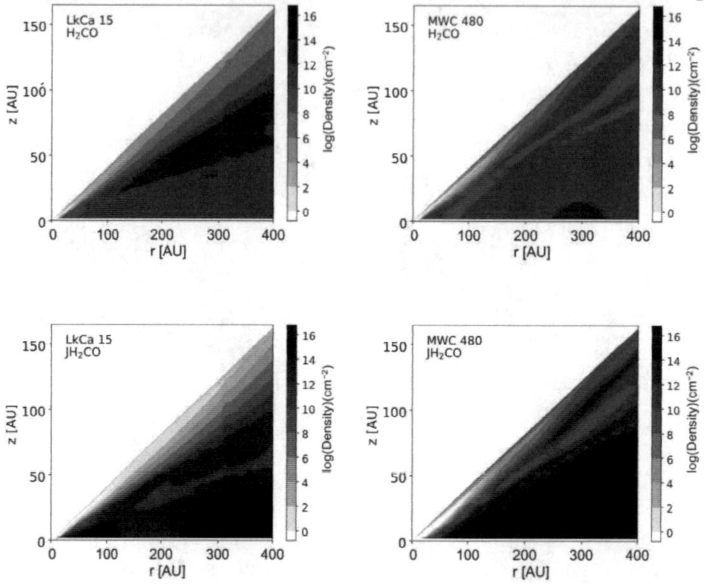

Figure 5. $H_2CO$ distribution in the LkCa 15 (left) and MWC 480 (right) disks.

### *3.4. $CH_3OH$*

$CH_3OH$ is the precursor of many complex organic molecules. However, the efficiency of forming $CH_3OH$ in the gas-phase is very low. Infrared Space Observatory observation showed that methanol was attached to the ice layer of dust particles. Therefore, Some studies proposed that $CH_3OH$ is formed on the surface of dust particles by CO hydrogenation.

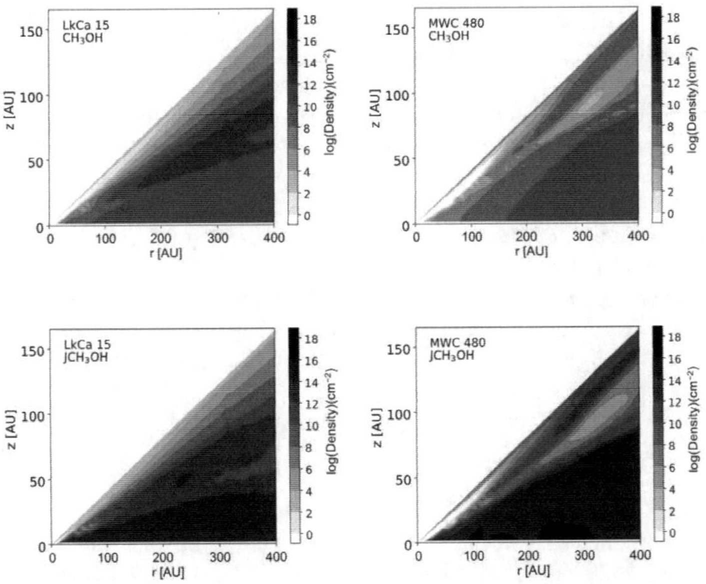


Figure 6. $CH_3OH$ distribution in the LkCa 15 (left) and MWC 480 (right) disks.

Our results show that compared with MWC 480, methanol in the gas phase is more abundant in the warm molecular region of LkCa 15. This result is similar to the above comparison between HD 163296 and TW Hya observations. In contrast, the column density of $CH_3OH$ on the ice surface in the LkCa 15 disk is two orders of magnitude lower than that in the MWC 480 disk in the intermediate region. In LkCa 15 disk, the high abundance of $CH_3OH$ in the warm molecular layer is because the chemical heat is released after radical recombination. And this can be partially used to desorb $CH_3OH$ to the gas-phase. In our model, that the result does not imply a stronger photodissociation effect on $CH_3OH$ by UV photons. The FUV in MWC 480 is stronger than in LkCa 15. This agrees with the laboratory results

## 4. Conclusions

In this paper, we calculate the chemical composition of protoplanetary disks around LkCa 15 and MWC 480, focusing on the distribution and formation of $H_2CO$ and $CH_3OH$. Also, we studied the CO and $N_2H^+$ that have been observed in these two disks, discussed the relationship between these molecules, and determined the overall trend when comparing these models with the observed column density. We compare LkCa 15 and MWC 480 to analyze the possible chemical differences between the disks. Our main results are the following:

(1) Except for $N_2H^+$, our calculated column densities observations are consistent with observed values. The average column density of CO in the MWC 480 disk was higher than that in LkCa 15 disk. The cylinder density of CO derived from the two objects seems to originate from different emission regions.

(2) The low abundance of $N_2H^+$ is due to its inhibition by the three-phase model, which is consistent with the simulation results. In future research, we need to model the spatial resolved $N_2H^+$ observation of the disk to distinguish the contribution of X-ray from the ionization of the disk, and to study the distribution of $N_2H^+$ in the disk. At present, this kind of observation on the disk is still rare, but it may become more common soon due to the significant progress of modern (submillimeter) interferometers, especially in terms of spatial resolution.

(3) We found that $N_2H^+$ and CO have a negative correlation, and $N_2H^+$ reaches the peaks at the position of CO depletion. This is due to the reaction competition between $N_2H^+$ and CO in the gas phase.

(4) The observed $H_2CO$ emission may come from the warm molecular layer in LkCa 15. Our model shows that the column density of the $H_2CO$ in LkCa 15 is higher than that in MWC 480, which is consistent with the observations. The results of our model show that $H_2CO$ is usually formed through the gas-phase and the surface of particles in LkCa 15 disk. And $H_2CO$ is formed by gas phase reaction in MWC 480. The low density of Herbig AE disk column relative to T Tauri disk may be caused by the different formation paths of $H_2CO$.

(5) We predict that $CH_3OH$ may be observed in LkCa 15 disk with a column density of $10^{12}$ $cm^{-2}$, and MWC 480 has a lower column density of $CH_3OH$ than LkCa 15 disk. The column density of $CH_3OH$ in of Herbig Ae/Be disk may be lower than that of T Tauri disk. $CH_3OH$ is usually produced by active ice chemistry in the protoplanetary disk, which is then desorbed into the gas phase by surface chemical exothermic reactions. Thermal desorption and photodissociation from UV are not effective for methanol ice.

**ACKNOWLEDGMENT**

This work was supported by the Science and Technology Projects in Jilin Province Department of Education grants JJKH20200607KJ and JJKH20210666KJ, Topic Foundation of Changchun Institute of Technology grants 320210016 and 320200040, Young people Foundation of Changchun Institute of Technology grants 320210033, and Doctor Foundations of Changchun Institute of Technology 2021.

**REFERENCES**

[1] Rivière-Marichalar, P., et al., “AB Aur, a Rosetta stone for studies of planet formation. I. Chemical study of a planet-forming disk,” Astronomy and Astrophysics, vol. 642, pp. A32, October 2020.

[2] Mordasini, C., Alibert, Y., and Benz, W., “Extrasolar planet population synthesis. I. Method, formation tracks, and mass-distance distribution,” Astronomy and Astrophysics, vol. 501, pp. 1139, July 2009.

[3] van 't Hoff, M. L. R., Walsh, C., Kama, M., Facchini, S., and van Dishoeck, E. F., “Robustness of $N_2H^+$ as tracer of the CO snowline,” Astronomy and Astrophysics, vol. 599, pp. A101, March 2017.

[4] Qi, C., et al., “Imaging of the CO Snow Line in a Solar Nebula Analog,” Science, vol. 459, pp. 239–242, May 2009.

[5] Dutrey, A., Guilloteau, S., and Guelin., “Chemistry of protosolar-like nebulae: The molecular content of the DM Tau and GG Tau disks,” Astronomy and Astrophysics, vol. 317, pp. L55, January 1997.

[6] Aikawa, Y., Furuya, K., Nomura, H., and Qi, C., “Analytical Formulae of Molecular Ion Abundances and the $N_2H^+$ Ring in Protoplanetary Disks,” The Astrophysical Journal, vol. 807, pp. 120, July 2015.

# According transformer parameters to design a high voltage power

**Junling Zhang[1], Renbo Jin[2]**

[1] China State Shipbuilding Corporation, Jiangsu Nanjing, China

[2] China State Shipbuilding Corporation, Jiangsu Nanjing, China

*First author e-mail: china723zjl@163.com

**Abstract**. When the load is empty pulse width modulation high frequency and high voltage power' stabilization depth can easily cause oscillation, and when the full load of the voltage regulation is too shallow the voltage regulation effect is not good, the reason for this phenomenon is because the equivalent ratio and load correlation. The first thing you need to do is to find out why you can avoid the problem early in the design. To find out the cause, should to know the influence of the high voltage rectifier circuit on the total gain of the circuit and the dispersed parameters of the high frequency and high voltage transformer.

## 1. Overview

Design of the high frequency transformers has always been a difficult point in design of the high frequency high voltage power supplies. The leakage inductance and distributed capacitance of the transformer become apparent at high frequencies and high ratios. The dispersed parameters are mostly the dispersed leakage inductance and capacitance.

This paper proposes a broad design approach: The distribution parameters of the transformer are used as resonant parameters, The method can ensure that the whole gain of the high voltage power is suitable for all range, which has engineering application value. Finally, experiments and simulation verify the effectiveness of the proposed method.

## 2. The basic methods

### *2.1. Obtain the dispersed parameters*

First, you need to get the dispersed parameters accurately of the high frequency transformers, the transformer's dispersed parameter ideal model is shown in Figure 1.

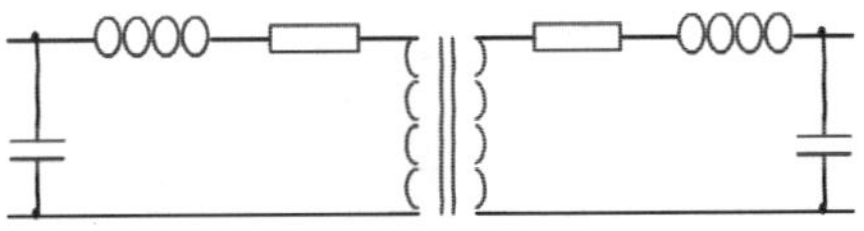

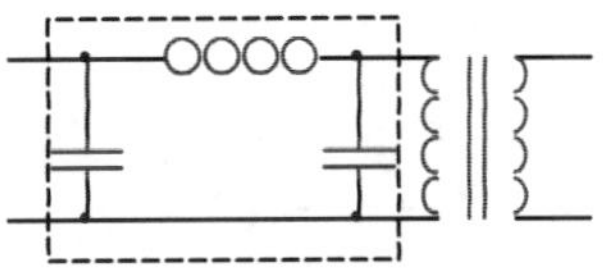

Figure 1. Transformer distribution parameter model Figure 2. Simplified transformer model

Figure 1 shows an air core transformer ideal model with no dispersed parameters. L1 and L2 is the leakage inductor of the primary side of it, C1 means the equivalent dispersed capacitor of the primary; C2 means the equivalent dispersed capacitor for the second, R1and R2 means the equivalent resistance of the primary and secondary. The secondary current of the transformer is much smaller than the primary, so R2 and L2 are ignore. The turns of primary are relatively smaller to the secondary, so R1 is ignore. Converting the equivalent dispersed capacitance of the secondary; you'll get a simpler model as Figure 2.

Due to same winding craft, the sub-side's turns are n times of the primary side turns, which can be obtained:

$$C2' = n^2 \times C2 = n \times C1 \tag{1}$$

The type two-port network within the dotted box in FIG. 2 may be equivalent to a T type and dual port network, shown in Figure 3.

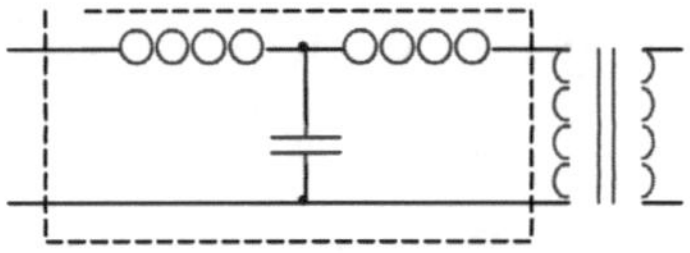
Figure 3. Two-port network

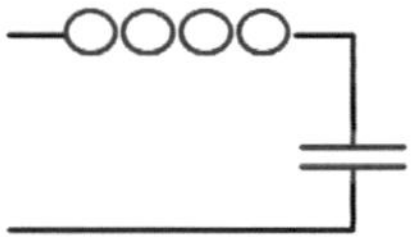
Figure 4. Empty equivalent circuit

The ideal air core transformer's electromagnetism is very large, the excitation current is ignoring, and Figure 3 is simplified to Figure 4.

### 2.2. Obtain the ideal model

To design high voltage power, you need to reduce the output ripple amplitude and ensure the output voltage stability, the secondary needs to use a large capacitor as a filter. Therefore, the output voltage in the steady state remains unchanged. Due to the large capacitance on the output side, the rectifier diode is turned on only when the voltage is greater than the primary voltage, and the voltage is clamped at the primary voltage and turned off when the inductor current is zero.

This steady-state analysis method of nonlinear properties has been studied very well in many documents, and is applied directly below: [1,2,3,4]

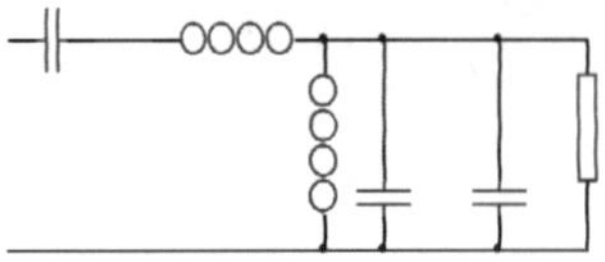
Figure 5. Transformer model

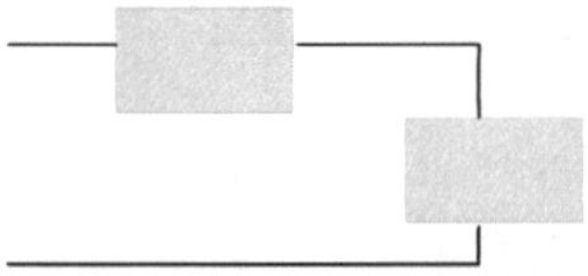
Figure 6. Transformer equivalent network model

In Figure 5, L1 is the leakages induct, C1 is the series capacitor, Cp is the secondary capacitor, Lm is the primary inductor, Re is the secondary output resistors, and Cp is the secondary output capacitors. Figure 6 is the simpler model:

This allows the rectifier diode's angle to be α:

$$R_e = \frac{\lambda_1^2}{2n^2} R_o \tag{2}$$

$$C_e = \frac{2n^2}{\omega R_o \lambda_1^2} \tan|\theta| \tag{3}$$

$$\lambda_1 = \frac{2}{\pi}\sqrt{(1+\cos\alpha)^2 - \left(\frac{\sin\alpha+\alpha-\pi}{1+\cos\alpha}-\sin\alpha\right)^2} \tag{4}$$

$$\theta = \arctan\frac{\frac{\sin\alpha+\alpha-\pi}{1+\cos\alpha}-\sin\alpha}{1-\cos\alpha} \tag{5}$$

$$Z_{12} = \frac{1}{j\omega C_1} + jwL1 \tag{6}$$

$$Z_{22} = \frac{1}{\frac{1}{j\omega L_2} + j\omega C_2 + \frac{1}{R_e}} \tag{7}$$

$$M_{21} = \frac{Z_2}{Z_2 + Z1} \tag{8}$$

$$K = \frac{1}{\lambda_1}|M_{21}|\frac{4}{\pi} \tag{9}$$

### 2.3. *The calculation of selected circuit parameters*

When the circuit is resonant, the secondary waveform's high-voltage spike is small and the reliability is large, so the design in this article is based on resonant state. The equivalent network of resonances has zero virtual parts, the circuit manifests as only resistance, and the circuit resonates when the input frequency is the same as the natural one.

After obtain an expression for the equivalent model, the required parameters can be evaluated. Select C1 and L1 through untie the system of equations: K=1, and the imaginary equals 0.

## 3. The exam design

To verify the design method, we design a power supply in the laboratory. The power supply is DC input 400V, output DC 6kV and 500mA.

In order to reduce the distributed capacitance and reduce the turns of the secondary, the method of doubling the voltage rectification is selected. Considering the magnitude of the current, this example uses the circuit format of voltage doubling rectification .The circuit diagram is showed as Figure 7:

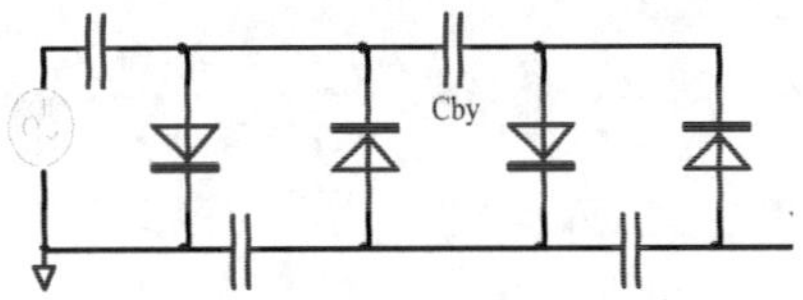


Figure 7. Circuit diagram of secondary rectifier

The calculation process is below:

$$Vout = 6kV\text{; } Iout = 0.5A\text{; } n = 5\text{; } \omega = 2\times\pi\times fs = 5.278\times10^5\frac{1}{s}\text{; } L_2 = Lm\text{; } C_W = 0.1\times10^{-6}\text{F}$$

$$Cp = 100\times10^{-9}\text{F}\text{; } Ro = \frac{Vout}{Iout} = 1.2\times10^4\Omega\text{; } C_1 = 150\times10^{-9}\text{F}$$

$$Z_2 = \frac{1}{\frac{1}{j\times\omega\times L_2}+j\times\omega\times C_2+\frac{1}{\text{Re}}} = 0.508-11.392i\ \Omega \tag{10}$$

$$M_1 = \frac{Z_2}{Z_1 + Z_2} = 0.877 + 4.797 \times 10^{-3} i \tag{11}$$

Because of above, so:

$$|M_1| = 0.877 \tag{12}$$

$$KDC = \frac{1}{\lambda_1} \times |M_1| \times \frac{4}{\pi} = 1.074 \tag{13}$$

## 4. Simulation

Here are some simulated waveforms:

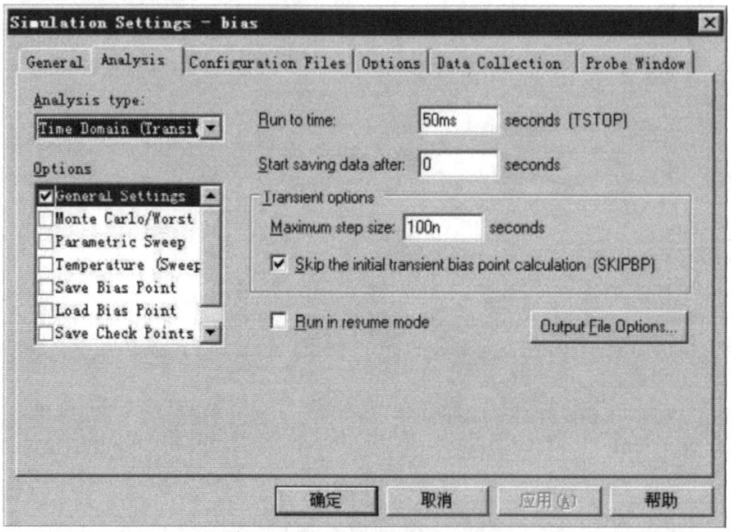


Figure 8. Simulation setup

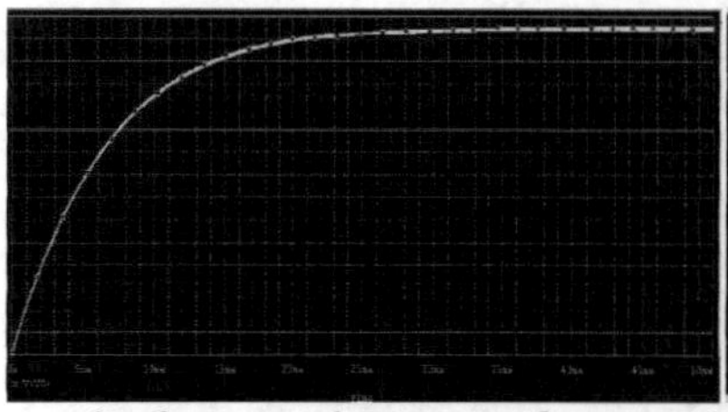

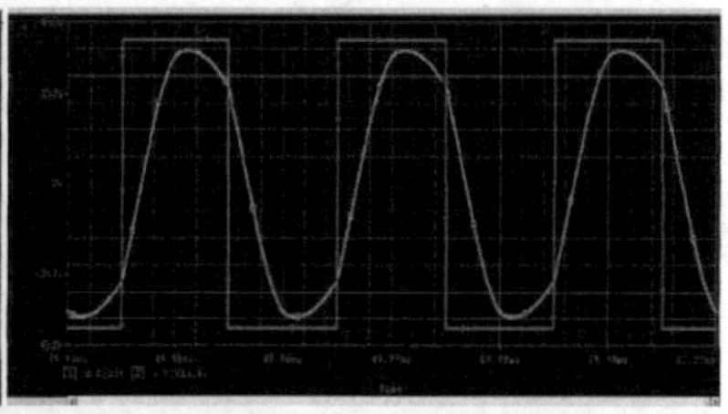

Figure 9. Output voltage waveform Figure 10. Input voltage and current waveform

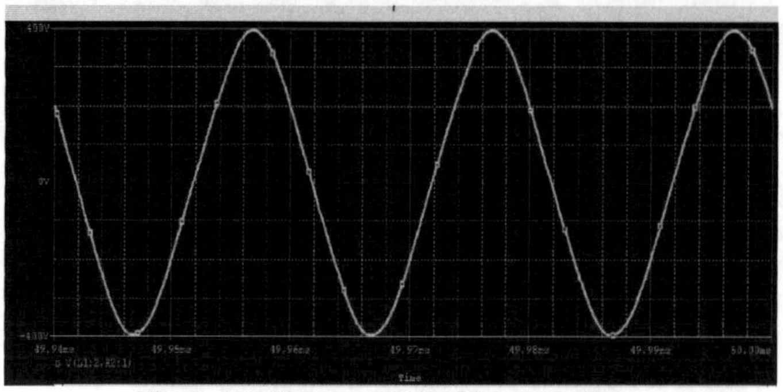

Figure 11. Resonant capacitive waveform

## 5. Conclusion

This paper first proposes a design method for resonant power supply considering transformer distribution parameters, and then establishes a simple and practical engineering model. Select the appropriate circuit parameters, and finally design a high voltage power supply by using the method, which proves the method's operability. This scheme not only solves the problem of large change in pulse width, but also ensures that the circuit works in a resonant state, greatly improving reliability and has significant engineering application prospects

## References

[1] Xinbo Ruan, Yanguang Yan. Soft switching technology for DC switching power supplies [M]. Science Press, 2000:262 -276.

[2] Weiping Zhang. Modelling and control of switching power supplies [M] China Power Electronics Press, 2005:35-52.

[3] Chunsheng Hu, Peng Hu. Integrated switching power supply design, commissioning and maintenance [M] People's Post and Telecommunications Publishing House, 1995:26-31.

[4] Ming Wei. Saber simulate a book [M]. China Power Electronics Press, 2017:55-70.

[5] Gaoping Hu. Three-phase three-level llc resonant DC converter suitable for high voltage and high power applications [J]. The 20th Annual Conference of the China Power Supply Society,2013:15-19

# Research on drag reduction characteristics of the surface of springtail cuticle

**Benshuai Fu[1,2*], Haiyan Xiao[1,2], Bingju Lu[2], Liping Qin[2], Zhu Zhu[2]**

[1]School of Mechanical and Electrical Engineering, Harbin Engineering University, Harbin 150001, PR China

[2]Henan Key Laboratory of Underwater Intelligent Equipment, Zhengzhou 450015, PR China

[*]Corresponding author's e-mail: fubenshaui@hrbeu.edu.cn

**Abstract.** The technology of underwater drag reduction is one of critical paths to break through the problem of ocean-going, and the bionic drag reduction is a very promising method. The bionic drag reduction uses sliding effect on the liquid-gas interface to reduce the drag, while the gas film is very unstable and the drag reduction function is easy to fail. The concave corner microstructure of springtail cuticle can make the air retention, which has very important significance to solve the problem of instability of gas film. In the paper, a model is established by referring to the concave corner microstructures of springtail cuticle. The k-ε turbulence model and volume of fluid multiphase flow model were used. The results show: the simulation model can be used to calculate the flow characteristics of the concave corner microstructure of springtail cuticle; as the area ratio of free shear increases, so does the drag reduction rate, which is beneficial for drag reduction; the drag reduction rate of different microstructure is slightly different, but it is mainly related to the area ratio of free shear of microstructure.

## 1. Introduction

We never stopped exploiting and scrambling for ocean resources. As the areas of ocean exploitation and ocean defense continue to extend to deep and far seas, breakthroughs in ocean-going technology become an urgent problem to be solved. Increasing energy reserves, improving propulsion efficiency and reducing vehicle resistance are the three main ways to achieve this goal. For underwater vehicle, the increase of energy reserve is limited by the volume of the shell, and the propulsion efficiency has basically reached the limit. Therefore, underwater drag reduction technology naturally becomes the focus of many researchers at home and abroad [1, 2].

Bionic drag reduction is a very promising method [3], among which bionic shark skin[4, 5] and bionic superhydrophobic surface [6, 7] have become hotspots in recent years. At present, the microstructure imitating shark skin has been widely used in aerospace, swimwear, pipeline transportation and other aspects. Inspired by the hydrophobicity of plant leaf surfaces such as lotus leaves and insect limb surfaces such as water striders, the application of hydrophobic surface in drag reduction has attracted more and more attention. The hydrophobic surface can seal micro-scale gas film on the wall surface, and the sliding effect on the liquid-gas interface can improve the fluid velocity and suppress the turbulence pulsation intensity, thus producing a macroscopic drag reduction effect. However, the small rib drag reduction technology based on shark skin groove has the problem of insufficient drag reduction performance, and the gas film drag reduction technology based on lotus leaf superhydrophobic principle has the problem of drag reduction function is easy to fail because of unstable gas film [8, 9].

In order to solve the stability problem of gas film, we turned our attention to the springtail [10, 11]. In order to maintain respiration in the humid environment, the springtail has evolved an excellent hydrophobic functional structure on the cuticle, namely the concave corner microstructure, so that the air is trapped in the concave corner microstructure. Therefore, we took springtail cuticle microstructure as the imitation object, established the concave corner microstructure model, and studied the drag reduction characteristics of the concave corner microstructure surface by numerical simulation.

## 2. Simulation calculation model

### *2.1. Geometric model and meshing*

The research object is the concave corner microstructure surface with regular morphology abstracted from the springtail cuticle. Considering that the microstructure has good periodicity, the flow field of the entire microstructure surface can be obtained only by calculating part of the microstructure in the computational domain. Therefore, the computational domain, Lx=15mm, Ly=5mm, Lz=0.66mm. The lower wall surface contains a microstructural surface with regular microscopic morphology, and the specific structural size is shown in Figure 1.

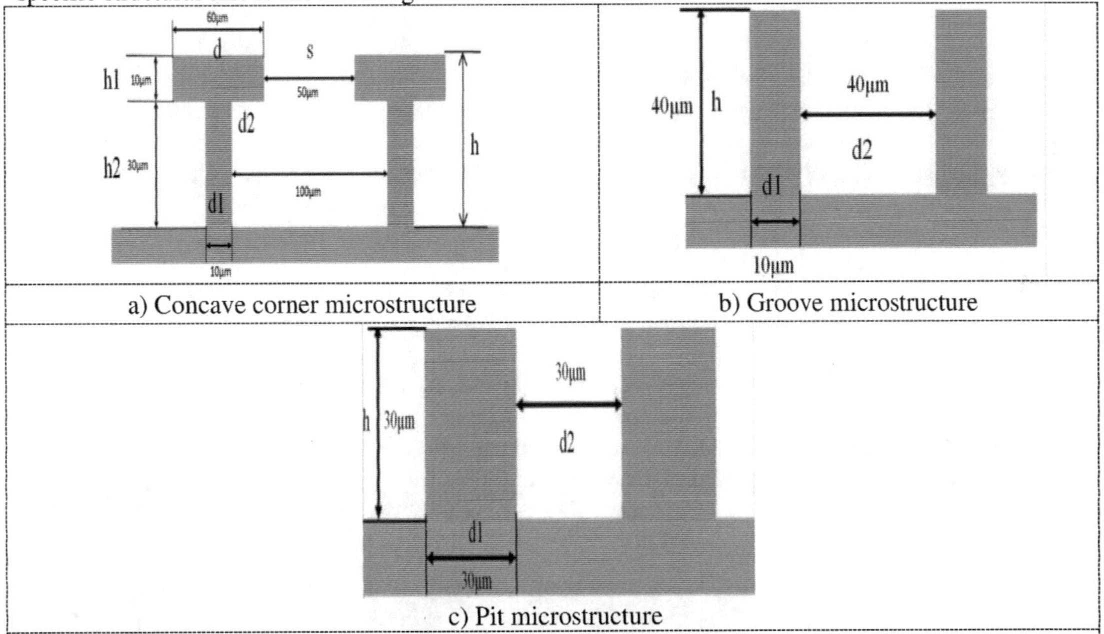


a) Concave corner microstructure

b) Groove microstructure

c) Pit microstructure

Figure 1. Microstructure size.

In order to improve the resolution of the meshes on the surface of concave corner microstructure without too much increase in the number of meshes, different density meshes were used at different positions in the computational domain to encrypt the concave corner microstructure area. This can not only meet the requirements of calculation accuracy but also reduce the amount of calculation to ensure the comprehensive flow field information. As shown in Figure 2, and the number is about 7 million.

a) Disconnected concave corner microstructure

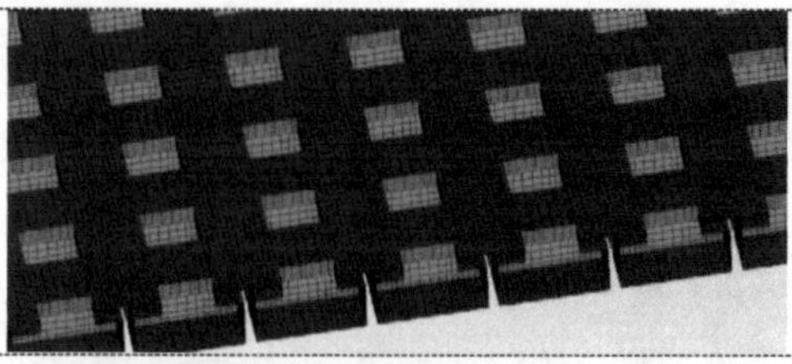

b) Connected concave corner microstructure

c) Groove microstructure

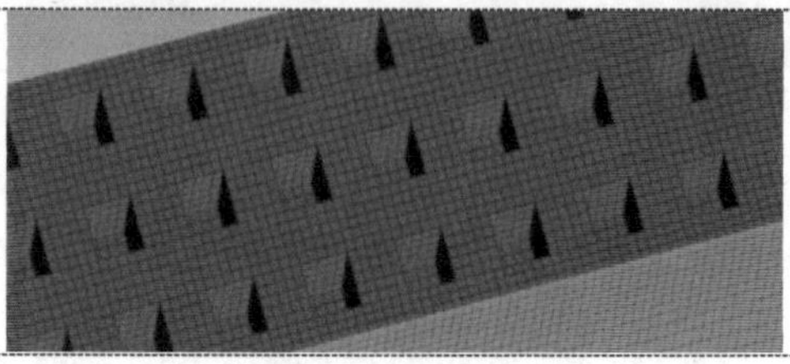

d) Pit microstructure

Figure 2. Meshing.

### 2.2. Other parameters

The governing equations were discretized by finite element method. Due to low flow rate and low compressibility in thc calculation condition, the solver based on pressure is selected. The discrete equations are solved by PISO algorithm in the separate solution method. The $k$-$\varepsilon$ turbulence model and VOF multiphase flow model were used for calculation. PRESTO! with moderate computational velocity and good convergence is selected for pressure dispersion, the relaxation factor was kept default, and the time step size was 1e-6s.

### 2.3. Boundary and initial conditions

In computing domain, the left side was the speed entry boundary; the right side was the pressure outlet boundary; the upper side was symmetrical boundary; the front and rear sides were periodic boundary; the lower side has the concave corner microstructure. In order to reduce the amount of computation, the udf program is compiled to assign the fully developed velocity distribution to the entrance boundary, as shown in Figure 3.

In initial state, the microstructure was filled with air (blue color), and the liquid-gas interface flushes with the top of microstructure, as shown in Figure 4.

Figure 3. Entry velocity distribution.

Figure 4. Initial liquid-gas interface.

### 2.4. Model verification

To check the reliability of the numerical calculation model established, numerical calculation was carried out by referring to the literature of Ou et al [12]. The pressure difference of the microstructure channel is calculated by averaging the pressure data of two sections at a certain distance. Then, the pressure drop ratio of dimensionless Π is calculated and compared with the pressure drop ratio of dimensionless obtained by Ou et al (Π is used to measure the effect of drag reduction, with expression: $\Pi = (\Delta P_s - \Delta P_{hy}) / \Delta P_s$, where $\Delta P_{hy}$ is the pressure reduction ratio of dimensionless in the microchannel of superhydrophobic surfaces, and $\Delta P_s$ is the pressure reduction ratio of dimensionless in the microchannel of smooth non-slip surfaces.), See Figure 5. From the figure, you can found the error between the calculation and the experiment results was smaller than 10%, showing the calculation model can calculate the flow on the microstructure surface well.

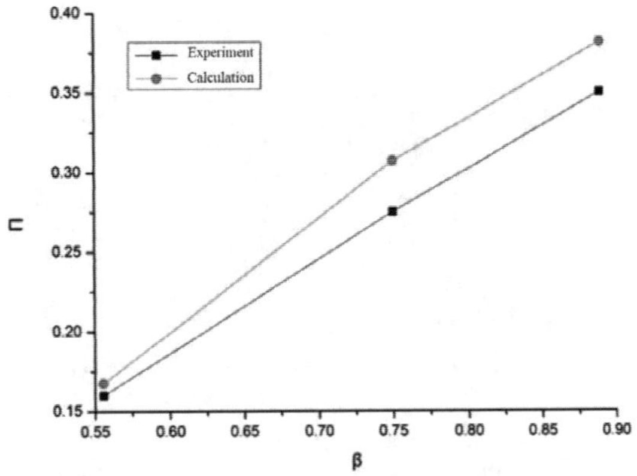


Figure 5. The calculation and experiment results.

## 3. Calculation results

### *3.1. Calculation conditions*

The calculation conditions are shown in Table 1.

Table 1. Calculation conditions.

| | Outlet pressure (Pa) | Microstructure | Free shear area ratio(β) |
|---|---|---|---|
| 3D-1 | 101325 | smooth non-slip surface | - |
| 3D-2 | 101325 | Disconnected concave corner microstructure | 0.70 |
| 3D-3 | 101325 | Connected concave corner microstructure | 0.21 |
| 3D-4 | 101325 | Groove microstructure | 0.80 |
| 3D-5 | 101325 | Pit microstructure | 0.75 |

Where, the area ratio of free shear $\beta$ is a very important structural characteristic parameter, which represents the proportion of the difference between the total area of the superhydrophobic surface and the non-slip surface area of microstructure in total area. Its expression can be written as $\beta = (A_{total} - A_{no\text{-}slip}) / A_{total}$.

### *3.2. Flow structure in surface of microstructure*

Figure 6 show the flow velocity distribution cloud of smooth non-slip surface and microstructure surface. On the whole, the velocity distribution on the microstructure surface is similar to that on the non-slip surface, indicating that the existence of the microstructure does not disturb the overall layered structure of the flow field, and its influence is limited to the near wall region.

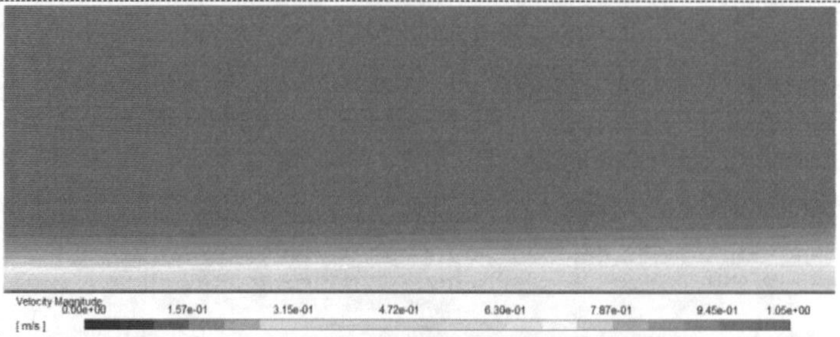

a) Non-slip surface

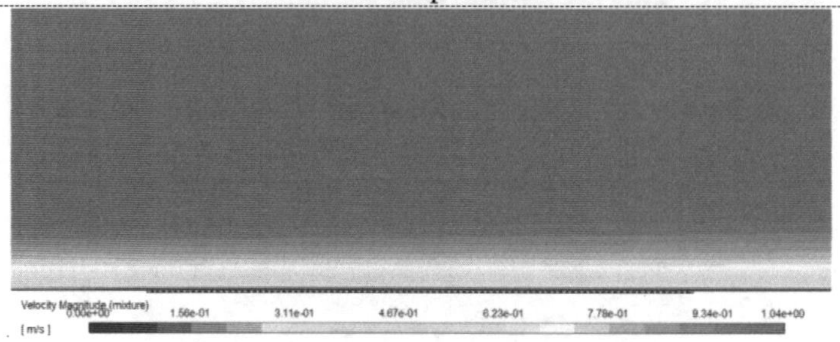

b) Concave corner microstructure surface

Figure 6. The flow velocity distribution cloud.

Figure 7 show the phase distribution cloud of microstructure surface. We can see that the liquid-gas interface is relatively clear and presents a certain radian, which is the same as the experiment result.

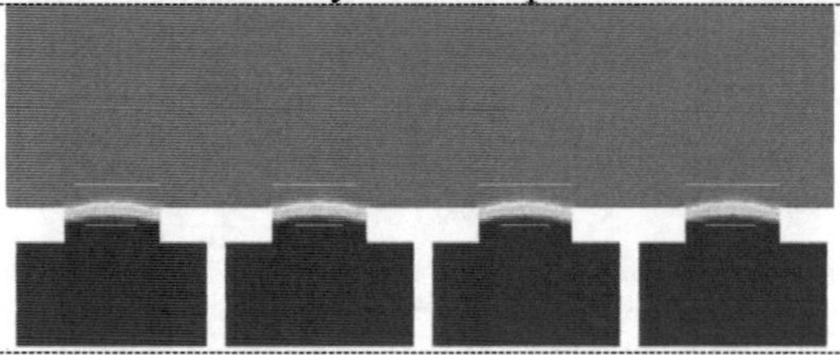

a) Concave corner microstructural surface

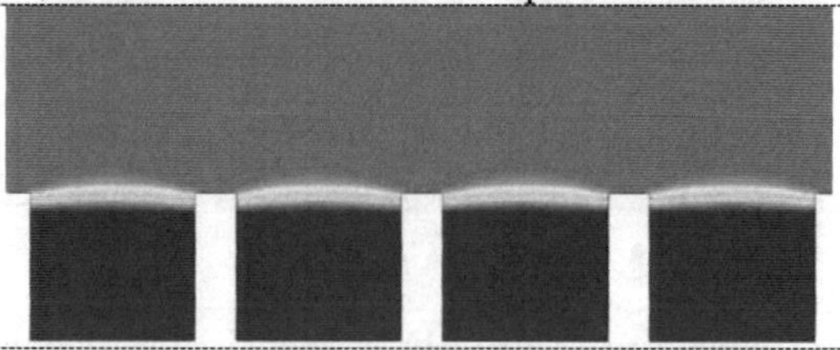

b) Groove microstructure surface

Figure 7. Phase distribution cloud of microstructure surface.

### 3.3. Velocity curves in surface of microstructure

In order to further display the velocity distribution on the microstructure surface in detail, the velocity distribution curves of boundary layer and near wall area at different positions on the microstructure surface are given below. The positions of velocity measurement lines are shown in Figure 8.

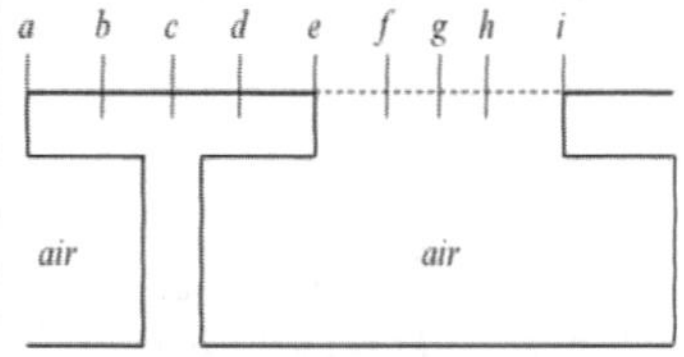


a) Concave corner microstructural surface

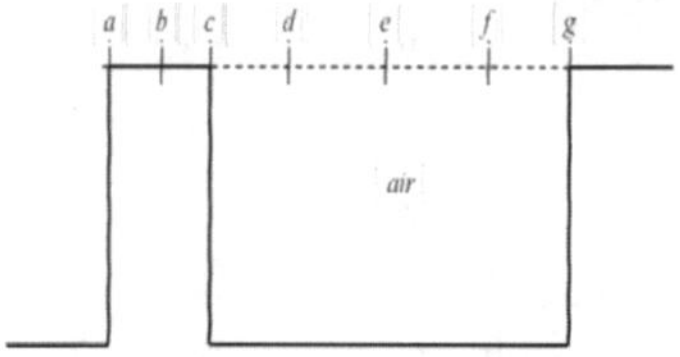


b) Groove microstructure surface

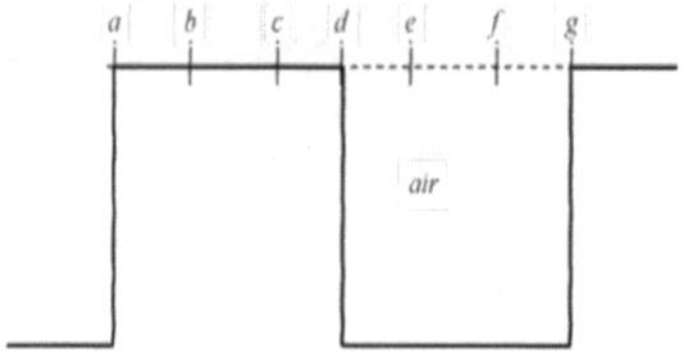


c) Pit microstructure surface

Figure 8. Location diagram of velocity measurement line on microstructure surface.

Figure 9 show the longitudinal velocity distribution of different position. It can be found, compared with the smooth non-slip surface, velocity distribution curve on different location appeared the phenomenon of rise, and on both sides of the microstructural lug and pit looked more apparent, this is caused by liquid-gas interface exists. but and thus the velocity is zero on the microstructural lug. Water flowing through liquid-gas interface, the slip caused water flow velocity increases of near surface, on the microstructural lug it still accord with no slip boundary, while in the microstructural pit it no longer meet the classic no slip boundary.

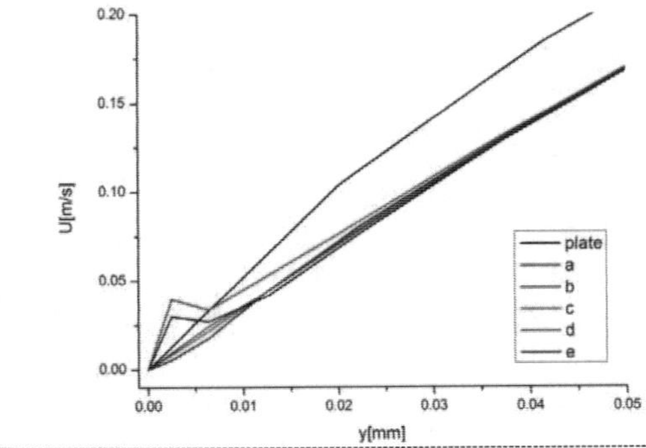


a）Velocity distribution of disconnected concave corner microstructure on the microstructural lug

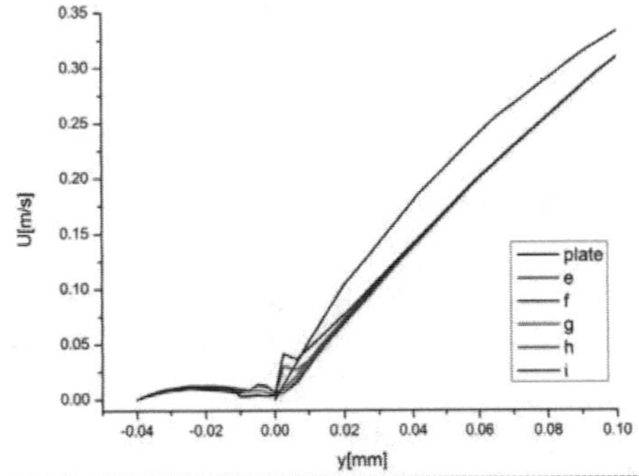


b）Velocity distribution of disconnected concave corner microstructure in the microstructural pit

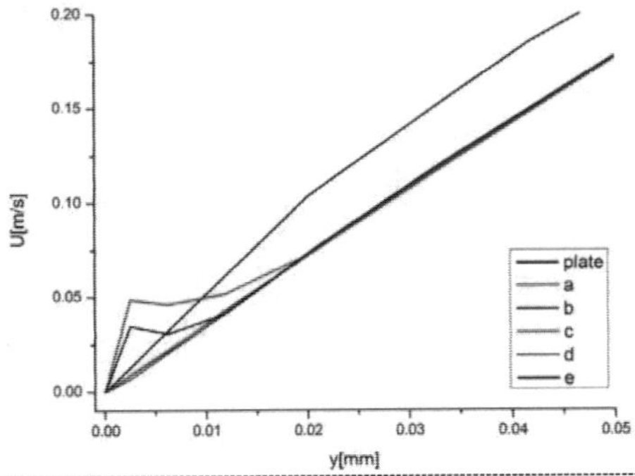


c）Velocity distribution of connected concave corner microstructure on the microstructural lug

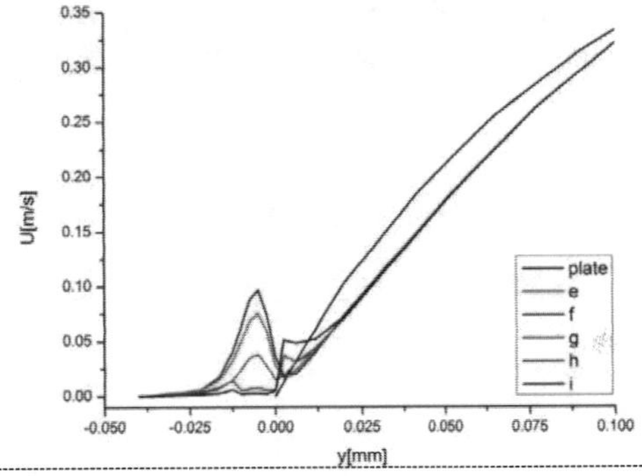


d）Velocity distribution of connected concave corner microstructure in the microstructural pit

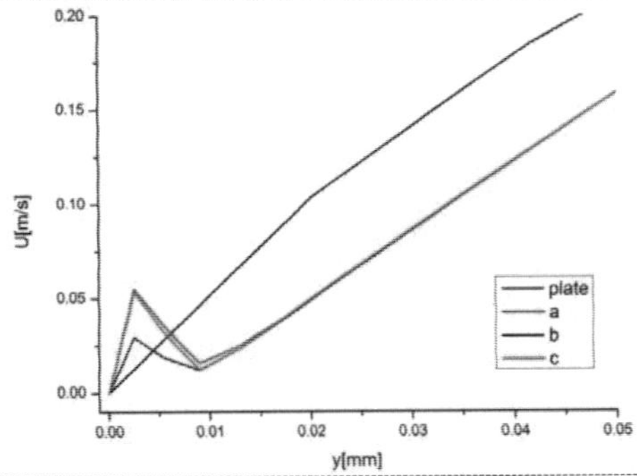


e）Velocity distribution of groove microstructure on the microstructural lug

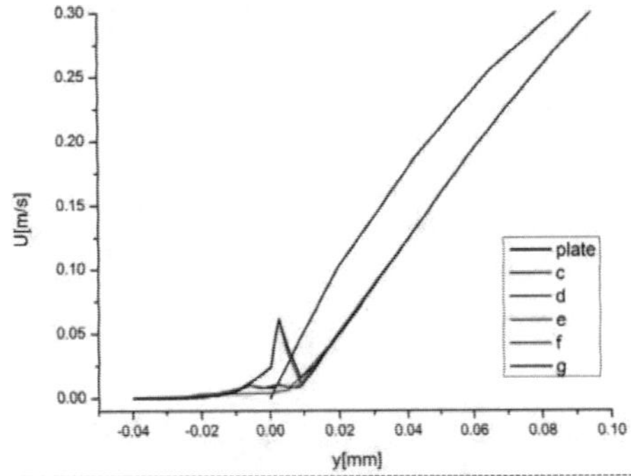


f）Velocity distribution of groove microstructure in the microstructural pit

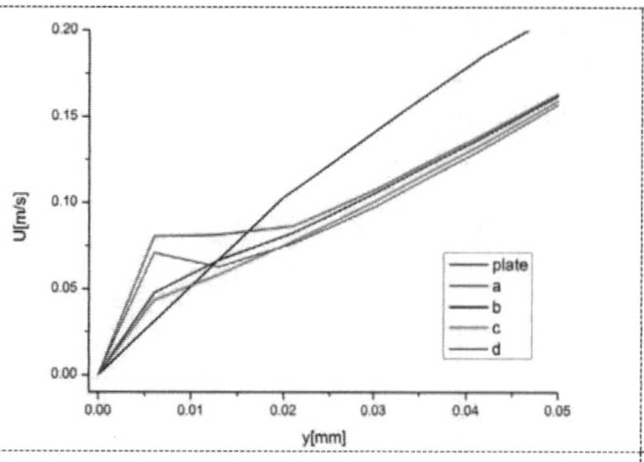


g) Velocity distribution of pit microstructure on the microstructural lug

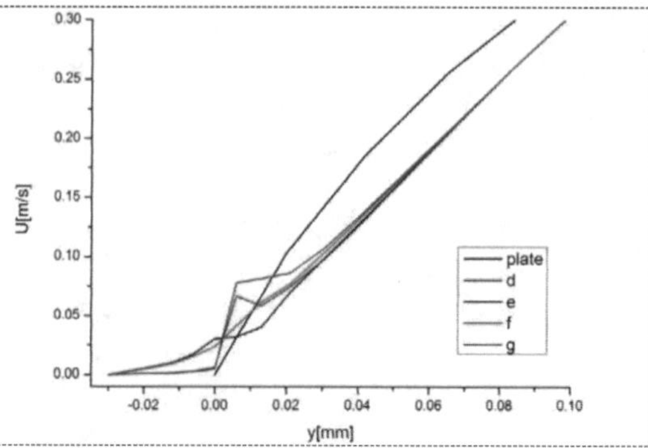


h) Velocity distribution of pit microstructure in the microstructural pit

Figure 9. Longitudinal velocity distribution curves of different position.

Figure 10 show the flow velocity distribution curve at microstructure surface. For liquid-gas interface in a single pit, the velocity fluctuates greatly along the flow direction, and the maximum velocity in pits on either side of the place. After the point of maximum velocity, the velocity steep fall in, this is the liquid-gas interface and wall on the flow direction staggered distribution. Liquid-gas interface has the effect of increasing the slip and the wall is the role of limited slip, at the liquid-gas interface, the slip along the flow direction should be increasing, but due to the limitation of no slip flow on the wall, the velocity drops sharply.

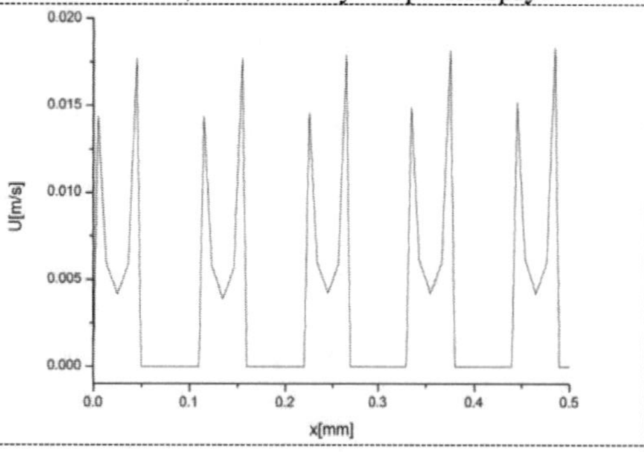


a) Disconnected concave corner microstructure

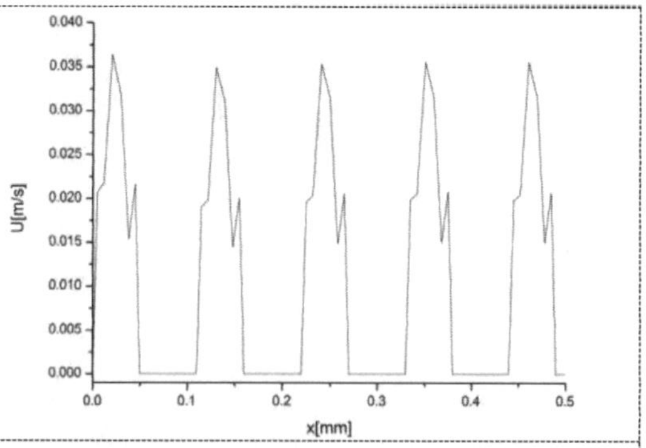


b) Connected concave corner microstructure

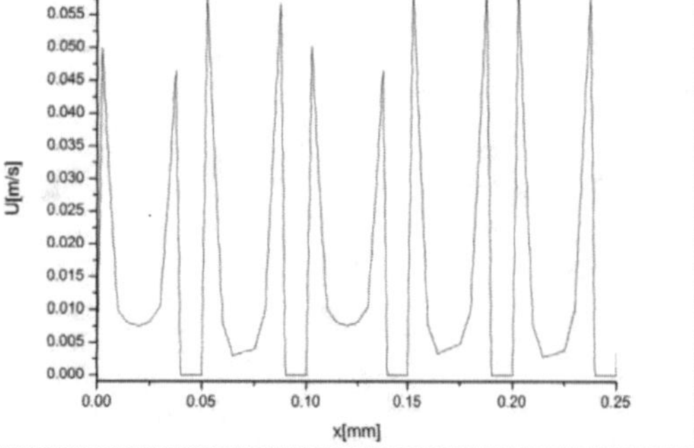


c) Groove microstructure

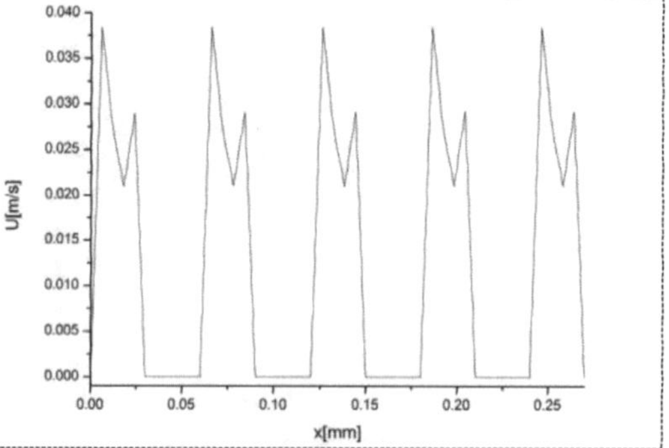


d) Pit microstructure

Figure 10. Flow velocity distribution curve at microstructure surface.

The velocity of flow field is not zero at the liquid-gas interface, which is called interface slip, which is the cause of drag reduction on the microstructure surface.

### *3.4. The characteristic of drag reduction*

Due to the presence of liquid-gas interface, the microstructure surface can reduce the resistance in flow, and the rate of drag reduction η (η is an important parameter to measure the drag reduction effect. Its expression is: $\eta=(F_s - F_{hy})/F_s$, where $F_s$ is the total resistance of no-slip surface and $F_{hy}$ is the total resistance of superhydrophobic surface.) showed in Table 2. You can found: as the area ratio of free shear increases, the rate of drag reduction of microstructure increases, which is conducive to drag reduction; the rate of drag reduction of different microstructure is slightly different, but it is mainly related to the area ratio of free shear of microstructure.

Table 2. The characteristic of drag reduction.

| | Free shear area ratio(β) | Pressure reduction ratio of dimensionless(△P) | Pressure drop ratio of dimensionless(Π) | Wall shear stress | Drag reduction rate(η) |
|---|---|---|---|---|---|
| 3D-1 | - | 1593.98Pa | - | 6.15Pa | - |
| 3D-2 | 0.70 | 1205.04Pa | 24.40% | 4.22Pa | 31.43% |
| 3D-3 | 0.21 | 1539.20Pa | 3.44% | 5.66Pa | 7.92% |
| 3D-4 | 0.80 | 1041.30Pa | 34.67% | 4.15Pa | 32.51% |
| 3D-5 | 0.75 | 1093.30Pa | 31.41% | 4.19Pa | 31.94% |

## 4. Conclusion

In the paper, the characteristics of flow field and drag reduction of typical microstructure were studied from the aspects of the velocity distribution, the shear stress of surface and the rate of drag reduction.

a) Compared with non-slip surfaces, water slips at the liquid-gas interface, resulting in velocity increasing. Therefore, the velocity of the liquid-gas interface is greater than zero.

b) The liquid-gas interface and the wall are staggered in the flow direction. The liquid-gas interface has the effect of increasing slip, while the wall has the effect of limiting slip.

c) The rate of drag reduction of microstructure increases with the increase of area ratio of free shear. The rate of drag reduction of different types of microstructure is slightly different.

## Acknowledgments

The work was supported by the National Natural Science Foundation of Henan Province (202300410003).

## References

[1] Liravi M, Pakzad H, Moosavi A, et al. A comprehensive review on recent advances in superhydrophobic surfaces and their applications for drag reduction[J]. Progress in Organic Coatings, 2020, 140: 105537.

[2] Ao M, Wang M, Zhu F. Investigation of the turbulent drag reduction mechanism of a kind of microstructure on riblet surface[J]. Micromachines, 2021, 12(1): 59.

[3] Liu G, Yuan Z, Qiu Z, et al. A brief review of bio-inspired surface technology and application toward underwater drag reduction[J]. Ocean Engineering, 2020, 199: 106962.

[4] Fu Y F, Yuan C Q, Bai X Q. Marine drag reduction of shark skin inspired riblet surfaces[J]. Biosurface and Biotribology, 2017, 3(1): 11-24.

[5] Martin S, Bhushan B. Modeling and optimization of shark-inspired riblet geometries for low drag applications[J]. Journal of colloid and interface science, 2016, 474: 206-215.

[6] Lyu S, Nguyen D C, Kim D, et al. Experimental drag reduction study of super-hydrophobic surface with dual-scale structures[J]. Applied Surface Science, 2013, 286: 206-211.

[7] Li Z, Marlena J, Pranantyo D, et al. A porous superhydrophobic surface with active air plastron control for drag reduction and fluid impalement resistance[J]. Journal of Materials Chemistry A, 2019, 7(27): 16387-16396.

[8] Xiang Y, Huang S, Lv P, et al. Ultimate stable underwater superhydrophobic state[J]. Physical review letters, 2017, 119(13): 134501.

[9] Zhang J, Yao Z, Hao P. Drag reductions and the air-water interface stability of superhydrophobic surfaces in rectangular channel flow[J]. Physical Review E, 2016, 94(5): 053117.
[10] Schmüser L, Zhang W, Marx MT, et al. Role of Surface Chemistry in the Superhydrophobicity of the Springtail Orchesella cincta (Insecta:Collembola). ACS Appl Mater Interfaces. 2020, 12(10):12294-12304.
[11] Filippov A E, Kovalev A, Gorb S N. Numerical simulation of the pattern formation of the springtail cuticle nanostructures[J]. Journal of The Royal Society Interface, 2018, 15(145): 20180217.
[12] Ou J, Perot B, Rothstein J P. Laminar Drag Reduction in Microchannels Using Ultrahydrophobic Surfaces[J]. Physics of Fluids, 2004, 16(12): 4635-4643.

# Dynamic pressure-based rocket PID parameters adaptive adjustment method and application

**1stMeng Zhao[a], Co-1stYifan Zhang[b], 2ndZhengxin Tao[c], 3rd Yiyue Liu[d], 4thRunqi Li[e*], 5th Ke Peng[f]**

Organization: College of Aerospace Science and Engineering, National University of Defense Techenology, Changsha,Hunan, China

[a]e-mail: 1290185302@qq.com; [b]e-mail: 1228613509@qq.com; [c]e-mail: 872679253@qq.com; [d]e-mail: 1073496213@qq.com; [e]e-mail: 2629128842@qq.com; [f]e-mail: pengke.nudt@nudt.edu.cn

*Corresponding Author: baixibin@hotmail.com

**Abstract**—This paper establishes a dynamic pressure-based PID parameter adjustment method, which uses the function of the ratio of the maximum dynamic pressure of the rocket to the actual dynamic pressure of the current flight to correct the rocket control parameters. Compared with the traditional time-based PID parameter gain scheduling method, the number of PID parameters to be adjusted is significantly reduced from nearly 50 to 10, and the adjustment complexity is significantly reduced. Based on the method in this paper, the PID parameter debugging of the attitude control of a model-guided rocket is completed. The six-degree-of-freedom ballistic simulation results show that the control law obtained by the method in this paper meets the requirements of the attitude control of the model-guided rocket, and is superior to the traditional time-based PID parameter gain adjustment method in terms of response rate.

## 1. INTRODUCTION

The traditional PID method of rocket attitude control adjusts PID parameters by taking characteristic points at different flight times and interpolating calculation. The calculation is more complicated and the anti-interference ability is poor. The response characteristics of attitude control during rocket flight are closely related to dynamic pressure, which changes significantly with time. Considering the influence of dynamic pressure in the process of adjusting PID parameters of rocket attitude control is an important aspect to improve the control effect[1]. Reference [1] established the actual model of the pneumatic load simulator through Amesim, and adopted the control methods of PID and PID + dynamic pressure feedback respectively, and concluded that the dynamic pressure feedback is better than PID regulation. In the ballistic design of the literature [5], a secondary downward pressure ballistic design method based on dynamic pressure changes is used to reduce the dynamic pressure while satisfying the flight constraints, and the rationality and effectiveness of the proposed method are verified through mathematical simulation. In the literature [6] the method based on dynamic pressure is also used for PID parameter adjustment. Aiming at the attitude control problem of model-guided rockets, this paper establishes an adaptive adjustment method of PID parameters based on dynamic pressure, completes the design of the rocket attitude control law and carries out simulation verification.

## 2. Adaptive adjustment method of PID parameters for attitude control based on dynamic pressure

The actual flight of the rocket is affected by various disturbances. As a result, the actual attitude angle deviates from the commanded attitude angle to produce attitude deviation, then the attitude controller calculates the rudder declination command based on the attitude deviation and transmits it to the rudder system (actuator module), finally the rudder blade turns the corresponding rudder deflection angle to generate control torque to change the rocket's attitude. The block diagram of the rocket attitude control system is shown in Fig. 1.

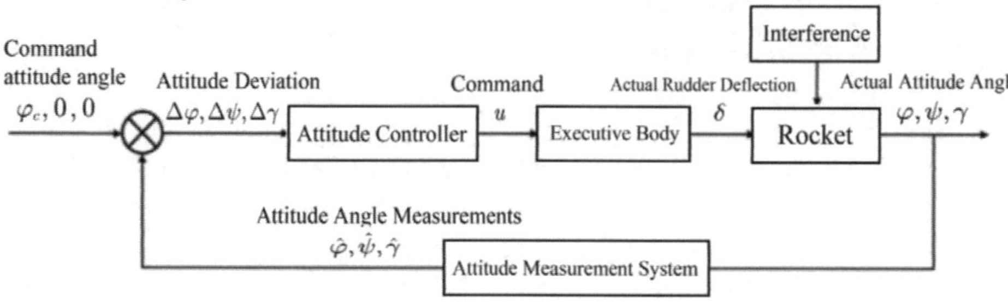


Figure 1. Block diagram of rocket attitude control system

The rocket closed-loop control process is shown in Fig. 2.

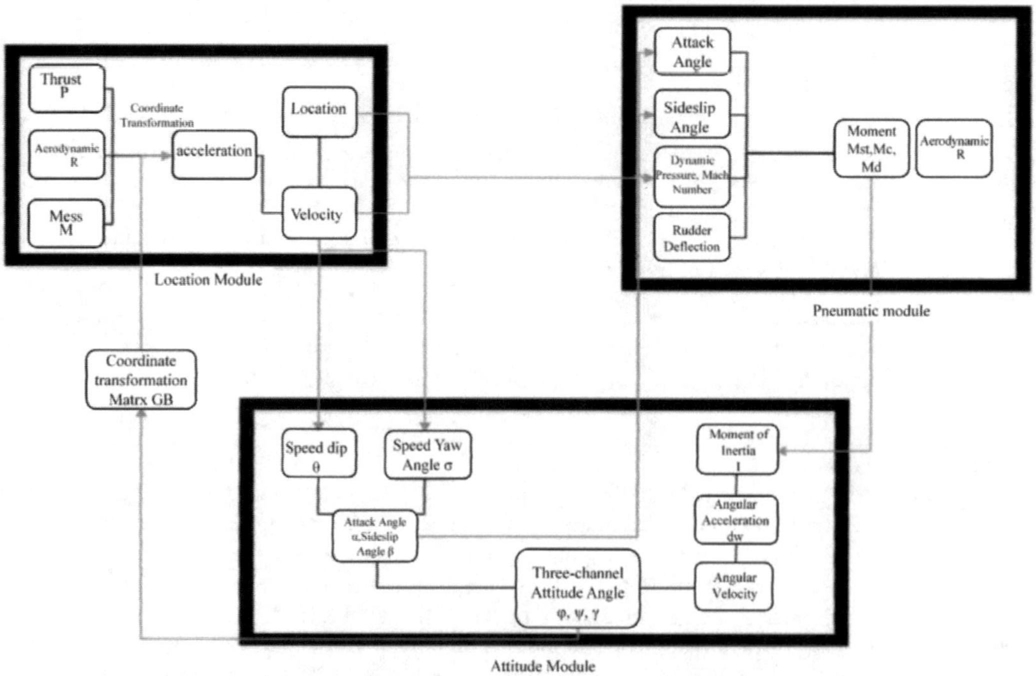


Figure 2 Rocket closed-loop control schematic

This paper applies a PID controller to control the rocket attitude:

$$u(t) = K_p e(t) + K_i \int_0^t e(t)dt + K_d \frac{de(t)}{dt} \tag{1}$$

In the rocket attitude control system, $e$ in the above equation (1) indicates the attitude angle error (the difference between the commanded attitude angle and the actual attitude angle). In this paper, the differential term is improved by using the direct feedback of the angular velocity of the rocket, which can be measured directly by the combined navigation system, and it is easy to build the control system. The controller of three channels is designed as (2).

$$\begin{cases} \Delta\delta_{\varphi} = K_{p\varphi}\Delta\varphi + K_{i\varphi}\frac{1}{s}\Delta\varphi - K_{d\varphi}s\Delta\varphi \\ \Delta\delta_{\psi} = K_{p\psi}\Delta\psi + K_{i\psi}\frac{1}{s}\Delta\psi - K_{d\psi}s\Delta\psi \\ \Delta\delta_{\gamma} = K_{p\gamma}\Delta\gamma + K_{i\gamma}\frac{1}{s}\Delta\gamma - K_{d\gamma}s\Delta\gamma \end{cases} \tag{2}$$

In this paper, a certain model rocket is selected as the controlled object. Its dynamic pressure varies significantly with time during flight and the dynamic pressure in the active section rapidly reaches its maximum value under the action of engine thrust and gradually becomes smaller before reaching the ballistic apex, and the dynamic pressure in the descending section gradually increases as the altitude decreases. The curves of dynamic pressure versus time during the flight of the model-guided rocket are shown in Fig. 3.

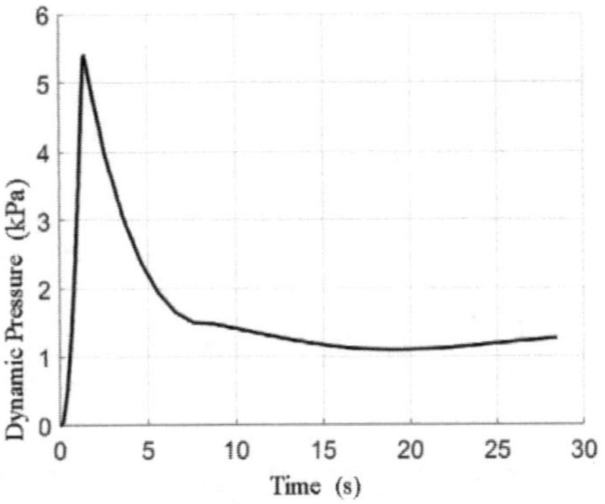


Figure 3 Curve of dynamic pressure versus time during the flight of model-guided rocket

The control effect of the rocket's rudder is closely related to the dynamic pressure of the rocket, with a smaller rudder effect required for higher dynamic pressure and a larger rudder effect required for lower dynamic pressure. This paper uses the function of the ratio of the maximum dynamic pressure $q_{\max}$ of the rocket to the actual dynamic pressure $q$ of the current flight to correct the PID parameters of the rocket (3).

$$K(q) = f(\frac{q_{\max}}{q})K \tag{3}$$

where $K$ is a constant control parameter value. Select $f(x) = (x)^{\frac{1}{n}}$ as the control function to regulate the PID parameters based on dynamic pressure, where $n$ is a constant adjustment factor, and should be taken as $(1 \le n \le 3)$. In case that the integral term in PID control tends to accumulate too large a value over time, the integral term should be limited, which is, when the angle deviation is greater than 30°, the integral term is set to 0.

In summary, the rocket's dynamic pressure-based PID regulation function is given in (4).

$$\begin{cases} K_p(q) = \left(\frac{q_{\max}}{q}\right)^{\frac{1}{n}} K_p \\ K_i(q) = \left(\frac{q_{\max}}{q}\right)^{\frac{1}{n}} K_i, \quad (|\delta_{\varphi/\psi/\gamma}| \ge 30°, K_i(q) = 0) \\ K_d(q) = \left(\frac{q_{\max}}{q}\right)^{\frac{1}{n}} K_d \end{cases} \tag{4}$$

When carrying out simulation tests in the simulation program for different values of $n$, it turns out that the control effect is best when $n$ is 2. Compared with the time-based PID parameter gain adjustment method, the number of PID parameters to be adjusted by dynamic pressure-based PID regulation is significantly reduced from nearly 50 to 10, which significantly reduces the adjustment complexity.

## 3. ROCKET FLIGHT CONTROL PARAMETERS DESIGN AND SIMULATION ANALYSIS

### *3.1. Rocket performance parameters*

The performance parameters of the rocket are shown in Table 1.

TABLE 1. ROCKET PERFORMANCE PARAMETERS

| Parameter Name | Parameter Value |
|---|---|
| Total length of rocket(mm) | 1255 |
| Maximum outer diameter(mm) | 90 |
| Rocket takeoff mass(kg) | 2.8 |
| Total engine impulse(N·s) | 296.14 |
| Launch inclination(°) | 55.6 |
| Maximum dynamic pressure(Kpa) | 5.41 |
| Rocket full-load center of mass position(mm) | 779.31 |
| Rocket no-load center of mass position(mm) | 725.53 |
| Launcher sliding length(m) | 1.2 |
| Rocket full-load rotational inertia(kg·m$^2$) | [0.0052 0.483 0.483] |
| Rocket no-load rotational inertia(kg·m$^2$) | [0.0047 0.418 0.418] |

### *3.2. PID parameters design*

Take the launch angle of the rocket as 55°, keep flying at zero angle of attack in the ascending section, and glide with a positive angle of attack after passing the trajectory vertex. The simulation results of the three-degree-of-freedom trajectory are shown in the Fig. 4.

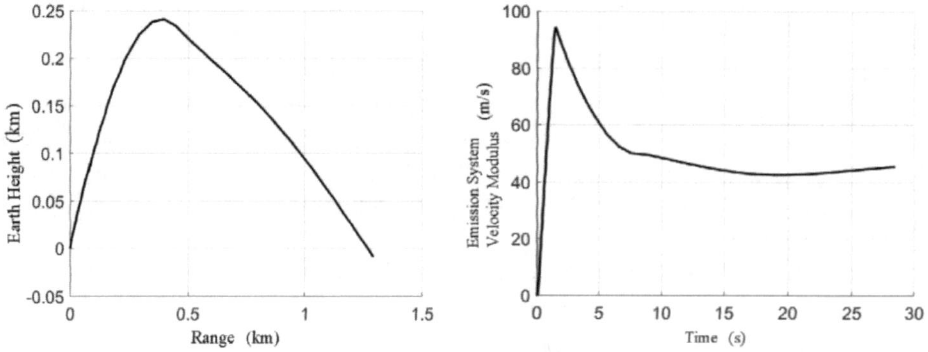


Figure 4 Simulation results of rocket three-degree-of-freedom trajectory

The range is 1.29 kilometers, the ballistic vertex height is 240m, the total flight time is 28.45 seconds, and the maximum speed is 93.6m/s.

The pitch program angle is set based on the calculation result of the three-degree-of-freedom ballistic simulation pitch angle, and the roll angle and yaw angle are kept at 0°. According to the mission requirements, the rocket is set to hit the target at a distance of 1050m and a height of 80m. After flying for 20s, the terminal guidance is turned on, and proportional guidance method is applied for the terminal guidance.

After repeated adjustment and analysis, the PID parameter settings based on dynamic pressure are shown in Table 2.

TABLE 2. MODEL-GUIDED ROCKET ATTITUDE CONTROL PID PARAMETER VALUES

| Control parameter | Value |
|---|---|
| Pitch proportional term coefficient | 20.5 |
| Pitch integral term coefficient | 8.2 |
| Pitch differential term coefficient | 4.7 |
| Yaw proportional term coefficient | 14.5 |
| Yaw integral term coefficient | 6.6 |
| Yaw differential term coefficient | 3.4 |
| Roll proportional term coefficient | 6.2 |
| Roll integral term coefficient | 2.7 |
| Roll differential term coefficient | 1.1 |
| Dynamic pressure adjustment term $n$ | 2 |

*3.3.Simulation analysis*

Import data such as guided rocket PID control parameters, rocket technical parameters, launch pad sliding length and engine thrust into the 6-DOF rocket simulation model to simulate the rocket flight process and calculate the speed, position and Euler angle during the rocket flight process. The control effects of the dynamic pressure-based PID parameter adjustment method and the traditional time-based PID gain adjustment method are analyzed and compared. The simulation results of the rocket's 6-DOF trajectory are shown in Fig. 5.

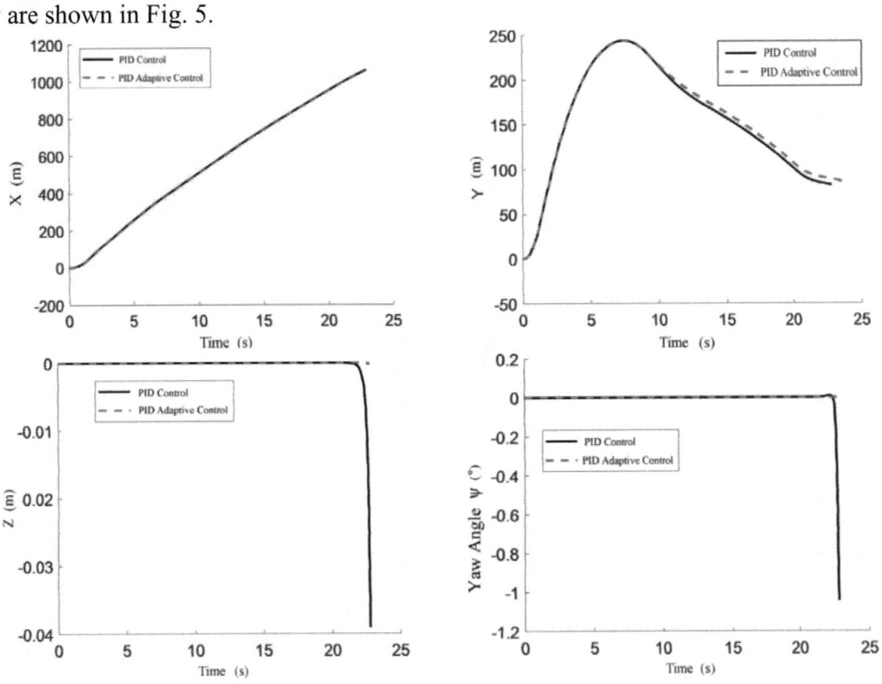

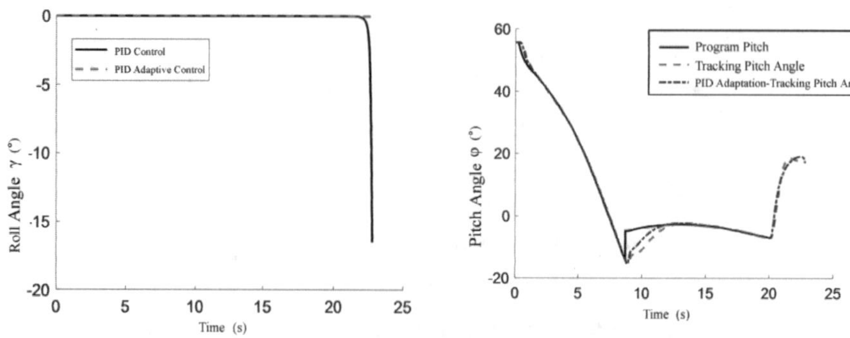


Figure 5 Simulation results of rocket 6-DOF ballistics

The results show that the control law obtained based on the method in this paper meets the requirements of the attitude control of the model-guided rocket, and is superior to the traditional method in terms of response rate.

## 4. CONCLUSION

According to the dynamic pressure variation law as well as the relationship between PID control law and dynamic pressure during rocket flight, a PID parameter adjustment method for rocket attitude control based on dynamic pressure is established. Using the function of the ratio of the maximum dynamic pressure $q_{max}$ of the rocket to the actual dynamic pressure $q$ of the current flight to adjust and correct the PID parameters of the rocket control adaptively. Compared with the traditional time-based PID parameter gain adjustment method, the number of PID parameters to be adjusted is greatly reduced from nearly 50 to 10, and the adjustment complexity is significantly reduced. Based on the method in this paper, the PID parameters adjustment of the attitude control of a model-guided rocket is completed. The six-degree-of-freedom ballistic simulation results show that the control law obtained by the method in this paper meets the requirements of the attitude control of the model-guided rocket, and is superior to the traditional time-based PID parameter gain adjustment method in terms of response rate.

## REFERENCES

[1] Feng Shize. Research on control strategy of pneumatic load simulator based on dynamic pressure feedback [D]. Harbin Institute of Technology, 2012.

[2] Luo Zhe, Li Xinsan, Wang Lixin, Liu Jieyu, Shen Qiang, Xue Liang. Maneuver penetration guidance strategy of hypersonic vehicle based on dynamic pressure constraints [J]. Chinese Journal of Inertial Technology, 2021, Vol. 29(1): 112-118.

[3] Xu Shihao. Research on adaptive augmented disturbance rejection control for launch vehicles [D]. Harbin Institute of Technology, 2020.

[4] Liu Junhui, Shan Jiayuan, Rong Jili, Zheng Xiong. Incremental reinforcement learning flight control with adaptive learning rate [J]. Chinese Journal of Astronautics, 2022, Vol. 43(1): 111-121.

[5] Wang Zhi, Ma Baohai, Zhang Jie, Xiong Wei, Song Jianshuang. A Ballistic Design Method of Rocket Booster Stage Based on Dynamic Pressure Change [J]. Missile and Space Launch Technology, 2018, (2): 65-68.

[6] Chen Guangshan, Zhang Tiejun, Xia Bin, Feng Hao, Liao Huanian, Zhu Wenwen. A Control Parameter Adjustment Algorithm Adapting to Pneumatic Nonlinearity of Rolling Channels [P]. Chinese Patent: CN112363521A, 2021. 02. 12.

[7] Ye Saixian, Zhong Kewei, Chen Guangshan, Zheng Ziyuan, Chen Liang, Yu Shuaixian, Liu Lu. A Stable Control Method Applicable to Low Rudder Efficiency Separation State [P]. Chinese Patent: CN112327908A, 2021. 02. 05.

[8] Peng Xiaobo, Zhang Yong, Huang Pei. A rocket attitude control system [P]. Chinese patent: CN111336868A, 2020. 06. 26.

[9] Cai Guangbin, Xu Hui, Huang Yueping, Yang Xiaogang, Xi Jianxiang, Hu Laihong. A Reentry Trajectory Optimization Method for Hypersonic Vehicles Based on Sequential Convex Programming [P]. Chinese Patent: CN111930145A, 2020. 11. 13.

# Dynamic load identification based on piecewise fitting trend term and smooth curve

**Wei Gao[1], Zhuoyu Ren[1*]**

[1]School of mathematics and statistics, Northeast Petroleum University, Daqing 163000, China

[*]Corresponding author's e-mail:[*]1248992926@qq.com

**Abstract.** $wilson-\theta$ back analysis method is a good load identification method. However, there are some problems, such as unsatisfactory load identification results caused by acceleration measurement errors. By deriving the recurrence formula of cumulative acceleration error of velocity and displacement, the different situations of displacement and velocity error are summarized. This paper introduces an algorithm for identifying the extreme points of trend items. The correction speed and displacement are modified by subsection fitting. The midpoint smoothing fitting of the load data after the initial identification is proposed which can better improve the accuracy of the load identification $wilson-\theta$ back analysis method. Finally, a simulation example shows that the displacement, velocity and the identified load error level are well controlled.

## 1. Introduction

Dynamic load identification technology is widely used in aerospace, bridge technology, machinery and other fields. After more than ten years of development, dynamic load identification technology is mainly divided into frequency domain identification and time domain identification. [1-2] Frequency domain methods commonly used include: frequency response function matrix inversion method [3], generalized orthogonal polynomial frequency domain identification method [4]; In the time domain, there are time domain identification method based on modal filter[5] and back analysis method.

$wilson-\theta$ method is a kind of good method to solve forward problems in structural dynamics, which has been widely used. Based on the wide application of $wilson-\theta$ method in forward problems, $wilson-\theta$ back analysis method has also received extensive attention in dynamic load identification. However, there are still many problems in $wilson-\theta$ back analysis method. ZHAO Fengyao[6] has better solved the problem of sensitive initial value of $wilson-\theta$ back analysis method by introducing polynomial fitting algorithm. Jiang Jinhui [7] introduced the golden section method to improve the cumulative error of $wilson-\theta$ back analysis model to a certain extent.Pu Yuxue[8] used the dichotomy iterative correction to effectively improve the problem of cumulative error of $wilson-\theta$ back analysis method.

## 2. Influence of acceleration error on wilson - θ back analysis model

In order to explore the influence of acceleration data error on speed, displacement and load identification results, it is assumed that the values of real acceleration at each time are $\ddot{u}_0,\ddot{u}_1\ldots\ddot{u}_t$, the actual input acceleration with noise at each time is $\ddot{u}_0^{\#},\ddot{u}_1^{\#}\ldots\ddot{u}_t^{\#}$ .At the same time, set the difference between the noisy acceleration and the real acceleration as $\delta_0,\delta_1\cdots\delta_t$ .

Namely $\ddot{u}_t^{\#}-\ddot{u}_t=\delta_t$ . The acceleration with error is substituted into $wilson-\theta$ back analysis model to calculate the influence on the load identification results.

According to the recursive formula (1) (2) (3) of $wilson-\theta$ method, the velocity and displacement in the whole time range can be calculated by acceleration.

$$\ddot{u}_{t+1}=\frac{6}{\theta^3\Delta t^2}[u(t+\theta\Delta t)-u_t]-\frac{6}{\theta^2\Delta t}\dot{u}_t+(1-\frac{3}{\theta})\ddot{u}_t \tag{1}$$

$$\dot{u}_{t+1}=\dot{u}_t+\frac{\Delta t}{2}(\ddot{u}_{t+1}+\ddot{u}_t) \tag{2}$$

$$u_{t+1}=u_t+\frac{\Delta t^2}{6}(\ddot{u}_{t+1}+2\ddot{u}_t) \tag{3}$$

Replace the acceleration with noise at each time into equation (2).

$$\dot{u}_{t+1}^{\#}=\dot{u}_t^{\#}+\frac{\Delta t}{2}(\ddot{u}_{t+1}^{\#}+\ddot{u}_t^{\#}) \tag{4}$$

$\dot{u}_t^{\#}$ is the velocity with noise. The error value at each time is obtained by making a difference between the theoretical real velocity value (2) and the noisy velocity value (4), i.e

$$\varepsilon[\dot{u}_t^{\#}]=\dot{u}_{t+1}^{\#}-\dot{u}_{t+1} \tag{5}$$

Sorting:

$$\varepsilon[\dot{u}_{t_n}^{\#}]=\frac{\Delta t}{2}(\delta_0+2\sum_{i=1}^{n-1}\delta_i+\delta_n) \tag{6}$$

Replace the acceleration with noise at each time into equation (3).

$$u_{t+1}^{\#}=u_t^{\#}+\Delta t\dot{u}_t^{\#}+\frac{\Delta t^2}{6}(\ddot{u}_{t+1}^{\#}+2\ddot{u}_t^{\#}) \tag{7}$$

Make a difference between the theoretical real velocity value (3) and the noisy velocity value (9) to obtain the error value at each time, i.e

$$\varepsilon[u_t^{\#}]=u_{t+1}^{\#}-u_{t+1} \tag{8}$$

Sorting:

$$\varepsilon[u_t^{\#}]=\frac{\Delta t^2}{6}(2\delta_0+3\sum_{i=1}^{n-1}\delta_i+\delta_n)+\frac{\Delta t^2}{2}[(n-1)\delta_0+(2n-3)\delta_1+\cdots+\delta_{n-1}] \tag{9}$$

From the above derivation, it can be seen that the acceleration makes the velocity and displacement produce random nonlinear trend terms. Thus, the accuracy of load identification results is poor. Therefore, the correction of velocity and displacement algorithm needs to be selected according to different situations. Comparing formulas (6) and (9) shows that the displacement error is more sensitive to the random noise generated by acceleration at each time.

## 3. Simulation example

The finite element model of real steel cantilever beam structure and cantilever beam structure. The cantilever beam structure and node position are shown in Figure 1. The specification of cantilever beam structure is $0.9m \times 0.05m \times 0.009m$, density $7.8 \times 10^3 kg / m^3$, elastic modulus of $200GPa$ [9]. The finite element model of cantilever beam is shown in Figure 2. Apply a simple harmonic load of $p(t) = 40 * \sin(40\pi)$ at node 8, time interval $\Delta t = 0.001, \theta = 1.42$.

Fig 1. Node location of the cantilever beam

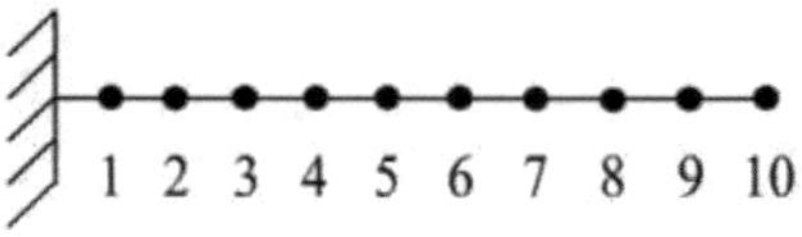


Fig 2. Finite element model of cantilever beam

### 3.1. Correction of displacement and velocity by extreme point fitting algorithm

In order to simulate the real environment, add noise to the acceleration response. Add a noise level of 5% to the acceleration signal.

The displacement discrete points with errors are calculated by the $Wilson - \theta$ back analysis model $(t_i, u_{t_i}^{\#})$. Calculate all the extreme points $k_i$. Since the maximum point and the minimum point appear alternately, calculate the midpoint of the two adjacent extreme points $(x_i, mid_i)$.

Connect all the midpoint $(x_i, mid_i)$ to fit a curve. There are still small fluctuations in the fitted curve, which will interfere with the identification of extreme points of trend items. Therefore, it is necessary to smooth the curve first fitted at the midpoint.

Calculate all extreme points $k_i^{'}$ of points $(x_i, mid_i)$ again Set an error tolerance $d$ (avoid smoothing the extreme point of the real trend item). When the difference between the abscissa of adjacent extreme points meets $x_{i+1}^{'} - x_i^{'} > d$, the extreme point $k_i^{''}$ is retained.

Then the displacement curve with error is segmented by the abscissa of these extreme points $k_i^{''}$. The displacement curve with errors after segmentation is fitted by polynomial least square method, and the trend term is removed.

Figure 3 and Figure 4 show the comparison of velocity and displacement before and after correction. Table 1 shows the comparison of error levels before and after speed and displacement correction. It can be seen that this method can better improve the noise resistance of velocity displacement.

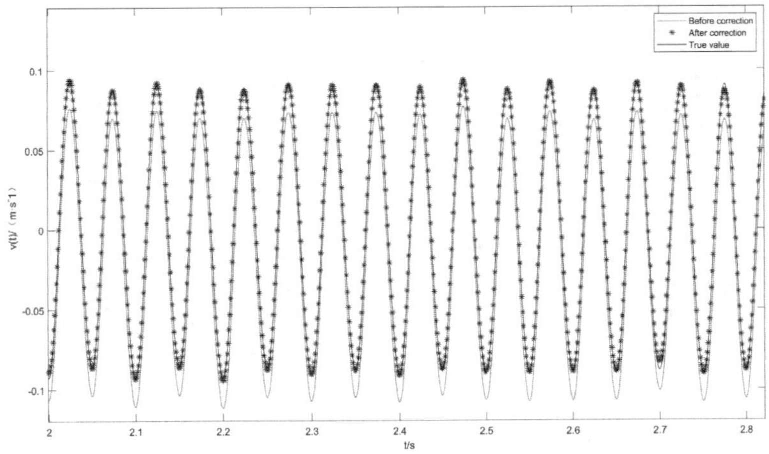


Fig 3. Before and after speed correction and true value

### 3.2. *Median smoothing fitting correction load*

In order to better evaluate the data correction results, calculate the percentage of the square root of the ratio of the square sum of the difference between the corrected value and the real value to the square sum of the actual real value as the evaluation data correction level, as shown in formula (10), which is defined as the error level.

$$\varepsilon = \left(\frac{\sum (f(i)_{fix} - f(i)_{real})^2}{\sum (f(i)_{real})^2}\right)^{1/2} \times 100\% \tag{10}$$

Through the above correction of displacement and velocity, the identification result of load is improved to a certain extent, but the random noise generated by acceleration still has a certain impact on the identification result. In order to further correct the load data, the median point correction data is used below.

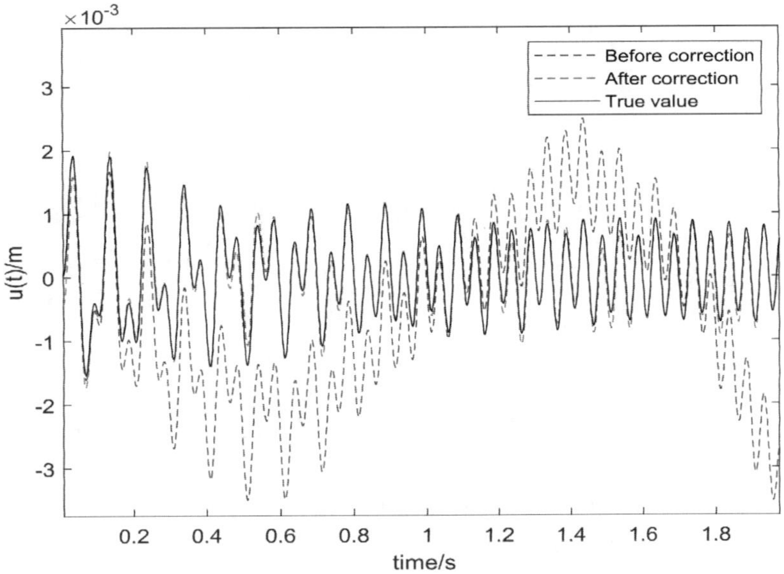


Fig 4. Before and after displacement correction and true value

A series of discrete points $(t_i, p^{\#}_{t_i})$ of the load are calculated from the corrected displacement, velocity and acceleration with noise. Calculate the midpoint of all adjacent points $(t_{fix}, p_{fix_1})$. Connecting all the midpoint $(t_{fix}, p_{fix_1})$ is called the first smoothing curve. Repeat the above steps for the curve after the initial smoothing until the required accuracy is achieved. Figure 5 shows the comparison before and after dynamic load correction. Table 1 shows the comparison of error levels before and after dynamic load correction and after multiple smoothing. It can be seen that this method can better improve the accuracy of load identification.

Table1. Displacement velocity noise level /%

| node | Before displacement correction | After displacement correction | Before speed correction | After speed correction |
|---|---|---|---|---|
| 2 | 1517.55 | 27.82 | 16.12 | 3.79 |
| 3 | 3443.83 | 19.13 | 22.42 | 3.61 |
| 4 | 724.23 | 23.76 | 10.24 | 4.45 |
| 6 | 593.53 | 17.23 | 15.26 | 3.85 |

## 4. Conclusion

Through the back analysis of the $wilson-\theta$ back analysis model, it is found that the trend terms of displacement and velocity have nonlinear and monotonic uncertainties. Displacement error is more sensitive. The method of fitting the trend term by segments at the extreme points is proposed, which effectively improves the errors of velocity and displacement in the back analysis method. A method of calculating the midpoint of adjacent points is proposed to smooth the curve and further improve the accuracy of load identification. The noise resistance of the model is improved.

Table2. Load noise level after multiple smoothing/%

| node | Before correction | After speed displacement correction | First smoothing | Second smoothing | Third smoothing | Fourth smoothing | Fifth smoothing |
|---|---|---|---|---|---|---|---|
| 2 | 6.25 | 4.99 | 3.56 | 3.10 | 2.84 | 2.73 | 2.60 |
| 3 | 9.55 | 4.98 | 3.53 | 3.07 | 2.82 | 2.72 | 2.59 |
| 4 | 5.09 | 4.82 | 3.42 | 3.00 | 2.74 | 2.68 | 2.56 |
| 6 | 5.16 | 4.95 | 3.56 | 3.11 | 2.86 | 2.75 | 2.63 |

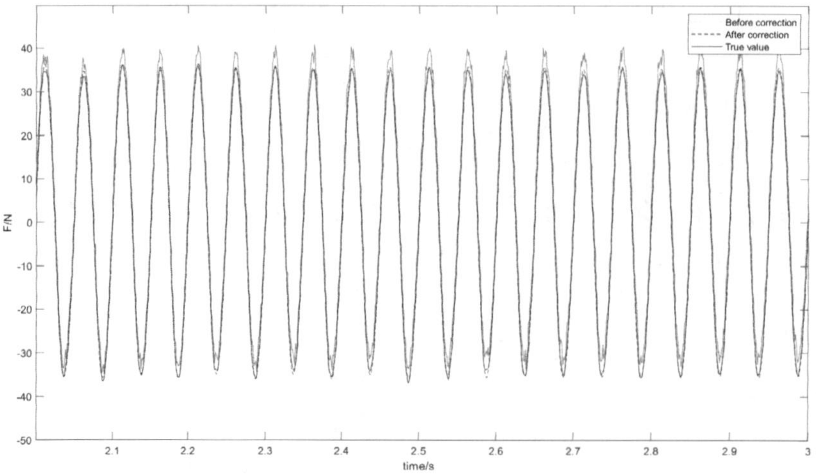


Fig.5. Before and after load correction and theoretical value

## References

[1] Xu Feng,Chen Huaihuai,Bao Ming.Force identification for mechanical vibration:state-of the art and prospect [J].China Mechanical Engineering,2002,13(6):526-531.(in Chinese)

[2] Tadeusz U.The inverse identification problem and its technical application[J].Arch Appl-Mech,2007,77:325-337.

[3] Desanghere G,Snoevs R.Indirect identification of excitation forces by modal coordinate transformation [C]//Proceedings of the 3rd IMAC.Florida,USA:[s.n.],1985:685-690

[4] Zhang Fang,Qin Yuantian,Deng Jihong.Research of identification technology of dynamic load distributed on the structure [J].Journal of Vibration Engineering,2006,19(1):81-85. (in Chinese)

[5] Xu Feng,Chen Huaihuai,Bao Ming.Force identification on model improved based on model filter[J]. Acta Aeronautica Et Astronautica Sinica,2002,23(1):51-54. (in Chinese)

[6] ZHAO Fengyao,GE Wei,et al.Dynamic load identification correction algorithm based on polynomial fitting initial value [J].Journal Of Vibration And Shock,2021,40(11):211-219.

[7] Jiang Jinhui,Xu Jing, et al.Accuracy of the load identification based on inverse analysis of $wilson-\theta$ method [J].Journal of Vibration,Measurement & Diagnosis, 2013, 33(5):782-788.

[8] XU Jing,ZHANG Fang,JIANG Jinhui,et al.An algorithm of dynamic load identification based on numerical iteration [J].Journal of Vibration Engineering,2014,27(5):702-707.

[9] GAO Wei,YU Kaiping, et al.Application of $L_\infty$ norm fitting regularization method in dynamic load identification of space-crafts [J].Journal of Vibration and Shock,2017,36(9):101-107.

# Numerical Calculation of Added Mass of Stratospheric Airship with X-Shaped Tail

**Sheng-zhe SHI, Ting JIANG, Fan WANG, Xiao-long ZHENG**

AVIC Special Vehicle Research Institute, Key Aviation Scientific and Technological Laboratory of High Speed Hydrodynamic, JingMen, CHINA

Shishengzhe05011232@126.com

**Abstract**— For solving added mass of stratospheric airship, take the method of planar motion mechanism numerical simulation, calculate main diagonal added mass of standard ellipsoid model, and then calculate main diagonal added mass and other coupling added mass of stratospheric airship with X-shaped tail. Error between calculated value and theoretical value is 2.94%~4.29%, that validates the method is feasible and effective; stratospheric airship added mass is expressed by matrix. Comparison with wind tunnel test, the calculation error of stratospheric airship with X-shaped tail added mass $\lambda_{55}$ and $\lambda_{66}$ are -25.8% and -19.7%. The result shows that the method of planar motion mechanism numerical simulation can solve added mass matrix of arbitrary shape stratospheric airship, not just main diagonal added mass; Compared with the added mass on the main diagonal, the coupling added mass at other positions is small.

## 1. Introduction

The calculation of the actual airship shows that the added mass value has a great impact on the airship movement, and the added mass value cannot be assumed arbitrarily[1]. At present, there are two main methods for calculating the added mass of objects with complex shapes. ye Ma[2], Huiping FU[3] and Changrun XIAO[4] calculate the added mass by calculating the force/torque of constant speed/ variable speed translational motion and constant speed/ variable speed rotational motion, and Kunlun HUANG[5], Gang LI[6], He ZHANG[7] and Zhao wei LIN[8] simulate the translational or rotational micro amplitude sinusoidal motion, the added mass is calculated by means of plane motion mechanism simulation. These calculation take submarine, underwater robot and other underwater vehicles as examples, because the added mass of underwater vehicle is large, and it is verified by the test of hydrodynamic plane motion mechanism, meanwhile the numerical calculation of airship added mass is relatively few. Moreover, the numerical calculation of additional mass only calculates the value on the main diagonal of the additional mass matrix, not all non-zero terms in the additional mass matrix. In this paper, the additional mass of the standard ellipsoid model is calculated by simulating the plane motion mechanism. After verifying that the method is feasible and effective, the additional mass on the main diagonal of the X-shaped tail stratospheric airship and the coupling additional mass at other positions are calculated and expressed in the form of square matrix.

## 2. Added Mass of Airship

When the airship makes unsteady motion in the ideal fluid, its aerodynamic force is proportional to the acceleration of the airship motion, and the direction is opposite to the acceleration direction, and its proportional constant is called the added mass, which is expressed by $\lambda_{ij}$ .The added mass is always positive, and it is specified that the movement along the x, y and z directions is represented by 1, 2 and 3; The rotation in the x, y and z directions is represented by 4, 5 and 6.When an object of any shape moves, there are 36 additional masses, which can be listed into a square matrix as follows:

$$\begin{bmatrix} \lambda_{11} & \lambda_{12} & \lambda_{13} & \lambda_{14} & \lambda_{15} & \lambda_{16} \\ \lambda_{21} & \lambda_{22} & \lambda_{23} & \lambda_{24} & \lambda_{25} & \lambda_{26} \\ \lambda_{31} & \lambda_{32} & \lambda_{33} & \lambda_{34} & \lambda_{35} & \lambda_{36} \\ \lambda_{41} & \lambda_{42} & \lambda_{43} & \lambda_{44} & \lambda_{45} & \lambda_{46} \\ \lambda_{51} & \lambda_{52} & \lambda_{53} & \lambda_{54} & \lambda_{55} & \lambda_{56} \\ \lambda_{61} & \lambda_{62} & \lambda_{63} & \lambda_{64} & \lambda_{65} & \lambda_{66} \end{bmatrix} \quad (1)$$

The added mass only depends on the shape of the airship and the choice of coordinate axis, and has nothing to do with the motion of the airship.

According to potential flow theory, it can be proved that:

$$\lambda_{ij} = \lambda_{ji} \quad (i,j=1,2,\ldots,6) \quad (2)$$

The items down the main diagonal are equal to the up items in the square matrix (1), only 21 items are independent. And the dimensions of each item: $\lambda_{11}, \lambda_{22}, \lambda_{33}, \lambda_{12}, \lambda_{13}, \lambda_{23}$ have mass dimensions, which is called added mass. $\lambda_{14}$、$\lambda_{15}$、$\lambda_{16}$、$\lambda_{24}$、$\lambda_{25}$、$\lambda_{26}$、$\lambda_{34}$、$\lambda_{35}$、$\lambda_{36}$ have the dimension of mass static moment, it is called added mass static moment. $\lambda_{44}, \lambda_{55}, \lambda_{66}, \lambda_{45}, \lambda_{46}, \lambda_{56}$ have the dimension of mass moment of inertia, which is called added moment of inertia[9].

If the airship has a plane of symmetry oxz, all terms in which subscript i + J = odd are zero.

If the airship has another plane of symmetry oxy, there is: $\lambda_{13} = \lambda_{15} = \lambda_{24} = \lambda_{46} = 0$ .

## 3. Numerical Calculation of Plane Motion Mechanism Test

For the numerical solution of added mass, numerical calculation of planar motion mechanism (PMM) is commonly used, which is obtained by analyzing the time history curve in the calculation.

### *3.1.Numerical Calculation of Pure Heave Motion*

Taking pure heave motion as an example, the law of linear motion meets the requirements[7]:

$$\begin{cases} \zeta = a\sin\omega t \\ \theta = \dot{\theta} = 0 \\ w = \dot{\zeta} = a\omega\cos\omega t \\ \dot{w} = -a\omega^2 \sin\omega t \end{cases} \quad (3)$$

In formula (3), $\zeta$ is the vertical displacement of airship; $a$ is the amplitude of heave motion of airship; $\omega$ is the circular frequency of airship pure heave motion; $\theta$ and $\dot{\theta}$ are the angle and angular velocity of the airship around the y-axis; $w$ and $\dot{w}$ are the vertical velocity and acceleration of airship; $t$ is time.

The linear equation of pure heave motion is:

$$\begin{cases} Z = Z_{\dot{\omega}}\dot{\omega} + Z_{\omega}\omega + Z_0 \\ M = M_{\dot{\omega}}\dot{\omega} + M_{\omega}\omega + M_0 \end{cases} \quad (4)$$

In formula (4), Z is the vertical force at the center of gravity of the airship; M is the moment around Y axis at the center of gravity of airship; $Z_0$ and $M_0$ are the vertical zero lift and zero moment around the y-axis; $Z_\omega$, $M_\omega$, $Z_{\dot{\omega}}$ and $M_{\dot{\omega}}$ are the linear velocity derivative and acceleration derivative.

Substitute formula (3) into formula (4) to obtain:

$$\begin{cases} Z = -a\omega^2 Z_{\dot{\omega}} \sin\omega t + aZ_\omega \cos\omega t + Z_0 \\ M = -a\omega^2 M_{\dot{\omega}} \sin\omega t + aM_\omega \cos\omega t + M_0 \end{cases} \tag{5}$$

The time history force and moment of airship during pure heave motion are decomposed into fluid inertia force (moment) in the same phase with heave, damping force (moment) orthogonal to heave and constant part. The subscript in is used to represent the in-phase component and out is used to represent the orthogonal component. Then:

$$\begin{cases} Z_{\dot{\omega}} = -\dfrac{Z_{in}}{a\omega^2} \\ Z_\omega = \dfrac{Z_{out}}{a\omega} \\ M_{\dot{\omega}} = -\dfrac{M_{in}}{a\omega^2} \\ M_\omega = \dfrac{M_{out}}{a\omega} \end{cases} \tag{6}$$

The added mass is equal to the value of the acceleration derivative, and the sign is opposite, $\lambda_{33}$=-$Z_{\dot{\omega}}$, $\lambda_{35}$=-$M_{\dot{\omega}}$.

*3.2.Numerical Simulation of Pure Pitch Motion*

Taking pure pitch motion as an example, the law of rotational motion meets the requirements[7]:

$$\begin{cases} \theta = \theta_0 \sin\omega t \\ q = \dot{\theta} = \theta_0 \omega \cos\omega t \\ \dot{q} = -\theta_0 \omega^2 \sin\omega t \\ w = \dot{w} = 0 \end{cases} \tag{7}$$

In formula (7), $\theta_0$ is the amplitude of pure pitch motion of airship; $q$ and $\dot{q}$ are the pitch angular velocity and angular acceleration of the airship

The linear equation of pure pitch motion is:

$$\begin{cases} Z = Z_{\dot{q}}\dot{q} + Z_q q + Z_0 \\ M = M_{\dot{q}}\dot{q} + M_q q + M_0 \end{cases} \tag{8}$$

In formula (8), $Z_q$, $M_q$, $Z_{\dot{q}}$ and $M_{\dot{q}}$ are the derivatives of angular velocity and angular acceleration.

Substitute formula (7) into formula (8) to obtain:

$$\begin{cases} Z = -\theta_0\omega^2 Z_{\dot{q}} \sin\omega t + \theta_0 Z_q \cos\omega t + Z_0 \\ M = -\theta_0\omega^2 M_{\dot{q}} \sin\omega t + \theta_0 M_q \cos\omega t + M_0 \end{cases} \tag{9}$$

The time history force and moment of airship during pure pitch motion are decomposed into fluid inertia force (moment) in the same phase with pitch, damping force (moment) orthogonal to pitch and constant part. The subscript in is used to represent the in-phase component and out is used to represent the orthogonal component. Then:

$$\begin{cases} Z_{\dot{q}} = -\dfrac{Z_{in}}{\theta_0 \omega^2} \\ Z_q = \dfrac{Z_{out}}{\theta_0 \omega} \\ M_{\dot{q}} = -\dfrac{M_{in}}{\theta_0 \omega^2} \\ M_q = \dfrac{M_{out}}{\theta_0 \omega} \end{cases} \tag{10}$$

The added mass is equal to the value of the acceleration derivative, and the sign is opposite, $\lambda_{35}$=- $Z_{\dot{q}}$, $\lambda_{55}$=- $M_{\dot{q}}$.

*3.3. Other Plane Motion*

The aerodynamic derivatives (velocity derivative and acceleration derivative) of pure sway motion are $Y_{\dot{v}}, Y_v, N_{\dot{v}}, N_v$, and the corresponding additional mass are $\lambda_{22}, \lambda_{26}$;

The aerodynamic derivatives (velocity derivative and acceleration derivative) of pure yaw motion are $Y_{\dot{r}}, Y_r, N_{\dot{r}}, N_r$, and the corresponding additional mass are $\lambda_{26}, \lambda_{66}$;

For pure surge motion, the aerodynamic derivatives (velocity derivative and acceleration derivative) are $X_{\dot{u}}, X_u, K_{\dot{u}}, K_u$, and the corresponding additional mass are $\lambda_{11}, \lambda_{14}$;

For pure rolling motion, the aerodynamic derivatives (velocity derivative and acceleration derivative) are $X_{\dot{p}}, X_p, K_{\dot{p}}, K_p$, and the corresponding added mass are $\lambda_{14}, \lambda_{44}$;

In the calculation of the added mass of stratospheric airship, due to the symmetry relationship, $\lambda_{22}=\lambda_{33}$, $\lambda_{55}=\lambda_{66}$, $\lambda_{26}=\lambda_{35}$.

## 4. Verification of Numerical Calculation Method

The ellipsoid ( $2a = 20m$ , $2b = 5.333m$ , $\rho = 1.225kg/m^3$ ) to be verified has been calculated by different methods in several papers [1] [2] [6], which can be regarded as a standard model. The ellipsoid grid calculated in this paper is shown in Figure 1, with a total of 609000 body grids, and the size of the first layer of boundary layer grid is 4.7MM.

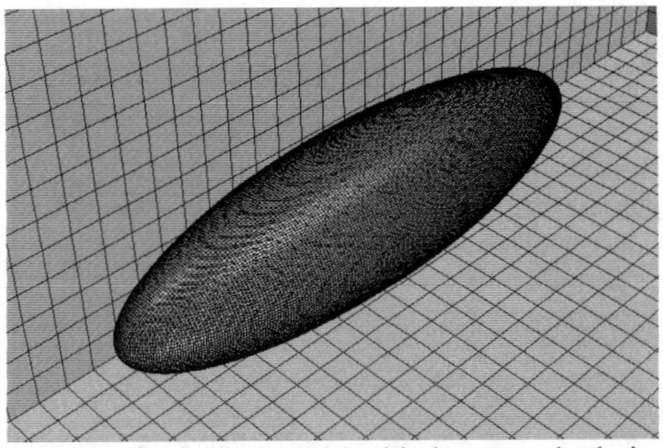

Figure1. Grid of ellipsoid added mass calculation

In paper 1, Hess Smith method is used to calculate the added mass of ellipsoid; In paper 2, the numerical simulation methods of translational acceleration and rotational acceleration are used to calculate the added mass of ellipsoid; In paper 6, viscous flow model and inviscid model are used to calculate the added mass of ellipsoid. The results of papers 1, 2 and 6 are compared with the results of this paper in Table 1.

TABLE 1. ADDED MASS OF ELLIPSOID

| ***added mass*** | $\lambda_{11}$ | $\lambda_{22}$ | $\lambda_{33}$ | $\lambda_{44}$ | $\lambda_{55}$ | $\lambda_{66}$ |
|---|---|---|---|---|---|---|
| Theoretical value | 32.84 | 302.82 | 302.82 | - | 4378.14 | 4378.14 |
| Hess-Smith method[1] | 32.79 | 313.17 | 313.17 | - | 4556.71 | 4556.71 |
| Translational acceleration and rotational acceleration methods[2] | 33.58 | 317.39 | 317.38 | - | 4366.41 | 4366.41 |
| Inviscid model[6] | 32.2 | 300.3 | 300.3 | - | 4376 | 4376 |
| Viscous flow model[6] | 32.3 | 299.8 | 299.8 | - | 4393 | 4393 |
| Calculated value in this paper | 34.25 | 311.72 | 311.72 | - | 4516.13 | 4516.13 |
| Error between and theoretical value(%) | 4.29 | 2.94 | 2.94 | - | 3.15 | 3.15 |

It can be seen from table 1 that the calculated value in this paper is close to the calcula·ted value of the calculation method in [1] [2] [6] in other literature, and the error with the theoretical value is 2.94% ~ 4.29%, which proves that the calculation method in this paper is effective.

## 5. ADDED MASS OF STRATOSPHERIC AIRSHIP WITH X-SHAPED TAIL

The computational grid of stratospheric airship with X-shaped tail in this paper is shown in Figure 2, with a total of 1556000 body grid, and the size of the first layer of boundary layer grid is 5.3mm.

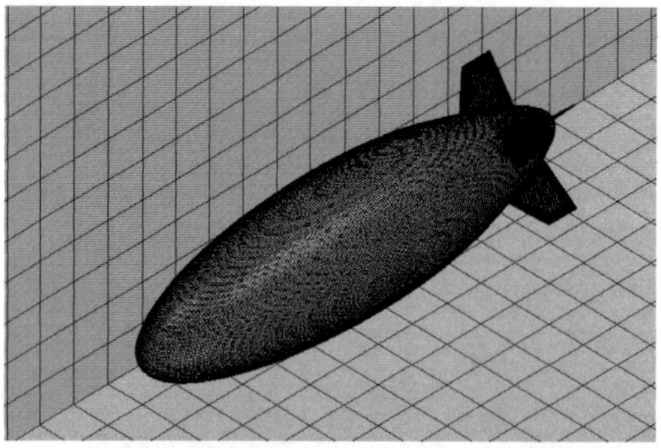

Figure2. Computational grid of additional mass of stratospheric airship with X-shaped tail

Taking the numerical simulation of pure pitching motion of stratospheric airship with X-shaped tail as an example, the pitching amplitude is 0.025rad, the pitching frequency is 0.3HZ, and the initial phase is 0rad; The air density is 0.0889 $kg/m^3$; Calculation time step $\Delta t$= 0.006s, that is, a motion cycle is set as 555.5 time steps, and the computational accuracy can meet the requirements.

Figure 3 shows the force time history curve of the stratospheric airship with X-shaped tail. The solid line is the vertical force on the pure pitching motion, and the dotted line is the pitching moment on the pure pitching motion. It can be seen from the figure that after the second cycle, the time history curves of the vertical force and pitching moment enter stable oscillation. The time history curve of pitching moment is sinusoidal, and the sinusoidal of vertical force is poor, but it still has obvious periodicity.

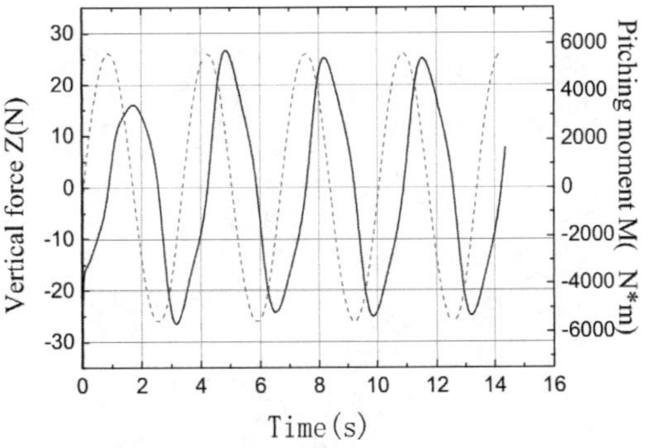


Figure3. Vertical force and pitching moment of stratospheric airship pure pitching motion

The sinusoidal fitting of vertical force and pitching moment after entering stable oscillation as shown in Figure 3 is as follows:

$$Z = 24.10\times\sin(1.885t - 1.5670)$$
$$= 0.0915\times\sin(1.885t) - 24.0998\times\cos(1.885t)$$
$$M = 5650.00\times\sin(1.885t - 0.0318)$$
$$= 5623.9345\times\sin(1.885t) - 510.0083\times\cos(1.885t)$$

So

$$\lambda_{35} = -Z_{\dot{q}} = 0.0915/\left(0.025\times1.885^2\right)$$
$$= 1.03 kg\cdot m^2$$

$$\lambda_{55} = -M_{\dot{q}} = 5623.9345/\left(0.025\times1.885^2\right)$$
$$= 63347.52 kg\cdot m^2$$

Thus, through pure heave motion, pure surge motion and pure roll motion, the additional mass on other main diagonals and the coupling additional mass at other positions can be obtained. Finally, the added mass of the stratospheric airship with X-shaped tail can be written in the form of a square matrix, that is:

$$\begin{bmatrix} 50.28\text{kg} & & & & & \\ & 489.80\text{kg}\cdot\text{m} & & & & 0.91\text{kg}\cdot\text{m} \\ & & 489.80\text{kg}\cdot\text{m} & & 1.03\text{kg}\cdot\text{m} & \\ & & & 44.83\text{kg}\cdot\text{m}^2 & & \\ & & 1.03\text{kg}\cdot\text{m} & & 63347.52\text{kg}\cdot\text{m}^2 & \\ & 0.91\text{kg}\cdot\text{m} & & & & 63347.52\text{kg}\cdot\text{m}^2 \end{bmatrix} \quad (11)$$

It can be seen from formula (11) that the numerical Calculation of plane motion mechanism can solve the additional mass square matrix of stratospheric airship with arbitrary shape, not just the additional mass on the main diagonal, especially for appendages with complex shape, which is easier to realize than the theoretical solution; Due to the asymmetry between the head and tail of the stratospheric airship, the added mass $\lambda_{35}$ and $\lambda_{26}$ are not zero. Compared with the added mass on the main diagonal, the coupling added mass at other positions is small.

## 6. Comparison of Wind Tunnel Test, Engineering Estimation and Numerical Calculation

The stratospheric airship was tested in the NH-2 low-speed wind tunnel of Nanjing University of Aeronautics and Astronautics (as shown in Figure 4). The scale ratio of the model is 1:45, the wind speed V is 30m / s, the reference length L is 14.67m, the reference area s is 300.2m$^2$, and the air density at an altitude of 20km is 0.0889kg/m$^3$.

Figure4. Dynamic derivative test of stratospheric airship with X-shaped tail in low speed wind tunnel

TABLE 2. PARTIAL RESULTS OF WIND TUNNEL TEST AND ENGINEERING ESTIMATION OF AIRSHIP DYNAMIC DERIVATIVES

| | $N'_{\dot{r}}$ | $M'_{\dot{q}}$ |
|---|---|---|
| Engineering estimation | -0.1954 | -0.3915 |
| Wind tunnel test | -0.2625 | -0.2424 |

Added mass $\lambda_{55}$ is obtained from pure yaw motion at $\alpha$=0°and β=0°:

$$\lambda_{55} = -M'_{\dot{q}} \times \frac{1}{2}\rho V^2 SL$$

Added mass $\lambda_{66}$ is obtained from pure pitch motion at $\alpha$=0°and β=0°:

$$\lambda_{66} = -N'_{\dot{r}} \times \frac{1}{2}\rho V^2 SL$$

TABLE 3. COMPARISON OF ADDITIONAL MASS TEST AND NUMERICAL CALCULATION OF STRATOSPHERIC AIRSHIP WITH X-SHAPED TAIL

| project | $\lambda_{55}$ | $\lambda_{66}$ |
|---|---|---|
| Engineering estimation kg*m$^2$ | 75867.1 | 37865.7 |
| Wind tunnel test kg*m$^2$ | 46973.6 | 50868.7 |
| Numerical calculation kg*m$^2$ | 63347.5 | 63347.5 |
| Error between numerical calculation and wind tunnel test % | -25.8 | -19.7 |

Table 3 compares the added mass $\lambda_{55}$ and $\lambda_{66}$ engineering Estimation, wind tunnel test and numerical calculation, The numerical calculation errors of airship added mass $\lambda_{55}$ and $\lambda_{66}$ are - 25.8% and - 19.7% respectively, and it can be used in practical engineering.

## 7. Conclusions

1) For verifying calculation of the standard ellipsoid, the calculated value of the added mass in this paper is close to that of the calculation method in other papers, and the error between the calculated value and the theoretical value is 2.94% ~ 4.29%, which proves that the calculation method is feasible and effective.

2) Compared with the wind tunnel test results, the numerical errors of the added mass $\lambda_{55}$ and $\lambda_{66}$ of stratospheric airship with X-shaped tail are - 25.8% and - 19.7% respectively.

3) The added mass matrix of stratospheric airship with arbitrary shape can be solved by numerical simulation of plane motion mechanism, not just the added mass on the main diagonal; Compared with the added mass on the main diagonal, the coupling added mass at other positions is small.

## References

[1] Dan LIU, Xiao-liang WANG, Xue-xiong SHAN. Added Mass to Stratospheric Airship and Its Effect on Motion.Computer Simulation, 2006, 23 (6): 52-56(in Chinese)

[2] Ye MA, Xue-xiong SHAN. A New Numerical Computation Method for Added Masses of Complicated Object. Computer Simulation, 2007, 24 (5) :75-78,113(in Chinese)

[3] Hui-ping FU, Jie LI.Numerical Studies of Added Mass Based on the CFD Method.Journal of Harbin Engineering University,2011,32(2):148-152(in Chinese)

[4] Chang-run XIAO, Rui-jie LIU, Yang LIU, et al. Calculations of Submarine's Inertial Hydrodynamics Coefficient.Journal of Naval University of Engineering,2014,26(5):44-47(in Chinese)

[5] Kun-lun HUANG, Yong-jie PANG, Yu-min SU,et al.Research on Linear Hydrodynamic Coefficients Calculation Method of Submergible Vehicle. Journal of Ship Mechanics,2008,12(5):697-703(in Chinese)

[6] Gang LI, Wen-Yang DUAN, Zhi-bin GUO. Added Mass of Submerged Vehicles with Complex Shape. Journal of Harbin Institute of Technology,2010,42(7):1145-1148(in Chinese)

[7] He ZHANG, Yong-jie PANG, Ye LI. Numeral Simulation of Hydrodynamic Tests Using FLUENT. Journal of System Simulation,2010,22(3):566-569(in Chinese)

[8] Zhao-wei LIN, Sheng MENG, Hong YIN, et al.Numerical Prediction of Maneuverability of Underwater Vehicle.Shipbuilding of China,2016,57(1):59-68(in Chinese)

[9] Sheng-da SHI. Submarine Maneuverability. BEI JING : Defense Industry Press,1995:23-29(in Chinese)

[10] Hui-ting WANG, Yi BI.Numerical Simulation on Planar Motion Mechanism of KCS Ship model with a body-force Propeller Model. Chinese Journal of Ship Reasearch,2016,11(4):29-37,66(in Chinese)

# The Numerical Simulation of Cavitation Phenomenon in a Venturi Tube

**Zibo Zhou[1,a], Qinglian Li[1,b*], Tao Liang[1,c], Shaotian Gong[1,d]**

[1]College of Aerospace Science and Engineering, National University of Defense Technology, Changsha, Hunan, 410073, China

[a]email: zhouzibo19@nudt.edu.cn, [c]email: 1362651023@qq.com, [d]email: 494679588@qq.com

[b]*Corresponding author's e-mail: lisiwei19@nudt.edu.cn

**Abstract:** Venturi tubes are an important component of the liquid rocket engine to achieve extensive thrust regulation. On the basis of the existing venturi tube model, the relevant numerical simulation is carried out by the Mixture model and the $k-\varepsilon$ turbulence model. Elaborated the typical cavitation flow field structure, cavitation area length and critical pressure ratio and analyzed the influence of outlet pressure on cavitation flow field in cavitation tubes. It was found that cavitation is always produced against the wall, and there is a core area of Cavitation. Cavitation in the core area is the most severe and the temperature is the lowest. From the core area outward, the gas phase area shows a patchy distribution, and there exists a gradient. Also, the critical pressure ratio that can produce cavitation is calculated. It is concluded that under the condition that cavitation can occur, the decrease of outlet pressure will lead to the lengthening of the cavitation area and the larger maximum temperature difference in the Venturi tube.

## 1. Introduction

Cavitation refers to the phenomenon that a liquid produces a large amount of steam during the flow process due to the pressure lower than the saturated vapor pressure of the liquid at that temperature.

Venturi tubes are widely used in the pipelines of liquid rocket engines to achieve the purpose of controlling flow and adjusting the thrust of liquid rocket engines. Its configuration is a convergencing - divergencing pipeline, and cavitation usually occurs in the larynx.

The dimensions of the venturi tube used in this article are shown in Figure 1. The diameter of the inlet basin channel is 20 mm, the diameter of the throat is 8 mm, and the diameter of the outlet is 15 mm. The convergence angle and diffusion angle are 60°and 8°, respectively.

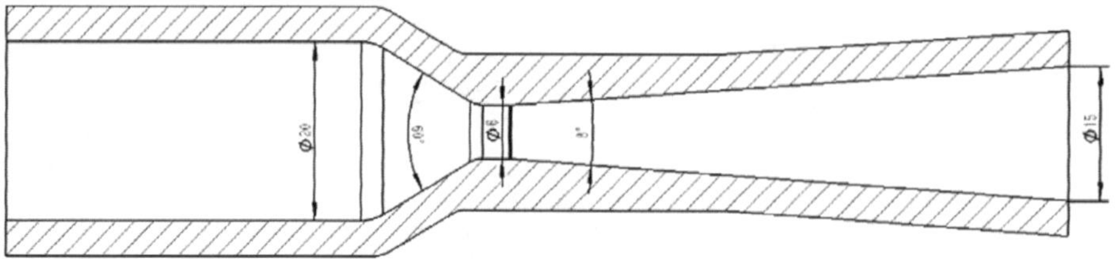


Figure 1 The configuration and dimensions of the venturi tube

The basic theory of cavitation phenomenon states that after the formation of cavitation in the throat, as long as the pressure of the outlet is unchanged, the liquid flow through the Venturi tube is not affected by the change of the outlet pressure of the Venturi tube within a certain range, maintaining the stability of the flow of liquid rocket engines. At the same time, it can suppress the propagation of downstream pressure disturbances upstream and ensure the stability of the system[1].

At present, most of the research on the phenomenon of Venturi tube cavitation is to study the influence of pressure ratio on the evolution process of cavitation zone. Define the pressure $P_r$ as follows:

$$P_r = \frac{P_{inlet}}{P_{outlet}}$$

Scholars have carried out a lot of research on the design of Venturi tube and the phenomenon of cavitation in Venturi tube. Qizhi Chen and Hanru Gao[2] gave the design method of the length of the cylindrical segment of the throat of the flow-adjustable Venturi pipe through analysis. Chibing Shen and Jiping Wu[3] solved the problems of ball screw self-locking, motor adjustment accuracy and repeatability encountered in the test process through the experimental study of two kinds of adjustable cavitation Venturi tube. Fei Tang, Yong Li[4] and others studied the design of linear adjustable cavitation venturide, and realized the purpose of large-scale linear adjustment of flow rate in Venturi tube. Ziyong Zhu[1] et al. calculated and compared the vapour biphasic flow in the Venturi tube with the numerical calculation and experimental comparison, and obtained the law that the gas phase region increases with the increase of the pressure ratio. Yujing Tao[5] obtained the changing law of the flow coefficient, and pressure recovery coefficient through a large number of cold experiments, studying the low temperature flow control system with the use of water. Xinping Long[6] studied the length of the three-dimensional Venturi tube cavitation area based on the standard deviation method, and found that the length of the cavitation area has a linear relationship with the pressure ratio, and there is a turning point. Abdulaziz[7] experimentally found that when the pressure ratio increases, the area of the cavitation zone increases, and the cavitation zone of the upper and lower walls will intersect. It can be seen that the pressure ratio is one of the important factors affecting the morphology of the cavitation zone, and the change of the pressure ratio will cause the cavitation zone to oscillate, and the length of the cavitation zone to change.

## 2. The main research object and content

This paper adopts the numerical simulation method to take the room temperature water at 300K as the working condition, keeping the inlet static pressure unchanged at 5.8MPa, and studies the flow phenomenon in the Venturi tube by changing the outlet pressure. This paper mainly investigates the influence of pressure ratio change on the length of the cavitation are, pressure field distribution, and temperature field distribution, and at the same time, by measuring the dimensionless length of the cavitation area under different pressure ratios, the changing curve of the dimensionless length of the cavitation area with different pressure ratios is obtained, and the critical pressure ratio of cavitation can also be found.

## 3. Numerical calculation model

### *3.1. The Mixture Model*

The internal flow field is a flow field that changes drastically when cavitation occurs in the Venturi tube, and the existence of cavitation phenomenon makes the calculation need to consider the joint action of the gas and liquid phases. So far, there are 3 kinds of multiphase flow models that are relatively mature and commonly used: The VOF(Volume of fluid) model, the Mixture model and the Euler model. The VOF model is suitable for hierarchical two-point free surface flows, while the Mixture and Euler models are suitable for flows with interphase mixing and separation.

This study uses the Mixture model. The Mixture model can simulate multiple flows at different speeds in each phase, assuming the balance of multiphase flows in local regions

*3.2. The Turbulence Model*

Turbulence models in multiphase flow simulations are complex. As a preliminary study, this case selects the $k-\varepsilon$ model to simulate the flow of gas-liquid mixed phases.

$$\frac{\partial(\rho k)}{\partial t}+\frac{\partial(\rho k\mu_i)}{\partial x_i}=\frac{\partial}{\partial x_j}\left[\left(\mu+\frac{\mu_t}{\sigma_k}\right)\frac{\partial k}{\partial x_j}\right]+G_k-\rho\varepsilon-Y_M+G_b \tag{1}$$

$$\frac{\partial(\rho\varepsilon)}{\partial t}+\frac{\partial(\rho\varepsilon u_j)}{\partial x_i}=\frac{\partial}{\partial x_j}\left[\left(\mu+\frac{\mu_i}{\sigma_\varepsilon}\right)\frac{\partial\varepsilon}{\partial x_j}\right]+G_k\frac{\varepsilon C_{1\varepsilon}}{k}-C_{2\varepsilon}\rho\frac{\varepsilon^2}{k} \tag{2}$$

$$G_k=\mu_t\{2[(\frac{\partial u}{\partial x})^2+(\frac{\partial v}{\partial y})^2+(\frac{\partial w}{\partial z})^2]+(\frac{\partial u}{\partial y}+\frac{\partial v}{\partial x})^2+(\frac{\partial u}{\partial y}+\frac{\partial w}{\partial x})^2+(\frac{\partial v}{\partial y}+\frac{\partial w}{\partial x})^2 \tag{3}$$

Where, $G_k$ is the turbulent flow energy due to the average velocity gradient, $G_b$ is the turbulent flow energy due to buoyancy, $Y_M$ is the fluctuation due to the diffusion of the transition in compressible turbulence and $C_{1\varepsilon}$, $C_{2\varepsilon}$, $\sigma_k$ and $\sigma_\varepsilon$ are constants, namely $C_{1\varepsilon}=1.44$, $C_{2\varepsilon}=1.92$, $\sigma_k=1.0$, $\sigma_\varepsilon=1.3$.

*3.3. Cavitation Dynamic Model*

The cavitation model is described in the literature [8], and his starting point is based on the fact that the flow field contains non-condensable gas nuclei in a dissolved or mixed state in the liquid. Since the mass fraction of the non-condensable gas is given in the incoming flow, the content of the non-condensable gas throughout the flow area is constant, but its density changes with the change of local pressure. Thus, the relationship between the mixing density function $\rho$ and the vapor mass fraction $f_v$ is as follows:

$$\frac{1}{\rho}=\frac{f_v}{\rho_v}+\frac{f_g}{\rho_g}+\frac{1-f_v-f_g}{\rho_1} \tag{4}$$

In the above equation, the subscript $v, g, l$ are steam, gas and liquid states, respectively; which can be solved by the continuity equation, the steam transport equation, and the momentum equation.

$$\frac{\partial(vVf_v)}{\partial t}+\nabla(\rho Vf_v)=\nabla(\Gamma\nabla f_\varepsilon)+R_e-R_c \tag{5}$$

In the above equation, the $\Gamma$ is the effective transformation coefficient; $R_c$ is the phase transition rate of compression and rupture of bubbles; $R_e$ is the phase transition rate of bubble generation and expansion, which can be derived from the Rayleigh-Plesset equation:

1) When $P_{st}>P$:

$$R_e=C_e\frac{V_{ch}}{\sigma}\rho_l\rho_v\sqrt{\frac{2(P_{st}-P)}{3\rho_l}}(1-f_v-f_g) \tag{6}$$

2) When $P_{st}<P$:

$$R_e=C_c\frac{V_{ch}}{\sigma}\rho_1\rho_v\sqrt{\frac{2(P_{sat}-P)}{3\rho_l}}f_v \tag{7}$$

where $C_e=0.02$, $C_c=0.01$, $V_{ch}$ is the local characteristic velocity (approximately equal to the local turbulence intensity); $\sigma$ is the surface tension coefficient of saturated liquids; $P_{st}$ is the saturated vapor pressure of the liquid; $P$ is the actual measured local pressure.

## 4. Numerical Calculation Methods

### *4.1. The Settings of Calculation Conditions and Working Conditions*
Due to the axially symmetrical nature of the Venturi tube, it can be reduced to a two-dimension problem. The mesh used for calculation is a structural mesh, which is easier to index, more accurate to calculate.

Figure 2 Two-dimension Venturi tube meshing

Numerical calculations are simulated by Fluent, a computational fluid dynamics software based on the finite volume method. The numerical simulation medium is water, the upper boundary is a non-slip boundary, the lower boundary adopts symmetrical boundary conditions, the inlet pressure is maintained at 5.8MPa, and the cavitation simulation calculation under different pressure ratios is carried out by changing the outlet pressure. The relevant calculated boundary parameters are shown in Table 1 below:

Table 1 Numerical calculation parameters for Venturi tubes with water

| Working conditions | Inlet Total Pressure (MPa) | Inlet Static Pressure (MPa) | Inlet Static Temperature (K) | Outlet Pressure (MPa) |
|---|---|---|---|---|
| Water | 6.0 | 5.8 | 300 | 3.0 |
| | 6.0 | 5.8 | 300 | 3.3 |
| | 6.0 | 5.8 | 300 | 3.8 |
| | 6.0 | 5.8 | 300 | 4.1 |
| | 6.0 | 5.8 | 300 | 4.4 |
| | 6.0 | 5.8 | 300 | 4.8 |

### *4.2. Mesh Independence Verification*
To ensure the correctness of the numerical calculations, the mesh should be independently verified.

Defines $\delta_N$ as a percentage of mesh encryption and $\delta_m$ as the changing rate of the mass flow:

$$\delta_N = \frac{N - N_o}{N_0}$$

$$\delta_m = \frac{\dot{m} - \dot{m}_o}{\dot{m}_0}$$

where N is the total number of encrypted meshes and $\dot{m}$ is the mass flow after encryption; $N_o$ is the total number of meshes of the original mesh, and $\dot{m}_o$ is the mass flow of the original mesh.

Table 2 Comparison of mass flow of the encrypted mesh with the original mesh

| $\delta_N$ | $\delta_m$ |
|---|---|
| 25% | 0% |
| 50% | 0% |
| 75% | 0.73% |
| 100% | 0% |

From the above table, the change in the mass flow of the encrypted mesh is little, so it means that the original mesh satisfies the grid independence condition and can be used for numerical calculation.

## 5. Calculation Results and Discussion
According to the numerical calculation results, this article will mainly elaborate from three aspects: the structure of the typical cavitation field, the length of the cavitation area and the critical pressure ratio, and the influence of outlet counterpressure on the cavitation field.

Journal of Physics: Conference Series **2364** (2022) 012051 doi:10.1088/1742-6596/2364/1/012051

### *5.1. Typical Cavitation Field Structure*

a)-c) in Figure 3 shows the distribution of the gas phase region, pressure field, and temperature field in the vapor erosion flow field with a pressure ratio of 1.41.

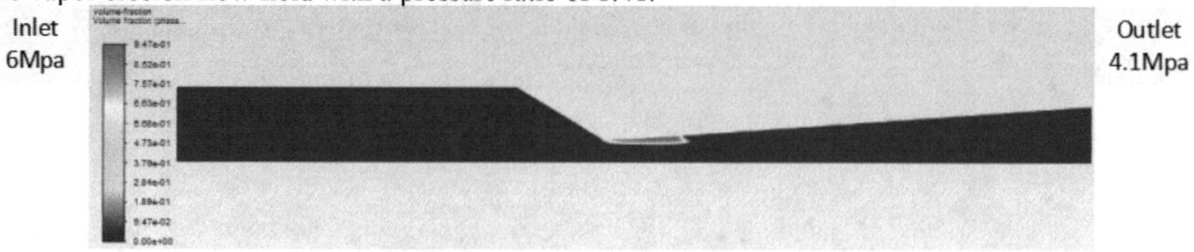


a) The distribution of the gas phase region

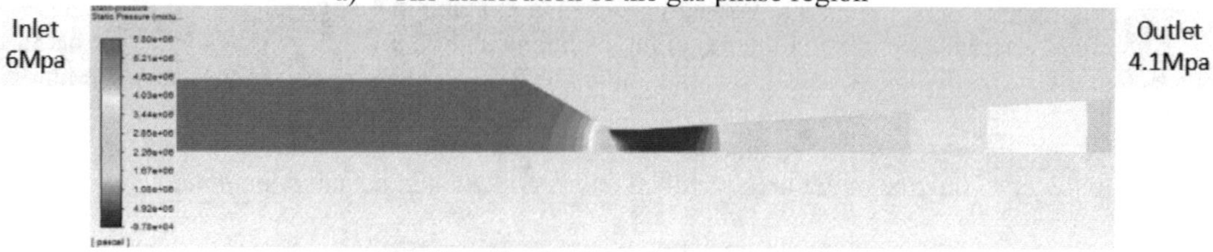


b) The distribution of pressure field

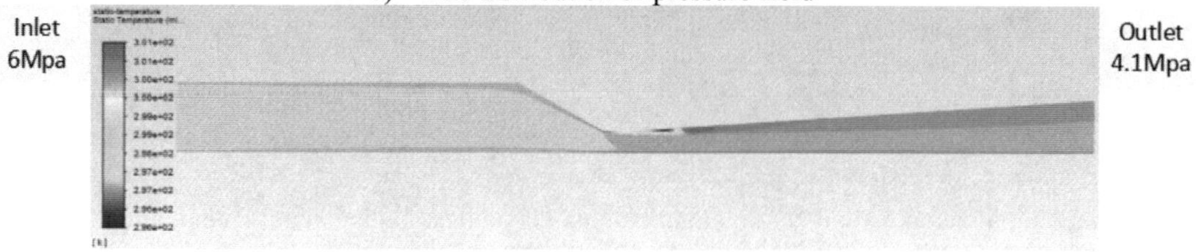


c) The distribution of temperature field

Figure 3 Cavitation flow field structure with the pressure ratio of 1.41

As can be seen from a) in Figure 3, the cavitation zone is a two-phase flow of water and gas, and the cavitation is always produced against the wall, showing a flaky distribution, and there is a clear gradient.

As can be seen from the pressure field cloud diagram in Figure b) above, and the static pressure curve in Figure 4 below. The static pressure in the Venturi tube drops rapidly from the inlet of the convergence section to the larynx, reaching the saturated vapor pressure at this temperature in the larynx, and then as the pipe expands, the cavitation weakens until it disappears, and the pressure returns to the outlet pressure.

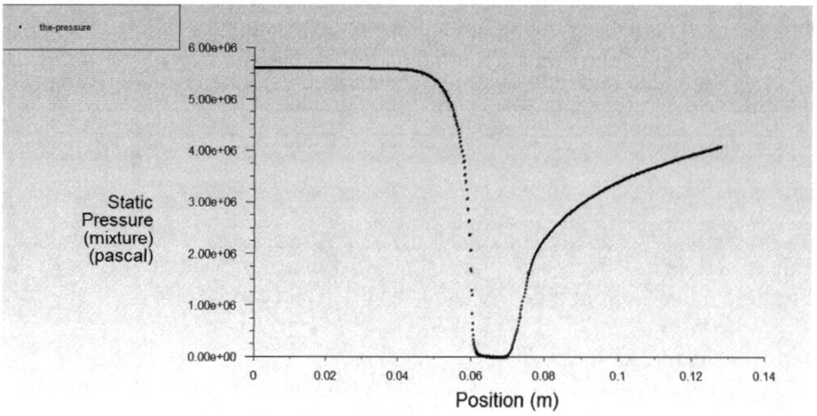


Figure 4 The static pressure distribution curve

It can be found that the fluid temperature distribution occurs after the cavitation zone, which is evident in the temperature field distribution in c) of Fig. 3 above.

Through analysis, it is speculated that the reason for this phenomenon is that there is a violent non-isotropic flow in the cavitation zone, and after the cavitation zone, a non-isotropic slip line appears between the fluid passing through the cavitation zone and the fluid that does not pass through the cavitation zone, and this slip line is matched by pressure. Therefore, the longitudinal distribution of the pressure field in Figure 3b) is relatively uniform, while the longitudinal distribution of the temperature field in Figure 3c) is layered.

Combined with the fractional characteristics of the cavitation area, the pressure field distribution, and the temperature field distribution, we can see that there is obviously a cavitation core area on the wall surface of the Venturi tube, and the closer it is to the cavitation core area, the more severe the cavitation phenomenon is and the lower the temperature will be.

### *5.2. The Length of the Cavitation zone and the Critical Pressure Ratio*

Define the area with the largest gradient of gas phase fractions is the tail of the cavitation zone, namely, the area with the most drastic color changes in the cloud map. As shown in Fig. 5, define the length of the cavitation zone divided by the length of the throat to the outlet is the dimensionless length $L_x$:

$$L_x = \frac{X}{L_0}$$

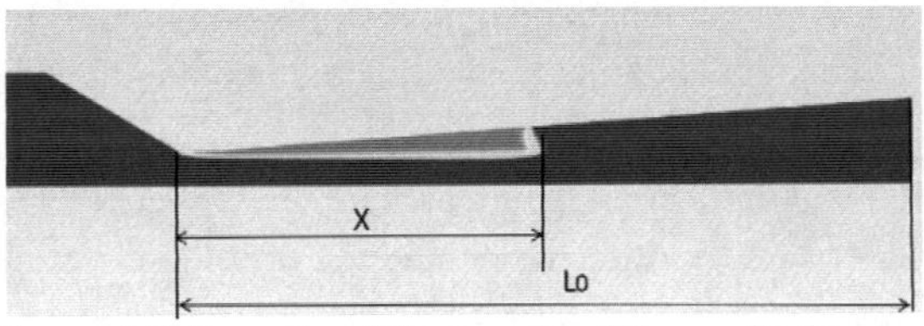


Figure 5 Schematic of the length of the cavitation area

Taking the pressure ratio $P_r$ as the independent variable and the dimensionless length $L_x$ as the dependent variable, the curve of the dimensionless length and pressure ratio of the cavitation area was made.

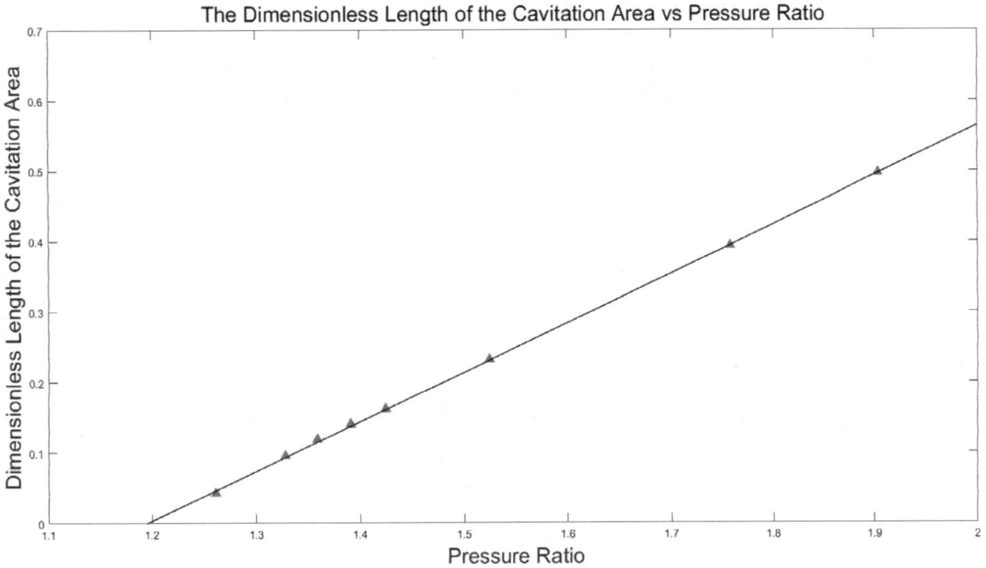


Figure 6 The Dimensional Length of the Cavitation Aera vs Pressure Ratio

It can be found that the dimensional length of the cavitation area changes as a straight line with the pressure ratio. According to the fitting relationship, it is found that when the pressure ratio is about 1.195, the cavitation phenomenon no longer occurs in the Venturi tube.

### *5.3. Influence of Outlet Pressure on Cavitation Field*

#### *5.3.1. Influence on the Pressure Field*

Figure 7 shows the pressure field distribution under different outlet pressure of the Venturi pipe. When the inlet pressure is 6Mpa, if cavitation occurs, the pressure of the water at 300K in the throat section is sharply reduced to the saturated vapor pressure of the water (3.54KPa). When the outlet pressure in the Venturi tube drops to 4.6-4.8MPa, the pressure ratio is lower than the critical ratio that produces cavitation phenomenon, and the cavitation phenomenon in the throat gradually disappears.

a) The outlet pressure is 3MPa

b) The outlet pressure is 3.3MPa

c) The outlet pressure is 3.8MPa

d) The outlet pressure is 4.1MPa

e) The outlet pressure is 4.4MPa

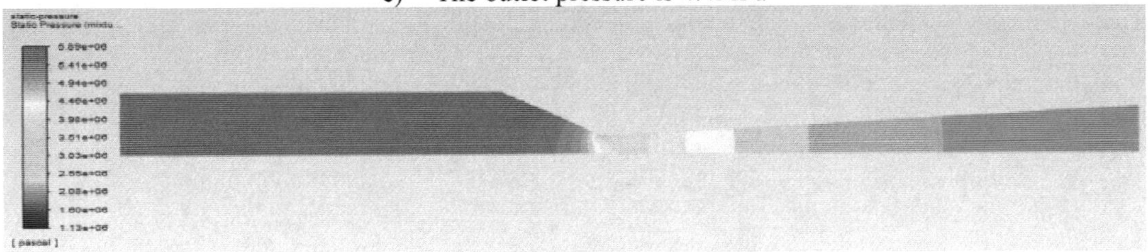

f) The outlet pressure is 4.8MPa

Figure 7 Pressure field distribution at different outlet pressure

From a)-f) in Fig. 7, it can be seen that when the inlet pressure is constant, as the outlet pressure decreases, the cavitation area spreads downstream, and the low-pressure area becomes longer.

### 5.3.2. *Influence on the Temperature Field*

Fig. 8 shows the images of the temperature field distribution at different outlet pressures in a Venturi tube. The distribution characteristics of the temperature field once again confirm the existence of the cavitation core area-It can be found that the lowest temperature in the Venturi tube is always present in the cavitation area. The reason is that the cavitation zone is mainly a vapor phase zone, and the liquefaction absorbs heat, so the cavitation zone and its surroundings are a low temperature zone. It can be obtained that the temperature drops most in the middle and lower reaches of the cavitation zone, which also shows that the cavitation phenomenon in this area is the most intense.

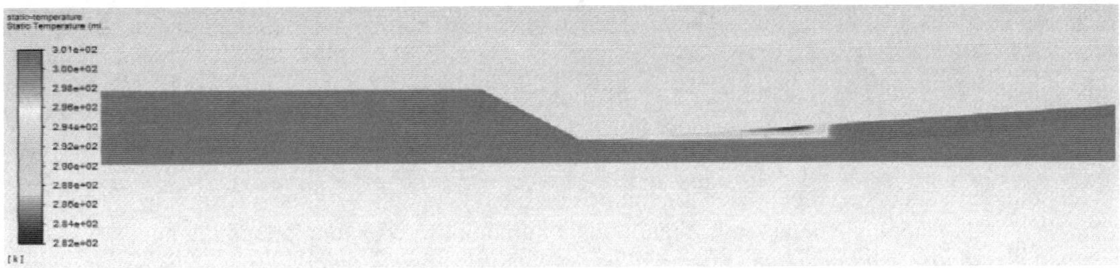

a) The outlet pressure is 3MPa

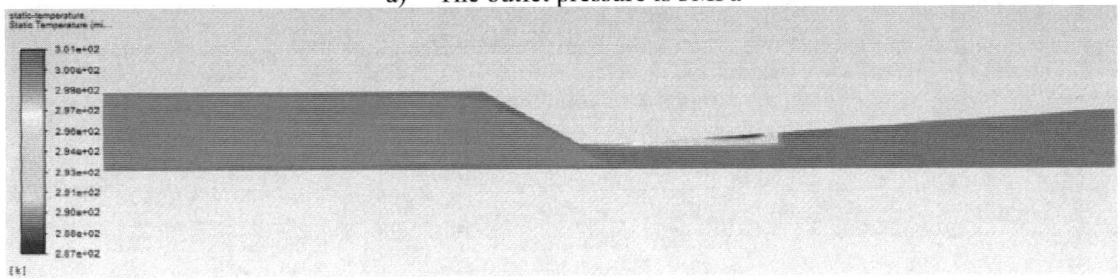

b) The outlet pressure is 3.3MPa

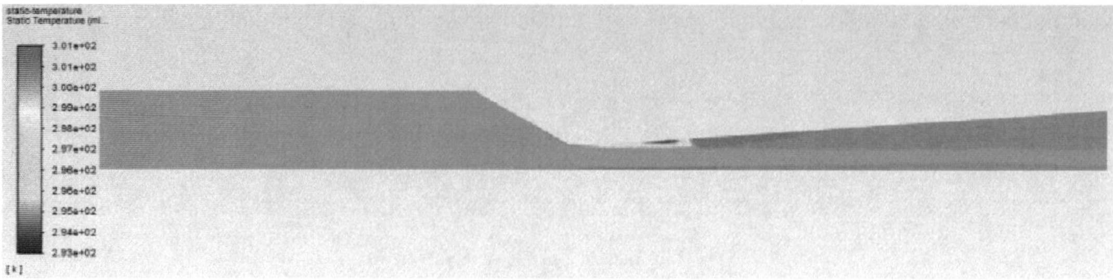

c) The outlet pressure is 3.8MPa

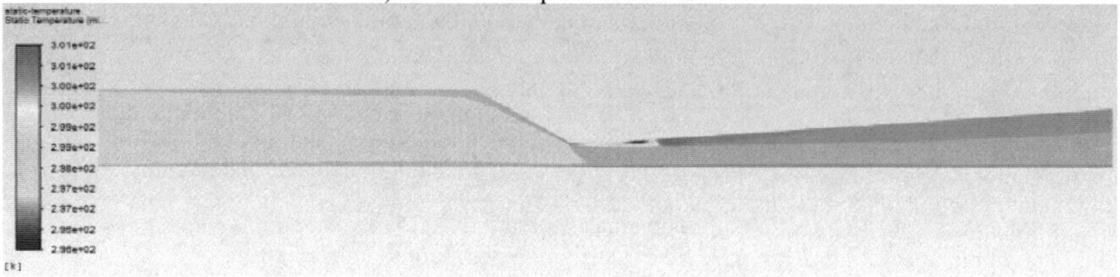

d) The outlet pressure is 4.1MPa

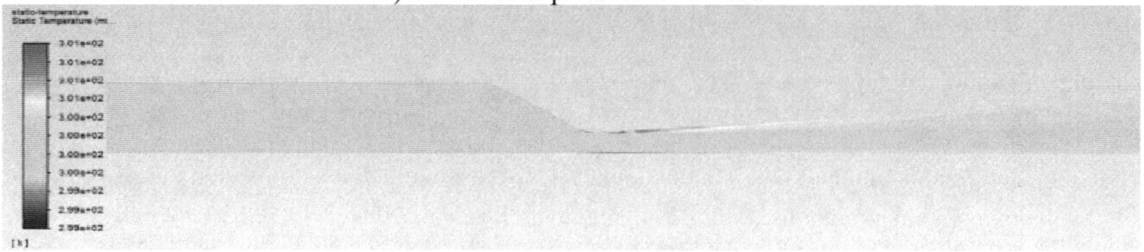

e) The outlet pressure is 4.4MPa

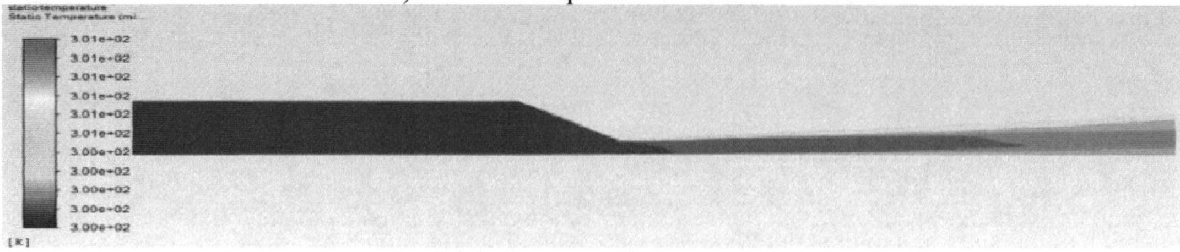

f) The outlet pressure is 4.8MPa

Figure 8 Temperature field distribution at different outlet pressures

Table 3 gives the temperature difference in the Venturi tube under different outlet pressure. It is found the larger the pressure ratio is, the greater the temperature difference in the Venturi tube will be.

Table 3 The Results of the Temperature Field

| Outlet Pressure(MPa) | Max Temperature(K) | Min Temperature(K) | The Difference(K) |
|---|---|---|---|
| 3.0 | 301.4657 | 282.1602 | 19.3055 |
| 3.3 | 301.4657 | 286.8944 | 14.5713 |
| 3.8 | 301.4657 | 293.0647 | 8.4010 |
| 4.1 | 301.4657 | 295.6285 | 5.8372 |
| 4.4 | 301.4651 | 298.7623 | 2.7028 |
| 4.8 | 301.1936 | 299.9232 | 1.2704 |

## 6. Conclusion

Through the CFD to simulate the evolution of the physical parameter fields of the cavitation region under different pressure ratios in the Venturi tube, some conclusions are gained as followings:

1) The cavitation phenomenon in the Venturi tube is always attached to the upper wall near the throat, and there is a cavitation core area. The degree of cavitation in the core area is the most intense, and the temperature is the lowest. From the core area outward, the gas phase area shows a flaky distribution, and there existing a gradient;

2) There is a critical pressure ratio, which is 1.192 in this calculation. When the pressure ratio is less than the critical pressure ratio, cavitation no longer occurs in the Venturi tube.

3) When the cavitation phenomenon can occur in the Venturi tube, keeping the inlet pressure unchanged, the outlet pressure has an impact on the length of the cavitation area, and the cavitation area gradually evolves downstream with the decrease of the outlet pressure. The maximum temperature difference in the Venturi tube also increases with the decrease of the outlet pressure.

**References**

[1] Ziyong Zhu, Hongtong Liu, Wanmin Sun (2014) Comparison Between the Numerical Simulation and Test Result for Two-Phase Flow in the Internal Cavity of Cavitating Venturi. J. Missile and Space Launch Technology, 01:75-81.
[2] Zhiqi Chen, Hanru Gao (1987) The Designing Method for the Length of the Cylindrical Segment of the Throat of the Adjustable Cavitating Venturi. J. Journal of Aerospace Power, (3):235-238+283.
[3] Chibing Shen, Jiping Wu (2004) Experimental Investigation on the Controllable Cavitating Venturi. J. Journal of Propulsion Technology, 05:473-476.
[4] Fei Tang,Yong Li, Yongbing Geng, Wen Li, Changjie Wen(2013) On Design and Numerical Simulation for Linear Throttling Cavitation Venturi. J. Space Control Technology and Applications, 39:12-16.
[5] Yujing Tao, Zhangfu Tian, Zhenguo Wang (2006) Experiment Study of the Mass Flow Control System. J. Journal of National University of Defense Technology, (4):6-9+21.
[6] Xinping Long, Jiong Wang, Dan Zuo, Junqiang Zhang, Bin Ji (2017) Experimental Investigation of the Global Cavitation Dynamic Behavior in a Venturi Tube with Special Emphasis on the Cavity Length Variation. J. International Journal of Multiphase Flow, 89:290-298.
[7] Abdulaziz A M. (2014) Performance and Image Analysis of a Cavitating Process in a Small Type Venturi. J. Experimental Thermal and Fluid Science, 53:40-48.
[8] Singhal Ashok K, Athavale Mahesh M, Huiying Li (2002) Mathematical Basis and Validation of the Full Cavitation Mode. J. Journal of Fluids Engineering,124:617-624.

# The category of black holes and state-of-art detection scenarios

**Zhenyu Peng***

San New high school, Wuhan, 43000, China

*18401036@masu.edu.cn

**Abstract**—The existence of black holes has been proposed for more than two centuries. Contemporarily, plenty of milestones have been achieved in this field, e.g., the first image of the black holes and the launch of the James Webb Space Telescope. This paper focuses on the organization of recent achievements in this field including three main parts: the history and fundamental concepts of the black holes; the classification of different black holes through their physical qualities; and progresses in this decade. These results shed light on the technical developments of the ability to detect black holes, which is beneficial to further research in the future.

## 1. Introduction

The black holes (BHs) are first suggested by Michell and Simon in the 18th century that celestial bodies, whose surface gravitational force is large enough to make their escape velocity faster than the speed of light [1]. In the 20$^{th}$ century, Einstein's theory of general relativity predicts that massive celestial bodies will curve space-time. If their mass reaches a certain level, they will collapse to become BHs [2, 3].

In the 21st century, new methods, new projects, and new topics about the BHs are proposed by a large number of scholars. The first image of BHs is captured by NASA in 2019 by the Event Horizon Telescope (EHT) [4]. Besides, the James Webb Space Telescope could give the specific information about the third stellar group, which is the first stars formed after The Big Bang [5].

In this paper, the classification system, state-of-art achievements, and suggestions for future research prospects about the BHs will be discussed. The rest part of the paper is organized as follows. In Section 2, a brief introduction of the history is presented, including basic formulas and concepts of BHs. Based on the physical properties of BHs (e.g., mass, charges, angular momentum), one can get two ways of classification. The first way is whether BHs have charges and angular momentum, which is mentioned in Section 3. In addition, BHs with different mass-level can be considered as the other way of classification, which will be discussed in Section 4. In Section 5, the limitations and drawbacks of current studies will be demonstrated and guidelines for further research are given accordingly. Eventually, a brief summary will be given in Section 6.

## 2. Basic description of BHs

The history of BHs can be traced from the 18th century, in that era scholars come up with the idea of BHs, they derive equations and formulae to describe the BHs. In Section 2.1, the BHs in Newton kinematics is mentioned; in Section 2.2 BHs after the general relativity is summarized; in Section 2.3, brief information about the physical properties of BHs is given; in Section 2.4, some general terms of BHs are explained.

### *2.1. BHs under Newton kinematics*
Michell should be considered as the first one who predicts the existence of BHs [1]. In his letter to Cavendish, he mentioned that if a star with the same mass as the sun, as its radius reaches 500 times compare of the sun, its surface escape velocity will be equivalent to the speed of light. According to the analysis, he suggested that BHs can be detected through gravitational phenomena from nearby stars. In this period, they call those stars the Dark Stars.

### *2.2. BHs under general relativity*
Just a few months after the publication of general relativity, Schwarzschild found one of the solutions to Einstein's field equation, which can describe the mass point and spherical volume [6]. The idea of Schwarzschild BHs is similar to Michell's idea, in which the surface escape velocity of the BHs is the speed of light. Besides, the radii of this kind of BHs can be described as the following equation:

$$r_s = \frac{2GM}{c^2} \tag{1}$$

where $r_s$ is Schwarzschild radius, $c$ is the speed of light in vacuum, $G$ is the gravitational constant, and $M$ is the mass of the BH. Lorentz gives the same solution, a few months later, and a more general description in Ref. [7]. Chandrasekhar calculated the Chandrasekhar limit [8], white dwarfs (WDs) with a mass higher than Chandrasekhar limit collapse to become neutron stars (NSs) [9], which led to further calculation in this area. In 1939, Oppenheimer and his co-workers predict NSs which are more massive than the Oppenheimer limit (TOV) will collapse to become BHs [10]. In 1958, Finkelstein described the inside of the Schwarzschild radius as Event Horizon (EH) [11]. In 1974, Hawking predicted the phenomenon of Hawking radiation [12].

BHS is hard to determine, even in the 21st century only a few numbers of BHs practically determined. One of the most famous BH is captured by EHT by NASA in 2019 which is shown in Figure 1 [4].

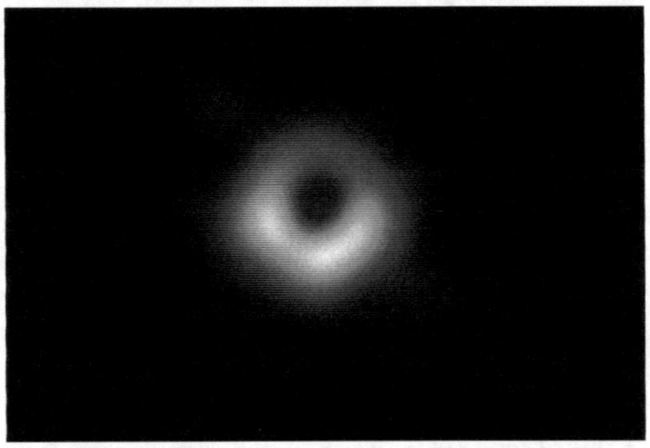

Figure 1. BH is directly imaged by EHT [4].

### *2.3. Three physical qualities of BHs*
In the 1970s, Israel [13], Carter [14], Robinson [15], and many other researchers contribute to the establishment of the *No Hair theorem*, which suggests that the stable BHs can be described by three conservable physical qualities: mass (M), charge (Q), and angular momentum (J). Those three qualities can be related to this following inequation.

$$Q^2 + \left(\frac{J}{M}\right)^2 \leq M^2 \tag{2}$$

### 2.4. *Terms in the theory of BHs*

#### 2.4.1. *Event Horizon*

Event Horizon (EH) means that the section inside the Schwarzschild radius. On the surface of the EH, time will flow slower than outside, according to general relativity. Therefore, substances that fall into the BHs seem slow down and need an infinity of time to reach the EH. On the other hand, the substance will not notice any difference. Larry suggested for stable BHs, EH always is spherical; for spinning BHs, EH is tilted [16].

#### 2.4.2. *Singularity*

Singularity is the center point of BHs. For spinning BHs the singularity shows the ring shape. Singularity can be defined as infinity small volume, infinity large density, and infinity large space curvature. In the process the substance from the edge of EH to the singularity, the substance will be extended by the tidal force, which is known as Spaghettification.

#### 2.4.3. *Thermodynamics for BHs*

The thermodynamics was proposed and constructed by Jacob [17]. According to the second law of thermodynamics, entropy will not decrease in a closed system. Then, Jacob pointed out that entropy must exist in BHs. Based on further calculation, it is found that the entropy of BHs is directly related to its surface area.

## 3. Category in terms of spin and charge

In retrospect, scholars suggest, theoretically, four kinds of BHs can be distinguished by two of their physical qualities: charges and angular momentum. BHs with no spin and no charge are Schwarzschild BHs, which will be discussed in Section 3.1. BHs with no spin but charge are R-N BHs, which will be introduced in Section 3.2. BHs with no charge but spin are Kerr BHs, which will be described in Section 3.3. BHs with both spinning and charges are K-N BHs, which will be demonstrated in Section 3.4. Among the four different BHs, Schwarzschild BHs were the first practical BHs that can be described by equations; R-N BHs and Kerr BHs are two special BHs; the K-N BHs are the most general BHs and the BHs that are detected in our universe.

### 3.1. *Schwarzschild BHs*

Schwarzschild's solution is the spherical solution for Einstein's field equation by Schwarzschild in 1915. This solution indicates the presence of Schwarzschild BHs, which have neither charge nor spin. Schwarzschild solution divergent in $r$=0 and $r$=$r_s$, where $r_s$ is Schwarzschild radius. EH is in $r$=$r_s$, which separates the space-time into two-part. Exterior Schwarzschild solution, when $r$>$r_s$, can be used to describe the external structure of the BHs. The Interior Schwarzschild solution is located in the area of 0<$r$<$r_s$, which is a different coordinate than the exterior. So, the divergence in $r=r_s$ is caused by an impropriate coordinate system [18].

In contrast, the divergent in r=0 shows a singularity, also known as gravitational singularity. One can use Kretschmann scalar to check, the Kretschmann scalar for Schwarzschild solution is:

$$K = \frac{12{r_s}^2}{r^6} = \frac{48G^2M^2}{c^4r^6} \tag{3}$$

It is also divergent in $r$=0, thus there should be a real singularity.

### 3.2. *R-N BHs*

Reissner–Nordstrom solution is another static solution for the Einstein field equation for BHs with charges and no spin, which is one of the advanced Schwarzschild solutions. In the following equation, $r_Q$ is a characteristic length scale, which is defined as:

$$r_Q^2 = \frac{Q^2 G}{4\pi\varepsilon_0 c^4} \tag{4}$$

Here, $Q$ is the charge of BH, $\varepsilon_0$ is the electric constant, $G$ is the gravitation constant, and c is the speed of light. For charged BHs with $r_Q \ll r_s$, they are similar to Schwarzschild BHs, but there are two horizons inside of them: EH and Cauchy horizon [19].

### *3.3. Kerr BHs*

Kerr BHs are predicted by Einstein's field equation and proposed by Kerr [20], which are a group of spinning BHs. In 2017, Chou, Yu-Ching suggested Ellipsoid coordinate transformation depends on Kerr matric.

### *3.4. K-N BHs*

Gravitational field around the Kerr-Newman BHs, BHs with both spin and charge, can be described by Kerr-Newman matric, which can be mathematically represented by the following equation:

$$c^2 d\tau^2 = -\left(\frac{dr^2}{\Delta} + d\theta^2\right)\rho^2 + (cdt - a\sin^2\theta\, d\varphi)^2 \frac{\Delta}{\rho^2} - \left((r^2 + a^2)d\varphi - acdt\right)^2 \frac{\sin^2\theta}{\rho^2} \tag{5}$$

The spherical coordinate system (r, θ, φ) is used in this equation, Q represents charge, c is the speed of light, and $t$ is time. The α, ρ, and Δ can be defined as the following equations:

$$a = \frac{J}{Mc} \tag{6}$$

$$\rho^2 = r^2 + a^2\cos^2\theta \tag{7}$$

$$\Delta = r^2 - r_s s + a^2 + r_Q^2 \tag{8}$$

where $J$ is the angular momentum of BHs, $r_s$ is Schwarzschild radius shown in equation 1, and $r_Q$ is a characteristic length scale of R-N BHs shown in Eq. (4).

## 4. Category in terms of mass

Generally, it is agreed that most of the BHs are K-N BHs, which are also known as normal BHs or general BHs. Thus, through the third physical property of the BHs, the mass, researchers also classify BHs by different mass-level, which is the more practical and commonly used way of distinguishing different BHs. In following subsections, the discussions for micro BHs, Stellar-mass BHs, Massive BHs and Super-massive BHs (SMBHs) will be discussed in the Section 4.1~4.4, respectively.

### *4.1. Micro BHs*

Micro BHs, also known as quantum mechanical black holes, are relatively much smaller than BHs that be observed. This idea is theoretical because micro BHs are never observed or made. According to Hawking radiation, one of the difficulties to find micro BHs is that BHs will evaporate continuously and the rate of evaporation is inversely related to the radius of the BHs [12]. Therefore, micro BHs would evaporate in a very short period.

Primordial BHs (PBHs), also known as proto BHs, is a theoretical BHs predicted to form in the very early universe. They formed under high pressure and density in the early universe. According to the Big Bang theory, in a very short time after the Big Bang, the universe is very dance and hot to give the environment for PBHs' formation. PBHs are also considered as a candidate for dark matter.

### *4.2. Stellar-mass BHs*

When the nuclear fusion in massive stars is stopped by gravitational force, stars with mass more than the Oppenheimer limit (TOV) will collapse to become new BHs [10]. They can be detected by Gamma-Ray Burst or Supernova explosion. The smallest BH that was detected has a mass = 2.73 ± 0.56 $M_\odot$

[21], and the largest stellar-mass BH is $70M_\odot$ detected by the Large Sky Area Multi-Object Fibre Spectroscopic Telescope (LAMOST) in China [22].

The core of BHs detection is in galactic centres or clusters, because those BHs are easy to determine by analysing the motion of adjacent celestial bodies. However, it is predicted that there should be around $10^8$ BHs in our galaxy in 1990, hence they make the hypothesis of isolated stellar-mass BHs. In 2022 Sahu and hundreds of co-workers determined the first isolated stellar-mass BH [23] (as shown in Fig. 2), by using the method of astrometric microlensing which is a phenomenon of astrometric shift by the curvature of space-time [24].

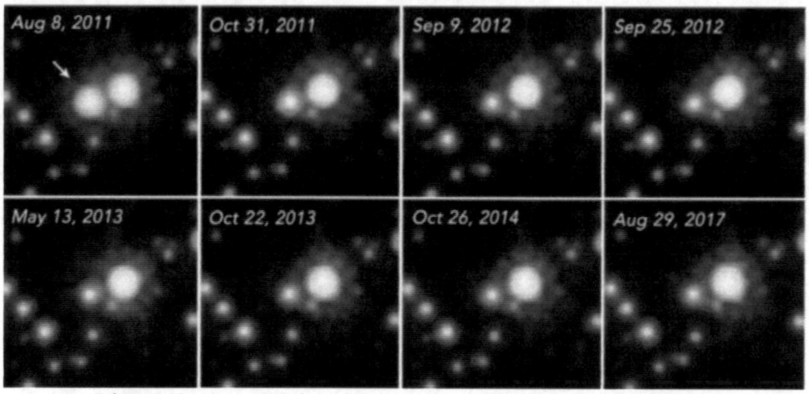

Figure. 2: astrometric microlensing of the isolated BH [23].

### *4.3. Massive BHs*
Massive BHs (MBHs), also known as intermediate-mass BHs, are BHs with mass from $10^2$ to $10^6$ $M_\odot$, which massive than stellar-mass BHs but much lighter than SMBHs. This kind of BHs is still theoretical as the current understanding of BH formation led to a gap in the BHs with intermediate-mass. Nevertheless, some scholars still hold the view that MBHs will form because they propose that BHs can combine to form new BHs. The detection process of gravitational waves can be the evidence, in 2015 a pair of binary BHs combined and their gravitational waves arrive on the Earth. These two BHs are $36M_\odot$ and $29M_\odot$, the resultant BH is $62M_\odot$. According to the results, BHs can be formed by the combination of different BHs. Thus, the hypothesis that stellar-mass BHs can form heavier BH can be made, i.e., there should be MBHs formed by the accretion of many smaller BHs.

### *4.4. Super-massive BHs*
SMBHs are BHs that have $10^5M_\odot$ to $10^9M_\odot$, it is believed that every galactic centre has at least one SMBHs contemporarily. SMBHs have unique characteristics compared to other BHs, i.e., they have very low density. Since density is mass divided by volume, and mass is directly related to their $r_s$, volume is related to the cube of $r_s$. In addition, SMBHs have low tidal force than stellar-mass BHs, because they have much larger $r_s$. Most of SMBHs are formed by a long process of accretion or by the collapse of nebulas or clusters.

Quasars (quasi-stellar radio sources) are active galactic nuclei (AGNs) in the early universe, they contain SMBHs in their core to form accretion disk. As the speed of light is constant, the distance from those quasars is related to their age, so further quasar means they formed earlier. The furthest quasar is found is PSO J0309+27, which is $1.33*10^{11}$ light-years away [25].

In 2020, Chinese astrophysicists capture a stellar tidal disruption by a candidate SMBH binary in the OGLE16aaa, which will contribute to future work in the field of the formation of gravitational waves [26]. In 2022, a group of astrophysicists reported an ultraviolet compact object, GNz7q. It is very dim in the X-ray frequency, which implies that GNz7q is the former part of luminous quasars at the later epochs. The Fig. 3 is the near-infrared images and spectrum of GNz7q, which were captured by the Hubble Space Telescope [27].

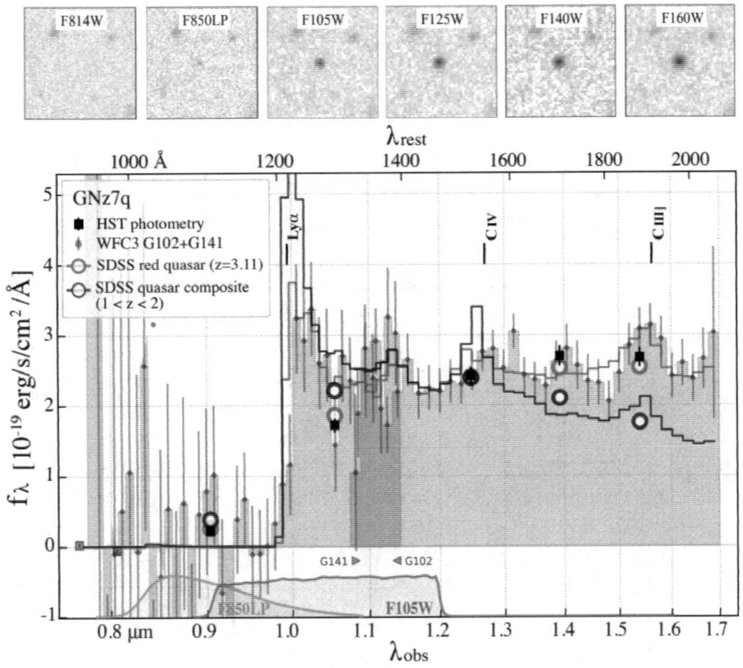


Figure. 3: the near-infrared images and spectrum of GNz7q [27].

## 5. Limitations and future expectations

The detection approaches for BHs are still limiting, and the obtained data is also inaccurate (i.e., low signal-noise ratio). For example, identical intermediate BHs were never detected. Hubble Space Telescope's ability for receiving inferred frequency electromagnetic waves is limited, which is unable to detect the very early stage of the universe.

Therefore, it is expected to derive more information about the BHs through the James Webb Space Telescope and other effective satellites or telescopes in the future. Especially in the field of BHs detection and analysis, it is necessary to determine the gap of the intermediate BHs, to make the theory of BHs' formation more complete, find the third-star group which only contains hydrogen and helium, and have a deeper understanding of thermodynamics in BHs.

## 6. Conclusion

In summary, this paper discusses the history, two ways to classify the BHs, and lists some accomplishments in this decade. Specifically, this paper gives fundamental concepts of BHs and the process of the theory of BHs development. Scholars use the charge, spin, and mass to distinguish BHs, and some state-of-art studies of those different BHs are presented in this paper. In this brand-new era, novel telescopes with better detection abilities will be constructed and operated to obtain information with higher accuracy, which will bring new insight into BHs in this century. Overall, these results offer a guideline for future progress in the study of BHs.

## References

[1] Montgomery, C., Orchiston, W., Whittingham, I. (2009) Michell, Laplace and the origin of the black hole concept. Journal of Astronomical History and Heritage, 12, 90-96.

[2] Wald, R. M. (1999) Gravitational collapse and cosmic censorship. In Black holes, gravitational radiation and the universe (pp. 69-86). Springer, Dordrecht.

[3] Overbye, Dennis. Black Hole Hunters (2015) The New York Times. https://www.nytimes.com/2015/06/09/science/black-hole-event-horizon-telescope.html
[4] Loff, S. (2019) Black Hole Image Makes History; NASA Telescopes Coordinated Observations. NASA. gov, 8.
[5] Gardner, J. P., Mather, J. C., Clampin, M., Doyon, R., Greenhouse, M. A., Hammel, H. B., ... Wright, G. S. (2006) The james webb space telescope. Space Science Reviews, 123(4), 485-606.
[6] Schwarzschild, K. (1999) On the gravitational field of a sphere of incompressible fluid according to Einstein's theory. arXiv preprint physics/9912033.
[7] Kox, A. J. (1992) General relativity in the Netherlands. Studies in the history of general relativity, 39-56.
[8] Venkataraman, G. (1992) Chandrasekhar and his Limit. Universities Press.
[9] Harpaz, A. (1993) Stellar evolution. AK Peters/CRC Press.
[10] Oppenheimer, J. R. Volkoff, 1939 Phys. Rev, 55, 374.
[11] Finkelstein, D. (1958) Past-future asymmetry of the gravitational field of a point particle. Physical Review, 110(4), 965.
[12] Hawking, S. W. (1974) Black hole explosions?. Nature, 248(5443), 30-31.
[13] Israel, W. (1967) Event horizons in static vacuum space-times. Physical review, 164(5), 1776.
[14] Carter, B. (1971) Axisymmetric black hole has only two degrees of freedom. Physical Review Letters, 26(6), 331.
[15] Robinson, D. C. (1975) Uniqueness of the Kerr black hole. Physical Review Letters, 34(14), 905.
[16] Smarr, L. (1973) Surface geometry of charged rotating black holes. Physical Review D, 7(2), 289.
[17] Bekenstein, J. D. (2020) Black holes and entropy. In JACOB BEKENSTEIN: The Conservative Revolutionary (pp. 307-320).
[18] Hughston, L. P., Tod, K. P. (1990) An introduction to general relativity (Vol. 5). Cambridge University Press.
[19] Chandrasekhar, S., Chandrasekhar, S. (1998) The mathematical theory of black holes (Vol. 69). Oxford university press.
[20] Kerr, R. P. (1963) Gravitational field of a spinning mass as an example of algebraically special metrics. Physical review letters, 11(5), 237.
[21] Orosz, J. A., McClintock, J. E., Remillard, R. A., Corbel, S. (2004) Orbital parameters for the black hole binary XTE J1650–500. The Astrophysical Journal, 616(1), 376.
[22] Liu, J., Soria, R., Zheng, Z., Zhang, H., Lu, Y., Wang, S., Yuan, H. (2020) Reply to: On the signature of a 70-solar-mass black hole in LB-1. Nature, 580(7805), E16-E17.
[23] Sahu, K. C., Anderson, J., Casertano, S., Bond, H. E., Udalski, A., Dominik, M., et al. (2022) An Isolated Stellar-Mass Black Hole Detected Through Astrometric Microlensing. arXiv preprint arXiv:2201.13296.
[24] Nucita, A. A., De Paolis, F., Ingrosso, G., Giordano, M., Manni, L. (2017) Astrometric microlensing. International Journal of Modern Physics D, 26(05), 1741015.
[25] Belladitta, S., Moretti, A., Caccianiga, A., Spingola, C., Severgnini, P., Della Ceca, R., et al. (2020) The first blazar observed at z> 6. Astronomy Astrophysics, 635, L7.
[26] Shu, X., Zhang, W., Li, S., Jiang, N., Dou, L., Yan, Z., et al. (2020) X-ray flares from the stellar tidal disruption by a candidate supermassive black hole binary. Nature communications, 11(1), 1-9.
[27] Fujimoto S., Brammer B., Watson D., Magdis E., Kokorev V., Greve R., et al. (2022) A dusty compact object bridging galaxies and quasars at cosmic dawn Nature volume 604, pages261–265.

# The Principle and State-of-Art Approach for Black Hole Detection

**Lingyan Guan[1, a†], Xianzhe Tang[2, †], Jialing Tian[3, *†], Jiayi Wu[4, c†]**

[1]Department of Physics, University of Strathclyde, Glasgow, United Kingdom

[2]Department of Physics and Astronomy, Stony Brook University, Stony Brook, NY, USA

[3]Department of Physics and Astronomy, The University of Manchester,Manchester, United Kingdom

[4] Department of Physics, Sichuan University, Chengdu, China

[a]lingyan.guan.2021@uni.strath.ac.uk; [b] xianzhe.tang@stonybrook.edu; [c]wujiayi777@stu.scu.edu.cn

*Corresponding author: jialing.tian@student.manchester.ac.uk;

[†]These authors contributed equally

**Abstract**—Black hole is a kind of special celestial subject whose density is so great that even the light cannot escape, which has always been a popular topic. This study will present the recent progress of observation related to three quantities of black holes. As a leading-in, we first introduced some of the most commonly discussed black holes by demonstrating their field equations, metric, and some other representative quantities. On this basis, the formation process of a black hole will be described. In addition, a picture of the detectors used in cosmology observation is discussed before getting any further into the methods applied in observation. Subsequently, we summarized observations regarding the three elements of a black hole (mass, charge and spin). For these three different characters of black hole, different methods and theories were put into use, including gravitational retro-lensing, twin-peak QPOs, accretion disks, continuum fitting method, and black body radiation. Eventually, possible limitations are evaluated and possible improving approaches on current observations are proposed. Overall, these results shed light on guiding further researches focusing on exploring the black holes.

## 1. Introduction

In general, black hole refers to the areas with extremely large density, it creates great gravitation potential, where no particles or electromagnetic can escape from their powerful tug of gravity. If matters (e.g., stars, planets, or spacecrafts) get too close, they will be stretched and compressed [1]. Black holes can be classified by three different properties: mass, charge and spin. According to the general relativity proposed by Einstein, the black holes ought to be existed in the universe, while the term is coined in 1967 by Wheeler [2]. In retrospect, the first possible discovery of the black holes was achieved in 1971, while the Interferometer Gravitational-Wave Observatory (LIGO) detected gravitational wave of the black holes merging in 2015 remarks key steps for further exploration of the black holes [2].

Black holes have been an intriguing topic ever since Einstein predicted their existence, people have gone from solving black hole solutions to attempting at observing relative phenomena. The essential feature of black holes that it could absorb even the light makes it a rather uneasy task to observe them, yet researchers were never hampered by this, they study phenomena specialized to black hole and made efforts trying to take use of these phenomena to show that black holes do exist. According to the observations, existence of black holes can be proved, meanwhile a better understanding of their properties can be gained. Actually, scholars have been reaching out for aspects including the three properties of black holes, namely mass, charge and the spin.

Researches on black hole are meaningful. Firstly, BH is a secret topic within the field of astrophysics, because it is hard to observe black hole with our eyes or simple equipment, also it cannot reflect any light, and there might be some special matters within the black hole. Meanwhile, there are lots of special matters in black holes, hence it is useful to get more information about it. Because of black holes, many objects and mater around them will be affect, it is a meaningful field to do researches and use these discusses to help other aspects in astronomy. The rest part of the paper is organized as follows. Sec. II will discuss basic descriptions of black holes. Sec. III will present several kinds of black holes. The Sec. VI will discuss the detection of three parameters black hole.

## 2. Basic descriptions of black holes

Black hole is an astronomical body with a great gravitation, where the spacetime around curved so much that even light can't escape from it. On this basis, the boundary of the black holes is what we call event horizon, because no information can be sent out through the event horizon since the strong gravity. For a non-rotate black hole, the basic equation of the event horizon can be described as:

$$r_S - \frac{2GM}{c^2} \tag{1}$$

Here, $G$ is gravitational constant, $M$ is the mass of the black hole, $c$ is the speed of light. However, almost all the black holes in the universe can be seen as rotating black hole, since nearly all the astronomical object that black holes generated from has a rotate spin. Furthermore, the collide of black holes and other astronomical body can generate angular momentum in black hole. In the theoretical research, a non-dimensional parameter $a$ is often given relating to the spin. It's relation with mass and angular momentum can be determined as:

$$a = \frac{cJ}{GM^2} \tag{2}$$

Here, $a$ is the non-dimensional parameter and $J$ is the angular momentum of the black hole. For a rotating black hole, as shown in Fig. 1, the event horizon can be determined as [3]:

$$\text{Outer event horizon:}\ r_+ = m + \sqrt{m^2 - a^2} \tag{3}$$

$$\text{Inner event horizon:}\ r_- = m - \sqrt{m^2 - a^2} \tag{4}$$

For a rotating black holes, the singularity will become a ring, which can be describe as $r = a$,

$$\begin{cases} x^2 + y^2 = a^2 \\ z = 0 \end{cases} \tag{5}$$

As a matter of fact, there are three quantities can be determined for a black hole, which are mass, electric charge and the spin. The value of the mass of a black hole has a wide range. For a supermassive black hole (e.g., Sgr A* in Fig.2), it can hold a mass about 0.1 million $M_\odot$. The primordial black holes generated soon after the big bang is thought to has $10^{-5}$g at minimum [4].

The electric charge for an astronomical black hole is likely far less than its theoretical maximum [6]. The universe is generally electrically neural; hence, the materials create the black holes and intake by the black holes usually will not take large electric. Therefore, black holes are not predicted to bring that extreme number of electric charges. Moreover, the reason that the electromagnetic repulsion is far larger than average gravitational force gives another strong certification of it.

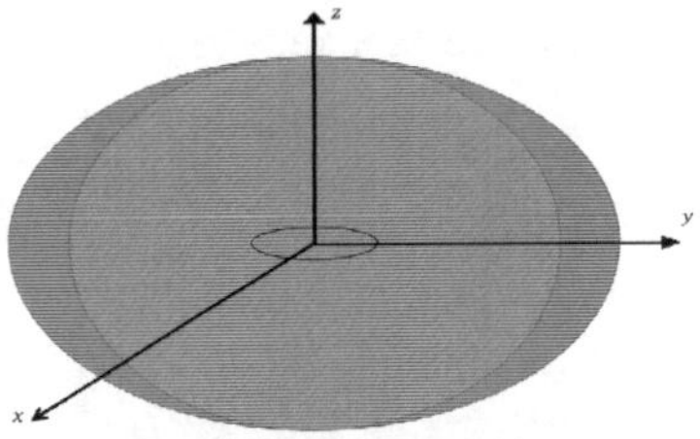


Figure 1. The inner and outer event horizon of the rotating Kerr black hole. The circular inside is the ring singularity

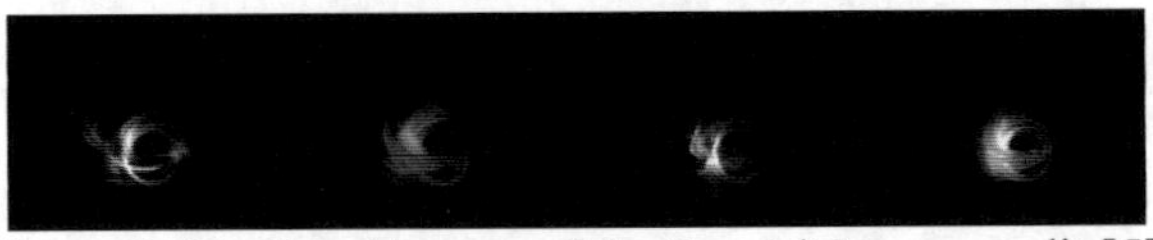

Figure 2. Predicted image of the Sgr A* Image credit [5].

## 3. Category of black holes

In general, we focus on 4 types of black holes: Schwarzschild, Reissner-Nordström, Kerr and Kerr-Newman black hole. The Schwarzschild geometry can be described as:

$$ds^2 = -\left(1 - \frac{2M}{r}\right)dt^2 + \left(1 - \frac{2M}{r}\right)^{-1} dr^2 + r^2 d\Omega_2{}^2 \quad (6)$$

where s is Schwarzschild metric, $0 < r < \infty$ is a radial coordinate, t is the time coordinate, and $\Omega$ is a point on the two spheres $s^2$, $d\Omega_2{}^2 = d\theta^2 + \sin^2\theta d\phi^2$ is the round metric on the two-sphere. From the Schwarzschild' geometry above, the least connected, massless scalar fields represent a spherically symmetric collapse. This solution depicts the generation of naked singularities and black holes depending on the values of specific parameters. Because black holes are known as Schwarzschild black holes, it's only reasonable to wonder how these black holes form when scalar fields collapse. Due to the mass M, the gravitation of a particle will become time independent at a point in the space. Fig. 3 gives a sketch of Schwarzschild black hole and the radial and angular direction in the Schwarzschild solution.

The escape velocity of the Schwarzschild black hole is equal to the velocity of light, which is:

$$v = \sqrt{\frac{2GM}{R}} \quad (7)$$

Where $v$ is the escape velocity of a Schwarzschild black hole, G is the Newton's gravitational constant, M is the mass, and R is the radius. Three classical representations of the Schwarzschild metric are given in Table. 1. The Schwarzschild metric is represented in six different ways as summarized in Table. 2.

With regard to the Reissner-Nordström (RN) black hole, it is a kind of charged non-rotating black hole with no angular momentum, which has geometry of according to Ref. [8]:

$$ds^2 = -\left(1 - \frac{2M}{r} + \frac{Q^2}{r^2}\right)dt^2 + \left(1 - \frac{2M}{r} + \frac{Q^2}{r^2}\right)^{-1} dr^2 + r^2(d\theta^2 + \sin^2\theta d\phi^2) \quad (8)$$

Here,

$$\mathrm{M} = \frac{Gm}{c^2} \quad (9)$$

$$Q^2 = \frac{q^2 G}{4\pi\varepsilon_0 c^4} \quad (10)$$

where q is the charge, $\varepsilon_0$ is the electric constant and the radii can be described as:

$$r_{\pm} = M \pm \sqrt{M^2 - Q^2} \tag{11}$$

As for Kerr black hole, the geometry can be written as

$$dr^2 = \left(1 - \frac{2M}{r}\right) dt^2 + \frac{4Ma}{r} dt d\phi - \frac{dr^2}{1 - \frac{2M}{r} + \frac{a^2}{r^2}} - \left(1 + \frac{a^2}{r^2} + \frac{2Ma^2}{r^3}\right) r^2 d\phi^2 \tag{12}$$

where r is a radial coordinate, J is the angular momentum of the black hole, a=J/M which also called Kerr parameter. Fig. 4 shows the structure of Kerr black hole. The three classical representations of the Kerr metric are listed in Table. 3.

A representation of the multiple horizons' spatial appearances and relationships, as well as the Kerr metric's singularity These are, expressively, from the outside to the inside (Seen from the Table. 4). Inner ergosurface and Cauchy horizon disappear at a = 0, while outer ergosurface and event horizon merge to form the Schwarzschild horizon [10].

TABLE 1. VELOCITY IN DIFFERENT COORDINATE SYSTEMS [7].

| | | Schwarzschild | Gullstrand-Painlevé |
|---|---|---|---|
| Particles | Coordinate velocity: $\frac{dr}{dt}$, Proper velocity: $\frac{dr}{d\tau}$ | $\pm(1-\frac{2GM}{r})\sqrt{\frac{2GM}{r}}$, $\pm\sqrt{\frac{2GM}{r}}$ | $-\sqrt{\frac{2GM}{r}}$, $\pm\sqrt{\frac{2GM}{r}}$ |
| Light rays | Coordinate velocity: $\frac{dr}{dt}$ | $\pm\left(1-\frac{2GM}{r}\right)$ | $\pm 1-\sqrt{\frac{2GM}{r}}$ |

TABLE 2. SIX WIDELY USED FORMS OF THE SCHWARZSCHILD METRIC [7].

| Schwarzschild metric in various coordinates | Coordinate transformation | Characteristic properties |
|---|---|---|
| **Schwarzschild:** $ds^2 = -(1-\frac{2m}{r})dt^2 + \frac{1}{1-\frac{2m}{r}}dr^2 + r^2 d\Omega^2$ | None | Area of spheres r = const is the 'Euchidean' $4\pi r^2$ |
| **Isotropic:** $ds^2 = -(\frac{1-\frac{m}{2\bar{r}}}{1+\frac{m}{2\bar{r}}})^2 dt^2 + (1+\frac{m}{2\bar{r}})^4(d\bar{r}^2 + \bar{r}^2 d\Omega^2)$ | $r = \bar{r}(1+\frac{m}{2\bar{r}})^2$ | Constant-curvature time slices |
| **Eddington-Finkelstein:** $ds^2 = -(1-\frac{2m}{r})dv^2 + 2dvdr + r^2 d\Omega^2$ | $v = t + r + 2m\ln\left\lvert\frac{r}{2m}-1\right\rvert$ | Ingoing light rays: $dv = 0$ |
| **Kerr-Schild:** $ds^2 = (\eta\alpha\beta + 2ml_a l_b)dx^\alpha dx^\beta$, $l_\alpha = \frac{1}{\sqrt{r}}(1, \frac{x}{\sqrt{r}}, \frac{y}{\sqrt{r}}, \frac{z}{\sqrt{r}})$ | $\bar{t} = v - r$, $r^2 = x^2 + y^2 + z^2$ | 'Cartesian' coordinates |
| **Lemaître:** $ds^2 = dT^2 + \frac{2m}{r}dR^2 + r^2 d\Omega^2$, $r = [\frac{2\sqrt{2m}}{3}(R-T)]^{\frac{2}{3}}$ | $dT = dt + \sqrt{\frac{2m}{r}}\frac{1}{1-\frac{2m}{r}}dt$ $dR = dt + \sqrt{\frac{r}{2m}}\frac{1}{1-\frac{2m}{r}}dr$ | Infalling particles: $dR = 0$ |
| **Gullstrand-Painlevé:** $ds^2 = -(1-$ | $dt = d\tilde{t} -$ | Infalling particles: $dr =$ |

| | | |
|---|---|---|
| $\frac{2m}{r})d\tilde{t}^2 + 2\sqrt{\frac{2m}{r}}d\tilde{t}dr + dr^2 + r^2 d\Omega^2$ | $\frac{dt}{\sqrt{\frac{r}{2m}}-\sqrt{\frac{2m}{r}}}$ | $-\sqrt{\frac{2m}{r}}d\tilde{t}$ |

TABLE 3. THE THREE CLASSICAL REPRESENTATION OF THE KERR METRIC [10].

| | | |
|---|---|---|
| **Kerr-Schild** $ds^2 = -dt^2 + dx^2 + dy^2 + dz^2 + \frac{2mr^2}{r^4+a^2z^2}(dt + \frac{r(xdx+ydy)}{a^2+r^2} + \frac{a(ydx-xdy)}{a^2+r^2} + \frac{z}{r}dz)^2$ | $x^2 + y^2 + z^2 = r^2 + a^2(1 - \frac{z^2}{r^2}), r = r(x,y,z)$ | $x = (r\cos\phi + a\sin\phi)\sin\theta$, $y = (r\sin\phi - a\mathrm{ocs}\phi)\sin\theta$ $z = r\cos\theta$ |
| **Boyer-Lindquist** $ds^2 = -(1 - \frac{2mr}{\rho^2})dt^2 - \frac{4mra\sin^2\theta}{\rho^2}dtd\phi + \frac{\rho^2}{\Delta}dr^2 + \rho^2 d\theta^2 + (r^2 + a^2 + \frac{2mra^2\sin^2\theta}{\rho^2})\sin^2\theta d\phi^2$ | $\rho^2 = r^2 + a^2\cos^2\theta$ $\Delta = r^2 - 2mr + a^2 = (r - r_+)(r - r_-)$ | $dv = dt + \frac{r^2+a^2}{\Delta}dr$ $d\varphi = d\phi + \frac{a}{\Delta}dr$ |
| **Kerr original** $ds^2 = -(1 - \frac{2mr}{\rho^2})(dv - a\sin^2\theta d\varphi)^2 + 2(dv - a\sin^2\theta d\varphi)(dr - a\sin^2\theta d\varphi) + \rho^2(d\theta^2 + \sin^2\theta d\varphi^2)$ | $r_{E\pm} = m \pm \sqrt{m^2 - a^2\cos^2\theta}$ $r_\pm = m \pm \sqrt{m^2 - a^2}$ | None |

TABLE 4. THE SPATIAL APPEARANCES AND RELATIONSHIPS OF THE KERR METRIC.

| | | |
|---|---|---|
| Outer | Ergosurface | $r_{E+} = m + \sqrt{m^2 - a^2\cos^2\theta}$ |
| Event | Horizon | $r_+ = m + \sqrt{m^2 - a^2}$ |
| Cauchy | Horizon | $r_- = m - \sqrt{m^2 - a^2}$ |
| Inner | Ergosurface | $r_{E-} = m + \sqrt{m^2 - a^2\cos^2\theta}$ |
| | singularity | r = 0 |

The Kerr-Newman black hole is a kind of charged rotating black hole with an angular momentum, whose geometry can be illustrated as following [11]:

$$ds^2 = \frac{\rho^2}{\Delta}dr^2 + \rho^2 d\theta^2 + \frac{\sin^2\theta}{\rho^2}[(r^2 + a^2)d\phi - acdt]^2 - \frac{\Delta}{\rho^2}[a\sin^2\theta d\phi - cdt]^2 \tag{13}$$

where,

$$\rho^2(r,\theta) = r^2 + a^2\cos^2\theta \tag{14}$$

$$\Delta(r) = r^2 - 2Mr + a^2 + Q^2 + P^2 \tag{15}$$

and,

$$a = \frac{J}{mc} \tag{16}$$

Similar to the Reissner-Nordström black hole,

$$M = \frac{Gm}{c^2} \tag{17}$$

$$Q^2 = \frac{q^2G}{4\pi\varepsilon_0 c^4} \tag{18}$$

The parameter P corresponds to the magnetic monopole, a is called the Kerr parameter. At the coordinate singularities $\Delta(r) = 0$, the Kerr–Newman black hole spacetime has two horizons represented by radii r , where outer and inner horizons are $r_+$ and $r_-$ [11].

$$r_{\pm} = M \pm \sqrt{M^2 - a^2 - Q^2 - P^2} \tag{19}$$

The inner and outer ergosphere is given by

$$r_{e_\pm} = M \pm \sqrt{M^2 - a^2 cos\theta - Q^2} \tag{20}$$

Figure 5 shows a charged rotating black hole with angular momentum J in spherical coordinates (r, , ). Figure 6 illustrates the inner structure of both Reissner–Nordström and Kerr–Newman black hole, where the near the origin of axis is the singularity ring, the area between singularity ring and the middle circle is the inner horizon, the area outside the inner horizon but inside the outermost circle is the outer horizon.

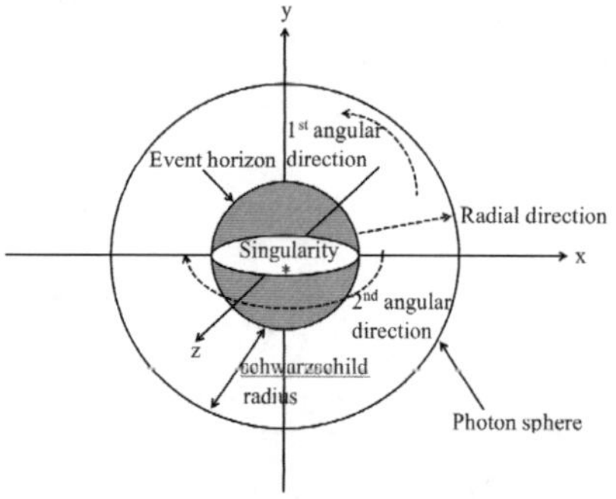


Figure 3. Anatomy of a Schwarzschild black hole.

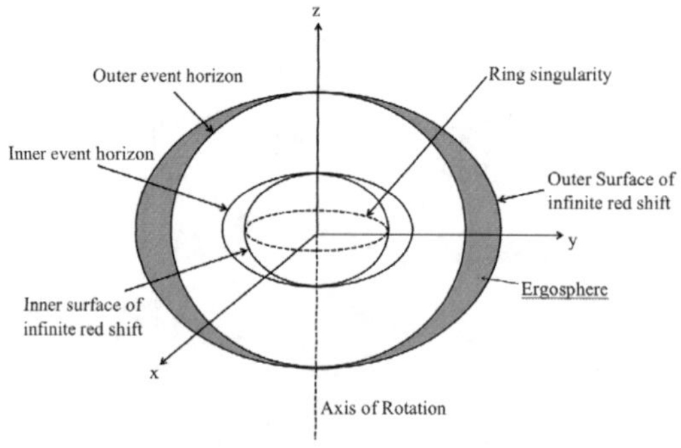


Figure 4. A sketch of Kerr black hole.

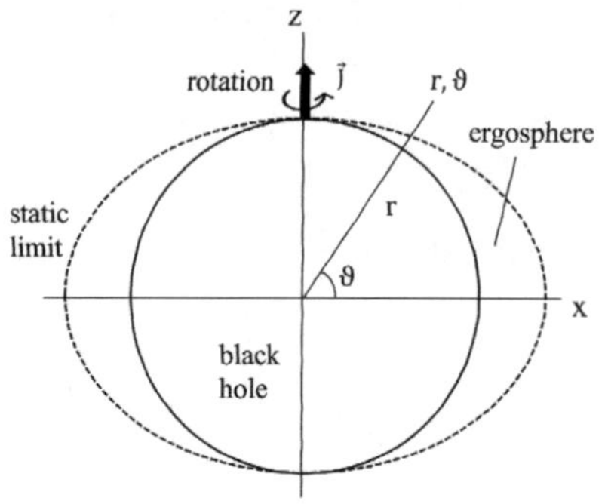


Figure 5. Kerr–Newman black hole [12].

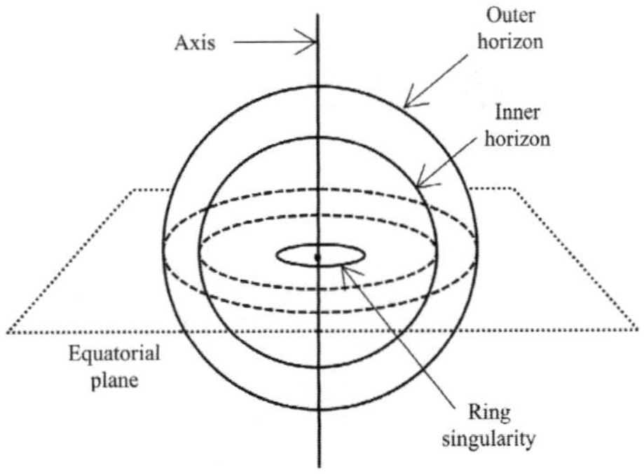


Figure 6. Structure of a charged black hole [13].

Table 5 lists the difference among these four types of black holes, the reference parameters are charge and angular momentum that describe whether the black hole is charged or uncharged and whether the black hole is rotated or not. The detail differences are demonstrated in Ref. [14].

TABLE 5. TABLE TYPE STYLES

| | Non-rotating (J = 0) | Rotating (J > 0) |
|---|---|---|
| Uncharged (Q = 0) | Schwarzschild | Kerr |
| Charged (Q ≠ 0) | Reissner–Nordström | Kerr–Newman |

## 4. DETECTION OF BLACK HOLE

### *4.1. Detector*

For any physical theory's validity to be tested, experimental evidence certainly holds high credit. Thus, researchers have been reaching out for observations into the universe to learn more about the universe, and to modify current theory on the cosmos. In addition, it's quite obvious that specific detectors are needed for cosmological purpose. Intended for different intentions, detectors also various from one another, including ground-based ones and space-based ones, one's probes actively and that detect passively. Despite these differences, none of the detectors can be built without proper technology, location, and enough fund supported, in fact, several detectors who are competent in such large-scale observations includes LIGO, LISA, Virgo, KAGRA, Cosmic Explorer (CE), Einstein Telescope (ET), Tianqin, dark matter detectors DARWIN, ARGO, and DAMPE, neutrino detectors Hyper-Kamiokande, JUNO and DUNE, and also Chandra X-ray Observatory CXO. What's more, different network is also applied of certain detector, e.g., LIGO-Hanford (H), LIGO-Livingston (L), Hanford–Livingston–Virgo (HLV), Hanford–Livingston (HL), Hanford–Virgo (HV) and Livingston–Virgo (LV) [15]. With tench, as well as academic advancing, researchers have been trying to improve the instruments to allow for further observations, increasing in configurations, distances and precision than previously allowed.

### *4.2. Detection of black holes (electric charge)*

The black hole charge can be determined by using the characteristic properties of gravitational retro-lensing images. The electromagnetic signal from a source located in front of a black hole may be significantly enhanced by the influence of strong gravitational lensing, this is the retro-lensing effect [16]. If we consider a photon with motion in the r-coordinate depends on the root multiplicity of the polynomial $\hat{R}(\hat{r})$ that satisfies

$$\hat{R}(\hat{r}) = \frac{R(r)}{M^4E^2} = \hat{r}^4 - \xi^2\hat{r}^2 + 2\xi^2\hat{r} - \hat{Q}^2\xi^2 \qquad (21)$$

The critical value of the impact parameter for a photon to be captured by a Reissner–Nordström black hole depends on the multiplicity root condition of the polynomial R(r). Introducing the notation $\xi^2 = l$, $Q^2 = q$,

$$\mathrm{R(r)} = r^4 - lr^2 + 2lr - qr \tag{22}$$

Thus, one derives

$$\Delta = 16l^3[l^2(1-q) + l(-8q^2 + 36q - 27) - 16q^3] \tag{23}$$

where the polynomial R(r) should obey the following expression for roots to exist,

$$l^3[l^2(1-q) + l(m(q)) - 16q^3] = 0 \tag{24}$$

where $m(q) = 8q^2 - 36q + 27$. Thereby, one obtains the critical value of the impact parameter,

$$l = \frac{m(q) + \sqrt{m^2(q) + 64q^3(1-q)}}{2(1-q)} \tag{25}$$

The detail discussions for the solutions are given in Refs. [16, 17].

*4.3. Measurement and estimation of mass*

Twin-peak QPOs: Variations in the accretion flow surrounding compact objects, (e.g., white dwarfs, neutron stars, and black holes), are thought to cause quasi-periodic oscillations (QPOs). In high-frequency QPO oscillations, the inverse mass scaling theory can be used to determine the system's mass. The unknown mass (MH) can be measured by using the formular:

$$M_H = \left[{v_{10}^{Hi}}/{v_H^{Hi}}\right] \times 10M_{\odot} \tag{26}$$

where H stands for the quantities of the unknown system, and $v_H^{Hi}$ is the higher peak value of the QPO frequency of an unknown object, $M_H$ is unknown mass, $M_{\odot}$is the solar mass, $v_{10}^{Hi}$ is High-frequency QPO.

The number of unknown systems is represented by the subscript H, while the peak of the unknown object's QPO frequency is represented by $v_H^{Hi}$. Fig. 7. is used to calculate the $v_{10}^{Hi}$ value of the $10M_{\odot}$ black hole. Two frequency curves are drawn based on the $v_{10}^{Hi}$ peak ratio. The peak ratio between 1 and 2 is = 0.999 on the curve, and the lower curve (ratio between extreme peaks 2 and 1.35) is $\in$ (1, −1) short extension curve (straight line) =0.999 from the peak ratio between 1.35 and 1. The $v_H^{Hi}$ frequency and ratio are used to describe the positions of different microquasars. A point can be determined on these curves to acquire the unknown frequency $v_{10}^{Hi}$ of a $10M_{\odot}$ candidate black hole using peak correlation ratio knowledge. The mass is computed this way using a formular.

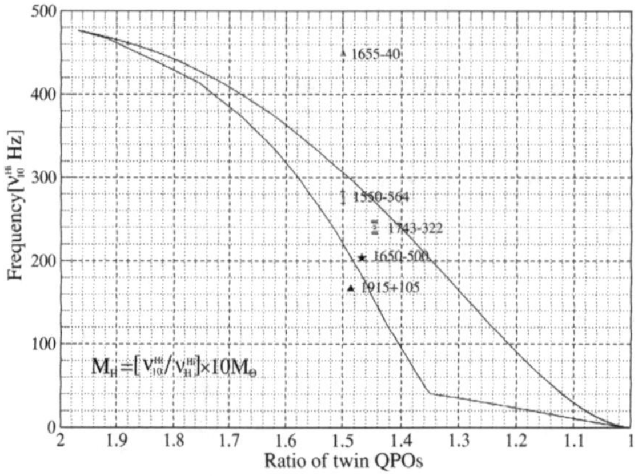


Figure 7. The peak ratios of a 10 M black hole's high-frequency ($v_{10}^{Hi}$) QPO are displayed versus their peak ratios [17].

As for accretion disks method, a flux variation from a point-like central source reaching a distant observer at time $t_0$ will be echoed after a span of time $t - t_0$ by the disc region created by omitting any general relativistic delays in light transmission.

$$t - t_0 = r(1 - sini \times sin\phi)/c \tag{27}$$

where I is the inclination angle from which the disk is observed. The highest blueshifts and redshifts in the echo front are produced in the weak-field approximation ($r \gg r_s$).

$$sin\phi_f = \{(sini)^{-1} - [(sini)^{-2} + 3]^{1/2}\}/3 \tag{28}$$

Where $\phi_f$ is the corresponding coordinate in the the disk plane.

$$r_f = c(t - t_0)(1 - sin\phi_f sini)^{-1} - r_s \tag{29}$$

Here, $r_s$ is Schwarzschild radii, $r_f$ is the corresponding radii, the last component on the right-hand side was added to roughly account for generic relativistic corrections at small radii:

$$(1+z)_f = \left(1 - \frac{3r_s}{2r_f}^{-1/2}\right)\left\{1 \pm \frac{\cos\beta_f}{[2r_f(1+\tan^2\xi_f)/r_s - 2]^{1/2}}\right\} \tag{30}$$

Where $(1+z)_f$ is the shifts of the drifting features, $cos\beta_f = cos\phi_f sini(1 - sin^2\phi_r sin^2 i)^{-1/2}$ and $\tan\xi_f = \tan\phi_f \cos\beta_f$. Over a broad range of disc settings, the drifting features' shifts $(1+z)_f$ are accurate to a few percent [18].

In Fig. 8, the solid line depicts our best power law fit for the data, whereas the dotted line depicts the thin disc theory's projected findings [19]. The predicted variance owing to the tendency is shown by the shadow band around the best fit. The solid line at the bottom of the illustration represents the tolerable radius range of the accretion disk's inner edge, representing stable circular orbits of the innermost layers of the greatest revolving Kerr black hole and Schwarzschild black hole.

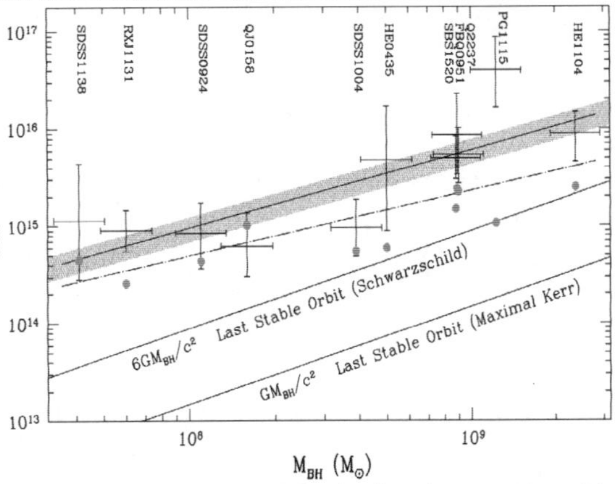


Figure 8. $R_{250}$ accretion disc size corrected for inclination vs $M_{BH}$ black hole mass [19].

### 4.4. Measurement and estimation of spin

QPOs (quasi-periodic oscillations) is a phenomenon that celestial body flickers at a particular frequency under the X-ray light range. Scholars gave the model of the QPOs of a black hole's system. As the reason for viscosity, the accretion flow is subsonic in the far distance and the position after the shock position while supersonic near the horizon. The oscillation shocks might be the reason of the QPOs [20]. As the double peak QPOs had been observed by X-ray variability of neutron stars, the models of resonance between epicyclic frequencies were made. Scholars have realized the orbit resonance and the ratio of the Quasi-periodic oscillation can help to solve questions about measurement of black hole's spin and others. The relationship of the Keplerian and epicyclic frequencies [21]. $\nu_{vert}$ and $\nu_{rad}$ with mass $M$ and parameter about the spin $a$:

 **2364** (2022) 012053 doi:10.1088/1742-6596/2364/1/012053 

$$\nu_K = \frac{1}{2\pi}\left(\frac{GM_0}{r_G^3}\right)^{\frac{1}{2}}\left(x^{\frac{3}{2}} + a\right)^{-1} \tag{31}$$

$$\nu_{vert}^2 = \nu_K^2\left(1 - 4ax^{\frac{-3}{2}} + 3a^2x^{-2}\right) \tag{32}$$

$$\nu_{rad}^2 = \nu_K^2\left(1 - 6x^1 + 8ax^{\frac{-3}{2}} - 3a^2x^{-2}\right) \tag{33}$$

where $\nu_{rad}$ is radial epicyclic frequency, and x can be determined as $x = \frac{c^2r}{GM}$. Those different types of resonance can help to get the formula of spin. The resonance constant could be $n{:}\,m$, which means:

$$n\nu_{rad} = m\nu \tag{34}$$

where $\nu$ can be the vertical orbital frequency or the Kepler frequency, and $m$,$n$ are thought to be positive integers. For observed resonance number, the spin $a$ can be calculated with a known mass and frequencies. As illustrated in Fig.9, the Török concluded that for the measurement of spin of Sgr A* showed the ratio of 3:2:1. By using 3:2 parametric resonance model, he gave the prediction about the spin range of Sgr A*, despite the 3:2 data is needed to be further confirmed.

Based on Fe-Kα line, a rotating black hole's Innermost stable circular orbit can be determined as [22]:

$$r_{isco} = \frac{GM}{c^2}\left[3 + \alpha + \sqrt{(3-\beta)(3+\beta+2\alpha)}\right] \tag{35}$$

Where

$$\begin{cases} \alpha = \sqrt{3\tau^2 + \beta} \\ \beta = 1 + \sqrt[3]{1-\tau^2}\left(\sqrt[3]{1-\tau} + \sqrt[3]{1+\tau}\right) \\ \tau = \dfrac{cJ}{M^2G} \end{cases} \tag{36}$$

If the accretion disk flows through the innermost stable circular orbit, it will make the innermost circular (ISCO) detected (seen from Fig.10 [23]). Inside the ISCO, the X-ray radiation cannot be observed since the accretion material that can be ionized is relative rare compared to outside of the ISCO. The Fe − Kα line is the most remarkable emission line which can be detected outside the ISCO, and the Fe − Kα line can gave us the details about the range of ISCO. For example, XTE JI550-546, was measured by this method, getting the result of the parameter $a \approx 0 \cdot 55^{+0.15}_{-0.22}$ [24].

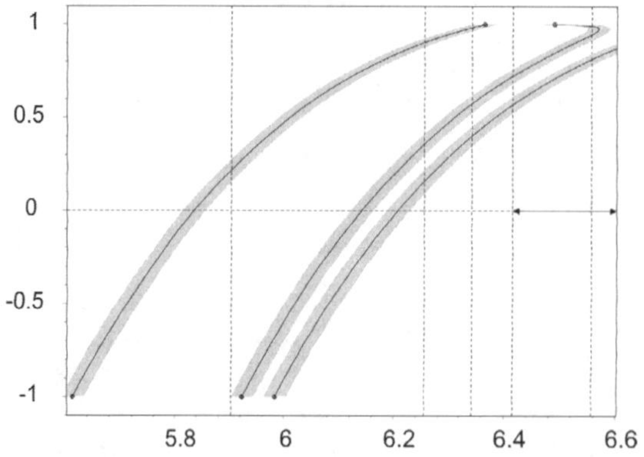


Figure 9. The relation for spin parameter $a$ and $log\left(M\ /\ M_{\odot}\right)$ when the resonance constant be 3:2 parametric, 3:1 and 2:1 for Sgr A*, and the shadows represent the degree of uncertainty [21].

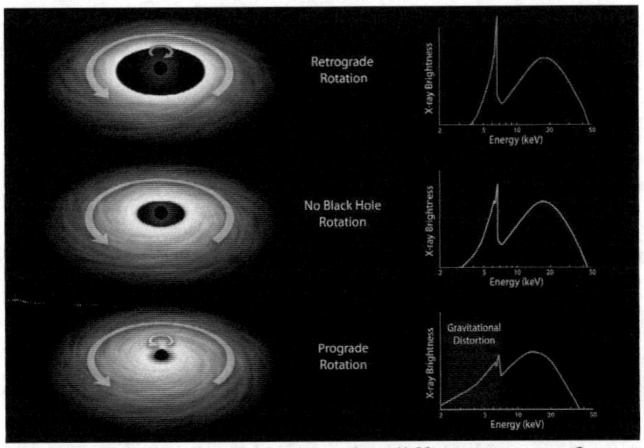


Figure 10. The X-ray's brightness distribution under different energy for retrograde rotation black holes, non-rotating black holes and prograde rotating black holes [23].

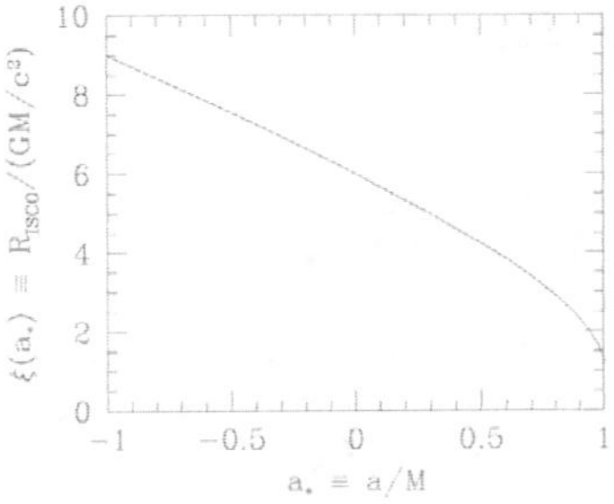


Figure 11. The dependence of $\frac{R_{isco}}{GM/c^2}$ on the spin parameter $a_*$ [27].

Regarding to continuum fitting method, it includes analyzing the data from the thermal continuum spectrum of the black holes' accretion disk and fitting with theoretical models. According to the model of thin accretion disc black holes developed by Novikov and Thorne which they improved from Shakura and Sunyaev's model, one can get the expression of the $R_{isco}$ with the thermal spectrum parameters [24].

$$\frac{3GM\dot{M}}{8\pi R^3}\left(1-\sqrt{\frac{R_{isco}}{R}}\right)=\sigma T_{eff}^4(R) \tag{37}$$

$$T_{eff}(R)=T^*\left(\frac{R_{isco}}{R}\right)^{\frac{3}{4}}\left(1-\sqrt{\frac{R_{isco}}{R}}\right)^{\frac{1}{4}} \tag{38}$$

Here, $T_{eff}$ is the effective temperature related to the black body radiation, as the accretion disk's similarity with the black body. $\sigma$ is the Stefan-Boltzmann constant. With given geometric parameters, by analyzing the data from the emission spectrum of it, researchers can get the $R_{isco}$. Furthermore, as shown in Fig.11, researchers can produce the spin parameter $a$ by its monotone relation with $\frac{R_{isco}}{M}$ [26, 27]. Some researchers have discussed about the deviation of the disc model developed by Novikov and Thorne. As a reason of the model using an extreme thin accretion disc model, the deviation of spin parameter $a$ may reach 0.2 when the accretion disk thickness parameter $\frac{h}{r}\sim 0.04-0.08$ [28].

Therefore, the condition for using the Novikov-Thorne model is the optically and geometrically thin of the disc.

## 5. Limitation and future prospect

When applying the twin-peak QPOs theory for measuring the mass, the relation between 'frequency' and 'ratio of twin QPOs' is plotted in Fig. 7 so that the unknown frequency can be derived. Yet, the curve itself is a modified one to be smooth, which could lead to a certain degree of inaccuracy. Additionally, as the QPOs theory suggests, the mass we get using the tangent line, relevant formation is the smallest possible mass of micro quasars rather than the practical one. As for the accretion desk method, apart from the certain condition needed to fit the weak-field approximation, this method is also currently restricted to certain types of black holes including Schwarzschild and Kerr ones. When making use of quasi-periodic oscillations to measure the spin, the mass should be calculated as a previous step, thus allow the possibility of accumulated error. Moreover, as mentioned above, the 3:2 data applied here need further confirmation. As apropos of continuum fitting method, disc has to be optically and geometrically thin, and this sets a limit the spin detectable under this circumstance. When it comes to detecting electric charge, type of black hole observable is also limited.

For future observation to hold more exactitude and farther space allowed, the possibility lies in more advanced instruments supported by improved technique as well as other theories discovered allowing less error and more types of black hole to be observed.

## 6. Conclusion

In conclusion, this paper investigates characteristic of different types of black holes and measurements of the three parameters of black holes. We study the four types of black holes (Schwarzschild black hole, Reissner-Nordström black hole, Kerr black hole, Kerr-Newman black hole) and demonstrates the space time metrics behind it. We mainly discuss about the measurements of the spin, mass and charge of a black holes. Specifically, the state-of-art detectors which made effort in detecting black holes are demonstrated. For the measurement of the mass, twin peaks QPOs and accretion disks methods are introduced. For the measurement of the spin, we discussed about double peaks QPOs method, the Fe-Kα line and Continuum fitting method. For the measurement of electric charge, the characteristic properties of gravitational retro-lensing images are mentioned. The limitations of those methods are discussed, similar to the inaccuracy and the selective black holes types from the twin-peak quasi-periodic oscillations method, the model limitation for long term continuum etc. According to the analysis, one can raise the prospect of higher accuracy detection with coming instruments and advanced theories. Overall, these results offer a guideline of detection for black holes.

## References

[1] B. Brügmann, A. Ghez, J. Greiner, J, "Black holes," Proceedings of the National Academy of Sciences, vol. 98(19), 2001, pp.10525-10526.
[2] B. Willke, "Observation of gravitational waves from a binary black hole merger – dawn of a new astronomy," Symmetry: Culture and Science, vol. 29(2), 2018, pp.257-264.
[3] M. Visser, "The Kerr spacetime: A brief introduction," arXiv preprint arXiv:0706.0622, 2007.
[4] B. Carr, et al., "Constraints on primordial black holes," Reports on Progress in Physics, vol. 84(11), 2021, 116902.
[5] A. Chael, et al., "The role of electron heating physics in images and variability of the Galactic Centre black hole Sagittarius A," Monthly Notices of the Royal Astronomical Society, vol. 478(4), 2018, pp. 5209-5229.
[6] M. Zajaček, et al, "Constraining the charge of the Galactic centre black hole," Journal of Physics: Conference Series, Vol. 1258, No. 1, 2019, pp. 012031, IOP Publishing.
[7] K. V. Kuchař, "Geometrodynamics of Schwarzschild black holes." Physical Review D, vol. 50.6, 1994, 3961.

[8] A. Hamilton, J. Lisle, "The river model of black holes." American Journal of Physics, vol. 76(6), 2008, pp.519-532.
[9] I. Petkov, "The Kerr Metric of Spinning Black Holes,", 2015 (), available at: https://www.researchgate.net/publication/276857998
[10] C. Heinicke, F. Hehl, "The Schwarzschild black hole," The Galactic Black Hole, vol. 12, 2002, pp. 3–34.
[11] H, Vargas-Rodríguez, et al., "Types of electromagnetic fields around black holes in Schwarzschild, Kerr and Kerr-Newmann spacetimes," arXiv preprint arXiv:2202.11707, 2022.
[12] Retrieving from: https://astronuclphysics.info/Gravitace4-4.htm
[13] Retrieving from: https://www.researchgate.net/figure/Sketch-of-a-Kerr-black-hole-with-its-two-horizons-and-the-ring-singulatrity_fig2_265603056
[14] Retrieving from: https://heasarc.gsfc.nasa.gov/docs/objects/heapow/archive/compact_objects/spinning_bh_chandra.html
[15] C. Cahillane, G. Mansell, "Review of the Advanced LIGO gravitational wave observatories leading to observing run four," Galaxies, vol. 10(1), 2022, 36.
[16] A. Zakharov, et al., "Direct measurements of black hole charge with future astrometrical missions," Astronomy & Astrophysics, vol. 442(3), 2018, pp.795-799.
[17] S. Mondal, "Black Hole Mass and Spin from the 2: 3 Twin-peak QPOs in Microquasars," The Astrophysical Journal, vol. 708(2), 2009, 1442.
[18] L. Stella, "Measuring black hole mass through variable line profiles from accretion disks," Nature, vol. 344(6268), 1990, pp. 747-749.
[19] C. W. Morgan, et al., "THE QUASAR ACCRETION DISK SIZE-BLACK HOLE MASS RELATION," The Astrophysical Journal, vol. 712(2), 2010, pp. 1129–1136.
[20] S. Mondal, "October. Estimation of spin of the black hole from QPO frequency," In AIP Conference Proceedings Vol. 1053, No. 1, 2008, pp. 349-352, American Institute of Physics.
[21] G. Török, "QPOs in microquasars and Sgr A.: measuring the black hole spin," Astronomische Nachrichten: Astronomical Notes, vol. 326(9), 2005, pp. 856-860.
[22] J. M. Bardeen, W. H. Press, S. A. Teukolsky, "Rotating black holes: locally nonrotating frames, energy extraction, and scalar synchrotron radiation," The Astrophysical Journal, vol. 178, 1972, pp. 347-370.
[23] Retrieving from: https://photojournal.jpl.nasa.gov/catalog/PIA16696
[24] J. F. Steiner, et al. "The spin of the black hole microquasar XTE J1550−564 via the continuum-fitting and Fe-line methods," Monthly Notices of the Royal Astronomical Society, vol. 416(2), 2011, pp. 941-958.
[25] I. D. Novikov, K. S. Thorne, "Astrophysics of black holes," In Black holes, 1973.
[26] S. L Shapiro, S. A. Teukolsky, "Black holes, white dwarfs, and neutron stars: The physics of compact objects," John Wiley & Sons, 2008..
[27] Shafee, R., McClintock, J. E., Narayan, R., Davis, S. W., Li, L. X., & Remillard, R. A. (2005). Estimating the spin of stellar-mass black holes by spectral fitting of the X-ray continuum. The Astrophysical Journal, 636(2), L113.
[28] A. K. Kulkarni, et al., "Measuring black hole spin by the continuum-fitting method: effect of deviations from the Novikov–Thorne disc model," Monthly Notices of the Royal Astronomical Society, vol. 414(2), 2011, pp. 1183-1194.

# The principle and state-of-art applications of Hawking radiation

**Chunyu Guo***

line 1 (of *Affiliation*): dept. name of organization, Shenyang foreign language school, Shenyang, China

* Corresponding author: 18409495@masu.edu.cn

**Abstract**—Hawking radiation, firstly discovered in 1974 by Stephen Hawking, is a crucial quantum phenomenon, which proves that a black hole has been losing its mass since its formation. In this paper, the method of information retrieval and literature analysis are fully applied to introduce the principle and state-of-art applications of Hawking radiation. Three parameters used to describe the evaporation of black holes are analyzed. Furthermore, the tunnelling effect of entangled pairs near event horizon is described to explain Hawking radiation from a microscopic perspective. Then, the evaporating black holes are observed under wave optical conditions by the Fourier transformation of the spatial correlation function and detected in Laboratory condition by equivalenting entangled pairs to surface wave packets with sum-zero energy. Finally, a prediction is made that the widely used of extra-dimensional theory and high-energy particle technology as well as numerical researches in laboratory conditions will be strongly pushing the observation of Hawking radiation. Overall, these results shed light on further exploring the universe in terms of Hawking radiation.

## 1. Introduction

Hawking radiation is an important quantum phenomenon proposed by British physicist Stephen Hawking in 1974, which makes a prediction that black holes produce thermal radiation [1]. According to general relativity, a black hole absorbs everything but emits nothing. In other words, it is a region of space that nothing (even the photons), can escape from its interior because of the intensive gravitational field [2]. However, this hypothesis cannot be explained by conventional quantum mechanics theories. As Hawking discovered, black holes should behave similar to black bodies and thus emit thermal radiation with temperatures proportional to their surface gravities on the event horizons. According to Hawking's calculations, thermal radiation is only significant for so-called quantum black holes, whose size is smaller than that of an atomic nucleus [3].

In order to improve the Hawking radiation theory, plenty of approaches are proposed and many appealing research results have been demonstrated. Gibbons and Hawking's research gives the temperature formular for Hawking radiation calculation. However, they questioned that it is unknown whether an event horizon associated with some dynamical spacetime at a given time exists or not, which poses a great difficulty in discussing Hawking radiation for dynamical black holes [4]. Hayward derived a local Hawking temperature for dynamical black holes and defined the Killing value to reflect measurements at infinity. In Newtonian limit, the geometrical surface gravity is converted to the Newtonian surface gravity, which may settle a longstanding problem about the Hawking temperature of the charged stringy black hole for the extremal limit [5].

Although the existence of Hawking radiation has not been observed yet, Hawking's quantum gravity theoretical model has become the basis of many studies in physics and cosmology. Singh's research revealed the Generalized Uncertainty Principle (GUP) effect of Baados-Teitelboim-Zanelli (BTZ) black holes by using Hamilton-Jacobi method and Dirac equation. Damour and Ruffini method, tortoise coordinate transformation and modified Klein-Gordon equation were used to modify Hawking radiation of massless particle from a BTZ black hole. The relation between the modified Hawking temperature and the energy of the emitted particle is deduced [6]. Zeng discussed the information missing puzzle in Hawking radiation. It proved that Hawking's neglecting or implicit averaging over the inner structure microstate of black holes leads to the information missing puzzle. Meanwhile, a semi-classic fuzzy ball picture for microstate of black holes and a Hamiltonian thus explicitly unitary formulation for Hawking radiations are proposed to make the various information missing argument in Hawking radiations easy to resolve [7].

Hawking's quantum black holes theoretical model resolves the contradictions existing in the black hole thermodynamics, reveals the connection between gravity theory, thermodynamics, and quantum theory. The rest part of the paper is organized as follows. The Sec. II gives basic descriptions of Hawking radiation, introducing a calculation model of Hawking temperature and the meaning of variables used in the model. The Sec. III lists several applications of Hawking radiation, especially in cosmology and thermodynamics. The Sec. IV points out the limitations of Hawking radiation theory in quantum mechanics while gives a preview of researches in this field. Eventually, a brief summary is given in Sec. V.

## 2. Basic Descriptions of Hawking Radiation

Hawking radiation theory explains the reason that black holes have continued to losing mass since their formations. There are three important parameters to describe this process as discussed following.

### *2.1.Hawking temperature*

Hawking temperature is an important parameter to calculate thermal radiation. It represents the surface temperature of a black hole. In Unruh effect, a relationship between vacuum temperature $T$ and observer's acceleration $a$ can be derived. By placing the particle trajectory in Schwarzschild horizon into geodesic equation, the Hawking temperature $T_H$ in Rindler coordinates is deduced. Eq. (1) shows the way to calculate the Hawking temperature $T_H$ of a black hole massed $M_{BH}$. All the parameters are constants, i.e., makes this formular so concise, which clearly indicates that the mass of black holes and the radiation temperature are in inverse proportion. The mathematical description can be described as follows:

$$T_H = \frac{hc^3}{8\pi kGM_{BH}} \tag{1}$$

where $h, c, k, G$ represent the Planck constant, the speed of light in vacuum, Boltzmann constant and gravity constant, respectively [8]. According to Eq. (1), the Hawking temperature is proportion to the mass of the black hole.

### *2.2.Bekenstein-Hawking Luminosity*

Bekenstein-Hawking luminosity is used to calculate thermal radiation power. Based on Eq. (1), the hawking temperature on the surface of the sun can be calculated. Subsequently, in terms of Wien displacement law, the maximum radiation wavelength of sun surface can be deduced:

$$P = \frac{hc^6}{15360\pi G^2 M_{BH}^2} \tag{2}$$

In Eq. (2), parameters are the same as Eq. (1) which holds on blackbody assumption [9].

*2.3.Evaporation Time*

The evaporation time of a black hole can be calculated by Eq. (3),

$$t_e = \frac{5120\pi G^2 M^3}{hc^4} \tag{3}$$

in which parameters have the same numerical value as Eq. (1) and Eq. (2). It could be estimated that the evaporation time of a black hole with the same mass as the Sun ($M_{sun} = 2 \times 10^{30} kg$) is about $10^{67}$years [10].

**3. Application of Hawking Radiation in Observing Black Holes**

Hawking radiation is thermal radiation, which proves that black holes are releasing energy continuously. This process is called the black hole's evaporation. It can be inferred that black holes evaporate away completely eventually. However, this process is so slow that a black hole with mass of $10^{12}$kg, has been evaporating since the early universe. Fig.1 describes different stages of process of the black hole's evaporation.

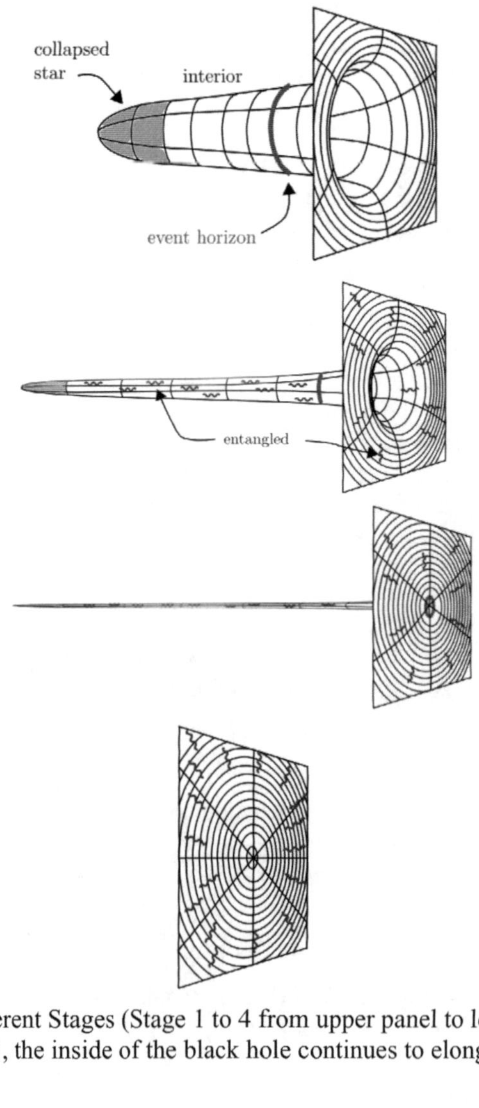


Figure 1. A sketch of different Stages (Stage 1 to 4 from upper panel to lower panel) of black hole's evaporation [11]. Stage.1, the inside of the black hole continues to elongate when stellar collapse.

Stage.2, one's escaping to infinity is observed as radiation after generation of entangled pairs. Stage.3 the angular directions shrink to zero size, becoming the singularity eventually. Stage4, the black hole's evaporation is completed at the end.

The process of evaporation can be described as 4 stages as illustrated in the Fig. 1. At the first stage, the space within in a certain distance from a collapsed star, which is called event horizon, continues to elongate in one direction, while the space out of event horizon is nearly stationary. At the second stage, according to Heisenberg Uncertainly Principle, entangled pairs continue to be created in vacuum. Under the great gravity in this space, one particle escapes to infinity while the other is trapped inside event horizon, which is observed as blackbody radiation. At the third stage, when the black hole almost evaporates to one point, the angular directions shrink to almost zero size. At the last stage, the black hole completely evaporates and disappears.

Investigation of Hawking radiation can help deepen human's understanding of black holes. In general relativity, a black hole is a bounded spacetime, where the gravity is so great that almost nothing captured by the gravitational of black hole can escape to the null infinity, which makes black holes difficult to be observed in optical field. The bound is called the event horizon. For a distant observer, the space distortion makes a set of photon orbits are observed as a distorted disk on a far observer's view plane. By detecting a black hole's Hawking radiation from a far distant, the wave optical images of the evaporating black hole can be reconstructed.

Assuming that there are two detectors placed near the cosmological horizon in a spherical coordinate system, the reflection and transmission coefficients in the upper detector are demonstrated in the three pictures above of Fig. 2. While the same parameters of the lower detector in Schwarzschild case are shown in the lower panel of Fig. 2 [12]. A superimposed image of black hole with the vacuum fluctuation in the UP mode is exhibited in Fig. 3. The photon sphere in geometric optics has a finite width shape in wave optics. The black hole has the appearance of a shining star and with a surface coinciding with the photon sphere.

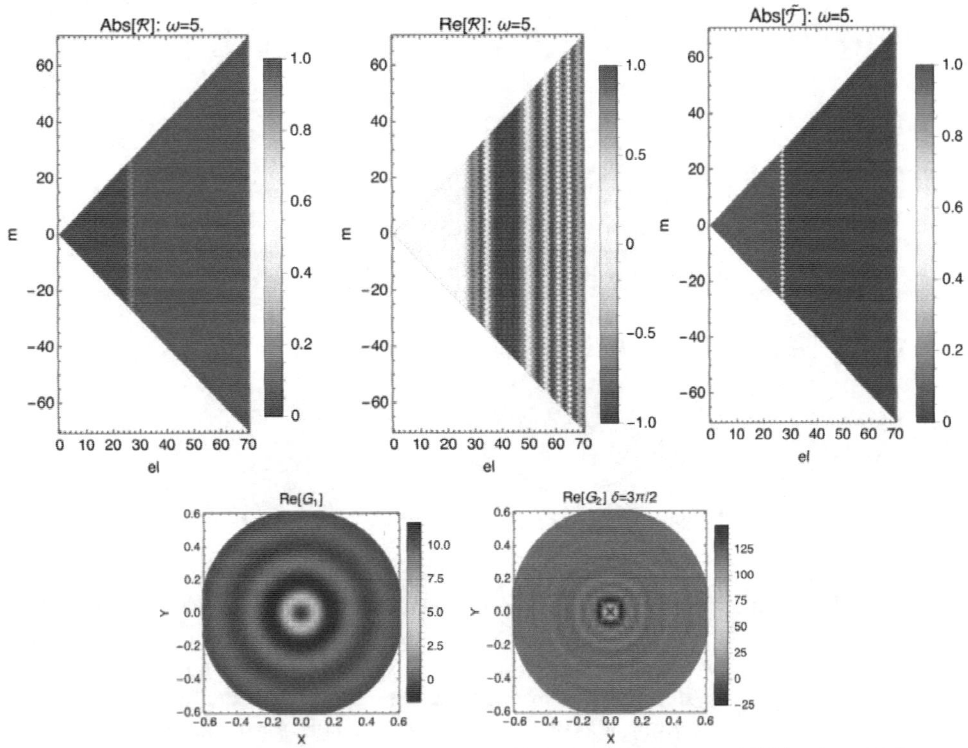


Figure 2. Reflection and transmission coefficients in Schwarzschild horizon [12].

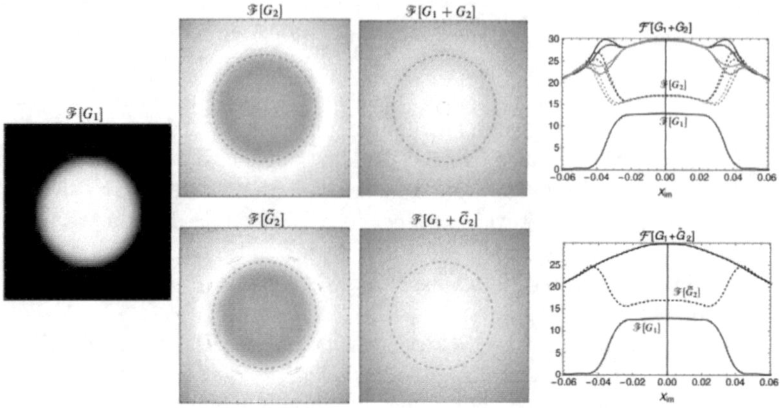

Figure 3. Images obtained by the Fourier transformation of G [12].

## 4. Detection of Hawking Radiation in Laboratory Analog

To detect some key properties of surface Hawking radiation of non-rotating black holes under laboratory conditions, two surface gravity wave pockets are selected, which are commonly used in the study of particle scattering problems. The surface gravity wave pockets with opposite energy but the same amount of wave action is used to simulate entangled pairs. Created out of near horizon vacuum fluctuations, entangled pairs consist of two particles. The virtual particle pairs with opposite energy are simulated by a minimal water wave model. In vacuum fluctuations near the black hole, the particle with positive energy escape to infinity, while the other falls in to the black hole, leading to black hole evaporation.

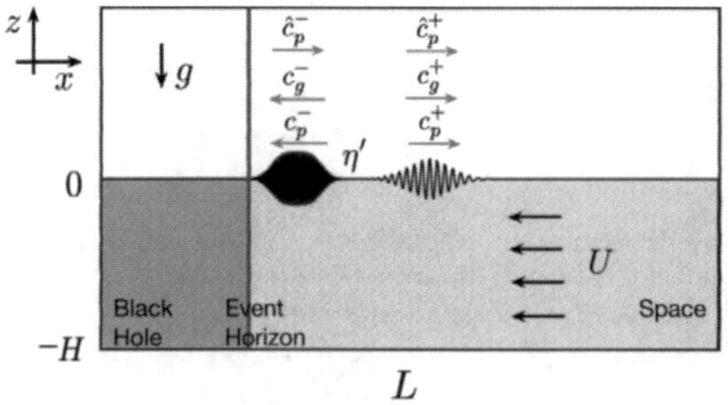


Figure 4. Schematic diagram of the entangled pairs simulated by surface gravity wave pockets [13].

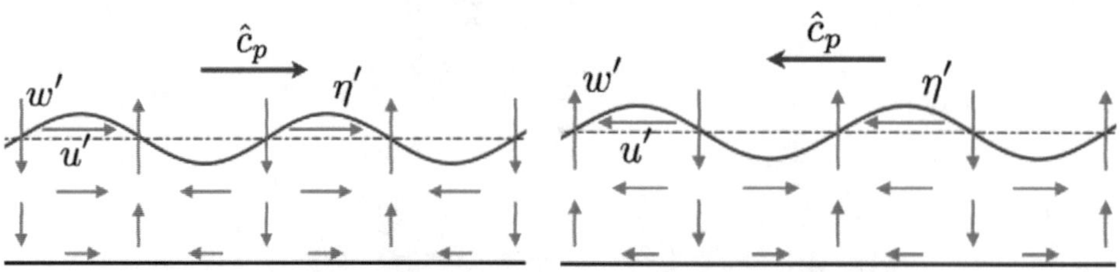

Figure 5. Schematic diagram of the pseudo-momentum of the pairs of wave packets [13].

Superposition pairs of wave packets with opposite values of pseudo-energy are chosen to ensure that the entangled pairs can be represented by them. Fig. 5 depicts that the two paired packets have opposite pseudo-momentum when the system has a tiny disturbance. It indicates that the two paired packets have opposite moving direction, performing the same with entangled pairs formatting in space near a black hole (seen from Fig.6).

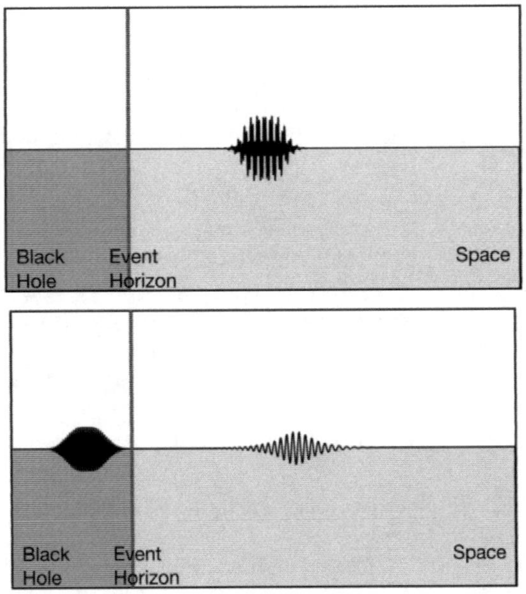


Figure 6. Simulation of sum-zero pseudo-energy packet [13].

## 5. Limitation & Future prospects

The existence of Hawking radiation has not been actually observed yet. As a great curved space-time quantum field theory, it ignored the internal contradiction of general relativity and quantum mechanics forcibly integrated together. However, there will be some irreconcilable limitations in Hawking radiation theory when the horizon gradually narrowed to a certain scale.

Nevertheless, Hawking radiation theory is a reasonable explanation of quantum gravity theory, explaining the path that the mass of a black hole is reduced and the reason for black holes' evaporation. Meanwhile, the reliability of detecting Hawking radiation by using surface gravity wave pockets will be demonstrated further through numerical simulation on the nonlinear effects of wave dissipation and wave mean flow interaction in laboratory conditions. In addition, the applications of extra-dimensional theory and high-energy particle collision experiments may create temporary micro black holes emitting radiation of specific wavelengths, which may allow Hawking radiation to be observed firstly in laboratory in the future.

## 6. Conclusion

In summary, this paper discusses the principle and state-of-art applications of Hawking radiation. Based on the analysis, Hawking's quantum black holes theoretical model builds the connection between gravity theory, thermodynamics, and quantum theory, which firstly proves that black holes keep losing mass since their formation. Hawking radiation theory is widely used in researches in cosmology and thermodynamics. In the future, abundant studies in theoretical and experimental will promote human's proof of Hawking radiation theory. Overall, these results offer a guideline for understanding the basic law of Hawking radiation theory and its significance to the study of black holes.

## References

[1] S. W. Hawking, "Black hole explosions?" Nature, vol. 248(5443), 1974, pp. 30-31.
[2] V. P. Frolov, A. Zelnikov, A., Introduction to black hole physics. Oxford University Press, USA, 2011.
[3] S. W. Hawking, "Particle creation by black holes. Communications in Mathematical Physics," 1976, pp. 167-188.

[4] G. W. Gibbons, Cosmological event horizons, thermodynamics, and particle creation, In Euclidean quantum gravity (pp. 281-294), 1977).
[5] S. A. Hayward, et al. "Local temperature for dynamical black holes. In AIP Conference Proceedings, vol. 1122, No. 1, 2008, pp. 145-151.
[6] T. I. Singh, Y. K. Meitei, I. A. Meitei, "Effect of gup on hawking radiation of BTZ black hole," International Journal of Modern Physics A, vol. 35.05 2020, 2050018.
[7] D. F. Zeng, "Information missing puzzle, where is hawking's error?" Nuclear Physics B, vol. 941, 2019, pp. 665-679.
[8] S. A. Hayward, R. D. Criscienzo, M. Nadalini, L. Vanzo, S. Zerbini, "Local hawking temperature for dynamical black holes," Classical and Quantum Gravity, vol. 26(6), 2009, 062001.
[9] S. Andrew, V. Cumrun, "Microscopic origin of the bekenstein-hawking entropy. Physics Letters B., vol. 379(1-4), 1996, pp. 99-104.
[10] W. G. Unruh, "Notes on black-hole evaporation," Physical review D: Particles and fields, vol. 14(4), 1976, 870.
[11] A. Almheiri, et al. "The entropy of Hawking radiation," Reviews of Modern Physics, vol. 93.3, 2021, 035002.
[12] Y. Nambu, S. Noda, "Interferometry of black holes with Hawking radiation," Physical Review D, vol. 105(4), 2022, 045022.
[13] A. Guha, E. Heifetz, A. Gupta, A. "Pairs of surface wave packets with zero-sum energy in the hawking radiation analog," Physical Review D, vol. 102(10), 2020, 104061.

# Design of INS/*GNSS* Precise Posture Measurement and Positioning System for Complex System

**Rong Han**[1,a]**, Jianfen Yang**[2,b]**, Yumin Tang**[3,c] **and Liben Yang**[1*]

[1]School of Automation and Electrical Engineering, Lanzhou Jiaotong University, Lanzhou City, Gansu Province, 730070, China

[2]School of Automation and Electrical Engineering, Lanzhou Jiaotong University, Lanzhou City, Gansu Province, 730070, China

[3]School of Automation and Electrical Engineering, Lanzhou Jiaotong University, Lanzhou City, Gansu Province, 730070, China

[a]email: 11210400@stu.lzjtu.edu.cn, [b]email: jfyang@mail.lzjtu.cn, [c]email: 951413516@qq.com,

[*]Corresponding author's e-mail: lbyang@mail.lzjtu.cn

**Abstract.** In this paper, aiming at the robustness of the UAV control system and the accuracy of positioning and mapping in complex situations, the visual odometry is used as the front end of the positioning and mapping system, and the data of the IMU inertial sensor is fused on the basis of the monocular camera. The analysis completes the loosely coupled initialization process; the back-end constructs the cost function to realize the tightly coupled optimization, and uses the loop closure detection relocation and the four degree of freedom graph optimization to realize the nonlinear optimization. It is verified by simulation that the algorithm can improve the visual robustness in time when the UAV moves quickly or the scene has wrong results, so that the trajectory captured by the camera and the real-time construction map maintain global consistency, effectively eliminating the integral drift of the IMU, and correcting The noise of the IMU improves its positioning accuracy.

## 1. Introduction

Inertial navigation systems (INS) are widely used in estimating the 6-DOF attitude of autonomous vehicle sensing platforms, especially in environments such as underwater, indoors, canyons or other planets where GPS signals are lacking[1]. Most inertial navigation systems rely on a 6-axis inertial measurement unit or IMU. In recent years, with the continuous advancement of hardware design and manufacturing capabilities, such as AR mobile augmented reality to VR technology, to autonomous driving, low-cost and lightweight micro electro mechanical IMU have become common, or mobile devices have micro air vehicles play an important role in countless emerging applications[2].

The VINS-Mono solution is to add an IMU pre-integration algorithm on the basis of optimization and sliding window visual odometry to build a tightly coupled processing model, which can realize automatic initialization, calibration of camera external parameters, relocation, closed-loop detection, and global Pose graph optimization and other functions[3]. This algorithm is further optimized on the basis of the original theory, and finally a fusion algorithm with good effect is formed.

## 2. Model establishment and algorithm analysis of UAV

### 2.1. Velocity of feature points on image plane

Instead of moving the entire camera or IMU sequence, we move the observation point of the feature on the timeline. The characteristic velocity is introduced for this purpose, for modeling and compensating for time offsets. The camera motion can be regarded as a uniform motion within a short period of a few milliseconds[4].

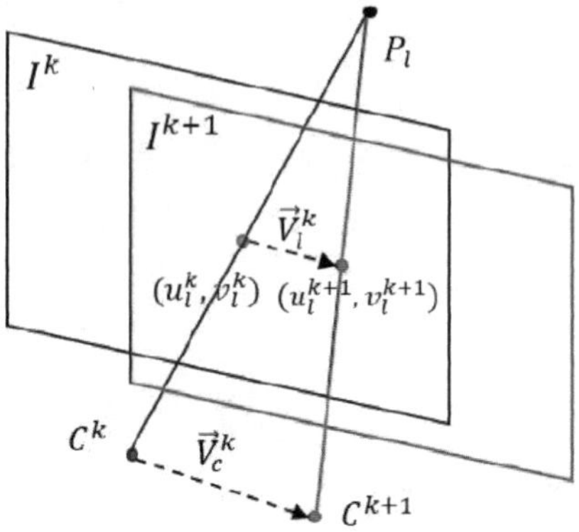


Figure 1. Feature point information map between two consecutive frames

Based on this assumption, we compute feature velocities on the image plane. For Figure 1., $I^k$ and $I^{k+1}$ are two consecutive image frames, assuming that the camera moves to $C^{k+1}$ at a constant speed $C^k$ in a short time $[t^k,t^{k+1}]$. We approximate that the feature L also moves on the image plane at a uniform speed in a short time. The speed is:

$$V_l^k = \left( \begin{bmatrix} u_l^{k+1} \\ v_l^{k+1} \end{bmatrix} - \begin{bmatrix} u_l^k \\ v_l^k \end{bmatrix} \right) / \left( t_{k+1} - t_k \right) \tag{1}$$

where $[u_l^k, v_l^k]$ and $[u_l^{k+1}, v_l^{k+1}]$ are the 2D observation points of the features in the image planes $I^k$ and $I^{k+1}$.

### 2.2. IMU pre-integration model

The inertial measurement unit IMU performs pre-integration between two consecutive frames to obtain the pose change between adjacent frames[5]. IMU pre-integration is proposed, and it is hoped that the relative measurement of the IMU can be processed to decouple it from the absolute pose, or it can be corrected as long as the linear operation is performed, so as to avoid repeated integration when the absolute pose is optimized. The IMU measurement model is:

$$\hat{a}_t = a_t + b_{a_t} + R_w^t g^w + n_a \tag{2}$$

$$\hat{\omega}_t = \omega_t + b_{\omega_t} + n_\omega \tag{3}$$

The rotation in all formulas adopts the right multiplication form, which can directly use the coordinates of the angular velocity in the body coordinate system, while the left multiplication form needs to convert the angular velocity to the world coordinate system. The pre-integration formula is:

$$\alpha_{b_{k+1}}^{b_k} = \iint_{t\in[t_k,t_{k+1}]} R_t^{b_k} \left( \hat{a}_t - b_{a_t} \right) dt^2 \tag{4}$$

$$\beta_{b_{k+1}}^{b_k} = \int_{t\in[t_k,t_{k+1}]} R_t^{b_k} \left( \hat{a}_t - b_{a_t} \right) dt \tag{5}$$

$$\gamma_{b_{k+1}}^{b_k} = \int_{t\in[t_k, t_{k+1}]} \frac{1}{2}\Omega\left(\hat{\omega}_t - b_{\omega_t}\right)\gamma_t^{b_k} dt \tag{6}$$

After correcting the offset we get:

$$\alpha_{b_{k+1}}^{b_k} \approx \hat{\alpha}_{b_{k+1}}^{b_k} + J_{b_a}^{\alpha}\delta b_{a_k} + J_{b_\omega}^{\alpha}\delta b_{\omega_k} \tag{7}$$

$$\beta_{b_{k+1}}^{b_k} \approx \hat{\beta}_{b_{k+1}}^{b_k} + J_{b_a}^{\beta}\delta b_{a_k} + J_{b_\omega}^{\beta}\delta b_{\omega_k} \tag{8}$$

$$\gamma_{b_{k+1}}^{b_k} \approx \hat{\gamma}_{b_{k+1}}^{b_k} \otimes \begin{bmatrix} 1 \\ \frac{1}{2} J_{b_\omega}^{\gamma}\delta b_{\omega_k} \end{bmatrix} \tag{9}$$

In the initial motion, in order to reduce the influence of the motion acceleration on the gravitational acceleration measured by the IMU, the vision and IMU values are measured separately, and the vision and IMU measurements are not optimized together[6].

*2.3. Nonlinear Tightly Coupled Optimization*

The main idea of VIO tight coupling scheme is to form a joint optimization problem by putting the residual term based on visual construction and the residual term based on IMU construction together, so the optimal solution of the whole optimization problem can be considered as a more accurate state estimation value[7].

$$\chi = \left[x_0, x_1, \cdots, x_n, x_c^b, \lambda_0, \lambda_1, \cdots, \lambda_m\right] \tag{10}$$

$$x_k = \left[p_{b_k}^{\omega}, v_{b_k}^{\omega}, q_{b_k}^{\omega}, b_a, b_g\right], k\in[0,n] \tag{11}$$

$$x_c^b = \left[p_c^b, q_c^b\right] \tag{12}$$

where $x_k$ represents the IMU state corresponding to time k when the camera obtains the picture; n is the number of key frames in the sliding window; m is the number of 3D visual features in the sliding window; P, V, Q are the pose, Speed, rotation information; $b_a$ is the accelerometer bias; $b_g$ is the gyroscope bias; $\lambda_m$ is the inverse depth of the mth 3D visual feature, which is the value calculated when it is first observed. The cost function is obtained as:

$$\min_{\chi}\left\{\left\|r_p - H_p\chi\right\|^2 + \sum_{k\in B}\left\|r_B\left(\hat{z}_{b_{k+1}}^{b_k}, \chi\right)\right\|_{P_{b_{k+1}}^{b_k}}^2 + \sum_{(l,j)\in C}\rho\left(\left\|r_C\left(\hat{z}_l^{c_j}, \chi\right)\right\|_{P_l^{c_j}}^2\right)\right\} \tag{13}$$

$$\rho(s) = \begin{cases} 1, s\geq 1 \\ 2\sqrt{s}-1, s<1 \end{cases} \tag{14}$$

The three terms in the above objective function are the marginalization prior error, the IMU measurement and the visual re-projection error. The residuals are weighted for each variable by the inverse of the covariance, the information matrix. ρ(s) is the Huber kernel function[8]. According to the Gauss-Newton method, the most important incremental equation in the optimization process is obtained. The optimization of these three objective functions together can explain why the system in this paper is a tightly coupled system.

## 3. Experiment and data analysis

The main indicator of this experiment is the positioning error between the final map and the actual geographic location. In the Ros-kinetic software on the virtual machine, the MH_01_easy data set of EuRoC is used for simulation, and the estimated data obtained is compared with the real path using the evo drawing tool. Analyze after comparison. The experiments in this chapter mainly have the following three aspects:

- Increasing the max_cnt value to 300, and keep the min_dist unchanged at 30 and comparing it with the real path provided in the data set, analyze the influence of the extraction radius and amplitude on the final mapping accuracy. The final generated estimated path is compared with the real value as shown in Figure 2.:

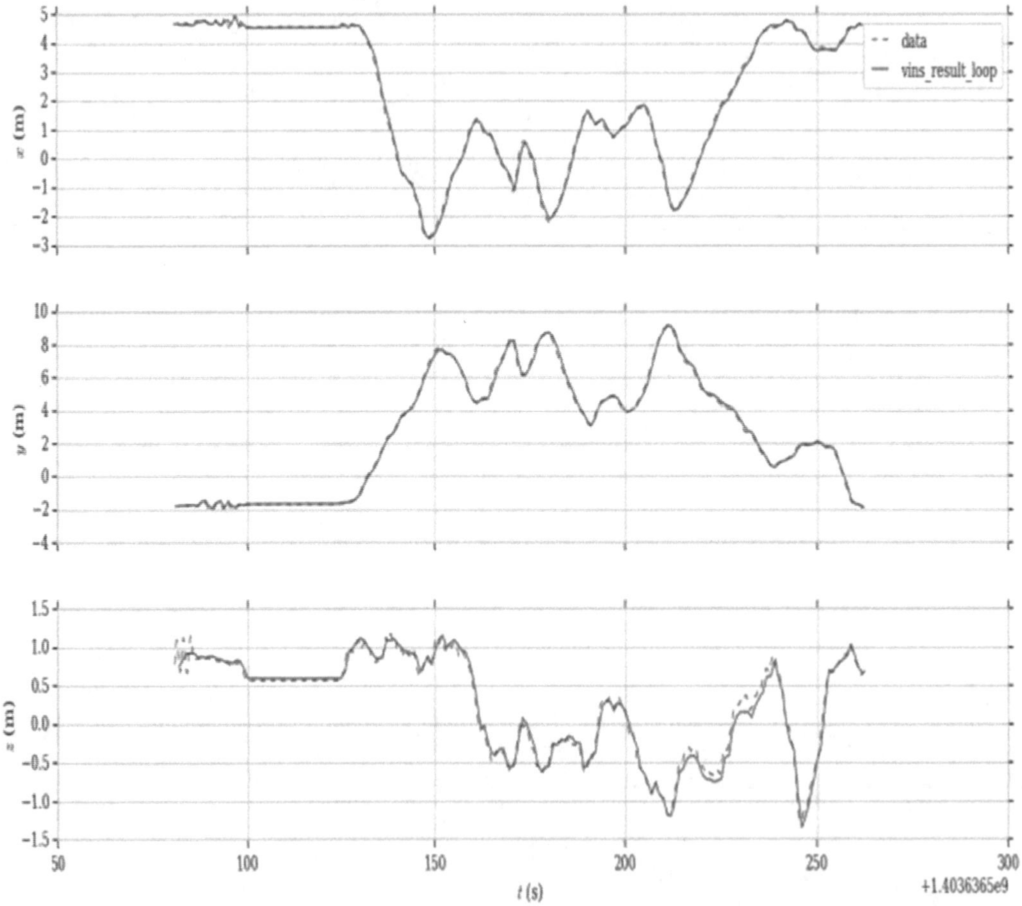


Figure 2. x_y_z three-axis coordinate diagram

- By changing the setting method of the camera external parameters, three modes are set in the simulation program in this paper, and the error range between the estimated path and the real path can be observed and compared, so that the back-end optimization capability of the algorithm can be intuitively understood. Modify the flag bit estimate_extrinsic in the program to 2, and change the camera external parameter calibration method to real-time calculation and update the estimated value, and the comparison between the generated trajectory map and the before modification is shown in Figure 3.:

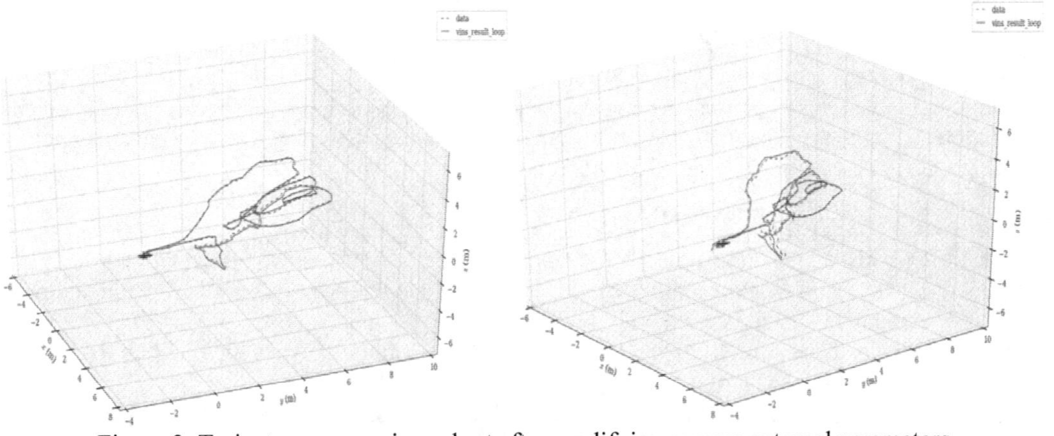


Figure 3. Trajectory comparison chart after modifying camera external parameters

- Loop detection is used for back-end optimization, and then loop_ The closure is changed to 0, and the angle value obtained after removing the loop relocation at the back end of the system is shown in Figure 4[9]. It can be seen that there is little difference between whether there is a loop in the conventional environment.

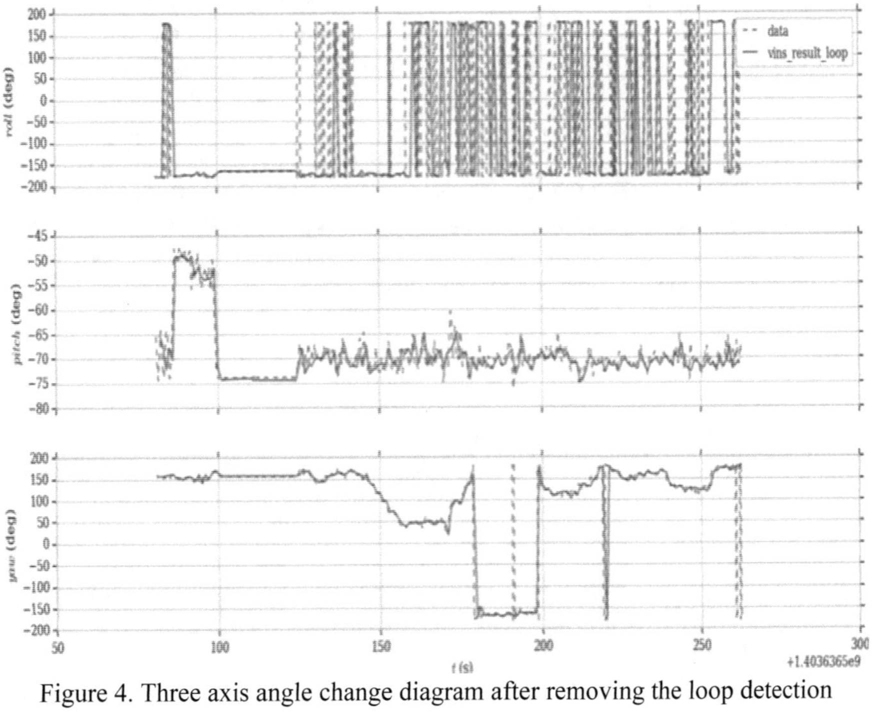


Figure 4. Three axis angle change diagram after removing the loop detection

## 4. Conclusions

This paper mainly designs and builds a complete monocular positioning and mapping system by researching and improving the algorithm of the monocular camera and IMU fusion navigation system, using the public data set to test, the front-end visual odometer can use loose The coupling method performs a preliminary processing on the data values collected by the sensors in different environments. After estimating the scale factor, the initialization is completed through the algorithm process, and the back-end performs nonlinear optimization. The use of four iteration equations can reduce the real-time error value to At last, the global optimization of the pose graph is performed to obtain a robust localization graph.

**Acknowledgments**
This article is one of the phased achievements of the general project "National Natural Science Foundation of China" (61863023) and "Natural Science Foundation of Gansu Province" (21JR1RA235) of the National Social Science Fund.

**References**

[1] He,Y.,Sun,W.,Huang,H.(2020)Pvn3d: A deep point-wise 3d key points voting network for 6dof pose estimation. In: Proceedings of the IEEE/CVF conference on computer vision and pattern recognition. Virtual Online. 11632-11641.
[2] Bian, J.W., Lin, W.Y., Liu, Y.(2020) GMS: Grid -Based Motion Statistics for Fast,Ultra-robust Feature. International Journal of Computer Vision,128:1580-1593.
[3] Li, D., Shi, X., Long, Q., (2020) DXSLAM: A robust and efficient visual SLAM system with deep features. arXiv preprint, arXiv:2008. 05416.
[4] Yang, N, Stumberg, L.V., Wang, R., (2020) D3vo: Deep depth, deep pose and deep uncertainty for monocular visual odometry. In: Proceedings of the IEEE/CVF Conference on Computer Vision and Pattern Recognition. Virtual Online,1281-1292.
[5] Liu, Q., Qin, S., Yu, B., (2020) T-B A: Bundle Adjustment Hardware Accelerator Based on Distribution of 3D-Point Observations. IEEE Transactions on Computers. 69(7): 1083- 1095.
[6] Chen, Y., Zhao,L., Lee, K.M., (2020) Broadcast your weaknesses: cooperative active pose-graph SLAM for multiple robots. IEEE Robotics and Automation Letters. 5(2):2200-2207.
[7] Ortiz, J., Pupilli, M., Leutenegger, S., (2020) Bundle adjustment on a graph processor. In: Proceedings of the IEEE/CVF Conference on Computer Vision and Pattern Recognition. Virtual Online, 2416-2425.
[8] Zhang, T., Zhang, H., Li, Y., (2020) Flowfusion: Dynamic dense rgb-d slam based on optical flow. In: IEEE Inter national Conference on Robotics and Automation. Paris. 7322-7328.
[9] Engel, J., Koltun, V., Cremers, D., (2017) Direct sparse odometry, IEEE transactions on pattern analysis and machine intelligence. 40(3): 611-625..

# A mission planning method for sunlight avoidance period of GEO optical remote sensing satellite

**Xiaoxia Shi [1*], Xindong Li [1] and Yu Wang[1]**

[1]Beijing Institute of Control Engineering, Beijing, CHINA

*Email： shixx331@163.com

**Abstract:** GEO optical remote sensing satellites need to avoid sunlight entering the camera optical system at midnight hours. For the existing evasion methods such as thermal doors and posture planning, nearly 15% of the satellite in-orbit life cannot be reasonably utilized due to the passive evasion methods. To solve this problem, a method based on mission planning is proposed. When evasion begins, the satellite changes the orbital coordinate system into the inertial coordinate system and maintains the safe attitude in inertial space. If space missions are planned such as camera calibrations and celestial bodies observations, the algorithm can determine the compliance of the missions with attitude constraints and energy constraints, and the eligible missions are executed. When evasion ends, the satellite maneuvers toward the Earth proceeding with Earth observation missions. The mathematical simulation indicates the method is feasible, ensuring the safety of the optical camera and executing the space missions meanwhile. The mission planning method extends relatively the service life of the satellite and makes the expansion of the space observation field possible.

## 1. Introduction

Due to GEO orbital characteristics, GEO optical remote sensing satellites can track the observed object continuously and response the mission command secondly. Especially in recent years, the payload resolution and specifications have increased by orders of magnitude, and the superiority of GEO optical remote sensing satellites in the field of earth observation becomes more and more obvious.

The payloads of GEO optical remote sensing satellites are optical cameras. Near midnight the camera is exposed to the sunlight. If the inner of the lens hood or the optical system is exposed, the performance and reliability of the camera are affected and seriously lead to failure of the whole satellite mission. Because of this problem, GEO optical remote sensing satellites have taken corresponding measures to avoid sunlight invasion. Based on personalized characteristics of the design for camera hoods and torque output capability of the control system, methods of the satellite evasion are diverse, for instance, thermal door protection, an attitude maneuver back to the sun, posture planning around the rolling axis, posture planning around the pitch/yaw axis.

Using thermal doors on lens hoods, the field of view of cameras are blocked and cameras cannot work. GEO-Africa satellite has maneuvered the attitude for 180 degrees and implements no more mission-related work. Algorithms by posture planning around the rolling or pitch/yaw axis do not consider the necessary requirements such as platform maintenance and payload calibrations. Included angles between the sunlight and the optic axis along the maneuver path are small, and the evasion path is set in advance, so the satellite cannot change the attitude according to payload demands. In addition, because the satellite has been maneuvering during the evasion period, the attitude stability cannot meet the calibration requirements. Although the above methods can complete the function of avoiding sunlight, one thing they have in common is that they can only achieve the most basic evasion protection. In the period, working loads of the satellite lay dormant, so nearly 15% of the satellite in-orbit life cannot be reasonably utilized.

In order to make full use of the evasion period, this paper improves the current sunlight avoidance method. For the requirements of on-board subsystem maintenance and payload services, the paper analyzes the constraints of sunlight evasion, and proposes a sunlight planning method for evasion period, ensuring the completion of space missions outside the earth which are meeting the constraints and improving the efficiency of the satellite in orbit.

## 2. Constraints Analysis

### *2.1. Space Constraints*

Set $\alpha_{SV}$ as the sun avoidance angle, which is determined by the design of the camera system。Define $\alpha$ as included angle between the satellite +Z-axis and the projection vector of the solar vector $S_O$ in XOZ coordinate plane of the satellite body coordinate system. The angle ($\alpha$) changes as the satellite moves in orbit. Since 0:00 midnight, angle $\alpha$ increases gradually from 0°. By 12:00 noon, angle $\alpha$ decreases gradually from 180°. If α is less than $\alpha_{SV}$, the satellite enters the sunlight avoidance area; otherwise, the satellite exits the sunlight avoidance area. The space constraints of the sunlight avoidance area as follows:

$$S_b = C_{bo} * S_o \tag{1}$$

$$\alpha = tan^{-1} \ (S_{bx} / S_{bz}) \tag{2}$$

Starting point of the avoidance period: $\alpha \leq \alpha_{SV}$
Ending point of the avoidance period: $\alpha \geq \alpha_{SV}$

### *2.2. Attitude Constraints*

The satellite runs orbiting to the starting point of the evasion period, ending the earth observation and turning into the sunlight avoidance mode. At this moment, the satellite attitude is critical but safe. Take this inertial attitude quaternion *q0* as the benchmark for the next attitude control. Take the geocentric equatorial inertial coordination as the reference coordinate system. The Euler parameters are as follows:

$$\text{Euler axis:}\{0,1,0\},\ \text{Euler angle:}\ \omega$$

The Euler angle $\omega$ represents the range of attitude maneuvers performed in inertial space during evasion, which is related to the design of the lens hood and the optical system of the camera. According to current satellite design status, the lens hoods can be divided into two categories: flat lens hoods and oblique cutting lens hoods. For two different types of lens hoods, the attitude constraints are also different.

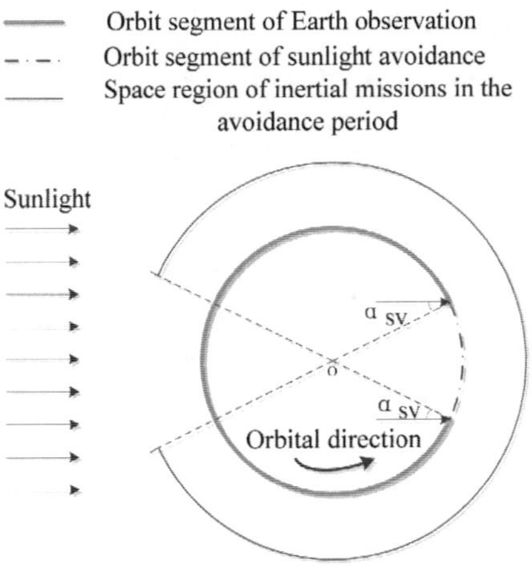


Figure 1 Attitude constraints of the flat lens hood

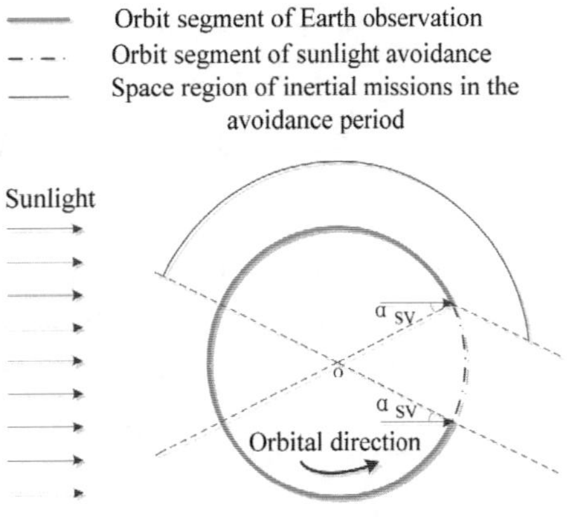


Figure 2 Attitude constraints of the oblique cutting lens hood

The value range of the Euler angle are as follows:

For flat lens hoods: $\omega \in (0, \pi-2\alpha_{SV})$

For oblique cutting lens hoods: $\omega \in 0, -\pi)$

### 2.3. Energy Constraints

In recent years, the technical level of satellites payloads has developed rapidly. The satellite index of ultra-high precision leads to more energy needs, especially from thermal system and payloads. Considering overall factors such as technology and cost, the influence of satellite attitude on the solar array should cause enough attention. Especially for GEO optical remote sensing satellites, the sunlight avoidance lasts for about 3.5 hours with large angular offset, and the energy efficiency of the solar array decreases seriously. The sunlight avoidance method around the rolling axis becomes unacceptable because energy damping in the evasion.

According to the requirement of the satellite power and the design of power system, the energy constraint during the sunlight avoidance can be proposed based on the analysis. Set $\beta$ as the allowed output threshold of the solar array. The Euler parameters are as follows:

Euler axis:{1,0,0}, Euler angle: $|\omega^{'}| \leq cos^{-1}(\beta)$

Rolling maneuver parameters of the satellite must conform to the above constraints for the power requirement.

### 2.4. TM/TC Constraints

Inertial missions during the avoidance period are executed automatically by satellite subsystems without interaction between the satellite and the ground. TM/TC Constraints can be ignored during the period.

## 3. Mission Design

The satellite performs missions of earth observation in orbital coordinate system. On the starting point of the avoidance period, the on-board system changes the orbital coordinate system into the inertial coordinate system and marks the initial attitude *q0*.

For the requirements of the on-board subsystem maintenance and payload services, the control system executes the missions in sequence and controls the attitude in a series of inertial space targets $q_{ti}$ (i=1~N).

Before missions, the control system gets the maneuver parameters:

$$q_m = q_0^{-1} * q_{ti} \quad (3)$$

According to the attitude constraints, the range of pitch maneuvers is determined as follows:

$$\theta = tan^{-1}\ (-L_{13}(q_m)/L_{33}(q_m)) \tag{4}$$

According to the energy constraints, the range of rolling maneuvers is determined as follows:

$$\varphi = sin^{-1}(L_{23}(q_m)) \tag{5}$$

If the constraint is touched, this mission would be refused and the next mission is executed. The process is shown in the following figure.

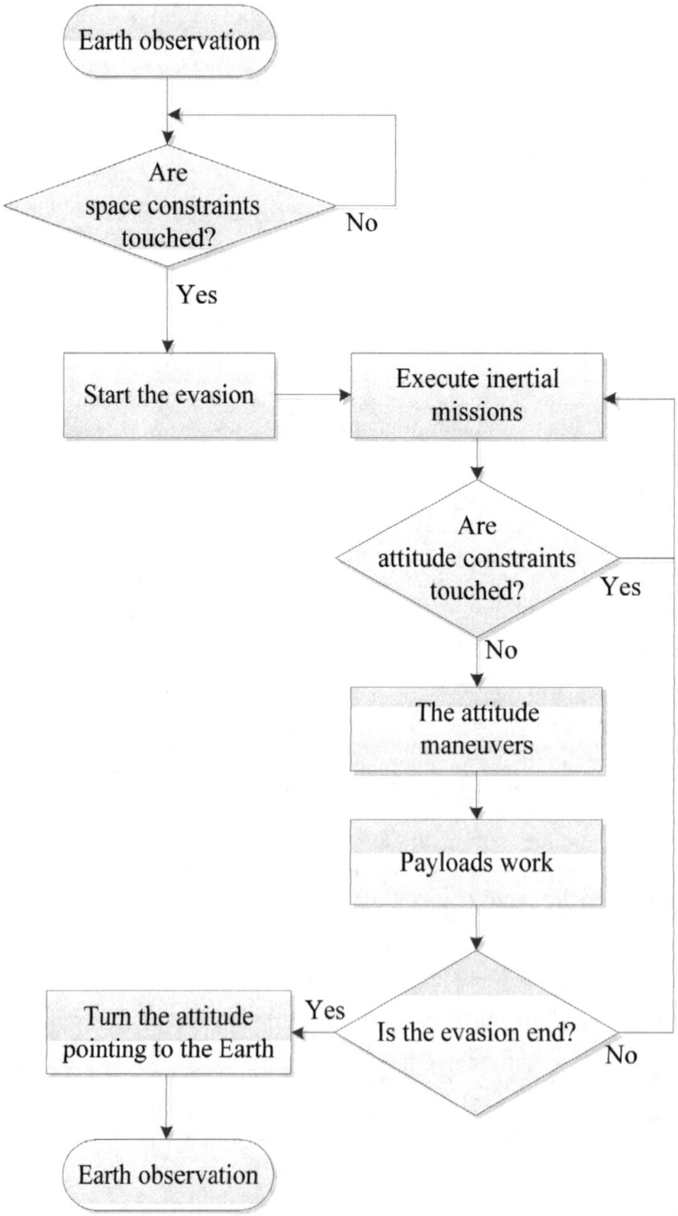


Figure 3 Mission process

## 4. Results

Using some satellite mathematical simulation platform, the mission algorithm during the sunlight avoidance is operated. The architecture of the simulation system is shown in the figure below.

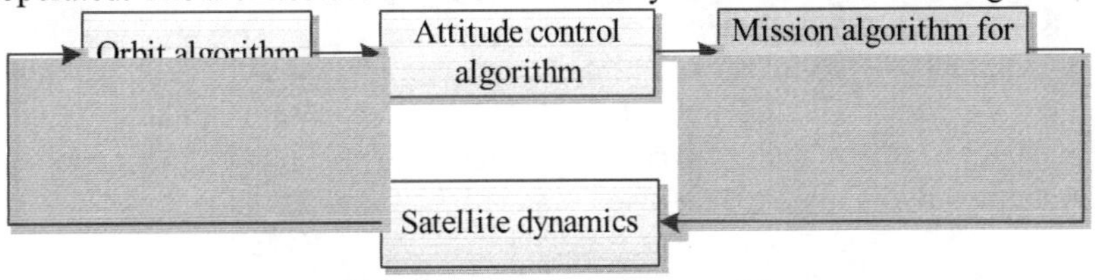


Figure 4 The simulation system architecture

The orbit parameters in the simulation: 2022-06-20 22:30 to 2022-06-21 1:30.

Simulation step: 1s.

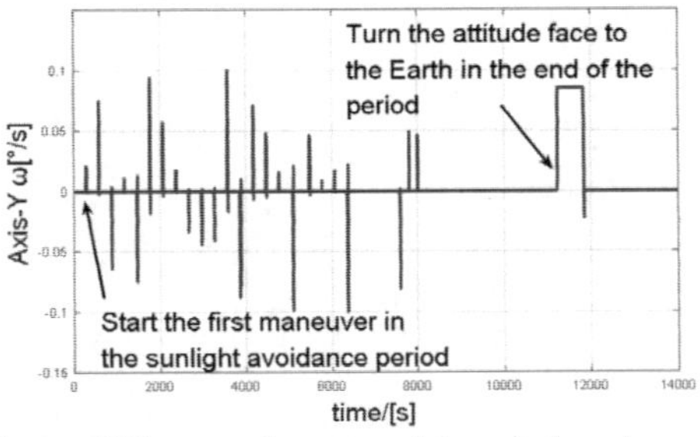


Figure 5 The angular rate of the pitch axis in the sunlight avoidance period

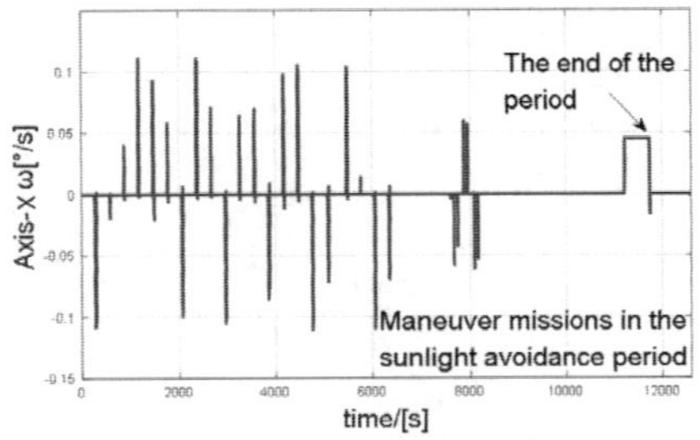


Figure 6 The angular rate of the roll axis in the sunlight avoidance period

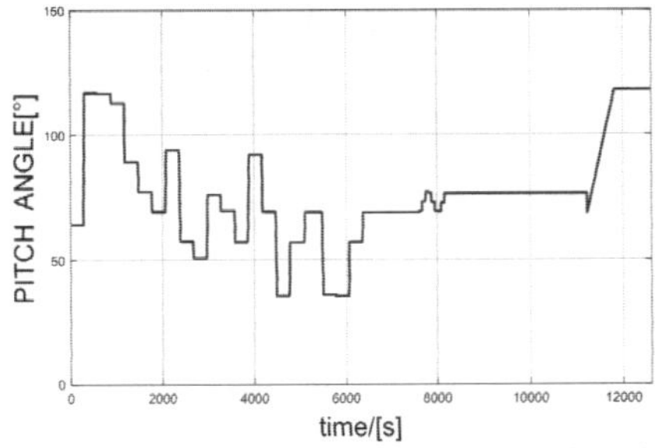


Figure 7 The angular rate of the pitch axis in the sunlight avoidance period

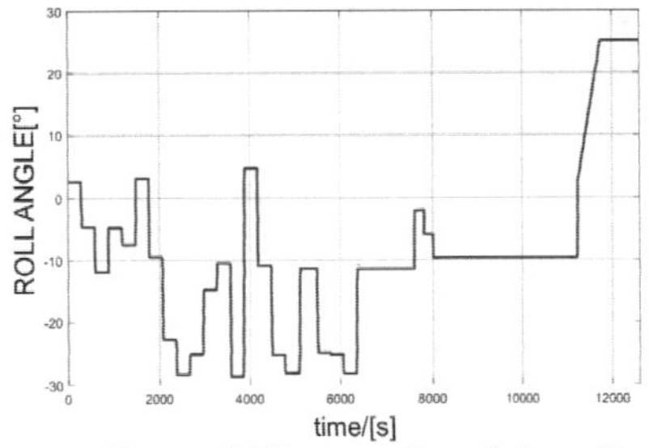


Figure 8 The angular of the roll axis in the sunlight avoidance period

In the whole avoidance period, included angle between the satellite +Z-axis and the projection vector of the solar vector $S_O$ in XOZ coordinate plane of the satellite body coordinate system is below.

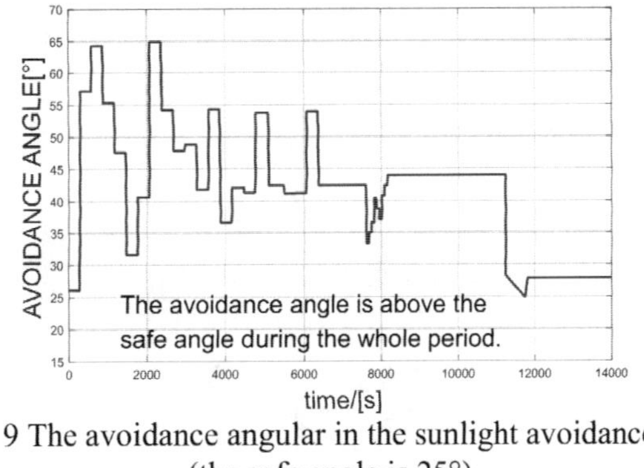


Figure 9 The avoidance angular in the sunlight avoidance period (the safe angle is 25°)

## 5. Conclusions

This paper compares the advantages and disadvantages of various avoidance methods. In the view of the fact that the current evasion method cannot carry out payload calibrations or other space observation missions, the paper proposes a mission planning method for sunlight avoidance period of GEO optical remote sensing satellite, analyzing space constraints, attitude constraints and energy constraints. The mathematical simulation indicates the method is feasible and the attitude curves are in line with expectations. This method provides a reference for the design of the sunlight avoidance period of GEO optical remote sensing satellite.

## References

[1] Z Peng, Z Li, G Qiao, X Liu. (2015) Sunlight avoidance design for camera of geostationary orbit remote sensing satellite. Chinese space science and technology, Co.35(2):57-62.
[2] Y Liu, F Liu, L Yu. (2014) Sunlight invasion avoidance method of GEO optical remote sensing satellite. Spacecraft engineering, Vol.23(6):24-29.
[3] D Gallagher, J Bergstrom, J Day, B Martin, M Tommeraasen. (2005) Overview of the optical design and performance of the high-resolution science imaging experiment (HIRISE). SPIE, 5874:1-10.
[4] X Cheng, H Cui, R Xu. (2012) Attitude maneuver control of spacecraft under geometric constraints[J]. Control and decision, Vol.27(5):724-730.
[5] Q. Dou, J. Li. (2013) Research on thermal door of agile remote sensing satellite camera. Spacecraft engineering, Vol.22(2):55-59.
[6] G Li, X Liu. (2021) High-precision and high-stability control technology for geostationary earth orbit optical telemetry satellite. Aerospace control and application, Vol.47(5):24-32.
[7] S Liu, B Liu. (2016) Impact of midnight solar intrusion on temperature of satellite-borne laser communication antenna and design of sunlight avoidance. Chinese journal of lasers, Vol.43(11): 197-203.

# Habitability of Exoplanets around Black Hole

**Zhe Hu[1,a†], Qingyao Jiang[2,*, †]Miao Li[3,b†]**

[1]Zhengzhou Foreign Language School, Henan, China

[2]West Buckland School, North Devon, United Kingdom:

[3]Department of physics, Inner Mongolia Normal University, Hohhot, China

[a]1432887464@qq.com; [*] 23jiangc@westbuckland.com; [b]2276356925@qq.com

[†]These authors contributed equally.

**Abstract**—Contemporarily, investigation for habitability of exoplanets around black hole remains a hot topic. In this paper, the state-of-art experimental results relevant to the topic as well as the basic principles will be summarized. To be specific, basic information about black holes will be introduced first, including the 4 categories and detection method. Based on several observations for the effect of black hole towards near-by planets, this paper mainly focuses on the effect of black hole radiation and infrared echo on the habitability of exoplanets and the subsequent impact from radioactive power on biological features. These result shed light on guiding further study about relationship between exoplanets and black holes.

## 1. Introduction

Generally, black hole is a predicted dense dark matter, which is verified by various astrophysics observations. It can be divided into some different types, and properties of these types of black holes are also very different, which can significantly affect the habitability around the black holes. Firstly, habitability around black holes can be determined through the black holes mass because properties of black holes can be classified by their mass. Secondly, there are three types of black holes in the mass classification method: Solar mass black holes, intermediate-mass black holes, and supermassive black holes [1]. Finally, the black holes which the exoplanets orbit around must be massive enough so that the planet can have a fit orbit and will not fall into the event horizon. There is another reason why the black holes must be "supermassive enough": the tidal effect, the radius at which tidal disruption will occur. If we take an exoplanet at earth's mass and radius into consideration, the mass of the central black hole has to be approximately M～1.63×108 M [2]. As a result, only the exoplanet must orbit around a large supermassive black hole to have the probability of being habitable.

The motivation for researching this subject is because the science fiction movie interstellar. Finding a habitable planet has become a popular and active areas of research in the exoplanet field nowadays. When we guess that if earth were orbiting a black hole or supermassive black holes instead of Sun. Whether we can continue survive becomes an interesting question. Whether we can find a habitable planet is also a challenge. Therefore, a specific way has been found that energy of the black hole can be extracted through the magnetic layers of Kerr black holes by building a special machine: The dual unipolar batteries [3]. As a result, spinning blackholes can provide the life on the exoplanets around black holes with extra energy to use especially when their planets go through the egrosphere [4].

The rest part of the paper is organized as follow. The section II will introduce the basic description of black hole including category (characteristics and differences) and detection (the gravitation waves produced by collision and the X-rays produced by the material). The section III will explain detailly on radiation effects which is divided into three parts (atmospheric loss, limitation and results, biological consequence). The section IV will research infrared echo from tidal disruption event. It is mainly through an experimental data and experimental phenomenon to explore. The limitations and future prospects for current studies are demonstrated in section V. Eventually, a brief summary is given in Sec. VI.

## 2. Basic description of black hole

### *2.1. Category of black hole*

In consideration of the mass of black holes, they can be categorized into 4 types, including the supermassive black hole (SMBH), intermediate mass black hole, stellar mass black hole and mini black hole. In consideration of other specific properties of black holes like charges and spin, another category method can be chosen.

The first type is Schwarzschild black hole with no charge or spin. Its center is regarded as a singularity surrounded by event horizon (Schwarzschild radius). The idea of Schwarzschild radius was first raised up by Karl Schwarzschild. By deviating from the escape velocity of a celestial body, the Schwarzschild radius can be calculated as follow:

$$R = \frac{2MG}{v^2} \quad (1)$$

where G stands for the universal gravitational constant, M is the celestial mass and v represents the speed of object escaping from the black hole. The second type of black hole is the Reissner-Nordström (R-N black hole) or charged black hole with 2 event horizons. It can be considered as the wormhole of charged black hole because of the fact that this property can be taken advantaged of for the "traversing" in universe. The third type of black hole is Kerr black hole with momentum and spin. It is considered as the black hole symmetrical about its axis and keeps rotating around the axis in a motion not varying with time. At the center of this type of black hole is an odd ring, with an inner and outer event horizon. The inner horizon is the limit of singularity of black hole, while the outer horizon is the limit of inescapability. The final type is suggested to be the Kerr-Newman black hole. It has mass, electromagnetic property and angular momentum. When Q=J=0, it can be reduced to the simplest Schwarzschild black hole. Kerr-Newman black hole is solved by Kerr-Newman metric to the Einstein-Maxwell equation of general relativity.

### *2.2. Detection technique*

To detect the black holes, the detectors must have ultra-high resolution to recognise the gravitational wave or other minor signals generated from the universe. On this basis, it usually has large scale for enhancing the signal-noise ratio (e.g., the EHT) [5]. The scientific goal of the EHT array is to observe SMBHs with spatial and temporal resolution sufficient for imaging [6]. The most well-known results are that a black hole surrounded by an optically thin bright plasma should have a 'shadow' morphology, with a dim core region outlined by the lensed photon orbit [7]. The size and shape of the shape of the resulting shadow are mostly determined by the mass of the black hole, with its spin and observer orientation playing just a minor role.

The principal EHT targets are the luminous radio core and Sgr A*, as the estimated mass and vicinity ensures the two most promising sources for examining SMBH with abundant resolution [8]. VIBL observations at a wavelength of 1.3mm have proven the existence of structure on the scale of the shadow in both the nucleus of M87 and Sgr A*. M87 is predicted to be home to a black hole with 3.3-6.2$\times 10^9$ solar mass ($M_\odot$), when scaled to a 16.8 Mpc distance [9]. With 1.3 mm VLBI, structure in M87 measuring 40 as was resolved [10]. The 1.3 mm emission from Sgr A*, a black hole candidate at the Galactic centre with a presumed mass of $4\times 10^6$ has been measured with 3.7 $R_{s.}$ [11]. Full polarimetric VLBI studies at 1.3 mm wavelength have recently revealed ordered and time-variable magnetic fields on horizon sizes within Sgr A* [12].

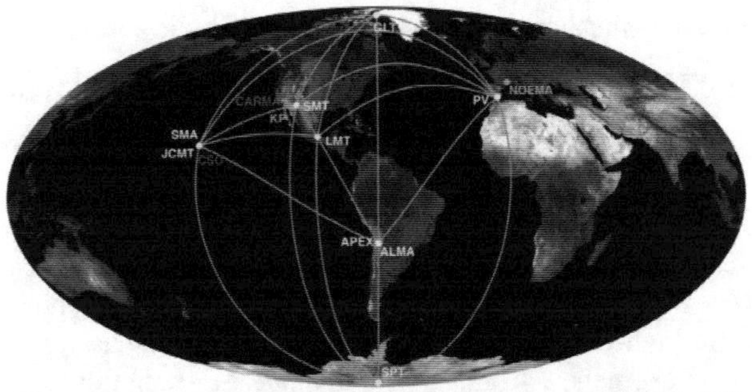


Figure 1. A sketch of the distributions for EHT. [6]

The first detections of horizon-scale structure are achieved attributed to the advanced new digital backend and the state-of-art recoding hardware, which was realized for mm-wavelength observations and had a recording rate of 4 gigabits per second1 (Gbps) [11]. A map of the EHT array is shown in Figure 1. The EHT used an array of eight telescopes to conduct global observations in April of 2017. The collecting area of 37 12 m diameter ALMA dishes was electronically combined by a custom-built system. These observing campaigns yielded preliminary scientific findings that prompted the development of an imaging system. Technology maturation was primarily focused on increasing the bandwidth of observations by increasing digitization and recording rates, which was fuelled by industry-driven trends. The development of phased array technologies in tandem with investments in site-specific infrastructure has resulted in a global EHT array capable of achieving its science objectives.

## 3. Radiation effects

Black holes can affect habitability of near-by planets in numerous aspects, amongst which radiation becomes one of the dominating factors. In the following session, radioactive effect on the planetary environment will be covered, leading to a subsequent catastrophic event referred to as atmospheric loss. The possibility of planetary formation in accretion disks has been already confirmed in the unified model of AGNs [13], in which thick torus made up of gas and dust can become sites for planetary formation. Further, terrestrial planets as long as Neptune-like icy giant planets are possible products in such areas [14]. Targeted at the central Supermassive Black Hole (SMBH) hosted in our Milky Way, known as Sagittarius A*(Sgr A*), existence of some planets around it is also not excluded. At 10 to 30 light-years from the nucleus, fatal effects from strong gravitational fields and tidal forces are alleviated to relatively safe extend [15]. In addition, the thermodynamic models of exoplanets due to the interaction of SMBH and cosmic microwave background (CMB) in such orbits supports the occurrence of carbon-based organisms [16].

In requirements of the growth of black holes, they undergo the accretion of gaseous emission. Several past analyses have already raised up hypothesis that photo-evaporation effect caused by XUV photons may be subjected to ranges of planetary outcomes, such as the primordial characteristics of Kepler planets [17]. Recent studies suggest that activity of SMBH can erode the planetary hydrogen–helium atmospheres and extend of the effect is mostly contributed by SMBH in its quasar phase for the reason that at that period the XUV radiation reaches its maximum value [18].

Several models have been created to study the evolution of Sgr A* and the effect of XUV flux on the near-by exoplanets in structure. Based on setup of previous models [19, 20], MESA planet model is constructed to target [21]. Considering exoplanets in orbital of stars in solar scale of properties life mass and metallicity, the calculated energy-related escape is formed as follow.

$$\dot{M} = -\frac{\epsilon_{XUV}\pi F_{XUV} R_{XUV}^3}{G M_p K_{tidal}} \tag{2}$$

where $\varepsilon_{XUV}=0.1$ is the mass-loss efficiency parameter, $F_{XUV}$ is the extreme ultraviolet energy flux impinging on the planet atmosphere, Mp is the mass of the planet, $R_{XUV}$ is the distance from the center of the planet to the point where the atmosphere is optically thick, the coefficient Ktidal corrects for tidal forces. Assuming an average luminosity one derives:

$$M_{lost} = \frac{3\epsilon}{12\pi} \frac{L_{XUV} \Delta t}{G \rho_p D^2} e^{-\tau} \tag{3}$$

In general, despite of the fact that the accretion of material from the supermassive blackhole is continuous, quasar phase still dominates in its contribution, playing a significant role in the growth of supermassive blackhole. Additionally, the radioactive effect on planets around also depends on the XUV flux. As a consequence, the high-energy radiation due to Sgr A* in its quasar phase can impart significant heat to the atmosphere of exoplanets from its vicinity and the radiation received due to the quasar dominates as well [22]. Typical simulation results are illustrated in Figure 2. Results can be inferred from the two figures in order to illustrate some implications for the final impact on habitability. According to Figure 2, the results can be significantly decided by Rgc. Changing the value from equal or less than 20 pc, the resulted planetary condition changes from the habitable evaporated cores (HECs) [22] to a complete evaporation by star-planet separations of 1.0 au. For those at galactocentric distances Rgc equal or less than 100 pc, the effect of high stellar XUV will last for a short period time (~1000 yrs).

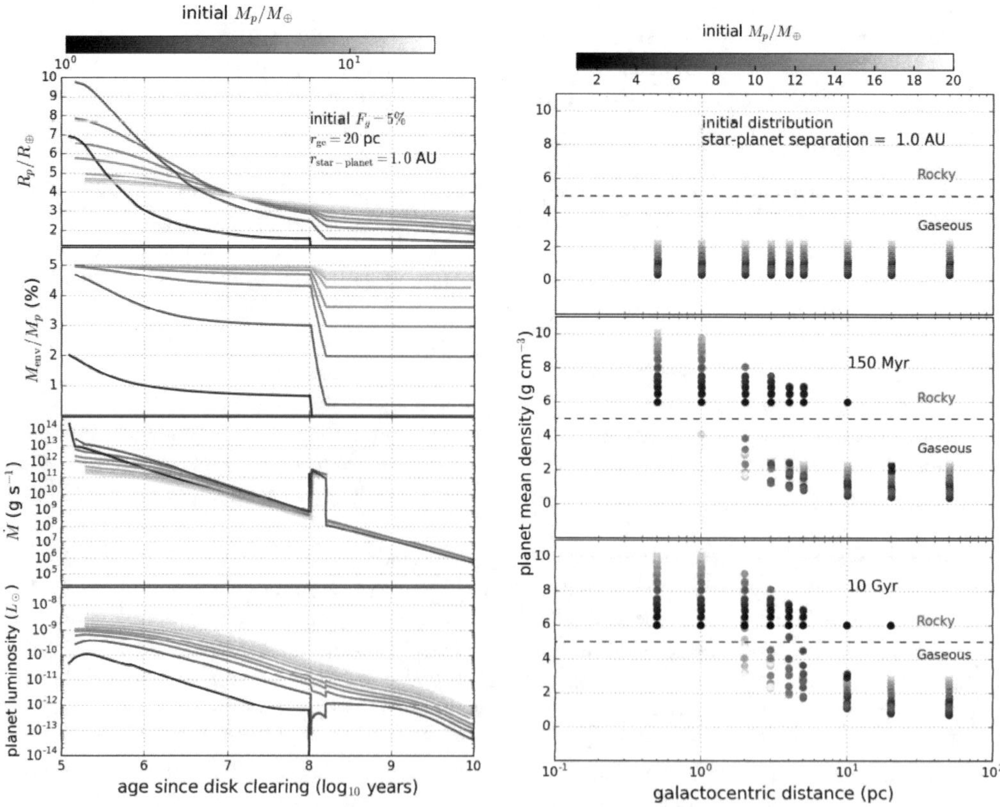


Figure 2. Typical simulation results [22].

To further discuss the habitability of HECs and initially-terrestrial planets, although the atmospheric mass loss caused be further XUV radiation of the central Sgr A* is still worth-noticing, several caveats still must be taken into account. First, considering energy-limited escape equation, which is the method used to calculate this resulted value, some uncertainty may exit for it becomes unsure whether the equation fits well to nitrogen-dominated atmosphere. Moreover, whether the method can adapt in atmosphere with higher mean-molecular weights still need further confirmations [23, 24].

The most luminous objects in the Universe are Active Galactic Nuclei (AGNs), which form when matter accretes onto supermassive black holes (SMBHs) and can emit electromagnetic radiation at rates close to the maximum equilibrium value, known as the Eddington luminosity. X-rays and ultraviolet photons account for a significant portion of the radiation emitted by AGNs [25]. This research also suggested that potential extra-terrestrial species could be harmed by XUV radiation at sub-kpc sizes. AGNs could have a positive impact at distances of 20 pc by stripping sub-Neptunes of their thick hydrogen-helium atmospheres, transforming them into rocky super-Earths [26]. AGNs have twofold impacts on the habitability. In this section, we mainly focus on some of the advantages of AGNs (during the quasar phase) to produce primordial substances. Furthermore, we will discuss some adverse impacts of AGNs on biota.

It has been suggested that the essential distance across which AGNs can promote prebiotic chemistry [27], and then we show the timescales over which these events may occur. The bolometric luminosity (L) of AGNs is modelled as Eq. (4), where $M_{BH}$ denotes the mass of the SMBH and $\varepsilon_{Edd}$ is the Eddington ratio:

$$L \approx 1.3 \times 10^{38}\, erg\, s^{-1}\, \epsilon_{Edd} \left(\frac{M_{BH}}{M_{\odot}}\right) \tag{4}$$

After simplification, we arrive to the following:

$$d_O \approx 2.2 \times 10^{-2}\, pc\, \epsilon_{Edd}^{\frac{1}{2}} \sqrt{\left(\frac{M_{BH}}{M_{\odot}}\right)} \tag{5}$$

Based on the mass of Sagittarius A* [28], we obtain $d_O \approx 44$ pc, if $\varepsilon_{Edd} \approx 1$ and $M_{BH} \approx 4\times10^6\, M_{\odot}$. The Salpeter time, appears to be a good approximation of the typical lifetime in theoretical models

$$t_s \approx 4.5 \times 10^7 yrs \left(\frac{1}{\epsilon_{EDD}(1-\epsilon_{BH})}\right)\left(\frac{\epsilon_{BH}}{0.1}\right) \tag{6}$$

It's hardly unexpected that creatures evolved photosynthesis early in Earth's evolutionary history because light is a plentiful source of energy [22]. Given the abundance of electromagnetic radiation emitted by AGNs, it's only natural to wonder if they can power photosynthetic processes. This type of radiation is referred to as photosynthetically active radiation (PAR) The "photosynthesis zone" is defined by the cut-off distance (dP) [22]

$$\Psi_{PAR} \approx \frac{\eta_{PAR} L}{4\pi d^2} \tag{7}$$

$$d_P \approx 0.17\, pc\, \epsilon_{EDD}^{\frac{1}{2}} \sqrt{\frac{M_{BH}}{M_{\odot}}} \tag{8}$$

We get $dP \approx 0.34$ kpc, by substituting the Milky Way parameters. This distance is considered "large" as it roughly equals the scale height of our Galaxy's thin disc. We show the critical distance at which UV radiation can harm Earth-like ecosystems, as well as the effects of X-rays and gamma rays. Because of the higher energy per photon, the shorter the photon wavelength, the greater the harmful impact on biota [30]. UV radiation has several negative effects, including the suppression of photosynthesis and the destruction of DNA and other macromolecules. The susceptibility of biota to UV fluxes from AGN, on the other hand, is determined by the organisms' physiology. Following that, we consider worlds with biospheres and atmospheres similar to our own. One can calculate the distance (dB) at which the threshold is satisfied:

$$d_B \approx 9.6 \times 10^{-4} pc\, \epsilon_{EDD}^{\frac{1}{2}} \sqrt{\frac{M_{BH}}{M_{\odot}}} \tag{9}$$

As a result, when we apply this formula to the Milky Way, one obtains dB ≈ 2 pc. It is implied that the dB level will be raised to

$$d_B \approx 1.5 \times 10^{-2} pc\, \epsilon_{EDD}^{\frac{1}{2}} \sqrt{\frac{M_{BH}}{M_\odot}} \tag{10}$$

Using Eq. (10) to calculate dB 30 pc for the Milky Way, we get dB 30 pc. Regardless of the host star's spectral classification, we can see that this distance is much smaller than the Milky Way's scale length and height. As previously stated, UV radiation from AGNs, even at d dB, will not always cause extinction-level event. As shown in Figure 3, the "zone" where negative effects become dominant appears to be smaller than the corresponding zones for powering prebiotic chemistry and photosynthesis; this can also be verified by comparing above equations [27].

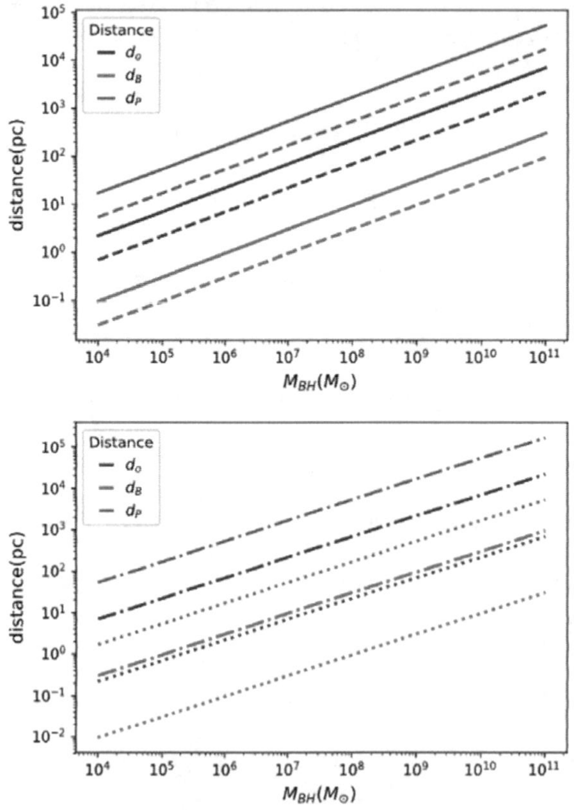


Figure 3. Distance as a function of the mass of the SMBH [27].

## 4. Infrared echo from tidal disruption event

Contemporarily, there have been several studies using TDE to trace the interstellar medium around quiet galaxies or weakly active galactic nuclei. ASAS-SN found a new flare ASAAAN-14Li. The flash occurs no more 0.04 from the center of the galaxy. It suggests that it was probably created by a black hole tearing apart a star at the galaxy's center. After a series of confirmations, asassn-14Li was found to be the closest TDE to us. It has obvious variable signal light and it's the first confirmed example of infrared reverberation from dust around TDE. Using NEOWISE-R data, infrared emission of ASASSN-14Li was investigated. WISE conducted an imaging photometric survey of the entire sky simultaneously in four wavelengths (wavelengths of 3.4, 4.6, 12, 22 um, labelled W1, W2, W3, and W4) from February to August 2010. Besides, W3 and W4 use solid hydrogen as coolant. Data from ALLWISE and

NeowISe-R were collected for all periods before and after the WISE satellite coolant depletion.obtaining the data that shown in it [31].

First, in each period, no short time changes were detected. There is no significant change in the first four periods. It can be thought of as being in a calm state and its infrared flow come from the host galaxy. But in the fifth period, it has obvious enhancement. Exactly it is 36 days after ASASSN-14Li is found. The period after the tidal rip flare burst is divided into E1 and E2. In E1, the infrared flux is brightened by 0.12and 0.16 magnitude in the W1 and W2 bands [31].

The enhanced infrared emission signal, closely associated with ASassn-Li14, can explain the re-emission of dust by absorption of ultraviolet photons produced by tidal rip flares. Because the distance from the black hole is farther than that from the ultraviolet radiation region, the infrared photo variation signal is delayed relative to ultraviolet signal. Assuming the mass of the black hole at the center of ASASSN-14Li, the distance of dust from the black hole center can be roughly estimated according to the time of infrared signal delay to ultraviolet photometry (~110).

It is assumed that the dust satisfies the MRN particle size distribution. Given a dust particle with radius A and distance R from the central heating source, its own thermal radiation is in balance with the radiation from ultraviolet/optical photons absorbed by the dust particle to heat it. Thus, the total thermal luminosity in uv/optical band can be calculated (it is simple black-body radiation).

$$\frac{L_{bol}}{4\pi R^2}\pi a^2 = 4\pi a^2 \sigma T^4 \tag{11}$$

Considering dust absorption efficiency, dust temperature can be fitted, and it can be calculated by the following formular [32]:

$$\frac{L_{bol}}{16\pi^2 R^2} \approx 1.47 \times 10^{-6} a T^6 \tag{12}$$

The results are related to the distribution of dust particle size. It is assumed that the dust satisfies the MRN particle size distribution. Then Lbol=$7.8\times10^{42}$ erg/s. If an average particle radius A= 0.1 μm, Then Lbol=$1.0\times10^{44}$ erg/s um, the total dust mass is about 10 times higher than the simple black-body radiation hypothesis.

## 5. Limitation & future prospect

To further discuss the habitability of HECs and initially terrestrial planets, although the atmospheric mass loss caused be further XUV radiation of the central Sgr A* is still worth noticing, several caveats still must be taken into account. First, considering energy-limited escape equation, which is the method used to calculate this resulted value, some uncertainty may exit for it becomes unsure whether the equation fits well to nitrogen-dominated atmosphere. Moreover, whether the method can adapt in atmosphere with higher mean-molecular weights still need further confirmations.

## 6. Conclusion

In summary, this paper summarizes the recent progress of exploring the habitability of exoplanets around black hole. With regards of the radiation effect of black hole, we discuss the consequent atmospherical loss event leading to the deterioration of the environment of near-by exoplanets. The research chooses the Sgr A* as study target, whose growth depends on the accretion of gaseous emission including XUV photons. As a result, the emitted radioactive substances can erode the H-He atmosphere In the MESA planet model, the energy-related escape rate is formed. Consequently, it is figured out that the major influence of this effect mainly comes from the quasar phase of the SMBH. According to the observations results, evolution of exoplanets is significantly decided by Rgc. For those at galactocentric distances Rgc equal or less than 100 pc, the effect of high stellar XUV will last for a short period time (~1000 yrs). AGNs can power UV-mediated prebiotic reactions that lead to the synthesis of biomolecular building blocks, according to existing research. In addition, because they emit large amounts of visible radiation, AGNs may be used to power photosynthesis. Thus, the positive impact may outweigh the negative effects of UV radiation on biota. However, we found that at distances of ≳1 kpc, neither positive nor negative phenomena are likely to matter in the case of a hypothetical AGN in the Milky Way. UV-induced extinction-level events have a zone that extends to ~2–30 pc, but prebiotic chemistry and photosynthesis have corresponding zones of ~44 pc and ~340 pc. Overall, these results offer a guideline for further exploration of the habituality in the universe.

## References

[1] M. Mezcua, "Observational evidence for intermediate-mass black holes." International Journal of Modern Physics D, vol. 26.11, 2017, 1730021.
[2] P. Bakala, J. Dočekal, and Z. Turoňová, "Habitable zones around almost extremely spinning black holes (black sun revisited)," The Astrophysical Journal, vol. 889.1, 2020, 41.
[3] J. D. Schnittman, "Life on Miller's Planet: The Habitable Zone Around Supermassive Black Holes", arXiv preprint arXiv:1910.00940, 2019.
[4] R. D. Blandford, R. L. Znajek, "Electromagnetic extraction of energy from Kerr black holes." Monthly Notices of the Royal Astronomical Society, vol. 179.3, 1977, pp. 433-456.
[5] S. Doeleman, E. Agol, D. Backer, "Astro2010: The astronomy and astrophysics decadal survey." Science White Papers vol. 68, 2009.
[6] K. Akiyama, et al., "First M87 event horizon telescope results. II. Array and instrumentation." The Astrophysical Journal Letters, vol. 875.1, 2019, L2.
[7] H. Falcke, F. Melia, E. Agol, "Viewing the shadow of the black hole at the galactic center." The Astrophysical Journal 528.1 (1999): L13.
[8] Johannsen, Tim, et al. "Masses of nearby supermassive black holes with very long baseline interferometry." The Astrophysical Journal, vol. 758.1, 2012, 30.
[9] J. L. Walsh, et al, "The M87 black hole mass from gas-dynamical models of space telescope imaging spectrograph observations." The Astrophysical Journal vol. 770.2, 2013, 86.
[10] K. Akiyama, et al, "230 GHz VLBI Observations of M87: Event-Horizon-Scale Structure During an Enhanced Very-High-Energy-Ray State in 2012." The Astrophysical Journal, vol. 807(2), 2015, 150.
[11] R. Abuter, et al., "Detection of orbital motions near the last stable circular orbit of the massive black hole SgrA." Astronomy & Astrophysics vol. 618, 2018, L10.
[12] M. D. Johnson, et al., "Resolved magnetic-field structure and variability near the event horizon of Sagittarius A." Science, vol. 350.6265, 2015, pp. 1242-1245.
[13] R. Antonucci, "Unified models for active galactic nuclei and quasars." Annual review of astronomy and astrophysics, vol. 31.1, 1993, pp. 473-521.
[14] H. Netzer, "Revisiting the unified model of active galactic nuclei." Annual Review of Astronomy and Astrophysics, vol. 53, 2015, pp. 365-408.
[15] A. Balbi, M. Hami, A. Kovačević, "The Habitability of the Galactic Bulge." Life, vol. 10.8, 2020, 132.
[16] K. Wada, Y. Tsukamoto, E. Kokubo, "Planet formation around supermassive black holes in the active galactic nuclei." The Astrophysical Journal, vol. 886.2, 2019, pp. 107.
[17] P. Bakala, J. Dočekal, Z. Turoňová, "Habitable zones around almost extremely spinning black holes (black sun revisited)." The Astrophysical Journal, vol. 889.1, 2020, 41.
[18] L. Zeng, S. B. Jacobsen, and D. D. Sasselov, "Exoplanet Radius Gap Dependence on Host Star Type." arXiv preprint arXiv:1712.05458, 2017.
[19] J. C. Forbes, A. Loeb, "Evaporation of planetary atmospheres due to XUV illumination by quasars." Monthly Notices of the Royal Astronomical Society, vol. 479.1, 2018, pp. 171-182.
[20] H. Chen, L. A. Rogers, "Evolutionary analysis of gaseous sub-Neptune-mass planets with MESA." The Astrophysical Journal, vol. 831.2, 2016, 180.
[21] J. E. Owen, Y. Wu, "Kepler planets: a tale of evaporation." The Astrophysical Journal, vol. 775.2, 2013, 105.

[22] H. Chen, J. C. Forbes, A. Loeb, "Habitable evaporated cores and the occurrence of panspermia near the Galactic center." The Astrophysical Journal Letters, vol. 855.1, 2018, L1.
[23] R. A. Murray-Clay, A. Loeb, "Disruption of a proto-planetary disc by the black hole at the milky way centre." Nature Communications, vol. 3.1, 2012, pp. 1-5.
[24] J. E. Owen, A. P. Jackson, "Planetary evaporation by UV and X-ray radiation: basic hydrodynamics." Monthly Notices of the Royal Astronomical Society, vol. 425.4, 2012, pp. 2931-2947.
[25] A. Balbi, F. Tombesi, The habitability of the Milky Way during the active phase of its central supermassive black hole." Scientific Reports 7.1, 2017, pp. 1-6.
[26] M. Lingam, I. Ginsburg, S. Bialy, "Active galactic nuclei: boon or bane for biota?." The Astrophysical Journal, vol. 877.1, 2019, 62.
[27] F. E. Reinhard, S. Gillessen, "The Galactic Center massive black hole and nuclear star cluster." Reviews of Modern Physics, vol. 82(4), 2010, 3121.
[28] Y. Shen, "The mass of quasars." arXiv preprint arXiv:1302.2643, 2013.
[29] D. Brock, "Life on a Young Planet: The First Three Billion Years of Evolution on Earth." Journal of College Science Teaching, vol. 33(4), 2004, 46.
[30] L. R. Dartnell, "Ionizing radiation and life." Astrobiology, vol. 11.6, 2011, pp. 551-582.
[31] L. Dou, et al. "Long fading mid-infrared emission in transient coronal line emitters: Dust echo of a tidal disruption flare." The Astrophysical Journal, vol. 832.2, 2016, 188.
[32] W. Yuan, et al., "Chandra and MMT observations of low-mass black hole active galactic nuclei accreting at low rates in dwarf galaxies." The Astrophysical Journal, vol. 782.1, 2014, 55.

# Experimental study on ignition and lean blowout characteristics of high temperature rise combustor

**Rui Luo[1a*], Yi Fu[1b], Jintao Jiang[1c], Pengyu Shi[1d], Penghui Liu[1e], Zhisheng Wang[1f], Wei Fan[1g*]**

[1]School of power and Energy, Northwestern Polytechnical University, Xi'an 710129, China

[a*]luorui@mail.nwpu.edu.cn, [b]fuyi@nwpu.edu.cn, [c]npu_jt@mail.nwpu.edu.cn, [d]shidande163@163.com, [e]2015301489@mail.nwpu.edu.cn, [f]jason_wang@mail.nwpu.edu.cn, [g*]Weifan419@nwpu.edu.cn,

**Abstract:** Ignition and lean blowout characteristics are an important performance index of the Aeroengine combustor, which is directly related to the reliability of the whole engine. The ignition process of the combustor was filmed by a CCD camera, and three stages of successful ignition were obtained. By changing the intake conditions, the variation of the minimum ignition energy in the fuel-air ratio (FAR) is tested. With the decrease of the FAR, the minimum ignition energy increases and there is a sudden increase process. As the room temperature and pressure conditions, the lean blowout (LBO) FAR is roughly 0.014, and Increasing kerosene temperature can improve the LBO performance in the combustor.

## 1. Introduction

Ignition and extinguishing characteristics are important performance indicators of aero-engine main combustion chambers, and mastering the effective engine operating range is an inevitable requirement for component matching and design optimization in the pre-research process [1] The new generation of advanced aero-engine combustion chambers have significant structural differences from the traditional annular combustion chambers, such as the elimination of the main combustion orifice to ensure sufficient air intake to the head. These changes have given rise to new research questions, typical of which is the point flameout problem.

The ignition performance of the combustion chamber is directly related to the reliability of the whole engine operation. Based on Lefebvre's theory [2], Mastorakos [3] described the process of spray ignition in an engine as a three-step process: (1) initial fire nucleus generation; (2) fire nucleus growth and propagation; and (3) flame stabilization on a head nozzle. A successful ignition event usually starts with an ignition nucleus, i.e., a region with extremely high reactivity and heat release, before the flame builds up. Ahmed [4,5] studied the effect of cyclonic flow on ignition success and found that the limit of stable combustion was close to the ignition limit under cyclonic flow and that the mixture distribution in the central return zone was highly sensitive to changes in the inlet gas velocity.

The ignition characteristics of cyclonic combustion chambers are a hot topic in the research field, and within the last decade, Professor Mastorakos at the University of Cambridge has carried out extensive studies on the ignition of cyclonic combustion chambers. Bach [6] found that in cyclonic combustion chambers, the two conditions for a high ignition probability are a low Karlovitz number and a negative mean velocity, respectively. Decreasing the equivalence ratio or decreasing the flow rate increases the time required for ignition. Chakraborty [7-9] used the DNS method for forced ignition in turbulent mixed layers and found that both local forced ignition and subsequent self-sustained combustion were affected by the fuel Lewis number, with the flame more likely to burst when the Lewis number was greater than 1; in addition spark ignition and subsequent Neophytou [10,11] developed a spark ignition model that can be used for both gases and liquids, and is able to calculate the evolution of the trajectory of spark generation and fire nuclei development, and obtain the ignition probability distribution and ignition time to evaluate the ignition performance of the combustion chamber. Pedro [12] studied the high-altitude re-ignition performance of kerosene gas turbine combustion chambers using a low-order ignition model and obtained the law of the effect of spray particle size on the ignition probability, but the final prediction of the ignition probability was low because the computational model could not take into account the inhomogeneity of the kerosene spray particle size distribution. In addition, they also studied and discussed the ignition probability, ignition time, and joint flame process of the full-ring combustion chamber [13-16].

The empirical model of Lefebvre [2] was proposed to provide a concise method for the prediction of the lean oil limit during engine pre-study and is used by researchers to this day for generalizing the experimental data and deriving the lean oil quenching relationship equation for a specific structure of combustion chamber. However, in high-temperature liter combustion chambers, the combustion chamber structure has been significantly different from that of the 1970s to 1980s in order to obtain higher combustion performance. Although these empirical models take into account part of the physicochemical processes in the real combustion chamber, they cannot capture the influence of the local flow field on the combustion process and are very specific to the basic structure of the combustion chamber, and cannot be directly used for the prediction of the lean oil limit in the high temperature rise combustor. Therefore, experimental means is still ideal for obtaining the ignition boundary of high-temperature lift combustion chambers.

In order to obtain the point-quench characteristics of the high-temperature lift combustion chamber, the point-quench test of a single-head high-temperature lift combustion chamber was conducted in this paper, and the point-quench characteristics of a configuration of high-temperature lift combustion chamber were obtained.

## 2. Experimental setup and method

### *2.1 Single head combustor*

The schematic diagram of single-head high-temperature liter combustor model is shown in Figure 1. In order to ensure sufficient air entering from the head of the combustor, the main combustion hole and doping hole structure on the traditional annular combustor are eliminated, and the flame tube adopts the form of air film cooling, with the designed cooling air volume accounting for 20%, and the rest of the air entering from the head. The head is mainly composed of two axial cyclones, the first/second cyclone number is 0.78 and 0.88 respectively; the upper wall projection of the flame tube is the igniter mount, and its position is on the radial section in the center of the main reflow area; in order to facilitate the high-speed camera to capture the ignition process in the combustor, the side of the flame tube has an annular sleeve view window.

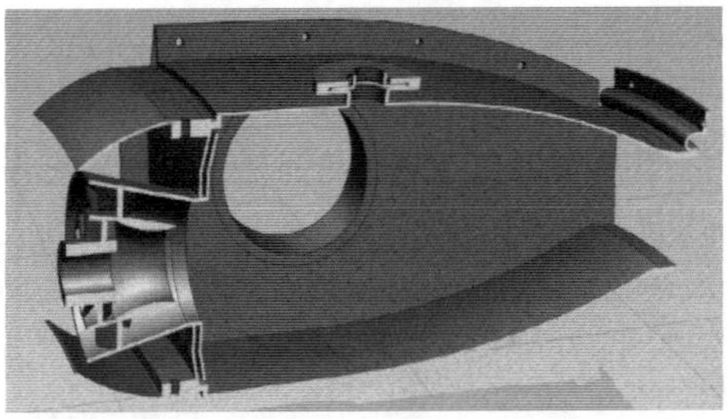

Figure 1 Schematic diagram of single head combustor model

Figure 2 shows the schematic diagram of the single-head combustor experiment system. The kerosene heating module can transfer the room temperature kerosene in the tank to the kerosene heater line through the high-pressure oil pump, and heat it to the required temperature to provide fuel for the combustor. The air in the high-pressure tank is introduced into the air heater by the high-pressure pipeline and heated to the required temperature to enter the combustor inlet section. The combustor experiment section is a single-headed high-temperature liter combustor with several pressure and temperature measurement points arranged along the flow direction to meet the experimental experiment requirements. The ignition module is a high-energy spark ignition system that provides a fixed ignition source for the combustor. The control and acquisition module controls the opening and closing of the gas source and oil circuit, and can also be used to output signals to trigger ignition and high-speed camera shooting, synchronous acquisition of experiment data, etc.

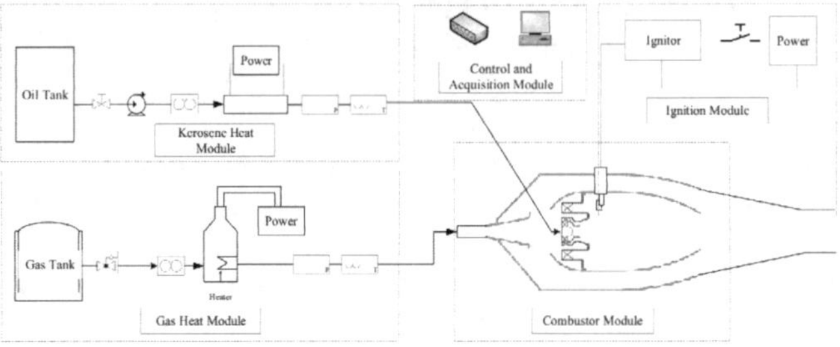


Figure 2 Schematic diagram of the single head combustor experiment system

*2.2 Experiment working conditions*

*2.2.1 Ignition experiment*

The ignition experiment was conducted at room temperature and pressure, and the fuel used in the experiment was RP-3 aviation kerosene. The minimum ignition energy for each successful ignition was recorded and the variation of the minimum ignition energy with the oil to gas ratio was obtained by controlling the kerosene flow rate in a single working condition and conducting the ignition experiment in the combustor by changing the air flow rate. Table 1 shows the set ignition conditions. The temperature inside the combustor is ensured to drop to room temperature before each ignition experiment, and the self-luminous signal of the ignition process is captured by a high-speed camera to capture the typical ignition process inside the combustor while the ignition experiment is conducted.

Table 1 Ignition experiment parameter settings

| Kerosene flow rate（g/s） | Air flow rate（g/s） | Ignition energy（J） |
|---|---|---|
| 1.92、2.09、2.26 | 20~90 | 0.05~20 |

*2.2.2 LBO experiment*

In order to investigate the LBO performance of the combustor at high temperature, the combustor LBO experiment was carried out, under the condition that the single experiment air flow rate was kept constant, the igniter was started until the combustor ignited successfully, the kerosene flow rate was gradually reduced until the combustor was extinguished, the kerosene flow rate at the time of extinguishment was recorded, and the experiment conditions were repeated five times, and the experiment conditions were shown in Table 2. The effect of kerosene temperature on the LBO performance of the combustor was investigated.

Table 2 LBO experiment conditions

| Case | Kerosene flow rate（g/s） | Air flow rate（g/s） |
|---|---|---|
| 1~7 | 3.50 | 120、140、160、180、200、220、240 |

## 3. Results and analysis

*3.1 Ignition experiment results*

According to the selected ignition conditions in Table 1, the ignition experiment was carried out at room temperature and pressure, and Figure 3 shows the trend of minimum ignition energy with FAR after converting the inlet air conditions to FAR. It can be seen from the graph that the minimum ignition energy tends to increase as the air flow increases and the FAR decreases, and there is a sudden increase in the minimum ignition energy after the FAR is less than 0.035. This is due to the poor atomization effect of liquid kerosene after entering the combustor environment at room temperature and pressure through the nozzle, which requires a large enough energy ignition source to promote the fuel atomization and evaporation process, ignite the combustible mixture to form the initial nucleus, and make the nucleus develop smoothly to the latter two stages of the ignition process to ensure the successful ignition in the combustor. Ignition is successful.

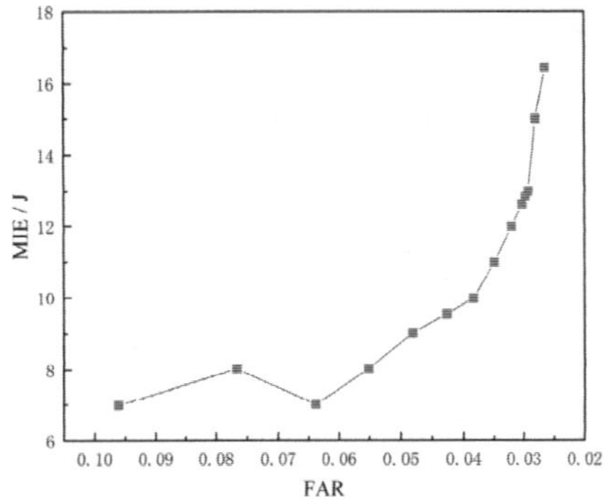


Figure 3 Trend of minimum ignition energy with FAR

*3.2 LBO experiment results*

Table 3 shows the results of the LBO experiment, in which five times of lean oil quenching kerosene flow values under single working condition are recorded, and the quenching FAR is calculated by geometric average. The FAR of lean oil quenching in the combustor at high temperature and pressure is about 0.014, which is relatively different from the required value of lean oil quenching in the combustor of not more than 0.005. The reasons for the large FAR of lean oil quenching are analyzed by the relationship between the equivalent ratio in the main combustion zone and the total FAR.

$$main\ combustion\ zone\ equivalent\ ratio=\frac{FAR_{total}}{0.068*The\ head\ of\ the\ combustor\ air\ intake\ ratio}$$

The head of the combustor air intake ratio of 80%, blowout FAR is 0.014, the main combustion zone equivalent ratio is calculated to 0.257, while the general quenching equivalent ratio of 0.2. This indicates that the use of a large amount of air into the head of high-temperature liters combustor close to the poor oil quenching experiment research when the main combustion zone is close to very poor combustion.

Table 3 Experimental results of LBO in high temperature rise combustor

| Air flow rate（g/s） | Extinguish kerosene flow（g/s） | Blowout FAR |
|---|---|---|
| 120 | 1.68、1.67、1.70、1.66、1.68 | 0.014 |
| 140 | 1.96、1.98、1.96、2.00、1.94 | 0.014 |
| 160 | 2.09、2.07、2.08、2.10、2.06 | 0.013 |
| 180 | 2.53、2.52、2.50、2.52、2.54 | 0.014 |

| | | |
|---|---|---|
| 200 | 3.00、3.05、2.97、2.98、3.02 | 0.015 |
| 220 | 3.07、3.06、3.04、3.10、3.08 | 0.014 |
| 240 | 3.36、3.38、3.35、3.35、3.36 | 0.014 |

Next, the effect of kerosene temperature on the LBO performance of the high-temperature liter combustor was investigated. The kerosene flow rate was selected as 3.25 g/s, and when the combustor was successfully ignited, the air volume was gradually increased until the flame out occurred in the combustor, and the air flow rate at the time of flame out was recorded, and the flame out FAR was obtained by calculation. Figure 4 shows the change trend of blow off flow rate and oil to gas ratio with kerosene temperature, from the figure can be seen that with the increase of kerosene temperature, the flame out flow rate in the combustor is increasing, at the kerosene temperature of 200 °C, the flame out flow rate is 302 g/s, which is converted to 0.01 oil to gas ratio of poor oil quenching, indicating that increasing the kerosene temperature is conducive to optimizing the LBO performance of high temperature liter combustor.

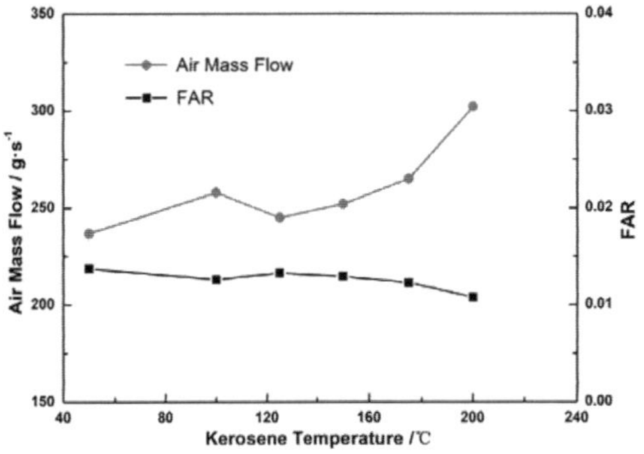


Figure 4 Variation of blow-out flow rate and oil to gas ratio with kerosene temperature

## 4. Conclusion

The ignition process of single-head high-temperature liter combustor was firstly photographed by high-speed camera under normal temperature and pressure, and three stages of successful ignition were obtained, and the variation law of minimum ignition energy with FAR was obtained by changing the inlet gas conditions. After that, a single-head combustor LBO experiment was carried out to obtain the combustor lean oil quench FAR and analyze the reasons for the large quench FAR, and it was found that increasing the kerosene temperature was beneficial to optimize the combustor LBO performance, and the main conclusions were as follows.

(1) The high-speed camera captured the three stages of a typical successful ignition and captured the development trend of the ignition process.

(2) Under the experiment conditions, the minimum ignition energy of single-head high-temperature liter combustor was above 7 J during cold start, and the minimum ignition energy showed a rising trend with the decrease of FAR, and there was a sudden increase after the FAR was less than 0.03.

(3) Under normal temperature and pressure environment, the lean oil quench FAR of single-head high-temperature liter combustor is about 0.014, and raising the kerosene temperature helps to optimize the LBO performance of the combustor.

## Acknowledgments

The authors acknowledge gratefully the financial support from National Natural Science Foundation of China Project (NO.51876179,52176133) and the National Science and Technology Major Project of China (No. 2017-III-0005-0030).

**References**

[1] Wang huiru, Jin jie. Overview of Lean Blowout Limit Prediction Methods for Aeroengine Combustors [J]. Aeroengine, 2014, 40(05):72-78.

[2] Lefebvre A, Ballal D. Gas Turbine Combustion (Alternative Fuels and Emissions, Third Edition) || Alternative Fuels[J]. 2010, 10.1201/9781420086058:443–511.

[3] Mastorakos E. Forced ignition of turbulent spray flames[J]. Proceedings of the Combustion Institute, 2017, 36(2):2367–2383.

[4] Ahmed S F, Balachandran R, Marchione T, al et. Spark ignition of turbulent nonpremixed bluff-body flames[J]. Combustion and Flame, 2007, 151(1):366–385.

[5] Marchione T, Ahmed S F, Mastorakos E. Ignition of turbulent swirling n-heptane spray flames using single and multiple sparks[J]. Combustion and Flame, 2009, 156(1):166–180.

[6] Bach E, Kariuki J , Dawson J R, et al. Spark ignition of single bluff-body premixed flames and annular combustors[C]// 51st AIAA Aerospace Sciences Meeting including the New Horizons Forum and Aerospace Exposition. 2013.

[7] Chakraborty N , Hesse H , Mastorakos E . Numerical Investigation of Edge Flame Propagation Behavior in an Igniting Turbulent Planar Jet[J]. Combustion Science & Technology, 2010, 182(10-12):1747-1781.

[8] Chakraborty N , Hesse H , Mastorakos E . Effects of Fuel Lewis Number on Localised Forced Ignition of Turbulent Mixing Layers[J]. Flow, Turbulence and Combustion, 2010, 84(1):125-166.

[9] Chakraborty* N , Mastorakos E , Cant R S . Effects of Turbulence on Spark Ignition in Inhomogeneous Mixtures: A Direct Numerical Simulation (DNS) Study[J]. Combustion Science & Technology, 2007, 179(1-2):293-317.

[10] Neophytou A , Richardson E S , Mastorakos E. Spark ignition of turbulent recirculating non-premixed gas and spray flames: A model for predicting ignition probability[J]. Combustion & Flame, 2012, 159(4):1503-1522.

[11] Neophytou A , Mastorakos E , Cant R S . DNS of spark ignition and edge flame propagation in turbulent droplet-laden mixing layers[J]. Combustion and Flame, 2010, 157(6):1071-1086.

[12] PMD Oliveira, Sitte M P , Zedda M , et al. Low-order modeling of high-altitude relight of jet engine combustors[J]. International Journal of Spray and Combustion Dynamics, 2021.

[13] Ciardiello R , Pathania R S , Allison P M , et al. Ignition Probability and Lean Ignition Behaviour of a Swirled Premixed Bluff-Body Stabilised Annular Combustor[J]. Journal of Engineering for Gas Turbines and Power, 2021, 143(4).

[14] Rc A , Pmdo A , Aws A , et al. Effect of spark location and laminar flame speed on the ignition transient of a premixed annular combustor[J]. Combustion and Flame, 2020, 221:296-310.

[15] Machover E , Mastorakos E . Experimental investigation on spark ignition of annular premixed combustors[J]. Combustion and Flame, 2017, 178:148-157.

[16] Edouard, Machover, Epaminondas, et al. Spark ignition of annular non-premixed combustors[J]. Experimental Thermal and Fluid Science, 2016, 73:64-70.

# The State-of-art Analytical Models and Detectors for Black Holes

**Zehua Yu***

Ulink College of Suzhou Industrial Park, Suzhou, China

*Corresponding author: Zehua.Yu@sz-alevel.com

**Abstract**—Since the 19th century, human beings have a much better understanding on the black holes with the fast progress on the optical detection technology. In this paper, the recent progress and the state-of-art detection scenarios for black holes are discussed, especially for the primordial black holes. With the result of evidence and theory, the current status of detecting black holes is detailly demonstrated. Specifically, five parts will be presented, the formation of the black holes, the growth of the black holes, the types of the black holes, the detection of the black holes and the relationship between primordial black holes and black matter. Overall, these results shed light on the astrophysics, the interaction between the black holes and the surrounding heavenly bodies and the universe with deeper insight.

## 1. Introduction

In the 18th century, Pierre-Simon Laplace asked a famous question that whether there is any heavenly body which escaping velocity is as same as the velocity of the light, which made the came out firstly. In the 19th, Albert Einstein created the Einstein gravitational field, which is a complex partial differential equation [1]. However, this is just a corner of the ice berg. Taking one step further, Karl Schwarzschild found a solution of Einstein gravitational field at the year of the 1916, which represents a black hole without self-rotation [2-4], which predicted the exist of the black hole. Meanwhile, Julius Robert Oppenheimer and his students claimed that the giant stars would have the possibility to collapse and form a singular point, which indicates that space time equals to zero [5, 6]. By the way, they also measured the minimum mass, 3.5 times as solar mass, to form black holes. In the year of 1963, Roy Kerr found out another solution, which represents the black holes with self-rotation [7]. Next year, the first black hole was claimed to be found by NASA.

Black holes are heavenly bodies which exist in the universe. There are many different types of black holes based on different standards. For example, a star that mass is 1.5 times bigger than the mass of the sum can be able to form black holes. Generally, considering the scale of mass, black holes can be classified as supermassive black holes, supermassive binary black holes, massive black holes, intermediate-mass black holes and stellar black holes. In addition, according to three physical quantities -mass, momentum and charge. The black holes which do not do self-rotation and have no charge are called Schwarzschild black holes. The black holes which do not do self-rotation but have charge is known as R-N black holes. The black holes which do self-rotation but have no charge are the Kerr black holes. The black holes which have self-rotation and have charge are named as the Kerr-Newman black holes.

Primordial black holes are the black holes which are produced during the period of the very early universe [1]. Since the very early universe cannot be homogenous, it means some specific regions would be sufficiently compressed. Based on the inflation theory, this makes the formation of the primordial black holes become possible as for the gravitational collapse. The conclusion is that many gravitationally collapsed objects are formed to $10^{-5}$g as the early universe fluctuates. Depends on the theory, all black holes would emit hawking radiation [8], which leads to the loss of mass. At the very early universe (smaller than 1 second), those primordial black holes with small mass would lose mass due to continuous emission, which means they finally face evaporation and release a massive energy out [9]. The emission of particles, this kind of the process is called the Hawking effect. This always undergoes at the edge of the horizon, the virtual pair of particles is broken which left one particle out of the horizon, and another one is absorbed by the black hole. In addition, the formation of the primordial black holes is not from the collapsing stars, so primordial black holes can be any size. Additionally, the primordial black hole is considered as one of the possible candidates to produce the black matter [10]. It is mainly because Hawking radiation plays a pivotal role of the evaporation of the primordial black holes. For this reason, the black matter is produced during this process. Besides, the concept of the primordial black holes is given out by Zeldovich, Novikov in the 1996s and Steven Hawking at the year of 1971 [11]. The first relevant paper about the primordial black holes was submitted on the year of 1974 by Carr Bernard and Stephen Hawking. So far, Mr. Carr still continues doing the further research on this topic.

The rest part of the paper is organized as follows. The Sec. II will describe the basic principles and mathematical equations for black holes. The Sec. III will demonstrate the state-of-art detections facilities and observations. Afterwards, the Sec. IV will discuss the current limitations and defects. Eventually, a brief summary is given in Sec. V.

## 2. Basic description and mechanism

The mass function for the stellar-mass function:

$$f(M) = \frac{k_c^3 T}{2_\pi G_N} = \frac{M^3 sin^3 i}{(M + M_2)^2} \tag{1}$$

where $k_c{=}v_c sini$ is the maximum line-of-sight Doppler velocity of the companion star, $V_c$ is the velocity of the companion star, *i* is the inclination angle of the orbital plane, T is the orbital period, and M and $M_C$ are the mass of the compact object and the stellar companion.

From this formation, we can get the minimum mass to cause the formation of the black hole which should be three times bigger than the solar mas, so any heavenly body which follow this way would be classified as the candidates of the black hole.

For primordial black holes, the necessary condition for collapse is [12]

$$uR^2 > \sim \frac{u_0}{u} \log \frac{u}{u_0} \tag{2}$$

Here, $u$ is energy density, $u_0$ is initial energy density. In fact, random turbulent motions might stop further collapse. Mass of the primordial black holes is [12]

$$M \sim \frac{t}{I + \frac{t}{t_0\left(\frac{t_o}{M_0} - I\right)}} \tag{3}$$

Here, $M_0$ is the initial mass of the black hole, $t_0$ is the time of formation. Regarding to radiation of the black holes, as for the effect of the quantum gravity in curved space time, this leads to the emission of particles of the black holes. The spectrum is a thermal spectrum which is connected to the temperature, because black holes with mass M is quasi-static. The period is

$$T = \frac{k}{2\pi} \tag{4}$$

With

$$k = \frac{1}{2} \partial_r \gamma(r) | r = R_H \tag{5}$$

where k is the surface gravity of the black hole

$$T = \frac{1}{8\pi}\frac{C^2 m_p^2}{K_B}\frac{1}{M} \approx 10^{-6} M_{\odot}[k] \tag{6}$$

Here, $M_{\odot} \approx 10^{54}$ TeV represents the solar mass. From this formula, it is able to get the regularity, the smaller the mass of the black temperature, the higher is its temperature.

The equation to describe the radius of orbit of black holes

$$r_{\gamma} = 2M\left[1 + \cos\left(\frac{2}{3}\arccos\left(\pm\frac{a}{m}\right)\right)\right] \tag{7}$$

where a is the dimensionless spin parameter of the spacetime. +a means it belongs to the event horizon, -a means it belongs to the inner horizon. A sketch of the disk accretion is given in Fig. 1, which shows accretion disk is thin and optically thick from the perspective of the geography. The accretion does radiate like a blackbody. Hot Corona is made of the group of the hot electron. In addition, the temperature of the disk accretion depends on the mass and accretion rate of its black holes. The center of the disk is the place where x-rays are produced majorly as for the Compton scattering of thermal photons. The fraction of the primordial black hiles where will form binary before evaporation [13].

$$f_{PBH,binary} \sim \frac{N_{PBH}^2}{InN_{PBH}}\left(\frac{10^{-8}pc^2}{r_c^2}\right)\left(\frac{M_{PBH}}{30M_{\odot}}\right)^2\left(\frac{v_d}{1kms^{-1}}\right)^{-\frac{18}{7}} \tag{8}$$

The binary formed will be absrobed into the center of the black hole to form a black hole with large scalein the early universe.The isothermal core of radius $r_c$ and $N_{PBH}$ Primordial black holes inciuded in the gigayear-timescale for the a core-halo structure which is due to stellar evaporation. Primordial black holes become the binary when GW emission eceeds their total kinetic energy. Fig. 2 is framed on Aprial 10, 2019, which is recorded by the Event Horizon Telescope (EHT). This black hole is M87, a supermassive black hole, is 55000 thousand far from the Earth. This picture illustrates the appearance of black hole is similar to a bright ring and the shadow of this "ring'' is the centre of black hole. With the further discovery of ETH, the astronomers found that the light around the black hole is polarized majorly. This evidence is very meaningful as the matter may ejected or absobed for the outer stage of the back hole, which is the first milestone of this rearch.One of the grandphenomenas is the jet of thematter and energy from the core has been transferred to the outer space at least 5000 light years. Fig. 3 depicts the bending effect of the light rays in the vicinity of the diatant quasar, which can be attributed to the warp fabric of space.

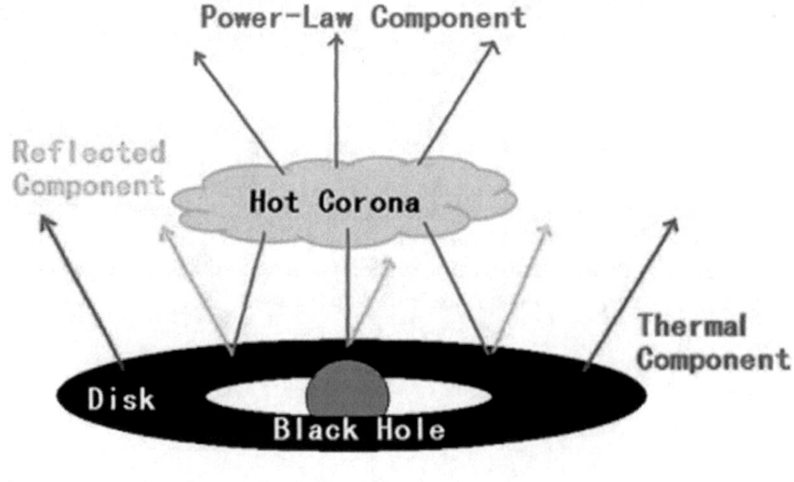


Figure 1. The structure of the disk accretion.

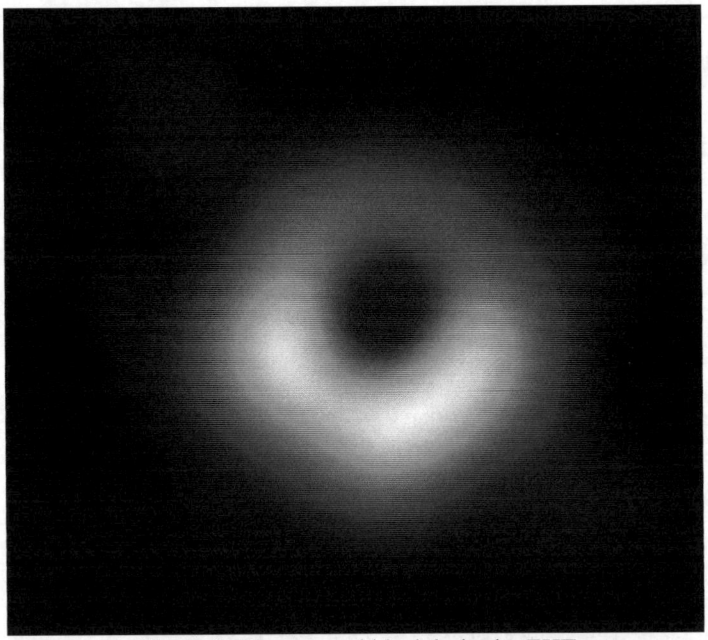

Figure 2. The recored black holes by EHT

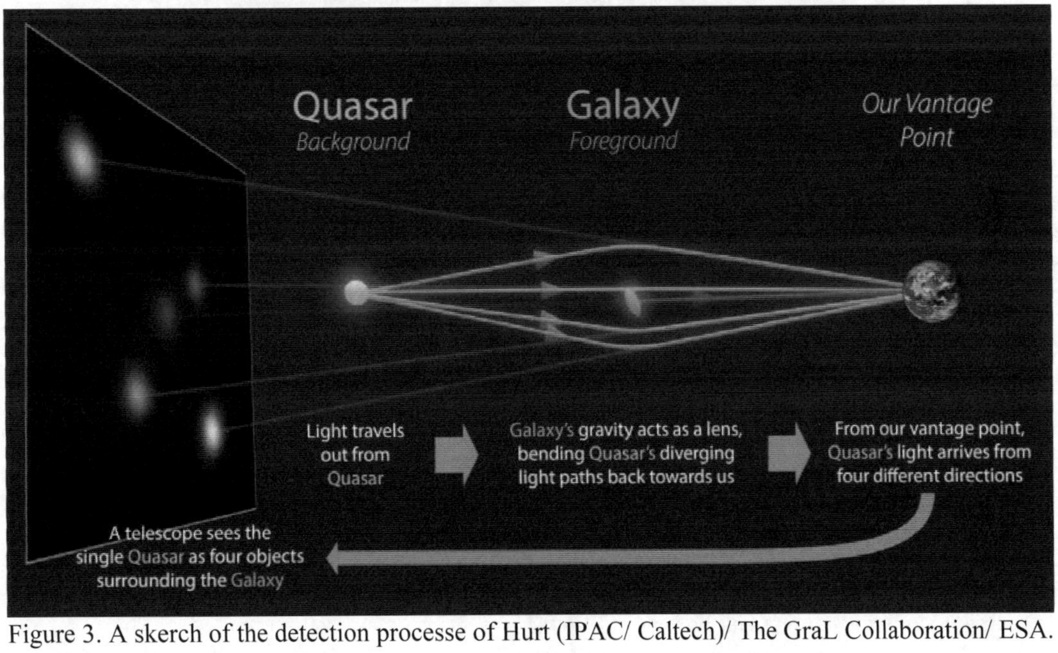


Figure 3. A skerch of the detection processe of Hurt (IPAC/ Caltech)/ The GraL Collaboration/ ESA.

## 3. Detection

With the development of technology, plenty of technique ways are applied to detect the exist of the black holes. Scholars have created five ways to do search the exoplanets. These includes transit, radical velocity, direct image, gravitational microlensing and astrometry.

Transit used to discover the largest number of exoplanets; 3759 planets has been discovered so far. It uses the change of level of light which changes to be dimmer to work. Although the change is extremely small, it is enough for astronomers to realize the presence of exoplanets. For instance, Kepler Space

Technique from NASA. Moreover, some candidates may be proved to be “false positives”. If they are proved to be found by any other two scopes, then this can be the sign to ensure the presence of exoplanets.

This detecting way is taking photographs directly. The advantage is to find out the size of the star detected and doing spectrum is also available. However, this could extremely depend on the situation. It is hard to take photographs on star directly from the earth because of the existence of the atmosphere, satellite is the best option to do this task.

As substance can bend the spacetime according to the general relativity, it causes any light rays which move in the gravitational field to be bended, especially for the moons around a star would cause a fluctuation in the light intensity. This kind of similar ideas can be used to detect and prove the existence of stars.

Any systems of heavenly body, detecting any change in the movement of the giant star. This requires a long-term detection; mechanism has to be accurate enough and the star required to be heavy enough to detect this effect.

Primordial black holes have ability to emit X-ray because of the effect of the accretion [14]. All the X-rays emitted can be found majorly in the background. Therefore, with the help of the specific satellite to detect the X-rays from the universe, it is accessible to deduce the location of the primordial black hole.

In addition, lensing can also be an effective way to find where are the primordial black holes [14]. For example, Hawkins once claimed evidence which was about a critical density of Jupiter-mass primordial black holes based on the observation of the quasar microlensing. By the way, Mediavilla et al. also had found evidence. These two findings show that 20 percent of the total mass of compact objects in the mass range 0.05-0.45 of the solar mass. Based on the function of the primordial black hole can have ability to describe this event perfectly. Hence the existence of primordial black holes can be proved.

Contemporarily, there is no direct way to detect the black matter, but indirect ways can be useful. For example, the gravitational effect which is caused by the black matter can be discovered through the gravitational lens as the light rays which pass around the back matter would be bended because of the nonuniform space-time.

## 4. Limitation and Future Prospects

Inflation theory was firstly posed by Alan Guth at the year of 1979. It is one of the pivotal bases for the formation of the primordial black holes in the very early universe, between $10^{-33}$ to $10^{-32}$ s from the beginning of the singular point. However, there is no direct evidence to support the inflation theory. In addition, this theory has limits, e.g., the monopole still remains to be determined. Furthermore, the process of LIGO-type black holes may need to be modified if primordial black holes do not be ones which produce the dark matter all or majorly.

Primordial Black holes will still be expected to be the candidate to produce the black matter, Astronomers seek for a more advanced machine to detect the black hole in order to confirm some theories or to have a new understanding.

Moreover, the formation of the middle-mass black holes is still a doubt. Although most physicists believe it is caused by two small black holes merging. There are still many other ways to explain this type of formation since the merging process of the black holes has still not been proved by direct evidences so far. Furthermore, lensing has its limit also. Trimble and Thorne pointed out that gravitational lens effect would be too small to detect for a collapsed star or neutron star since the bending of light is not obvious enough at the year of 1969 [8].

## 5. Conclusion

In summary, this paper investigates the general topics of black holes and primordial black holes based on the results observed nowadays. Specifically, this paper focuses on the formation of the primordial black holes and how primordial black holes can be possible to produce black matter based on the inflation theory. On this basis, some basic functions and pictures are also used to describe the structure of black holes, which can do further explanation. In the coming tomorrow, scholars ought to find a more advanced detector in order to find direct evidence to prove the reality of the inflation theory and the process to produce the black matter or any ruins of the black matter. Overall, these results off era brief guideline of the black holes and primordial black holes, which is the root of the development of the black holes and primordial black holes.

## REFERENCES

[1] S. Hawking, "Gravitationally collapsed objects of very low mass." Monthly Notices of the Royal Astronomical Society, vol. 152.1, 1971, pp. 75-78.
[2] J. Franchi, and Y. L. Jan, "Relativistic diffusions and Schwarzschild geometry." Communications on Pure and Applied Mathematics: A Journal Issued by the Courant Institute of Mathematical Sciences, vol. 60.2, 2007, pp. 187-251.
[3] S. Chandrasekhar, "Dynamical instability of gaseous masses approaching the Schwarzschild limit in general relativity," Physical Review Letters, vol. 12(4), 1964, 114.
[4] Y. Hagihara, "Theory of the relativistic trajeetories in a gravitational field of schwarzschild." Japanese journal of astronomy and geophysics vol. 8, 1930, 67.
[5] J. R. Oppenheimer, G. M. Volkoff, "On massive neutron cores," Physical Review, vol. 55(4), 1939, 374.
[6] J. R. Oppenheimer, and H. Snyder, "On continued gravitational contraction." Physical Review vol. 56.5, 1939), 455.
[7] R. P. Kerr, "Gravitational field of a spinning mass as an example of algebraically special metrics." Physical review letters, vol. 11.5, 1963, 237.
[8] P. Hajicek, "Origin of Hawking radiation." Physical Review D vol. 36.4, 1987, 1065.
[9] S. Hossenfelder, "What black holes can teach us." arXiv preprint hep-ph/0412265, 2004.
[10] B. Carr, and F. Kühnel. "Primordial black holes as dark matter: recent developments." Annual Review of Nuclear and Particle Science, vol. 70, 2020, pp. 355-394.
[11] I. D. Novikov, et al., "Primordial black holes." Astronomy and Astrophysics, vol. 80, 1979, pp. 104-109.
[12] B. J. Carr, and S. W. Hawking. "Black holes in the early Universe." Monthly Notices of the Royal Astronomical Society, vol. 168.2 1974, pp. 399-415.
[13] A. Kashlinsky, "LIGO gravitational wave detection, primordial black holes, and the near-IR cosmic infrared background anisotropies." The Astrophysical Journal Letters, vol. 823.2, 2016, L25.
[14] A. M. Green, and B. J. Kavanagh, "Primordial Black Holes as a dark matter candidate." Journal of Physics G: Nuclear and Particle Physics vol. 48.4, 2021, 043001.

# Simulation research on positioning acquisition of field rescue equipment based on LoRa

**Pengfei Kou [1,a], Mingyao Chen [1,b] and Fei Wang [1,c]**

[1] College of Information and Communication, National University of Defense Technology, Wuhan, Hubei, 430000, China

[a]email: k450437065@163.com, [b]email: 752119037@qq.com,

[c]email: 450437065@qq.com

**Abstract.** Timely and accurate acquisition of the position information of people in distress in the field is an important prerequisite for successful rescue. This paper analyzed the main positioning means of the current call equipment and the research status of LoRa technology in the position information acquisition of the current call equipment. The author proposed a method to obtain the position information of field rescue equipment based on LoRa, and simulated it in NS3 network simulator. This scheme can meet the requirements of high positioning accuracy, long communication distance and long standby time of field rescue equipment, and has high value on practical application.

## 1. Introduction

With the country's economic development and the improvement of people's living standards, more and more people are engaged in field research and exploration. The hikers in the field are not clear about the route and do not have a comprehensive grasp of the road condition, they cannot accurately locate and timely communicate with the outside world in distress, resulting in the occurrence of accidents in the field. Therefore, it is necessary to develop a reliable field rescue equipment. GPS/ Beidou satellite navigation technology is commonly used to obtain the position information of traditional distress equipment. However, due to the harsh LOS condition of jungles, canyons and hills, satellite positioning technology cannot be applied in full time, and satellite positioning has the limitations of high energy consumption, high cost and weak signal. Therefore, a supplementary positioning method with strong applicability and stable effect is urgently needed. LoRa (Long Range), as a representative of emerging Internet of Things technology, not only has the characteristics of wide coverage, low power consumption, low cost, but also has the advantages of simple network structure, flexible deployment and indoor and outdoor positioning, which can be well applied to the field environment. However, the current research mainly focused on the transmission of position information using the wide coverage performance of Lora technology, which was rarely used directly for positioning, and its positioning accuracy had not yet reached the practical application level. Therefore, this paper proposes a LoRa-based scheme for obtaining the position information of field rescue equipment. The LoRa gateway is carried out by the UAV for general positioning based on TDOA (Time Difference of Arrival) algorithm; According to the general positioning results, the UAV is maneuvered to the search and rescue region and the gateway deployment is adjusted; Kalman filter is used for precise positioning to reduce positioning error and narrow the search and rescue area, so as to achieve better search and rescue effect.

## 2. Research status of LoRa technology in field rescue equipment

LoRa, as an emerging Internet of Things technology, adopts a typical star network architecture in its application process. It not only features wide coverage, large connections, low power consumption and low cost, but also has the advantages of simple network structure, flexible deployment and large regional positioning. Launched at Mobile World Congress in March 2015, the LoRa Alliance has been dedicated to maintaining and promoting LoRa technology standards for low-power, wide-area position business applications in the Internet of Things. Driven by the LoRa Alliance, the LoRa network is being piloted and deployed in a large number of positions around the world, and LoRa positioning technology is gradually becoming an integral part of the emerging Internet of Things positioning technology. Due to its extensive coverage, LoRa can greatly reduce the density of gateway deployment to reduce the cost of positioning. LoRa equipment can realize both front-end acquisition and rear transmission of position information, which greatly reduces the cost and complexity of the system. The LoRa terminal provides adaptive data transfer rate and periodic sleep and wake up functions, enabling low-power positioning and long battery life. The end-device is small in size, light in weight and durable, so that it has good portability. Therefore, LoRa positioning technology is very suitable to be applied to the position information acquisition of field rescue equipment.

At present, LoRa technology has been well used in the search and rescue field. The search and rescue system designed in literature [1][2][3] made full use of LoRa technology's advantages of wide coverage and low energy consumption. In those papers, it was used as the main information transmission method, which solved the problem of lack of basic communication facilities in the field environment. However, its positioning means was still limited to a single satellite positioning technology, which was difficult to ensure the reliability of positioning when the GPS signal was weak or disturbed. The system designed in literature [4][5][6] used LoRa technology to achieve positioning, but the gateway position was relatively fixed and the coverage of the system was small, so the advantage of long-distance transmission of LoRa technology was not given full play. In addition, the field search and rescue area was often far away from the rear camp, so the system mentioned above was difficult to meet the requirements of field application.

In general, LoRa technology has achieved some research results in the field of position information acquisition of rescue equipment, but has not yet reached the conditions of practical application. This paper will give full play to the advantages of LoRa technology, and use it directly for positioning, and improve positioning accuracy by adjusting gateway deployment and Kalman filtering, so as to make it more in line with field application requirements and improve practical value.

## 3. Research status of LoRa technology in field rescue equipment

The realization architecture of LoRa positioning of field rescue equipment is shown in Figure 1.Firstly, the UAVs carrying LoRa gateway in the rear camp are used to conduct general positioning of the distress end-device, so as to quickly determine the reference coordinates and obtain the general range of search and rescue. Then, the UAVs are maneuvered near the reference coordinate and the deployment strategy is changed to improve the positioning accuracy. Finally, the high-precision position estimation coordinates is achieved so as to improve the success rate of search and rescue through multiple precise positioning, combined with real-time Kalman filter processing data to further reduce the search and rescue scope.

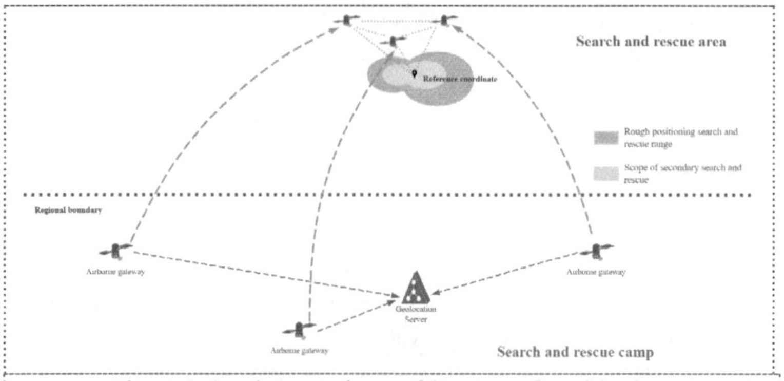


Figure 1. Implementation architecture of positioning.

In order to improve the practicality of the device and enhance the adaptability to multiple scenarios, the architecture is processed as follows:

### *3.1. Expanding the deployment spacing of gateways during general positioning*

As the search and rescue area is usually 3 kilometers away from the rear camp, the first round of general positioning coverage depth is more than 5 kilometers. By expanding the three gateways at an obtuse Angle, with the bottom facing the search and rescue area and increasing the spacing, the search and positioning range can be expanded to cover a broader area in the search and rescue direction.

### *3.2. Adopting the equilateral triangle gateways deployment strategy during precise positioning*

After maneuvering to the general positioning reference coordinates, it is necessary to narrow the gateway spacing and change the deployment strategy to create conditions for precise positioning. According to LoRa's positioning white paper, when the number of gateways was 3, the optimal positioning effect can be achieved by adopting equilateral triangle deployment centered on reference coordinates[7]. Therefore, UAVs are further adjusted and deployed according to this method.

### *3.3. Adopting the positioning algorithm based on TDOA*

At present, there are two widely used LoRa positioning technologies: Received Signal Strength Indication (RSSI) and Time Difference of Arrival (TDOA). The positioning accuracy of RSSI is at the level of 1000 meters, which is difficult to meet the requirements of positioning accuracy. Moreover, the received signal strength is easily affected by the complex shadowing environment in the field, which has great limitations. TDOA positioning is based on transmission delay that is less affected by shadowing, which can provide higher positioning accuracy and is more suitable for field applications. Therefore, in the process of implementation, TDOA based positioning algorithm is used for each positioning.

### *3.4. Processing Real-time data by Kalman filtering*

Kalman filter is an algorithm that using linear system state equation to achieve the optimal estimation through the system input and output observation data. The real-time processing of the observed values in the precise positioning process by Kalman filter can obviously reduce the positioning error caused by the gateway clock synchronization noise, thus improve the positioning accuracy.

## 4. Simulation and analysis of position information acquisition method of field distress equipment based on LoRa

### *4.1. NS3 network simulator*

This paper will use NS3 network simulator to simulate the operation process of field rescue equipment. NS3 is a highly open source network simulator designed and developed by Professor Thomas

R.Henderson and his research team at the University of Washington. The underlying code is implemented by C++ language, and is well compatible with Python language, providing convenience for users of different programming languages. The basic network architecture of NS3 is composed of Node, Protocol, Channel, Net Device and topology Helper. It restores the real state of network operation as much as possible through the combination of network components and module applications.

The scalability of NS3 is mainly reflected in its modular characteristics. Its modules are similar to functions in MATLAB, which are developed by researchers in various fields and added with the constant change of NS3 version. Its common modules include core, mobility, Propagation and point-to-point, which are respectively used to realize the basic mechanism of NS3, node mobility function, definition of propagation model and point-to-point communication. Through the comprehensive application of the model, researchers can carry out simulation research in the same field more conveniently on the basis of predecessors.

In order to realize the simulation of LoRa technology on NS3, Davide Magrin of Federico II University in Naples, Italy, built LoRaWAN module in 2017, which gave a detailed simulation establishment process for throughput, reception success probability, gateway coverage evaluation and other aspects[8-9], and standardized the contents of LoRa end-device, gateway physical layer, MAC layer and channel transmission and so on. Based on this module, this paper realizes the configuration of gateway, end-device, server and channel environment and the simulation of interaction process in communication network.

### *4.2. The design of application scenario simulation*

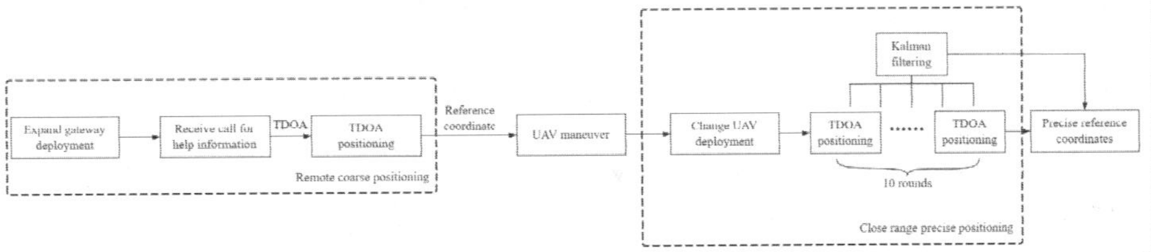


Figure 2. Flowchart of simulation implementation.

Figure2 shows the specific process of field rescue equipment simulation, which is expressed as follows:

(1) Remote general positioning

The line Y=0 is used as the dividing line between the rear camp and the search and rescue area. Three UAVs carrying LoRa gateway are deployed at an isosceles obtuse angle with the bottom facing the search and rescue direction. Three gateway nodes and one terminal node are created in NS3, and the gateway coordinates are set as (-2000, 0), (2000,0) and (0, -1700) using the mobility module. In order to meet the practical application of remote search and rescue in the field, the coordinates of the call terminal are set as (0,5000). After that, the gateway, terminal and channel parameters are configured. Gaussian white noise with standard deviation of 0.3 microseconds is introduced into the gateway receiving time to simulate the main error source of the gateway clock synchronization error. The OneShotSenderHelper class is used to realize the call terminal to send a single message to each gateway. Finally, each gateway sends the recorded receiving time to the position server through the uplink data frame. The server calculates the time difference according to the extracted receiving time and calculates the general positioning coordinate by Chan algorithm.

(2) UAV maneuvering and change of deployment

After the general positioning reference coordinates are obtained, the gateway is deployed as approximately equilateral triangle. A set of equilateral triangular coordinates centered on the origin of coordinate axis (-1000,570), (1000,570), (0, -1140) is preset in the simulation script. By adding the coordinates above with the reference coordinates, the modified three gateway coordinates are determined to simulate the process of UAVs maneuvering to the reference position and deployment.

(3) Accurate positioning at close range

After the gateway is deployed, PeriodicSenderHelper is used to periodically send messages to the gateways from the end-device, and 20 rounds of positioning are performed. The X and Y coordinates are processed in real time with Kalman filter, and the filtering value of the 20th round is finally obtained as the positioning result. In order to realize the iterative process of Kalman filter more conveniently, the filtering process is defined as a function in the simulation, and the filter parameters are stored in a structure.

The struct of filter parameters and the Kalman filter function are shown in Figure 3 and Figure 4.

```
typedef struct{
        double filterValue;   // Filter value at time K-1
        double A;             // x(n)=A*x(n-1)+u(n),u(n)~N(0,Q)
        double H;             // z(n)=H*x(n)+w(n),w(n)~N(0,R)
        double Q;             // Predicted noise variance
        double R;             // Measuring noise deviation
        double P;             // Estimation error covariance
        double kalmanGain;    // Kalman gain
} KalmanInfo;
```

Figure 3. Struct of filter parameters.

```
double KalmanFilter(KalmanInfo* kalmanInfo, double lastMeasurement)
{
  double predictValue = kalmanInfo->A* (kalmanInfo->filterValue);

  kalmanInfo->P = kalmanInfo->A*kalmanInfo->A*kalmanInfo->P + kalmanInfo->Q;

  kalmanInfo->kalmanGain = kalmanInfo->P*kalmanInfo
          ->H / (kalmanInfo->P*kalmanInfo->H*kalmanInfo->H + kalmanInfo->R);

  kalmanInfo->filterValue = predictValue + (lastMeasurement - predictValue)*kalmanInfo->kalmanGain;

  kalmanInfo->P = (1 - kalmanInfo->kalmanGain*kalmanInfo->H)*kalmanInfo->P;

  return  kalmanInfo->filterValue;
}
```

Figure 4. Kalman filter function.

The main functions used to achieve the process above are shown in the following table:

Table 1. Main NS3 functions and their interpretation.

| **Function** | **Explanation** |
|---|---|
| CreateObject<LogDistancePropagationLossModel> () | Set logarithmic distance propagation model |
| CreateObject<ConstantSpeedPropagationDelayModel> () | Set propagation delay model |
| CreateObject<LoraChannel> (loss, delay) | Set channel condition |
| mobilityEd.SetPositionAllocator () | Set the initial coordinates of the ED node |
| mobilityEd.SetMobilityModel ("ns3::ConstantPositionMobilityModel") | Set ED node as fixed-point model |
| macHelper.SetDeviceType (LorawanMacHelper::ED_A) | Set the working mode to Class A |
| PeriodicSenderHelper.SetPeriod (Seconds ()) | Set packet sending period |

| | |
|---|---|
| GatewayLorawanMac::Receive () | Receive packets |
| Simulator::Now().GetNanoSeconds() | Get nanosecond time of the simulator |

### *4.3. Simulation results and analysis*

Simulation results are shown in Figure 5.

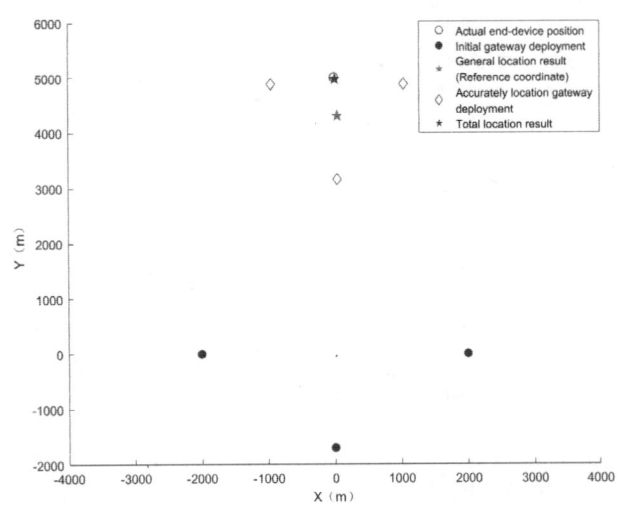


Figure 5. Simulation results.

The result of remote rough positioning is (49.1, 4302.9), and the error is 698.8m, so the new gateway deployment coordinates are determined as (-950.9, 4872.9), (1049.1, 4872.9), and (49.1, 3162.9). After 20 rounds of positioning, the X and Y coordinates are processed with Kalman filter, and the last filtering value (14.44, 4967.6) is taken as the result, with an error of 35.5m. The filtering effect is shown in Figure 6 and Figure 7.

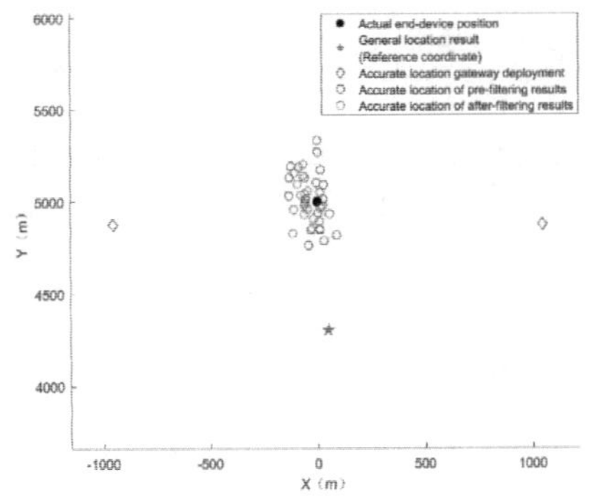


Figure 6. Scatter diagram.

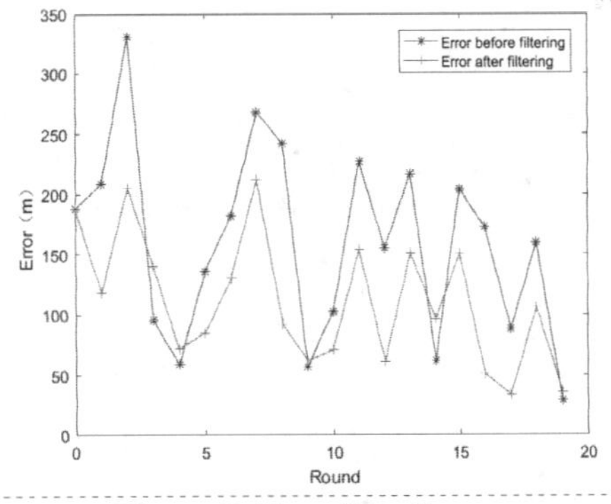


Figure 7. Error line diagram.

The analysis shows that after kalman filter processing, the positioning errors of most rounds are significantly reduced, and the overall error shows a convergence trend. The final result is reduced from 698.8m in rough positioning to 35.5m, and the amplitude of error fluctuation decreases after filtering, and the search and rescue range is further narrowed, which is in line with the expectation. It can be seen that the equipment can meet the needs of field search and rescue operations 5km away. To sum up, the field rescue equipment and its positioning scheme have good practical value.

## 5. Conclusions

This paper proposes a LoRa-based scheme to obtain the position information of field rescue equipment, and simulates it in NS3 network simulator. The simulation results show that the positioning scheme can cover the field search and rescue area 5 kilometers away from the camp, and can quickly respond to the call for help signal. The whole process, including UAV maneuver, can be completed in 10 minutes, and the call source can be within the visible range of the final positioning result, which improves the success rate of search and rescue. The scheme and the corresponding equipment meet the requirements of high positioning accuracy, long communication distance and long standby time of field rescue equipment, and have high value on practical application.

## References

[1] Ma Shuai. Field operation personnel search system research and design [D]. University of north, 2020. The DOI: 10.27470 /, dc nki. GHBGC. 2020.000300.

[2] Li Lu, Hu Nairui, Wang Junyi. Design of field search and rescue monitoring system based on Beidou positioning and LoRa wireless technology [J]. Electronic Devices, 201,44(02):428-433.

[3] Liu Shuo, Peng Bo, Li Ji, He Zhen. Design and implementation of chest belt search and rescue system based on beidou and LoRa [J]. Medical equipment,2020,41(12):35-39.DOI:10.19745/j.1003-8868.2020271.

[4] Zeng Yibin. Tunnel construction personnel positioning based on the technology of LoRa research [D]. Hangzhou university of electronic science and technology, 2019. The DOI: 10.27075 /, dc nki. GHZDC. 2019.000272.

[5] Wu Yaqin, Shi Lanlan. Research on personnel position algorithm of fire rescue based on LoRa [J]. Computer applications and software,2020,37(06):70-75.

[6] A. Mackey and P. Spachos, "LoRa-based Localization System for Emergency Services in GPS-less Environments," IEEE INFOCOM 2019 - IEEE Conference on Computer Communications Workshops (INFOCOM WKSHPS), 2019, pp. 939-944, doi: 10.1109 / INFCOMW. 2019.8845189.

[7] LoRa Alliance™ Strategy Committee.LoRa Alliance Geoposition Whitepaper[M], 2018,1:6-15.

[8] Vangelista L, Centenaro M, Magrin D. Performance evaluation of LoRa networks in a smart city scenario[C]// 2017 IEEE International Conference on Communications (ICC). IEEE, 2017.

[9] Magrin D, Capuzzo M, Zanella A, et al. Performance Analysis of LoRaWAN in Industrial Scenarios[J]. IEEE Transactions on Industrial Informatics, 2020, PP (99):1-1.

# Design and Simulation of a Position and Average Velocity Closed-loop Control Method of a Hydraulic Servo System

**Xuyao Mao[1*], Wei Duan[2], Chao Xiang[1], Chao Wu[1], Ze Wang[1]**

[1] Wuhan Second Ship Design and Research Institute, Wuhan, Hubei, 430205, China

[2] Wuhan Maritime Communication Research Institute, Wuhan, Hubei, 430079, China

[*]Corresponding author's e-mail: dreamforty@126.com

**Abstract.** A concise position and average velocity closed-loop control method based on PID of valve controlled hydraulic systems is proposed for both positional precise control and average velocity precise control. First, set the target position and the target average velocity; secondly, the average velocity closed-loop control is preferentially carried out at start-up; thirdly, switch to the position closed-loop control circuit when the ratio of the actual position to the target position reaches a critical value; at last, adjust the PID parameters to a proper configuration. The simulations show that this method can acquire both the position and the average velocity control with sufficient accuracy at the same time, in which the position deviation can be less than 0.001% and the average velocity deviation is 2.08%. In addition, the simulations indicate that the system can maintain almost the same precision of the position control and the average velocity control by adding an external load as a disturbance item.

## 1. Introduction

Electro-hydraulic servo elements (valves or variable pumps) are used as electro-hydraulic conversion and amplification components to achieve a certain control law. The output can change rapidly with the input. It not only has fast and easy-to-tune and high-precision response capabilities, but also has the advantage of controlling large inertia to actualize high-power output. Therefore, electro-hydraulic servo control system has been widely studied and used in various technical fields. Precise position closed-loop control with single signal feedback of position or angular displacement was realized by simply adjusting the PID controller parameters [1~3]. For electro-hydraulic servo systems with low damping ratio, it is necessary to improve the natural frequency and damping ratio with correction methods. The stability and rapidity of the system can be significantly improved by increasing the damping ratio to 0.4~0.8 [4]. Commonly used correction methods include velocity and acceleration feedback correction, pressure feedback correction, dynamic pressure feedback correction, etc. Velocity feedback or force feedback were added to the position feedback control at the same time, which effectively improved the position accuracy and running stability of the system [5~7]. Reference [8] put forward a double-loop limiting conversion strategy for a full hydraulic leveller. When one loop was in a working state, the other loop was always in a saturated state, and finally a smooth conversion of the double loop was achieved. In this paper, based on scenarios in which both position and average velocity precise control were needed, a concise position and average velocity closed-loop control method based on PID of valve controlled hydraulic system was proposed. And the effectiveness of the method was verified by simulation analysis.

## 2. Conceptual Design

Figure 1 shows the principle block diagram of a typical position closed-loop control hydraulic servo system. The deviation (*Es*) between the displacement output (*S*) of the actuator and the target position (*S*0) is amplified by the PID controller to adjust the opening of the servo valve. In the process of PID parameter matching and adjusting, it is found that the average velocity (*V*) of the actuator is difficult to control. In this paper, the average velocity refers to the average position change velocity from the time when the position command is issued to the time when the position of the actuator is stable, that is, $V \approx S0/t$, where *t* is the process time mentioned above. Obviously, an input parameter needs to be added to the control loop of the system to regulate the average velocity, which is not reflected in Figure 1.

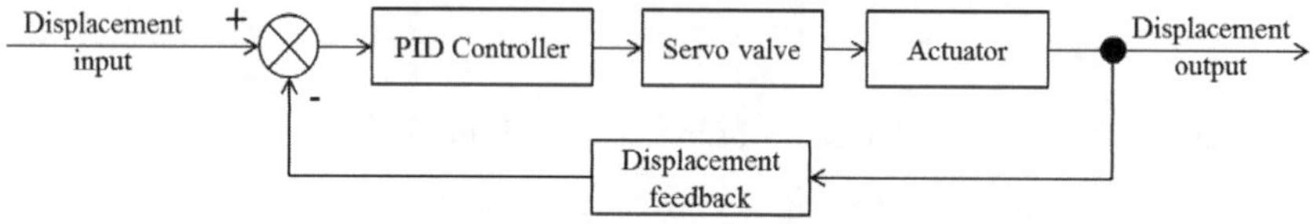


Figure 1. Block diagram of a typical position feedback control

A similar closed-loop control circuit of average velocity was brought out (as shown in Figure 2). The difference is that a clock is introduced for the average velocity calculation to record the process time since the system starts running, and the instantaneous average velocity can be obtained by the ratio of *S*/*t*. The closed-loop control of average velocity with sufficient accuracy could be realized theoretically.

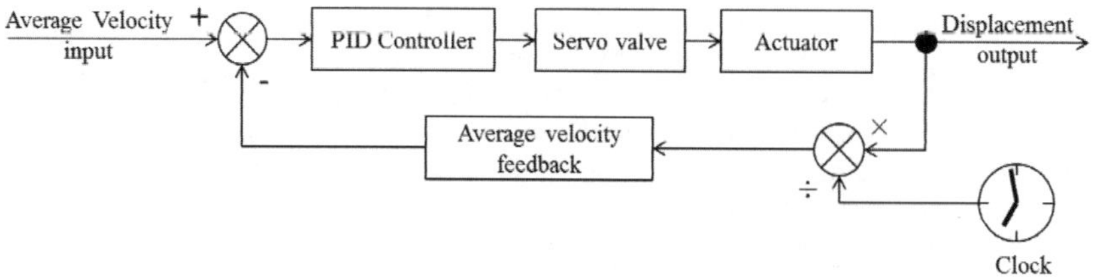


Figure 2. Block diagram of average velocity feedback control

Then a concise switching strategy was designed to fuse the position closed-loop control circuit shown in Figure 1 with the average velocity closed-loop control circuit shown in Figure 2. The details are as follows: (1) Define a process parameter, $R=Es/S0$. (2) Set a critical value of the process parameter as *Rc*. (3) Switch control circuits. When $R>Rc$, the average velocity closed-loop control is put into operation, and when $R \leq Rc$, the position closed-loop control is switched into operation. The comprehensive feedback block diagram can be obtained (as shown in Figure 3), where the position deviation gain $k1=1/S0$.

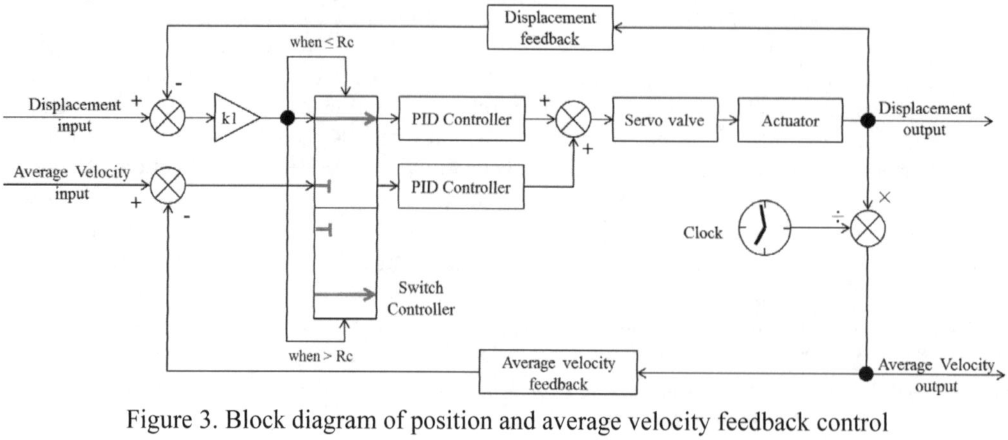


Figure 3. Block diagram of position and average velocity feedback control

## 3. Characteristic analysis

### *3.1. AMESim modeling*

AMESim was adopted for the simulations (Figure 4). Some assumptions and settings were made to simplify the analysis without affecting the simulation results.

- Positive position control and positive average speed control were simulated, and the negative control could follow this schematic diagram by setting more thoughtful boundary conditions. The latter would not be analysed for the time being in this paper.
- The hydraulic actuator was subject to a constant stiction force and a constant Coulomb friction force, and other viscous forces were ignored. The stiction force was converted according to the starting pressure of the actuator 0.2MPa, and the Coulomb friction was calculated by setting the ratio of stiction force/Coulomb friction force to 1.5.
- Major non-linear factors such as servo valve dead zone were not considered, the frequency of the servo valve was set to 50Hz, and the flow capacity was 100L/min when the pressure difference at each valve port was 3.5MPa.
- The flowrate provided by the hydraulic power unit must be greater than the flowrate required at the target average velocity.

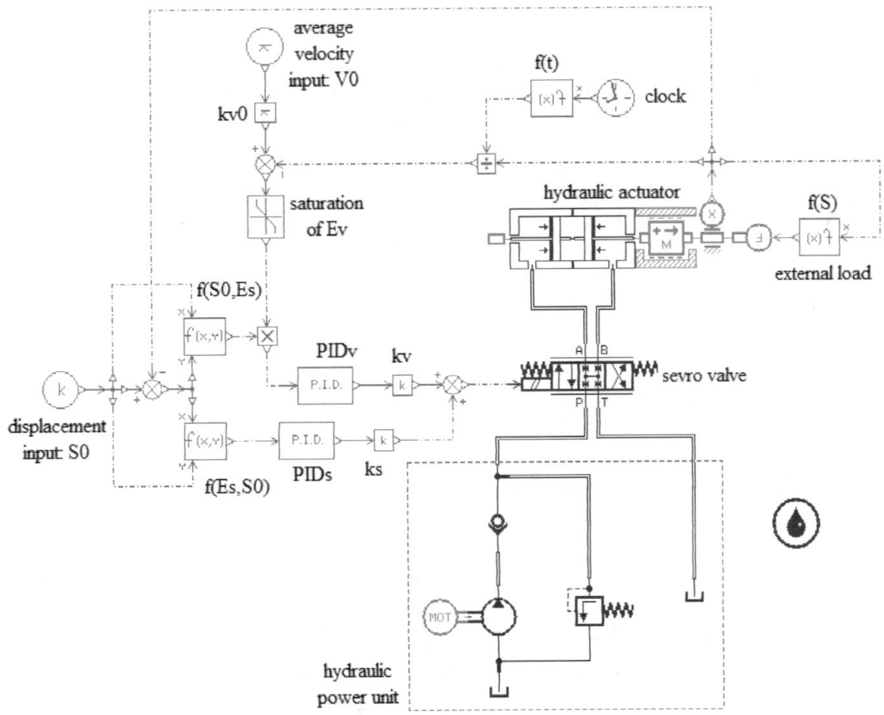


Figure 4. AMESim diagrammatic of the closed-loop control system.

Some main calculation parameters and their values were displayed in Table 1. Symbols or functions shown in Figure 4 were described in Table 2.

Table 1. Parameters and values for the analysis

| Parameter | Value | Unit | Parameter | Value | Unit |
|---|---|---|---|---|---|
| Flowrate of the power unit | 100 | L/min | Relief valve cracking pressure | 10 | MPa |
| Pipeline diameter of each port of the servo valve | 20 | mm | Pipeline length of each port of the servo valve | 10 | m |
| Piston diameter of the actuator | 200 | mm | Rod diameter of the actuator | 120 | mm |
| Piston mass of the actuator | 100 | kg | *Rc* | 0.01 | Null |

Table 2. Settings displayed in Figure 4

| Symbol or function | Value or formula | Usage |
|---|---|---|
| $f(t)$ | $f(t)=\begin{cases}\varepsilon=10^{-12} & \text{when } t=0\\ t & \text{when } t>0\end{cases}$ | Prevent calculation overflow when t=0 |
| $kv0$ | —— | Adjust the gain of average velocity input |
| Saturation of $Ev$ | $Ev=\begin{cases}0 & \text{when } Ev<0\\ Ev & \text{when } Ev\geq 0\end{cases}$ | Ensure that the actuator does not move back when $Ev < 0$ |
| $f(S0,Es)$ | $f(S0,Es)=\begin{cases}0 & \text{when } \frac{Es}{S0}\leq Rc\\ 1 & \text{when } \frac{Es}{S0}> Rc\end{cases}$ | Switch control principle |
| $f(Es,S0)$ | $f(Es,S0)=\begin{cases}0 & \text{when } \frac{Es}{S0}> Rc\\ 1 & \text{when } \frac{Es}{S0}\leq Rc\end{cases}$ | Switch control principle |
| $PIDv$ | —— | Adjust the PID parameters of average velocity closed-loop control |
| $PIDs$ | —— | Adjust the PID parameters of position closed-loop control |
| $kv$ | —— | Adjust the gain of average velocity closed-loop control |
| $ks$ | —— | Adjust the gain of position closed-loop control |
| $f(S)$ | $f(S)=\begin{cases}0 & \text{without load}\\ 10000\times\sin\left(\frac{S}{0.8}\times\frac{\pi}{2}\right) & \text{with load}\end{cases}$ | Load calculation input or exit function |

### *3.2. Simulation and analysis*

The simulation duration was set to 20 seconds. Figure 5 shows the time-varying curves of the actuator displacement of different target positions but the same target average velocity, and Figure 6 shows the time-varying curves of the actuator displacement of different target average velocities but the same target position. It can be seen that the displacement changes smoothly with almost no overshoot, and the final displacement can be stabilized at S0×(1±0.01%). Therefore, for the convenience of calculation, another setting was made by taking the average velocity as $V=S0\times 0.9999/t_{99.99\%}$, where $t_{99.99\%}$ means the time that the actual position ($S$) reaches $S0\times 0.9999$. The results of Figure 5 and Figure 6 are summarized in the following Table 3. It indicates that the maximum position simulation deviation is 0.0007% and the maximum average velocity simulation deviation is 2.08% under various simulation conditions, and there exists a trend that the smaller the target position, the larger the average speed deviation. This is because the smaller the target position, the greater the influence of control fluctuation when the average velocity control starts, and the smaller the time left for PID adjustment, resulting in a large control deviation (as seen in Figure 7). The decrease of the back end of the curve in Figure 7 is due to the decrease of the calculated average velocity caused by the continuous increase of time when the position is holding.

Table 3. Results summarized from Figure 5 and Figure 6

| Simulation conditions | Position control (m) | | Average velocity control (m/s) | |
|---|---|---|---|---|
| | Target | Simulation (deviation) | Target | Simulation (deviation) |
| S0=0.5m V0=0.05mm/s | 0.5 | 0.499999~0.500002 (0.0004%) | 0.05 | 0.0496342(0.73%) |
| S0=0.3m V0=0.05mm/s | 0.3 | 0.299999~0.300002 | 0.05 | 0.0489613 (2.08%) |

| | | | | |
|---|---|---|---|---|
| | | (0.0007%) | | |
| S0=0.7m V0=0.05mm/s | 0.7 | 0.699999~0.700002 (0.0003%) | 0.05 | 0.0499501(0.10%) |
| S0=0.5m V0=0.03mm/s | 0.5 | 0.499999~0.500000 (0.0002%) | 0.03 | 0.0302580 (0.86%) |
| S0=0.5m V0=0.07mm/s | 0.5 | 0.499999~0.500000 (0.0002%) | 0.07 | 0.0699712(0.04%) |

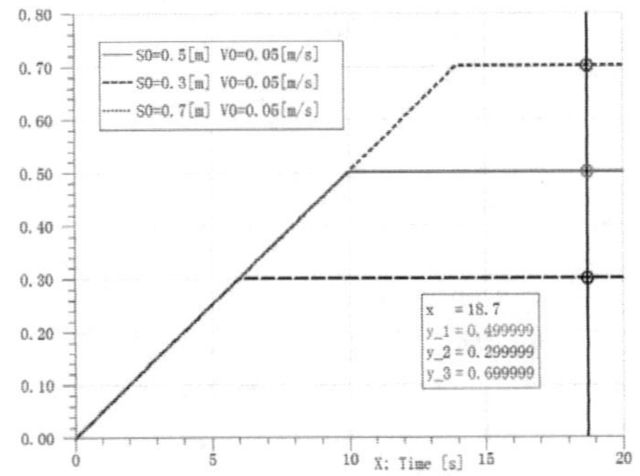


Figure 5. displacement of the actuator (m)

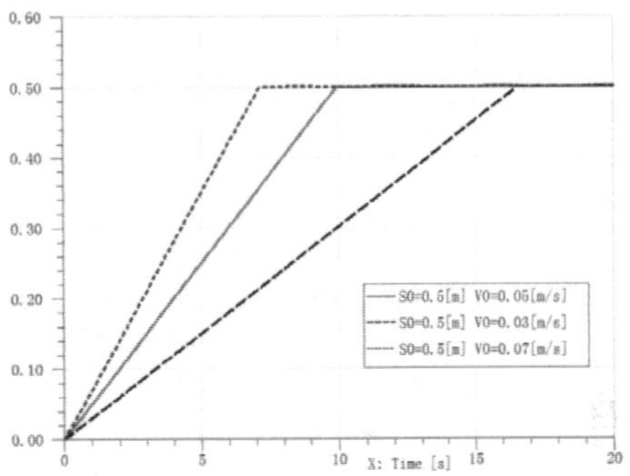


Figure 6. displacement of the actuator (m)

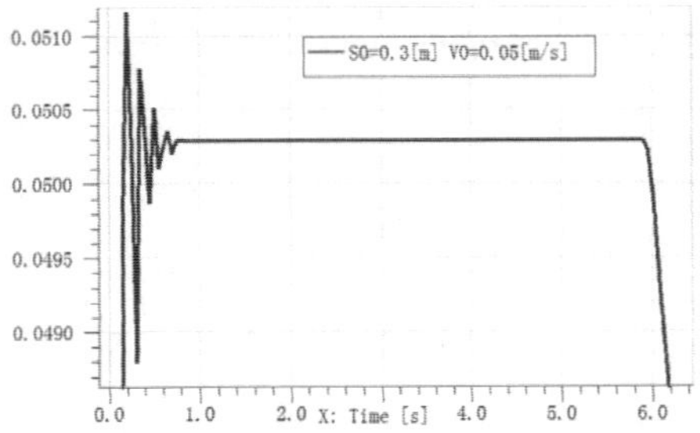


Figure 7. average velocity of the actuator (m/s)

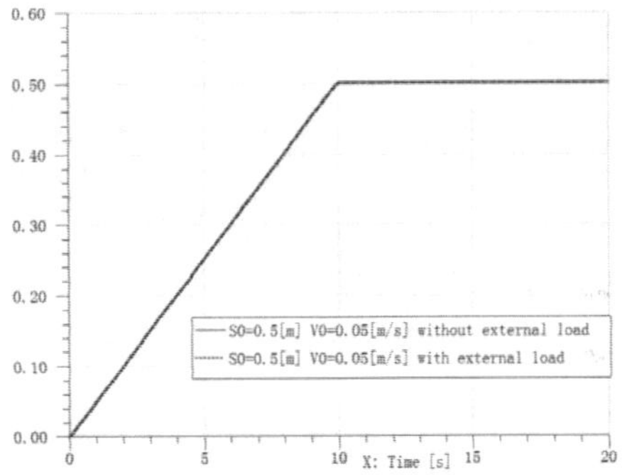


Figure 8. displacement of the actuator (m)

### *3.3. Effect of external load*

Figure 8 shows the comparison of actuator displacements with and without external load. The load can be obtained in the last row of Table 2, which is a sine function expression related to the actuator displacement. It is found that under the action of dynamic load, the curves of the two simulation conditions almost perfectly overlap, and the system position control accuracy and average velocity control accuracy with external load are basically the same as those without external load, that is to say, this closed-loop control method can effectively withstand the load interference and maintain the control accuracy.

## 4. Conclusions

Generally, the following conclusions could be obtained through simulation analysis above.

- The closed-loop control method of position and average velocity of the hydraulic servo system involved in this paper has simple logic and strong enforceability, and can effectively realize the high-precision control of position and average velocity synchronously.
- The smaller the target position, the larger the average velocity deviation. And the deviation should be reduced by adjusting the PID parameter separately.

- The closed-loop control can well compensate the influence of external load and other disturbances on the system, and realize that the system response is "unrelated" of external load and other disturbances to a certain extent.

**References**

[1] X.J. Chen, X.D. Wu. (2009) Design and Implementation of Digital PID Based on Hydraulic Proportional Position Control, *Mechanical Engineering & Automation* , (06): 126-128 (in Chinese)
[2] M.J. Qiu, X.L. Guo, B. Ning, etc. (2017) Modeling, simulation and experimental research on the closed-loop system of hydraulic cylinder position control by common directional valve based on Simulink, *Hydraulics Pneumatics & Seals*, 37(01): 50-53 (in Chinese)
[3] J.J. Zhao, M.Y. Mu. (2021) Research on hydraulic angular position closed-loop control system based on AMEsim, *Mechanical Engineer* , (11): 69-74 (in Chinese)
[4] M.L. Yi, S.P. Cao, Y.S. Liu. (2010) Electro-hydraulic control technology. *Huazhong University of Science & Technology Press*(in Chinese)
[5] Z.G. Qiao, X.G. Zhang, B. Zhao, etc. (2019) Simulation Research on Speed-Position Composite Control Characteristics of Fast Forging Hydraulic Press, (08): 13-20 (in Chinese)
[6] W.N. Huang, L. Quan, H. Cheng, etc. (2020) Speed and position compound control characteristics of hydraulic excavator slewing system based on handle integral. Transaction of Beijing Institute of Technology, 40(02): 193-197 (in Chinese)
[7] H. Liu, C.J. Zhao, Y.S. Ning, etc. (2022) Research on the position and pressure compound control of the hydraulic servo system of the spinning machine. *Chinese Hydraulics & Pneumatics*, 46(07): 11-16 (in Chinese)
[8] D.P. He, H.Y. Han, Q.X. Huang, etc. (2016) Research on position closed-loop and force closed-loop conversion strategy of full hydraulic leveller, *Journal of Taiyuan University of Science and Technology*, 37(04): 302-307 (in Chinese)

# Research on Common Mode Analysis Method in System Safety Analysis

**Dawei LI[1*], Jianguo Gao[1], Zhongyuan ZHANG[2], Jihui XU[1]**

[1] Equipment Management and Unmanned Aerial Vehicle Engineering College, Air Force Engineering University, Xi'an 710051, China

[2] Unit 93220 of PLA, Harbin 150049, China

[*]Corresponding author's e-mail: lidaweitg@163.com

**Abstract.** In system safety analysis, the existence of Common Mode Failure (CMF) greatly undermines the independence of components and poses great threat to safety of the system. In Common Mode Analysis (CMA), whether CMF needs to be considered for basic events under different types of logic gates in Fault Tree, it lacks strict reasoning and comprehensive consideration. This paper takes CMF of the same and different levels under "AND gate" and "OR gate" in fault tree as an example, and uses reasoning analysis and mathematical calculation methods to compare the influence of the existence of CMF on failure rate of the system. The results show that whether CMF need to be considered for the component depends on the type of logic gate correspond to the higher-level component. The research results have certain reference value for the analysis of CMF in CMA.

## 1. Introduction

In system safety analysis of aircraft, an important goal is to eliminate single point of failure and ensure safety level of the system. Common Cause Failure (CCF) refers to simultaneous failure of multiple components in the system caused by the same cause or event. The corresponding safety analysis method is called Common Cause Analysis (CCA). Existence of CCF greatly increases the probability of single point of failure.

In SAE ARP4761 [1], CCA includes three methods: Zonal Safety Analysis (ZSA), Particular Risks Analysis (PRA) and Common Mode Analysis (CMA). PRA focuses on the effect of common cause other than the system or components themselves, such as bird strikes, icing, rotor blasting of engine, etc. ZSA evaluates the potential mutual influences between each sub-system and equipment by the method of regional division. CMA mainly analyzes whether the redundant system is independent, and evaluates whether the redundant system can achieve the expected safety level.

In many literatures, it is usually concluded that CMF must be considered for basic events under "AND gate"(i.e., parallel system) in Fault Tree Analysis (FTA), while basic events under "OR gate" (i.e., a series system), CMF does not need to be considered. However, few literatures provide a rigorous proof for this conclusion. At the same time, in the specific analysis, if two basic events are at different levels of fault tree, there is no relevant conclusion whether CMF need to be considered. Therefore, this paper aims to evaluate the influence of CMF from the perspective of reasoning analysis and mathematical calculation, and give a theoretical analysis of the influence of CMF.

## 2. Common Mode Analysis

Common Mode Analysis is a method of CCA, which can be used to verify the independence of basic events under "AND gate" in FTA, which runs through the whole process of aircraft development, using and maintenance, and can be used to ensure rationality of the design. The point of CMA is to identify multiple failures due to a single cause, which may come from the same design, manufacturing defects, human error, or external factors [2].

CMA can be applied to any type of system. However, in actual safety analysis, it is mainly used to analyze safety-critical system with redundancy design [3-5]. CMA should be carried out in the early stage of equipment development to identify key systems that need to be considered in the design. Detailed component-level CMA needs to be carried out in detailed design stage.

## 3. Causes of CMF

Causes of CMF can be divided into engineering factors, usage factors and environmental factors [6-8].

### *3.1 Engineering factors*

Engineering factors refer to the fact that designers have not taken appropriate measures due to lack of understanding to causes of failure, which usually include the following aspects:

(1) Normative conditions. Incomplete or conflicting norms and conditions.

(2) Design. For example, the same component is used in working system and backup system, if it fails, the same failure mode may occur in each channel; several independent components of hydraulic system are connected together, if lack of isolation, they can fail at the same time due to external impacts; several channels of a computer are designed with the same software code, if a design flaw is existed in the code, several channels of the computer may fail at the same time.

(3) Manufacturing. Equipment or components produced in the same batch may have defection during manufacturing process, because they are not produced according to the specified process requirements, or incorrect assembly.

### *3.2 Usage factors*

(1) Storage. For example, aircraft spare parts are stored in warehouse all the year round. If storage environment is not suitable or maintenance is not proper, the spare parts may be deteriorated.

(2) Maintenance. Such as maintenance error, electric generator is failed to incorrectly adjust by maintenance personnel for insufficient technology.

### *3.3 Environmental factors*

Multiple aircraft systems and components can be failed for changes of natural environment in aircraft operation, such as icing, lightning strikes, bird strikes, etc. For CMF caused by environmental factors, PRA is mainly used to analyze the failure due to their big influence.

Experiences show that in CMF caused by the above factors, engineering factor is the most important, and it is also the factor that can be controlled in design and development, and is the focus in CMA.

## 4. CMF impact analysis

### *4.1 Series system*

For a series system composed of two components, fault tree is established taking the system failure as top event as shown in Figure 1. Among them: $S$ represents system failure. $A$, $B$ represents failure of component $A$ and $B$. $A'$, $B'$ represents independent failure of component $A$ and $B$. $C_{AB}$ represents CMF of $A$ and $B$. Figure 1(a) shows the fault tree when there is no CMF in $A$ and $B$. Figure 1(b) shows two drawing methods of the fault tree when there is CMF in $A$ and $B$. It can be drawn as the left figure, and can also be simplified to the right figure.

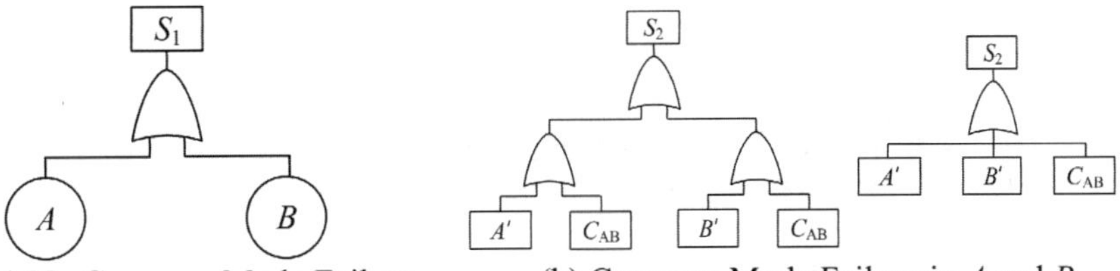


(a) No Common Mode Failure (b) Common Mode Failure in *A* and *B*

Figure 1. Fault tree of system failure in series system

In the case of no CMF, minimum cut set of the system is $\{A\}$ and $\{B\}$. In the case of CMF, minimum cut set of the system is $\{A'\}$, $\{B'\}$ and $\{C_{AB}\}$. Failure rate of the system can be obtained from the minimum cut set as follows:

$$\begin{aligned} P(S_1) &= P(A \cup B) = P(A) + P(B) - P(A)P(B) \\ &= P(A') + P(C_{AB}) + P(B') + P(C_{AB}) \\ &\quad -[P(A') + P(C_{AB})][P(B') + P(C_{AB})] \\ &= P(A') + P(B') + 2P(C_{AB}) - P(A')P(B') \\ &\quad -P(C_{AB})[P(A') + P(B')] - P(C_{AB})^2 \end{aligned} \tag{1}$$

$$\begin{aligned} P(S_2) &= P(A' \cup B' \cup C_{AB}) = P(A') + P(B') \\ &\quad +P(C_{AB}) - P(A')P(B') - P(A')P(C_{AB}) \\ &\quad -P(B')P(C_{AB}) + P(A')P(B')P(C_{AB}) \end{aligned} \tag{2}$$

Subtracting equation (1) from equation (2):

$$P(S_2) - P(S_1) = P(C_{AB})[P(A')P(B') + P(C_{AB}) - 1]$$

In general, assuming $P(A')$, $P(B') < 0.1$, we can get $P(S_1) > P(S_2)$. Therefore, when there is CMF in *A* and *B*, failure rate of the top event is much lower. CMF does not need to be considered at this time.

### *4.2 Parallel system*

For a parallel system composed of two components, fault tree is established taking the system failure as top event as shown in Figure 2. Figure 2(a) shows the fault tree when there is no CMF in *A* and *B*. Figure 2(b) shows two drawing methods of the fault tree when there is CMF in *A* and *B*. It can be drawn as the left figure, and can also be simplified to the right figure.

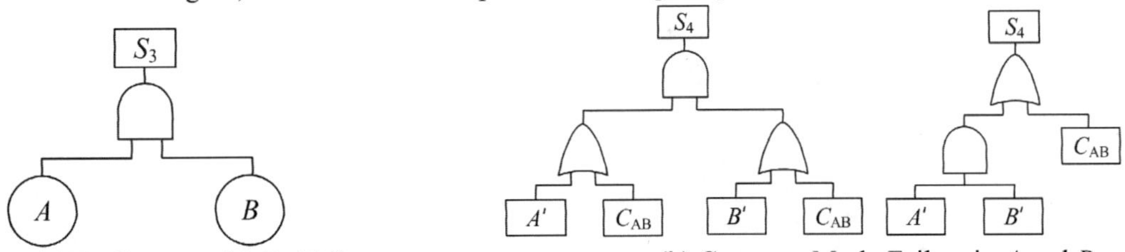


(a)No Common Mode Failure (b) Common Mode Failure in *A* and *B*

Figure 2. Fault tree of system failure in parallel system

In the case of no CMF, minimum cut set of the system is $\{A, B\}$. In the case of CMF, minimum cut set of the system is $\{A', B'\}$ and $\{C_{AB}\}$. Failure rate of the system can be obtained from the minimum cut set as follows:

$$\begin{aligned} P(S_3) &= P(A \cap B) = P(A)P(B) \\ &= [P(A') + P(C_{AB})][P(B') + P(C_{AB})] \\ &= P(A')P(B') + P(C_{AB})[P(A') + P(B')] + P(C_{AB})^2 \end{aligned} \tag{3}$$

$$\begin{aligned} P(S_4) &= P(C_{AB} \cup A'B') \\ &= P(C_{AB}) + P(A'B') - P(C_{AB})P(A'B') \\ &= P(C_{AB}) + P(A')P(B') - P(C_{AB})P(A')P(B') \end{aligned} \tag{4}$$

Subtracting equation (3) from equation (4):

$$P(S_4)-P(S_3)=P(C_{AB})[1-P(A')-P(B')-P(A')P(B')-P(C_{AB})]$$

In general, assuming $P(A'),P(B')<0.1$, we can get $P(S_4)>P(S_3)$. Therefore, when there is CMF in *A* and *B*, failure rate of the top event is much higher. CMF must be considered at this time.

*4.3 Bottom event not at the same level of the fault tree*

(1) CMF of basic events at different levels under "OR gate"

For the fault tree shown in Figure 3, when there is no CMF in the system, it is shown in Figure 3(a); when there is CMF in *B* and *C*, it is shown in Figure 3(b). Among them: *S* represents system failure. *A*, *B* and *C* represents failure of component *A*, *B* and *C*. $B'$, $C'$ represents independent failure of component *B* and *C*. $C_{BC}$ represents CMF of *B* and *C*.

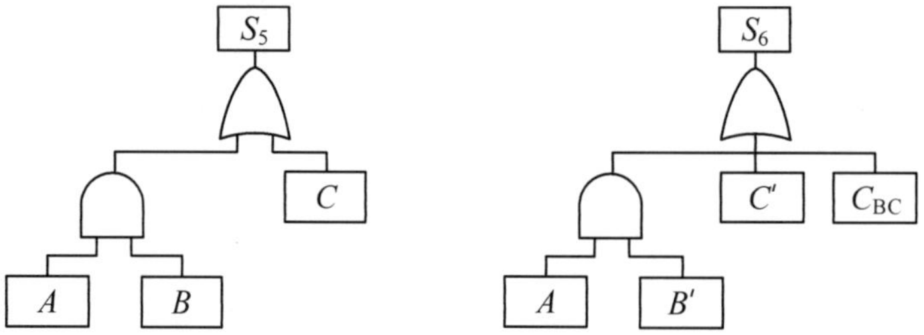


(a)No CMF (b)CMF in *B* and *C*

Figure 3. Fault tree of system failure under "OR gate"

Assuming that the failure rate of *A*, *B*, *C* is $P(A)=P(B)=P(C)=0.01$. Independent failure rate of *B* and *C* is $P(B')=P(C')=0.009$. The probability of CMF of *B* and *C* is $P(C_{BC})=0.001$.

Calculation result shows that failure rate of the system is $P(S_5)=0.0101$ without counting CMF, and failure rate of the system is $P(S_6)=0.01008$ while CMF of *B* and *C* is taken into account. *B* and *C* are at different levels of the fault tree, they are bottom events whose upper layer belong to "OR gate". It can be seen that when there is CMF in *B* and *C*, failure rate of the top event is smaller than there is no CMF. Therefore, there is no need to consider CMF in *B* and *C* at this time.

(2)CMF of basic events at different levels under "AND gate"

When basic events with CMF are not at the same level of fault tree, and they are bottom events whose upper layer belong to "AND gate", it is shown in Figure 4.

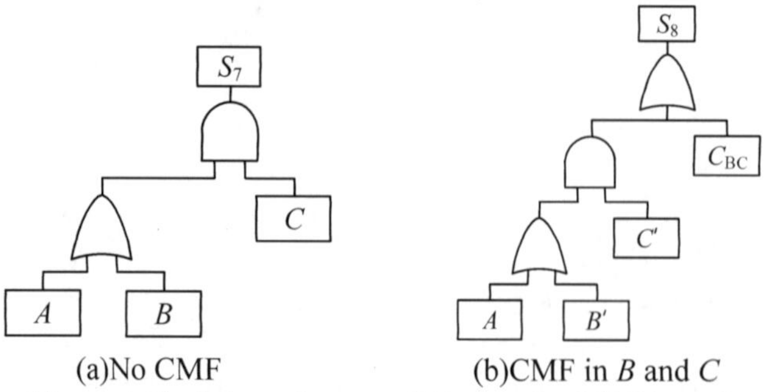


(a)No CMF (b)CMF in *B* and *C*

Figure 4. Fault tree of system failure under "AND gate"

Meanings of each symbol and failure rate of basic events are the same as above. Calculation result shows that failure rate of the system is $P(S_7)=0.000199$ without counting CMF, and failure rate of the system is $P(S_8)=0.00177$ while CMF in *B* and *C* is taken into account. *B* and *C* are at different levels of the fault tree, they are bottom events whose upper layer belong to "AND gate". It can be seen that when there is CMF in *B* and *C*, failure rate of the top event is higher than there is no CMF. Therefore, influence of CMF in *B* and *C* on the analysis results must be considered at this time.

*4.4 Result analysis*
Based on the above reasoning and calculations, the following conclusions can be drawn:

(1)CMF in the system can undermine the assumption of independence of basic events and alter the results of the safety analysis.

(2)Analysis results in Section 3.1 and 3.2 show that for basic events under the "OR gate" of fault tree, failure rate of top event is much lower after considering CMF, so the influence of CMF does not need to be considered. For basic events under the "AND gate" of fault tree, failure rate of top event is much higher after considering CMF, therefore, influence of CMF must take into account.

(3) Calculation results in Section 3.3 show that for basic events at different levels of fault tree in the system, analysis results depend on the type of the logic gate correspond to high-level basic event. If the high-level basic event corresponds to an "OR gate", the effect of CMF does not need to be considered, and if it corresponds to an "AND gate", the effect of CMF must take into account.

## 5. Conclusions
By the method of reasoning analysis and mathematical calculation and under certain assumptions, it is proved that the influence of CMF must be considered for basic events which are under "AND gate", and the influence of CMF does not need to be considered for basic events which are under "OR gate" in CMA. At the same time, an example is carried out on whether CMF need to be considered for basic events at different levels of fault tree. Valuable conclusions are obtained, which have certain reference value for engineers and researchers to analyze CMF in CMA.

## Acknowledgments
This paper is one of phased achievements of NSFC (The mission safety modelling and sensitivity analysis of "liveware-equipment-environment" complex systems for an uncertainty perspective, 71701210).

## References
[1] SAE ARP4761. Guidelines and Methods for Conducting the Safety Assessment Process on Airborne Systems and Equipments [S]. America: The Engineering Society For Advancing Mobility Land Sea Air and Space, 1996

[2] Zhao Tingdi, Jiao Jian. Safety Design Analysis and Verification[M]. Beijing: National Defense Industry Press, 2011

[3] QI Jiajun, ZHAO Xinwen, ZHANG Yongfa, et al. Common Cause Failure Analysis Method of Safety Injection System Based on BDD[J]. Journal of Ordnance Equipment Engineering, 2017, 38(9): 143-147

[4] Wang Mengmeng, Sun Fan. Methodology of Common Mode Analysis and Utilization in Civil Aircraft Design[J]. Civil Aircraft Design & Research, 2015(2): 43-46

[5] Zhang Yongxian, Liao Zhao-cong, Wang Yueqian, et al. Common cause failure analysis method for double two out two system[J]. Computer Engineering and Design, 2019, 40 (1): 285-289

[6] Ni Dajiang. Research on Civil Airplane Safety Work System and Methods of Common Mode Analysis[D]. Nanjing: Nanjing University of Aeronautics and Astronautics master thesis, 2007

[7] Kong Fanfan, Sun Youchao, Wang Wei. Research on Common Mode Failure Analysis Method for Civil Airplane[J]. Journal of Civil Aviation University of China, 2008, 26(5): 49-51

[8] Kong Fanfan. Research on Common Mode Failure Analysis Method and Its Engineering Application for Civil Airplane[D]. Nanjing: Nanjing University of Aeronautics and Astronautics master thesis, 2008

# Numerical study of air-breathing two-phase rotating detonation engine under Ma 6 flight conditions

**Chenglong Yan[1], Wei Lin[1*], Chen Shu[1], Yue Zhi[2] and Wei He[1]**

[1]Department of Aerospace Science and Technology, Space Engineering University, Beijing, 101416, China

[2]College of Environment and Ecology, Chongqing University, Chongqing, 400044, China

[*]Corresponding author's e-mail: linweiqy@163.com

**Abstract.** Air-breathing two-phase rotating detonation engines possess high thermodynamic cycle efficiency and have attracted extensive attention in domain of wide range flight aircraft. In this study, an engine configuration is proposed, and the corresponding numerical model is established using the Euler-Lagrange method. The engine type is suitable for flying at an altitude of 28 kilometers and a flying speed of Ma 6. Our data show that the engine operates primarily in chaos in this flight state. The peak pressure of the detonation wave is about 0.85 MPa, the peak temperature of the detonation wave is about 4500 K, the average thrust of the engine is 2637 N, and the average fuel-specific impulse is about 2989 s.

## 1. Introduction and experimental configuration

The detonation engine has the advantages of high thermal cycle efficiency, fast heat release and compact structure. Among them, two-phase rotating detonation ramjets are considered to be suitable for aerospace applications. Maintaining stable operation of detonation engines at high flight speed is difficult. Previous research on air-breathing rotating detonation engines has mainly concentrated on gas-phase fuel [1]-[9], and the engines for two-phase detonation concentrated on rocket-based rotating detonation engines [10]-[15]. However, due to challenges associated with testing and other complications, few studies have been reported publicly to date regarding air-breathing two-phase rotating detonation engines. To date, no numerical model of a two-phase rotating detonation ramjet has yet been publicly reported. To fill the knowledge gap, we proposed a three-dimensional air-breathing two-phase rotating detonation numerical engine. It is established for the flight altitude of 28 km and the flight velocity of Ma 6 in this study. We characterized the flow field of the air-breathing two-phase rotating detonation engine at the condition of high flight Mach number.

The configuration is shown in Figure 1. Under supersonic inflow, the air-breathing rotating detonation engine adopts a three-dimensional annular combustion chamber, and the detonation combustion chamber includes three parts: isolation section, expansion section, and combustion chamber. The expansion angle of the expansion section is 15°. This study injects kerosene into the engine in the isolation section. The injection position is located 35mm from the inlet. Three hundred sixty square jet orifices of 0.3mm×0.3mm are used. The jet orifices are divided into two groups, 180 in each, placed on the inner and outer walls, respectively.

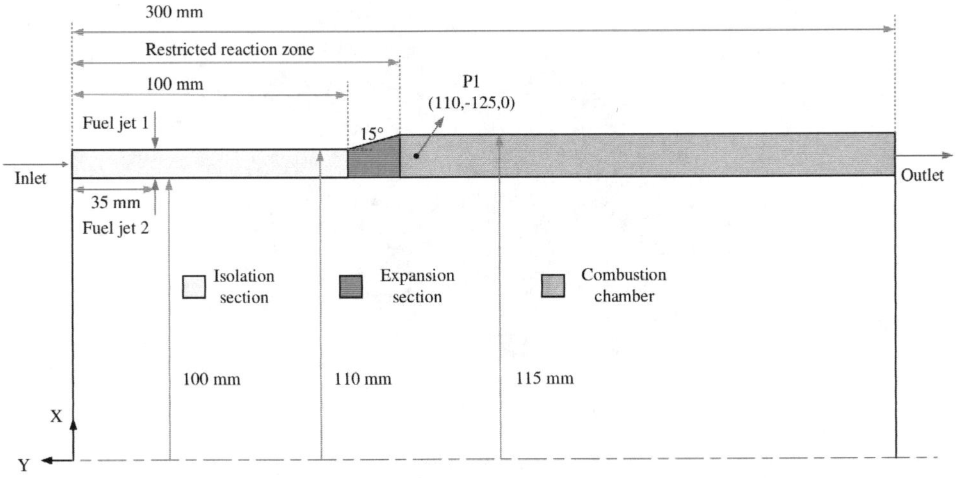

Figure 1. Simulation configuration and monitoring points.

## 2. Simulation method

The numerical model established in this study is derived from the Euler-Lagrangian method. The gas phase adopts the three-dimensional reaction flow equation in the Cartesian coordinate system and integrates the particle force differential equation in the Lagrangian coordinate system to solve the trajectory of the discrete phase droplet. In this study, the laminar finite-rate model was used, and the Arrhenius formula was used to calculate the chemical source term, ignoring the effect of turbulent pulsation. Using $C_{12}H_{23}$ instead of kerosene, the chemical reaction equation is:

$$C_{12}H_{23}+17.75O_2=12CO_2+11.5H_2O$$

The total pressure recovery coefficient is 0.4, and the supersonic flow conditions at the inlet of the combustion chamber are shown in Table 1. $Y_{O2}$ and $Y_{N2}$ represent the mass fraction of oxygen and nitrogen, respectively.

Table 1. The supersonic flow conditions at the inlet of the combustion chamber.

| Flight velocity | Total temperature /K | Total pressure /Pa | Static temperature /K | Static pressure /Pa | $Y_{O2}$ | $Y_{N2}$ |
|---|---|---|---|---|---|---|
| Ma 6 | 1842 | 1001640 | 658 | 27268 | 0.232 | 0.768 |

The pressure inlet was adopted in the simulation. With the equivalence ratio of 0.9, the mass flow of fuel was 0.09 kg/s. The finite volume method is used for numerical solution. The gradient was computed using the least squares cell-based method, the convection flow was calculated by the second-order upwind method. The first-order upwind method was used to calculate the turbulence equation and specific dissipation rate. The structured mesh was drawn using ICEM software, and the locations of the jet orifices are densified. After processing, the minimum grid is 0.3 mm, and the total number of grids is 2.6676 million. To better organize the combustion while ensuring the stability of the calculation. It is generally assumed that the fuel before the expansion section does not participate in the reaction. Therefore, the chemical reaction is limited in both the expansion and isolation sections. The structured grid of the computational domain is shown in Figure 2.

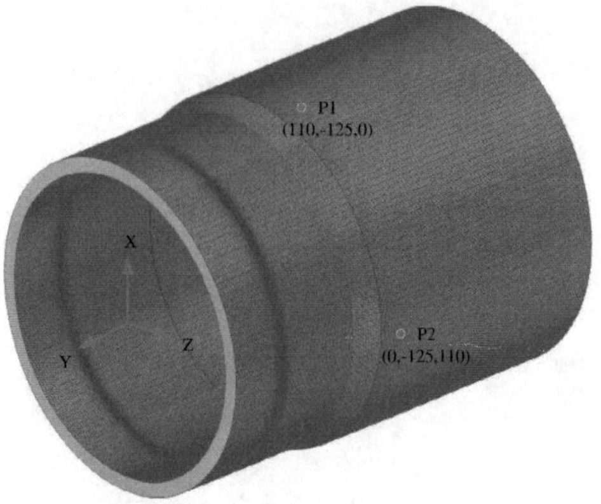


Figure 2. Computational domain described by structured grid.

## 3. Results and Discussions

After the detonation flow field is established, the flow field of 3.2 ms is analyzed, and the pressure contour and kerosene droplet distribution are shown in Figure 3. It can be seen that the detonation wave presents a chaotic state in the combustion chamber. For example, the propagation directions of several detonation waves are inconsistent, and there is no definite distribution law. In the combustion chamber, the detonation waves collide or propagate in the same direction.

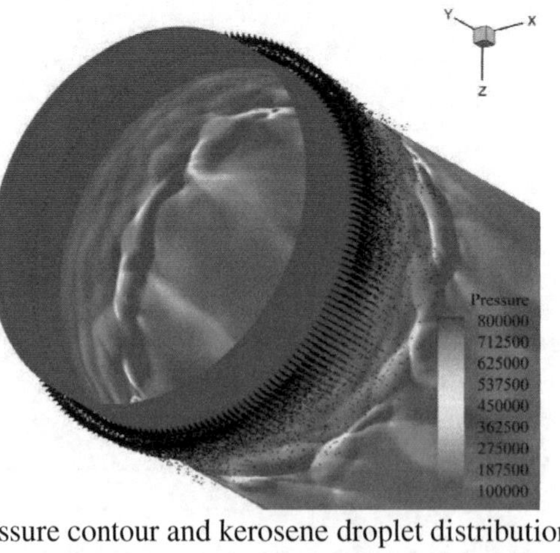


Figure 3. Pressure contour and kerosene droplet distribution (3.2 ms).

Figures 4 and 5 are the pressure and temperature monitoring curves for monitoring P1 and P2, respectively. In Figure 4, the pressure monitoring curves of P1 and P2 are significantly different. Under low Mach number (under Ma 3) flight conditions, due to the low total temperature (about 606 K) of the incoming flow, the detonation wave will propagate in the same direction, and the propagation direction will not change. Even in the multi-wave head state, the propagation directions of several detonation waves in the combustion chamber will tend to be consistent over a while. In this situation, the fluctuation trend of the pressure monitoring curves at points P1 and P2 should be similar, and there is only one phase difference between the functions of different monitoring points. However, under the high Mach number flight conditions described in Table 1, hot spots are easily formed in the combustion chamber due to the high total temperature, resulting in new detonation waves. The uncertainty of the propagation direction of the new detonation wave results in a chaotic state under the current conditions. Therefore, the pressure monitoring curves of points P1 and P2 are pretty different.

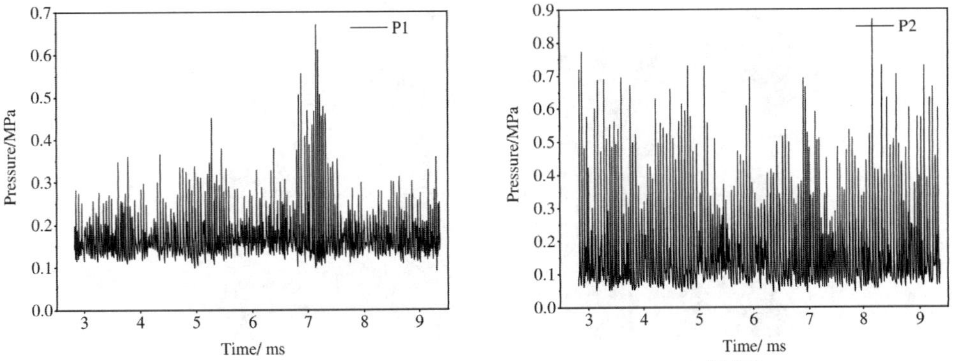


a) P1 b) P2

Figure 4. Pressure monitoring curve.

As can be seen from the monitoring curves of the pressure and temperature of the detonation wave (Fig. 4 and Fig. 5), the peak pressure and temperature of the detonation wave are ~0.85 MPa and ~4500 K, respectively. Under these flight conditions, the static temperature in the engine is high. Therefore, the thermal protection of the engine needs to be focused on under high Mach number flight conditions rather than the pressure resistance.

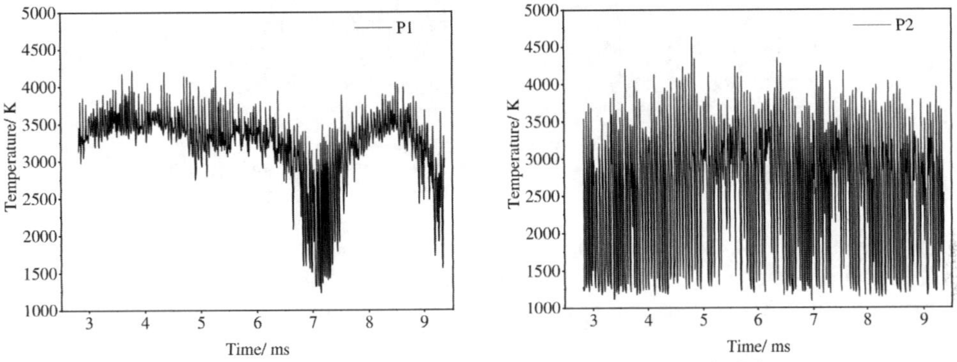


a) P1 b) P2

Figure 5. Temperature monitoring curve.

As detailed in Figure 3, the collision and pursuit of detonation waves will cause the annihilation and regeneration of detonation waves in the engine. Therefore, the chaotic state of engine detonation waves is temporary. Under certain conditions, the propagation directions of several chaotic detonation waves may converge to form an ordered state of detonation wave propagation. For example, as shown in Figure 6: at 4.65 ms, the detonation wave propagation in the combustion chamber presents a chaotic state, and with the development of a specific time, the detonation wave develops into an orderly propagation state at 7.35 ms. However, the time occupied by the ordered state is relatively short under the current flight conditions. It will immediately transform into a chaotic state propagation after the ordered state is established, as shown in Figure 6(d).

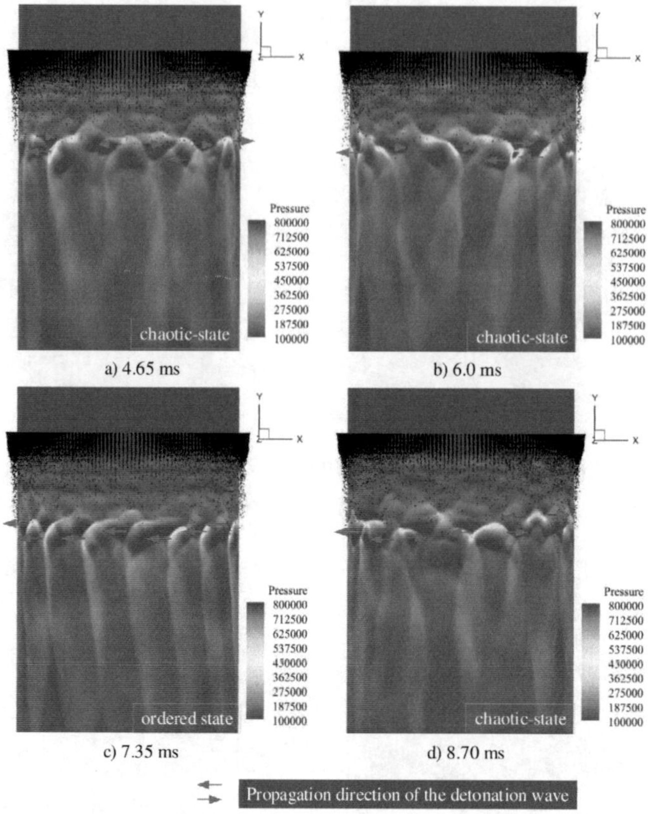


Figure 6. Transition between chaotic and ordered states.

In this study, the propulsion performance of the current engine is predicted without considering the effect of the engine's tail nozzle. The thrust monitoring curve is shown in Figure 7. The average thrust of the engines was 2637 N. The maximum thrust of the engine is about 2725 N, the minimum thrust of the engine is about 2525 N, and the engine's thrust is relatively stable. At the same time, the fuel-specific impulse of the engine is about 2989 s. Therefore, the propulsion performance of the engine is excellent.

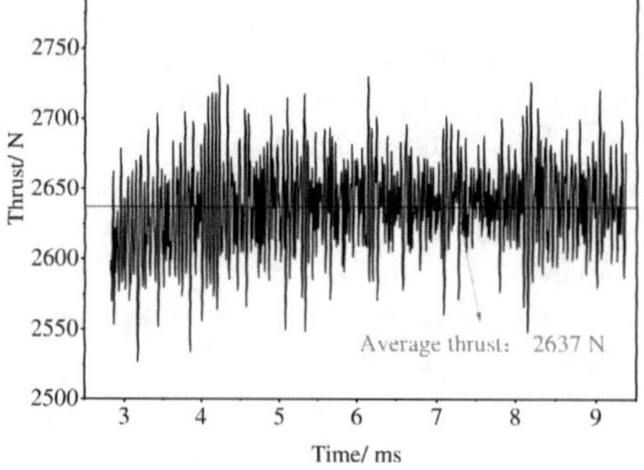


Figure 7. Thrust monitoring curve.

## 4. Conclusions

This study designed an air-breathing rotating detonation engine configuration suitable for Ma 6 flight conditions. The structure in the flow field is obtained by numerical simulation, and the two states of chaotic propagation and ordered propagation of the detonation wave are analyzed. The propulsion performance of the detonation engine is predicted, which can provide theoretical basis for the design of the detonation engine.

**Acknowledgments**
The authors gratefully acknowledge support from the National Natural Science Foundation of China (Grant No. 11902361).

**References**

[1] Ivanov, V., Frolov, S., Zangiev, A., Zvegintsev, V., & Shamshin, I. (2021a). Hydrogen fueled detonation ramjet: Conceptual design and test fires at Mach 1.5 and 2.0. Aerospace Science and Technology, 109, 106459.
[2] Wu, K., Zhang, S., Luan, M., & Wang, J. (2020). Effects of flow-field structures on the stability of rotating detonation ramjet engine. Acta Astronautica, 168, 174–181.
[3] Braun, E. M., Lu, F. K., Wilson, D. R., & Camberos, J. A. (2013b). Airbreathing rotating detonation wave engine cycle analysis. Aerospace Science and Technology, 27(1), 201–208.
[4] Frolov, S., Zvegintsev, V., Ivanov, V., Aksenov, V., Shamshin, I., Vnuchkov, D., Nalivaichenko, D., Berlin, A., Fomin, V., Shiplyuk, A., & Yakovlev, N. (2018). Hydrogen-fueled detonation ramjet model: Wind tunnel tests at approach air stream Mach number 5.7 and stagnation temperature 1500 K. International Journal of Hydrogen Energy, 43(15), 7515–7524.
[5] Frolov, S., Aksenov, V., Ivanov, V., Shamshin, I., & Zangiev, A. (2019). Air-breathing pulsed detonation thrust module: Numerical simulations and firing tests. Aerospace Science and Technology, 89, 275–287.
[6] Smirnov, N., Nikitin, V., Stamov, L., Mikhalchenko, E., & Tyurenkova, V. (2019). Three-dimensional modeling of rotating detonation in a ramjet engine. Acta Astronautica, 163, 168–176.
[7] Wang, C., Liu, W., Liu, S., Jiang, L., & Lin, Z. (2015). Experimental verification of air-breathing continuous rotating detonation fueled by hydrogen. International Journal of Hydrogen Energy, 40(30), 9530–9538.
[8] Zhdan, S. A., & Rybnikov, A. I. (2014). Continuous detonation in a supersonic flow of a hydrogen-oxygen mixture. Combustion, Explosion, and Shock Waves, 50(5), 556–567.
[9] Dubrovskii, A. V., Ivanov, V. S., Zangiev, A. E., & Frolov, S. M. (2016). Three-dimensional numerical simulation of the characteristics of a ramjet power plant with a continuous-detonation combustor in supersonic flight. Russian Journal of Physical Chemistry B, 10(3), 469–482.
[10] Bykovskii, F. A., Zhdan, S. A., & Vedernikov, E. F. (2019). Continuous Detonation of the Liquid Kerosene—Air Mixture with Addition of Hydrogen or Syngas. Combustion, Explosion, and Shock Waves, 55(5), 589–598.
[11] Hayashi, A. K., Tsuboi, N., & Dzieminska, E. (2020a). Numerical Study on JP-10/Air Detonation and Rotating Detonation Engine. AIAA Journal, 58(12), 5078–5094.
[12] Xu, G., Wu, Y., Xiao, Q., Ding, C., Xia, Y., Li, Q., & Weng, C. (2022). "Characterization of wave modes in a kerosene-fueled rotating detonation combustor with varied injection area ratios." Applied Thermal Engineering, 212, 118607.
[13] Han, J. X., Bai, Q. D., Zhang, S. J., Wang, F., & Weng, C. S. (2022). Experimental study on propagation characteristics of rotating detonation wave with kerosene fuel-rich gas. Defence Technology.
[14] Zhao, M., Wang, K., Zhu, Y., Wang, Z., Yan, Y., Wang, Y., & Fan, W. (2022). Effects of the exit convergent ratio on the propagation behavior of rotating detonations utilizing liquid kerosene. Acta Astronautica, 193, 35–43.

[15] Ding, C., Wu, Y., Xu, G., Xia, Y., Li, Q., & Weng, C. (2022). Effects of the oxygen mass fraction on the wave propagation modes in a kerosene-fueled rotating detonation combustor. Acta Astronautica, 195, 204–214.

# Research on Road Extraction Algorithm Based on Improved Otsu's Thresholding Method

**Yanmei Wang[1*], Mingyu Lu[1], Pengfei Feng[1]**

[1] Traffic Engineering College, Anhui Sanlian University, Hefei, Anhui, 230601, China

[*]Corresponding author's e-mail: lumingyu@mail.slu.edu.cn

**Abstract.** In this paper, the widely used method of Otsu's thresholding method is studied. On the basis of the traditional Otsu's thresholding method, according to the distribution characteristics of the target and background in the actual image, a new method based on traditional Otsu's thresholding method is presented.and the road extraction experiment is carried out with GF-2 remote sensing image.The experimental results show that the improved method has the advantages of high segmentation accuracy and strong anti noise ability compared with the traditional Otsu's thresholding method.

## 1. Introduction

With the development of remote sensing technology in China, the spatial resolution of remote sensing image is getting higher and higher, which can reach sub meter level.The high spatial resolution of remote sensing image makes people use remote sensing image to study macro region to micro region.At present, it has been widely used in the fields of land and resources, intelligent transportation and environmental protection.Automatic and semi-automatic extraction of urban road has become a basic problem in remote sensing image application[1-2].However, for high spatial resolution remote sensing images, there are a lot of interference information, such as building shadows, vehicles and pedestrian crossing lines on the road, pedestrian bridges, which make the road network become discontinuous and the edges become blurred.In addition, the existence of road like features produced by parking lots, rivers and building tops also brings difficulties to road extraction.At present, the main technologies of road information extraction based on remote sensing image, such as edge detection[3], image segmentation[4], artificial neural network[5], mathematical morphology[6], support vector machine[7], deep learning[8]etc.But still have some limitations in the degree of automation of extraction, operation speed, applicability, accuracy and data collection of the algorithm,Generalization and robustness.

In the past 40 years, image segmentation technology has been widely concerned by scholars at home and abroad. The traditional Otsu's thresholding method proposed by Japanese scholar QSTU in 1979. It has been widely used because of its simple calculation, stability and effectiveness[9].The traditional Otsu's thresholding method is very sensitive to noise and target size, and it only produces good segmentation effect for images with single peak between class variance.When the inter class variance is bimodal or multi-modal, the effect is not ideal, and the segmentation threshold will fall into local optimal rather than global optimal.As a fast and effective threshold selection technique[10], genetic algorithm can make up for the defects of the traditional Otsu's thresholding method.The main research content of this paper is to extract the road information with better greening rate and less interference information based on improved Otsu's thresholding method.

## 2. Algorithm principle

*2.1. Traditional Otsu's thresholding method*

The traditional Otsu's thresholding method is based on the maximum inter class variance to obtain the global threshold. It is better for the threshold segmentation method with bimodal histogram structure. The basic principle of the traditional Otsu's thresholding method is to divide the image into target and background according to the threshold. The specific algorithm is as follows: suppose the gray level range of the image is$[1,L]$, the gray value is $i$ , the number of pixels is recorded as $n_i$, the total number of pixels is

$$N = \sum_{i=1}^{L} n_i \tag{1}$$

The probability of occurrence of each gray level is

$$P_i = \frac{n_i}{N} \tag{2}$$

If $\bar{\mu}$ is the average gray level of the image, and its value is

$$\bar{\mu} = \sum_{i=1}^{L} i \cdot P_i \tag{3}$$

Set threshold $T$ divide an image into A, B two types, then the gray value range of $A$ *is* $[1, T]$, the gray value range of $B$ *is* $[T, L]$. The probabilities of the two classes are

$$\omega_A = \sum_{i=1}^{T} P_i \tag{4}$$

$$\omega_B = \sum_{i=T+1}^{L} P_i \tag{5}$$

Order $\omega_A = \omega(T)$, the average gray levels of the two types are respectively

$$\bar{\mu}_A = \frac{1}{\omega(T)} \sum_{i=1}^{T} i \cdot P_i \tag{6}$$

$$\bar{\mu}_B = \frac{1}{1 - \omega(T)} \sum_{i=T+1}^{L} i \cdot P_i \tag{7}$$

The average gray level of the whole image $\bar{\mu}$ can also be expressed as

$$\bar{\mu} = \omega_A \bar{\mu}_A + \omega_B \bar{\mu}_B \tag{8}$$

Order $\bar{\mu}(T) = \bar{\mu}_A$ , then the inter class variance of the two classes is

$$\sigma^2 = \omega_A (\bar{\mu}_A - \bar{\mu})^2 + \omega_B (\bar{\mu}_B - \bar{\mu})^2 = \frac{[\bar{\mu}\omega(T) - \bar{\mu}(T)]^2}{\omega(T)[1 - \omega(T)]} \tag{9}$$

The maximum threshold value satisfying the above formula T is the best segmentation threshold obtained by the traditional Otsu's thresholding method. The formula is aimed at target and background segmentation, that is, single threshold selection.When three kinds of objects are segmented, it can be extended to double threshold selection.

### *2.2. Improved Otsu's Thresholding Method*

As a basic method of image segmentation, the traditional Otsu's thresholding method has a good effect when there is a significant difference between the target and the background. When the ratio of target to background is very large, and the variance function between classes may be bimodal or multimodal, the effect is not good. As an effective search algorithm, genetic algorithm can well meet the needs of improving the traditional Otsu's thresholding method. Because the traditional Otsu's thresholding method is to find the optimal solution when the variance between the target and background is the largest, the genetic algorithm can quickly and stably find the best threshold value, which makes the image achieve a good segmentation effect. Therefore, this paper proposes a genetic algorithm based on the traditional Otsu's thresholding method to extract road information from remote sensing images.

The application of genetic algorithm in image processing is to extract the target by taking the gray value of the image as the search factor. The specific process is shown in Figure 1 below.In the process of calculation, the main steps are as follows:

- Firstly, the initial population required for random generation in the image is regarded as the first generation. Mutation probability, crossover probability, population size and individual length are fixed initial values.
- The fitness value of the parent population was calculated.
- Compare with the termination condition. If the termination condition is satisfied, the fourth step is executed. if not, the selection, crossover and mutation operations are carried out to form the offspring population, and the obtained offspring population is taken as the parent population of the next iteration, and then the fitness value of the parent population was calculated once again.
- The output of the best threshold.

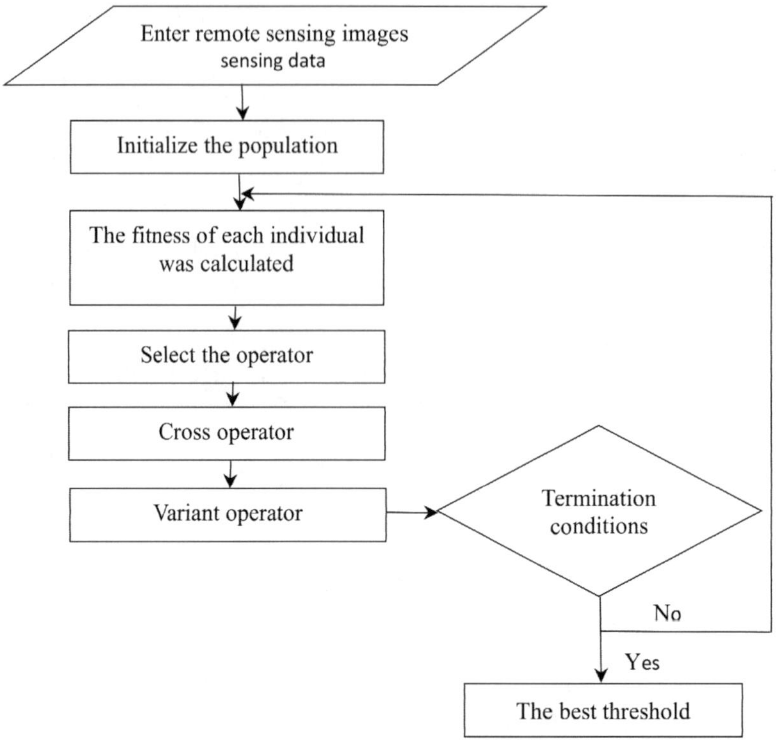


Figure 1. Flow chart of obtaining optimal threshold value by genetic algorithm

The improved Otsu's thresholding method based on genetic algorithm is used to extract road information from GF-2 remote sensing image.The specific steps are as follows:

- The gray value of road image is encoded to generate chromosome unit.
- The initial population is generated randomly to determine the initial value of the solution.
- Decode each chromosome. The variance can be set as the fitness function of each chromosome. The larger the variance of chromosome, the more likely it is to approach the optimal solution.
- The best threshold is obtained by genetic operation.

## 3. Experimental results and analysis

In order to verify the superiority of the improved Otsu's thresholding method based on genetic algorithm compared with the traditional Otsu's thresholding method. In this experiment, the traditional Otsu's thresholding method and the improved Otsu's thresholding method based on genetic algorithm are used to verify the experiment. The remote sensing data of GF-2 are selected, as shown in figure 2. Figure (a) is experiment one, the image size is 231×238. Figure (e) is experiment 2, the image size is 243×230. The difference of gray value between road and background is large, so the method based on threshold segmentation is suitable.

In the experiment, the initial population number is set as 8 and the maximum evolutionary iteration is 100. In experiment one, the best threshold of the traditional Otsu's thresholding method and the improved Otsu's thresholding method based on genetic algorithm are 295 and 304 respectively. In experiment two, the best threshold of the traditional Otsu's thresholding method and the improved Otsu's thresholding method based on genetic algorithm are 310 and 316 respectively. Figure (b) and figure (c) are the road information extracted by the traditional Otsu's thresholding method and the improved Otsu's thresholding method based on genetic algorithm respectively. Figure (e) and figure (f) are the road information extracted by the traditional Otsu's thresholding method and the improved

Otsu's thresholding method based on genetic algorithm respectively from the remote sensing data of experiment two.

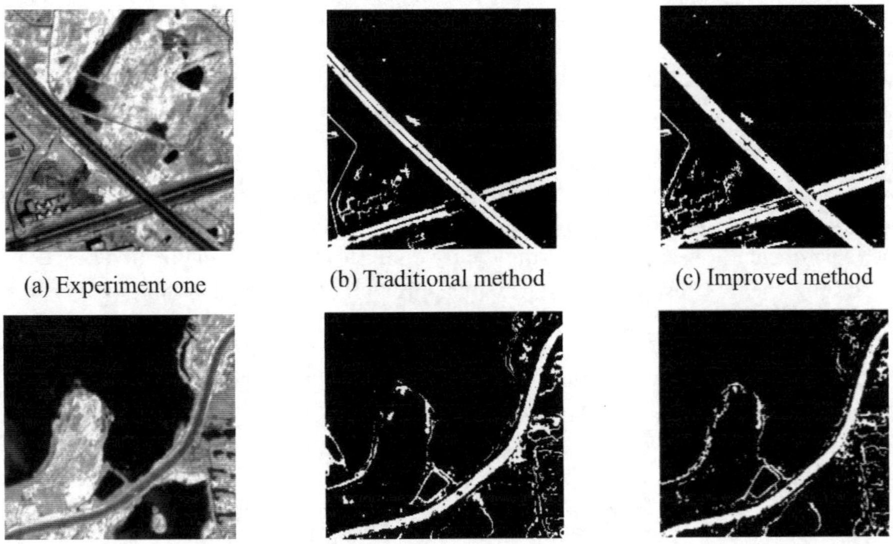

(a) Experiment one (b) Traditional method (c) Improved method

(d) Experiment two (e) Traditional method (f) Improved method

Figure 2. Comparison of road extraction

From figure (b), (c), (e) and (f), it can be seen that the traditional method and the improved method can better extract the road information and retain the characteristics of the original image. Compared with the traditional method, it can not only separate the image from the background more clearly and effectively, but also can distinguish the target from the background more clearly and effectively.

In order to evaluate the accuracy of the two methods more accurately, data analysis method can be used to evaluate the accuracy of road extraction. According to the geometric length of the extracted road, the accuracy, redundancy error and omission error can be used to evaluate the extracted road results.

$$\text{accuracy}=\frac{\text{correct extracted road target length}}{\text{total length of the road target}}\times 100\% \tag{10}$$

$$\text{redundancy error}=\frac{\text{excess road target length}}{\text{total length of the road target}}\times 100\% \tag{11}$$

$$\text{omission error}=\frac{\text{total length of the missed road target}}{\text{total length of the road target}}\times 100\% \tag{12}$$

According to the results of road information extracted in the experiment, the accuracy, redundancy error and omission error of road information extraction in experiment one and experiment two are listed in Table 1. From the evaluation results, the accuracy, redundancy error and omission error of the improved method are better than those of the traditional method. Moreover, the accuracy of the two experiments is more than 90%, which shows that the method is effective for extracting road information from remote sensing data.

Table 1. Accuracy statistics of road target extraction results (unit:%)

| Experiment | Road extraction method | Accuracy | Redundancy error | Omission error |
|---|---|---|---|---|
| Experiment one | Traditional method | 91.32 | 5.85 | 8.68 |
| | Improved method | 94.95 | 5.74 | 5.05 |
| Experiment two | Traditional method | 92.15 | 7.19 | 7.85 |
| | Improved method | 93.03 | 5.63 | 6.97 |

## 4. Conclusions

In this paper, based on the traditional Otsu's thresholding method, an improved Otsu's thresholding method based on genetic algorithm is proposed. And the method is verified by GF-2 remote sensing image. The experimental results show that the improved Otsu's thresholding method based on genetic algorithm can make up for the deficiency of the traditional Otsu's thresholding method to a certain extent, which can significantly improve the accuracy of road extraction, reduce the redundancy error and omission error significantly, and improve the effect of road information extraction. In the next step, we will study the difference between the target and the background in the process of road target extraction to enhance the generalization ability and robustness of the algorithm.

## Acknowledgments

This paper is one of the research achievements of the natural science key projects of Anhui Provincial Department of education, "Research on traffic information extraction technology of high spatial resolution satellite data based on machine learning" (KJ2019A0902) and "Research on control and guidance of multi Lane micro traffic under vehicle road coordination" (KJ2021A1187).

## References

[1] Dai L.G, Wang Y etc. (2020)Overview of road extraction methods from optical remote sensing images, Remote Sensing Journal, 24 (07): 804-823.
[2] Abdur Raziq, Xu A.G, Li Y.(2016)Automatic Extraction of Urban Road Centerlines from High-Resolution Satellite Imagery Using Automatic Thresholding and Morphological Operation Method，Journal of Geographic Information System, 8:517-525.
[3] Yan K, Wang F.C etc.(2013)Edge Detection of Composite Insulators Hydrophobic Image Based on Improved Canny Operator，Energy and Power Engineering, 5:593-596.
[4] Min L, Gao K, etc.(2020)A Review of the Optical Remote Sensing Image Segmentation Technology，Spacecraft Recovery & Remote Sensing, 41(6):1-13.
[5] Nima Ghasemloo, Mohammad Reza Mobasheri etc.(2013)Road and Tunnel Extraction from SPOT Satellite Images Using Neural Networks，Journal of Geographic Information System ,5:69-74.
[6] Ali Mohammadzadeh, Ahad Tavakoli etc.(2006)Road extraction based on fuzzy logic and mathematical morphology from pan-sharpened IKONOS images，Journal of Geographic Information System , 21(113):44-60.
[7] Yang T, Zhang L.L etc.(2014)Road Vector Map Change Monitoring Based on High Resolution Remote Sensing Images，Advances in Remote Sensing, 3:272-279.
[8] Hu f, Xia G.S etc. (2015)Transferring Deep Convolutional Neural Networks for the Scene Classification of High-Resolution Remote Sensing Imagery，Remote Sens, 7:14680-14707.
[9] OTSU N. (1979)A threshold selection method from gray—level histogra. IEEE Transactions on Systems，Man and Cybernetics，9: 62—66.
[10] Tai X.L. (2012)Research on road information extraction from remote sensing images, Kunming: Kunming University of science and technology.

# Numerical Simulation of Jet Interaction Flow Field with Different Flow Rates

**Yuwei Liu[1], Xinchen Wang[1, *], Kangkang Li[1] and Yaofeng Liu[1]**

[1] China Academy of Aerospace Aerodynamic Technology, Beijing, 100074, China

[*]Corresponding author's e-mail: 1285017483@qq.com

**Abstract.** In this paper, the complex disturbance flow caused by the interaction of jets with different flow rates and hypersonic incoming flow on a cone-cylinder-skirt of revolution is numerically studied, the typical shock wave structure and the influence range caused by the jet interaction are given, and the effect of the jet flow rate on the interaction flow field is discussed. The results show that the interaction between the jet flow and the incoming flow increases with the rise of the jet flow rate, and the influence range of the wrapping effect enlarges. If the jet flow rate is equivalent to the incoming flow rate, the leading edge separation line forms a closed loop.

## 1. Introduction

The jet direct force control technology can supplement the control capability of aerodynamic control surfaces at high altitude, quickly change the trajectory and attitude of the aircraft, and can be used to achieve a variety of different control purposes such as braking and separation of the aircraft. In order to enhance the maneuverability and penetration ability and improve the control accuracy of the new hypersonic vehicle in the future, there is an urgent need for this technology.

The interaction between the high-speed incoming flow and the jet flow of the control engine forms more complex high-order induced vortices and waves around the nozzle, upstream and downstream flow fields, including flow separation and reattachment, shock wave/shock wave interaction, shock wave/boundary layer interaction, shear layer and the interaction between them, while coupling complex multi-physics effects such as real gas thermal jet chemical reaction, unsteady and so on. The flow is very complex.The influence of jet interaction on the aerodynamic/thermal environment of aircraft is restricted by many factors, such as flight altitude, speed, flight attitude, gas parameters, jet direction, nozzle layout and so on, which makes it very difficult to study.

In recent years, in the low altitude continuous flow region, a large number of mechanism studies have been carried out for the small flow control jet interaction[1-6], but the mechanism studies for the large flow and large thrust jet strong interaction are relatively weak[7]. Compared with the small flow rate and small thrust attitude control engine, the large flow rate and large thrust engine has stronger interference between the jet and the incoming flow, larger coverage of the impact area, strong shock wave, large-scale separation flow and so on, which has stronger interference on the aerodynamic/thermal characteristics of the whole aircraft and its components (such as wing, rudder, body flap, etc.) And more complex interference law.For the future development of new hypersonic vehicles towards faster, higher and more accurate direction, the strong interference problem caused by the use of large flow and large thrust jet control has become a bottleneck restricting the development of new hypersonic vehicles in the future.It is necessary to deeply grasp the aerodynamic/thermal interference mechanism and law to meet the requirements of low redundancy design and high accuracy control of hypersonic vehicle.

In this paper, the jet interaction mechanism of a cone-cylinder-skirt shape of revolution is studied numerically, and the flow characteristics of jet strong interaction and the influence of parameters are obtained.The results can give references to the design of aircraft with lateral jet control.

## 2. Research shape and calculation condition

### *2.1. Research shape*

The cone-cylinder-skirt configuration consists of a nose cone, a cylinder, a nozzle, and a tail skirt, as shown in Figure 1. The overall length of the profile is 2250 mm and the nozzle exit is centered at 1177 mm from the nose.

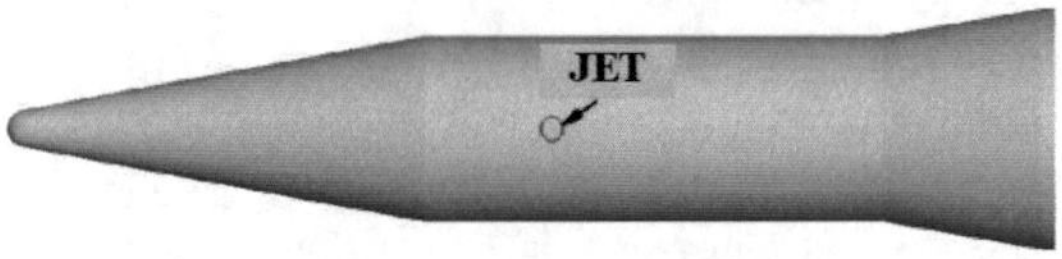


Figure 1. Cone-cylinder-skirt configuration.

### *2.2. Incoming flow conditions and jet parameters*

The free stream conditions and jet parameters corresponding to the configurations studied in this paper are given in Table 1 and Table 2, respectively.

Table 1. Incoming flow conditions.

| **H(km)** | **Ma** | **α(°)** | **β(°)** |
|---|---|---|---|
| 30 | 6 | 0 | 0 |

Table 2. Jet parameters.

| **Parameter Name** | **Cold spray parameter values** |
|---|---|
| Outlet diameter(mm) | 78 |
| Throat diameter(mm) | 44.77675 |
| Length of expansion section(mm) | 42 |
| Outlet Mach number(Mj) | 2.65 |
| Outlet pressure(MPa) | 0.1333 |
| Outlet temperature(K) | 121.9435 |
| Outlet gas specific heat ratio | 1.4 |
| Total temperature(K) | 293.15 |
| Total pressure(MPa) | 2.87 |
| Gas Constant(J/(kg*K)) | 287 |
| Thrust(N) | 6800 |

## 3. Numerical Simulation Methods and Computational Grid

### *3.1. Numerical method*

A finite volume method based on the multi-block patched grid technique is applied to the discretization of the three-dimensional Reynolds-averaged Navier-Stokes equations. The Roe's scheme is adopted and the accuracy is improved to the second order by MUSCL interpolation. The viscous term is discretized according to the conventional central difference method and the implicit LU-SGS method is used to discretize the time term. The turbulence model is S-A model[16].

*3.2. boundary condition*
Inlet boundary: Supersonic incoming flow with given static temperature, Mach number and static pressure.

Outlet boundary: For supersonic outflow, extrapolate all values of the parameters.

Solid wall boundary: The velocity is under the condition of no slip, and the temperature is treated as adiabatic wall or isothermal wall.

Symmetric boundary: On the symmetric plane, the normal velocity is zero, and the normal gradient of all variables is zero.

Jet boundary: The pressure, temperature and inflow velocity of the plenum of the nozzle are given by nozzle-external flow field coupling calculation.

Butt Boundaries: Based on the mesh topology, the data at the mesh points corresponding to the boundary points is used directly.

*3.3. Computational Grid*
The calculation grid of the jet interaction flow field is generated by the block butted grid technology, the grid quantity is 13 million, and the first layer of normal grid spacing is $10^{-5}$m. Figure 2 gives the overall computational grid for the cone-cylinder-skirt configuration jet interaction and the surface grid diagram.

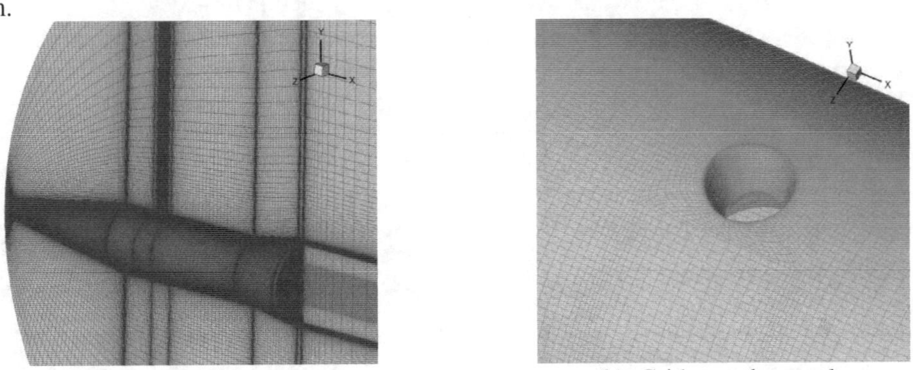

(a). Integral grid structures. (b). Grid near the nozzle.

Figure 2. Computational grid of jet interaction for cone-cylinder-skirt configuration.

## 4. Results and analysis of numerical simulation

*4.1. Basic characteristic of jet interaction flowfield*
A nose shock is formed at the nose of the hypersonic vehicle by the hypersonic incoming flow, and a strong bow shock of the jet is generated by the strong interference between the incoming flow and the lateral jet, and is deflected to the downstream step by step.The high pressure upstream of the nozzle propagates to the upstream through the boundary layer and interferes with the outflow, thus leading to the separation of the boundary layer, and a strong separation shock is formed at the upstream of the jet.

Figure 3(a) is the equal pressure cloud chart of the symmetry plane of the jet interaction flow field calculated under the conditions of Ma=5, Mj=3, $\alpha$=-2°, $P_j/P_\infty$=29.4. The color transition from blue to red in the chart indicates that the pressure level is from low to high. According to the pressure distribution in the interaction area and the pressure mutation characteristics before and after the strong shock wave, not only the typical shock wave position can be distinguished, It can also show the detailed distribution of high and low pressure areas in the interference area.Figure 3 (b) is a wind tunnel schlieren photograph taken under the same conditions, which reflects the superposition of the spanwise density gradient of the flow field and clearly shows the interference characteristics of the bow shock, bow shock and near the nozzle. By comparing the two figures, it can be seen that the interference characteristics of the main shock structures such as bow shock, bow shock and nozzle are consistent with the experimental results.

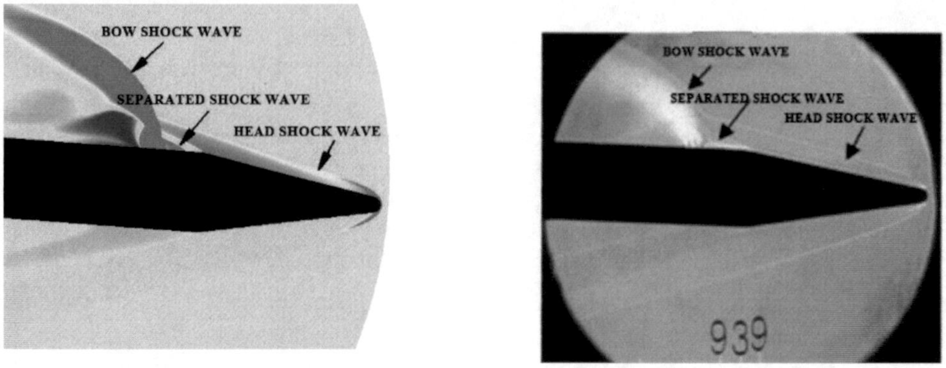


(a) Pressure nephogram of flow field symmetry plan (b) Wind tunnel schlieren photograph

Figure 3. Typical shock structure in jet interaction flow field.

The jet will cause complex shock wave/boundary layer interaction separation flow. Figure 4 (a) shows the surface limit streamlines of the jet interaction flow field calculated under the conditions of Ma = 5, Mj = 3,$\alpha$=-2°, and $P_j/P_\infty$ = 29.4, which can show the flow characteristics of the separation flow on the surface, and can clearly distinguish the main separation line at the leading edge, the reattachment line, and the wake separation line downstream of the nozzle. Figure 4 (b) gives the wind tunnel test surface oil flow pattern for the same conditions, showing that the leading edge main separation line and reattachment line are in agreement with the calculated results.

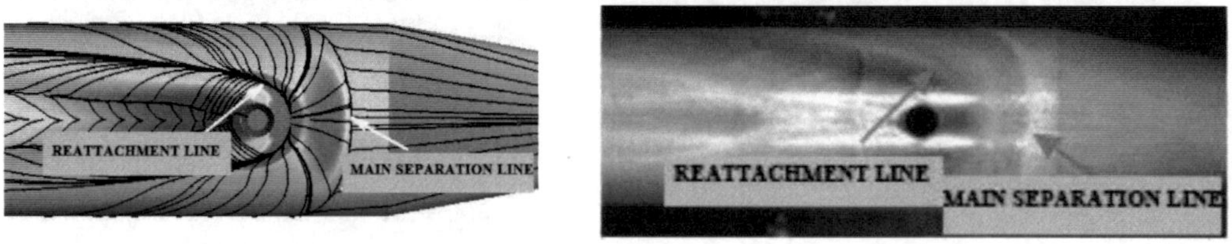


(a) Surface Limiting Streamlines Obtained by Numerical Simulation (b) Surface oil flow patterns from wind tunnel test

Figure 4 . Flow separation characteristics of jet interaction flowfield.

## *4.2. The effect of jet flow rate on the interaction flow field*

In practical engineering applications, different thrust control engines are usually installed on the aircraft to achieve different control purposes and overload capacity requirements, such as orbit control, attitude control and so on. Different thrust control engines often correspond to different exit areas, that is, different jet flow rates. Therefore, this paper focuses on the influence of the mass flow rate change caused by changing the nozzle exit area on the jet interaction flow field.

The influence of the jet flow rate on the jet interaction flow field is studied under the same inflow conditions of Ma = 6, H = 30 km, $\alpha$=0°. By changing the nozzle outlet diameter to achieve different jet flow, while maintaining the nozzle outlet Mach number, outlet pressure, outlet temperature unchanged. For ease of handling, the nozzle was removed and the jet conditions were applied directly to the nozzle orifice as shown in Figure 5.

Jet flow:

$$\dot{m}_j = \rho_j V_j A_j = \rho_j V_j \frac{\pi d^2}{4} \tag{1}$$

Incoming flow:

$$\dot{m}_\infty = \rho_\infty V_\infty A_\infty = \rho_\infty V_\infty \frac{\pi D^2}{4} \tag{2}$$

d is that diameter of the nozzle and D is the diameter of the column section of the cone-cylinder-skirt profile.

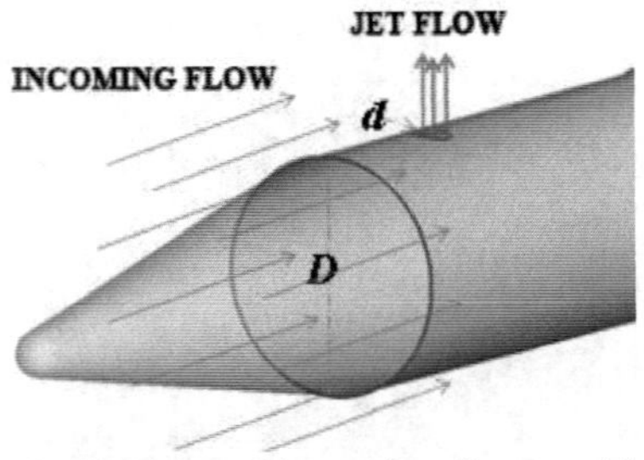


Figure 5. Schematic diagram of incoming flow and jet flow direction.

The corresponding jet parameters for different nozzle exit diameters are given in Table 3. d = 78mm is that exit diameter of the original nozzle, and the other four exit diameter are 63mm, 47. 25mm, 40mm, 32mm and 16mm. It can be seen that, with other parameters unchanged, the jet to stream flow ratio is proportional to the square of the diameter ratio d/D, where the jet to stream flow ratio is 1 at d/D = 0.12.

Table 3 Jet parameters for different nozzle exit diameters.

| Outlet diameter(mm) | 78 | 63 | 47.25 | 40 | 32 | 16 |
|---|---|---|---|---|---|---|
| Outlet Mach number(Mj) | 2.65 | 2.65 | 2.65 | 2.65 | 2.65 | 2.65 |
| Outlet pressure(MPa) | 0.13 | 0.13 | 0.13 | 0.13 | 0.13 | 0.13 |
| Outlet temperature(K) | 122 | 122 | 122 | 122 | 122 | 122 |
| specific heat ratio | 1.4 | 1.4 | 1.4 | 1.4 | 1.4 | 1.4 |
| Gas Constant(J/(kg*K)) | 287 | 287 | 287 | 287 | 287 | 287 |
| Diameter ratio d/D | 0.2 | 0.16 | 0.12 | 0.1 | 0.08 | 0.04 |
| Flow ratio | 2.73 | 1.78 | 1.0 | 0.7 | 0.46 | 0.11 |

Figure 6 gives the pressure distributions on the jet interaction flowfield surface and symmetry plane for different nozzle exit diameters obtained at Ma = 6, H = 30 km, $\alpha$=0°. The figure shows the basic flow field characteristics of the symmetry plane of the jet interaction flow field, including the spatial interaction range, shock wave structure and vortex structure.It can be seen that as the diameter ratio d/D increases gradually, that is, the jet flow rate increases gradually, the interaction between the jet flow and the incoming flow increases gradually, and the strength and influence range of the bow shock and separation shock of the jet increase.

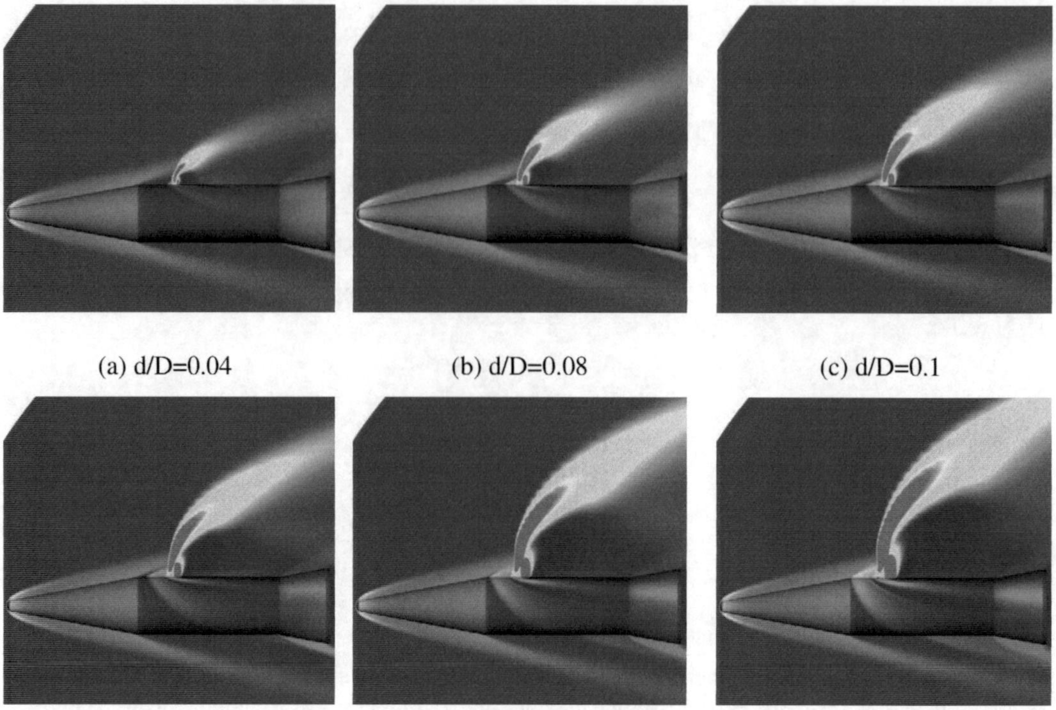

(a) d/D=0.04 (b) d/D=0.08 (c) d/D=0.1

(d) d/D=0.12 (e) d/D=0.16 (f) d/D=0.2

Figure 6. Pressure distribution on the surface and symmetry plane of the jet interaction flow field for different nozzle exit diameters.

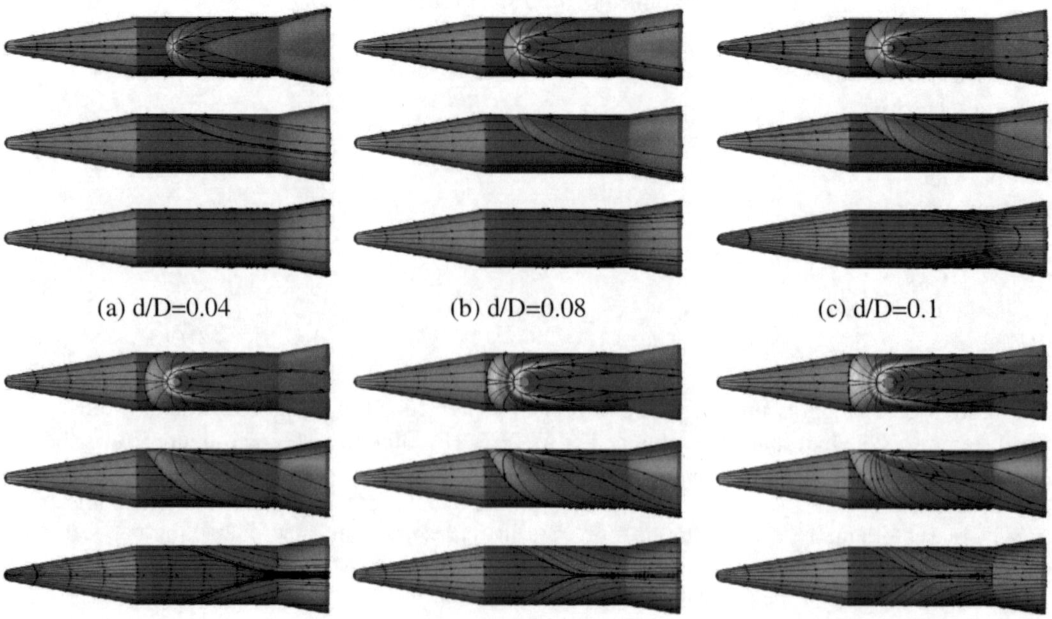

(a) d/D=0.04 (b) d/D=0.08 (c) d/D=0.1

(d) d/D=0.12 (e) d/D=0.16 (f) d/D=0.2

Figure 7. Surface pressure contours and limiting streamlines of jet interaction flow field for different nozzle exit diameters.

Figure 7 shows the surface pressure contours and the limit streamlines of the jet interaction flowfield for different nozzle exit diameters. It can be seen that as the diameter ratio d/D gradually increases, the leading edge main separation line gradually moves away from the nozzle toward the upstream, and meanwhile, the range of the lower surface annulus gradually increases, that is, the so-called wrapping effect gradually strengthens, which indicates that the jet interference area gradually increases.

Note that when d/D = 0.12 (where the flow ratio is 1), the leading edge main separation line extends from the front sides of the nozzle to the center line of the lower surface, thus forming a closed ring envelope, along which the incoming flow is completely cut off and in a completely separated state. With the increase of d/D, the scope of this ring becomes larger and larger, indicating that the jet interaction region becomes larger and larger.

Figure 8 shows the spatial flow field pressure contours obtained by superimposing 10 cross sections of the jet interaction flow field for different nozzle exit diameters (in the range of X = 977 mm to 1877 mm, one cross section is taken every 100 mm). From the comparison of these spatial pressure contours, it can be seen more intuitively that with the increase of the diameter ratio d/D, the influence range of the complex shock wave formed by the interaction between jet and incoming flow becomes larger and larger, and gradually wraps up to the lower surface to form a closed influence area.

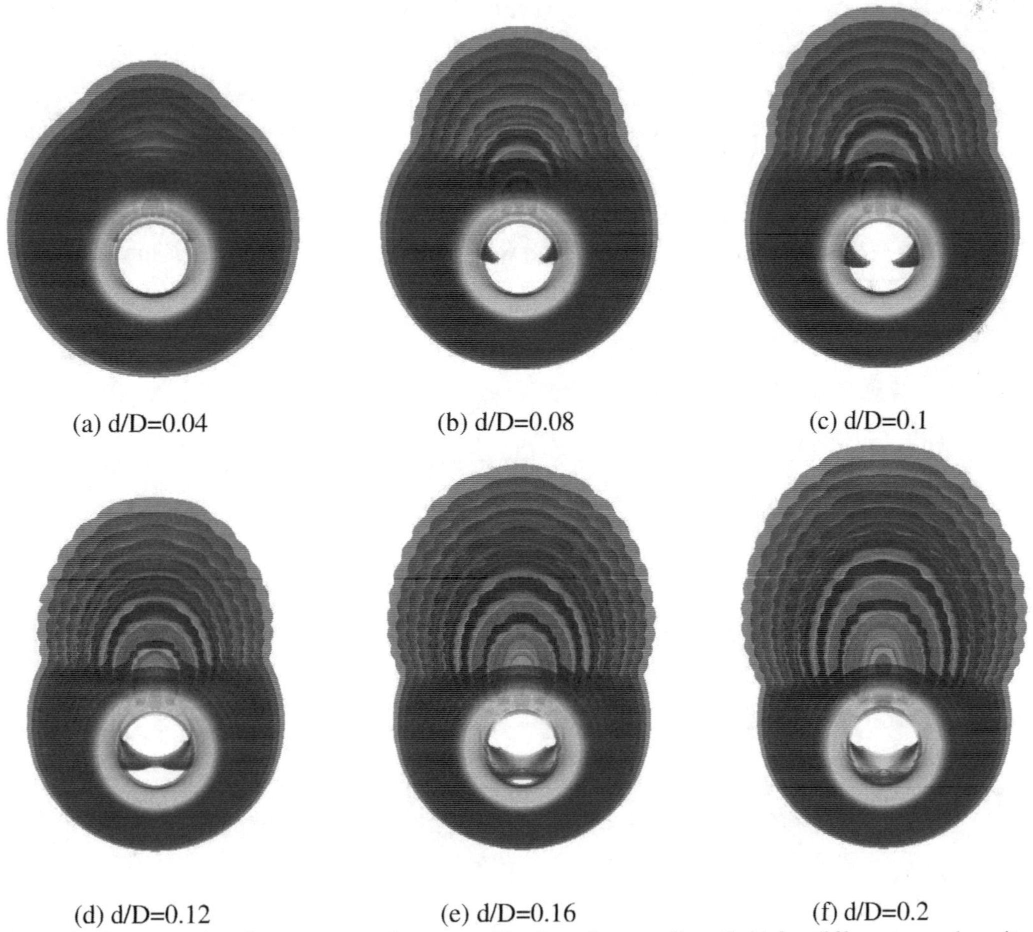

(a) d/D=0.04 (b) d/D=0.08 (c) d/D=0.1

(d) d/D=0.12 (e) d/D=0.16 (f) d/D=0.2

Figure 8. Cross-sectional pressure nephogram of jet interference flow field for different nozzle exit diameters.

Journal of Physics: Conference Series **2364** (2022) 012065 doi:10.1088/1742-6596/2364/1/012065

## 5. conclusion

In this paper, the complex interaction flow caused by the interaction of lateral jet flow and hypersonic incoming flow on a cone-cylinder-skirt of revolution is numerically studied, the typical shock structure caused by jet interaction and the influence range are given, and the effect of jet flow rate on the interaction flow field is discussed. Through the study of this paper, within the scope of the conditions of this study, the following conclusions can be drawn:

- With the increase of the jet flow rate, the interaction between the jet and the incoming flow is enhanced, and the strength and influence range of the bow shock and separation shock are increased;
- The wrapping effect increases with the increase of the flow rate of the jet. When the flow rate of the jet is equal to that of the incoming flow, the main separation line at the leading edge forms a closed loop.

## References

[1] Brandeis J, Gill J. Experimental Investigation of Side Jet Steering for Missiles at Supersonic and Hypersonic Speeds. AIAA 95-0316.
[2] Brandeis J, Gill J. Experimental Investigation of Super- and Hypersonic Jet Interaction on Configurations with Lifting Surfaces. AIAA 97-3723.
[3] Fan H, Bowersox R. Gaseous Injection Through Diamond Orifices at Various Incidence Angles into a Hypersonic Freestream. AIAA 2001-1050.
[4] Guelhan A, et al. Experimental Study on Aerothermal Heating Caused by Jet-Hypersonic Crossflow Interaction. Journal of Spacecraft and Rockets, 2008, 45(5):891~899.
[5] Beresh S J, et al. Crossplane Velocimetry of a Transverse Supersonic Jet in a Transonic Crossflow. AIAA 2006-906.
[6] Clark S W, Chan S C. Effects of Turbulence Models and Adaptive Grids on the Jet Interaction Flowfield. AIAA 93-3524.
[7] John M S, Bernard J J. Acoustic Associated with Fluid/Structure Coupling for Interceptor Missile Fow Fields. AIAA 2000-2035.

# Study of Combustion Mode Identification Method on Dual-Mode Solid Rocket Scramjet

**Libei Zhao [1, a], Zhixun Xia [2, b], Likun Ma [3, c*], Chaolong Li [4,d], Yunchao Feng [5,e], Binbin Chen[6,f], Pengnian Yang[7,g]**

[1]College of Aerospace Science and Engineering, National University of Defense Technology, Changsha, Hunan, China

[2]College of Aerospace Science and Engineering, National University of Defense Technology, Changsha, Hunan, China

[3]College of Aerospace Science and Engineering, National University of Defense Technology, Changsha, Hunan, China

[4]College of Aerospace Science and Engineering, National University of Defense Technology, Changsha, Hunan, China

[5]College of Aerospace Science and Engineering, National University of Defense Technology, Changsha, Hunan, China

[6]College of Aerospace Science and Engineering, National University of Defense Technology, Changsha, Hunan, China

[7]College of Aerospace Science and Engineering, National University of Defense Technology, Changsha, Hunan, China

[a]e-mail: zhaolibei@nudt.edu.cn; [b]e-mail: zxxia@nudt.edu.cn;
[*] corresponding author: [c]malikun@nudt.edu.cn; [d]e-mail: lichaolong13@nudt.edu.cn;
[e]e-mail: yunchaofeng@nudt.edu.cn; [f]e-mail: chenbinbin11@nudt.edu.cn; [g]e-mail: yangpengnian19@nudt.edu.cn

**Abstract**: A Direct connection experiment study of combustion mode transition was completed under the air conditions with Mach 2.2, total pressure 1.1 MPa and total temperature 1660 K, and a real-time identification method of combustion mode on dual-mode solid rocket scramjet was proposed. Based on the adiabatic assumption, the Mach number at the isolator exit could be calculated and used as the criterion for identifying the combustion mode. The pressure of the upper engine wall was measured during the test. It was found that the pressure of the isolator changes significantly during the combustion mode transition by analyzing the pressure changes in the isolator. The ratio of pressure between the exit and inlet $P_i/P_0$ is below 2 when the combustion works in scram-mode, while the value of $P_i/P_0$ is over 2 when it is in ram-mode.

## 1. Introduction

With the pursuit of propulsion technology for wide-speed hypersonic flight, the research on dual-mode scramjet engine systems have received extensive attention in recent years [1-4]. The dual-mode scramjet could operate in a wide range of flight speeds, can translate the combustion mode through thermal adjustment and geometric adjustment to match the incoming flow under different flight conditions, so that it can work stably in scram-mode and ram-mode [5-6]. The method of realizing mode transition by geometric adjustment will increase the quality and complexity of the engine, most people mainly study the mode transition through thermal adjustment. By adjusting the equivalence ratio, the distribution of heat in the combustor is changed to realize the transition of the combustion mode in most conditions. In ram-mode, the lower inlet airflow velocity can be fully mixed with the fuel, and the larger heat release generated by the fuel combustion can form a thermal throat. In the scram-mode, the higher airflow velocity is not conducive to the combustion of fuel, and the heat released by the fuel is not enough to keep the thermal throat.

It is important to accurately determine the combustion mode, because there are significant differences in the combustion regularity and performance of dual-mode scramjets under different combustion modes [7-8]. The various criterion to recognize combustion modes are based on different ground test data, as well as numerical simulation and theoretical analysis [9-11], but no widespread consensus has been formed due to the limitations of experimental data acquisition and observation means [12]. The method of directly measuring the velocity of the exit of the isolator proposed by Heiser[13]is not applicable in the actual test process because of the limitation of the test measurement methods. In different combustion modes, the shock train structures and Mach number distribution of isolator are different, resulting in different wall pressure distributions. Based on this characteristic, Cabell [11]studied the influence of various injection methods on the different work modes, and took the increase of the pressure before the fuel injector as the mode identification criterion. Liu [14]obtained the internal flow structure of the engine through numerical simulation, and combined the shock train with the Mach of the combustor to recognize the operation mode. Pan [15]adopted the mass-averaged one-dimensional Mach number as the criterion for identifying the combustion mode. Xiao [16]analyzed the influence of different combustion modes on the isolator of the engine, and selected the change rate of the wall pressure ratio of the isolator as the criterion for distinguishing combustion modes. However, the current research on combustion mode identification methods is based on the dual-mode scramjet using liquid fuel, and the accuracy of combustion mode distinguish for solid scramjets needs to be discussed.

In this paper, the ground direct-connect experiment of the combustion mode on dual-mode solid rocket scramjet using boron-based propellant as fuel was carried out. The structure of the dual-mode solid rocket scramjet is shown in Fig. 1. Under the adiabatic assumption of the isolator, the parameters of the engine inlet are known, and the Mach number can be got by the pressure data of the isolator exit. The Mach is used as the basis for identifying the combustion mode to distinguish the combustion modes, and the pressure changes of the isolator under different combustion are analyzed.

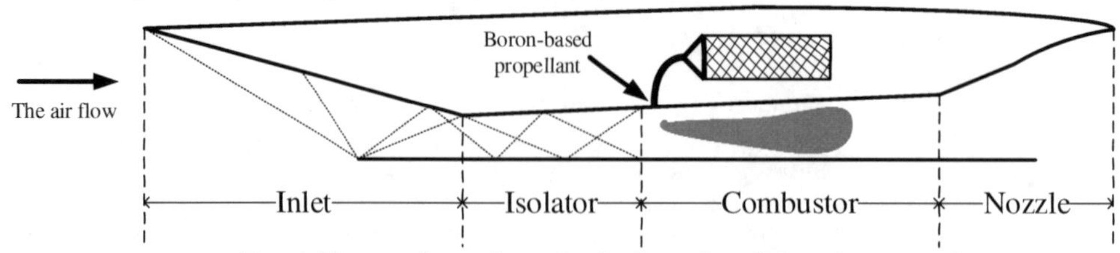

Fig. 1 The configuration of a dual-mode solid rocket scramjet

## 2. Testing System

### *2.1. Test Modal*

The dual-mode solid rocket scramjet configuration shown in Fig. 2 is used for the ground direct-connect experiment investigation. The width of the engine inlet is $H$, the width of the engine outlet is $2H$, and the length of the engine is $27.4H$. The isolator could balance the pressure change in heater outlet and the combustor. The fuel injector adopts an equal-area configuration. And a step is used to resist the upstream movement of the combustor pressure in the fuel injector. It has an angle of 60° to the direction of the incoming flow. The internal size of the combustor expands along the flow direction, which is used to match the heat release of the engine when the work modes change. In addition, there are many cavities in the combustor.

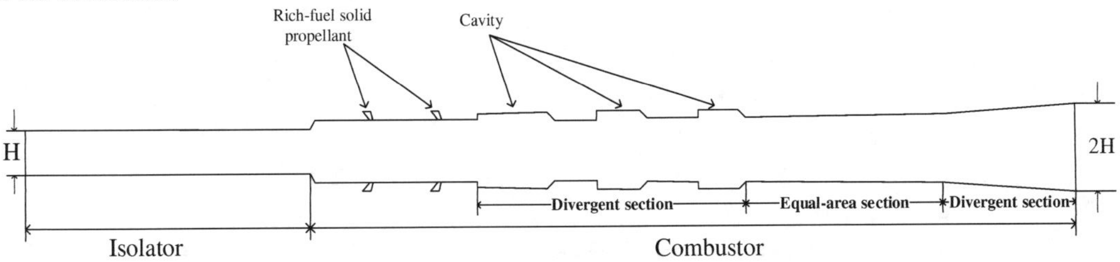


Fig. 2 Schematic diagram of the scramjet

### 2.2. Test Design Parameters

The ground direct-connect experiment simulates the inflow conditions of the isolator inlet on Mach number 6.0 and flight altitude 25 km. It can be seen from table 1 that the gas is a high-enthalpy gas heated by alcohol and air, and the airflow parameters at the inlet of the isolator.

Table 1 Parameters of the air at isolator inlet

| Parameter | Value |
|---|---|
| Mach number | 2.2 |
| Total pressure/MPa | 0.8 |
| Total temperature/K | 1660 |
| Air mass flow rate/(kg/s) | 1.00 |

To facilitate the analysis of the engine combustor, the test is divided into five stages, denoted as TI-T5, according to the fuel equivalence ratio and the working time of the gas generator, as illustrated in Table 2. The fuel equivalence ratio is changed by the working state of the gas generator.

Table 2 The description of the phases

| Experimental status | Time period | Title |
|---|---|---|
| Air heater ignite — Gas generator 1 ignite | 2.57s – 9.19s | T1 |
| Gas generator 1 ignite — Gas generator 2 ignite | 9.19s – 14.10s | T2 |
| Gas generator2 ignite — Gas generator 1 flameout | 14.10s – 22.19s | T3 |
| Gas generator 1flameout — Gas generator 2 flameout | 22.19s – 27.61s | T4 |
| Gas generator 2 flameout — Air heater flameout | 27.61s – 42s | T5 |

## 3. Results and Discussion

The Mach at the isolator exit is used as the criterion for identifying the combustion mode. And the outlet Mach number is below 1, it is a ram-mode, and when it is over 1, it is a scram-mode. According to the known airflow pressure parameters of the isolator inlet during the experiment, the Mach of the isolator exit can be calculated to distinguish the combustion mode. During the test, the pressure data is obtained by the pressure scanning valve and the pressure sensor. The analysis process is as follows:

The operating conditions of the experimental air heater during the test are shown in Fig. 3, including the mass flow of the measurement system and the air pressure. All parameters remain unchanged during the test, that is, and the air heater can maintain a stable operation.

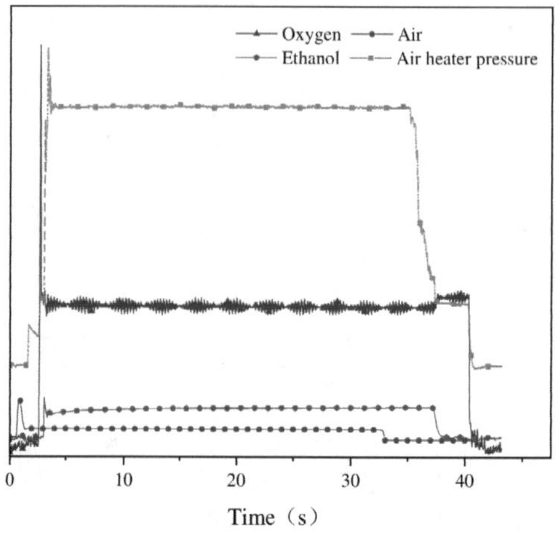

Fig.3 The air heater condition

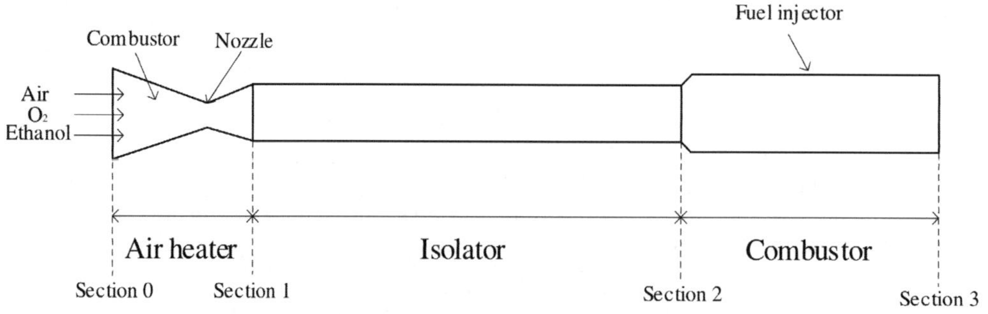

Fig. 4 The experimental system

The isolator inlet is connected to the air heater outlet in the ground direct connection experiment.

The air, oxygen and alcohol mass flow, combustor and nozzle outlet pressure, and cross-sectional area in the air heater are obtained by measurement. The parameters of the total temperature and enthalpy in the combustor could be got by the ratio and mass flow of air, oxygen and alcohol components. The total pressure is related to the internal geometry of the combustor. Therefore, the pressure $P_{t,1}$ and Mach $Ma_1$ of the isolator inlet can be obtained from the design parameters of the air heater. In addition, the size of the isolator is constant, and there is no chemical reaction inside in the test. According to the adiabatic assumption[18], the total enthalpy of the isolator is constant, as following:

$$h_1 = C_{p,1}T_{t,1} = C_{p,2}T_{t,2} = h_2 \tag{1}$$

Where $C_p$ represents the constant pressure specific heat capacity. It is defined as:

$$C_p = \frac{\gamma R}{\gamma - 1} \tag{2}$$

Where $\gamma$ represents the specific heat ratio, and $R$ is the gas constant of the mixture. $R$ is expressed as:

$$R = \frac{\bar{R}}{\bar{M}_{gas}} \tag{3}$$

Where $\bar{M}_{gas}$ represents the average molar mass of the mixture. And $\bar{R}$ represents the universal gas constant whose value is 8.314 J/(mol K).

The mass equation at the entrance of the isolator can be expressed as:

$$\dot{m} = \dot{m}_{Air} + \dot{m}_{O_2} + \dot{m}_{C_2H_6O} = \rho V A \tag{4}$$

Where $\rho$, $V$, $A$ represent density, velocity, and area, respectively. These can be obtained directly.
The state equation of mixture gas can be expressed as:

$$P = \rho PT \tag{5}$$

Where $P$ could be measured by the experiment, and $T$ represents the static temperature in the experiment.

These, such as the $T$, $P$, $V$ and $Ma$, are related as follows:

$$\frac{T_t}{T_0} = 1 + \frac{\gamma - 1}{2} Ma \tag{6}$$

$$\frac{P_t}{P_0} = (\frac{T_t}{T_0})^{\frac{\gamma}{\gamma-1}} = (1 + \frac{\gamma-1}{2} Ma)^{\frac{\gamma}{\gamma-1}} \tag{7}$$

$$V = Ma \times \sqrt{\gamma RT} \tag{8}$$

Under the condition that the $T$, $P$, $V$ and Mach of the isolator inlet are known, the Mach $Ma_2$ at the isolator exit can be calculated according to the above formulas. The Mach at the isolator outlet is calculated by measuring the centerline pressure of the isolator outlet in Fig.5. When $t_1$ = 17.9 s and $t_2$ = 22.3 s, the Mach number $Ma_2$ is equal to 1, which is the critical point of the two combustion modes. Therefore, it can be considered that at $t = 17.9 - 22.3\ s$, the combustor operates in the ram-mode.

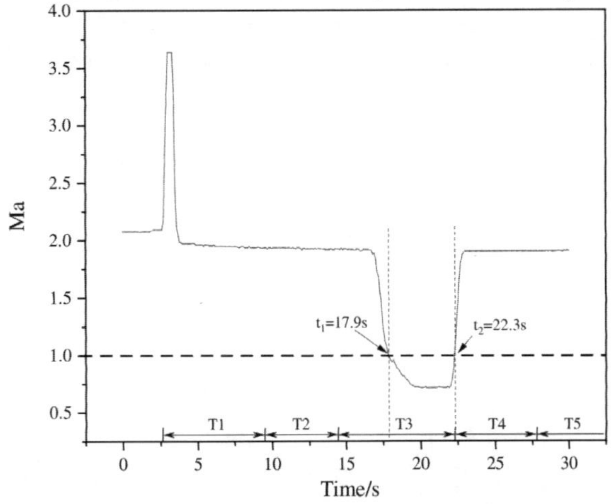


Fig.5 Schematic of the distribution of Mach numbers in section 2

The Mach number at the isolator outlet obtained by theoretical calculation realizes the real-time identification of the combustion mode. By the pressure change of the isolator, the influence of the position change of the shock train and the mode transition on the pressure change of the isolator was analyzed.

The wall pressure ratio during the test is shown in Fig. 6. Comparing the pressure ratio data at 16, 18, 20, and 23 s, it can be seen that the downstream pressure of the isolator changes significantly, and there is a trend of forwarding to the upstream of the isolator. At 16 and 23 s, there is no significant pressure change at $x/h$=0.95. At 18 and 20 s, there is a significant increase in pressure at $x/h$=0.95, while at $x/h$=0.67 there is no significant pressure fluctuation. Therefore, it can be inferred that the position of pressure increase is between $x/h$=0.67 and 0.95.

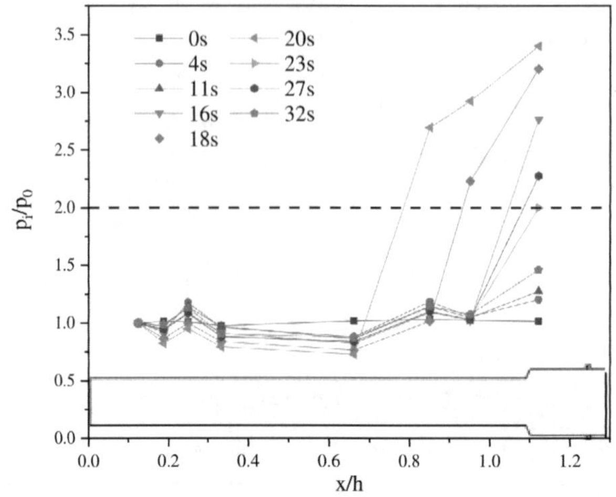


Fig.6 Pressure comparison of the isolator on the upper wall during the test

The change of Mach number and the pressure ratio is shown in Fig. 7. And the relationship of the Mach number and the pressure ratio can be intuitively represented in Fig. 7. When the Mach $Ma_2$ at the exit of the isolator is below 1, the engine operates in ram-mode, and the pressure ratio $P_i/P_0$ is over 2. When the Mach number $Ma_2$ is over 1, the engine works in scram-mode, and the $P_i/P_0$ is below 2.

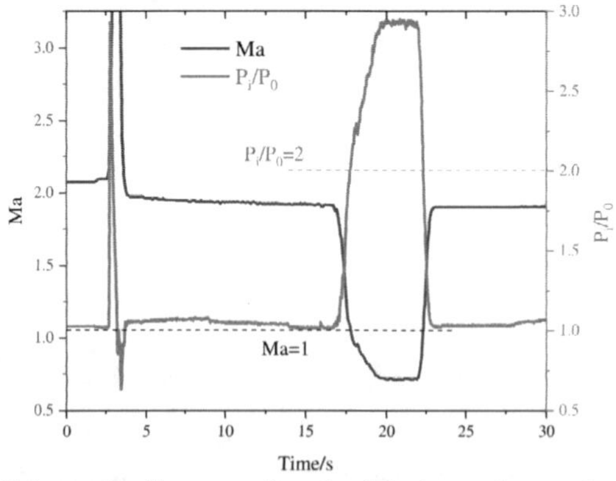


Fig. 7 Schematic diagram of outlet Mach number and pressure

## 4. Conclusion

The ground direct-connect experiment of a dual-mode solid rocket scramjet is carried out, and the pressure change process of the isolator is analyzed and compared under the condition of Mach number 2.2.

1) Based on the experiment pressure data, the Mach number can be calculated. The Mach number is used as the criterion for identifying the combustion mode, and the real-time identification of operation mode is realized. When the $Ma$ is over 1, the combustor operates in scram-mode, and the Mach number is below 1 in ram-mode.

2) Under the condition studied in this paper, the variation range of the pressure ratio $P_i/P_0$ corresponding to the two combustion modes is different. When the engine operates in scram-mode, the $P_i/P_0$ between the outlet and inlet of the isolator is below 2, and the pressure ratio $P_i/P_0$ is over in ram-mode.

In the next step, the pressure variation characteristics and flow field variation characteristics of the combustor and the inner wall under multiple equivalence ratio and fuel injection methods will be studied to verify the universality of the mode identification method and form a more accurate and specific distinguishing method of the combustion mode.

**Acknowledgments**
This research was funded by the National Natural Science Foundation of China grant number No. 52006240, Changsha Science and Technology Project No. kq2107001 and Hunan Provincial Innovation Foundation for Postgraduate No. QL20210022.

**References**
[1] Zhao, X., Xia, Z., Ma, L., Li, C., Fang, C., Natan, B., Gany, A. (2022) Research Progress on Solid-Fueled Scramjet. Chinese Journal of Aeronautics., 35:398-415.
[2] Cao, D., Brod, H. E., Yokev, N., Michaels, D. (2021) Flame Stabilization and Local Combustion Modes in a Cavity-Based Scramjet Using Different Fuel Injection Schemes. Combustion and Flame., 233:111562.
[3] Chen, Y., Fan, Y., Xu, L., Shan, X., Han, Q., Bai, X. (2022) Experimental and Numerical Study of Flow and Ignition and Lean Blowout Characteristics of Jet-cooled Wall Flameholder in a Dual-Mode Combustor. Aerospace Science and Technology., 122:107403.
[4] Thillai, N., Thakur, A., K., S., J., D. (2021) Analysis of flow-field in a dual mode ramjet combustor with boundary layer bleed in isolator. Propulsion and Power Research., 10: 37–47.
[5] Chen, J., Liu, S., Liu, W., Bai, H. (2017) Sensitivity Analysis of Combustor Incoming Parameters on Dual-Mode Scramjet Performance Potential at Ma6. Journal of Propulsion Technology., 38:2422-2430.
[6] Tian, Y., Xiao, B., Zhang S., Xing, J. (2016) Preliminary Study on Combustion Performance of Dual-Mode Scramjet Engine Journal of Aerospace Power., 31: 2921-2927.
[7] Yang, P., Xia, Z., Ma, L., Chen, B., Feng, Y., Zhang, J. (2021) Experiment study on the influence of the injection structure on solid scramjet performance. Acta Astronautica., 188:229-238.
[8] Yang, P., Xia, Z., Ma, L., Chen, B., Feng, Y., Li, C., Zhao, L. (2021) Influence of the Multicavity Shape on the Solid Scramjet. 2021:9718537.
[9] Waltrup, P., J., Billig, F., S. (1973) Structure of Shock Waves in Cylindrical Ducts. AIAA Journal., 11: 1404-1408.
[10] Liu, H., Song, W., Wang, D. (2011) Investigation of Mode Transition in A Dual Mode Supersonic Combustor with Hydrogen Fuel. Journal of Aerospace Power., 6: 576-581.
[11] Hass, N., Cabell, K., Storch, A., Gruber, M. (2011) HIFiRE Direct-Connect Rig (HDCR) Phase I Scramjet Test Results from the NASA Langley Arc-Heated Scramjet Test Facility. In: 17th AIAA International Space Planes and Hypersonic Systems and Technologies Conference. San Francisco. 2011-2248.
[12] Zhang, Y., Zhu, S., Liu, G., L, X., Xu, X. (2013) An Overview on Mode Transition in Dual Mode Ramjet. Journal of Propulsion Technology., 34: 1719-1728.
[13] Heiser, W., Pratt, D., Daley, D., Mehta, U. (1994) Hypersonic Airbreathing Propulsion. AIAA Press, Reston.
[14] Liu, H., Song, W., Wang, D. (2011) Investigation of mode transition in a dual mode supersonic combustor with hydrogen fuel. Journal of Aerospace Power., 26: 576-581.
[15] Yu, P., Wei-Dong, L., Zhen-Guo, W. (2007) Investigation on the Combustion Mode Translation of Thermal Choked Scramjet. In: 43rd AIAA/ASME/SAE/ASEE Joint Propulsion Conference and Exhibit. Cincinnati. 2007-5387.
[16] Xiao, B., Yan, Z., Tian, Y., Xing, J. (2015) Preliminary Study on Criterion of Identifying Combustion Mode for Scramjet. Journal of Propulsion Technology., 36: 1121-1126.
[17] Li, C., Xia, Z., Ma, L., Zhao, X., Chen, B., Yang, P. (2021) Performance Evaluation for Scramjet Based on Ground Direct-Connected Test: A Method Investigation. Aerospace Science and Technology., 117:106895.

# Numerical simulation study on the optimization of the central ejector intake tube

**Qi Jing[1,a], Wanwu Xu[2,b*], Wei Ye[3,c], Zhenkang Zhang[4,d]**

[1]College of Aerospace Science and Engineering, National University of Defense Technology, Changsha, Hunan, China

[2]College of Aerospace Science and Engineering, National University of Defense Technology, Changsha, Hunan, China

[3]College of Aerospace Science and Engineering, National University of Defense Technology, Changsha, Hunan, China

[4]College of Aerospace Science and Engineering, National University of Defense Technology, Changsha, Hunan, China

[a]e-mail: 1454261606@qq.com; [b*]e-mail: cfdxww@nudt.edu.cn;

[c]e-mail: 510519265@qq.com; [d]e-mail: zhangzknudt@163.com

**Abstract**—A CFD numerical simulation study was conducted for the primary flow non-axial inlet L/Y intake tube ejector. The effects of different configurations of the inlet piping on the secondary flow of the ejector were analyzed in terms of total pressure loss, separation zone volume, and velocity uniformity. The results showed that: the volume of the separation zone decreases with the increase of secondary flow total pressure. The L intake tube causes a larger separation zone under the same working condition. When the total pressure of secondary flow is 20kPa, the volume of the separation zone caused by the L intake tube reaches 15 times that of the Y intake tube, and the total pressure loss is 2.3 times that of the Y intake tube. A smaller separation zone has less effect on the flow field structure, resulting in better velocity uniformity in the mixing chamber inlet section of the Y intake tube ejector.

## 1. Introduction

From the invention of the world's first ejector in 1858 to the present, several types have been invented. The main ones are the central ejector, ring ejector, multi-nozzle ejector, and multi-branch plate ejector. It has the advantages of low cost, simple processing, low friction loss during operation, and a large entrainment ratio at small pressure ratio. These advantages have led to a wide range of applications in aerospace, chemical, and new energy fields.

The performance of an ejector is determined by the entrainment ratio and the ratio pressure. Normally, it is impossible to meet both of them at the same time. In recent years, numerous scholars have carried out a series of studies mainly based on its performance. Keenan and Neuman [1] were the first to propose two basic configurations of the ejector. After the practical experience, it is generally agreed that Constant Area Mixing ejector can obtain a higher entrainment ratio, while the Constant Pressure Mixing ejector has more stable performance in a wide range of back pressure. In addition, the primary flow nozzle exit position (NXP) has a significant impact on the performance of the central ejector. Pianthong [2] et al. found that the NXP excess would substantially reduce the entrainment ratio. Chunnanond and Aphornratana [3] showed that the expansion wave will face stronger compression if NXP gradually increases. This would produce a smaller expansion angle and a larger effective mixing area. However, too large NXP can make the boundary layer separation more severe [4]. It will not allow the mixed fluid to overcome back pressure to exit the ejector. The literature [5-7] indicated that there is an optimal value of NXP and discussed the factors influencing it. Therefore, the location of the primary flow nozzle needs to be fully considered when designing the ejector. In terms of structural parameters, the entrainment ratio is also influenced by other factors. The mixing area ratio has a strong influence on the ejecting performance, and an optimum value exists when other conditions are certain [8]. But the effect of mixing chamber contraction angle on the ejecting performance is inconclusive. Chong [9] found that the critical back pressure increased with the increase of the mixing section aspect ratio. Ouzzane and Aidoun [10] showed that the mixing chamber length greatly affects the mixing of primary and secondary flow by one-dimensional calculations.

The effect of ejector mixing chamber configuration on ejecting performance has been studied a lot in the past studies. However, the influence of the intake tube in front of the mixing chamber on the ejector is still less studied. Figure 1(a) shows the primary flow axial inlet central ejector. At this point, the secondary flow is only affected by the friction of the nozzle and mixing chamber wall surface. It can be assumed that the velocity at the inlet interface of the mixing chamber is circumferentially and uniformly distributed. Figure 1(b) shows the secondary flow axial inlet central ejector. In this configuration, the secondary flow will be influenced by the intake tube and flow separation will occur. The intake tube also causes a total pressure loss in the secondary flow. A large separation zone leads to poor velocity uniformity in the mixing chamber inlet section.

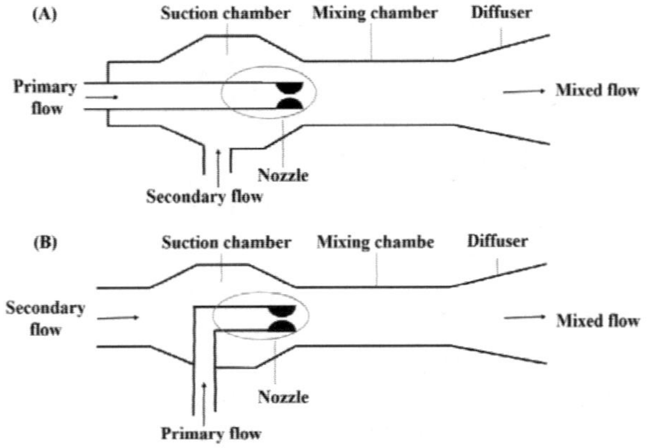


FIGURE 1 (A) PRIMARY FLOW AXIAL INTAKE CENTER EJECTOR
(B) SECONDARY FLOW AXIAL INTAKE CENTER EJECTOR

Based on this, a Y intake tube central ejector was designed (hereinafter referred to as Y-ejector). Compared to the L intake tube central ejector (hereinafter referred to as L-ejector), the blockage ratio and other structural parameters are the same. And the central ejector of the L/Y intake tube was analyzed through numerical simulations. The main evaluation parameters are the volume of the flow separation zone, total pressure loss, and cross-sectional velocity uniformity due to the intake tube. The magnitude of these parameters has an impact on the performance of the ejector. Therefore, this work will provide some basis for the improvement of the ejector inlet tube. And we were able to select the intake tube configuration for the ejector more easily.

## 2. NUMERICAL STUDIES

### *2.1. Computational model and boundary conditions*

Two types of ejectors are shown in Figure 2 for the numerical simulation. The secondary flow suction chamber, primary flow intake tube, primary flow nozzle and constant area mixing chamber are included.

Four different operating conditions are set as shown in Table 1. The L-ejector is a symmetrical structure but the Y-ejector is a periodic structure. To save computing resources, 1/2 of the L-ejector and 1/3 of the Y-ejector are set as the calculation domain. And the mixing chamber of constant area is 5D long (D is the diameter of the mixing chamber). The solid wall is set up as an adiabatic no-slip boundary. Again, the inlet and outlet are set as pressure boundary conditions. The working fluid of primary flow is Air, and the secondary flow is nitrogen. Two fluids satisfy the ideal gas law. The total temperature of both primary and secondary streams is 300 K, which is not listed in the table 1. In Table 1, $P_{0s}$ and $P_{0p}$ correspond to the total pressure of secondary flow and primary flow, and $P_{out}$ is the outlet pressure.

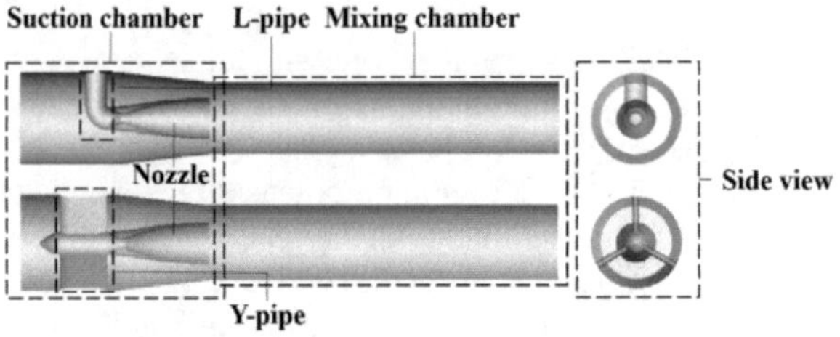


FIGURE 2 L/Y-EJECTOR CONFIGURATION

TABLE 1 DETAILED PARAMETERS OF BOUNDARY CONDITIONS

| Ejector | case | $P_{0s}$(kPa) | $P_{0p}$(MPa) | $P_{out}$(kPa) |
|---|---|---|---|---|
| L/Y | 1 | 5 | 3.6 | 0 |
| | 2 | 10 | | |
| | 3 | 15 | | |
| | 4 | 20 | | |

*2.2. Solution method and turbulence model*

In this study, the flow field characteristics of different configurations of the ejector are investigated by solving Reynolds Averaged Navier-Stokes equations system in three-dimensional integral form. The SST k-ω turbulence model with compressibility correction is used to deal with turbulence-related parameters. The model calculates the shear flow process with high efficiency and accuracy. And it is especially suitable for the calculation of complex flow fields with shock/expansion wave disturbances. And the numerical calculations were carried out in Fluent 18.0.

*2.3. Meshing generation and grid sensitivity analysis*

ANSYS ICEM was used to mesh the computational fluid domain, and the mesh was encrypted at all near-wall locations. The Y-ejector fluid domain was preliminarily divided into 2,449,440 structured hexahedral meshes.

To exclude the influence of grids number on the calculation results, a grid sensitivity analysis was carried out. Taking the case4 as an example, four sets of mesh models with the same topology and a continuously varying number of nodes are set up. The number of grids and the comparison of calculation results are shown in Table 2 and Figure 3. When the number of grids reached 3937400 (Grid3) and continued to increase to 6331000 (Grid4), the primary and secondary flow rates change very little and stabilize. As shown in Figure 2, the overall trend of the velocity profile of the model center axis is consistent at different grid numbers. However, Grid1 and Grid2 are slightly lacking in the ability to capture flow details. Therefore, Grid3 was chosen as the final grid volume for numerical simulation to ensure accuracy while saving computational resources. In Table 2, $m_p$ and $m_s$ represent primary and secondary flow rates.

TABLE 2 GRID SENSITIVITY ANALYSIS RESULTS

| Grid label | Grid number | $m_p$(kg/s) | Error-p(‰) | $m_s$(kg/s) | Error-s(‰) |
|---|---|---|---|---|---|
| 1 | 2449440 | 7.49476 | -- | 1.06032 | -- |
| 2 | 3073160 | 7.49583 | 0.10 | 1.06056 | 0.23 |
| 3 | 3937400 | 7.49681 | 0.13 | 1.06157 | 0.95 |
| 4 | 6331000 | 7.49681 | 0.00 | 1.06159 | 0.02 |

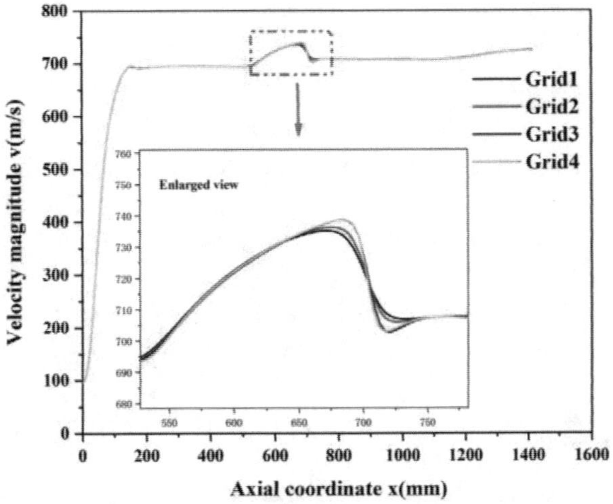


FIGURE 3 CENTER AXIS LINE SPEED FOR DIFFERENT NUMBER OF GRIDS

Similarly, the L-ejector mesh is finally divided into 6287306 structured hexahedral meshes.

## 3. ANALYSIS OF RESULTS AND DISCUSSION

Under the same incoming flow conditions, the L/Y intake tube of the ejector lead to a large impact on the flow field. The resulting differences are mainly in the size of the separation zone volume and the total pressure loss of the secondary flow. The numerical simulation results of the L/Y-ejector are analyzed and discussed with respect to the above differences.

Theoretically, since the windward areas of L and Y intake tube are different, the resulting separation zones must also be different. As shown in Figure 4, the trend of the volume of the separation zone and $V_s$ caused by the L and Y intake tube at different working conditions. It can be obtained that when the $P_{0s}$ increases, $V_s$ also increases gradually. $V_s$-L and $V_s$-Y are very close to each other. This is caused by the same blockage ratio of L/Y-ejector. The volume of the separation zone caused by the L/Y intake tube gradually decreases as the secondary flow total pressure $P_{0s}$ increases from 5kPa to 20kPa. Because when the shape of the intake tube is constant, the larger the airflow velocity, the smaller the volume of the separation zone. During the increase of the $P_{0s}$, the separation zone caused by the L intake tube is always larger than that of Y intake tube. When the total secondary flow pressure is 5 kPa, the volume of the separation zone caused by the L intake tube is 15 times larger than that of Y intake tube. This value will increase to 15 times when $P_{0s}$ reaches 20 kPa. The above results show that the Y intake tube can largely reduce the flow separation. Figure 5 shows the three-dimensional diagram of the separation zone caused by the L/Y intake tube when $P_{0s}$ is 5kPa. It is obvious from the figure that the two conformations bring different sizes of separation zones.

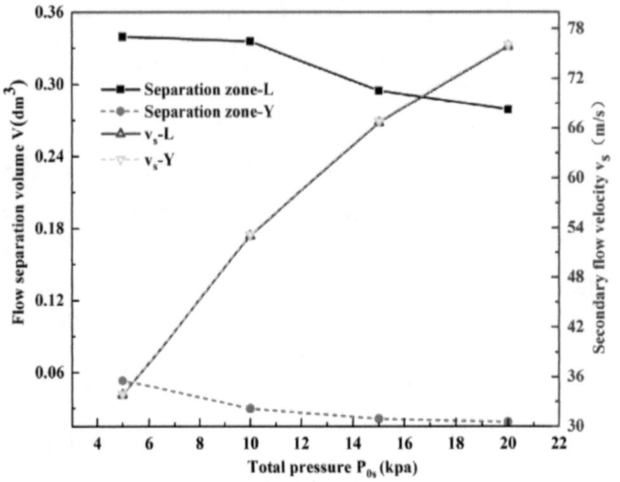


FIGURE 4 SEPARATION ZONE VOLUME AND SECONDARY FLOW VELOCITY VARIATION

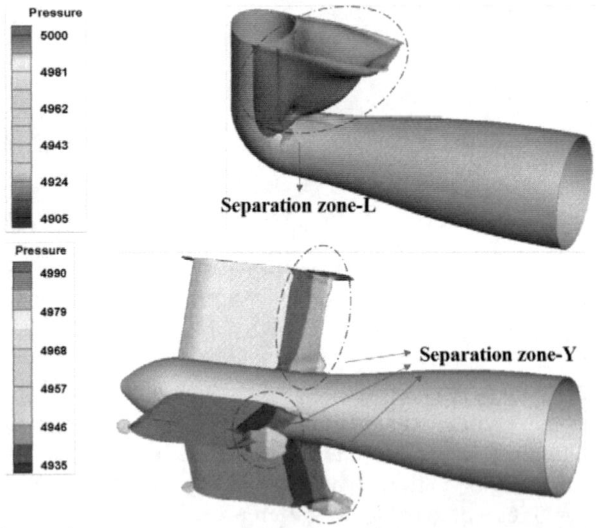


FIGURE 5 VOLUME EQUIVALENCE SURFACE IN THE SEPARATION ZONE, $P_{0S}$=5KPA

Figure 6 shows the velocity variation of the secondary flow and the total pressure loss after passing through the intake tube. It is possible to get that there is also a difference in the total pressure loss of the L/Y-ejector.

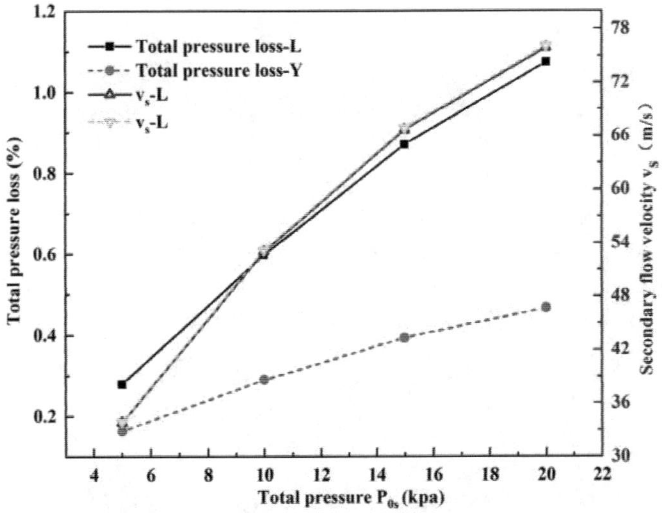


FIGURE 6 TOTAL PRESSURE LOSS AND SECONDARY FLOW VELOCITY VARIATION

As shown in Figure 6, the total pressure loss due to the intake tube increases with the higher $P_{0s}$. And it has the opposite trend to the volume change of the separation zone. Because an increase in total pressure brings about a higher secondary flow velocity, and high speed brings about larger pressure loss. When $P_{0s}$ reaches 20kPa, the total pressure loss caused by the L-intake tube will be the largest, which is 2.3 times of the Y-intake tube. This indicates that the Y intake tube can reduce the total pressure loss of the secondary flow to some extent.

The velocity uniformity of the inlet section of the mixing chamber is also very different due to the influence of the intake tube configuration. As Figure 7 shows the velocity clouds of the L/Y-ejector at this cross-section for different working conditions.

Comparing the velocity clouds of the mixing chamber inlet section for each operating condition, significant differences can be found. When $P_{0s}$ increases, the velocity heterogeneity of the L/Y-ejector in the inlet section of the mixing chamber gradually decreases. The main reason is that the volume of the separation zone decreases with the increase of $P_{0s}$. The reduced volume of the separation zone makes it possible to reduce the range of influences on the flow. So the velocity uniformity of this cross-section is improved. The velocity uniformity of the Y-ejector at the inlet section of the mixing chamber seems to be better for the same operating conditions. The reason for that is similar to the above; smaller separation zone volume brings better velocity uniformity.

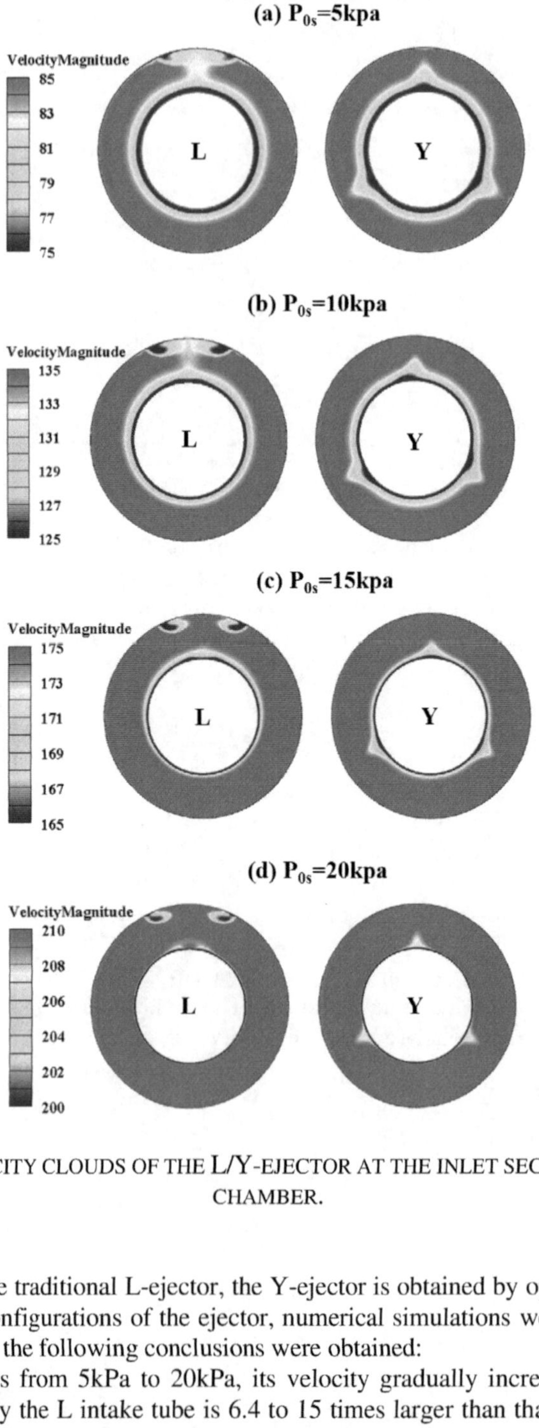


FIGURE 7 THE VELOCITY CLOUDS OF THE L/Y-EJECTOR AT THE INLET SECTION OF THE MIXING CHAMBER.

## 4. CONCLUSIONS

In this study, based on the traditional L-ejector, the Y-ejector is obtained by optimizing the intake tube. For two different inlet configurations of the ejector, numerical simulations were performed for several operating conditions, and the following conclusions were obtained:

1) When $P_{0s}$ increases from 5kPa to 20kPa, its velocity gradually increases. The volume of the separation zone caused by the L intake tube is 6.4 to 15 times larger than that of the Y intake tube. It shows that the Y intake tube can reduce the flow separation to a large extent.

2) When $P_{0s}$ is 20kPa, the total pressure loss caused by the L intake tube is the largest, reaching 2.3 times of the Y intake tube. So the Y intake tube can reduce the total pressure loss of the secondary flow to a certain extent.

3) Under the same working condition, the Y-ejector has better velocity uniformity in the inlet section of the mixing chamber. The improved Y intake tube can improve the uniformity of the secondary flow.

**ACKNOWLEDGMENT**
This study did not receive any specific funding.

**REFERENCES**

[1] Keenan, J.H., E.P. Neumann. A Simple Air Ejector[J]. Journal of Applied Mechanics-Transactions of the Asme, 1942, 64: A75-A81.
[2] Pianthong K, Seehanam W, Behnia M, et al. Investigation and improvement of ejector refrigeration system using computational fluid dynamics technique[J]. Energy Conversion & Management, 2007, 48(9):2556-2564.
[3] Chunnanond K, Aphornratana S. An experimental investigation of a steam ejector refrigerator: the analysis of the pressure profile along the ejector[J]. Applied Thermal Engineering, 2004, 24(2-3):311-322.
[4] Han Y, Wang X, Sun H, et al. CFD simulation on the boundary layer separation in the steam ejector and its influence on the pumping performance[J]. Energy, 2018.
[5] Yapc R, Ersoy H K, Aktoprako Lu A, et al. Experimental determination of the optimum performance of ejector refrigeration system depending on ejector area ratio[J]. International Journal of Refrigeration, 2008, 31(7):1183-1189.
[6] Yana J, Caia W, Li Y. Geometry parameters effect for air-cooled ejector cooling systems with R134a refrigerant[J]. Renewable Energy, 2012, 46:155-163.
[7] Tashtoush B, Alshare A, Al-Rifai S. Performance study of ejector cooling cycle at critical mode under superheated primary flow[J]. Energy Conversion & Management, 2015, 94(apr.):300-310.
[8] Yana J, Caia W, Li Y. Geometry parameters effect for air-cooled ejector cooling systems with R134a refrigerant[J]. Renewable Energy, 2012, 46:155-163.
[9] Chong D T, Yan J J, Wu G S, et al. Structural optimization and experimental investigation of supersonic ejectors for boosting low pressure natural gas[J]. Applied Thermal Engineering, 2009, 29(14-15):2799-2807.
[10] Hakkaki-Fard A, Aidoun Z, Ouzzane M. A computational methodology for ejector design and performance maximisation[J]. Energy Conversion & Management, 2015, 105(NOV.):1291-1302.

# Rocket Recovery Cabin Section Passive location based on whale optimization algorithm

**Chao Lu[1], Pei Wang[2,a], Zhiyuan Dong [2], Zhihui Ye [2]**

[1]Beijing Institute of Astronautical Systems Engineering, Beijing, China

[2]School of Astronautics, Northwestern Polytechnical University, Xi'an, China

[a]jimmy5579@nwpu.edu.cn

**Abstract**—The accurate location of the rocket recovery cabin section is one of the key technologies to realize the reliable recovery rockets. Based on the multi-station time difference of arrival (TDOA) passive location principle, a technical solution for the recovery and positioning of the rocket recovery cabin section is designed. Through the active radiation signal of the recovery cabin, the whale optimization algorithm (WOA) is used to locate the position of the rocket recovery cabin section. The recovery process trajectory of the rocket's stage 1 cabin was adopted to verify the feasibility of the designed recovery positioning technical scheme.

## 1. INTRODUCTION

Rockets are the main means of delivery for space activities all over the world. At present, the engines of most rockets are disposable. Research shows that sending 1kg of material into space costs up to 10000 ~ 20000 US dollars [1,2]. Such a high cost has greatly limited the process of developing Aerospace Engineering in various countries. With the increasing complexity of the missile and arrow system, the recovery of some or all of the debris of the module after the missile and arrow launch has become a problem worthy of discussion [3,4]. The recovery of the debris of the rocket can not only reduce the launch cost by reusing the recyclable part of the rocket, but also the most direct and effective method to obtain the flight technical data of the rocket, which will provide reliable original data for analysis and troubleshooting, It is helpful to test the design effect and speed up the development progress. Therefore, it is necessary to study the recovery technology of rocket recovery cabin section.

There are some problems in the recovery of rocket cabin section, such as wide landing area, complex terrain and landform, limited recovery time window under special tasks and so on. Therefore, the high-precision trajectory monitoring of the rocket recovery cabin section is one of the key technologies to realize the reliable recovery of the rocket compartment. If 1~3 types of active radiation sources are arranged on the rocket recovery cabin section to transmit positioning signals, and the passive detection technology with long action distance and strong concealment of electromagnetic radiation is used to realize the rapid positioning of the rocket recovery cabin section in the process of high-speed landing, it will provide effective positioning information support for the reliable recovery of the rocket recovery cabin section. At the same time, the technology can also be applied to the recovery of returnable spacecraft and missile test devices, and has a certain application prospect in the field of civil aerospace and deep space exploration.

The passive location method owns better anti-interference capability, can apply around the clock. Before the UAV takes off, the method can achieve the positioning of the UAV and the operator. In this paper, the problem of passive location of rocket recovery cabin section is discussed, and the passive location algorithm based on TDOA method is applied to the position of the rocket recovery cabin section at any time. By calculating the difference of interval among the signal of the rocket recovery cabin section to main station with each auxiliary station, it's enabled that the passive location method based on TDOA method is completed, associating the hyperbolic equations composed by the coordinate of every base station. The method has been widely studied and applied, with the characteristics of low environmental requirements, low complexity and high positioning accuracy.

Currently the TDOA location problem can be solved by many ways. The Taylor series expansion method is a kind of the traditional method [5,6]. Nevertheless, to solve the evaluated location of the target through iterative computations is the content of the kind of solution. When the initial value is not initialized properly, the algorithm will not converge easily. The least squares algorithm is a kind of popular algorithm solving the positioning problem, too. This kind algorithm's advantage is that it needn't iterative calculation[7-8]. But the improved algorithms which are based on least squares exist two evident disadvantages. First, the calculation process of the least squares method requires the inversion of the matrix. Second, when the observation error is small, the least square method is more suitable. The algorithm has evident threshold effect when the error increases. The positioning accuracy will decrease rapidly when the difference is bigger than the threshold value. If the arrival time difference is accurate, the Chan algorithm can obtain more accurate positioning, otherwise, the positioning accuracy is significantly reduced.

In 2007, Particle-Swarm-Optimization was raised for finishing off the TDOA position issue by Kenneth et al. [9]. Compared to least square method, the algorithm can calculate position result without initial value and can achieve greater positioning result. In 2010, an enhanced adaptive PSO algorithm applied for TDOA positioning was raised by Yuanyuan WEI et al. [10]. In terms of finishing off the seriously nonlinear problem of TDOA localization, the precision of algorithm is also greater than that of the least square method. In 2018, Salvia algae group algorithm was raised to finish off the TDOA positioning by Chen Tao et al. [11]. The intelligent optimization algorithm abandons the complex calculation procedure. The kind of method generates lots of random points in the observation scope. It assesses each stochastic point by establishing a function of fitness among the stochastic points with the measured actual values. Then it finds the best target estimated position using the iterative approach while updating the position of the stochastic point. The algorithm need not to inverse the matrix. It overcomes the defects of least squares algorithm which is multiple observation stations need to be arranged.

The current intelligent optimization algorithm used in TDOA positioning has problems such as slow convergence in the later stage and easy to fall into local extreme values. In addition, each intelligent optimization algorithm has unique algorithm control parameters, and excessive parameters will not only influence the speed of calculation, but also cause parameter setting problems.

Consequently, to finish off the actual problems of the algorithm presently applied concurrently, it is crucial to look for a more appropriate intelligent optimization algorithm for the TDOA positioning model.

In 2016, Australian scholar Mirjalili S, based on the predation behavior of humpback whales to this end, a new swarm intelligence algorithm, the whale optimization algorithm, is proposed. Optimization algorithm, WOA) [12]. The advantages of this global search algorithm is that the structure is simple, the parameters to be adjusted are few, the calculation speed is fast, and it is also indirectly reduced. parameter influence problem. Consequently, this paper applies WOA to improve the TDOA location problem, then the algorithm proved its own validity with better performance within the multiple stations TDOA localization problem range.

## 2. MULTI-STATION TIME DIFFERENCE POSITIONING MODEL OF ROCKET RECOVERY CABIN SECTION

The passive positioning system based on multi station can be split into multifarious types on the basis of the known quantities applied. It mainly includes: arrival time difference positioning technology derived from time information [13], arrival angle positioning technology obtained from angle information, as well as Doppler detection technology obtained from frequency information. There are also stroke positioning technology [14] and some node positioning technology [15].Among above, the arrival time difference positioning technology is to find the location of the target radiation signal according to the time difference between the arrival of the pulse signal to multiple base stations. The positioning technology is simple to use, has a wide detection range, and has the ability to locate the target radiation source instantaneously. A TDOA positioning system based on multi station is used to position the rocket recovery cabin section. The location system is shown in Figure 1. The positioning criteria is to calculate the time difference between the main station and each auxiliary station when receiving the positioning signal after obtaining the time when the main station and each auxiliary station receive the radiation source signal, so as to realize the time difference positioning of the target.

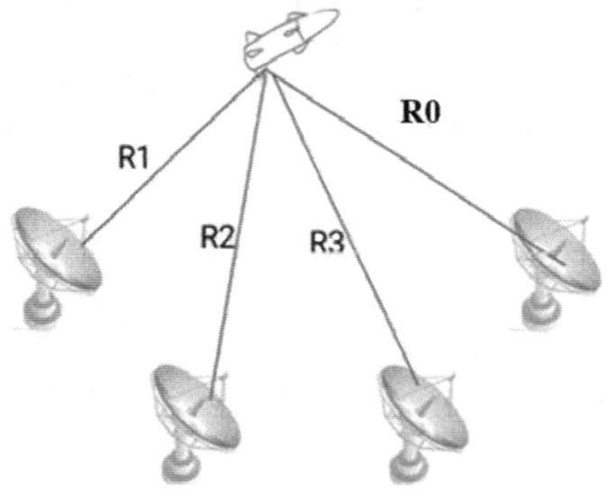


Figure 1. TDOA location system for rocket recovery cabin section
（This image has been authorized to use）

The designed ground radar station is composed of 4 base stations, including 1 primary station and 3 secondary stations. Their accurate position information can be obtained by GPS or other measurement means, and recorded as $(x_0, y_0, z_0)$ for the primary station and$(x_i, y_i, z_i), i = 1,2,3$for the secondary station, the location of rocket recovery cabin section is unknown, denoted as$X = (x, y, z)$.

The time stamp of the recovery cabin section of the rocket is recorded as $t$, and when the ground station receives this time stamp, the time at this time is recorded as$t_i$.The propagation delay time $t - t_i$can be determined by comparing the time when the signal is sent from the wreckage compartment of the missile and the time when the signal reaches the ground stations. The delay time multiplied by the speed of light is the pseudo range between the two, and expressed as the speed of light, the following formula can be obtained:

$$R_i = \sqrt{(x_i - x)^2 + (y_i - y)^2 + (z_i - z)^2} = (t - t_i)C \tag{1}$$

By gauging time difference among the stations with the rocket recovery cabin section, the relative distance between the stations and the rocket recovery cabin section can be calculated. It is known as pseudo range owing to errors of atmospheric propagation and clock. It expresses the distance between the rocket recovery cabin section and the master station, expresses the distance between the rocket recovery cabin section and every secondary station, and expresses the distance difference between the rocket recovery cabin section, the master station and the secondary station. The positioning equation is as follows:

$$\begin{cases} r_0 = \sqrt{(x_0 - x)^2 + (y_0 - y)^2 + (z_0 - z)^2} = (t - t_0)c \\ r_i = \sqrt{(x_i - x)^2 + (y_i - y)^2 + (z_i - z)^2} = (t - t_i)c \\ \Delta r_i = r_i - r_0 = (t_0 - t_i)c = \Delta t_i c \end{cases} \tag{2}$$

In the formula,$r_0$ means the distance of the target object to the primary station,$r_i$means the distance of the target object to every ancillary station. Finally $\Delta r_i$means the difference among the distance from the target object to the primary station with the distance from the target object to the secondary station.

Because it is inevitable that there is measurement between the ancillary station and the primary station, when the signal that is gauged by the observation station arrives between every ancillary station and the primary station, the TDOA measurement value can be expressed as:

$$\Delta t_{ci} = t_0 - t_i + e_i \tag{3}$$

In the above formulation, the TDOA measurement error $e_i$ follows the Gaussian distribution with mean 0 and variance $\sigma_t^2$.

The actual time that the signal reaches at the master station, the actual time that the signal reaches at the ancillary station, as well as the TDOA measurement value which is among the time that the signal reaches at the mater station with the time that the signal reaches at every ancillary station are all recorded as N −1 dimensional vectors, and the form is as follows:

$$\begin{aligned} \vec{t}_0 &= [t_0 \quad t_0 \quad \cdots \quad t_0]^T \\ \vec{t} &= [t_1 \quad t_2 \quad \cdots \quad t_N]^T \\ \Delta\vec{t} &= [\Delta t_{c1} \quad \Delta t_{c2} \quad \cdots \quad \Delta t_{cN}]^T \end{aligned} \tag{4}$$

Maximum likelihood method can be used for evaluating the concrete coordinates of the objective position$(x, y, z)$,in which case$t_1$follows Gaussian distribution, whose mean is$t_0 - t_i$and whose variance is$\sigma_t^2$.We assume that the measured results are uncorrelated with others, then the likelihood function can be deduced as follows.:

$$L = \prod_{i=2}^{N}\left[\frac{1}{\sqrt{2\pi}\sigma_t} exp\left(-\frac{(\Delta\tilde{t}_{ci} - t_0 + t_i)}{2\sigma_t^2}\right)\right] \tag{5}$$

Select

$$\begin{aligned} (x, y, z) = \arg\{max[\,[\frac{1}{\sqrt{2\pi}\sigma_t}]^{N-1} \\ exp\left(-\frac{(\Delta\vec{t} - \vec{t}_0 + \vec{t})^T(\Delta\vec{t} - \vec{t}_0 + \vec{t})}{2\sigma_t^2}\right)]\} \end{aligned} \tag{6}$$

From the likelihood function which is educed from equation (3), it can be seen that finding the maximal coordinate result of the likelihood function is the same as finding the goal function related to equation (5):

$$(\tilde{x}, \tilde{y}, \tilde{z}) = \arg\{min[(\Delta\tilde{t} - t_0 + t)^T(\Delta\tilde{t}_i - t_0 + t)]\} \tag{7}$$

In the formula, $\tilde{X} = (\tilde{x}, \tilde{y}, \tilde{z})$is the calculation result of passive positioning of carbon recovery section.$\Delta\tilde{t}_i$ is the corresponding arrival time difference of $\tilde{X} = (\tilde{x}, \tilde{y}, \tilde{z})$.

$$\begin{cases} \tilde{t}_0 = \sqrt{(x_0 - \tilde{x})^2 + (y_0 - \tilde{y})^2 + (z_0 - \tilde{z})^2}/c \\ \tilde{t}_i = \sqrt{(x_i - \tilde{x})^2 + (y_i - \tilde{y})^2 + (z_i - \tilde{z})^2}/c \\ \Delta\tilde{t}_i = \tilde{t}_i - \tilde{t}_0 \end{cases} \tag{8}$$

Therefore the arrival time difference vector $\Delta\vec{\tilde{t}}$that is of $\tilde{X} = (\tilde{x}, \tilde{y}, \tilde{z})$is:

$$\Delta\vec{\tilde{t}} = [\Delta\tilde{t}_1 \quad \Delta\tilde{t}_2 \quad \cdots \quad \Delta\tilde{t}_N]^T \tag{9}$$

It is difficult to finish off these problems by typical analytical methods. As a result, to obtain the optimal solution of the objective function, we can use WOA. And then the fitness of individual can be derived in the evaluation intelligent algorithm, through the following formula (5):

$$J(\tilde{X}) = (\Delta\vec{\tilde{t}} - \Delta\vec{t})^T \left(\Delta\vec{\tilde{t}} - \Delta\vec{t}\right) \tag{10}$$

In this way, this problem has changed from the solution of passive positioning equation that is of the rocket recovery capsule to the optimization problem with the recovery capsule coordinates as the design variables. This problem is showed as equation (11):

$$\begin{aligned} min\,i\,mize \quad & J(\tilde{X}) = (\Delta\vec{\tilde{t}} - \Delta\vec{t})^T(\Delta\vec{\tilde{t}} - \Delta\vec{t}) \\ & \tilde{X} = (\tilde{x}, \tilde{y}, \tilde{z}) \end{aligned} \tag{11}$$

## 3. Whale optimization Algorithm

Whale optimization algorithm(WOA) is a kind of special predation line simulating humpback whale Swarm intelligence optimization algorithm for. The algorithm was proposed by mirjalili, a scholar at Griffith University, Australia, in 2016. WOA simulates hunting behavior by using a random or optimal search agent and simulates the bubble-net attack mechanism of humpback whales by using a spiral. The whale algorithm consists of three behaviors: encirclement, hunting and search.

### *3.1Surround prey*

Humpback whales choose to surround their prey when hunting. Equations (12) simulate this process:

$$D = |CX^*(t) - X(t)| \qquad X(t) = X^*(t) - AD \tag{12}$$

Where $A$and $C$are coefficients,$t$is the current iteration times, $X^*(t)$ is the current greatest whale position vector, and$X(t)$ is the current whale position vector. $A$ and $C$ are calculated from equations (13):

$$\begin{aligned} A &= 2ar_1 - a \\ C &= 2r_2 \\ a &= 2 - \frac{2t}{T_{max}} \end{aligned} \tag{13}$$

Where $r_1$and $r_2$ are random numbers on the (0,1) interval, the value of$a$decreases linearly from 2 to 0, $t$is the current number of iterations, and $T_{max}$ is the maximum number of iterations.

### *3.2 Hunting behaviour*

In the hunting behavior of humpback whale, it approaches its prey with a spiral trajectory, and its mathematical model is shown in formula (14):

$$X(t+1) = X^*(t) + D_p e^{bl} \cos(2\pi l) \tag{14}$$

Where, $D_p = |X^*(t) - X(t)|$is the distance between individual whales and prey. $b$ is a constant used to define the shape of the spiral.

Whales swim in a spiral path to their prey while shrinking the enclosure. We can assume that whales have a 50% probability to choose the contraction encirclement mechanism and a 1% ~ 50% probability to choose the spiral model to update the position. The mathematical model is shown in formula (15), and P is a random number between [0,1]:

$$X(t+1)=\begin{cases} X^*(t)-AD & P<0.5 \\ X^*(t)+D_p e^{bl}\cos(2\pi l) & P\geq 0.5 \end{cases} \tag{15}$$

*3.3 Search for prey*

The model of whale hunting prey is shown in equations (16):

$$D=|CX_{rand}(t)-X(t)|X(t)=X_{rand}(t)-AD \tag{16}$$

Where $X_{rand}(t)$is the randomly selected whale location. The algorithm stipulates that when$A\geq 1$, a random agent is randomly selected from the population, and the positions of other individuals are updated according to the randomly selected individual positions of whales, so as to make whales deviate from the current target, find a more suitable prey, and improve the global search ability of WOA algorithm.

*3.4 position update*

The WOA algorithm begins with a series of random values. In each iteration, either a randomly selected search agents or currently obtained best solutions are used to update positions of search agents. Considering exploration and development, the parameter reduces from two to zero respectively. To update the positions of the search agents, when $|A|\geq 1$,one random search agent is adopted, in the mean time the best solution is adopted when $|A|<1$. According to the value of $p$, WOA can switch between either spiral or circular movement. Finally, the WOA algorithm is stopped until the termination condition is reached.

Extract random numbers and update $a$, $A$, C, $l$ and $P$. Update the location in accordance with the following formula:

```
if (p < 0.5)
   if (|A| < 1)
        Update the position of
        the current searchagent by the Eq.(10)
   else if (| A |≥ 1)
        Select a random searchagent (X_rand)
        Update the position of
        the current searchagent by the Eq.(14)
   end if
else if (p ≥ 0.5)
        Update the position of
        the current searchby the Eq.(12)
end if
```
(16)

where X_"rand" is a random position vector selected from the current population, that is, a random whale.

*3.5 TDOA passive location algorithm based on WOA*

The steps of location algorithm are:

Step 1: decide the population size of whales$N$, initialize algorithm parameters including $a$ and $b$ and the maximum of iterations $T_{max}$;

Step 2: Obtain the arrival time difference measurement vector $\Delta\vec{t}(g)$ of the time$g$;

Step 3: Current iterations$l=1$, Initialize whale swarm status.

Step 4: Compute the individual fitness value of every whale. Compare it to decide the individual whale with the best currently fitness result, which is recorded as$X^*(t)$;

Step 5: enter the main loop of the algorithm and update the whale position according (17);

Step 6: re-evaluate the entire whale population to seek out the best individual whale globally and its location;

Step 7: when the maximum number of iterations of the algorithm is reached, the cycle ends; otherwise go to step 4 and continue the algorithm iteration.

Step 8: output the global optimal individual$X^*(t)$.

Step 9: Whether the positioning is finished. If it is finished, exit; Otherwise, return to step 2 and wait for the arrival of the new measured value.

The algorithm flowchart is shown as Figure 2.

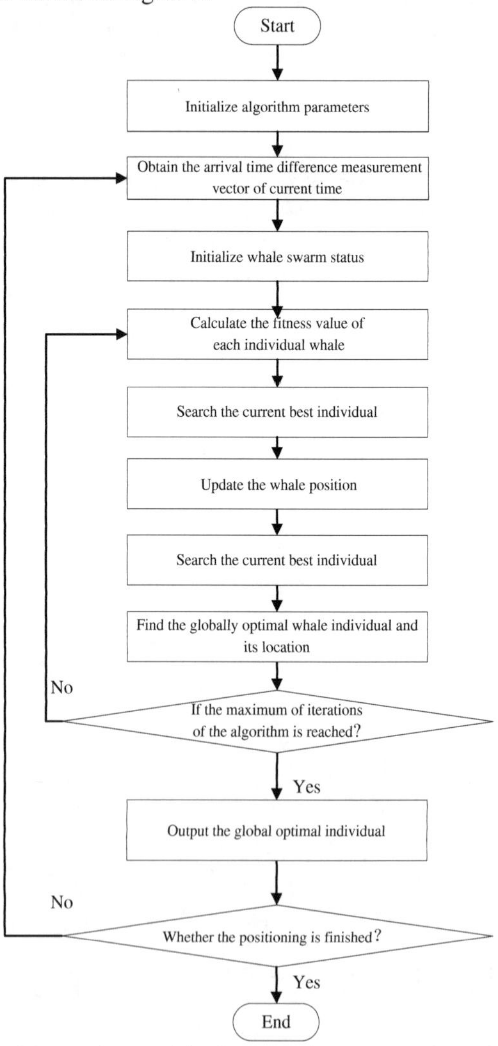


Figure 2. TDOA passive positioning solution process based on WOA

## 4. Simulation experiment and RESULT analysis

In this section, through simulation experiments, the feasibility of positioning the rocket recovery cabin section by WOA under the three-dimensional multi station TDOA positioning model is verified. The movement trajectory of rocket recovery cabin section is shown in Fig 2.

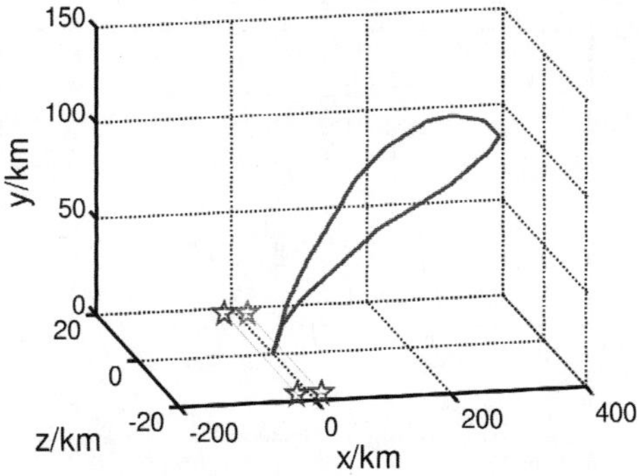


Figure 3. rocket recovery cabin section trajectory

Here, our configuration scheme is to use a square base station, as shown in Fig 3. The central base station is the master station, and that base stations which are at the four corners are secondary stations. The distance between each secondary station and the primary station is RKM.

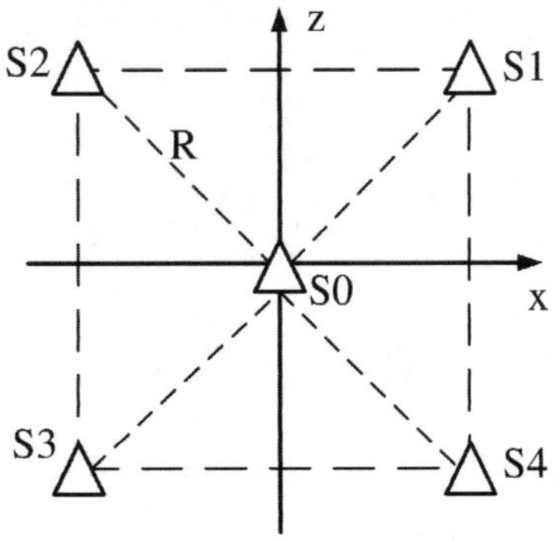


Figure4.Base station layout

We set the position of each base station as the following equation:

$$S_0 = [0,50,0](m),$$

$$S_1 = [R\cos 45°, 50, R\sin 45°](m)$$

$$,S_2 = [-R\cos 45°, 20, R\sin 45°](m),$$

$$S_3 = [-R\cos 45°, 20, -R\sin 45°](m).$$

$$S_4 = [R\cos 45°, 20, -R\sin 45°](m)$$

It is assumed that there is an error when measuring time difference of ground station, and the error follows the normal distribution error, whose mean value is 0 s and whose variance is 20ns.

The formation spacing is r = 25km, 50km and 75km. Chan algorithm and the solution method designed in this paper are used to locate the flight trajectory of rocket recovery cabin section respectively. The comparison of positioning errors is shown in Fig 5-10.

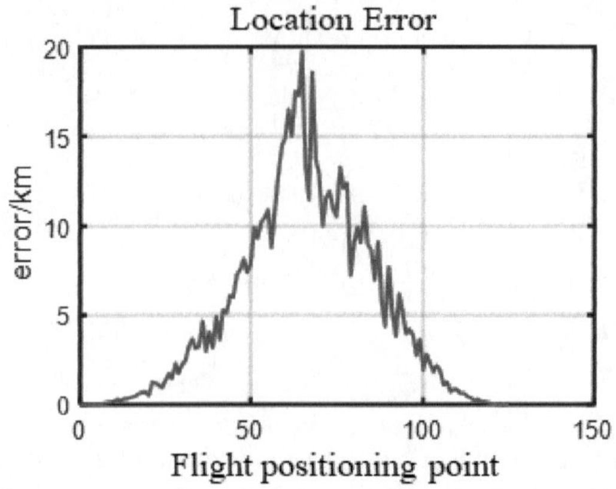


Figure 5. Chan algorithm location error when square base station configuration spacing R=25km

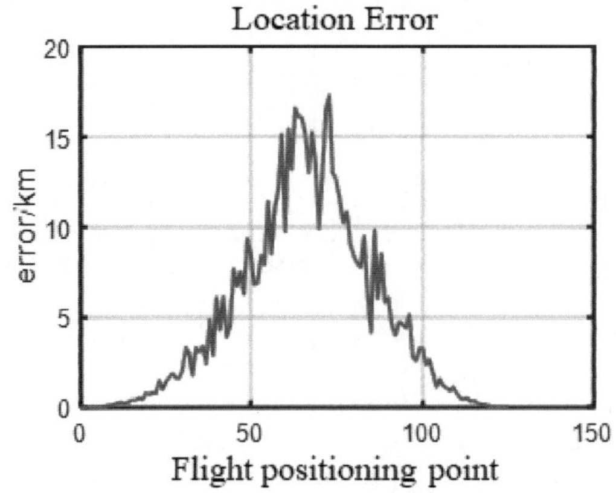


Figure 6. WOA algorithm location error when square base station configuration spacing R=25km

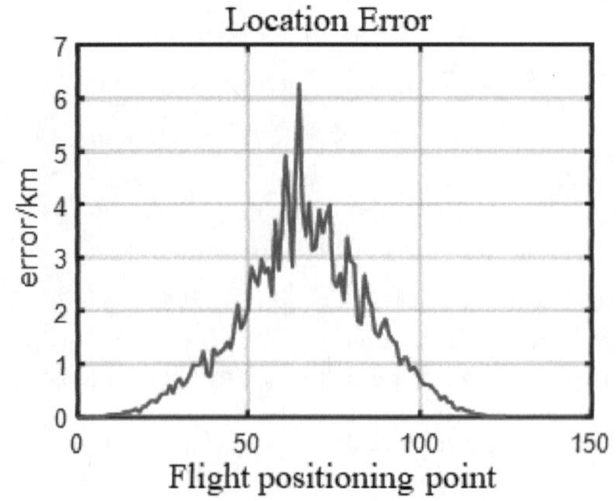


Figure 7. Chan algorithm location error when square base station configuration spacing R=50km

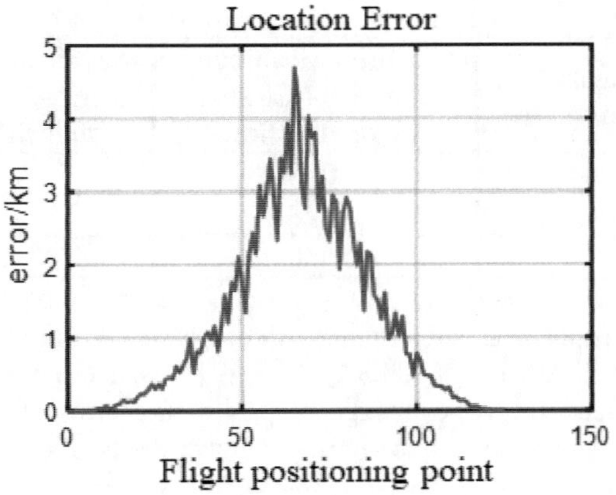


Figure 8. WOA algorithm location error when square base station configuration spacing R=25km

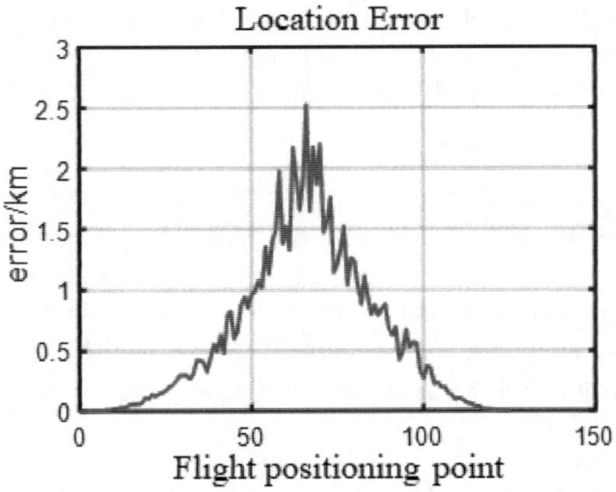


Figure 9. Chan algorithm location error when square base station configuration spacing R=75km

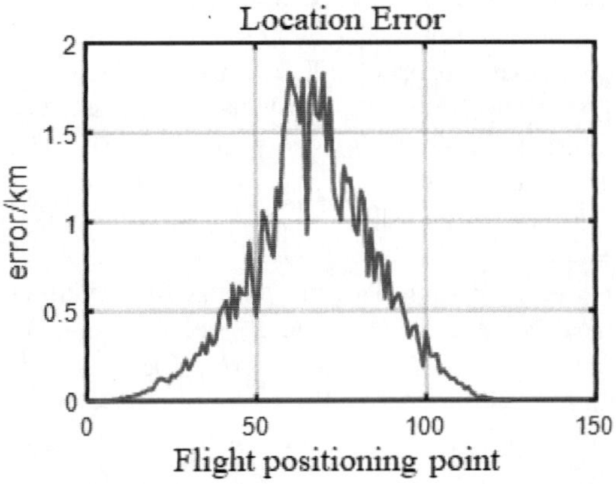


Figure 10.WOA algorithm location error when square base station configuration spacing R=75km

TABLE 1 COMPARISON OF LOCATION ERROR BETWEEN CHAN ALGORITHM AND WOA ALGORITHM

| **Square configuration spacing distance (km)** | **maximum positioning error of Chan algorithm (km)** | **maximum positioning error of WOA algorithm (km)** |
|---|---|---|
| 25 | 19.8 | 17.2 |
| 50 | 6.23 | 4.71 |
| 75 | 2.58 | 1.86 |

It can be seen from Figure.5-10 and Table 1 that the accuracy of the passive location method based on WOA designed in this paper is greatly improved compared with the traditional Chan algorithm under different array spacing, with 13.1%, 24.3% and 27.9% respectively under the three spacing. At the same time, it can be seen from the change of positioning error of the two algorithms under different spacing that the larger the array spacing is, the higher the positioning accuracy is.

## 5. CONCLUSIONS

In this paper, The WOA is applied to solve the nonlinear problem of TDOA positioning. The results show that WOA has the advantages of simple structure, high calculation efficiency, high positioning accuracy and good stability. It has a certain reference value for the actual use of base station positioning. The simulation results verify the feasibility of WOA in rocket recovery cabin section passive location solution.

## REFERENCES

[1] Wang Xiaowei, Zhang Puzhuo, Wu Shengbao, Shen Lin. Recovery Technology of Launch Vehicle Stage [J]. Spacecraft Recovery & Remote Sensing,2016, 37(03):19-28.
[2] Chen Zhangrui. Thinking and Analysis on the recovery technology of launch vehicle [J]. Public Communication of Science & Technology, 2019, 11(6):140-142.
[3] Pang Zhihao. Significance and key technology of recovering rocket [J]. Science & Technology Review, 2016,34(01):15-19.
[4] Pual Tattabini, Brett Starr. Ares I-x Separation and Reentry Trajectory Analyses[C]. AIAA Atmospheric Flight Mechanics Conference, AIAA 2011-6462, Portland, Oregon, 2011.
[5] Wade Foy. Position-location solutions by Taylor-series estimation[J]. IEEE Transactions on Aerospace & Electronic Systems, 2007, AES-12(2): 187-194.
[6] Fang Jiaqi, Feng Dazheng, and Li Jin. Research on modified Newton and Taylor-series methodsin TDOA[J]. Journal of Xidian University, 2016, 43(6): 27-33.
[7] Deng Bing, Sun Zhengbo, Yang Le, et al. Geo location of a known altitude object using TOA measurements[J]. Journal of Xidian University, 2017, 44(3): 133-137.
[8] Feng Qi, Qu Changwen, and Li Tingjun. Closed-form solution for passive location based on constrained weighted lease squares[J]. System Engineering and Electronics, 2017, 39(2): 263-268.
[9] Kenneth Lui, Jun Zheng. et al. Particle swarm optimization for time-difference-of-arrival based localization[C]. Signal Processing Conference, Wielkopolskie, Poland, 2007, pp.414-417.
[10] Wei Yuanyuan and Yao Jinjie. Application on target localization based on adaptive particle swarm optimization algorithm[C]. 2010 6th International Conference on Wireless Communications Networking and Mobile Computing. Chengdu, China, 2010, pp. 1245-1254.
[11] Wang Chentao, Meng xin, Huang Xiangsong. Time Difference of Arrival Passive Location Based on Salp Swarm Algorithm[J]. Journal of Electronics & Information Technology, 2018, 40(07): 72-78.
[12] Mirjalilis Lewisa. The whale optimization algorithm[J]. Advances in Engineering Software, 2016, 95:51-67.

[13] Lu Hongtao, Ma Fei. Three-station Passive Locating Technology Based on The Doppler Frequency Difference[J], SHIPBOARD ELECTRONIC COUNTERMEASURE, 2008, 31(1):29-31

[14] Li Jinzhou, Pang Hongwei, Guo Fucheng, et al. Localization of multiple disjoint sources with prior knowledge on source locations in the presence of sensor location errors[J]. Digital Signal Processing, 2015, 40(C): 181-197.

[15] Li Jinzhou, Guo Fucheng, Le Yang, et al. On the use of calibration sensors in source localization using TDOA and FDOA measurements[J]. Digital Signal Processing, 2014, 27(1): 33-43.

# AUTHOR INDEX